ENGINEERS AT WAR

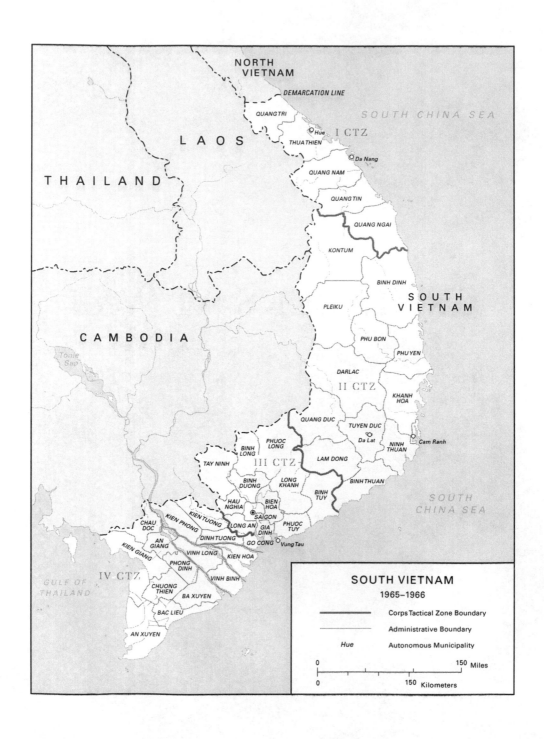

NORTH VIETNAM

DEMARCATION LINE

QUANG TRI

SOUTH CHINA SEA

L A O S

○ Hue   I CTZ

THUA THIEN

○ Da Nang

QUANG NAM

T H A I L A N D

QUANG TIN

QUANG NGAI

KONTUM

BINH DINH

SOUTH
VIETNAM

PLEIKU

C A M B O D I A

PHU BON

PHU YEN

Tonle
Sap

DARLAC

II CTZ

*KHANH
HOA*

QUANG DUC   *TUYEN DUC*

○ Da Lat   ○ Cam Ranh

*NINH
THUAN*

*PHUOC
LONG*

*BINH
LONG*

TAY NINH   III CTZ   *LAM DONG*

BINH THUAN

*BINH
DUONG*   *LONG
KHANH*

*HAU
NGHIA*   *BIEN
HOA*   *BINH
TUY*

KIEN TUONG   ◎ SAIGON

*CHAU
DOC*   *KIEN PHONG*   *LONG AN*   *GIA
DINH*   SOUTH
CHINA SEA

*PHUOC
TUY*

*AN
GIANG*   *DINH TUONG*   *GO CONG*   ○ Vung Tau

KIEN GIANG   *VINH LONG*   *KIEN HOA*

*PHONG
DINH*

GULF OF
THAILAND   IV CTZ   *VINH BINH*

*CHUONG
THIEN*   *BA XUYEN*

BAC LIEU

AN XUYEN

SOUTH VIETNAM

1965–1966

―――― Corps Tactical Zone Boundary

·········· Administrative Boundary

Hue Autonomous Municipality

0 ―――――――――― 150 Miles

0 ―――――――――― 150 Kilometers

United States Army in Vietnam

ENGINEERS AT WAR

by

Adrian G. Traas

MILITARY INSTRVCTION

Center of Military History
United States Army
Washington, D.C., 2010

Library of Congress Cataloging-in-Publication Data

Traas, Adrian George, 1934–
  Engineers at war / by Adrian G. Traas.
    p. cm. — (The United States Army in Vietnam) 1. Vietnam War, 1961–
1975—Engineering and construction. 2. Military engineering—Vietnam—
History—20th century. 3. Military engineers—Vietnam—History—20th
century. 4. Military engineers—United States—History—20th century. 5.
United States. Army. Engineer Command, Vietnam—History. I. Center of
Military History. II. Title.
  DS558.85.T73 2010
  959.704'34—dc22

                                                                    2010036741

                                    CMH Pub 91–14–1

                                    First printing

Cover: Jungle Clearing *by Sp4c. William E. Shuman, a construction draftsman
for Company B, 815th Engineer Battalion (Construction), at Pleiku.*

United States Army in Vietnam

Richard W. Stewart, General Editor

Advisory Committee

(*As of 9 July 2010*)

U.S. Army Center of Military History

Jeffrey J. Clarke, Chief of Military History

. . . to Those Who Served

Foreword

The military engineers who supported the U.S. Army in Vietnam wrote a proud record of achievement that spanned nearly two decades of war. Starting with a handful of advisers in the mid-1950s, Army engineers landed in force with U.S. ground units in 1965 and before long numbered more than 10 percent of the U.S. Army troops committed to the fight. Working in one of the world's harshest undeveloped regions, and under constant threat from an elusive and determined foe, the engineers met every test that came their way. They built ports and depots for a supply line that reached halfway around the globe, carved airfields and airstrips out of jungle and mountain plateaus, repaired roads and bridges to clear the advance for the combat infantryman, and constructed bases for an army whose communications grew in complexity with each passing year. They were often found in the thick of the fighting and fought as infantrymen as part of a long tradition of fighting while building. When the U.S. involvement in the Vietnam War began to wind down, the engineers were given another demanding mission, imparting to the South Vietnamese Army their specialized skills in construction and management. They left in place a robust infrastructure to support the South Vietnamese as they vainly struggled for survival against the armored spearheads of the North Vietnamese Army.

Engineers at War is the eleventh volume published in the United States Army in Vietnam official series. Like its companion volumes, it forcibly reminds us that the American soldier in Vietnam was courageous, infinitely adaptable, and tireless in pursuit of the mission. For the engineers, that mission and their comrades sustained them, in the best engineer tradition, even as the political and popular will to sustain the fight diminished. Their story and dedication should inspire all soldiers as they face a future of sustained operations around the world.

Washington, D.C.
30 September 2010

RICHARD W. STEWART
Chief Historian

The Author

L t. Col. (Retired) Adrian G. Traas, U.S. Army Corps of Engineers, was born in Milwaukee, Wisconsin, and received his education at schools in Milwaukee and Delafield, Wisconsin, completing high school at St. John's Military Academy in Delafield, commission and bachelor of arts degree in history from Marquette University, and a master of arts degree in history from Texas A&M University. His military education includes the engineer basic and advanced courses, the U.S. Army Command and General Staff College, and the Air War College.

Colonel Traas served in a variety of command and staff positions as an officer in the Corps of Engineers from 1957 to 1989. He was company commander of engineer units in Korea and at Fort Belvoir, Virginia; post engineer in Verona, Italy; executive officer and commanding officer of the 64th Engineer Battalion (Base Topographic) with headquarters in Leghorn, Italy, and mapping projects in Ethiopia, Liberia, and Iran; assistant professor of military science at Texas A&M; chief of the Combat Support Branch and staff officer in the Concepts and Studies Division at the U.S. Army Engineer School, Fort Belvoir; and professor of military science at Marquette. He served two tours in Vietnam: the first tour as S–3 and executive officer with the 19th Engineer Battalion (Combat) and contract liaison and installation master planning officer with the 45th Engineer Group; the second tour as an adviser with the South Vietnamese 30th Engineer Combat Group, a deputy engineer with the Third Regional Assistance Command, and later deputy region engineer in I Corps. Prior to his military retirement, he served in administrative positions and as a historian at the U.S. Army Center of Military History.

He is the author of *From the Golden Gate to Mexico City: The U.S. Army Topographical Engineers in the Mexican War, 1846–1848* and contributed to *The Story of the Noncommissioned Officer Corps* and *The United States and Mexico at War: Nineteenth-Century Expansionism and Conflict*. He currently holds the title of visiting professor at the Center of Military History.

Preface

Engineers at War describes the experiences of engineers in support of combat operations and carrying out construction in a distant theater. "The performance of United States Army Engineers in Vietnam, "wrote General Harold K. Johnson, the Army chief of staff, "adds another brilliant chapter to their history." The building effort in South Vietnam from 1965 to 1968 allowed the United States to deploy and operate a modern 500,000-man force in a far-off undeveloped region. Although the engineers faced enormous construction responsibilities, the Army's top priority remained providing combat support to tactical operations. As a result, ground combat troops with their supporting engineers were able to fight the enemy from well-established bases, which gave U.S. and allied forces the ability to concentrate and operate when and where they wanted. Although most of the construction was temporary, more durable facilities—including airfields, port and depot complexes, headquarters buildings, communications facilities, and an improved highway system—were intended to serve as economic resources for South Vietnam.

In the course of research and writing an engineering history, I have used many statistics. During the conflict, reports and correspondence used both the English and Metric Systems of measurement such as kilometers and miles and meters, yards, and feet. Engineers frequently used kilometers to measure road distances and feet to measure bridge spans. For consistency, I used the English System of measurement.

In researching and writing this book, I have received guidance and help from many people. At the Center of Military History much help and encouragement came from my colleagues in the Histories Division. Since combat operations involved a considerable part of the book, I relied on the knowledge and resources of the historians working on the combat volumes. These include George L. MacGarrigle, John M. Carland, Dale W. Andrade, and Erik B. Villard. Help was also gleaned from the extensive files of Vincent H. Demma and the work done by Charles R. Anderson, Andrew J. Birtle, Graham A. Cosmas, Richard A. Hunt, William M. Hammond, and David W. Hogan. Much credit goes to Joel D. Meyerson who helped me put the manuscript in final form. Thanks also go to John Schlight and Jeffrey J. Clarke who helped me early in the project.

I would like to recognize those in the Center's chain of command. The efforts and encouragements by a succession of branch chiefs are appreciated: John Schlight, Jeffrey Clarke, Joel Meyerson, and Andrew Birtle. My thanks also go to several History Division chiefs: Col. James W. Dunn,

Lt. Col. Richard O. Perry; Cols. Robert H. Sholly, William T. Bowers, and Clyde L. Jonas; Richard W. Stewart, and Joel Meyerson. I also appreciate the support given by several chiefs of military history: Brig. Gens. Douglas Kinnard, William A. Stofft, Harold W. Nelson, John W. Mountcastle, and John S. Brown; and Jeffrey Clarke.

Others at the Center of Military History contributed to this work. Staff at the Center under the leadership of Keith R. Tidman, Beth F. MacKenzie, and Diane S. Arms guided the manuscript to publication. Hildegard J. Bachman dedicated her editing talents to put a final touch to the book, cartographer Sherry L. Dowdy prepared the maps, and Gene Snyder designed the layout of the book. Contractor Anne Venzon created the index. Much credit for gathering research material for the book goes to Hannah M. Zeidlik and Robert K. Wright, former chiefs of the Historical Resources Branch; Frank R. Shirer, the current branch chief; James B. Knight, former librarian; and Rebecca C. Raines and Jennifer A. Nichols of the Force Structure and Unit History Branch.

Very helpful recommendations came from the panel review convened by Richard Stewart, the Center's chief historian and panel chair. I wish to thank the members of the panel: Brig. Gen. Gerald E. Galloway (USA, Ret.), University of Maryland and former Corps of Engineers officer; John C. Lonnquest, Corps of Engineers History Office, who provided extensive written comments; Donald A. Carter, Histories Division; Keith Tidman, Publishing Division; and Joel Meyerson, then acting chief of the Histories Division.

Those very helpful outside the Center of Military History include Richard L. Boylan, National Archives and Records Administration; Richard J. Sommers, U.S. Army Military History Institute; William C. Heimdahl and Wayne W. Thompson, Office of Air Force History; Jack Shulimson, History and Museums Division of the U.S. Marine Corps; Edward J. Marolda, U.S. Naval Historical Center; Vincent A. Transano, Lara D. V. Godbille, and Gina L. Nichols, U.S. Navy Seabee Museum, Port Hueneme, California; Martin K. Gordon and Michael J. Brodhead, Corps of Engineers History Office; Larry D. Roberts, U.S. Army Engineer School, Fort Leonard Wood, Missouri; and the numerous employees at various research facilities, especially the staffs at the Library of Congress and the Command and General Staff College library.

Of course, I alone remain responsible for the interpretations and conclusions expressed in this book. The hope is that my service as an engineer officer who served two tours in Vietnam will lend some credibility to the engineer story.

Washington, D.C.   ADRIAN G. TRAAS
30 September 2010

Contents

PART ONE
Engineers Enter the War

PART TWO
Supporting the Offensive

PART THREE
Changing Course

Tables

Charts

Maps

Illustrations

Illustrations courtesy of the following sources: cover, United States Army Engineer Command; pp. 4, 400, United States Operations Mission to Vietnam, *Annual Report for Fiscal Year 1961*; 7, 324, Corbis; 8, 116, 277, 290, 437, 453, U.S. Navy Seabee Museum, Port Hueneme, California; 9, 155, 177, 196, 461, 535, Department of Defense; 23, 149, 248, Dunn, *Base Development in South Vietnam*; 24, 30, 44, 62, 90, 184, 239, 315, 533, Office of the Chief of Engineers (OCE) History; 21, 29, 52, 63, 88, 113, 123, 127, 169, 171, 212, 213, 214, 215, 222, 227, 228, 286, 289, 299, 370, 372, 375, 377, 392, 399, 429, 439, 444 (*bottom*), 448, 454, 475, 476, 493, 494, 497, 498, 501, 508, 575, U.S. Army Center of Military History; 60, 191, National Archives and Records Administration; 69, Sharp and Westmoreland, *Report on the War in Vietnam (As of 30 June 1968)*; 74, Associated Press; 85, Engineer Command, Special Operational Report– Lessons Learned, 31 January 1967; 100, 106, 255, 256, 266, 275, 346, 347, 349, 410, 420, 433, 555, 557, 561, 562, Author's Collection; 159, 309, 381, U.S. Army Procurement Agency, *Procurement Support in Vietnam, 1966–1968*; 201, 1st Engineer Battalion, *Always First*, 1965–1967; 204, U.S. Army Heritage and Education Center; 139, 242, 268, 282, 305, 307, 444 (*top*), 455, OCE Briefing, "Military Engineering in South Vietnam," 5 January 1968; 245, Marolda, *Illustrated History of Vietnam*; 259, Michael McCabe; 382, Vinnell Corporation (Northrop Grumman); 335, Donald G. Lindberg; 342, *Hurricane,* II Field Force, Vietnam, Magazine; 364, 1st Cavalry Division Yearbook, August 1965–1969; 387, 395, 416, 422, 432, 485, 505, *Kysu'*, Engineer Command Magazine; 525, 531, 536, 538, Col. Kenneth E. McIntyre.

PART ONE
Engineers Enter the War

The Path to War

The decision to commit American troops to Vietnam, first brigades, then divisions, launched the engineers, and the rest of the support establishment, on a rapid buildup for war. At the start of 1965, as the Communist insurrection neared victory, South Vietnam had about 23,000 U.S. advisers and support troops, including some one hundred military engineers who, along with contracting firms, managed facilities and performed construction on a limited scale. Base development planning had barely scratched the surface, hampered by a dearth of staff engineers and the absence of a broadly accepted concept of operations that could serve as a point of departure for future work on ports, airfields, and roads. When the first large contingent of Army engineers went ashore in June with the combat forces, the construction troops faced immense demands for preparation and coordination in order to shift the development of the battlefield into high gear on short notice. The lateness of the hour and the steeply mounting scale of engineer effort required left unsettled whether the combat units could be adequately supported in the early months of the war.

Early Years

American planners had been concerned for years about the state of South Vietnam's defenses, but until 1965 improvements had been hard to come by, as successive administrations had enforced limits on the degree of U.S. involvement in that country's affairs. When the First Indochina War ended in 1954, and the United States supplanted France as South Vietnam's protector against Communist attack, international prohibitions set by the 1954 Geneva Agreements against establishing new military bases in the South shaped the next five years of American contingency planning for Southeast Asia. Thailand became the cornerstone of American strategy in that part of Asia, and the Republic of Vietnam a secondary holding action. Almost all operational preparations were aimed at launching an expedition into the Laotian panhandle and points farther north, and readying northeast Thailand with stockpiles, airfields, and forward bases through a robust program of military construction while South Vietnam would be thinly defended along its borders.[1]

---

[1] Ronald H. Spector, *Advice and Support: The Early Years, 1941–1960*, United States Army in Vietnam (Washington, D.C.: U.S. Army Center of Military History, 1983), pp. 357–58; Richard Tregaskis, *Southeast Asia: Building the Bases: The History of Construction in Southeast Asia* (Washington, D.C.: Government Printing Office, 1975), pp. 13, 16.

If intervention came to Southeast Asia, construction support would fall to the Army and Navy engineering corps. Throughout most of its history, the Army Corps of Engineers had served as a combat support branch supporting troops in the field, as a construction agency for the Army and, later, the Air Force, and as an office to carry out civil works in the United States. Likewise, the Bureau of Yards and Docks carried out major building programs for the Navy and the Marine Corps. To avoid duplication of effort, the Department of Defense had divided overseas construction responsibilities between the Corps of Engineers, the Bureau of Yards and Docks, and, to a lesser degree, the Air Force. The Navy was given responsibility for Southeast Asia, including South Vietnam. In December 1955, the Navy established an engineering office in Bangkok, Thailand, initially to supervise contractors building air bases in that country. Because the military advisers in Vietnam had little engineering capability, they asked the Bangkok office to design and supervise any military assistance construction projects in South Vietnam.[2]

As long as the peace lasted in the 1950s, American construction projects in Vietnam remained modest and generally out of the spotlight and were financed with nonmilitary aid. A road network in the Central Highlands designed to open the remote area to Vietnamese settlement kept U.S. contractors and South Vietnamese Army and civilian engineers busy through the end of the decade. The project had the additional benefit of connecting the highlands with strategic terrain across the border in southern Laos. Runway construction in Vietnam fit the contingency plans as well. Worried that the state of the country's airfields would slow troop deployments by air in case of war, the U.S. Military Assistance Advisory Group in Saigon drew up plans to improve airfield facilities at key locations. But since the United States continued to abide by the restrictions under the Geneva Agreements against building new military bases, a logical recourse was to use economic assistance to build commercial airfields as a subterfuge. In 1956, a contractor hired by the U.S. Operations Mission, the agency charged with carrying out economic assistance in developing countries, completed a new north–south concrete runway at Saigon's international airport, Tan Son Nhut. Further improvements followed, and Saigon soon had an airfield capable of handling military jet aircraft paid for by foreign aid.[3]

---

[2] Erwin N. Thompson, *Pacific Ocean Engineers: History of the U.S. Army Corps of Engineers in the Pacific, 1905–1980* (Honolulu: Pacific Ocean Division, n.d.), p. 211; Lt. Gen. Carroll H. Dunn, *Base Development in South Vietnam, 1965–1970*, Vietnam Studies (Washington, D.C.: Department of the Army, 1972), pp. 16–17; Tregaskis, *Building the Bases*, pp. 13, 15, 19; Edward J. Marolda and Oscar P. Fitzgerald, *From Military Assistance to Combat, 1959–1965*, United States Navy and the Vietnam Conflict (Washington, D.C.: Naval Historical Center, Department of the Navy, 1986), p. 24. For an overview of the Corps of Engineers, see *The U.S. Army Corps of Engineers: A History* (Alexandria, Va.: Office of History, U.S. Army Corps of Engineers, 2007). Army engineer responsibilities during this period are outlined in U.S. Army Field Manual (FM) 5–1, *Engineer Troop Organizations and Operations*, May 1961, pp. 38–41.

[3] Spector, *Early Years*, pp. 306–08, 360; Robert F. Futrell, *The Advisory Years to 1965*, United States Air Force in Southeast Asia (Washington, D.C.: Office of Air Force History, United States Air Force, 1981), p. 52; Ray L. Bowers, *Tactical Airlift*, United States Air Force in Southeast Asia (Washington, D.C.: Office of Air Force History, United States Air Force, 1983),

In 1957, the South Vietnamese president, Ngo Dinh Diem, on a state visit to the United States, received a pledge of additional assistance, including a package for the rehabilitation of his coastal highways laid waste during World War II and the Indochina War. Underpinning the pledge lay the belief in Saigon and Washington that the new nation was on the road to political stability and, with generous financial support assured from the United States, the economic future was bright. As before, the Operations Mission took the lead on the highway project. One segment of road, a twenty-mile stretch from Saigon northeast to Bien Hoa, became a model for observers of what a well-run economic aid program in Vietnam could accomplish. "The Bright Spot in Asia," an article in an American weekly magazine, described the positive developments in South Vietnam. The country's bright future was based on interviews with leading aid officials and senior members of the military advisory group. Other reports spoke of Diem's enormous popularity and effectiveness. All failed to heed the embers of insurrection igniting in the countryside.[4]

Initially, the scattered and sporadic nature of the violence misled almost every professional. As late as 1959, the chief of the advisory group was playing down the internal security threat represented by the Communist guerrillas, or Viet Cong, and declined to recommend that South Vietnamese Army combat units interrupt their training to join in pacification. But even as military officials were publicly affirming the point of view that the menace was overdone, the United States was shifting toward a progressively stronger counterinsurgency policy and the logistical wherewithal to support it. In 1960, at the direction of the State and Defense Departments, the embassy and military mission in Saigon drew up a comprehensive plan of political and military action that they hoped would help Diem reverse the trend on the battlefield. One year later, certain that the situation was worsening and that U.S. policy to contain Communist expansion was under attack in Southeast Asia, the administration of John F. Kennedy raised the level of American commitment to the Diem government, and the first U.S. military units entered South Vietnam. With this deployment, Special Forces teams, Army and Marine helicopter units, Air Force transports, Navy patrol boats, and more advisers descended on South Vietnam in 1961, quadrupling the American presence in the theater to 3,200 men. This number more than tripled again in 1962. A full military field headquarters, the U.S. Military Assistance Command, Vietnam (MACV), headed by a four-star Army general, took up station in Saigon in February 1962. While the advisory group headquarters stayed on as a subordinate element

pp. 42, 161–62; S. Sgt. Bob Reid, U.S. Air Force (USAF), "Vietnam Engineers in War," Military Engineer Field Notes, *Military Engineer* 55 (1963): 243; United States Operations Mission (USOM), *Annual Report for Fiscal Year (FY) 1962* (Saigon, 1962), pp. 22–23, 28–29, Historians files, U.S. Army Center of Military History (CMH), Washington, D.C.

    [4] Robert G. Scigliano, *South Vietnam: Nation Under Stress* (Boston: Houghton Mifflin Co., 1963), pp. 102–29; Spector, *Early Years*, pp. 304, 315–16; USOM, *Activity Report of the Operations Mission to Vietnam, 30 June 1954 to 30 June 1956* (Saigon, n.d.), pp. 38–40, USOM, *Annual Report for FY 1961* (Saigon, 20 November 1961), pp. 23–26, both in Historians files, CMH.

The Dong Nai River Bridge on the Saigon–Bien Hoa Highway was built with U.S. economic assistance.

until 1964, the winds of change were tending in one direction: American units were entering the war.[5]

The Defense Department paid close attention to all facets of the deployment, including construction. Under the auspices of the Defense Department, theater-planning conferences met regularly in Honolulu at U.S. Pacific Command headquarters starting in December 1961. Almost everything short of combat troops, said Secretary of Defense Robert S. McNamara at the first meeting, would be offered to the South Vietnamese, with the survival of that nation the highest priority. During the conference, McNamara approved construction of a new airfield at Pleiku City in the western highlands, fuel storage tanks at Qui Nhon on the central coast, ammunition storage bunkers at several sites, and a refueling station at Tan Son Nhut. Other airfield projects were already under way at Bien Hoa, and Da Nang on the northern coast, and these too received support. What McNamara refused to authorize was the use of

[5] Leslie H. Gelb with Richard K. Betts, *The Irony of Vietnam: The System Worked* (Washington, D.C.: Brookings Institution, 1979), pp. 69–80; Spector, *Early Years*, pp. 333–35, 362–65; Admiral U. S. G. Sharp and General William C. Westmoreland, *Report on the War in Vietnam (As of 30 June 1968)* (Washington, D.C.: Government Printing Office, 1969), pp. 77, 80; Maj. Gen. George S. Eckhardt, *Command and Control, 1950–1969*, Vietnam Studies (Washington, D.C.: Department of the Army, 1974), pp. 20–23, 27–31, 41–42. For more on Special Forces, see Col. Francis J. Kelly, *U.S. Army Special Forces, 1961–1971*, Vietnam Studies (Washington, D.C.: Department of the Army, 1973), pp. 3–18.

U.S. engineer troops on construction projects. The entire building effort would be undertaken by civilian contractors under Navy management. This would have the effect of keeping U.S. troop ceilings in Vietnam within authorized limits, and would leave the Army and Navy engineer units serving on active duty in the Far East free to continue their projects elsewhere.[6]

This approach, relying on civilian construction contractors in a war zone, continued for the next three years, and seemed to most observers a worthwhile solution to the limited demands of a counterinsurgency campaign. The biggest contractor was the joint venture of two large firms: Raymond International of New York City and Morrison-Knudsen of Asia, Inc., based in Boise, Idaho. Known as RMK, the two firms had a proven record of handling construction projects overseas. The initial contract was of a fixed-price type, in which the consortium's profit depended upon the efficiency of its operations. After several false starts, missed deadlines, and changed requirements because of the war's innumerable vagaries, the Navy shifted to a cost-plus-fixed-fee contract, assuming most of the risk of construction and furnishing RMK with materials and equipment and paying its transportation costs. By the end of 1962, RMK employed more than 3,000 workers, of whom 2,900 were Vietnamese. The contract was expected to last a couple of years, and the firm would be responsible for the demobilization of its workforce when work neared completion. Little was it realized that this contract would continue almost to the end of the American involvement in the conflict, nearly eleven years later.[7]

To oversee the civilian construction projects, the Navy's Officer in Charge of Construction, Southeast Asia, located in Bangkok had established a branch office in downtown Saigon in February 1961. The first contracts it administered were to Thomas B. Bourne Associates, a Washington, D.C. company, which designed plans for new and improved airfields at nine locations, including Bien Hoa, Tan Son Nhut, and Da Nang. The Saigon office also awarded a contract in October to the Tudor Engineering Company and Pacific Architects and Engineers, Inc., a joint venture of two California firms, to design air control and warning stations at Da Nang and Tan Son Nhut. At the end of 1961, when the RMK contract began and the scope of work expanded dramatically, the Bureau of Yards and Docks elevated the title of the Saigon branch from the Resident Officer in Charge of Construction, Republic of Vietnam, to the Deputy Officer in Charge of Construction, Republic of Vietnam. In addition to managing the U.S. contracts, the Navy's construction office in Saigon hired some thirty local contractors, mostly Vietnamese firms, to join in building facilities for incoming units. Some of these projects were delayed by funding

[6] Marolda and Fitzgerald, *From Military Assistance to Combat*, pp. 164–65; Headquarters (HQ), Commander in Chief, Pacific, "CINCPAC Command History, 1961" (Honolulu: Deputy Chief of Staff for Military Assistance, Logistics and Administration, 1962), p. 103, CMH (hereafter cited as CINCPAC History, date); CINCPAC History, 1962, pp. 198–99; CINCPAC History, 1963, p. 106; Rpt, HQ CINCPAC, Secretary of Defense Conference, 16 Dec 61, pp. 1–2, 45, 48, Historians files, CMH.

[7] CINCPAC History, 1962, p. 198; Tregaskis, *Building the Bases*, pp. 28–31; Capt. Charles J. Merdinger, U.S. Navy (USN), "Civil Engineers, Seabees, and Bases in Vietnam," in *Vietnam: The Naval Story*, Frank Uhlig Jr., ed. (Annapolis: Naval Institute Press, 1986), pp. 228–53.

restrictions on peacetime contracts, such as the need to gain congressional approval for jobs exceeding $175,000, and a requirement mandating the use of American-made construction supplies that caused some projects to be redesigned mid-stride. Once the contractors shifted their work on facilities into higher gear, however, U.S. advisers and support troops began to move from their temporary quarters, which were often little more than tent cities, into new semipermanent cantonments.[8]

Between 1962 and mid-1964, South Vietnam experienced a mini-building boom as construction projects funded by the United States sprang up in a dozen cities and towns across the country. During this period, RMK completed most of its assigned projects, including a pair of jet-capable airstrips at Bien Hoa and Da Nang and two smaller all-weather runways at Pleiku and Can Tho, a town in the Mekong Delta. The work gave the small but expanding South Vietnamese Air Force greater capability to fly tactical support missions and to transport soldiers and materiel from region to region. With the completion of the runways, each of the country's four military zones—I Corps, II Corps, III Corps, and IV Corps—had a modern and centrally located airfield. Those sites also supported the growing number of U.S. aviation companies that were arriving in South Vietnam. To improve the country's limited port facilities, the firm built a new deep-draft pier at Cam Ranh Bay in II Corps, a sandy peninsula on the south-central coast that sheltered one of the world's fine natural harbors. The aim during these years was still limited—to support an advisory effort that, by most projections, was expected to peak in 1964 and then slowly decline in size as the South Vietnamese armed forces gained in skill and confidence. Hardly anyone paid attention to the engineering requirements that would be needed if the United States changed its policy and decided to commit large numbers of troops over a relatively short period of time.[9]

As the number of U.S. troops climbed, the need for Army engineers to support them grew apace. The steady influx of new units, particularly the helicopter companies that required a high degree of engineer support, put a strain on

[8] Tregaskis, *Building the Bases*, pp. 22–23, 27, 33, 35; Sharp and Westmoreland, *Report*, p. 77; Merdinger, "Civil Engineers, Seabees, and Bases in Vietnam," p. 234; Info Sheet, Office Deputy Chief of Staff for Logistics (DCSLOG), Department of the Army (DA), 6 Jun 62, sub: Status of Military Construction, South Vietnam, Incl. in Information Book for Chief of Staff, U.S. Army, Visit to MACV (Military Assistance Command, Vietnam), 10–17 Jun 62, Historians files, CMH; Joint Logistics Review Board, *Logistic Support in the Vietnam Era*, Monograph 1, *Advanced Base Facilities Maintenance* (Washington, D.C.: Department of Defense, ca. 1970), p. 12 (hereafter cited as JLRB, Monograph #, title); JLRB, Monograph 6, *Construction*, p. 51.

[9] Sharp and Westmoreland, *Report*, p. 81; Tregaskis, *Building the Bases*, pp. 32–33, 36, 39–47, 64–65, 66, 149; CINCPAC History, 1962, p. 198, ibid., 1963, pp. 108–10, and ibid., 1964, fig IV-11; Ltr, RMK (Raymond, Morrison-Knudsen) to Brig Gen Daniel A. Raymond, Dir of Construction, MACV, 23 May 67, sub: Observations of the Deputy Chairman, Raymond, Morrison-Knudsen, Brown and Root, and J. A. Jones (RMK-BRJ) Operating Committee, Incl. in Rpt, Brig Gen Daniel A. Raymond, sub: Observations on the Construction Program, RVN [Republic of Vietnam], 1 Oct 65–1 Jun 67, 1 Jun 67, Historians files, CMH (hereafter cited as Raymond, Observations on the Construction Program). For more on RMK's accomplishments during this period, see Diary of a Contract, NBy (Navy Bureau of Yards and Docks) 44105, January 1962–June 1967, RMK-BRJ, Jul 67, pp. 1–76, U.S. Navy Seabee Museum, Port Hueneme, Calif. (hereafter cited as Diary of a Contract).

Generals Lyman L. Lemnitzer, chairman of the Joint Chief of Staff, and Taylor, Secretary of Defense McNamara, and President Kennedy, 1961

the limited number of specialists such as plumbers, electricians, refrigeration technicians, and water supply experts. By the middle of 1962, the permanent engineer element in the Army component in Vietnam, the U.S. Army Support Group, Vietnam, stood at only four officers and thirty-six enlisted men, not enough to engage in planning or any degree of support.[10] There had been a high-level request just a few months before to increase the number of engineers, but it had not come to fruition. Following the visit of General Maxwell D. Taylor, the personal military adviser to President Kennedy, to South Vietnam in October 1961, the general had recommended sending a 6,000- to 8,000-man flood relief task force, primarily engineer troops, to the monsoon-ravaged Mekong Delta. While the engineers fulfilled their humanitarian mission, Taylor also expected them to improve the military infrastructure in the delta where the insurgency was at its most active. In the end, President Kennedy opted not to send the flood relief force because he wished to keep the American troop commitment as small as possible.[11]

---

[10] Bowers, *Tactical Airlift*, pp. 109–10; Futrell, *Advisory Years to 1965*, p. 111; Rpt, Feb–Jun 62, Office of Engr, pt. I, pp. 1, 4, in Staff Office Rpt, Jan–Jun 62, n.d., U.S. Army Support Group, Vietnam, Historians files, CMH.

[11] *United States-Vietnam Relations, 1945–1967: Study Prepared by the Department of Defense,* 12 vols. (Washington, D.C.: Government Printing Office, 1971), bk. 2, ch. IV.B.1, pp. 58–101, 114–37, bk. 11, pp. 327–28, 331–44, 359–66, 410–21; Marolda and Fitzgerald, *From Military*

Beginning in 1963, teams of Army and Navy engineers carried out building projects such as this dispensary by the Seabees.

A second and more modest request for Army engineers came in early 1964 when the U.S. Operations Mission asked for three engineer officers with civil engineering backgrounds and 112 engineers, organized into eight advisory teams, to carry out a variety of civic action projects in the delta. Four small teams of Army engineers had arrived a year earlier for four-month tours and had succeeded in making some headway in the countryside. Three teams had gone to Can Tho where they had advised South Vietnamese officials on jobs ranging from equipment maintenance to well-digging. The fourth team had deployed to the highlands, building and repairing Special Forces camps, and working alongside several U.S. Navy Seabee (construction battalion) Technical Assistance Teams, which had a more robust presence in Vietnam than the Army. But the Operations Mission soon reevaluated its civic action plan and withdrew the troop request following the poor results of the government's Strategic Hamlet and New Life programs, two controversial schemes to concentrate rural

Assistance to Combat, pp. 122–29; Maxwell D. Taylor, *Swords and Plowshares* (New York: W. W. Norton and Co., 1972), pp. 216–40, 245–47; Graham A. Cosmas, *MACV: The Joint Command in the Years of Escalation, 1962–1967,* United States Army in Vietnam (Washington, D.C.: U.S. Army Center of Military History, 2006), p. 20.

South Vietnamese engineers undergo bridge construction training

people into small defended enclaves. An expansion of Army civic action would have to await better days.[12]

So, apparently, would support of the South Vietnamese Army. Despite their small numbers, the engineers did their best to train their Vietnamese counterparts in such areas as equipment maintenance, staff planning, and the management of supplies and equipment. Two advisers were stationed at the Vietnamese Engineer School at Phu Cuong, fifteen miles north of Saigon, to improve its standards of education. Still, the number of engineer advisers through 1963, one per infantry division and a few at corps level, was woefully inadequate to give the South Vietnamese Army the help it needed. Deadline rates for South Vietnamese engineer equipment hovered near 50 percent, and, a forty-mile stretch of highway between the cities of Pleiku and Kontum took a South Vietnamese engineer construction battalion nearly three years to finish, even though the United States funded most of the heavy subgrading.[13]

[12] Kelly, *U.S. Army Special Forces,* p. 194; Marolda and Fitzgerald, *From Military Assistance to Combat*, p. 355; Headquarters, U.S. Army, Pacific (USARPAC), "History of U.S. Army Operations in Southeast Asia, 1 January–31 December 1964" (Honolulu: Military History Division, Office of the Asst CofS, G–3, 1965), pp. 227, 229–30.

[13] Spector, *Early Years*, pp. 260–62; Lt. Col. Ralph F. Lentz, "Little Fort Belvoir of Southeast Asia," *Army Information Digest* 19 (August 1964): 17; Maj. Gen. Robert R. Ploger, *U.S. Army Engineers, 1965–1970*, Vietnam Studies (Washington, D.C.: Department of the Army, 1974),

With no hope of meeting its expanding requirements with Army engineers, the U.S. command once again turned to contractors. In 1963, under a cost-plus-fixed-fee arrangement with U.S. Army Support Group, Pacific Architects and Engineers took over facilities engineering services at U.S. installations. The initial contract provided for support at Tan Son Nhut, Pleiku, Qui Nhon, Nha Trang on the central coast, and Soc Trang, a helicopter base in the delta. The company's first main office consisted of a six-man squad tent in the Air Force headquarters section of Tan Son Nhut. Early priorities were to resolve maintenance problems plaguing electrical generators, water purification plants, and other utilities equipment. Over time, Pacific Architects and Engineers also began to supervise local contractors in the construction of new cantonments as part of the country's mini-building boom. Within a year, the firm had grown to 762 employees, more than double its original labor force, and had extended its reach to four additional sites as the number of American support troops continued to rise.[14]

In fact, the entire contracting effort—RMK, Pacific Architects and Engineers, and the local firms—was outgrowing its makeshift beginnings and, by the summer of 1963, was recording significant headway on every front. In July, a new cantonment with wooden barracks opened at Pleiku capable of housing an aviation battalion. Named Camp Holloway for CWO2 Charles E. Holloway, an Army helicopter pilot who was killed in action in Phu Yen Province in December 1962, it featured a new mess hall, officers and enlisted men's clubs, a barber shop, a laundry, and even an indoor theater. Flight line and other operational facilities were either completed or in progress. Meanwhile, cantonments and other facilities were going up for other aviation units at Bien Hoa, Vung Tau on the coast southeast of Saigon, and Vinh Long in the delta. Projects for the 8th Field Hospital at Nha Trang were expected to be completed by 1 April 1964 and for the 39th Signal Battalion's station at Phu Lam just south of Saigon by 30 May 1964.[15]

But though the construction effort was showing signs of progress, the military and political situation continued to deteriorate. With each passing month, the Viet Cong gained strength and expanded their foothold in the countryside, while American aid could not reverse the trend. The enemy was proving more resilient than expected; the South Vietnamese forces less effective than hoped. Even though U.S. helicopter companies were replacing their old and slow H–21 Shawnees with the more powerful UH–1 Iroquois, or "Huey," the Viet

pp. 26–27; Comments of Col William J. Parsons, Ofc Asst DCSLOG (Materiel Readiness), DA, 16 Jan 63, p. 5, on Briefing, Brig Gen Frank A. Osmanski, Asst CofS J–4, MACV, for Paul R. Ignatius, Asst Sec Army (Installations and Logistics), 11 Dec 62, Historians files, CMH.

[14] Ploger, *Army Engineers*, pp. 27–28; JLRB, Monograph 1, *Advanced Base Facilities Maintenance*, pp. 12, 18; Dunn, *Base Development*, pp. 89–90; Eric D. Johns, *History, Pacific Architects and Engineers Incorporated: Repairs and Utilities Operations for U.S. and Free World Military Forces in the Republic of Vietnam, 1963–1966*, [Saigon, n.d.], pp. 5–14, copy in CMH (hereafter cited as PA&E History, date).

[15] CINCPAC History, 1963, p. 106; Rpt, Jul–Dec 63, Office of Engr, pt. I, pp. 2–6, in Staff Office Rpt, Jul–Dec 63, n.d., U.S. Army Support Group, Vietnam, Historians files, CMH; John D. Bergen, *Military Communications: A Test for Technology*, United States Army in Vietnam (Washington, D.C.: U.S. Army Center of Military History, 1986), pp. 82–83.

Cong were adapting to the threat. At the battle of Ap Bac in January 1963, a Viet Cong unit defeated a much larger South Vietnamese force supported in part by the newer helicopters.

The status of the South Vietnamese government was hardly more encouraging. In May 1963, a Buddhist uprising in Hue, Saigon, and other cities threatened the stability of the government and the increasingly unpopular president, Ngo Dinh Diem. The crisis worsened in August when Diem authorized a violent crackdown on the Buddhist monks who had originally organized the rebellion. With the overall trends looking bleak, the way forward pointed to an even larger advisory force and a greater demand for engineering and construction support. And, for the first time, U.S. planners began to examine in a serious way the logistical and engineering obstacles the United States would face if it chose to intervene in Vietnam on a larger scale.[16]

Intervention

The situation in Vietnam continued to worsen as 1963 drew to a close. On 1 November 1963, a coterie of South Vietnamese generals engineered a coup that overthrew the government and led to the death of President Diem. Three weeks later, President Kennedy met his death from an assassin's bullet while on the campaign trail in Dallas, Texas. While the twin shocks did not immediately alter the U.S. approach to the war, they set the stage for a deeper American commitment to Vietnam. The military junta that came to power after the death of Diem proved to be fractious and inept, giving the Viet Cong even greater opportunity to spread their influence. A second coup in January 1964 only served to increase the instability of the South Vietnamese government and degrade the effectiveness of its armed forces. In the United States, meanwhile, Kennedy's successor as president, Lyndon B. Johnson, pledged to continue and, if necessary, to increase U.S. military support for the embattled South Vietnamese state. As the frequency and intensity of Viet Cong attacks grew with each passing month, so did the pressure for greater American intervention in the conflict.

The inadequacy of the logistical base in South Vietnam, and the inability of the base development plans to keep pace with even a small U.S. buildup became evident by the summer of 1964. In August, General William C. Westmoreland, the MACV commander, advised Admiral Ulysses S. G. Sharp, the head of Pacific Command, and the Joint Chiefs of Staff of problems accepting a 4,700-man buildup any faster than the nine months called for in earlier planning. Advancing the deployment of additional advisers and aviation troops to 30 September, Westmoreland noted, would cause particular trouble at airfields already saturated with aircraft and troops, and further deployments would result in overtaxed and overcrowded bases. The MACV commander added it would take five months to construct facilities for the new troops, and his entire logistical and administrative base was already operating

---

[16] Gelb with Betts, *Irony of Vietnam*, pp. 80–89; Dave R. Palmer, *Summons of the Trumpet: U.S.-Vietnam in Perspective* (San Rafael, Calif.: Presidio Press, 1978), pp. 22–44.

11

on a shoestring. Furthermore, real estate would have to be acquired through local channels, sometimes a slow process, because land and facilities could not simply be commandeered.[17]

With little or no planned increase in the number of military engineers assigned to Vietnam, it became evident that a large contractor force would have to be kept in the country. Anticipating the deployment of more forces to South Vietnam, the Army, Navy, and Air Force commands developed a broad spectrum of project requirements that, in turn, greatly expanded RMK's labor force and scope of work. By the end of 1964, the firm's strength had grown to nearly 4,000 and the company was building $1.2 million worth of projects every month, a figure projected to increase to $12 million per month by early 1965.[18] Pacific Architects and Engineers also expanded its operations as the war heated up. In early August, a clash between U.S. Navy destroyers and North Vietnamese patrol boats in the Gulf of Tonkin led President Johnson to authorize one-time air strikes against North Vietnam. Immediately afterward, MACV called on the firm to improve air base defenses, a task typically done by troops. In short order, the contractor erected emergency shelters, sand bag bunkers, security fences, and perimeter lighting at Tan Son Nhut, Bien Hoa, and other airfields. By September, the firm was devoting approximately 80 percent of its construction effort to this work.[19]

Despite the expanded contractor effort, a prodigious amount of work remained to be done before South Vietnam could support even a modest increase in U.S. forces. MACV OPLAN (Operation Plan) 32–64, published in mid-1963, provided a detailed outline of South Vietnam's logistical facilities and highlighted the many improvements that would be needed to support a large U.S. expeditionary force. The airfield directory listed 251 usable and abandoned fields, most of which were small and unimproved. The commercial fuel storage facilities were barely adequate for South Vietnamese needs and could not hope to support an expanded U.S. presence. In evaluating ports, the MACV plan noted that Saigon and Da Nang were the country's only primary points of entry. Da Nang, however, lacked deep-draft piers, which meant that the United States would have to employ over-the-beach techniques. While Cam Ranh Bay on the central coast had tremendous potential as a port, and Pacific Command had successfully lobbied to have the deep-draft pier built under the Military Assistance Program, years of work would be required before the sleepy fishing village became a modern naval harbor. In the meantime, OPLAN 32–64 identified smaller ports that could be used, as well as coastal landing sites suitable for landing cargo in over-the-beach operations.

[17] JLRB, Monograph 6, *Construction*, p. H-8; Ploger, *Army Engineers*, p. 24; Msg, COMUSMACV (Commander, U.S. Military Assistance Command, Vietnam) MACJ3 7738 to CINCPAC, 11 Aug 64, sub: Additional Support, RVN; Msg, CINCPAC to Joint Chiefs of Staff (JCS), 12 Aug 64, sub: Additional Support RVN; Sharp and Westmoreland, *Report*, p. 93. Messages in Historians files, CMH.

[18] Tregaskis, *Building the Bases*, pp. 65, 71, 77–78; Observations of the Deputy Chairman, RMK-BRJ Operating Committee, 23 May 67; JLRB, Monograph 6, *Construction*, p. 96; Ltr, RMK to Raymond, 23 May 67.

[19] PA&E History, 1963–1966, pp. 14–21.

Recognizing the vast amount of work to be done and the limits on his engineering resources, General Westmoreland set his construction priorities for the future in the following rank order: airfields, roads, railroads, ports, and logistics bases.[20]

In late 1964, General Westmoreland recommended the deployment of an Army logistical command and an engineer construction group to support the growing needs of his command and the expanding war. The MACV staff under Brig. Gen. Frank A. Osmanski, the J–4 (assistant chief of staff for logistics), had foreseen the need for a centralized logistical organization as early as 1962, but a proposal had not made it out of the headquarters. General Osmanski revived the idea in 1964 with a plan that included a robust construction capability. Justifying the need for engineer troops, Osmanski explained that RMK faced an increasing backlog, and the deployment of a 2,700-man engineer group of three battalions would help close that gap. Although RMK was in the process of increasing its monthly work in place, Osmanski pointed out that the firm would take twenty months just to complete its current backlog. MACV estimated that the construction program by late January 1965 would exceed $130 million. This information served as the basis for requesting the engineer group.[21]

The Navy disagreed with Osmanski, arguing that RMK could satisfactorily carry out an increased workload. In Honolulu, R. Adm. James R. Davis, the chief of the Pacific Division of the Bureau of Yards and Docks, and Pacific Command's adviser on construction in Vietnam, declared that the firm's "mobilization and rate of construction accomplishment can and will be promptly expanded as required by further program expansion." In Washington, R. Adm. Peter Corradi, the head of the Bureau of Yards and Docks, backed this position. Admiral Corradi added that there were no major constraints in the way to prevent the contractor from expanding operations provided plans and procurement actions were done in advance. He also cautioned the Defense Department to defer consideration to have an engineer group coordinate construction in Vietnam because any shift of responsibility from the Navy's construction office in Saigon would be disruptive.[22]

Notwithstanding these arguments, the Joint Chiefs of Staff endorsed Osmanski's plan on 15 January 1965 and recommended dispatching an

---

[20] Joint Tables of Distribution, MACV, 1 Jul 63, p. 7; Tregaskis, *Building the Bases*, pp. 45–46; JLRB, Monograph 6, *Construction*, pp. 22, H-5; JLRB, Monograph 12, *Logistics Planning*, pp. 27–28; COMUSMACV OPLAN (Operation Plan) 32–64, Phase II, 1 Jul 63, with four changes through 16 Nov 67. Joint Tables of Distribution and OPLAN 32–64 in Historians files, CMH.

[21] Msg, CINCPAC to JCS, 24 Dec 64, sub: Improvement of U.S. Logistic System in RVN, Historians files, CMH; Msg, Osmanski MAC 370 to Lt Gen Richard D. Meyer, Dir of Logistics, JCS, 26 Jan 65, sub: Engineer Const Gp for RVN, Westmoreland Message files, CMH; Lt. Gen. Joseph M. Heiser, *Logistic Support*, Vietnam Studies (Washington, D.C.: Department of the Army, 1974), p. 9; Joel D. Meyerson, "War Plans and Politics: Origins of the American Base of Supply in Vietnam," in *Feeding Mars: Logistics in Western Warfare From the Middle Ages to the Present*, ed. John A. Lynn (Boulder, Colo.: Westview Press, 1993), pp. 281–87.

[22] JLRB, Monograph 6, *Construction*, p. 107 (quotation); Msg, CINCPAC to JCS, 24 Dec 64, sub: Improvement of U.S. Logistic System in RVN; Msg, Meyer JCS 163 to Westmoreland, et al., 14 Jan 65, Westmoreland Message files, CMH; Tregaskis, *Building the Bases*, p. 80.

advance party to Saigon to set up the logistical element. Defense Secretary McNamara approved the plan in principle but, in no hurry to commit ground troops, decided further justification was warranted, particularly for the engineer group. In late January, he dispatched a Defense Department task force to Vietnam to make an on-site inspection and to review the management of the construction program. By early February, the team concluded that an engineer group was not needed at the time, but it did agree to recommend the deployment of a scaled-down logistical command. The team also proposed augmenting the Navy construction office in Saigon and increasing its responsibility for technical engineering support for U.S. forces. As far as depending on RMK to carry out an expanded construction role in Vietnam, the team felt that the firm had virtually unlimited capacity for expansion and a proven capability to work in a combat theater. Besides, the team noted that Seabee battalions stationed in the Pacific could be called upon for rapid augmentation. On 12 February, McNamara's deputy, Cyrus R. Vance, approved the team's report. This decision proved shortsighted. One month later, the need for the engineer group became critical.[23]

On 7 February, and again on the eighth and eleventh, President Johnson ordered reprisal air raids against the North after guerrillas struck American installations at Pleiku and Qui Nhon, killing thirty-two soldiers and wounding more than a hundred. A few days later, dissatisfied with the uncertain trajectory of his reprisal policy, the president took the next step up the rung of pressure toward a wider war, authorizing the start of the bombing offensive called ROLLING THUNDER and inviting discussion on dispatching an expeditionary ground force of soldiers and marines. An engineer group was only one element in the administration's debate on escalation as the commitment of ground troops hung fire through February into March. But once the White House tilted the issue in favor of intervention, other decisions followed that clarified the course of construction policy, not quite as quickly as General Westmoreland and the Joint Chiefs of Staff believed was required, but with enough forward momentum to begin shifting the theater of operations to a wartime footing.

The need for military engineers received confirmation in the weeks that followed when officials in the theater and the Defense Department started weighing plans of action to deploy American combat units. Initially Westmoreland recommended an aggressive enclave approach, with infantry battalions deployed in beachheads along the coast and an infantry division rushed to the Central Highlands as a blocking force to counter a growing concentration of North Vietnamese regulars in the border area. He also laid out a more ambitious option taken from the theater's OPLAN 32–64 of a full corps force on the northern border and on into Laos in order to isolate the southern battlefield from enemy reinforcement. The Joint Chiefs after quickly endorsing the

---

[23] Heiser, *Logistic Support*, p. 9; JLRB, Monograph 6, *Construction*, p. 107; Msg, Westmoreland MAC 582 to Adm Ulysses S. G. Sharp, CINCPAC, 6 Feb 65, Westmoreland Message files, CMH; Memo, Cyrus R. Vance, Dep Sec Def, for JCS, 12 Feb 65, sub: COMUSMACV Plan for Introduction and Employment of a U.S. Army Logistical Command, Historians files, CMH.

enclave approach expanded it to three full divisions—two American and one South Korean—toward the end of March.[24]

By this time, the theater's construction shortfalls had begun to loom large in official Washington. The tactical requirements alone for engineers were giving pause to the top-level service and secretariat staffs that were puzzling over the use of force and its costs. Every scenario now seemed fraught with limiting logistical factors and other dangers. The deployment of a division to the highlands, officials now believed, would absorb an engineer group all by itself, and was simply unsustainable without a buildout at the port of Qui Nhon, repairs to Highway 19 (the main east–west artery), and expansion of the Pleiku airfield. As for pitching a corps-size force on the northern border and in the Laotian panhandle, it would take every engineer battalion in the active Army and several from the reserves, and would require massive concurrent construction projects in Vietnam and Thailand, none of which could begin, given the realities of logistical lead time, until the autumn at the earliest.[25]

But there were deeper problems with Vietnam, inadequacies endemic to the country and the American commitment that had been understood for years, and all through March the logistical and construction professionals attempted to grapple with them, laying out with increasing clarity and urgency the steps the administration needed to take immediately to prepare the theater for war. Their recommendations addressed fundamental sources of unreadiness: from the importance of developing direct sea access to Vietnam by dredging and constructing new ports along the coast, and pre-positioning lighterage and construction material within a few days' sailing time of Vietnam, to streamlining administrative procedures for funding construction, and deploying the full logistical command and engineer construction group, and even laying the political groundwork for mobilizing the Reserves and National Guard to ensure that the combat troops received enough support.[26]

These were the overriding issues for the military leaders, what they believed to be the perils facing them, when President Johnson in March took control of

[24] Msg, COMUSMACV MACJ41 6125 to CINCPAC, 27 Feb 65, sub: Southeast Asia Logistic Actions, Historians files, CMH; Cosmas, *Years of Escalation, 1962–1967*, pp. 202–03; William C. Westmoreland, *A Soldier Reports* (Garden City, N.Y.: Doubleday and Co., 1976), p. 126.

[25] Memo, Acting Ch of Engineers for DCSOPS (Deputy Chief of Staff for Operations and Plans), 26 Mar 65, sub: South Vietnam Engineer Estimate, Pleiku-Kontum, with Incl, Engineer Strategic Studies Group, South Vietnam Engineer Estimate, Pleiku-Kontum, Mar 65, box 8, 68A/5926, Record Group (RG) 319, National Archives and Records Administration (NARA), Washington, D.C.; Memo, DCSOPS for CofS Army, 26 Apr 65, sub: Engineer Estimate, U.S. Army Corps Force Southeast Asia, with Incl, Engineer Strategic Studies Group, Engineer Estimate, U.S. Army Corps Force Southeast Asia, Apr 65, box 9, 68A/5926, RG 319, NARA; William C. Baldwin, *The Engineer Studies Center and Army Analysis: A History of the U.S. Army Engineer Studies Center, 1943–1982* (Fort Belvoir, Va.: U.S. Army Corps of Engineers, 1985), p. 129.

[26] Memo, Office CofS Army, 22 Feb 65, sub: VN Discussions; Msg, Commander in Chief USARPAC to CINCPAC, 19 Mar 65, sub: Contingency Planning for Southeast Asia/Western Pacific; Msg, JCS to CINCPAC, 20 Mar 65, sub: Emergency Construction Requirements, Southeast Asia, all Historians files, CMH; Meyerson, "Origins of the American Base of Supply," pp. 282–83.

the debate on intervention, settling on a gradualist approach to troop deployments. His decisions offered the possibility of even larger commitments in the future while postponing potential trouble with Congress and the public. A first decision in early March put Marine units at Da Nang when the ROLLING THUNDER raids started over North Vietnam. Because the mission was limited to providing for air base defense, few officials regarded the troops as the beginning of an American ground war. Over the next two weeks, however, on 15 and 20 March, the president opened Vietnam to larger deployments, ordering the dispatch of dredges to theater harbors along with LSTs (landing ships, tanks), and approving two large construction projects—a new runway at Da Nang air base in an effort to handle the rising tempo of operations and a jet airfield farther south at Chu Lai to catch the theater's overflow—two projects Pacific Command had been pushing as construction priorities since the fall of 1964.[27]

The main objective of the support professionals, a decision for a full logistical command and engineer construction group, emerged during a meeting of the National Security Council on 1 April. After weeks of high-level controversy over the question of ground troops, the president deferred the three-division force that the military leaders recommended but authorized what amounted to its support train for deployment, plus two more battalions of marines and an expansion of their mission. If the decision represented a continuation of the policy of incrementalism, it still constituted a serious commitment of American ground troops, providing for a complete theater logistical command totaling 5,900 soldiers, plus 13,000 more spaces in over a hundred logistical and engineer units, among which was a second engineer group to help lay the base for any additional commitments of forces he might later make.[28]

The president would make three such commitments in the weeks that followed: a task force of the 173d Airborne Brigade that landed at Bien Hoa and Vung Tau in early May and two Army infantry brigades—the 2d Brigade, 1st Infantry Division, and the 1st Brigade, 101st Airborne Division—scheduled for enclaves on the central coast in early summer. The hope in military circles was that the engineers would arrive just as quickly or even more so, and pace the rate of arrival of the rest of the support package and the combat units. But already, according to officials at the Department of the Army, there were signs that the engineer deployment was slipping badly, reinforcing old fears, Westmoreland's among them, that the support base would never be ready to receive the infantry and properly sustain it. At the last possible moment,

---

[27] John M. Carland, *Combat Operations: Stemming the Tide, May 1965 to October 1966*, United States Army in Vietnam (Washington, D.C.: U.S. Army Center of Military History, 2000), pp. 15–18; Msg, JCS 007929 to CINCPAC, 26 Mar 65, sub: COMUSMACV Requirement for Additional LST Support, Historians files, CMH; "The Joint Chiefs of Staff and the War in Vietnam, 1960–1968," Part 2, "1965–1966" (Historical Division, Joint Secretariat, Joint Chiefs of Staff, 1970), ch. 20, pp. 12–14, CMH; Jack Shulimson and Maj. Charles M. Johnson, U.S. Marine Corps (USMC), *U.S. Marines in Vietnam: The Landing and the Buildup, 1965* (Washington, D.C.: History and Museums Division, Headquarters, U.S. Marine Corps, 1978), pp. 6–15, 29–30.

[28] Msg, Sharp to Gen Earle G. Wheeler, Chairman JCS, 27 Mar 65, sub: U.S. and ROK (Republic of Korea) Deployments, Westmoreland Message files, CMH; National Security Action Memorandum 328 to Sec State, et al., 6 Apr 65, box 13, 70A/5127, RG 330, NARA.

engineer units in the United States were preparing to deploy, but on a schedule that raised questions about the soundness of the support buildup for the fighting that lay ahead.[29]

[29] Sharp and Westmoreland, *Report*, pp. 108–09.

2

Engineers Cross the Pacific

The expanding engineer involvement came about more as a reaction to growing U.S. military strength rather than the execution of carefully drawn-up plans. When President Johnson directed the increase of support troops on 1 April 1965, the initial engineer increment consisted of one group headquarters, two battalions, three separate companies, a platoon, and three detachments. As more combat troops arrived in the summer and fall, additional increments of engineers filtered in. By November, three groups and ten battalions were building ports, airfields, and bases throughout the country and providing tactical support to combat units. And from the first, the demands placed upon the engineers were immediate and taxing.

The 35th Engineer Group (Construction) Deploys

On 10 April 1965, Col. William F. Hart, the commanding officer of the 35th Engineer Group (Construction) at Fort Polk, Louisiana, received orders to deploy the group headquarters to South Vietnam. The 84th Engineer Battalion (Construction) at Fort Ord, California, and the 864th Engineer Battalion (Construction) at Fort Wolters, Texas, were to accompany the group. Planning began at once to transport equipment and supplies to nearby ports to be put aboard cargo ships bound for Vietnam. On 3 May, advance parties from the 35th Engineer Group and 864th Engineer Battalion departed by air for Saigon. Lt. Col. Thomas C. Haskins, the group executive officer, headed the two advance parties. Lt. Col. James E. Bunch, the commanding officer of the 864th, and members of his staff made up the second advance party.[1]

The deployment of the 35th Group was not trouble-free. Although the ninety-eight-man headquarters company had a high manpower readiness rating, the Army's deployment criteria, which specified a six-month period of remaining service and other conditions, created immediate turbulence.

---

[1] Two assigned battalions, the 46th Engineer Battalion (Construction) at Fort Leonard Wood, Missouri, and the 168th Engineer Battalion (Combat) at Fort Polk, Louisiana, remained behind for the time being. Quarterly Cmd Rpt, 1 Apr–30 Jun 65, 35th Engr Gp, 8 Jul 65, p. 1, Historians files, CMH; Interv, Maj John F. Hummer, 16th Mil Hist Det, with Lt Col James E. Bunch, Commanding Officer (CO), 864th Engr Bn, 1 May 66, Vietnam Interview Tape (VNIT) 10, p. 1, CMH. Depending on tables of organization and equipment and modified tables of equipment, the construction battalions were authorized approximately 900 troops organized in a headquarters and headquarters company, an engineer equipment and maintenance company, and three engineer construction companies. For a summary of the types of engineer units, their functions, strengths, and types of equipment, see Ploger, *Army Engineers*, app. D, pp. 215–18. Also see FM 5–1, *Engineer Troop Organizations and Operations*, September 1965.

Ultimately, only four of the original complement of twenty officers and warrant officers deployed to Vietnam: Colonel Hart, the executive officer, the adjutant, and the commander of the aviation section. Replacements reported just before embarkation, and they did not get acquainted with the unit until aboard the troop ship on its way to Vietnam.[2]

The 864th Engineer Battalion's problems were on a much larger scale. Colonel Bunch had only nine days to get his equipment to the port at Beaumont, Texas, to meet the 21 April shipping deadline. The nearly 900-man battalion also lost about one-third of its strength, including many important officers and noncommissioned officers. Among the key staff officers only the adjutant remained. Because Bunch traveled with the advance party, he did not meet his new executive officer and operations officer until the main body arrived at Cam Ranh Bay.[3]

After reaching Saigon, Colonel Haskins met with Col. Robert W. Duke, the commander of the recently activated 1st Logistical Command. Haskins now learned that the 35th Group would report to the logistical command. Duke and his staff had evaluated anchorages, road nets, and security requirements and had selected sites for base areas. It was evident that expansion of port and airfield capacities at Saigon, Qui Nhon, and Vung Tau would be major tasks facing the group. The port of Saigon alone could not handle the number of ships coming over to unload troops, equipment, and supplies. A MACV survey had concluded that despite some construction problems, the sandy Cam Ranh peninsula and its well-protected natural harbor would make an excellent port and logistical complex. Haskins and his team began developing a plan to distribute the first increment of 2,300 engineer troops expected to arrive by the end of the month. On 10 May, Colonel Bunch and his thirty-seven-man advance party flew from Saigon to Cam Ranh Bay to prepare for arrival of the group. Five days later, General Westmoreland gave the go-ahead to develop bases at Cam Ranh Bay, Qui Nhon, and elsewhere.[4]

Concurrent with this planning, MACV initiated a series of requests for real estate. The basis for acquiring real estate was a provision in the 1950 Pentalateral Agreement between United States, France, and the three Indochina states, which stipulated that the host country would provide land at no cost. In South Vietnam, real estate was either requested from the Joint General Staff for government-owned land or leased from private owners. In some cases, the Vietnamese government would purchase private land, and MACV would reimburse the government. The United States also paid the cost of indemnification and relocation when squatters were on the property, but the title itself was obtained and held

[2] Ploger, *Army Engineers*, pp. 32, 38; Quarterly Cmd Rpt, 1 Apr–30 Jun 65, 35th Engr Gp, pp. 1–2; Table of Organization and Equipment (TOE) 5–112D, Engr Gp (Const), 9 Nov 61.

[3] Interv, Hummer with Bunch, 1 May 66, p. 1.

[4] Heiser, *Logistic Support,* pp. 9–11, 15; Quarterly Cmd Rpt, 1 Apr–30 Jun 65, 35th Engr Gp, p. 2; Interv, Maj John F. Schiller, 15th Mil Hist Det, with Col Robert W. Duke, CO, 1st Log Cmd, 3 Jan 66, VNIT 1, pp. 5–7, CMH; Interv, Hummer with Bunch, 1 May 66, p. 2; Headquarters, United States Military Assistance Command, Vietnam, "Command History, 1965" (Saigon, Vietnam: Military History Branch, Office of the Secretary, MACV, 1966), pp. 107–08, CMH (hereafter cited as MACV History, date).

When the 35th Engineer Group arrived at Cam Ranh Bay, the only deep-draft pier was the one built by RMK in 1964.

by the Vietnamese government. This system was satisfactory as long as small plots were involved and time was not a factor, as was usually the case during the advisory years. But the arrival of marines and Seabees to build a jet airfield at Chu Lai in May revealed problems with current practice. Only two weeks before the marines landed, MACV dispatched two Army captains, an engineer and a finance officer, to negotiate with government officials and land owners for "each fruit tree, each banana tree, rice paddy, thatched hut and grave" at the construction site. Because this involved payment to 1,800 different property owners, the transaction was barely completed in time. Time was also a consideration at Cam Ranh Bay. By the time the Vietnamese government approved U.S. entry onto the peninsula, advance elements were already ashore. The U.S. Command expressed little concern about these formalities given its sense of tactical urgency and the fact that the peninsula was virtually uninhabited. The State Department took a different position, however, by not pronouncing itself satisfied with the legality of the occupation until an investigation was completed during the summer.[5]

Upon reaching Cam Ranh Bay, the advance party got right to work. The current tenants of the peninsula, the South Vietnamese Navy, provided the engineers one of the old French military buildings, a fortuitous move because the Americans did not bring any tents. With only hand tools, Bunch and his party prepared a temporary landing beach and installed "dead-men" (buried logs used as anchors) to hold down the LSTs now beginning to frequent the nascent port.[6]

[5] MACV History, 1965, pp. 108, 124 (quoted words, p. 124); Interv, Schiller with Duke, 3 Jan 66, p. 7; Dunn, *Base Development*, p. 29.
[6] Interv, Hummer with Bunch, 1 May 66, pp. 1–3.

The main body of the 35th Engineer Group departed San Francisco on 13 May aboard the U.S. Navy Ship (USNS) *Eltinge*, a World War II Liberty ship just out of mothballs. Aboard the *Eltinge* were the group headquarters, the 84th and 864th Engineer Battalions, the 513th Engineer Company (Dump Truck), the 584th Engineer Company (Light Equipment), the 178th Engineer Company (Field Maintenance), and the 53d Engineer Company (Supply Point). The *Eltinge* was expected to arrive at Cam Ranh Bay around 30 May. The ship, however, suffered repeated mechanical breakdowns and ended up being towed to Midway Island. At Midway, the troops and cargo were transferred to the USNS *Barrett*. Meanwhile, three ships carrying equipment proceeded on, and they arrived weeks before the main body. The *Barrett* continued to its original destination, Subic Bay in the Philippines, before proceeding to Vietnam. Colonel Haskins flew to the Philippines from Saigon to brief commanders and staff officers. When plans were finished, he flew back to Saigon to complete preparations for landing and deploying the group.[7]

When the ships carrying the equipment arrived at Cam Ranh Bay on 22 May, the small group of engineers and transportation troops on the scene faced a daunting task. Dump trucks, bulldozers, graders, and the multitude of other construction equipment were slung off the ships and on to the peninsula's sole pier. It took four days to unload the first vessel. After sitting on board the ships for almost a month, most of the construction equipment and vehicles had dead batteries and flat tires. Maintenance and repairs could not be done because the operators were on the *Eltinge* and later the *Barrett*. Colonel Bunch and his party pitched in by using makeshift jumper cables to start the equipment and vehicles. Then the officers and senior noncommissioned officers, trying to recall how to operate the equipment, managed to drive the items off the crowded pier to a parking area in the sand. Finally, after about a week and a half of round-the-clock effort, they finished the unloading.

On 9 June, the USNS *Barrett* dropped anchor off Cam Ranh Bay. On hand to receive the disembarking troops were Brig. Gen. John Norton, the commander of the U.S. Army Support Command; the advance parties; and some Transportation Corps soldiers. The ship's captain had balked at entering the bay because his charts did not show an approved harbor there, so the troops had to debark over the side of the vessel into landing craft. It took two to three hours until the offloading began. After half the troops came ashore, a driving rainstorm struck the bay, soaking everyone, a welcome relief from the extreme heat. Once ashore, the group headquarters, the 864th Engineer Battalion, the four separate engineer companies, a finance detachment, and Company D, 84th Engineer Battalion, moved to preselected defense positions in the sand dunes. The *Barrett* and the rest of the 84th Battalion proceeded to Qui Nhon.[8]

---

[7] Ploger, *Army Engineers*, p. 38; Quarterly Cmd Rpt, 1 Apr–30 Jun 65, 35th Engr Gp, p. 1.

[8] Interv, Schiller with Duke, 3 Jan 66, p. 8; Interv, Hummer with Bunch, 1 May 66, pp. 2–3; Quarterly Cmd Rpt, 1 Apr–30 Jun 65, 35th Engr Gp, p. 1; Capt. Lindbergh Jones, "Operations at Cam Ranh Bay," *Military Engineer* 58 (July-August 1966): 10.

Engineers of the 35th Group transfer to a landing craft at Cam Ranh Bay, June 1965.

Digging In

After building a campsite and getting the equipment in working order, the group commenced the transformation of the sandy peninsula into a depot and port. The first task required removing a large sand dune in order to build a LST ramp on the south beach. Bulldozers and earthmoving scrapers leveled other areas and gradually a road network began to take shape. Initially, bulldozers cut the first trails through the deep sand, twenty to thirty feet in places. These first tracks were so unstable that soon bulldozers were positioned at various points to tow stalled vehicles. After a few weeks, vehicle and equipment operators developed techniques to maneuver in the sand. They reduced tire pressure below the minimum prescribed standard, and climbed the dunes by moving along the contours instead of attempting a direct approach. Still, sand was a never-ending problem. Its abrasive action on moving parts caused maintenance problems on nearly all equipment. Traveling by foot through the deep sand also sapped the men's energy. Finally, the gleaming granules of sand magnified the already intense heat inside the dark green tentage. When the occupants raised the tent sides for ventilation, onshore winds carried even more sand into offices and troop quarters. Sand also crept into kitchens, foodstuffs, and clothing. Only the constant breeze, albeit mixed with sand, seemed to offer some relief from the searing sun.[9]

[9] Interv, Hummer with Bunch, 1 May 66, pp. 3–5; Jones, "Operations at Cam Ranh Bay," pp. 10–11.

Because of high daytime temperatures, crews pave concrete pads for buildings at Cam Ranh Bay at night.

The daytime heat forced the engineers to adjust work schedule and construction procedures when they started work on the depot buildings and troop housing. Temperatures, intensified by the reflection from the white sand, rose to 120°F. To avoid the effects of the heat as much as possible, crews worked in two shifts to take advantage of the cooler night air. One shift worked from 0100 to 1100 and the second began at 1500, working until 0100. The 35th Group also adopted special procedures to prevent damage caused by the intense heat. Crews set forms during the daylight hours, leaving the heavier work of placing concrete slabs to the evening hours. This practice served both to protect the men and ensure that the concrete properly set before the heat of the day removed the hydration water. At least the sand could be used to make a low-grade concrete that proved adequate for floor slabs and hardstands. The abundant supply of sand also made it possible to fill the large numbers of sandbags used in bunkers, revetments, machine gun positions, and steps.[10]

Until infantry troops arrived, the 35th Group provided most of the security on the peninsula. During the first night, the engineers placed temporary machine gun positions and listening posts along the high dunes. Some nervousness prevailed. Colonel Bunch recalled: "I think every shadow got a hole

[10] Jones, "Operations at Cam Ranh Bay," p. 12; Interv, Hummer with Bunch, 1 May 66, pp. 2–4.

shot in and there were very few VCs [Viet Cong] in the area I'm sure, but my people didn't realize that so they were actually more of a hazard to themselves than they were to anyone else." Though the Viet Cong made their presence felt with some long-range, ineffective sniping, the 1,200 engineers ashore that first day seemed sufficient. While getting settled, the 35th Group worked out a mutual security plan with the local South Vietnamese naval commander, but diverting engineers to guard duty imposed a heavy burden on operations. Colonel Bunch used 160 to 170 men per day for manning defenses and patrolling. On 12 July, an infantry battalion from the 1st Infantry Division came ashore and took over the defense of the peninsula. In October, South Korean troops assumed security responsibility for Cam Ranh Bay peninsula and the surrounding area.[11]

Meanwhile, the 84th Engineer Battalion, less Company D, completed its sea voyage to Qui Nhon on 11 June. While Transportation Corps lighters moved equipment ashore, the construction battalion, under the command of Lt. Col. Joseph J. Rochefort, started setting up a base camp just west of the city near the intersection of Highway 1 and Route 440. After settling in, the battalion embarked on the long-term transformation of the port city and surrounding area into a large logistics base. Soon, permanent beach ramps for landing craft were completed, and landfill operations were under way for a supply depot. With the help of a platoon from the 497th Engineer Company (Port Construction), the 84th installed two 4-inch marine pipelines to pump petroleum products from tankers offshore. Just south of the city, the battalion began building ammunition storage pads and set up an improvised rock-crushing plant next to Highway 1. Here, too, sand was abundant, not as loose as at Cam Ranh, but still enough to cause equipment breakdowns.[12]

While the troops got situated at Qui Nhon and Cam Ranh Bay, the 84th Battalion's Company D transshipped from Cam Ranh to Vung Tau, some thirty-five miles southeast of Saigon. After setting up its tent camp on the Vung Tau peninsula, the construction company began turning the port town and the surrounding area into a combat support base designed to relieve Saigon of some of its offloading and storage burdens. The most urgent task facing the engineers was improvement of the existing shallow-draft port facilities, a major project for a company-size unit of two hundred men. Unlike the sandy and mountainous terrain at Cam Ranh and Qui Nhon, the Vung Tau peninsula was marshy. The company commander, 1st Lt. Reed M. Farrington, learned that the plans for the depot did not take the marshy soil into account. A redesign became necessary, and the staging depot complex fell behind schedule. Despite such setbacks, the officers and men of Company D carried out their assignment. Operating with a minimum of supervision and support from higher headquarters, getting whatever supplies they could from local vendors,

---

[11] Interv, Hummer with Bunch, 1 May 66, pp. 2–3; Jones, "Operations at Cam Ranh Bay," p. 10.

[12] Maj. Gerald E. Galloway, "Essayons: The Corps of Engineers in Vietnam" (Master of Military Art and Science thesis, U.S. Army Command and General Staff College, 1968), p. 35, copy in CMH; Quarterly Cmd Rpt, 1 Jul–30 Sep 65, 84th Engr Bn, 14 Oct 65, pp. 2–3, Historians files, CMH.

Company D fended for itself until the 46th Engineer Battalion (Construction) arrived in September.[13]

All the while, the war continued to go badly for the South Vietnamese. Viet Cong main forces and guerrillas, steadily increasing in numbers and effectiveness, were systematically bleeding Saigon's forces in large and small engagements. In April, Westmoreland and Sharp proposed hastening troop deployments, raising U.S. strength to 82,000 plus 7,200 troops from Australia and South Korea. On 5 May, the 173d Airborne Brigade was airlifted from Okinawa to Bien Hoa to relieve South Vietnamese Army forces of some of their security responsibilities and to free them to counter the enemy threat. In early June, MACV confirmed the presence of elements of a North Vietnamese division in northern II Corps. The command also suspected another division to be nearby in the Laotian panhandle. By mid-June, two new military leaders took over control of the government. Lt. Gen. Nguyen Van Thieu became the de facto chief of state, and Air Vice Marshal Nguyen Cao Ky assumed the premiership. Washington now realized that the new government would need even more U.S. combat troops to help stem the tide.

More men were on the way. On 10 July, President Johnson ordered the deployment of an additional 10,400 logistical and engineer troops. Four days later, five engineer battalions received orders to Vietnam. On the twenty-eighth, Johnson announced the deployment of the 1st Cavalry Division (Airmobile) to II Corps. The remainder of the 1st Infantry Division had also been alerted for Vietnam. By August, the add-ons would increase the number of soldiers committed to the war to 210,000, followed in September with a further requirement of 9,800 support troops, many of them engineers.[14]

One of the five engineer battalions alerted for deployment to Vietnam was the 87th Engineer Battalion (Construction) at Fort Belvoir, Virginia. Lt. Col. John J. McCulloch, who assumed command in June, faced problems that confronted other deploying units that summer. Most of the problems in themselves were not unusual. The difficulty stemmed from the number of problems arising simultaneously. Manpower shortages plagued the battalion immediately. While needing more soldiers with construction skills, the 87th lost 40 percent of its men because of reassignments, discharges, and other reasons. The Army alleviated the shortage to some degree by assigning combat engineers as substitutes. Colonel McCulloch also took measures to adjust his organizational equipment for service in Vietnam. He obtained more tents, some salvaged household refrigerators, and water distributors, and left behind unneeded items such as space heaters. On 26 July, McCulloch and his advance party of sixty-seven men departed for Saigon. Following several delays and a rough flight, McCulloch's party arrived in Saigon on 4 August only to find that the battalion's destination had changed. Instead of the Saigon region, the 87th Battalion would be assigned to the 35th Group at Cam Ranh Bay. While in Saigon, McCulloch made arrangements with the 1st Logistical Command

---

[13] Ploger, *Army Engineers*, pp. 49–50, 79.
[14] Cosmas, *Years of Escalation, 1962–1967*, pp. 200–56; Carland, *Stemming the Tide*, pp. 47–49.

for some missing items, including enough tentage for the entire battalion and small electrical generators. Most of the advance party then flew on to Cam Ranh Bay.[15]

The main body of nearly eight hundred men arrived at Cam Ranh Bay on 24 August and set up camp. During the next four months, the 87th Engineer Battalion built roads to the northern end of the peninsula and began building a 6,400-man cantonment, a petroleum tank farm, and a seven-and-a-half-mile fuel pipeline from the port to the new air base under construction by RMK. The 35th Group also asked the battalion to conduct tests to stabilize the sand. The 87th tried combinations of cement, crushed coral, asphalt, crushed rock, water, and decomposed granite, mixed in various degrees with sand, on the routes to the depot area and the ammunition supply point. For the time being, a mix of the decomposed granite and cement at an eight-inch thickness and moist cured for seven days afforded a substantial base course. Eventually, this method became the standard base course for all depot roads at Cam Ranh Bay.[16]

Also arriving on the same day as the 87th Engineer Battalion was the 497th Port Construction Company. This versatile unit of slightly more than two hundred men, previously based at Fort Belvoir and commanded by Capt. Paul L. Miles, proceeded to lay the ground work for the arrival of a DeLong pier, a prefabricated self-elevating barge pier developed by the DeLong Corporation of New York City. While the 87th Engineer Battalion worked on a combination rock fill causeway and panel bridge connecting the shore to the pier, the port construction company built a 550-foot sheet-pile bulkhead to protect the pier from beach erosion.[17]

Several other engineer companies arrived. The 102d Engineer Company (Construction Support), commanded by Capt. Jesse M. Tyson Jr., reached Cam Ranh Bay in late August with its quarrying, asphalt paving, and other specialized equipment. Captain Tyson and his lieutenants had no experience in asphalt production, but by using the knowledge of some of the noncommissioned officers, the 102d set up a plant. The unit began crushing rock in early November, and when rock production reached sufficient quantities, the roads around the peninsula received a topping of asphalt pavement. The 553d Engineer Company (Float Bridge), commanded by Capt. Richard L. Copeland, arrived a few days after the 102d. In October, the company put its M4T6 bridging equipment to good use, beginning regular ferry service to the mainland. Because of the heavy traffic, especially dump trucks hauling laterite—a soil rich in secondary oxides and used as a subgrade—to the peninsula, the 553d soon assembled a longer and swifter raft.[18]

---

[15] Ploger, *Army Engineers*, pp. 33–35; Quarterly Cmd Rpt, 1 Jul–30 Sep 65, 87th Engr Bn, 5 Oct 65, pp. 1, 3, Historians files, CMH.

[16] Ploger, *Army Engineers*, pp. 50, 53–54; Quarterly Cmd Rpt, 1 Jul–30 Sep 65, 87th Engr Bn, pp. 1–5; Galloway, "Essayons," pp. 42–43; Lt. Col. James M. Mueller, "Taming the Sands of Cam Ranh Bay," *Military Engineer* 58 (July-August 1966): 238–39.

[17] Ploger, *Army Engineers*, pp. 50–53; Dunn, *Base Development*, pp. 54–55.

[18] Ploger, *Army Engineers*, pp. 56–58; Galloway, "Essayons," pp. 46–47; Quarterly Cmd Rpt, 1 Oct–31 Dec 65, 35th Engr Gp, 19 Jan 66, pp. 1–4, Historians files, CMH.

The landing of Lt. Col. Paul D. Triem's 62d Engineer Battalion (Construction) (from Fort Leonard Wood, Missouri) at Cam Ranh Bay on 28 August marked the arrival of the fourth Army construction battalion and the second to be diverted from its original destination. Originally ordered to Qui Nhon to build an airfield, Triem received last-minute instructions to construct an airfield at Phan Rang farther down the coast. This change in plans illustrated the shifting priorities taking place during the hectic buildup that summer and the desperate need for another jet-capable airfield in South Vietnam. Studies at the Qui Nhon site had found it to be impractical from an engineering and security standpoint, which meant the proposed airfield at Phan Rang suddenly assumed the highest priority after that at Cam Ranh Bay. The advance party had already reached Qui Nhon, and the transport ship carrying the rest of the battalion approached the coast when MACV decided to shift the battalion. Colonel Triem sent the heavy equipment to Phan Rang by landing craft and moved the lighter equipment by road. After reaching Phan Rang the following month, the 62d Engineer Battalion started to build a 10,000-foot AM2 aluminum matting airstrip and cantonments. One construction company remained at Cam Ranh until November to help build the depot.[19]

By early September, the Cam Ranh peninsula began to take on the appearance of a bustling military facility. In the first month, the engineers dug in and tackled the environment. Within the first thirty days, the 35th Group built a tent and sandbag camp, lengthened the existing 1,100-foot airstrip to 1,400 feet, and built new roads or reinforced existing ones in the southern part of the peninsula. To relieve some of the congestion on the port's sole deep-draft pier, work crews expanded the LST unloading site on the beach and extended the existing 350-foot-long pier to 600 feet. Mid-August saw the completion of temporary motor pools, storage platforms, and storage areas for fifty-five-gallon petroleum drums. More engineers arrived, and the group headquarters refined plans for the building of a vast port and logistical complex. RMK was also hard at work in the northern part of the peninsula, building the country's sixth jet-capable airfield. By early September, with help from the contractor, the 35th Group completed more than thirty miles of all-weather road. In the port area, work proceeded feverishly to prepare the site for the arrival of the DeLong pier.[20]

Unfortunately, weather, as it often would in Vietnam, plagued progress or caused damage. In late 1965, the northeast monsoon rains washed out the unpaved roads crisscrossing the peninsula before they got paved. Heavy traffic and saturated subgrades combined to create a morass that could only be cured by removing the roadbed or mitigated by heavy applications of sand. The first monsoon that troops experienced in this part of Vietnam clearly showed the need for rapid paving and paying more attention to preserving the existing

---

[19] Ploger, *Army Engineers*, pp. 50, 55; Galloway, "Essayons," p. 45; Quarterly Cmd Rpt, 1 Oct–31 Dec 65, 62d Engr Bn, 8 Jan 66, p. 1, Historians files, CMH. AM2 matting consisted of 2-by-12-foot or 2-by-6-foot panels with honeycombed interiors that could be laid by hand. The latest in runway matting, it was designed to support landings of jet aircraft.

[20] Ploger, *Army Engineers*, p. 47; Tregaskis, *Building the Bases*, p. 142.

Deep-draft piers at Cam Ranh Bay in January 1966 showing the RMK pier built in 1964 at left and the recently completed DeLong pier and causeway on the right

vegetation. Strong winds caused the sands to drift like snow, a problem further exacerbated when unaware work crews cleared areas of their scanty scrub cover. This problem would be minimized later by planting grass and erecting snow fences.[21]

As for the arrival of the DeLong pier, the U.S. Army Materiel Command made arrangements to tow the only available pier stored at the Charleston Army Depot in South Carolina. The command obtained a contract tug, and on 11 August the tug, with the pier in tow, left Charleston for the long voyage via the Suez Canal and Indian Ocean. After eighty-one days at sea, the 300-by-90-foot steel pier arrived at Cam Ranh Bay on 30 October. By then, the engineers had completed the connecting causeway and sheet-pile bulkhead. DeLong engineers advised the 497th Port Construction Company on the emplacement and elevation of the pier. This work was completed in mid-December 1965. As 1966 dawned, Cam Ranh Bay boasted two deep-draft piers.[22]

The Delong pier had many advantages over conventional construction. The pier, essentially a barge supported by eighteen tubular steel caissons six feet wide and fifty feet long, could easily be moved into position. If harbor depths exceeded the caissons' lengths, then workers installed additional fifty-foot sections. Pneumatic jacks attached to large collars around the caissons

---

[21] Ploger, *Army Engineers*, p. 62; Galloway, "Essayons," p. 83.

[22] Dunn, *Base Development*, pp. 54–55; *Arsenal for the Brave: A History of the United States Army Materiel Command, 1962–1968* (Washington, D.C.: Historical Office, U.S. Army Materiel Command, 1969), p. 136; Quarterly Cmd Rpt, 1 Oct–31 Dec 65, 35th Engr Gp, pp. 3–4.

DeLong pier at Cam Ranh Bay nearing completion, December 1965

were used to jack the barge up on its legs to a usable height. Because of the mud conditions at Cam Ranh Bay, work crews joined three lengths of caissons totaling 150 feet for each leg. Although two sections could be joined before erection, the third had to be welded in place, a process that required twenty days. Most of the fittings and hardware of the barges and caissons arrived in poor condition, but the port construction crews succeeded in repairing or rebuilding the pier's vital components. The first DeLong pier at Cam Ranh Bay took forty-five days for construction by sixteen men. Estimates showed that constructing a timber-pile pier would have required at least six months by a forty-man construction platoon, plus supporting equipment and operators and a large number of hard-to-get timber piles and lumber.[23]

Such advantages inspired a demand for more of the unique piers. Logisticians and engineers recognized that the quickly assembled mobile piers could save valuable man-hours in readying the Vietnamese ports to receive the large influx of war materiel. The Army identified requirements for additional DeLong piers at Cam Ranh Bay, Qui Nhon, and Vung Tau. On 6 December, Secretary McNamara gave his approval for the Army Materiel Command to purchase more piers from the DeLong Corporation. Initially the contract called for delivery of twenty-one pier barges, seven measuring 300 by 80 by 13 feet (A-type barge or unit) and fourteen units measuring 150 by 60 by 10

[23] 1st Lt. David P. Yens and Capt. John P. Clement III, "Port Construction in Vietnam," *Military Engineer* 59 (January-February 1966): 20; Dunn, *Base Development*, p. 55; Ploger, *Army Engineers*, p. 52. For early background of the DeLong pier, see 2d Lt. Robin R. Forsberg, "Portable Piers and Packaged Ports," *Army Information Digest* 9 (September 1954): 55–59.

feet (B-type barge or unit). (One of the seven units was slated for Okinawa.) Manufacturing of the pier barges and caissons would be in Japan, thus cutting the towing time to eighteen days and reducing manufacturing costs. Procurement of jacks, compressors, valves, and other mechanical items from manufacturers continued in the United States. DeLong Corporation then married the components at the ports. In April 1966, McNamara authorized a further increase in the funding and number of DeLong units.[24]

During the summer and fall of 1965, RMK completed a 10,000-foot AM2 aluminum matting expeditionary runway at Cam Ranh Bay on an all-sand subgrade. Based on lessons learned from the Seabees at the Chu Lai airfield, RMK paid attention to stabilizing the loose, granular particles before placing the rectangular aluminum honeycombed panels. Workers flooded the sand with sea water and compacted it with rubber-tired rollers. Following the compacting and grading with equipment borrowed from the 35th Group, the contractor applied a bituminous sealer, and in September began laying airfield matting. By 1 November 1965, the firm's deadline date, a runway, parallel taxiway, high-speed turnoffs, and a parking apron (totaling some 2.2 million square feet of aluminum matting and 1.3 million square feet of pierced steel planking) were ready. South Vietnam now had five airfields (Tan Son Nhut, Bien Hoa, Da Nang, Chu Lai, and Cam Ranh Bay, with a sixth, Phan Rang, under construction) that could handle jet fighters. RMK completed other facilities at the Cam Ranh air base and began preparations to add a second runway, a 10,000-foot concrete runway, before moving its workforce to help the depot projects in February 1966.[25] (*Map 1*)

The Buildup Quickens

While construction accelerated, the increases in the U.S. troop commitment during the summer imposed immediate manpower burdens on the engineers. Army planners had assumed that any augmentation to the existing force structure would come from the reserve components. In order to maintain a high degree of readiness, the Army kept a large proportion of combat formations in its active forces. Nearly one-half of the engineers and engineer equipment were in reserve units. When President Johnson decided in July not to call up the National Guard or Reserve units, the burden to furnish the engineer units fell on the active-duty force. This meant that nearly all the active combat and construction battalions in the United States would deploy to Vietnam. The training base would have to be

---

[24] *Arsenal for the Brave*, pp. 136–38.

[25] Tregaskis, *Building the Bases*, pp. 143–46, 148; USAF Airfield Construction in South Vietnam, July 1965–March 1967, Historical Division, Directorate of Information, Headquarters, Seventh Air Force (AF), n.d., pp. 110–12, copy in Historians files, CMH; Memorandum for the Record (MFR), MACJ02, sub: J–4 Briefing 2 Feb, Base Development at Cam Ranh Bay, 3 Feb 66, Historians files, CMH; C. M. Plattner, "The War in Vietnam: U.S. Air Buildup Spurs Base Construction," *Aviation Week & Space Technology* 84 (14 March 1966): 76; E. T. Lyons, "Aluminum Matting Runways in Vietnam," *Military Engineer* 58 (July-August 1966): 245.

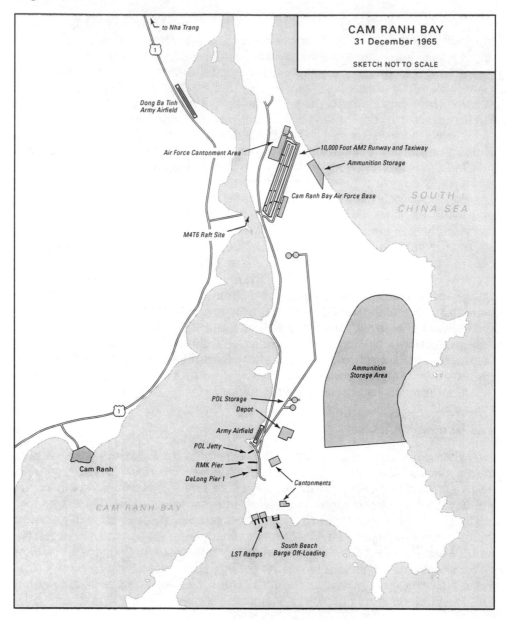

Map 1

expanded and new units organized, equipped, and trained to meet future Vietnam deployments and worldwide commitments.[26]

With the increase in troop levels under way that summer, General Westmoreland asked for more engineer battalions. Before a deployment

[26] Ploger, *Army Engineers*, pp. 6, 8. For a list of active-duty engineer units in the United States and those deployed to Vietnam, see ibid., pp. 12–16.

conference convened at Pacific Command in early August, the MACV commander stated that he needed ten engineer battalions, five combat and five construction, excluding the organic battalions assigned to the 1st Infantry and 1st Cavalry Divisions. This number increased to twelve during the conference to meet an emergency requirement for more facilities. The Joint Chiefs of Staff approved this number, but before the end of the month Westmoreland added three more construction battalions to his request. This brought the total to fifteen, all needed in South Vietnam before the end of the year. Westmoreland pointed out that engineer units in country and scheduled for deployment still could not complete the required facilities before 31 December.[27]

The Army did what it could to meet Westmoreland's requirements. On 21 August, the Department of the Army informed MACV that five engineer combat and five construction battalions should be in Vietnam by December. By then, the Army had no choice but to deploy combat battalions as substitutes for the construction battalions, since all but two of the active-duty construction battalions in the United States had been deployed or alerted for duty in Vietnam. The only other active Army construction battalions were committed in Europe, Korea, Thailand, and other parts of the world. Approved mobilization plans included another eight construction and three combat battalions, but these units still needed to be manned, equipped, and trained. Accordingly, the Army nominated two of the nine remaining deployable engineer combat battalions in the United States for duty in Vietnam, and two more combat battalions became candidates to meet MACV's latest request. Deploying one of the two remaining construction battalions would round out the total of fifteen, which would not be reached until September 1966. In addition, the Army promised a light equipment company and a construction support company for two of the combat battalions. These two companies gave the combat battalions some comparable capabilities of two construction battalions. This, however, left the Army short of these types of companies in the United States.[28]

For the last year, the 70th Engineer Battalion (Combat) at Fort Campbell, Kentucky, had been readying itself for deployment. Because of its prolonged alert status, the 70th managed to avoid many of the manpower problems that plagued the construction battalions ordered to Vietnam. Lt. Col. Leonard Edelstein, the battalion commander, had almost all his key men on hand when the word came to deploy this first nondivisional engineer combat battalion, some 580 troops, arrived in mid-June. New standard road graders and the multifuel series of trucks had also reached the battalion.[29]

[27] Fact Sheet, Deputy Chief of Staff for Military Operations to Army Chief of Staff, 23 Aug 65, sub: COMUSMACV Requirements for Army Engineer Battalions, Historians files, CMH; Army Buildup Progress Rpt, 25 Aug 65, p. 35, CMH.

[28] Army Buildup Progress Rpts, 25 Aug 65, p. 35, and 1 Sep 65, pp. 36–37, CMH; Msg, DA 729240 to COMUSMACV, 21 Aug 65, sub: Army Engineer Battalions for RVN; Status of Deployment Spreadsheets, Incl to Memo, Dep CofS for Mil Opns for Army CofS, 4 Aug 65, both in Historians files, CMH.

[29] Ploger, *Army Engineers*, pp. 32–33; Galloway, "Essayons," pp. 29–30, 32; Quarterly Cmd Rpt, 1 Jul–30 Sep 65, 70th Engr Bn, 12 Oct 65, pp. 1–3, Historians files, CMH.

The battalion's destination was Qui Nhon. Equipment shipped out of the port of Mobile, Alabama, in mid-July arrived in Qui Nhon well before the main body, making the 70th one of the few engineer battalions of any type to find its equipment waiting on the shore. Troops followed a few weeks later. The advance party reached Qui Nhon on 16 August, and with the help of the 84th Engineer Battalion it moved the equipment to an assembly area. When the main body arrived on the nineteenth, operators moved vehicles and equipment to a base camp outside the city. Two days later, Companies A and C, escorted by security forces of the 1st Brigade, 101st Airborne Division, moved up Highway 19 to An Khe. There, the two companies helped launch the construction of a huge base camp for the 1st Cavalry Division, then en route and expected to arrive by mid-September. The rest of the battalion moved to An Khe a few weeks later.[30]

When the USNS *Mann* carrying the 70th Engineer Battalion reached Qui Nhon, it also disembarked the headquarters of the 937th Engineer Group (Combat) commanded by Col. Roland A. Brandt. The 937th Group assumed responsibility from the 35th Group for all projects and nondivisional engineer units in the northern II Corps area. From August until December, the group concentrated its efforts on Qui Nhon and An Khe with a smaller force at Pleiku. In Qui Nhon, the major emphasis was the development of the port, depot, and fuel storage facilities and the construction of a 400-bed hospital. At An Khe, the group supervised the 70th Battalion's work on the 20,000-man division base camp. By October, work had also started there on a 140-bed hospital and a second access road to Highway 19. In December, the engineers launched a major self-help effort to build a cantonment of the tropical-type wooden buildings to house the division and its support units. By the end of the year, group units were also building facilities for U.S. units at Pleiku. The 937th Group also supported the development of the Republic of Korea (ROK) Capital Division's base camp in an area dubbed ROK Valley approximately twelve and a half miles west of Qui Nhon.[31]

September marked the arrival of the second nondivisional engineer combat battalion, the 19th from Fort Meade, Maryland. The battalion, commanded by Lt. Col. Amos C. Mathews, disembarked at Qui Nhon on 2 September and moved into the 70th Engineer Battalion's former base camp outside the city. Priority to offload the 1st Cavalry Division, which arrived at the same time, caused delays getting the battalion's equipment ashore. By the end of the month, the battalion began to work on several construction projects. These included the expansion and improvement of the ammunition depot and taking over the construction of a 50,000-barrel petroleum storage area and refueling facility from the 70th Engineer Battalion. The 19th also helped South Korean

---

[30] Ploger, *Army Engineers*, pp. 33, 75; Galloway, "Essayons," pp. 29–30, 32; Quarterly Cmd Rpt, 1 Jul–30 Sep 65, 70th Engr Bn, pp. 1–3.

[31] Quarterly Cmd Rpts, 1 Jul–30 Sep 65, 937th Engr Gp, 15 Oct 65, pp. 2–5, and 1 Oct–31 Dec 65, 937th Engr Gp, 15 Jan 66, pp. 2–5, both in Historians files, CMH; Ploger, *Army Engineers*, pp. 75–78; Galloway, "Essayons," pp. 25, 27, 29, 32; *The Logistics Review, U.S. Army, Vietnam, 1965–1969*, vol. 7, *Engineering Services System*, (U.S. Army, Vietnam, n.d.), pp. T-46, T-48 (hereafter cited as Logistics Review, USARV, vol. 7, Engineering Services System).

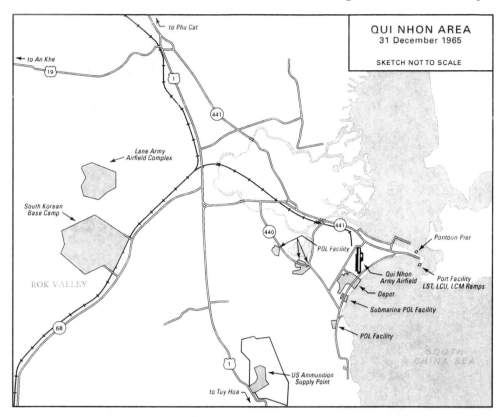

MAP 2

engineers build access and perimeter roads for the Capital Division's base camp. Before the end of the year, the battalion completed a macadam heliport for fifty UH–1 helicopters, a maintenance area, and a cantonment for two U.S. Army aviation companies at Lane Army Airfield near the Korean camp. Like the 70th Engineer Battalion at An Khe, the 19th found itself doing work typically done by a construction battalion. To help carry out its work, the 19th was supported by the 509th Engineer Company (Panel Bridge) from Fort Riley, Kansas, which arrived about the same time. When the bridge company's vehicles reached Qui Nhon the following month, the unit was used as a dump truck company.[32] (*Map 2*)

The fifth Army construction battalion ordered to Vietnam, the 46th from Fort Leonard Wood, had its destination changed from Qui Nhon to the Saigon area. Its primary mission would be to turn Long Binh Plantation, a formerly cleared area now overgrown with scrub jungle fifteen miles northeast of Saigon on Highway 1A, into a major logistics base. Upon disembarking at Vung Tau on 27 September, the battalion, commanded by Lt. Col. George G. Hagedon,

[32] Galloway, "Essayons," p. 36; Quarterly Cmd Rpts, 1 Jul–30 Sep 65, 19th Engr Bn, 15 Oct 65, p. 2, and 1 Oct–31 Dec 65, 19th Engr Bn, 31 Dec 65, pp. 2–3, both in Historians files, CMH.

took control of Company D, 84th Engineer Battalion, and the work in the port. The 46th's Company D proceeded to Qui Nhon to replace the 84th's Company D. Within a few weeks, the companies were exchanged—Company D of the 46th became Company D of the 84th and Company D of the 84th became Company D of the 46th. This exchange would recur in Vietnam to avoid moving two like units over long distances to their parent units. By 4 October, the 46th completed its move to Long Binh, built its base camp, and began its first major construction project, a 400-bed facility for the 93d Evacuation Hospital. Next, the battalion built a tactical operations center for the arriving 1st Division at Di An several miles to the west, several tropical buildings for the MACV flight detachment at Tan Son Nhut, and a large ammunition storage area at Long Binh covering some eight and a half square miles. To keep up the high pitch of construction, the battalion worked two 10-hour shifts, seven days a week with the men getting off an average of two and a half days per month. However, the departure of many experienced equipment operators and other troops due to separation from service and retirement made it difficult to maintain two shifts. Only a few replacements arrived, and by the end of the year the battalion was 164 men below its authorized strength of 895. Hiring 325 Vietnamese workers helped somewhat.[33]

Supply Deficits

From the outset, shortages of materials slowed the construction effort. The underdeveloped economy of Vietnam forced logisticians to import most construction items. Initially, lumber, airfield matting and membrane, DeLong piers, generators, and prefabricated buildings were all scarce. These items, except lumber, required long lead times. Although sand, rock, and gravel could be obtained in Vietnam, the initial paucity of quarries in secure areas and the scarcity of crushing and screening equipment had forced the engineers to consider even importing crushed rock. As early as December 1964, U.S. Army, Pacific, proposed pre-positioning lumber, barbed wire, airfield matting, and other construction stocks within a few days' sailing time of Vietnam. In Washington, the chief of engineers prepared several estimates of construction materials needed to support the buildup in Southeast Asia. The engineers recommended $70 million in obligating authority to buy and place additional materials in Vietnam. The Joint Chiefs of Staff passed the proposal on to Pacific Command, but the command requested only $10 million. This measure, authorized in April 1965, proved too little and too late when the 35th Engineer Group arrived at Cam Ranh Bay two months later. Although the 1st Logistical Command had submitted requisitions, the depots in Okinawa did not receive significant deliveries of material until December. Only small quantities of plumbing and electrical components arrived. Meanwhile, the 35th Group pressed the logistical command for urgently needed stores of timber and cement. Instead of waiting for supplies from the United States,

---

[33] Ploger, *Army Engineers*, pp. 79–81; Galloway, "Essayons," pp. 48–49, 51; Quarterly Cmd Rpt, 1 Oct–31 Dec 65, 46th Engr Bn, 13 Jan 66, pp. 1–4, Historians files, CMH.

the command placed orders with the procurement office at U.S. Army, Japan, for offshore acquisition in the Far East. When the requirement for electrical generators, ranging in size from 100 to 1,500 kilowatts, came up, these, too, were purchased in Japan.[34]

All units complained about not getting the right amount of construction materials. According to Colonel McCulloch, shortages of lumber, nails, plumbing and electrical fixtures, and reinforcing bar put his 87th Engineer Battalion at Cam Ranh Bay well behind schedule. In Qui Nhon, Colonel Mathews' 19th Engineer Battalion started work on the fuel storage facility on 7 September only to be hamstrung by shortages of culvert, steel tanks, piping, and fittings. The steel tanks did not arrive until early December. Shortages of culvert also held up roadwork and heliport construction at the South Korean camp outside Qui Nhon. Rain-damaged bags of cement were issued by the 82d Engineer Supply Point Company to the 19th with the admonition "take it or leave it." The 62d Engineer Battalion reported a loss of 294 platoon days at Phan Rang because of shortages of cement, culvert, and matting. At Long Binh, Colonel Hagedon complained of shortages of cement, nails, hinges, plywood, lumber, screening, and culvert. Projects, especially the 400-bed hospital, were held up because of the lack of cement.[35]

Under these conditions, engineers made use of whatever they could get their hands on. Maximum use was made of locally available materials. Expedients were common solutions. For example, the 46th Engineer Battalion substituted napalm cylinders for culvert, and the 19th Engineer Battalion welded fifty-five-gallon drums and also used napalm cylinders. The 84th Engineer Battalion resorted to applying boat oil to control dust. Some materials such as airfield matting came from different manufacturers and did not match. At Phan Rang, the 62d Engineer Battalion found that M8 airfield matting used in the parking areas had slight dimensional variations which, multiplied over several rows, caused improper seating of the bayonet lugs. This problem was solved by welding and banding the bayonet lugs under the locking slot.[36]

From both an equipment and construction material outlook, rock-producing quarries had become a critical concern. When the engineers arrived, they found only a few old French rock quarries in operation. It did not take long to realize that rock was a scarce and critical resource in a country where much of the terrain was sandy and marshy. Since construction of the base

[34] Interv, Schiller with Duke, 3 Jan 66, p. 29; JLRB, Monograph 6, *Construction*, pp. 161, 163, 165–66; Special Operational Report–Lessons Learned (ORLL), 1 Jan–31 Oct 65, DA, 23 Aug 66, incl. 3, pp. 3–4, Historians files, CMH; Msg, DA 709720 to CINCUSARPAC [Commander in Chief, United States Army, Pacific], 2 Apr 65, sub: Stockpiling Construction Materiel and Equipment, Historians files, CMH; Quarterly Cmd Rpt, 1 Apr–30 Jun 65, 35th Engr Gp, p. 2.

[35] Quarterly Cmd Rpts, 1 Jul–30 Sep 65, 87th Engr Bn, p. 6, 1 Jul–30 Sep 65, 84th Engr Bn, p. 2, 1 Oct–1 Dec 65, 19th Engr Bn, pp. 2, 4–5, 1 Oct–31 Dec 65, 62d Engr Bn, p. 8, and 1 Oct–31 Dec 65, 46th Engr Bn, p. 3.

[36] Quarterly Cmd Rpts, 1 Oct–31 Dec 65, 19th Engr Bn, p. 3, 1 Oct–31 Dec 65, 62d Engr Bn, p. 5, 1 Jul–30 Sep 65, 84th Engr Bn, pp. 1–2, and 1 Oct–31 Dec 65, 46th Engr Bn, 13 Jan 66, p. 3; Section III, Lessons Learned, 19th Engr Bn, 13 Mar 66, p. 3, Incl to Command Rpt for Quarterly Period Beginning 1 October 1965, 31 Dec 65, Historians files, CMH.

complexes and roads depended on adequate supplies of rock for base course foundations and concrete, one of the first things the engineers did was to find a source of rock and set up quarries and crusher sites. This the 864th Engineer Construction Battalion did at Cam Ranh Bay by reopening an abandoned quarry used earlier by RMK. Outside Qui Nhon, the 84th Engineer Construction Battalion found a quarry site but had to build three miles of access roads to get to it. To speed up production, the 84th built a makeshift plant to feed the crusher. Using dunnage from cargo vessels in the harbor, the resourceful quarrymen built a lumber retaining wall. Then they covered it with a salvaged 2½-ton dump truck bed with attached channel beams and pierced steel planks to provide a funnel onto the feeder of the crusher. This improvised plant became the highest volume producer of crushed granite in South Vietnam until contractor plants reached heavy production levels in late 1966.[37]

Meanwhile, the Army made an effort to get more rock crushers to Vietnam. On 19 October, fifteen sets (thirteen 75-ton-per-hour and two 225-ton-per-hour) of quarrying and crushing equipment were authorized for shipment. Both 225-ton-per-hour sets were undergoing rebuilding at the Granite City, Illinois, depot. The first set was expected to reach Vietnam by the end of January 1966 and the second by the end of March. Five of the 75-ton-per-hour sets were ready for shipment except for two 100-kilowatt generators per crusher. These generators were being purchased.[38]

The Navy's construction material stock position before the buildup fared somewhat better. Of the established pre-positioned war reserve requirements for forward bases, the Navy had 51 percent already available in stockage, known as the Advanced Base Functional Components System. Developed by the Navy in World War II to meet the needs of the island-hopping campaign in the Pacific, this functional components system provided deploying naval engineer units with material or prefabricated packages of components to build complete overseas facilities. In Vietnam, the 30th Naval Construction Regiment at Da Nang oversaw supplies and construction materials for four Seabee battalions and two Marine Corps engineer battalions.[39]

Like the Navy, the Army had a program known as the Engineer Functional Components System. Developed during the Korean War, the Army's system consisted only of data published in three technical manuals (5–301, 5–302, and 5–303), which provided staff guidance, plans, construction details, and bills of materiel for the specific buildings and facilities. The Army intended to use the system as a basic planning guide based on building blocks for constructing bases in a theater of operations. The data in the manuals could be used to order and construct individual buildings or facilities or even an entire installation in a single requisition.[40]

[37] Ploger, *Army Engineers*, pp. 44, 48, 123–24.
[38] Army Buildup Progress Rpt, 3 Nov 65, p. 56, CMH.
[39] For a full description of the Navy's functional components system, see JLRB, Monograph 6, *Construction*, pp. 26, 33, 163–64, 172, 183, and B-3 to B-6.
[40] Ibid., p. 33; Dunn, *Base Development*, pp. 47, 115–16.

The Army found the Engineer Functional Components System unsatisfactory in Vietnam and would end its use in late 1966. Although the engineer system saved a certain amount of preparation in compiling long, complicated bills of materials and thousands of separate requisitions, it did not include many required facilities. The design criteria also did not turn out to be compatible with requirements. All the structures in the system were wooden and designed for a temperate climate as opposed to tropical Vietnam. Modifications were allowed when requesting materials for individual structures. Corrugated steel roofing replaced the tar paper called for in the specifications. The locally adapted tropical-type buildings also needed more screening. However, if a single requisition for an entire installation went forward, an entire bill of materials including insulation, roofing felt, and tar paper arrived, for depots in the United States did not stock the materials as sets. Instead, some depots only carried certain components. Pallets of lumber, crates of nails, and electrical and plumbing fixtures arrived in bulk, making identification with a specific project a major problem. Material lists were not updated to meet current construction practices. Old designs still included World War II–vintage electrical wiring known as knob and tube wiring and incandescent light fixtures instead of fluorescent. Design standards also turned out to be lower than those adopted for Vietnam. This resulted in excesses of component items and shortages of requested modified items.[41]

The Navy appeared more satisfied with its system. During the buildup in Vietnam, the Seabees and contractor used many elements of the Advanced Base Functional Components System. Despite some degree of obsolescence, the components system provided the Navy the building blocks for the planning and construction of bases ashore. An example of construction with the Advanced Base Functional Components System was the Da Nang Hospital built almost entirely of Quonset huts.[42]

The Air Force looked to the Army and Navy to provide the bulk of materials and construction forces to build airfields in Southeast Asia. The Navy administered the procurement of materials for the contractor before the buildup under the Military Assistance Program and Air Force construction funds later. Since contingency plans did not call for additional major air bases in Vietnam, Army engineer units faced difficulties in getting sufficient quantities of airfield matting. Air Force civil engineers and construction squadrons deploying to Vietnam also required materials for base maintenance and construction. To meet the supply needs of the first two Red Horse squadrons (Rapid Engineering Deployment and Heavy Operational Repair Squadron,

[41] Dunn, *Base Development*, pp. 47, 116–18; JLRB, Monograph 6, *Construction*, pp. 166–67; Logistics Review, USARV, vol. 7, Engineering Services System, p. T-49. For an analysis of the Engineer Functional Components System, see Engr Studies Ctr Rpt no. 159, Analysis of the EFCS [Engineer Functional Components System], Jun 67, Historians files, CMH.

[42] Edwin B. Hooper, *Mobility, Support, Endurance: A Story of Naval Operational Logistics in the Vietnam War, 1965–1968* (Washington, D.C.: Naval Historical Division, Department of the Navy, 1972), pp. 37, 75, 170; Merdinger, "Civil Engineers, Seabees, and Bases in Vietnam," p. 244.

Engineering), due to arrive in early 1966, the Air Force Logistics Command assembled supply packages of lumber, cement, pipe, and other materials, which later deploying squadrons took with them. Although base supply offices requisitioned additional materials, they could not keep up with the demand.[43]

Although the Air Force depended on the Army and Navy to do major construction, the air service also developed construction support packages for its bases. The priority of effort went to operational facilities—runways, ramps, taxiways, and facilities to support weapons systems, followed by maintenance facilities. "Grey Eagle" supply kits airlifted to Vietnam contained minimum support equipment such as tents, electric generators, portable runway lights, and vehicles. In June 1965, the Air Force also contracted for inflatable shelters that could be erected quickly and easily to provide enough covered space for maintenance of fighter aircraft, critical components, ground equipment, ammunition, and war readiness materiel. When more permanent operational facilities were completed, base civil engineers turned their attention to building more permanent administrative and living facilities. They replaced tents with barracks and the portable generators with permanent generating plants.[44]

The key to much of the early construction in Vietnam was keeping engineer equipment in operation. It did not take long for breakdowns in equipment to occur under heavy use. Unlike other units in country, engineers typically operated on a two-shift schedule, approximately twenty hours a day. While a tank ran three or four hours a day, a bulldozer ran much longer, with little time out for maintenance. Heavy use—especially of bulldozers, scoop loaders, and five-ton dump trucks—led to breakdowns, with equipment remaining inoperative for long periods because of poor resupply of repair parts. By November 1965, the wear and tear, humidity, sand, and the lack of spare parts were affecting engineer operations. Of 1,218 pieces of Army construction equipment in country, 190 had been deadlined (down for repairs) for more than seven days.[45]

Units that had only fifteen days of repair parts in stock and no provision for automatic resupply soon faced critical shortages. The 510th Engineer Direct Support Maintenance Company arrived at Cam Ranh Bay in June 1965 without its stockage of repair parts. The unit remained ineffective until stocks began to arrive some ninety days later. The 70th Engineer Battalion, which arrived at Qui Nhon with only fifteen days' stockage plus a few extra parts, saw its readiness drop to a little more than 50 percent in forty-six days because of the lack of replacement parts and maintenance support units. Five of seven TD20 bulldozers were down as well as fifteen of the battalion's thirty-nine

[43] JLRB, Monograph 6, *Construction*, pp. 163–64, 172; Contemporary Historical Evaluation of Combat Operations (CHECO) Rpt, Pacific Air Forces (PACAF), Project RED HORSE, pp. 9–10, copy in Historians files, CMH.

[44] John Schlight, *The War in South Vietnam: The Years of the Offensive, 1965–1968*, United States Air Force in Southeast Asia (Washington, D.C.: Office of Air Force History, United States Air Force, 1988), p. 169; *Department of Defense Annual Report for Fiscal Year 1965* (Washington, D.C.: Government Printing Office, 1967), p. 336.

[45] Dunn, *Base Development*, pp. 120–21; Ploger, *Army Engineers*, pp. 50, 201–02; Quarterly Cmd Rpts, 1 Jul–30 Sep 65, 18th Engr Bde, 15 Oct 65, pp. 9–10, Historians files, CMH, and 1 Jul–30 Sep 65, 70th Engr Bn, p. 3.

dump trucks. When the 19th Engineer Battalion was informed that no spare parts were available in Vietnam for seven new HD16 bulldozers scheduled for issue to the unit before its departure from the United States, the 19th elected to take its old TD18 bulldozers. But three of the older bulldozers were inoperative when offloaded in Qui Nhon, and spare parts were unavailable. Due to their age, the TD18s became a perpetual maintenance problem. It did not help that the 578th Engineer Direct Support Company only reached Qui Nhon in late November, five months after the first two construction battalions. Worse, U.S. Army, Vietnam, did not have a heavy maintenance capability in the command or depot stocks to borrow equipment for use as "floats." The command did not expect the first two engineer heavy-equipment maintenance companies to reach Vietnam until the end of 1966.[46]

To compensate, Army Materiel Command began assembling special push packages of spares for airlift to Vietnam. A special sixty-day shipment was flown to Qui Nhon and Cam Ranh Bay in July when construction equipment began breaking down. Washington took a similar action to get electrical generators to Vietnam. Although a thirty-day shipment of engineer repair parts reached Saigon on 19 October, it proved to be only partially effective in removing equipment from deadline. Assemblies and components, such as torque converters, engine clutches, radiators, and axles, were in greater demand than repair parts.[47]

The rate of equipment down for repairs continued to worsen as 1965 drew to a close. In early December, the Army reported 36 percent or 49 of the 136 bulldozers in Vietnam were deadlined. The shortage of repair parts became so serious that in December, following one of his visits to Vietnam, Secretary McNamara authorized an emergency airlift system known as the Red Ball Express. The engineers took advantage of this rapid delivery system, and complete bulldozer tracks and other heavy items seemingly inappropriate for airlifting across the Pacific were soon being delivered. This around-the-clock, seven-day-a-week operation helped to get at least some equipment back into operation. Sometimes parts arrived within ten days after requisitioning. Still, by late February 1966, the engineers' deadline rate remained high—23.5 percent or 445 out of 1,852 pieces of equipment were inoperative.[48]

---

[46] Quarterly Cmd Rpts, 1 Jul–30 Sep 65, 18th Engr Bde, pp. 9–10, 1 Oct–31 Dec 65, 18th Engr Bde, 18 Jan 66, p. 9, 1 Jul 30 Sep 65, Historians files, CMH, 1 Jul–30 Sep 65, 19th Engr Bn, pp. 4–5, 1 Jul–30 Sep 65, 70th Engr Bn, p. 3; Special ORLL, 1 Jan–31 Oct 65, DA, incl. 3, p. 17; Situation Report of Engineer Equipment in South Vietnam, incl. 17, and Ltr, Commanding General (CG), 18th Engr Bde to CG, 1st Log Cmd, sub: Engineer Maintenance Support, 13 Jan 66, incl. 17, tab B, Office of the Chief of Engineers (OCE) Liaison Officer Trip Rpt, no. 1, 15 Mar 66, OCE Historical Ofc, Fort Belvoir, Va.; Msg, CG, USARV AVD-MD 50482 to CINCUSARPAC, 6 Feb 66, sub: High Mortality of Engineer Construction Equipment and MHE [materials handling equipment], box 11, 69A/702, RG 319, NARA.

[47] *Arsenal for the Brave*, pp. 209–11; Army Buildup Progress Rpt, 2 Mar 66, p. 56, CMH; Trip Rpt, Special Assistant to the Chief of Staff for Supply and Maintenance, 17 Dec 65, p. 2–1, Historians files, CMH.

[48] *Arsenal for the Brave*, pp. 253–54; Heiser, *Logistic Support*, pp. 50, 176; Ploger, *Army Engineers*, p. 69; Dunn, *Base Development*, p. 122; Special ORLL, 1 Jan–31 Oct 65, DA, incl. 3, p. 17; Situation Rpt on Engr Equipment in South Vietnam, incl. 17, tab A, OCE Liaison Officer

Among the most recalcitrant pieces of equipment were bulldozers. In January 1966, Army engineers in Vietnam had twelve makes and models, and less than 20 percent of the bulldozers had been removed from deadline status through Red Ball–furnished repair parts. The problem was this: at the beginning of the buildup, the Army had found it difficult to come up with enough engineer equipment to meet the requirement of its units. Consequently, bulldozers, scrapers, and cranes of several makes and varieties, including hard-to-support Korean War–vintage items, were drawn from depots and active and reserve units throughout the United States and shipped to Vietnam. The diversity of repair parts required by this equipment created a logistical nightmare. At the end of January, the deadline rate for bulldozers in Vietnam had risen to 47 percent. While there were plans to reduce the number of makes and models to three, with preference to the Caterpillar D7, the D7s would not be shipped before the summer. As an interim measure, the Army Materiel Command sent over 118 Allis Chalmers HD16Ms. Recognizing that another make and model complicated the repair parts problem, officials recommended that the Allis Chalmers be used only in the Cam Ranh Bay area. Requirements in Vietnam were such, however, that the HD16Ms had to be split between Cam Ranh Bay and Saigon. It was summer before the bulldozer deadline rate dropped below 25 percent.[49]

A New Engineer Brigade

Between June and September 1965, all nondivisional Army engineer units came under the command of Colonel Duke's 1st Logistical Command. Planning for Army construction centered in the command's small Engineer Office, which in May was transferred from the support command following the activation of the logistical command. During August, the Engineer Office, staffed with seven assigned officers augmented by temporary duty soldiers of the 539th Engineer Detachment (Control and Advisory) on Okinawa, concentrated on preparing the Army's construction program in Vietnam, acquiring real estate, coordinating construction work done for the Army by civilian firms, overseeing facilities engineering, and requisitioning construction materials. Lt. Col. Floyd L. Lien, who became command engineer in August, found himself in an unusual position of directing the work of the 35th and 937th Engineer Groups, both commanded by colonels.[50]

Colonel Hart also found himself in an unusual position soon after his 35th Engineer Group set up shop at Cam Ranh Bay. Colonel Duke charged him with the added responsibility of establishing the Cam Ranh Bay Logistical

Trip Rpt no. 1, 15 Mar 66; Msg, CG, USARV AVD-MD 50482 to CINCUSARPAC, 6 Feb 66, sub: High Mortality of Engineer Construction Equipment and MHE.

[49] *Arsenal for the Brave*, pp. 248–52; JLRB, Monograph 6, *Construction*, p. 181; MACV History, 1966, p. 282.

[50] Ploger, *Army Engineers*, pp. 5, 65; Galloway, "Essayons," p. 19; History, 539th Engr Det, 1965, p. 14; JLRB, Monograph 6, *Construction*, p. 23; Interv, Schiller with Duke, 3 Jan 66, p. 24; Interv, Maj John F. Schiller, 15th Mil Hist Det, with Lt Col Floyd L. Lien, Engineer, 1st Log Cmd, 7 Aug 66, VNIT 29, pp. 1–9, CMH.

Area and commanding all the logistical troops in the vicinity. Colonel Hart and his staff took on other functions ranging from mail, chaplain, and medical services; overseeing a military police detachment; and establishing and operating depots for all supplies. This arrangement lasted for the better part of three months until the 504th Quartermaster Depot arrived and took over logistical support for the area on 7 August.[51]

The decisions to send more American troops in the summer of 1965 included the newly activated 18th Engineer Brigade. A brigade headquarters seemed appropriate because of the growing number of engineer units. During the late summer of 1965, the 18th Engineer Brigade prepared for its move to Vietnam. Under its acting commander, Col. C. Craig Cannon, the brigade's Headquarters Company assembled at Fort Bragg. Most of the 34 officers and 110 enlisted men came from the 159th Engineer Group (Construction) alerted earlier for deployment. A small advance party led by Colonel Cannon reached Saigon on 3 September, and the brigade's main body followed on the twenty-first. The headquarters settled in several former U.S. Operations Mission buildings next to Tan Son Nhut Air Base. Nondivisional units placed under 18th Brigade control consisted of the 35th and 937th Groups, four construction (62d, 84th, 87th, and 864th) and two combat (19th and 70th) battalions, and nine separate companies—altogether over 6,200 men. Before the end of the month, the brigade assumed operational planning and supervision for forty-four construction projects at nine separate locations in II, III, and IV Corps.[52]

In the meantime, the Army selected an engineer general officer to command the 18th Brigade. General Harold K. Johnson, the Army chief of staff, advised Westmoreland that "in view of the monumental tasks confronting this brigade and to provide the requisite skill and experience I feel that a brigadier general should command it." Westmoreland readily agreed and placed the brigade directly under U.S. Army, Vietnam (USARV), the successor headquarters of the U.S. Army Support Command. Although Westmoreland also assumed command of USARV, General Norton, who became deputy commanding general, ran the command on a day-to-day basis. The MACV commander decided not to place the brigade under the 1st Logistical Command because he believed the latter already had enough problems associated with the logistical buildup. In addition, a large share of the brigade's effort would be supporting the Air Force, and Westmoreland wanted a general officer to coordinate and negotiate priorities at a senior level. Late in the evening of 12 August, Lt. Gen. William F. Cassidy, the chief of engineers, called the recently promoted

[51] Ploger, *Army Engineers*, p. 41; 1st Lt Wallace R. Wade, Spec Charles Miele, and Spec Edward C. Swab, A History of Cam Ranh Bay Through June 1966, Cam Ranh Bay Sub Area Cmd, Nov 66, p. 86, box 18, 70A/782, RG 334, NARA.

[52] Msg, DA 723668 to COMUSMACV, et al., 14 Jul 65, sub: Deployments to RVN, Historians files, CMH; Ploger, *Army Engineers*, pp. 63–69; Quarterly Cmd Rpt, 1 Jul–30 Sep 65, 18th Engr Bde, 15 Oct 65, pp. 1–6, Historians files, CMH; Interv, Lt Col Lewis C. Sowell Jr. with Maj Gen Robert R. Ploger, 21 Nov 78, sec. 8, pp. 2–3, Senior Officer Oral History Program, U.S. Army Military History Institute (MHI), Carlisle Barracks, Pa.; Ltr, Ploger to Office, Chief of Engineers, n.d., Historians files, CMH.

General Ploger
(shown as major general)

Brig. Gen. Robert R. Ploger and informed him that he would command the 18th Brigade.[53] The surprised Ploger, who had just become the New England Division Engineer, was given short notice to rendezvous with the advance party. He arrived in Saigon on 5 September.[54]

Not all nondivisional engineers were assigned to the 18th Brigade. When General Norton asked Ploger to prepare a letter of instruction to assign engineer units and missions, the question of engineer maintenance and supply units came up. Ploger did not argue against assigning these units to the 1st Logistical Command. He knew that the Army was undergoing a reorganization to transform technical service maintenance and supply units into functional organizations. Besides, he felt he already had enough to handle, and preferred to attack supply and maintenance problems from a position of control but not operational responsibility. The 1st Logistical Command Engineer retained facilities engineering responsibilities, mainly overseeing Pacific Architects and Engineers and utilities detachments, and served as the clearinghouse for the command's construction requirements.[55]

When General Ploger reported for duty, General Norton agreed that the 18th Brigade commander should hold the dual responsibility of 18th Engineer Brigade commander and U.S. Army, Vietnam, engineer. Such an arrangement eliminated the need for duplicate staffs at brigade and the USARV headquarters and conserved scarce manpower resources. Holding both offices gave Ploger procurement and management authorities and put him in an ideal

[53] General Ploger received his commission in the Corps of Engineers upon graduating from the U.S. Military Academy in 1939. During World War II, he served as commander of the 129th Engineer Battalion, 29th Infantry Division, which participated in the D-Day assault on Omaha Beach and in later operations in France and Germany. Subsequent assignments included staff positions in Washington, D.C.; liaison work with the Atomic Energy Commission; engineer district work on Okinawa; and command of an engineer group at Fort Lewis, Washington. Interv, Sowell with Ploger, 16 Nov 78, sec. 7, p. 32.

[54] Ploger, *Army Engineers*, p. 63; Msg, Gen Johnson WDC 6543 to Westmoreland, 2 Aug 65 (quotation); Msg, Lt Gen John L. Throckmorton, Dep COMUSMACV MAC 3994 to Westmoreland, 5 Aug 65; Msg, Westmoreland HWA 2069 to Gen Johnson, 6 Aug 65, all in Westmoreland Message files, CMH.

[55] Interv, Maj Roy Bower, 26th Mil Hist Det, with Maj Gen Robert R. Ploger, CG, Engr Cmd, Vietnam, 8 Aug 67, VNIT 89, pp. 5–6, CMH; Quarterly Cmd Rpt, 1 Oct–31 Dec 65, 1st Log Cmd, 19 Feb 66, p. 14, Historians files, CMH.

position to oversee the allocation of equipment and material resources. He transferred several officers from the brigade to the small USARV Engineer Office, which up until then consisted of Lt. Col. Andrew Gaydos and one or two enlisted men. The brigade staff absorbed the bulk of engineer staff matters. Ploger assigned Col. William W. Watkin, who had just arrived in country, as his deputy at USARV headquarters. Colonel Cannon, the deputy brigade commander, ran the day-to-day operations at the brigade. This dual-hatted role idea would be adopted by several other senior staff officers. In time the surgeon, the provost marshal, and the aviation officer commanded brigade-size troop units.[56]

General Ploger's first major act was to draft a blueprint to carry out the rapidly expanding construction load and other support requirements. Following a suggestion made by General Norton, Ploger on 7 October issued a one-page policy statement entitled "Our Objectives and Standards." The document emphasized the brigade's purpose: to serve the combat forces and support the "man with the rifle." Ploger also admonished his commanders and staff to make sure that drainage, roads, and associated utilities were included in project design and construction. Brigade units would devote their primary efforts to the needs of others. Improvements to engineer camps were never to be refined beyond those of the units they supported.[57]

General Norton also suggested that Ploger present the engineer picture to the MACV commander. On 4 November, Ploger briefed General Westmoreland on the brigade's construction program. He reported that Army projects for fiscal years 1965 and 1966 totaled some 170 battalion-months, a battalion-month comprising the expected outcome of either one construction battalion or one combat battalion plus a light equipment company. These figures did not include work to be done for the Air Force. At this time, the brigade consisted of 7,900 officers and men, or the equivalent of 8.4 battalions. This magnitude of work when compared to the brigade's capability meant that almost two years would be required to complete its projects, provided the brigade did not commit troops to the support of tactical operations, a very unlikely prospect.[58]

Ploger highlighted several factors affecting engineer priorities. Primary support would go to the tactical forces followed by building the logistical facilities. The construction effort would meet the minimum needs of units, then gradually improve the living conditions in the base camps. Besides the backlog of work, the engineers faced other problems, including shortages of construction material, breakdowns of equipment, complex real estate arrangements, and peacetime construction procedures in a war zone. "The inescapable conclusion," Ploger remarked, "is that engineer operations in the foreseeable future will be a continuous allocation of shortages."[59]

---

[56] Ploger, *Army Engineers*, pp. 86–87; Interv, Sowell with Ploger, 21 Nov 78, sec. 8, pp. 6–7; Quarterly Cmd Rpt, 1 Jun–31 Aug 65, USARV, 15 Nov 65, p. 12, Historians files, CMH.

[57] Ploger, *Army Engineers*, pp. 65, 67 (quoted words, p. 65). A copy of this statement is on p. 219.

[58] Ibid., pp. 67, 201. A copy of this briefing is on pages 199–204.

[59] Ibid., pp. 201–04 (quotation, p. 204).

In concluding his briefing, Ploger emphasized priorities and standards of construction. Since a substantial portion of the work dealt with building troop cantonments, he pointed out that commanders should follow six pre-scribed levels of physical improvement. These ranged from Standard 1, no site preparation, to Standard 5, modified, which would include buildings and waterborne sewage systems. Standard 2 called for leveling sites for the camps and building roads while troops put up their own tents. Standard 3 added floors and wooden frame buildings for kitchens, administration, storehouses, infirmaries, and bath houses; electrical distribution to the buildings; and, piped water from a storage tank. Standard 4, which Ploger recommended as the current objective for cantonment construction, improved the tents with the addition of wood frames and electrical distribution throughout the camp. If units expected to remain at the same location for more than twelve months, buildings could be added under Standard 5.[60]

Westmoreland concurred with the gradual upgrading to Standard 4 and a priority list that Ploger submitted during the briefing. The fifty-five (later fifty-nine) priority categories ranged from immediate tactical and operational requirements to less urgent but still required items. Essential and basic tasks, such as clearing and grubbing of troop areas, field fortifications, and clear-ing of fields of fire around defense perimeters, topped the list. Water supply points, ramps for landing craft, airstrips, roads and hardstands at ports, hos-pitals, ammunition storage areas, and communications facilities came next. At the bottom of the list were officer billets and chapels. His rationale regard-ing the inclusion of chapels was to make sure they were not left out. Initially mess halls and service clubs could be used for religious assemblies. He also requested that the MACV commander admonish all commanders that they must lower their expectations at least for the time being. Instead, the troops should be encouraged to do as much cantonment construction as possible through self-help on a gradual scale of upgrading and not to rely entirely on engineers. Westmoreland agreed and advised his senior commanders of these points during meetings.[61]

Ploger presented the same briefing to other commanders in late 1965 and early 1966. He briefed Maj. Gen. Stanley R. Larsen, commander of Field Force, Vietnam (the tactical headquarters set up at Nha Trang by General Westmoreland in September 1965 to command and control U.S. Army combat units in II Corps), and Maj. Gen. Jonathan O. Seaman, commander of the 1st Infantry Division, informing them where things stood and trying to gain their agreement to keep construction to the essentials. Since he began to travel extensively to check on progress and lend encouragement to engineer soldiers, Ploger checked for deviations. Sometimes he found slight variations from the standard when his troops helped a commander to carry out a pet project. "I was aware of it," he recalled, "but there was no point in hassling at it." He tried to point out that use of materials for nice-to-have projects deprived others of

---

[60] Ibid., pp. 67, 204–11; Interv, Sowell with Ploger, 21 Nov 78, sec. 8, p. 21.

[61] Ploger, *Army Engineers*, pp. 69, 204, 229–30; Interv, Sowell with Ploger, 21 Nov 78, sec. 8, pp. 22–25; MACV History, 1965, p. 127.

more essential requirements. Commanders always tried to better the lot of their soldiers, and the misuse of materials remained a pervasive problem.[62]

More Units on the Way

More engineer units arrived in the last three months of 1965 or were on their way. In October 1965, a third Army construction group and another combat engineer battalion arrived. Two combat battalions arrived in November, and on New Year's Day 1966 two more combat battalions joined the brigade. The 299th Engineer Battalion (Combat) from Fort Gordon, Georgia, and the 630th Engineer Company (Light Equipment) from Fort Bliss, Texas, landed at Qui Nhon on 22 October. Neither unit immediately operated at full capacity; the 299th had to wait nearly a month for its equipment to arrive. Offloading took time, and by the end of the year the light equipment company still had some of its equipment aboard a ship anchored at Cam Ranh Bay. Working under the direction of the 937th Engineer Group, the 299th joined the 19th, 70th, and 84th Engineer Battalions in northern II Corps. The 299th Battalion's initial projects in the Qui Nhon area included taking over the ammunition depot and roadwork. A reinforced platoon was also dispatched to Pleiku to work on base development projects.[63]

On 30 October, the headquarters of the 159th Engineer Construction Group arrived from Fort Bragg, set up headquarters at Long Binh, and took control of 18th Brigade units and operations in III and IV Corps. The group, commanded by Col. James H. Hottenroth, had been brought up to strength over the summer after losing troops to the 18th Brigade. The group inherited the 46th Engineer Construction Battalion at Long Binh, and by the end of the year it added two engineer combat battalions, two separate companies, and several detachments. In early November, the 588th Engineer Battalion (Combat) from Fort Lee, Virginia, arrived at Vung Tau followed later by its equipment on other ships. The estimated shipping time of twenty-seven days stretched to forty-four days for one ship and fifty-seven days for the other two ships. Initially the battalion worked on base development projects for the 1st Division at Phu Loi and Di An and the 173d Airborne Brigade at Bien Hoa. Shortages included lumber in all sizes, culvert, sandbags, barbed wire, dynamite, and fuses. In late November, the 168th Engineer Combat Battalion from Fort Polk arrived. Like other engineer units its equipment arrived later, 7 December. Early efforts also included base development projects for the 1st Division.[64]

On New Year's Day 1966, the 20th Engineer Battalion (Combat) from Fort Devens, Massachusetts, and the 39th Engineer Battalion (Combat), and the 572d Engineer Company (Light Equipment) from Fort Campbell,

[62] Interv, Sowell with Ploger, 21 Nov 78, sec. 8, pp. 24–26 (quotation, p. 26).

[63] Ploger, *Army Engineers*, p. 79; Quarterly Cmd Rpt, 1 Oct–31 Dec 65, 937th Engr Gp, pp. 2–3, 5; Galloway, "Essayons," p. 37.

[64] Ploger, *Army Engineers*, pp. 79–80, 83; Galloway, "Essayons," pp. 48–49, 51–52; Quarterly Cmd Rpts, 1 Oct–31 Dec 65, 159th Engr Gp, 14 Jan 66, pp. 1–4, 1 Oct–31 Dec 65, 588th Engr Bn, 13 Jan 66, pp. 1–4, and 1 Oct–31 Dec 65, 168th Engr Bn, pp. 1–3, all in Historians files, CMH.

Kentucky, joined the 35th Group at Cam Ranh Bay. Shortly after its arrival, the 39th Engineer Battalion, the 572d Light Equipment Company, and a South Vietnamese engineer float bridge company built a badly needed 1,115-foot M4T6 bridge across the bay. It took two days to assemble the rafts and one day to install the bridge between the peninsula and the mainland. When General Larsen flew over the site on 8 January, he could only exclaim "Where did that damn thing come from?" Just the day before a ferry offered the only crossing. Later in the month, the 20th Engineer Battalion, helped by light equipment and dump truck companies, renewed work on an Army aviation base at Dong Ba Thin on the mainland side of the bay. The arrival of the two combat battalions also enabled the group to accelerate the construction of warehouses, open storage hardstands, and a convalescent hospital in the northern part of the peninsula.[65]

By the close of 1965, the equivalent of some twenty Army, Navy, and Marine Corps engineer battalions; many smaller engineer units, including Air Force civil engineering teams; and civilian contractors were building the operational, logistical, and support facilities required for the expanding U.S. and allied commitment. By then U.S. troop strength had reached 184,300. The 18th Engineer Brigade had grown to over 9,500 men consisting of three groups, ten battalions, and an assortment of separate companies. (*Chart 1*) Several thousand Army engineers were also assigned to the two divisions (1st Infantry and 1st Cavalry), two separate airborne brigades (173d and 1st Brigade, 101st Airborne Division), one infantry brigade (3d Brigade, 25th Infantry Division), and the logistical command. Some 2,100 Seabees were assigned to the 30th Naval Construction Regiment's four construction battalions. Marine Corps engineers totaled nearly two thousand men in two battalions and smaller units. As early as August 1965, the Air Force deployed small specialized construction teams (dubbed Prime Beef—a combination nickname and acronym for Base Engineer Emergency Force) to Southeast Asia on temporary duty tours to augment the base engineers. Prime Beef teams built cantonments, installed utilities, and quickly erected revetments to protect aircraft from enemy shelling. Meanwhile, the Air Force was readying battalion-size Red Horse civil engineer squadrons to undertake larger projects at air bases. In August, the Navy's construction consortium added two more construction firms, J. A. Jones Corporation and Brown and Root, to form Raymond, Morrison-Knudsen, Brown and Root, and J. A. Jones, better known as RMK-BRJ. Before the end of the year, the RMK-BRJ's workforce jumped from 2,000 to 22,000. Pacific Architects and Engineers' workforce rapidly multiplied from its 2,000-person ceiling in June 1965. Plans were under way to construct several deep-draft ports with satellite shallow-draft ports; to build airfields, depots, and other facilities within these port complexes; to improve the roads inland from the ports; to build inland bases; and finally to expand the roads to these bases.[66]

---

[65] Ploger, *Army Engineers*, pp. 56–59, 84 (quotation, p. 58); Galloway, "Essayons," pp. 79–81; ORLL, 1 Jan–30 Apr 66, 35th Engr Gp, 15 May 66, pp. 1–2, 5–6, Historians files, CMH.

[66] Ploger, *Army Engineers*, pp. 84; Heiser, *Logistic Support*, p. 14; Tregaskis, *Building the Bases*, p. 183; Quarterly Cmd Rpt, 1 Oct–31 Dec 65, 18th Engr Bde, 21 Jan 66, p. 1; MACV

CHART 1—18TH ENGINEER BRIGADE, DECEMBER 1965

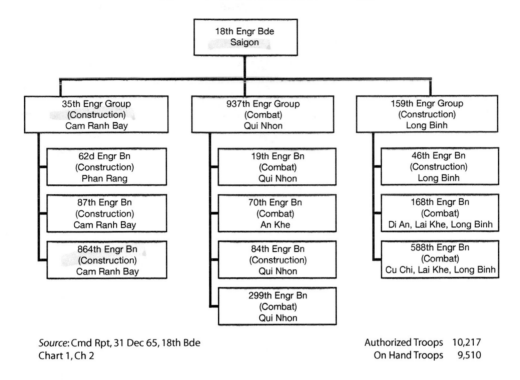

Source: Cmd Rpt, 31 Dec 65, 18th Bde
Chart 1, Ch 2

Authorized Troops   10,217
On Hand Troops   9,510

Concentration on base development would become much harder because of the increasing demand to support tactical operations. By the end of the year, the construction backlog reached 212 battalion-months. The pool of military engineer units, however, had just about reached bottom. With the arrival of the 20th and 39th Engineer Battalions in the first days of 1966, General Ploger could look forward to receiving only a few more units by summer. The Joint Chiefs of Staff and the Department of Defense had approved the deployment of up to twenty-five engineer battalions, but that number was simply not available without calling up the reserves. The Department of the Army in its annual report for 1965 noted that engineer resources in the United States "could not provide balanced support to current combat forces and that currently deployed engineer units fall considerably short of meeting balanced force requirements." Newly activated Army engineer construction battalions could not be expected for some time.[67]

History, 1965, p. 126; Shulimson and Johnson, *U.S. Marines in Vietnam, 1965*, app. F; PA&E History, 1963–1966, p. 30.

[67] MACV History, 1965, p. 128; JLRB, Monograph 6, *Construction*, p. 109; Army Buildup Progress Rpts for 6 Oct, 3 Nov, and 1 Dec 65, CMH; Special ORLL, 1 Jan–31 Oct 65, DA, incl. 3, p. 2 (quotation).

3

First Battles

By the summer of 1965, substantial numbers of U.S. Army and Marine ground forces, with supporting air and naval elements, had deployed to South Vietnam. Initially assigned to guard the air bases at Da Nang and Bien Hoa, American ground troops were soon committed by General Westmoreland to "the points of maximum peril on a 'fire brigade' basis." In the south, he built up U.S. forces—the 173d Airborne Brigade and the 1st Infantry Division—in a protective arc around Saigon. The evidence presented to the MACV commander also suggested that the North Vietnamese Army intended to cut South Vietnam in half along Highway 19 from Pleiku in the highlands to Qui Nhon on the central coast. To counter this threat, Westmoreland deployed the 1st Cavalry Division to the Central Highlands. While the cavalry division took on the mission to open—and to keep open—Highway 19, other reinforcements, including the 1st Brigade, 101st Airborne Division, stiffened South Vietnamese forces on the strategically critical highly populated coast.[1]

Organic engineer units—companies with separate brigades and battalions with divisions—accompanied the combat forces. Organized to help the movement of friendly troops and impede the movement of the enemy, combat engineers also found themselves assigned projects in the base camps. Still, by virtue of their mission the combat engineers bore the brunt of the support for the maneuver elements.

Early Operations

The first Army engineer unit of appreciable size to reach South Vietnam was the 173d Engineer Company, 173d Airborne Brigade. In early May 1965, the three-thousand-man brigade under the command of Brig. Gen. Ellis W. Williamson began arriving to protect Bien Hoa Air Base and the airfield and port at Vung Tau. Engineer elements consisted of a brigade engineer, Maj. Harold P. Austin, his small staff, and the 152-man 173d Engineer Company commanded by Capt. Thomas L. Morley. Immediately upon arrival, the company's 1st Platoon at Vung Tau and the 2d Platoon at Bien Hoa helped the infantry battalions dig in and set up water and shower points. By 12 May, the rest of the company, which included an equipment section and bridge platoon, reached Bien Hoa, where the airborne brigade set up headquarters and a base camp on high ground northeast of the airfield. The brigade quickly settled in and went to work building bunkers for protection against mortars, digging

---

[1] Sharp and Westmoreland, *Report*, pp. 98–99, 107 (quoted words p. 98); MACV Concept of Operations, 30 Aug 65, an. A (Intelligence), p. A-1.

*Engineers of the 173d Engineer Company, 173d Airborne Brigade, carry out a
river reconnaissance in one of their inflatable rafts.*

trenches, clearing fields of fire, and erecting barbed wire barriers. From the
bunkers and trenches, the paratroopers could see about 656 feet of open area,
which then gave way to thick brush and, finally, jungle farther north.[2]

While the 173d Engineer Company helped the brigade settle in, Major
Austin's office coordinated base camp development with the Navy's Officer in
Charge of Construction; ordered engineer equipment and supplies; and reconnoitered roads, helicopter landing zones, existing and potential airstrips, water
sources, and river crossing locations. This and information gathered through
other sources became the basis for selecting convoy routes used in operations.
In essence, Major Austin and his successor, Maj. Merritt R. Holcomb, who
replaced him on 7 July, operated like a division engineer, but did not command
the engineer company. Captain Morley, who was soon promoted to major and

[2] Hist Rpt of the 173d Engr Co (Abn), 173d Abn Bde (Sep), 1 Jan–31 Dec 65, n.d., p. 10;
Quarterly Cmd Rpt, 1 May–31 Jul 65, 173d Abn Bde, 15 Aug 65, p. 1; Annual Hist Supp, 1965,
173d Abn Bde (Sep), n.d., p. 36, app. C, all in Historians files, CMH; Galloway, "Essayons," p.
5; Carland, *Stemming the Tide*, p. 23.

replaced by Capt. Mark S. Sowell Jr. on 4 July, ran the company on a day-to-day basis.[3]

Although the 173d Engineer Company concentrated on base construction, elements of the company were dispatched to the field. During May, the Viet Cong ended a two-month lull and launched attacks against South Vietnamese forces. In northern III Corps, the enemy's Dong Xoai campaign began on 11 May with an attack on the capital of Phuoc Long Province, Song Be. Overrunning most of the town, the attackers held their ground into the next day. Two days following the attack, a twelve-man team from the engineer company, led by M. Sgt. Earnest F. Pena, arrived to boost the defenses of the joint Special Forces and MACV advisory camp at Song Be. The team quickly built and fortified bunkers, repaired perimeter fences, dug defensive emplacements, and helped guard the camp. Meanwhile, back at the Bien Hoa camp, a fifteen-man demolition team on 14 May joined a paratroop company patrolling outside the perimeter. As the brigade's two infantry battalions—the 2d Battalion (Airborne), 503rd Infantry, at Bien Hoa and the 1st Battalion (Airborne), 503d Infantry, at Vung Tau—moved out on independent operations, engineer demolition teams joined the infantry companies in the field. The Viet Cong offered little resistance to these initial forays.[4]

Then, during the night of 9 June, a paramilitary Civilian Irregular Defense Group (CIDG) camp and adjacent district headquarters compound just west of Dong Xoai in Phuoc Long Province came under attack. Most of Seabee Team 1104 had arrived from Tay Ninh City a few days earlier to improve the CIDG camp. For the next fourteen hours, Seabees, Special Forces troops, and advisers fought alongside their allies against an overwhelming enemy. Following an intense mortar barrage, the attackers breached the barbed wire defenses, overran the CIDG compound, and penetrated the district headquarters compound. The surviving Americans and some South Vietnamese held out in the district headquarters building, calling in air strikes. One wounded Seabee, Construction Mechanic Third Class Marvin G. Shields, carried the seriously wounded Special Forces commander from one of the defensive berms to the district headquarters building. Then Shields and a Special Forces officer, 2d Lt. Charles Q. Williams, second in command of the Special Forces detachment, manning a 3.5-inch rocket launcher, destroyed a Viet Cong machine-gun position. Shields was wounded again and later died aboard an evacuation helicopter. By then most of the Americans had been picked up by U.S. Army helicopters. Two of the Seabees, who had become separated during the fighting, were later found by a South Vietnamese Army relief force still holding out in the area. For a time, the 173d Airborne Brigade was poised to intervene, but the enemy withdrawal eliminated the need.[5]

[3] Annual Hist Supp, 1965, 173d Abn Bde, pp. 64–66.

[4] Hist Rpt of the 173d Engr Co, 1 Jan–31 Dec 65, pp. 10–11; Quarterly Cmd Rpt, 1 May–31 Jul 65, 173d Abn Bde, pp. 1, 3–15; Annual Hist Supp, 1965, 173 Abn Bde, pp. 36, 64.

[5] Shields and Williams received the Medal of Honor. Later the Navy awarded the Navy Unit Commendation Medal to members of the Seabee team present at the battle. COMCBPAC Reports, Special Edition Seabee Teams, October 1959–July 1968, "Helping Others Help Themselves," (Commander Naval Construction Battalions, U.S. Pacific Fleet, 1969), pp. 45–51, copy in

The Viet Cong shift to larger attacks led Westmoreland to request and get approval to deploy U.S. ground troops on unlimited offensive operations. Near the end of June, the 173d Airborne Brigade, now enlarged to three infantry battalions with the arrival of an Australian task force, and with all battalions at Bien Hoa, extended operations farther afield. In one operation, a battalion-size force of paratroopers penetrated Viet Cong–controlled War Zone D just beyond the Dong Nai River some six miles north of the brigade base camp. Then, on the twenty-eighth, the entire brigade and South Vietnamese Army troops ventured deep into the enemy redoubt. During the three-day sweep, demolition teams from the 173d Engineer Company uncovered and blew up mines and booby traps; cleared helicopter pads and landing zones; and destroyed Viet Cong tunnels, supply caches, and weapons emplacements. The rest of the engineer company joined with the gun jeeps of Troop E, 17th Cavalry, and M113 armored personnel carriers and M56 self-propelled antitank guns of Company D, 16th Armor, to form a composite battalion to patrol the roads and protect the brigade's command element and artillery.[6]

On 6 July, the 173d Engineer supported the brigade in another operation in War Zone D. The brigade's three battalions made airmobile assaults in an attempt to close a trap around the Viet Cong just north of the Dong Nai. While demolition teams moved out with the infantry, the rest of the company under Captain Sowell moved into an artillery firebase south of the river. At the firebase, the engineers' mission was to serve as a security force for the artillery and to run platoon-size patrols to clear the area to the north and east while the armor patrolled to the west and south. Besides inflicting casualties on the Viet Cong, the airborne brigade uncovered a large base complex consisting of mess halls, classrooms, latrines, and an extensive tunnel system. More than 150 booby traps were found during the operation. The engineers returned to Bien Hoa on the ninth and resumed base development work.[7]

The next operation took the 173d Airborne Brigade south to Phuoc Tuy Province. Leading off the operation on 28 July, the brigade's artillery battalion and the 173d Engineer Company moved by vehicle convoy along Highway 15 to an artillery firebase near Vung Tau. In the process, the task force cleared the highway for the first time in months, allowing the South Vietnamese Army to resupply a garrison at Binh Gia along the way. The following day, the brigade's

Historians files, CMH (hereafter cited as "Helping Others Help Themselves"); Tregaskis, *Building the Bases*, pp. 117–33; Shelby L. Stanton, *The Green Berets at War: U.S. Army Special Forces in Southeast Asia, 1956–1975* (Novato, Calif.: Presidio Press, 1985), pp. 120–23; After Action Rpt (AAR), Dong Xoai, Phuoc Long Province, 5th Special Forces (SF) Gp (Abn), 1st SF, 7 Jul 65, Historians files, CMH.

[6] Hist Rpt of the 173d Engr Co, 1 Jan–31 Dec 65, pp. 11–14; Quarterly Cmd Rpt, 1 May–31 Jul 65, 173d Abn Bde, p. 15, incl. 15; Lt. Gen. John J. Tolson, *Airmobility, 1961–1971*, Vietnam Studies (Washington, D.C.: Department of the Army, 1973), p. 64; *The First Three Years: A Pictorial History of the 173d Airborne Brigade (Separate)* (Brigade Information Office, n.d.), Section III, "Combat," 5 May 65–Jan 66, copy in CMH.

[7] Hist Rpt of the 173d Engr Co, 1 Jan–31 Dec 65, pp. 11–15; Quarterly Cmd Rpt, 1 May–31 Jul 65, 173d Abn Bde, p. 16, incl. 17; F. Clifton Berry Jr., *Sky Soldiers,* The Illustrated History of the Vietnam War (New York: Bantam Books, 1987), pp. 24–25; *Pictorial History of the 173d Airborne Brigade*, Section III, "Combat"; Tolson, *Airmobility*, p. 64.

MAP 3

two U.S. battalions began the operation with an airmobile assault. While the two battalions swept the area to sever a suspected Viet Cong supply route and other brigade forces set up blocking positions, the engineer and armor companies, now relieved of their security mission at the firebase, checked out reported Viet Cong locations and caches. The significance of the campaign was evident: the airborne brigade was no longer confining itself to the Bien Hoa bridgehead.[8] (*Map 3*)

Meanwhile, a second U.S. Army infantry brigade, the 2d Brigade, 1st Infantry Division, arrived in Vietnam. The 3,900-man unit, organized to operate on its own much like the 173d Airborne Brigade until the remainder of the division followed, left Fort Riley, Kansas, in June 1965. Company B, 1st

[8] *Pictorial History of the 173d Airborne Brigade*, Section III, "Combat"; Hist Rpt of the 173d Engr Co, 1 Jan–31 Dec 65, p. 15.

Engineer Battalion, commanded by Capt. Michael Volpe, deployed with the brigade. Plans called for the 2d Brigade to guard Qui Nhon and Cam Ranh Bay, but a change in plans sent the brigade to Bien Hoa, freeing the 173d Airborne Brigade for use as a mobile reserve. After a three-week sea voyage, the 2d Brigade reached Cam Ranh Bay. One battalion task force, including an engineer platoon, disembarked to guard the base in the early stages of development. The balance of the brigade proceeded to Vung Tau and from there to Bien Hoa where, under the operational control of the 173d Airborne Brigade, it would help protect the air base.[9]

After reaching Bien Hoa, the 2d Brigade moved into an area about two miles southeast of the flight line. Men of the brigade immediately set about clearing the elephant grass, thick brush, and ant hills that clogged the encampment site. Despite daily downpours and nightly sniping, the troops, sometimes working around the clock, cleared the area. Other tasks included building artillery gun emplacements and clearing fields of fire along the outer perimeters. As the brigade extended operations out from the encampment, Company B provided demolition teams, cleared areas of dud explosives, built bunkers, reconnoitered roads, and helped set up ambushes. Back at camp, the company built more than twelve and one-half miles of interior roads, drainage systems, buildings, and a tactical operations center.[10]

Company B provided a platoon to each infantry battalion during the operations of the 2d Brigade. The 1st Battalion, 18th Infantry, and Company B's 1st Platoon remained in the Cam Ranh Bay area for approximately three weeks. The engineer platoon worked on roads, bunkers, and helipads. In early August, the 2d Battalion, 18th Infantry, moved with the 173d Airborne Brigade to II Corps. The brigade's third infantry battalion, the 2d Battalion, 16th Infantry, carried out operations around Bien Hoa. In September the 2d Battalion, 16th Infantry, ran the first battalion-size operation in Long Thanh District, southeast of Bien Hoa, followed by a one-day operation in Tan Uyen District to the northwest. The presence of extensive Viet Cong activity in the area led to further battalion sweeps. In one encounter, the engineers used 800 pounds of explosives to destroy a Viet Cong tunnel. With the 2d of the 18th Infantry back from the Central Highlands, the brigade mounted a two-battalion operation, which resulted in the first encounter with the enemy and the destruction of well-defended, dug-in positions. On 29 September, the 1st of the 18th relinquished its mission at Cam Ranh Bay to the recently arrived 1st Brigade, 101st Airborne Division, and rejoined the 2d Brigade. For the first time since reaching Vietnam, the brigade's battalions and engineer company were together at one location.[11] (*Map 4*)

[9] *Vietnam, the First Year: A Pictorial History of the 2d Brigade, 1st Infantry Division*, (Tokyo, Japan: Brigade Information Office, n.d.), "Company B, 1st Engr Bn," and "Deployment to Vietnam" sections, copy in CMH.

[10] *Pictorial History of the 2d Brigade, 1st Infantry Division*, "Company B, 1st Engr Bn," section; *Always First: A Pictorial History of the 1st Engineer Battalion, 1st Infantry Division, October 1965–March 1967* (1st Engineer Battalion, n.d.), "Company B" section, copy in CMH.

[11] *Pictorial History of the 2d Brigade, 1st Infantry Division*, "Deployment to Vietnam" and "Company B, 1st Engr Bn," sections.

The 1st Brigade, 101st Airborne Division, followed an even more vaga-
bond road as a mobile brigade. Originally slated to relieve the 173d Airborne
Brigade at Bien Hoa so that the 173d could return to Okinawa, the 1st Brigade
instead was dispatched to II Corps where the strategically important areas
between Qui Nhon and Cam Ranh Bay were still wide open and at risk. Like
the 173d, the 3,600-man 1st Brigade consisted of qualified paratroopers, but
was bigger, having three airborne infantry battalions and a three-platoon engineer company, Company A, 326th Engineer Battalion. Upon reaching Cam
Ranh Bay on 29 July, the brigade set up camp on the mainland at Dong Ba
Thin where, unlike the soft sands of the Cam Ranh peninsula, the ground was
so hard that Company A engineers used road-cratering charges before digging
could commence on defensive positions. As the brigade set up a rough-and-
ready encampment, the paratroopers started on-the-job training from squad-
size patrols to a brigade-size operation west of Nha Trang. The brigade also
underwent intensive training in airmobile tactics. Starting in late August, the
brigade in Operation HIGHLAND provided security for the 1st Cavalry Division
by protecting the port of entry at Qui Nhon, clearing Highway 19, guarding
convoys, and initially securing the airmobile division's base area at An Khe,
allowing the division to move safely inland. In mid-September, the brigade
ran into a strong enemy force north of the An Khe Pass in the largest encounter with the enemy to date by U.S. Army forces. Between 2 October and 10
November, the 1st Brigade screened the arrival and deployment of the South
Korean Capital Division in the Qui Nhon area. It did the same for the Korean
2d Marine Brigade at Cam Ranh Bay. Company A engineers supported the
brigade in the field with minesweeping and demolition teams. Having completed its pathfinder assignment of opening up II Corps to follow-on reinforcements, the brigade moved on to Phan Rang, some twenty-five miles south
of Cam Ranh Bay.[12]

One of the airborne engineer company's tasks during Operation HIGHLAND
was to clear land mines around the An Khe airstrip inside the base. Inhabitants
were consulted to get an idea of minefield boundaries. Using mine detectors
and probing with bayonets at first, the 2d Platoon found two old unmarked
minefields containing U.S. antipersonnel mines. Mines found were destroyed
in place with one-half-pound explosive charges. The plastic M14 blast-type
mines were especially difficult to detect. One of the undetected M14 mines
was stepped on and detonated by a soldier who had just probed the area; he
lost part of his foot. Once reasonably assured there were no antitank mines, a
remote-controlled bulldozer traversed the area mainly to detonate undetected
M14 mines. Then the bulldozer pulled a sheepsfoot roller over the area several
times to compact the soil. Several M16 bouncing antipersonnel fragmentation
mines were found in the Song Ba riverbed inside the base just west of the

[12] Galloway, "Essayons," p. 53; Quarterly Cmd Rpts, 1 Jul–30 Sep 65, Field Force Vietnam
(FFV), 15 Oct 65, p. 6, and 1 Oct–31 Dec 65, FFV, 14 Jan 66, pp. 9–10; Quarterly Cmd Rpt, 1
Jul–30 Sep 65, 1st Bde, 101st Abn Div, 18 Oct 65, pp. 1–3, all in Historians files, CMH.

airfield. Most of those were pointed out by a Vietnamese boy who grazed his cattle in the area. Altogether Company A removed 225 mines.[13]

In November, MACV asked the 1st Brigade, 101st Airborne Division, to set up a new base camp and to protect Phan Rang Air Base, then undergoing transformation into a jet fighter field. In addition to building the air base, the 62d Engineer Construction Battalion was charged with building a cantonment for the paratroopers. Company A of the 326th helped when it could but spent most of the time in the field supporting the brigade's tactical operations. The brigade continued to spend most of its time away from its base camp. Shortly after the brigade's arrival at Phan Rang, two infantry battalions and supporting troops were airlifted to Bien Hoa to support operations in III Corps. The units returned to Phan Rang on 22 December.[14]

After the 173d Airborne Brigade returned from the highlands in early September, it resumed operations the following month in War Zone D and the enemy Iron Triangle, and provided security for units of the 1st Infantry Division as they moved into base camps near Saigon. The Iron Triangle region lay northwest of Saigon, and its features included fairly level terrain covered by patches of dense second-growth jungle mingled with an undergrowth of bamboo and other foliage. To soften up the Viet Cong redoubt, Air Force B–52 Stratofortress heavy bombers, in their first direct-support mission, bombed suspected enemy locations. Four task forces swept through the approximately sixty-square-mile area, but encountered only light opposition. Left behind, however, were large numbers of booby traps and command-detonated mines. Demolition teams, which landed with the heliborne assault forces, destroyed enemy camps, bunkers, and tunnels. At the close of the operation, General Williamson prematurely remarked: "The Iron Triangle is no more." Later in the month, the brigade cleared the nearby Phu Loi–Di An region for the 1st Division.[15]

A Base for the 1st Cavalry Division

While these operations were under way, the 16,000-man 1st Cavalry Division readied itself for the long sea voyage to South Vietnam. A 1,000-man advance party, which included Company C, 8th Engineer Battalion, arrived at the division's base camp at An Khe in late August. The An Khe area in western Binh Dinh Province, selected as a forward combat base by General Westmoreland, afforded the airmobile division a lodgment from which to operate in a wide radius. An Khe also lay within reasonable distance from Qui Nhon for ground resupply along Highway 19. While the 1st Brigade, 101st Airborne Division, guarded the region, the advance party led by Brig. Gen. John M. Wright Jr. laid out unit traces and fixed the approximate size of the

[13] AAR, Opn HIGHLAND, 1st Bde, 101st Abn Div, 5 Dec 65, incl. 13, p. 1, Historians files, CMH.

[14] Galloway, "Essayons," p. 53; Quarterly Cmd Rpt, 1 Oct–31 Dec 65, FFV, pp. 10–11.

[15] Hist Rpt of the 173d Engr Co, 1 Jan–31 Dec 65, pp. 17–18; Berry, *Sky Soldiers*, pp. 29–32; *Pictorial History of the 173d Airborne Brigade*, Section III, "Combat" (quoted words).

Troops clear the helicopter area at the 1st Cavalry Division's new base camp at An Khe.

camp. Wright then gathered all available officers and enlisted men in the party to clear trees and brush by hand. He placed the division's heliport near the center of the base and announced that the 3,000-by-4,000-foot field would "look exactly like a fine golf course." According to Wright, clearing by hand, instead of using engineer equipment, would preserve the turf and minimize the dust caused by the swirling blades of the division's helicopters. The division hired large numbers of civilian refugees to continue with the work. The heliport became known as the Golf Course.[16]

Meanwhile, Colonel Edelstein's 70th Engineer Combat Battalion arrived from Qui Nhon, where the unit had just disembarked, to work on the base camp. Tasks included building an access road from Highway 19 to the division base area, maintaining the existing pierced steel plank airstrip at An Khe, helping clear the large division heliport, and providing a company on standby to maintain Highway 19 from Qui Nhon to An Khe. Work on the eight-mile access road began on 28 August. The goal was to complete the road before the main body of the division began arriving on 11 September.[17]

---

[16] Interv, Lt Col David M. Fishback with Lt Gen John M. Wright Jr., 1 Mar 83, 2:382–92 (quoted words, p. 391), Senior Officer Oral History Program, MHI; Westmoreland, *A Soldier Reports*, p. 189; Maj. J. D. Coleman, ed., *1st Air Cavalry Division: Memoirs of the First Team, Vietnam, August 1965–December 1969* (Tokyo, Japan: Dai Nippon Printing Co., n.d.), pp. 25–26, 181; Galloway, "Essayons," p. 54; Rpt, 14th Mil Hist Det, 1st Cav Div (Ambl), 9 Jun 67, sub: Seven Month History and Briefing Data (September 1965–March 1966), p. 18, box 16, 74/053, RG 319, NARA.

[17] Quarterly Cmd Rpt, 1 Jul–30 Sep 65, 70th Engr Bn, 12 Oct 65, p. 2; AAR, Opn HIGHLAND, 70th Engr Bn, 5 Oct 65, p. 1. Both in Historians files, CMH.

It soon became apparent that more earthmoving equipment would be needed to complete the road by the required date. Initially, a grader and several bulldozers were borrowed from the 84th Engineer Battalion in Qui Nhon. This was followed a week later by two front loaders and two more graders from the 632d Engineer Company (Light Equipment) and a section containing twelve 5-ton dump trucks from the 513th Engineer Company (Dump Truck), both companies also at Qui Nhon.[18]

Since all entrances to the 1st Cavalry Division area were constricted by the Song Ba River on the east and Hon Cong Mountain on the west, a massive rock-tipped hill, and one or more villages along the way, a relatively narrow corridor remained through which to build the road. Although the engineers tried to follow the high ground to the camp, more than 30,000 cubic yards of fill had to be used to stabilize the sandy silt, which formed the subgrade of the major portion of the road. Removing the sandy silt would have entailed a prohibitive excavation effort within the limited time available. To make matters worse, a high water table would have made the soil completely unstable and virtually untrafficable to engineer equipment. Instead, just the top layer of soil and vegetation was removed, ditches cut, and the road topped off with four to six inches of laterite or decomposed granite. This method stabilized the soil and produced a suitable wearing surface in dry weather. Initially Edelstein's Company A hauled fill from an open-pit rock quarry on Highway 19 about two and one-half miles from An Khe, and Company C opened another pit at the base of Hon Cong Mountain on the western edge of the base little more than one-half mile from the road project. Later, the Highway 19 site was abandoned because of the longer haul distance, and both units used the Hon Cong Mountain pit. The most time-consuming part of the project involved the installation of numerous drainage facilities. A drainage survey revealed the need for nine bridges and thirty-one culverts, but bridge materials and culverts were not available at the time. Concrete culverts were procured locally, and supply channels managed to provide 540 feet of corrugated metal culvert. Bridging remained unavailable, and it became necessary to depend entirely on culverts. Unfortunately, culverts could not handle the flow of water during heavy rains. Working two 10-hour shifts with 2-hour maintenance breaks and using all available earthmoving equipment, the battalion completed the access road, dubbed Route 70, on the night of 11 September. Much work still remained, including widening sections of the road from a single lane to two lanes. The road also ran through a village and could not be widened until arrangements were made through civil affairs channels.[19]

During this period, security was a major concern, and soldiers had to be diverted from work sites and placed on guard duty. Although the 70th Engineer Battalion's bivouac areas and work sites were within the thinly held perimeter of the 1st Brigade, 101st Airborne Division, this provided little protection against isolated sniper fire and small guerrilla bands during darkness.

[18] AAR, Opn HIGHLAND, 70th Engr Bn, pp. 2–3.
[19] Ibid., pp. 3–6; Quarterly Cmd Rpt, 1 Oct–31 Dec 65, 1st Cav Div, 10 Jan 66, p. 24, Historians files, CMH.

Heavy-duty equipment such as this bulldozer from the 70th Engineer Battalion was used to build roads at the 1st Cavalry Division's Camp Radcliff.

Since the paratroopers were usually conducting operations requiring most of their men, isolated work sites such as the quarries and access road were usually protected by the engineers themselves. Security forces at each site usually consisted of four men with machine guns and M79 grenade launchers. Strict policies were adopted to avoid friendly casualties. Returning fire to disperse one or two snipers was restricted to small areas to avoid hitting the outer perimeter manned by the airborne brigade or in some other area occupied by friendly troops. Actually, only a few sniper rounds were received, and progress at the work sites was not impaired.[20]

In mid-September, the main body of the 1st Cavalry Division came ashore at Qui Nhon and moved inland to An Khe. Maj. Gen. Harry W. O. Kinnard, the commanding general, recognized the vulnerability of the highlands base camp. In a directive, he called for a small and compact camp that could be defended with few troops. Construction of Camp Radcliff (named in honor of Maj. Donald G. Radcliff, the first member of the division to lose his life in Vietnam) became a cooperative effort of all units on the base under the supervision of the division's 8th Engineer Battalion commanded by Lt. Col. Robert J. Malley. To provide the link between the division and other engineer organizations assigned to help build the base, Colonel Malley assigned one of his officers as Assistant Division Engineer, Base Development. This small

[20] AAR, Opn HIGHLAND, 70th Engr Bn, pp. 6–7.

The future Camp Radcliff shown here on 30 August 1965 would soon be transformed from a small tent city to a large base camp of wood frame and metal prefabricated buildings.

office clarified division priorities into guidance for the other engineers in the area.[21]

Because of an ever-increasing number of base projects, the engineers could only respond to the most urgent requests and frequently worked around the clock. One pressing assignment for the 8th Engineer Battalion was to level the top of Hon Cong Mountain for use as a signal relay station. With their light and modular equipment designed for transport by Army helicopters, which eliminated the need to cut an access road to the top, the airmobile engineers got right to work. Within three days, trees and vegetation had been cleared, a small bulldozer had been airlifted in to clear a helipad, and a larger bulldozer, a D6B, at 22,000 pounds, had been flown to the mountaintop in sections and reassembled to start moving dirt and rock. Over the next thirty days, the D6B, working with demolitions teams, cleared a 200-by 400-foot area and cut 12 to 14 feet off the top of Hon Cong Mountain. The engineers also built a security fence around the site and constructed bunkers for the radio vans. During their work on the hilltop, the engineers drew Viet Cong attention and exchanged weapons fire with the enemy for several nights. In the following months, more radio equipment,

---

[21] Galloway, "Essayons," pp. 54–55; Tolson, *Airmobility*, p. 72.

AN KHE BASE CAMP
31 December 1965

SKETCH NOT TO SCALE

Perimeter Barrier

Golf Course Heliport

2d Surgical Hospital

70

Hon Kong Mountain

Logistics Area

Bailey Bridge

70

to Qui Nhon

to Pleiku   19

An Khe Airfield

19

An Khe

Map 5

an Air Force weather station, artillery counterbattery detectors, and an Armed Forces Radio Service transmitter were placed on the hilltop.[22] (*Map 5*)

Gradually, the engineers transformed Camp Radcliff's tent city into a permanent combat base. In addition to the access road, the 70th Engineer Battalion also built a nine-mile perimeter road. After completing the roads, the 70th turned to building logistics facilities, protective bunkers, more roads within the camp, and Quonset huts for a 140-bed mobile surgical hospital. By early December, the 1st Cavalry Division began a major program to construct,

---

on a largely self-help basis, tropical-type building cantonment facilities for the twenty thousand men of the division and supporting units. Getting lumber through supply channels was very involved and slow, and only limited amounts were received. Timber was abundant in the area, and time-consuming efforts were soon under way to authorize local procurement. Pacific Architects and Engineers, which set up its facilities engineering shop in September, was able to get enough material to prefabricate kits to build latrines and showers.[23]

Another high-priority task was to build a strong perimeter defense. With the help of the 70th Engineer Battalion and some two thousand Vietnamese laborers, the 8th Engineer Battalion built a truly formidable obstacle system. Over time, the engineers built a defense perimeter one hundred yards in depth around the base camp. The perimeter consisted of four layers of barbed wire fences, a layer of claymore antipersonnel command-detonated mines, and inner and outer cattle fences to keep people and animals from straying into the mines that guarded the succeeding layers of wire. Vietnamese laborers worked for the most part outside the fence clearing vegetation. The claymore mine, developed by the Army before the war and introduced into combat in Vietnam, was designed as a one-shot, directional-fragmentation weapon that could be aimed to cover a specific area. Consisting of a plastic body, a fixed slit-type sight, four adjustable legs, electrical wire connections, and hand-held firing device, the claymore when fired spread steel fragments over a wide area. The weapon became quite popular for use in permanent and temporary defense positions and ambushes. In considering the terrain and the interlocking fields of fire, the division laid out long sections of the perimeter in a straight line to make it easier for helicopter gunships to help defend the base. The 70th Engineer Battalion further embellished the perimeter with 1,032 lights on concrete poles and sixty-eight guard towers, each equipped with a searchlight.[24]

To give the 70th Engineer Battalion at An Khe a greater construction capability, Colonel Brandt dispatched additional engineers and equipment from 937th Group resources. Although the 70th with its array of bulldozers, front loaders, and graders had more construction equipment than the lightly equipped 8th Engineer Battalion, it lacked the heavy equipment to do all of the required base camp construction. More equipment and operators from the Qui Nhon–based 632d Light Equipment Company were borrowed, increasing the light equipment element to a platoon. During the first months, the 70th Engineer Battalion had to make do with 80 to 160 Vietnamese hired to crush rock by hand, an unsophisticated but necessary means to obtain the all-important material. Not until the newly arrived 630th Engineer Company

[23] Quarterly Cmd Rpts, 1 Oct–31 Dec 65, 1st Cav Div, 10 Jan 66, p. 24, 1 Oct–31 Dec 65, 937th Engr Gp, 15 Jan 66, pp. 3–4; ORLL, 1 Feb–30 Apr 66, 70th Engr Bn, 7 May 66, pp. 2–6, Historians files, CMH; PA&E History, 1963–1966, p. 32.

[24] Rpt, 14th Mil Hist Det, 1st Cav Div, 9 Jun 67, sub: Seven Month History and Briefing Data (September 1965–March 1966), p. 19; Lt. Col. Albert N. Garland, ed., *Infantry in Vietnam* (New York: Jove Books, reprint of 1967 Infantry edition, 1985), pp. 257–59; Interv, Fishback with Wright, 1 Mar 83, 2: 397–98, 401–02; Quarterly Cmd Rpt, 1 Oct–31 Dec 65, 937th Engr Gp, pp. 4–5; ORLLs, 1 Jan–30 Apr 66, 937th Engr Gp, 30 Apr 66, p. 5, 1 Jan–30 Apr 66, 70th Engr Bn, p. 5, and 1 May–31 Jul 66, 70th Engr Bn, 15 Aug 66, pp. 3, 6, Historians files, CMH.

(Light Equipment) joined the 70th in early 1966 did the battalion get the benefit of a 75-ton-per-hour rock crusher.[25]

Maintaining the 4,200-foot airfield runway became a chronic problem for the 70th Engineer Battalion. Company C formally took over maintenance of the airstrip from South Vietnamese Army engineers on 29 August. Already on the scene was a platoon the 84th Engineer Construction Battalion had dispatched earlier to repair defects in the combination M8 and M6 pierced steel plank matting, which caused dangerous breaks in the airstrip's surface. Using an arc welder left behind by the 84th's platoon when it returned to Qui Nhon, the 70th's welding crew worked almost continuously to keep the airfield in operation. The condition of the strip under unusually heavy C–130 transport traffic deteriorated to the point that the welders could not keep up repairs. Despite all the effort, C–130 aircraft suffered frequent cuts and blowouts of landing gear tires. The Air Force complained bitterly, but the airfield had become so indispensable to 1st Cavalry Division operations that its closure for repairs could not be tolerated. This precluded engineer work crews from removing the old matting and repairing the base, a critically required task. A number of methods were considered, and the decision was simply to lay another layer of matting over the existing runway. By working only at night or during light traffic over a period of six nights, work crews placed some 1,500 linear feet of M8 matting. At the end of each night's work, welders fused the second layer's leading edges to the old surface. To compensate for the undulations in the existing pierced steel planking caused by base failures, a layer of crushed rock was placed on the low areas before placing the second layer of planking. This was only a temporary expedient, and plans were drawn up to build a new parallel runway.[26]

A Second Division Arrives

In October 1965, the rest of the 1st Infantry Division arrived to take up positions on the northern approaches to Saigon. While preparing for the deployment, General Seaman, acting on guidance from the Army chief of staff, General Johnson, began to reorganize the division from its mechanized configuration to a force considered more suitable for the counterinsurgency and jungle warfare in Vietnam. The result was a lighter organization, and units and equipment considered inappropriate to Vietnam stayed behind. These included the two armor battalions and units assigned to employ nuclear weapons, such as the Honest John missile battalion and the atomic demolitions platoon from the engineer battalion. To make up for the loss of the armor battalions, the Army transferred two infantry battalions from another division. In turn, the division's two mechanized infantry battalions were reorganized as standard dismounted infantry battalions, thus building the division around nine infantry battalions. The armored cavalry

---

[25] ORLLs, 1 Feb–30 Apr 66, 70th Engr Bn, pp. 4–5, 8–9, and 1 May–31 Jul 66, 70th Engr Bn, pp. 3, 8.

[26] Galloway, "Essayons," p. 30; Bowers, *Tactical Airlift*, p. 209; Quarterly Cmd Rpt, 1 Jul–30 Sep 65, 70th Engr Bn, p. 2; AAR, Opn HIGHLAND, 70th Engr Bn, pp. 1–2.

squadron, which did retain its M48A3 Patton diesel-powered tanks, replaced the older M114 armored reconnaissance vehicles with M113 armored personnel carriers, which had better overall reliability and mobility. Training geared to counterinsurgency warfare followed in the short time remaining before deployment. In mid-September, the division started to embark at ports along the west coast.[27]

The commander of the 1st Engineer Battalion, Lt. Col. Howard L. Sargent Jr., readied the battalion for its deployment. By the time the division began its move, the engineer battalion reached its authorized strength of over nine hundred men and made necessary adjustments in equipment. At first, there was a possibility that the battalion's float bridge company would remain behind to lighten the division. Considering the large number of rivers in Vietnam, the Army and MACV agreed to keep the bridge unit. To satisfy the division's lighter configuration, the heavier Class 60 floating bridge, which required a large crane to emplace the heavy steel deck bays, was replaced by the aluminum-decked M4T6 (Class 50–55) bridge. Manpower alone could easily place the twenty-two hollow sections, known as balk, in staggered pattern in each bay. The battalion's six 60-foot scissors-type armored vehicle launched bridges and four M48 tank launchers were left behind to lighten the division. Yet, the battalion was allowed to keep its four M48 tankdozers, the only tanks in the division outside the cavalry squadron brought to Vietnam, to provide armor protection to the engineers and infantry. Special training for the 1st Engineer Battalion included demolitions, jungle warfare, counter-ambush techniques, Viet Cong booby traps, and the M4T6 bridge.[28]

The vessels carrying the 1st Division reached Vietnam between 1 and 30 October. After disembarking at Vung Tau, the troops were flown to Bien Hoa and trucked to a nearby staging area. Supplies and equipment came through Vung Tau and the port of Saigon. After picking up its equipment, the division dispersed to four base camps north of the capital under the protection of the 2d Brigade and the 173d Airborne Brigade. Division headquarters and support troops moved to Di An, the 1st Brigade to Phuoc Vinh, the 3d Brigade to Lai Khe, and the division artillery to Phu Loi. These bases together with the previously established camps at Bien Hoa were positioned to guard the approaches to Saigon from the northwest, north, and northeast and to block enemy movement between War Zones C and D. Although the Viet Cong avoided a major fight, the 2d Brigade and the 173d Airborne Brigade encountered large numbers of mines and booby traps, mostly the command-detonated directional type. In all, the division succeeded in moving over 9,600

---

[27] Quarterly Cmd Rpt, 1 Oct–31 Dec 65, 1st Inf Div, n.d., pp. 1–2, 6, 12, Historians files, CMH; Carland, *Stemming the Tide*, pp. 63–66.

[28] *Pictorial History of the 1st Engineer Battalion*, Deployment to RVN; Quarterly Cmd Rpt, 1 Oct–31 Dec 65, 1st Inf Div, p. 11; 1st Engineer Battalion Unit Historical Rpt, 1965, 30 Mar 66, p. 2, Historians files, CMH (hereafter cited as 1st Engineer Battalion History, 1965); Msg, DA 729121 to CG, U.S. Continental Army Cmd, 20 Aug 65, sub: Float Bridge Requirements, Historians files, CMH.

troops and their equipment and supplies to their destinations in III Corps without loss of life or serious injury to anyone.[29]

After picking up their equipment at the Saigon port, elements of the 1st Engineer Battalion moved to the base camp sites that would house the units they supported. Company A moved north to Phuoc Vinh to support the 1st Brigade, and Company C proceeded up Highway 13—III Corps' main north-south artery—to Lai Khe to support the 3d Brigade. This left Headquarters and Headquarters Company and Companies D and E (the bridge company) for the clearing and development of the division's camps at Di An and Phu Loi just north of Saigon. At each camp, the engineer companies began constructing defenses and facilities. Initial tasks included clearing fields of fire and building roads, portable tent floors, wood buildings, security fences, bunkers, concrete floors, and helipads.[30]

At the same time, the 1st Engineer Battalion's intelligence section reconnoitered the area. Roads and bridges leading to Tay Ninh, Phuoc Vinh, and Vung Tau were surveyed. Wells were checked to determine their capacity for use as water points. Over twenty sites were located for possible laterite pits for use in road and hardstand construction. Reconnaissance teams also made soundings along the channel on the Dong Nai River for ferrying heavy equipment from Saigon to Tan Uyen north of Bien Hoa.[31]

At first, little base development support came from the 18th Engineer Brigade. The only other engineer battalion in the area, the 46th Construction Battalion, had arrived about the same time, and it was committed to projects at Long Binh and Vung Tau. Construction support from the engineer brigade began picking up toward the end of the year with the arrival of the 168th and 588th Engineer Combat Battalions. By then the 46th Engineer Battalion was able to complete the 1st Division's tactical operations center at Di An.[32]

In October, with some of the combat and support units still settling into their base camps, the 1st Division starting seeking out the Viet Cong. General Westmoreland had given General Seaman free rein to operate in a fan-shaped area opening northward from Saigon over a distance of some thirty-four miles in a region ranging from grasslands to rolling forested hills. Over the next three months, the 173d Airborne Brigade, under the operational control of the 1st Division, conducted many company-size and smaller operations east and northeast of the capital city. The 1st Division sought out the enemy in battalion-size or larger operations, mostly north of Saigon. Some of these operations resulted in major encounters with the Viet Cong.[33]

The 173d Airborne Brigade's biggest battle to date took place during Operation HUMP. In early November, the 173d again entered War Zone D to

[29] Quarterly Cmd Rpt, 1 Oct–31 Dec 65, 1st Inf Div, pp. 12, 23; Carland, *Stemming the Tide*, pp. 66–68; *Pictorial History of the 2d Brigade, 1st Infantry Division*, "Deployment to Vietnam" and "Company B" sections.

[30] Quarterly Cmd Rpt, 1 Oct–31 Dec 65, 1st Inf Div, p. 23; 1st Engineer Battalion History, 1965, p. 3.

[31] 1st Engineer Battalion History, 1965, pp. 1–2.

[32] Galloway, "Essayons," pp. 48–49, 51–52.

[33] Carland, *Stemming the Tide*, pp. 73–76.

Engineers use hand tools to clear a landing zone.

hunt down an enemy force reported massing near the confluence of the Dong Nai and Song Be Rivers. In one sharp battle, the enemy tried to encircle one of the brigade's battalions. The dense jungle and close fighting made resupply and evacuation impossible. On the early morning of 9 November, engineer teams were lowered from hovering helicopters to clear landing zones. With their power saws, they cut down trees, some 250 feet high and 6 feet in diameter, allowing helicopters to land and pick up casualties. The enemy, identified as a hard-core Viet Cong regiment, had been decimated and left behind more than 400 dead.[34]

In November and December, the 3d Brigade, 1st Division, teamed up with the South Vietnamese 5th Infantry Division in search-and-destroy operations in the area between Highway 13 and the Michelin Rubber Plantation to the west. During Operations BUSHMASTER I and II, the brigade encountered strong Viet Cong forces at Bau Bang, Trung Loi, and Ap Nha Mat in the fiercest fighting by units of the division thus far. The sharp battles resulted from chance encounters or enemy ambushes on returning convoys. During its sweeps, the brigade uncovered large caches of food, medicine, and supplies. Company C, 1st Engineer Battalion, provided a composite platoon and

---

[34] AAR, Opn HUMP, 173d Abn Bde, 19 Dec 65, pp. 1, 5–6, 9; 173d Abn Bde, Critique of Operation Hump, 19 Nov 65, pp. 14–15, both in Historians files, CMH; Hist Rpt of the 173d Engr Co, 1 Jan–31 Dec 65, pp. 18–19; Berry, *Sky Soldiers*, pp. 33–40; *Pictorial History of the 173d Airborne Brigade*, Section III, "Combat."

demolition teams, which helped the infantry destroy several camps, training facilities, tunnel complexes, and defense works.[35]

During November, the 2d Brigade continued clearing operations outside the 1st Division's base camps at Di An and Phu Loi. During Operation VIPER, which lasted until the end of the year, American troops located more than seventy Viet Cong tunnels and thirty-four camp sites. The division also found out how well the enemy had designed his defenses. Viet Cong camps were protected by perimeters of sharp wooden punji stakes, five-foot-deep trenches with firing ports, bunkers, and occasionally barbed wire. Often a system of tunnels branched out from the center of the camps. Company B engineers who ventured into the tunnel complexes also found compartments consisting of kitchens, dispensaries, classrooms, and living quarters, an indication of the many enemy complexes awaiting discovery.[36]

Before the year ended, other operations extended the reach of American forces in III Corps, but this time the action took place east of Saigon. During Operation NEW LIFE, which ran from 21 November to 17 December, the 173d Airborne Brigade moved by air to the air strip at Vo Dat in Binh Tuy Province and joined with South Vietnamese forces to prevent the rice harvest from falling into enemy hands. North of Vo Dat, in the La Nga Valley, a large group of refugees from North Vietnam had arrived in 1954 and transformed the once virgin jungle into the fifth highest rice-producing area in South Vietnam. In 1963, the Viet Cong moved in and with little resistance confiscated one half of the rice crop. They came back the following year and took the entire crop. During NEW LIFE, the airborne brigade saturated the area with patrols and sweeps. In the end, the farmers harvested their crops without interference and officials restored district-level government to the thirty thousand inhabitants of La Nga. During the operation, the 173d Engineer Company together with Australian engineers repaired over twenty-seven miles of local roads, built eleven bridges, and completed several civic action projects. For the first time, the 173d Engineer Company took along its bridging, assault boats, and construction equipment. Besides putting up a thirty-eight-foot dry span, the bridge platoon joined forces with Troop E, 17th Cavalry, to patrol the La Nga River. Using two assault boats, the twenty-man river patrol set out on 4 December, traveling about thirty miles. Along the way, the patrol discovered several caves, tunnels, and hidden sampans. At one point, the patrol came under fire, and the boats took several hits. While the engineer operators guided the boats out of range, artillery fire was directed on the enemy.[37]

---

[35] Quarterly Cmd Rpt, 1 Oct–31 Dec 65, 1st Inf Div, pp. 17–18; AARs, Opn BUSHMASTER I, 1st Inf Div, 21 Dec 65, p. 4, and Opn BUSHMASTER II, 1st Inf Div, 30 Dec 65, p. 5, both in Historians files, CMH; 1st Inf Div Info Release, Battle of Ap Nha Mat, 20 Feb 67, Vietnam Interview (VNI) 137, CMH. For more on the 1st Division's operations to the end of 1965, see Carland, *Stemming the Tide*, pp. 84–92.

[36] Quarterly Cmd Rpt, 1 Oct–31 Dec 65, 1st Inf Div, p. 17; *Pictorial History of the 2d Brigade, 1st Infantry Division*, "Deployment to Vietnam" and "Company B, 1st Engr Bn," sections.

[37] AAR, Opn NEW LIFE, 173d Abn Bde, 26 Jan 66, pp. 1–3, 12, 24, Historians files, CMH; *Pictorial History of the 173d Airborne Brigade*, Section III, "Combat"; Hist Rpt of the 173d Engr Co, 1 Jan–31 Dec 65, pp. 19–21.

From Operation NEW LIFE, the 173d Airborne Brigade moved directly to Operation SMASH, another search-and-destroy mission in a loosely coordinated operation with the 1st Division's 2d Brigade. Intelligence had indicated that the Viet Cong might be planning a major assault in the Xuan Loc area during the Christmas holiday period. The 173d's battalions patrolled west of Route 2, and the 2d Brigade worked east of Highway 15. During the week-long operation, demolition teams of the 2d Platoon, 173d Engineer Company, attached to the 2d Battalion, 503d Infantry, saw action. On 18 December, the airborne infantry and engineers successfully assaulted a strongly defended trench system manned with heavy machine guns. On the same day, the engineers escorted by Troop E, 17th Cavalry, moved south along Highway 15 toward Vung Tau to repair the road. In typical fashion, the Viet Cong had blocked the highway with ditches and other obstacles. Sniper fire caused the engineers to bring two squads for additional security. By the end of the operation, work crews filled some sixty ditches along fourteen miles of thoroughfare. On 22 December, the brigade returned to Bien Hoa to enjoy the Christmas holiday. It found a much improved base camp. Members of the engineer company who had remained behind not only upgraded the living quarters, but they also had built a new mess hall in time for Christmas dinner. As for the 2d Brigade, it saw no Viet Cong but found bases and destroyed supply caches. While the effects on enemy operations could not be determined, U.S. commanders believed that they had prevented a holiday offensive. More importantly, the presence of their units at some distance from the capital had served notice that the Americans were now here to stay in these contested parts of III Corps.[38]

The units under General Seaman's command gained valuable experience as they roamed throughout III Corps. Air transports and helicopters airlifted infantry task forces to airstrips and landing zones prepared and maintained by combat engineers. During these operations, combat engineers carried out a variety of tasks. The 1st Engineer Battalion provided mine detector and demolition teams, set up water points, maintained supply roads, and built bridges and rafts. The battalion's tankdozers proved very popular with the infantry by clearing paths through the jungle, exposing tunnel complexes, and detonating booby traps.[39]

Securing Highway 13, which ran through the enemy's well-entrenched strongholds of War Zones C and D, continued to be one of the priorities involving the divisional engineers. Like most Vietnamese national highways, Highway 13 had a bituminous surface treatment, adequate for the limited volume and weight of traffic before the U.S. buildup. The combination of an insufficient subgrade and poor drainage, heavy rains, Viet Cong sabotage, little maintenance by the South Vietnamese, and the pounding from heavy military convoys had resulted in a complete breakdown of the road. The 1st Division was determined to break Viet Cong control and open the road to a normal

[38] AAR, Opn SMASH, 173d Abn Bde, 26 Jan 66, pp. 1–2, 5–8, Historians files, CMH; *Pictorial History of the 173d Airborne Brigade*, Section III, "Combat"; Hist Rpt of the 173d Engr Co, 1 Jan–31 Dec 65, pp. 21–22.

[39] 1st Engineer Battalion History, 1965, p. 3.

flow of commerce. The 1st Engineer Battalion concentrated on clearing mines and making repairs. Since the 1st Division had arrived at the end of the rainy season, the engineers did not have to worry about the road being inundated in many places.[40]

To help the infantry destroy enemy tunnel complexes, the 1st Engineer Battalion organized a platoon made up of armored personnel carriers mounted with flamethrowers. Referred to by Colonel Sargent as the "Tunnel Killer Team," the platoon also used tankdozers, smoke pots, scoop loaders, and air compressors to carry out its mission. After the engineers inserted an air hose, the air compressor drove smoke through the tunnel exposing other holes in the dense jungle growth. Demolitions, scoop loaders, or tankdozers then sealed the holes. Still, the dense jungle could conceal other holes. Sargent reported that during the air compressor's first use, other holes could not be found. He concluded there must have been other holes because the compressor never idled, an indication of high pressure in the tunnel.[41]

The Ia Drang

While General Seaman extended his reach in III Corps, General Kinnard's immediate task in II Corps was to secure his base at An Khe and to hold open the highway to the coast. To do so, the 1st Cavalry Division on 6 October dispatched elements of the 2d Brigade to take control of the Vinh Thanh Valley, a Viet Cong–dominated rice-growing area located about twelve and one-half miles east of An Khe. Operation HAPPY VALLEY, which lasted until 19 November, involved air assaults, patrols, night ambushes, and civic action. Units of the 3d Brigade eventually replaced the 2d Brigade. While HAPPY VALLEY continued, the 1st Cavalry Division launched SHINY BAYONET, a five-day thrust into the Suoi Ca Valley in an attempt to find and destroy North Vietnamese units reported operating northeast of the division base. Little contact was made, but the cavalrymen gained experience operating in the field.[42]

During the two operations, the division's 8th Engineer Battalion dispatched elements appropriate for the mission. Companies B and C, respectively, supported the 2d and 3d Brigades. In addition, Company B constructed the Vinh Thanh Special Forces camp in the Vinh Thanh Valley complete with defensive fortifications, wire, underground bunkers, interconnecting trenches, quarters, and a dispensary. The engineers also did daily minesweeping and clearing along Highway 19 and Route 3A leading to the camp. Another task for Companies B and C was road maintenance. During this work, each company lost a 3/4-ton dump truck to land mines.[43]

---

[40] Ibid.

[41] Ibid.; Ltr, Lt Col Howard L. Sargent, CO, 1st Engr Bn, to Chief of Engrs, 14 Nov 65, Historians files, CMH.

[42] Quarterly Cmd Rpt, 1 Oct–31 Dec 65, 1st Cav Div (Ambl), 10 Jan 66, pp. 13–15, Historians files, CMH.

[43] *Memoirs of the First Team*, p. 27; Rpt, 14th Mil Hist Det, 1st Cav Div, 9 Jun 67, sub: Seven Month History and Briefing Data (September 1965–March 1966), pp. 18–19, 31–33, 36.

The assumption was that the 1st Cavalry Division's next operation would also be in Binh Dinh Province, but instead the division found itself increasingly oriented to the western highlands. In an uninhabited and rugged area between the Cambodian border and the Special Forces camp at Plei Me, some twenty-five miles south of Pleiku City, the North Vietnamese had concentrated three regiments. The Chu Pong Massif, with peaks rising to over 2,300 feet, dominated the area. A river, the Ia Drang, flowed along the mountain's northern edge, generally to the northwest. The dense vegetation provided excellent concealment for the enemy. On 19 October, Plei Me came under attack by one of the regiments. The South Vietnamese hesitated dispatching a relief force, fearing to leave Pleiku undefended. On 22 October, General Kinnard sent a battalion task force, including an artillery battery, to secure Pleiku upon departure of the relief force. Then, seeing an opportunity for combat, he obtained permission the following day to transfer the entire 1st Brigade to Pleiku. In a series of leapfrog moves by helicopter, the air cavalry's artillery kept the South Vietnamese column under cover of its 105-mm. howitzers. The relief column cautiously advanced, beating off a fierce attack before reaching Plei Me on the twenty-fifth.[44]

These events led to the Pleiku Campaign, the 1st Cavalry Division's first real combat test as a fighting force. On 27 October, General Westmoreland visited the division's 1st Brigade outpost at Landing Zone HOMECOMING south of Pleiku, where he gave General Kinnard the go-ahead to seek out and destroy the enemy in the western highlands. For the next two weeks, the 1st Brigade and the 1st Squadron, 9th Cavalry, combed the jungle near Plei Me, making occasional contact with the enemy.[45]

At that point, on 9 November, General Kinnard substituted the 3d Brigade. On the fourteenth, Lt. Col. Harold G. Moore's 1st Battalion, 7th Cavalry, made an airmobile assault into Landing Zone X-RAY, adjacent to the Chu Pong range. Without realizing it, the battalion had deployed almost on top of two North Vietnamese regiments. The North Vietnamese fell on the cavalry battalion. Over the next two days, in some of the fiercest fighting of the war in an area covered with high grass interspersed with ant hills nearly as high as a man's head, the battalion and reinforcements from the 3d Brigade beat back repeated assaults. Artillery at nearby Landing Zone FALCON and Air Force fighter-bombers pounded suspected enemy concentrations around the perimeter. The Air Force also unleashed B–52 raids on enemy positions and supply routes. At great risk, UH–1 Hueys flew in reinforcements, ammunition, medical supplies, rations, and water and carried off the dead and wounded. At dawn on the sixteenth, the Americans still held the landing zone. The North Vietnamese withdrew toward the sanctuary of Cambodia,

[44] AAR, Pleiku Campaign, 1st Cav Div, 4 Mar 66, pp. 10, 18–25, Historians files, CMH; Rpt, 14th Mil Hist Det, 1st Cav Div, 9 Jun 67, sub: Seven Month History and Briefing Data (September 1965–March 1966), p. 38; Carland, *Stemming the Tide*, pp. 99–104; Sharp and Westmoreland, *Report*, pp. 99, 110; Westmoreland, *A Soldier Reports*, pp. 189–90.

[45] Westmoreland, *A Soldier Reports*, p. 190; AAR, Pleiku Campaign, 1st Cav Div, pp. 27–67, 76; Tolson, *Airmobility*, pp. 74–75; Rpt, 14th Mil Hist Det, 1st Cav Div, 9 Jun 67, sub: Seven Month History and Briefing Data (September 1965–March 1966), pp. 37–44.

Troops of the 1st Battalion, 7th Cavalry, advance at Landing Zone X-RAY where attached engineers also fought as infantry.

but in the process bloodied the 2d Battalion, 7th Cavalry, in an ambush near Landing Zone ALBANY. The 2d Brigade moved in and chased the enemy to the border. General Kinnard sought permission to enter and finish off the North Vietnamese units, but officials in Washington rejected the request. President Johnson had decided that the United States sought no wider war and would fight the ground war solely within the borders of South Vietnam. Unable to pursue the enemy into Cambodia, American efforts west of Pleiku quickly petered out. On November 26, the troops returned to An Khe, satisfied in the knowledge that airmobility coupled with strong artillery and air support had dealt a blow to the North Vietnamese.[46]

The 8th Engineer Battalion provided continuous support throughout the Pleiku Campaign. Companies A, B, and C provided direct support to the 1st, 2d, and 3d Brigades, respectively. In turn, platoons were attached to the infantry battalions, and squads or demolition teams deployed with the infantry companies to clear landing zones and booby traps. The 3d Brigade deployed

---

[46] AAR, Pleiku Campaign, 1st Cav Div, pp. 69–132; Rpt, 14th Mil Hist Det, 1st Cav Div, 9 Jun 67, sub: Seven Month History and Briefing Data (September 1965–March 1966), p. 44–50. For more on the Pleiku Campaign, see Carland, *Stemming the Tide*, pp. 111, 113–47; John A. Cash, "Fight at Ia Drang, 14–16 November 1965," in *Seven Firefights in Vietnam* by John A. Cash, John Albright, and Allan W. Sandstrum (Washington, D.C.: Office of the Chief of Military History, U.S. Army, 1970), pp. 3–40; Joseph L. Galloway, "Fatal Victory," *U.S. News & World Report* (29 October 1990): 32–34, 36–37, 40–43, 44–45, 47–51; and Lt. Gen. Harold G. Moore and Joseph L. Galloway, *We Were Soldiers Once . . . and Young: Ia Drang—The Battle That Changed the War in Vietnam* (New York: Random House, 1992), pp. 59–327.

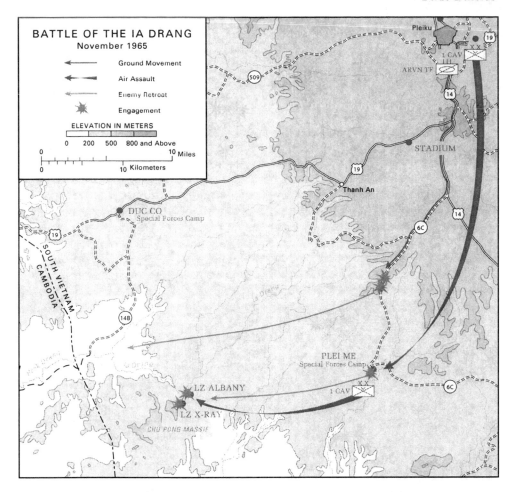

MAP 6

nine five-man demolition teams from Company C, 8th Engineer Battalion, to its three infantry battalions, three teams to each battalion. Besides overseeing the committed units, Colonel Malley and his staff reconnoitered the roads and bridges and areas for landing zones and airstrips. At STADIUM, a forward support base southwest of Pleiku astride Highway 19, the battalion set up a water point and provided the tactical units with a continuous supply of water throughout the operation. Water was a special problem at X-RAY because the nearby streambed was dry, making the cavalrymen totally dependent on the water hauled in by helicopters.[47] (*Map 6*)

---

[47] AAR, Pleiku Campaign, 1st Cav Div, pp. 1, 5; Rpt, 14th Mil Hist Det, 1st Cav Div, 9 Jun 67, sub: Seven Month History and Briefing Data (September 1965–March 1966), p. 18; AAR, Ia Drang Campaign, 8th Engr Bn, 16 Dec 65, p. 1, Historians files, CMH; Carland, *Stemming the Tide*, p. 126.

Seven engineers from Company C were at X-RAY clearing helicopter land-ing areas and fighting as infantry. The largest area available to land helicop-ters was vulnerable to enemy fire, and Colonel Moore noticed a smaller area just east of his battalion command post that could take two helicopters if some trees were removed. He turned to his engineer demolition team leader, Sgt. George Nye, to clear the trees. As the enemy fire intensified Nye recalled, "You could see the enemy, and suddenly we were part of the 1st Battalion, 7th Cavalry. It's tough to try to be an infantryman and a demolition specialist at the same time, but we did it. We blew those trees; no sawing. The intensity of fire made working with a saw tough, working without a weapon. By blowing the trees we could spend more time fighting." After the enemy withdrew from X-RAY, the remaining engineers collected and destroyed 100 enemy rifles and machine guns, 300 to 400 hand grenades, 7,000 rounds of ammunition, 3 cases of rocket-propelled grenades, and 150 entrenching tools.[48]

Throughout the forward areas, hasty building tasks characterized the 8th Engineer Battalion's work in the field. CH–47 Chinook helicopters lifted equipment, in sections if necessary, to job sites. After the attack on Plei Me, Company A, less the 3d Platoon and reinforced with equipment from Headquarters Company, moved to the camp to repair the airfield, destroyed unexploded shells, and buried enemy dead. A major effort included filling a large bomb crater, some fifty-five feet wide and twenty feet deep in the center of the runway. At STADIUM, Company A worked on field fortifications and extended the runway. As the brigades moved in and out of the area, Companies B and C took over engineering tasks. Company C continued maintaining the airstrip at STADIUM, filling ruts caused by CV–2 Caribou and C–123 Provider aircraft. By this time, it had become routine for the airmobile engineers to build landing zones, participate in reconnaissance missions and infantry patrols, clear mines, and repair roads.[49]

During the Pleiku Campaign, the 8th Engineer Battalion suffered its first battle deaths. While on a road reconnaissance near STADIUM, a jeep from the company headquarters hit a mine, killing two noncommissioned officers and wounding two other engineers. When the North Vietnamese struck the perim-eter at STADIUM in a mortar and ground attack, Company C defended an assigned sector, which the enemy failed to penetrate. Company C lost one sol-dier killed in action and three wounded. The demolition teams from Company C attached to the 1st Battalion, 7th Cavalry, at Landing Zone X-RAY suffered two deaths. By the end of the Pleiku Campaign, the 8th Engineer Battalion had lost six killed and thirteen wounded.[50]

[48] Moore and Galloway, *We Were Soldiers Once*, pp. 76, 116, 161–63, 199 (quoted words, pp. 76 and 162); AAR, Ia Drang Campaign, 8th Engr Bn, pp. 3–4.

[49] Rpt, 14th Mil Hist Det, 1st Cav Div, 9 Jun 67, sub: Seven Month History and Briefing Data (September 1965–March 1966), pp. 18–19; AAR, Ia Drang Campaign, 8th Engr Bn, pp. 2–4.

[50] Rpt, 14th Mil Hist Det, 1st Cav Div, 9 Jun 67, sub: Seven Month History and Briefing Data (September 1965–March 1966), p. 19; AAR, Ia Drang Campaign, 8th Engr Bn, pp. 3–4 and incl. 1; Galloway, "Fatal Victory," p. 45; Moore and Galloway, *We Were Soldiers Once*, p. 76.

Following the Ia Drang battles, General Westmoreland decided to move another brigade into the highlands to provide a permanent American presence. Between 18 December 1965 and 23 January 1966, the 3d Brigade, 25th Infantry Division, moved from Hawaii by sea and air to Pleiku. Two infantry battalions disembarked at Cam Ranh Bay and flew to Pleiku, while the third flew direct from Hawaii. The preparations begun in 1962 to build an airfield at Pleiku capable of handling large transport planes now proved well worth the effort.[51]

The Hawaii-based 25th Division, commanded by Maj. Gen. Frederick C. Weyand, had long prepared for operations in the Far East. Tapped as the Army's immediate reaction force in the Pacific, the Tropic Lightning Division often participated in deployments and exercises to Thailand, Taiwan, and Okinawa. The division specialized in jungle warfare and counterinsurgency operations. In January 1963, soldiers of the 25th Division responded to MACV's call by volunteering to serve as aerial door gunners to protect UH–1 helicopters. With the departure of the 3d Brigade, the rest of the division prepared itself for the long sea voyage leading to a different destination northwest of Saigon.[52]

As the 3d Brigade began to arrive and set up its base camp on the high ground northeast of Pleiku, elements of the 1st Cavalry Division guarded the surrounding area. The 3d Brigade soon took over its own defenses, placing priority on building a perimeter and an austere base camp at the same time. Accompanying the 3d Brigade was Company D, 65th Engineer Battalion. Attached to Company D were a bridge platoon from Company E, 65th Engineer Battalion, and two water supply teams, an engineer equipment section, and a medical section from Headquarters Company.[53]

As 1965 came to an end, elements of the 1st Cavalry Division directed their attention to Binh Dinh Province east of An Khe. Company C, 8th Engineer Battalion, supported the 3d Brigade in Operation CLEAN HOUSE, the division's last major operation of the year. Between 17 and 30 December, short but fierce encounters took place in the Suoi Ca Valley a few miles west of Highway 1 and north of Qui Nhon. The 1st Platoon attached to 2d Battalion, 7th Cavalry, and the 3d Platoon attached to the 1st Battalion, 8th Cavalry, served primarily as infantry, while the 2d Platoon attached to the 1st Battalion, 7th Cavalry, did extensive demolition work that included blocking caves believed to be used by the enemy. The balance of the company used its own and borrowed equipment from Headquarters Company to maintain roads leading to the brigade's forward support element.[54]

---

[51] Westmoreland, *A Soldier Reports*, p. 198; Quarterly Cmd Rpt, Oct–Dec 65, 25th Inf Div, n.d., pp. 4–6, and ORLL, 1 Jan–30 Apr 66, 3d Bde, 25th Inf Div, 23 Jun 66, pp. 1–2, both in Historians files, CMH.

[52] Quarterly Cmd Rpt, Oct–Dec 65, 25th Inf Div, pp. 8–9; *The 25th's 25th . . . in Combat, Tropic Lightning, 1 October 1941–1 October 1967*, 25th Infantry Division, 25th Div Public Affairs Ofc, n.d., pp. 129–67, 179–81, copy in CMH.

[53] Quarterly Cmd Rpt, Oct–Dec 65, 25th Inf Div, Part II, incl. 9; Msg, CG, 25th Inf Div to Asst Div Cmdr-2, 12 Dec 65, sub: HOLOKAI XI & XII, Historians files, CMH; ORLL, 1 Jan–30 Apr 66, 3d Bde, 25th Inf Div, p. 2.

[54] *Memoirs of the First Team*, p. 30; Rpt, 14th Mil Hist Det, 1st Cav Div, 9 Jun 67, sub: Seven Month History and Briefing Data (September 1965–March 1966), pp. 19, 51–56.

The 8th Engineer Battalion's Intelligence Section was also hard at work. Throughout December, the section's reconnaissance troops flying in a UH–1D helicopter at treetop level recorded data on all major roads in the II Corps area. The reconnaissance engineers also landed at the bridge locations to determine their types and vehicle-carrying capacities.[55]

By year's end, General Westmoreland believed that the emergency phase of the war had passed. Combat engineers assigned to the divisions and separate brigades had settled into a pattern of closely supporting infantry, armor, and artillery units in the field. Base development responsibilities remained secondary missions, only accomplished on a regular basis by troops remaining behind at the base camps. As for nondivisional combat engineer support, it remained under centralized command and concentrated on base development, providing operational support when necessary. Because the buildup proceeded so rapidly, construction had been done on a crash basis. As the engineer effort grew in size, the need for organizational and managerial changes soon became evident.

[55] Rpt, 14th Mil Hist Det, 1st Cav Div, 9 Jun 67, sub: Seven Month History and Briefing Data (September 1965–March 1966), p. 19.

4

Organizing the Construction Effort

With the arrival of contingents of Army engineers, Navy Seabees, Marine Corps engineers, Air Force engineers, and civilian contractors, U.S. construction strength mounted rapidly. Changing requirements for facilities from which to conduct or support combat operations and deployments interfered with the establishment of construction priorities, which, in turn, depended on the availability of labor, equipment, materials, and sites, for which there was intense competition among the services. Because of the rapid buildup, construction was initially accomplished on a crash basis under existing command and control arrangements. As the number of engineer units increased, the requirement grew to organize the construction effort.

Preliminary Proposals

Before 1965, several engineering offices in South Vietnam were involved in construction matters, but no single agency exercised supervision and control. Although the Navy's construction offices in Bangkok and Saigon provided some supervision of the construction program, the engineering offices within MACV planned and set priorities. MACV did this through three separate offices: the Base Development Branch, J–4; the Engineer Branch, Directorate of Army Military Assistance Program Logistics, J–4 (formerly the Engineer Advisory Branch, Military Assistance Advisory Group [MAAG]); and a small Engineer Branch within the Headquarters Commandant, which had the housekeeping responsibility for the headquarters. The Army's Support Command engineer, the Air Force's 2d Air Division civil engineer, and the Navy's Public Works provided engineering support to their respective services. Added to this were the contractors; Vietnamese Army engineers; Army, Navy, and Air Force engineer detachments; and Pacific Command and its subordinate service commands and their engineering staffs in Hawaii.[1]

Rules governing who would control a coordinated construction program in Vietnam, however, had yet to be worked out. In December 1961, the U.S. buildup had been expected to increase the tempo of construction, and the Military Assistance Advisory Group established an office to obtain real estate and monitor and coordinate construction matters for all U.S. military personnel in the country. Later these functions were passed to the MACV Base Development Branch. This small branch, however, had limited authority and concerned itself with other matters. It watched out for possible duplication of facilities or inequitable construction standards among the services, saw

---

[1] Joint Tables of Distribution, MACV, 1 Sep 64, pp. 47–50, Historians files, CMH.

that military assistance construction projects fitted into contingency and base development plans, prepared base development portions of MACV war plans, and acquired real estate. Funding and approval of projects continued to be processed through service channels, and the office did not oversee construction. The Base Development Branch was strictly a planning and coordinating office with "primary cognizance" over construction matters. Initially the branch was headed by a Navy commander, assisted by an Army major, who also served as real estate officer, and a Marine Corps major. The branch underwent an organizational change in September 1964 when an Army colonel was authorized as the chief, assisted by a Navy commander and an Air Force major.[2]

A loose organization known as the U.S. Construction Staff Committee composed of representatives from the various U.S. agencies met primarily to coordinate construction for the Vietnamese military forces and civilian agencies. Chaired by the MACV J–4, the committee included representatives of the Engineer Branch, Directorate of Army Military Assistance Program Logistics, J–4; the U.S. Operations Mission; the Deputy Officer in Charge of Construction, Republic of Vietnam; the U.S. Army Support Command; and other MACV and service agencies in South Vietnam. The modest requirements for the U.S. advisory teams, the MACV staff, and the small number of U.S. military units took up little of the committee's time.[3]

During the spring of 1965, the Engineer Branch, Directorate of Army Military Assistance Program Logistics, J–4, expanded its attention beyond military assistance projects. Col. Kenneth W. Kennedy took over the branch in November 1964, and under his tenure the branch assumed a larger role planning U.S. base requirements. Kennedy and members of his staff reconnoitered and prepared a major study for the development of Cam Ranh Bay as a logistics base. He also participated in the J–4 proposal to deploy an Army logistical command and engineer group to Vietnam in anticipation of a future military buildup. Throughout this time, Kennedy advocated the creation of a MACV command engineer office.[4]

Colonel Kennedy's proposal received strong backing from Lt. Gen. Walter K. Wilson, the chief of engineers, and Lt. Gen. Lawrence J. Lincoln, deputy chief of staff for logistics, also a Corps of Engineers officer. During the Department of Defense team's visit to Vietnam in January 1965 to investigate the logistical command and engineer group proposals, it also recommended that command engineering responsibilities should be passed to the Navy's construction office in Saigon. Generals Wilson and Lincoln strongly disagreed. Wilson mainly disagreed, not because the Deputy Officer in Charge

[2] Ltr, Chief, Military Assistance Advisory Group (MAAG), Vietnam, to CINCPAC, 8 Feb 62, sub: Report of Chief, MAAG, Vietnam for Period 2 Sep 1961 to 8 Feb 1962; Joint Tables of Distribution, MACV, 1 Jul 63, p. 7 (quoted words); Joint Tables of Distribution, MACV, 1 Sep 64, pp. 47–50; Interv, Capt E. Gregory, Mil Hist Br, MACV, with Col Kenneth W. Kennedy, MACV Engineer, 10 Nov 65, p. 1, box 6, 69A/702, RG 334, NARA.

[3] MFR, Col Kenneth W. Kennedy, 8 Feb 71, sub: Thoughts on Vietnam: Nov 1964–Nov 1965, pp. 6–7, OCE Hist Ofc.

[4] Interv, Gregory with Kennedy, 10 Nov 65, p. 1; MFR, Kennedy, 8 Feb 71, sub: Thoughts on Vietnam, pp. 1–2.

of Construction in Saigon lacked effectiveness in supervising contract construction, but, rather, because of the need for a command engineer with a sufficient staff to plan and advise on all military engineering matters for the Military Assistance Command, Vietnam. Lincoln supported Wilson and noted that the team's proposal considered the engineer mission only within the scope of construction, facility, and real estate matters. "Such an arrangement," he argued, "would require a Navy Civil Engineer to perform a role for which he has no background." In mid-March, after his visit to Vietnam, Lt. Gen. William F. Cassidy, the chief of engineer's designee, lent his support. Lt. Gen. Bruce Palmer Jr., the deputy chief of staff for military operations, who also looked into the proposal during his visit to the theater, agreed with the engineers when he reported his findings to the chief of staff, General Johnson. As a result, the MACV engineer, the Army contended, logically had to be a Corps of Engineers officer.[5]

On 7 April 1965, Colonel Kennedy was appointed MACV engineer. The appointment came when the J–4, with General Westmoreland's concurrence, notified him that in addition to heading the Engineer Branch, he would also be the director of Army Military Assistance Program Logistics and take over direction of the Base Development Branch, with the mission of planning and supervising construction of facilities needed for the buildup. A few days later, Kennedy was on his way to Hawaii to attend the deployment planning conference at Pacific Command headquarters. Within twenty-four hours after his arrival, he prepared a hasty base development plan estimated to cost $266 million. Troop construction requirements totaled 113 battalion-months for Army construction units and another 21 battalion-months in Seabee effort. The Army and Navy representatives at the conference made tentative agreements to deploy eight Army and three Navy construction battalions. They also recognized that construction would lag far behind the imminent buildup of combat forces, that construction would be austere, and that operations would have to be supported across the beach. But Kennedy later recalled: "Many forgot this as soon as they got ashore."[6]

Returning to Saigon, Kennedy was shocked and dismayed to find out that the Support Command engineer, Lt. Col. Edmond J. Cochard, had briefed General Taylor, now the U.S. ambassador to Saigon, without coordinating with the MACV engineer. Taylor, who initially resisted the U.S. buildup, only approved the deployment of five construction battalions, three less than the

[5] Ch of Engrs, Summary Sheet, 13 Feb 65, sub: Engineer Staff for MACV, and Addendum to Summary Sheet, Lt Gen Lawrence J. Lincoln, DCSLOG, to CofS Army, 19 Feb 65, sub: Engineer Staff for MACV (quoted words), Incls in Memo, Secretary of the General Staff for CofS Army, 27 Feb 65, sub: Engineer Staff for Hq MACV; Memo, Lt Gen Bruce Palmer, DCSOPS, for CofS Army, 25 Feb 65, sub: Engineer Staff for Hq MACV, all in Historians files, CMH; Interv, Lynn Alperin, OCE Hist Ofc, with Brig Gen Kenneth W. Kennedy, 23 Apr 84, pp. 297–98, OCE Oral History Interview Collection; MFR, Kennedy, 8 Feb 71, sub: Thoughts on Vietnam, pp. 1–2.

[6] MRF, Kennedy, 8 Feb 71, sub: Thoughts on Vietnam, pp. 2–3 (quoted words p. 3); Interv, Gregory with Kennedy, 10 Nov 65, pp. 1, 3; Interv, Alperin with Kennedy, 23 Apr 84, pp. 299–306.

CHART 2—MACV ENGINEER DIVISION, MAY 1965

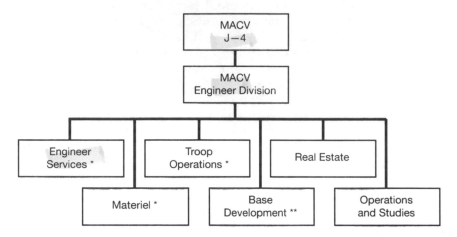

Source: Interv, Gregory with Kennedy, 10 Nov 65

* Original advisory sections of Engineer Branch, Director of Army Military Assistance Program Logistics, J–4
** Assumed control in April 1965

number Kennedy and the conferees had agreed upon in Hawaii. Fortunately, Kennedy found this out in time to change the message sent to the Department of the Army. The modified message read that five battalions were needed at once with the other three to be provided on call. In any case, Kennedy recognized that the engineers needed to improve coordination.[7]

In May, further reorganization affected Kennedy's MACV engineer operations. He was relieved of his duties as director of Army Military Assistance Program Logistics, and those functions were absorbed by the J–4. A separate MACV Engineer Division was formally established under J–4 supervision. Two new branches were added: Real Estate and Operations and Studies. With the arrival of U.S. military units now under way and with the need to acquire land for new bases, Kennedy had no recourse but to take the arriving 64th Engineer Detachment (Terrain Analysis) and assign the topographic unit to the Real Estate Branch. The U.S. Construction Staff Committee continued to meet with Kennedy, who served as the executive secretary.[8] (*Chart 2*)

As the buildup intensified so did the burden on the Engineer Division. Twelve officers and ten enlisted men were authorized; their workload increased by three to four times. Colonel Kennedy shuttled between Saigon and Honolulu, attending more planning conferences at Pacific Command headquarters. The

[7] MFR, Kennedy, 8 Feb 71, sub: Thoughts on Vietnam, p. 3; Interv, Alperin with Kennedy, 23 Apr 84, pp. 307–08.

[8] MFR, Kennedy, 8 Feb 71, sub: Thoughts on Vietnam, p. 4; Interv, Gregory with Kennedy, 10 Nov 65, p. 1; Interv, Alperin with Kennedy, 23 Apr 84, pp. 312–13; Misc files, MACV Real Estate Br, 15 Oct 65, Historians files, CMH.

planners struggled to develop a basic construction program and justify supplemental funds for all the services. As the scope of engineer activities expanded with the buildup, the J–4 on 15 November underwent another reorganization. This included elevating the MACV engineer to the J–4 deputy for engineering and finally authorizing additional personnel. But the engineer staff increased by less than half to eighteen officers and seventeen enlisted men.[9]

During this time, Kennedy faced practices that made coordinating and controlling construction difficult. Engineer offices of service headquarters continued to submit projects through their service chain of command for funding and approval and had a free hand in assigning work to troops. For instance, Army requirements proceeded through the U.S. Army, Pacific, which reviewed all Army construction projects in the Pacific and integrated them into a priority listing. The avalanche of work involved in base development planning overwhelmed the small MACV engineer staff, and more of this responsibility was delegated to the service commands. This meant the Army Support Command and its successor U.S. Army, Vietnam, would prepare base development plans for II, III, and IV Corps, the Air Force's 2d Air Division for specified air bases, and the III Marine Amphibious Force for I Corps.[10]

Without clear-cut control, the MACV Engineer Division began exerting more influence in construction matters. The division reviewed plans for new projects exceeding $25,000 and had authority to establish priorities. As a MACV staff office, the division issued regulations outlining base development procedures, the acquisition of real estate, and construction standards. When the subordinate service headquarters received approval and funding from their services, Kennedy had the option to do the work either by troops or contract. Projects considered more suitable for accomplishment by contract, usually larger and more complex projects in more secure areas, were passed on to the Deputy Officer in Charge of Construction in Saigon. This arrangement made sense because the Navy construction office in Saigon had increased its architect-engineering capability. It also helped that the Bureau of Yards and Docks had raised the Saigon office to an independent Officer in Charge of Construction to handle growing construction requirements. Projects not sent to the Officer in Charge of Construction were assigned to the Army and Navy commands in Vietnam. Those projects assigned to U.S. Army, Vietnam, were further delegated to the 1st Logistical Command and later the 18th Engineer Brigade. Although more projects were assigned to the growing number of Army engineer units, the logistical command also looked to Pacific Architects and Engineers to carry out much of the construction. In I Corps, the III Marine Amphibious Force assigned its troop projects to the 30th Naval Construction Regiment or a Marine Corps engineer battalion.[11]

[9] Raymond, Observations on the Construction Program, p. 9; Interv, Gregory with Kennedy, 10 Nov 65, pp. 2–3; Dunn, *Base Development*, p. 18.

[10] Interv, Gregory with Kennedy, 10 Nov 65, pp. 4–6; Interv, Alperin with Kennedy, 23 Apr 84, pp. 313–15; MFR, Kennedy, 8 Feb 71, sub: Thoughts on Vietnam, pp. 4–5; JLRB, Monograph 6, *Construction*, p. 86; MACV History, 1965, p. 127.

[11] Interv, Alperin with Kennedy, 23 Apr 84, pp. 314–15; Tregaskis, *Building the Bases*, p. 106; MFR, Kennedy, 8 Feb 71, sub: Thoughts on Vietnam, pp. 8–9; MACV History, 1965, p. 127; Army Buildup Progress Rpt, 22 Sep 65, p. 34, CMH.

As MACV engineer, Kennedy also used his position to call on help from other agencies. In his role as senior adviser to the South Vietnamese Army engineers, he exerted influence over the employment of those units. In several instances, the South Vietnamese engineers prepared beach unloading sites and campsites, and gave other help during the initial buildup. Help also came from the U.S. Operations Mission. Since Viet Cong harassment had stopped almost all aid projects, the mission made equipment and materials available for U.S. military use. For instance, nine hydraulic dredges were placed under the MACV engineer's control, and several metal prefabricated warehouse kits were transferred to the military. This enabled the engineers to build some badly needed warehouses at Da Nang, Cam Ranh, and the Saigon port.[12]

One of Kennedy's high-priority projects was Cam Ranh Bay. After General Westmoreland decided to transform the sandy peninsula into a major base in May 1965, interest by the three services began to snowball. Months later Kennedy remarked: "At first we couldn't get anybody to put anything there but now we can't keep people out and it's now one of the most overcrowded facilities that I know of." He advocated putting the air base on the mainland. Westmoreland, Kennedy, and Lt. Gen. Joseph H. Moore, commander of the 2d Air Division, boarded the MACV commander's helicopter and reconnoitered the area from the air. Westmoreland then ordered the aircraft to land at a spot on the northern part of the peninsula and made up his mind to put the airfield there. After listening to Kennedy's proposal for a mainland site, Westmoreland noted he understood the rationale, but his preference remained with the peninsula location primarily for security reasons. During the building of Cam Ranh Bay, Kennedy held weekly planning sessions with representatives of the agencies concerned to decide funding and location. They divided the peninsula functionally into four areas: a northern sector for the Army for later development into a convalescent hospital and a replacement and recreation center, a sector for the Air Force base, another sector for the Army port and depot complex, and a small sector on the southeast corner for a Navy facility.[13]

The arrival of two senior engineering officers in September placed Kennedy, still a colonel, in the unusual position of directing an admiral and a general. R. Adm. William F. Heaman, arrived to take over temporarily as Officer in Charge of Construction. General Ploger's arrival the same month to take command of the 18th Engineer Brigade added to Kennedy's predicament. As uncomfortable as this arrangement may have appeared, Kennedy's relations with the two were excellent. Besides, he figured the two flag officers had enough to handle. At the end of November, he ended his tour and turned over the office to Col. Daniel A. Raymond, a brigadier general designate.[14]

[12] MFR, Kennedy, 8 Feb 71, sub: Thoughts on Vietnam, pp. 7–8.

[13] Quote from Interv, Gregory with Kennedy, 10 Nov 65, p. 13; MFR, Kennedy, 8 Feb 71, sub: Thoughts on Vietnam, p. 11; Interv, Alperin with Kennedy, 23 Apr 84, pp. 318–22; Development of Cam Ranh Bay, 1st Log Cmd, 30 Oct 66, p. 1, box 13, 68A/4975, RG 319, NARA.

[14] Tregaskis, *Building the Bases*, pp. 106, 183, 187; Interv, Alperin with Kennedy, pp. 341–44; Msg, COMUSMACV MACJ4-EN to CINCPAC, 18 Aug 65, sub: Construction Effort, Historians files, CMH.

The 18th Engineer Brigade staff at its headquarters in Saigon.
General Ploger is fifth from the left in the front row.

The Construction Boss Concept

Back in Washington, high-level officials in the Department of Defense had come to the opinion that control of the construction program needed to be tightened. After his visit to South Vietnam in July 1965, Edward J. Sheridan, the deputy assistant secretary of defense for properties and installations, in his role as deputy overseer of Defense Department construction matters, recommended a central office within MACV other than the J–4 to coordinate the military building effort—a construction boss or "construction czar." During his visit to Cam Ranh Bay, he expressed concern over what appeared to him to be vague construction responsibilities between troops and contractors. General Westmoreland initially agreed but later indicated that the situation had improved to his satisfaction. In an exchange of messages with Admiral Sharp at Pacific Command, he pointed out that Sheridan's visit came at a time his command faced a herculean task to plan and formulate construction requirements. Westmoreland believed that Kennedy's office had been strengthened and the staff had gotten valuable experience. He also felt the impending arrivals of General Ploger and his brigade headquarters and Admiral Heaman as the new Officer in Charge of Construction would give him the organizations and expertise to satisfy construction requirements.[15]

Neither Sheridan nor Defense Secretary McNamara shared Westmoreland's views. McNamara felt that the job warranted an Army major general. By then, Sheridan seriously considered selecting Brig. Gen. Carroll H. Dunn, an Army

[15] JLRB, Monograph 6, *Construction*, p. 87; MFR, Lt Col John E. Gray, OCE, 27 Jan 66, sub: Conference of Deputy Assistant Secretary of Defense (Properties and Installations), Ofc Asst Sec Def (Installations and Logistics), with Brig Gen Carroll H. Dunn, Carroll H. Dunn Papers, CMH; Msg, COMUSMACV MACJ4-EN to CINCPAC, 18 Aug 65, sub: Construction Effort; Msgs, Sharp to Westmoreland, 26 Aug 65, and Westmoreland MAC 4356 to Sharp, 27 Aug 65, both in Westmoreland Message files, CMH.

engineer serving in Korea as deputy chief of staff of Eighth Army, as best qualified to run the theater's construction program. When McNamara visited Saigon
in November 1965, he was dismayed to find no progress on the construction czar
proposal. Upon his return to Washington, he directed General Earle G. Wheeler,
chairman of the Joint Chiefs of Staff, to examine the problem. In a series of messages to Sharp and Westmoreland, Wheeler presented the Defense Department's
case for a construction czar. He pointed out that construction projects in Vietnam
cut across all the services and were financed, controlled, and executed by several
agencies, including a civilian contractor. He also noted that the defense secretary
had departed Saigon with the realization that the construction program would
grow to the billion-dollar level. Besides the appointment of a construction chief,
Wheeler added that Washington would authorize a large expansion of the MACV
engineer office to about 150 men. The Army general heading this organization
would report directly to Westmoreland, not the J–4. Wheeler summed up by
noting that the chiefs of the services' engineering officers agreed.[16]

Westmoreland continued to support the status quo. He argued that his
new deputy chief of staff J–4 for engineering, soon to be General Raymond,
exercised sufficient control over the construction program, and that Admiral
Heaman, besides serving as the Officer in Charge of Construction, also helped
the J–4 as a special assistant for contract construction. He also felt satisfied
with his three construction agencies: the 18th Engineer Brigade, the 30th
Naval Construction Regiment, and the Officer in Charge of Construction.
What he really wanted, besides sufficient funding for the construction program, was more flexibility in using the construction funds allocated to him.
Wheeler responded that every effort would be made to allow more flexibility
and emphasized the best way to isolate any construction problems would
be to assign such responsibility to a single manager. This approach would
simultaneously reduce a large workload on the J–4. He also recognized that
Dunn, who would soon be a major general, might cause some rank disparity
with the J–4, a brigadier general. This could not be helped.[17]

On 8 December, Westmoreland answered that he could not solve logistical and construction problems by changes in organization. He had a
board of officers look at the construction czar proposal, and it reaffirmed
his position to keep the J–4 in charge of a strong engineering organization.
Westmoreland also independently solicited the views of Admiral Heaman,
who fully indorsed the board's findings. The MACV commander felt his
deputy assistant chief of staff J–4 for engineering did everything in the
Defense Department's proposal. During McNamara's visit in November,
he had outlined what he considered to be major construction problems.
These did not include organization or management within his command.

---

[16] Msg, Wheeler JCS 4658–65 to Sharp, 1 Dec 65; Msg, JCS 7886 to CINCPAC, 4 Dec 65,
both in Westmoreland Message files, CMH; MFR, Gray, 27 Jan 66, sub: Conference of Deputy
Assistant Secretary of Defense (Properties and Installations), Ofc Asst Sec Def (Installations
and Logistics), with Dunn, Dunn Papers, CMH. See also Cosmas, *Years of Escalation, 1962–
1967*, pp. 281–82.

[17] Msg, Westmoreland MAC 6176 to Sharp and Wheeler, 5 Dec 65; Msg, Wheeler JCS 4761–
65 to Westmoreland and Sharp, 7 Dec 65, both in Westmoreland Message files, CMH.

Instead, MACV's problems were the lack of construction units, the lack of spare parts for construction equipment, and the delays in the delivery and distribution of critically needed building materials. Westmoreland especially disliked the funding restrictions that precluded rapid responsiveness to new requirements. He agreed that an augmentation of twenty-four people for his engineering office would improve its effectiveness. He also pointed out that a separation of the construction from the overall logistics function could cause more problems than it would solve. "Aside from [the] proliferation of staff elements," Westmoreland argued that, "the close coordination between construction, movement control and transportation would be impaired."[18]

By this time Westmoreland believed that the Pentagon was trying to tell him how to do his job. "It is incredible to me," he recorded in his personal notes, "that higher headquarters would suggest how a subordinate command would be reorganized to do the job the commander is responsible for. Nevertheless, this proposal has been raised on several occasions, and despite my position, it is still an issue." Sensing he would lose the argument anyhow, he proposed to Sharp another way to use General Dunn. In a carefully crafted backchannel message, Westmoreland suggested raising the rank of his J–4 to major general and putting Dunn in that position. "It has become increasingly clear to me," he wrote, "that the magnitude of the logistics mission confronting the command calls for a J–4 of greater rank and more varied experience." He praised the effectiveness of his current J–4 but noted "that an Army engineer officer is most likely to combine general logistic experience with competence to manage the extensive construction program." Therefore, he concluded, the Defense Department should reconsider the construction czar proposal since the J–4 would be an Army engineer.[19]

On 18 December, Wheeler provided further guidance. He explained that the construction program—now approaching $1 billion—amounted to nearly one-half of the entire Corps of Engineers' construction effort, both civil and military. This fact had impressed the chairman and seemed to prove that MACV required "powerful organizational arrangements." Such a strong, centralized organization under a "construction boss," Wheeler added, would receive "quick, top level approval" to Westmoreland's appeals for more flexibility in the construction program. Without a construction boss in MACV, Wheeler predicted, "it is practically certain that procedures giving such flexibility will not receive approval, at least not at this time."[20]

Although Admiral Sharp had supported Westmoreland, he realized something had to be done. He identified the construction problems in Vietnam to be those concerned with funding, the lack of construction personnel and

---

[18] Quote from Msg, COMUSMACV 43885 to CINCPAC, 8 Dec 65, sub: MACV Construction Staff Augmentation, box 44, 71A/2351, RG 330, NARA; Westmoreland Historical Notes, no. 2, 8 Dec 65, Westmoreland History files, CMH; JLRB, Monograph 6, *Construction*, p. 87.

[19] Westmoreland Historical Notes, no. 2, 8 Dec 65; JLRB, Monograph 6, *Construction*, p. 87; Msg, Westmoreland MAC 6257 to Sharp, 8 Dec 65, Westmoreland Message files, CMH.

[20] Msg, Wheeler JCS 4934–65 to Westmoreland and Sharp, 18 Dec 65, Westmoreland Message files, CMH; Tregaskis, *Building the Bases*, p. 204.

General Westmoreland

equipment, and the long supply lines. If an Army engineer served as the J–4, then, according to Sharp, the MACV construction office should be headed by a Navy Civil Engineer Corps rear admiral. By 22 December, Sharp agreed with the points brought out in Wheeler's 18 December message and withdrew his suggestion for the Navy engineer construction boss if it would help settle the matter. He also wanted to make sure that they did not set up the construction agency outside the MACV staff, with the agency's operation left to "Westy's judgement." Realizing that further debate over a construction boss seemed fruitless, Sharp recommended that they "get on with setting him up."[21]

On 20 December, the Joint Chiefs of Staff recommended establishment of the position of construction boss, and on 6 January 1966 the deputy secretary of defense, Cyrus R. Vance, concurred. "It should be clearly understood," Vance stated in a memorandum to Wheeler, "that the 'engineer construction boss' has full authority to discharge the responsibilities placed on him, and that such authority rests in him and not in the MACV-J4." Sharp alerted Westmoreland, and Westmoreland promptly requested General Dunn for assignment as his J–4 with authority vested in him to coordinate and control all construction projects. Brig. Gen. John D. Crowley, his current J–4, would serve as deputy J–4 and concentrate on matters other than construction. Westmoreland was convinced this approach offered him the most effective and efficient arrangement. This would permit Dunn to arbitrate conflicting interests involving port operations, transportation, and related matters essential to support construction. In another message to Sharp the following day, Westmoreland reiterated his concern that his construction boss should have control over funding allocations and expressed "strong views as to how my staff should be organized."[22]

---

[21] JLRB, Monograph 6, *Construction*, p. 87; Msg, Sharp to Wheeler, 22 Dec 65, Westmoreland Message files, CMH.

[22] Memo, JCS JCSM–891–65 for Sec Def, 20 Dec 65, sub: Construction Management in Vietnam, quoted in Msg, JCS 1836 to COMUSMACV, 17 Jan 66, sub: Construction Management in Vietnam, Dunn Papers, CMH; Memo, [Cyrus R.] Vance for Chairman, JCS, 6 Jan 66, sub: Construction Management in Vietnam (first quotation), OCE Hist Ofc; Msg, Wheeler JCS 103–66 to Sharp and Westmoreland, 6 Jan 66; Msg, Westmoreland MAC 0179 to Sharp and Wheeler, 9 Jan 66, sub: Construction Czar; Msg, Westmoreland MAC 0215 to Sharp, 10 Jan 66 (second quotation), all in Westmoreland Message files, CMH.

Wheeler replied on 13 January that no one had intended to tell Westmoreland how he should organize his staff. Still, Defense Department officials expressed skepticism over the proposal to make Dunn the J–4 and whether one officer could handle both jobs. Wheeler pointed out that pursuing the J–4 proposal "will inevitably delay the decentralizing of functions and funds as you have requested." Reflecting on his experience as a general staff officer, the chairman did not see any problems of "rank inversion" between Crowley and Dunn. Westmoreland's proposal, Wheeler added, might be the best solution, but he frankly doubted it. Two days later, Wheeler reiterated these concerns to both Sharp and Westmoreland. In discussing the matter with Secretary Vance, he learned that any relaxation of controls by the Defense Department "over the construction effort was predicated upon the construction boss being neither the J4 nor reporting through the J4 to COMUSMACV."[23]

On 14 January, Wheeler informed Sharp and Westmoreland that Secretary McNamara had approved and issued a memorandum outlining interim procedures for construction management in Vietnam. In essence, Westmoreland's construction boss had supervision and directive authority over all military and civilian agencies involved in military construction in the country. To help carry out this vast task, the engineering staff would be increased by three to four times the number of additional people requested by MACV. McNamara's construction charter contained Defense Department policies dealing with approval and reprogramming of construction funds and standards of construction. The charter also gave to the MACV commander broad authority to issue construction directives within fifteen broadly defined functional categories (for example, cantonments, ports, maintenance buildings, hospitals, and airfield pavements). Westmoreland now had the authority to transfer funding from one category to another, provided the amount did not exceed 10 percent of the receiving functional category. The following month, McNamara amended this charter along with subsequent policy memos dealing with fiscal accounting and construction reporting procedures.[24]

Westmoreland decided to accept, at least for the time being, assigning Dunn solely as his construction chief. He still had reservations about the wording in Secretary Vance's 6 January memorandum. In a message to Wheeler on the eighteenth, he complained: "It is inconceivable that 'full' authority be vested in one of my subordinates and not in me and further that I be restricted in the authority I might choose to delegate to my J4 in the interest of carrying out my broad and demanding responsibilities." Still, he concluded "there is nothing to be gained at this juncture by making an issue of what I consider an arbitrary and unprecedented decision by Secretary Vance." One week later,

[23] Msg, Wheeler JCS 205–66 to Sharp and Westmoreland, 13 Jan 66; Msg, Wheeler JCS 227–66 to Sharp and Westmoreland, 15 Jan 66, both in Westmoreland Message files, CMH.

[24] Msg, Wheeler JCS 213–66 to Sharp and Westmoreland, 14 Jan 66, Westmoreland Message files, CMH; Tregaskis, *Building the Bases*, pp. 204–05; Memos, [Robert S.] McNamara for JCS and Secretaries of Military Departments, 14 Jan and 12 Feb 66, sub: Construction Approval Procedures for South Vietnam; JCS JCSM–39–66 for CINCPAC, 14 Jan 66, sub: Construction Management in Vietnam, all in MACV Dir of Const, Plans and Opns Div, Development of the Construction Directorate, 15 Jun 70, Historians files, CMH.

General Dunn

Westmoreland learned that *Time* magazine was preparing an article about construction in Vietnam. He decided to announce that General Dunn would be designated as his director of construction, not MACV engineer, since he would not have any troops under his command.[25]

In mid-January, Dunn left his assignment in Korea and proceeded to Washington to be briefed on the construction picture in Vietnam.[26] He met with Sheridan and members of the Southeast Asia Construction Group. This group, established the previous September, monitored and supported the military construction programs presented to Congress. Its members included people from the Defense Department's installations and logistics office and an Army colonel, an Air Force colonel, and a Navy captain from the services engineering offices. Dunn also met with Sheridan's chief, Paul R. Ignatius, the assistant secretary of defense for installations and logistics. Ignatius pointed out that Secretary McNamara had granted to General Westmoreland a broad construction charter, which gave wide managerial latitude to his construction chief. Dunn would be given ninety days to develop a detailed construction program, while minimizing the inflationary effect of a huge construction program on the South Vietnamese economy. He also considered Dunn to be his representative in Saigon. Dunn

[25] Msg, Westmoreland MAC 0459 to Wheeler, 18 Jan 66, Westmoreland Message files, CMH; Westmoreland Historical Notes, no. 3, 25 Jan 66, Westmoreland History files, CMH.

[26] Upon his graduation from the University of Illinois with a degree in mechanical engineering, Dunn was commissioned a second lieutenant in 1938 through the Reserve Officers' Training Corps and was selected to serve in the Regular Army. During World War II, he served as commander, 105th Engineer Battalion, 30th Infantry Division, which arrived in Normandy shortly after D-Day. He led the battalion across France to the Elbe River in Germany. Subsequent assignments included command of the 1153d Engineer Group (Combat) in France; Assistant Chief of Staff, G–4, 2d Infantry Division, Camp Swift, Texas, and Fort Lewis, Washington; instructor at the Engineer School, Fort Belvoir, Virginia; and the Engineer Section, General Headquarters, Far East Command, Tokyo, Japan, during the Korean War. After promotion to colonel, he served as Executive Officer to the Chief of Engineers in Washington, D.C.; Director of the Army Waterways Experiment Station, Vicksburg, Mississippi; student at the Industrial College of the Armed Forces in Washington, D.C.; Area Engineer, Thule, Greenland; and Director and Deputy Commander of the Titan II Missile System Construction Program. As a brigadier general, he served as Division Engineer, Southwestern Division, Dallas, Texas; and Deputy Chief of Staff, Eighth Army, Seoul, Korea.

took this comment as a hint of the detailed control that would be exerted by the Defense Department over his operations in the war zone.[27]

On 18 January, Dunn and Sheridan met with members of the Military Construction Subcommittee of the House Appropriations Committee. Both soon found themselves the recipients of an informal hearing before the entire subcommittee gathered in the office of Congressman Robert L. F. Sikes of Florida, the chairman. Sikes remarked that the subcommittee had recently toured South Vietnam and voiced concern about the lack of construction battalions and lighterage to unload ships. Congressman Charles D. Long of Maryland asked about the engineer's views on the call-up of reserve construction battalions. Dunn replied that deployment priorities should include units needed for the logistics base buildup, but he could not address the manner of providing such units. Long also came up with some interesting statistics. He foresaw a requirement for sixty construction battalions in Southeast Asia plus some 60,000 civilian laborers. South Vietnam had only 23,000 skilled workers, and the congressman prodded Sheridan for the Defense Department to make contractors run training programs raising the skill levels of Vietnamese workers.[28]

Dunn next met Air Force engineering officials. Brig. Gen. Guy H. Goddard, the deputy director for construction in the Air Force's Directorate of Civil Engineering, pointed out that air base construction, although approved and funded, did not receive high enough priority. The chief problem, according to Goddard, stemmed from the Army's failure to live up to its troop support agreement for construction in Vietnam and Thailand. Goddard also pointed out that the new Air Force Red Horse squadrons, organized to repair airfields and do light construction tasks, could not alleviate the airfield construction gap. The civilian head of the secretary of the Air Force's installations office expressed similar concerns.[29]

In his meeting with R. Adm. A. C. Husband, chief of the Navy's Bureau of Yards and Docks, Dunn was briefed on the problems faced by the Officer in Charge of Construction. First was the long lead times in letting contracts, and this tied up funds. The admiral pointed out the shortages of LSTs and lighterage and the backup of ships carrying construction materials. To make matters worse, the South Vietnamese government imposed enough red tape, which made it difficult to hire workers from other countries. It took a long time to get real estate. Equipment deliveries were frequently delayed. The

---

[27] MFR, Gray, 27 Jan 66; sub: Conference of Deputy Assistant Secretary of Defense (Properties and Installations), Ofc Asst Sec Def (Installations and Logistics), with Dunn; MFR, Gray, 27 Jan 66, sub: Assistant Secretary of Defense (Installations and Logistics) Meeting with Gen Dunn, MACV Engr Designate, Dunn Papers, CMH; Dunn, *Base Development*, p. 43; *Engineer Memoirs, Lieutenant General Carroll H. Dunn* (Alexandria, Va.: Office of History, U.S. Army Corps of Engineers, 1998), p. 102.

[28] MFR, Gray, 27 Jan 66, sub: Appearance of Gen Dunn Before Military Construction Subcommittee, House Appropriations Committee, Dunn Papers, CMH.

[29] MFRs, Gray, 27 Jan 66, sub: Briefing of Gen Dunn, by the Department of the Air Force Directorate of Civil Engineering, and Conference of Gen Dunn with Deputy for Installations, Office of the Assistant Secretary of the Air Force (Installations and Logistics), both in Dunn Papers, CMH.

Navy's office in Saigon needed more specific instructions by the using agencies and increased staffing to prepare plans. In several cases, contractors could not disengage from projects after completing most of the work because of add-on refinements by the customers. Refinements could be better done by military engineers. To obtain a better balance of troop and contract construction, the Navy engineers advised assigning construction troops in areas threatened by the enemy. Dunn also learned that the Navy's contracting effort, running at a rate of some $14 million per month, would gradually climb to $40 million by October 1966.[30]

Dunn's whirlwind visit to the capital included meeting more officials. He attended meetings with the Army chief of engineers, members of the Army Staff, and the Joint Chiefs of Staff J–4. The new construction chief wrapped up his stay by resolving some questions in a final conference with Ignatius and Sheridan. Dunn pointed out his concern with the fiscal accountability restrictions in McNamara's 14 January memorandum, which he felt, if not resolved, would impede management of the construction program. The construction charter called for funds to be apportioned to each service by Defense Department categories, altogether forty-five separate accounts. Dunn suggested a breakout of funds not further than the three services. Otherwise he would become a "chief accountant" at the expense of his construction management function. Ignatius agreed, and the Defense Department made this and several other administrative changes in a memorandum signed by McNamara on 12 February. In addition, the Officer in Charge of Construction would maintain fiscal accounting for the contracts it administered. This practical arrangement did not last long. Later, the Defense Department reinstated restrictive funding and accounting procedures.[31]

A New Directorate

On 15 February 1966, the MACV Directorate of Construction became operational. Besides overseeing the construction program, General Dunn as the director of construction also assumed control of the advisory effort for the South Vietnamese Army engineers. Structured along functional lines, the new directorate organized the former MACV engineering office into seven main divisions: Administrative, Engineering and Base Development, Construction Management, Real Estate, Program Management, Plans and Operations, and Engineer Advisory. The initial organization called for 144 people, about one-half Army, one-quarter Navy and Marine Corps, and one-quarter Air Force. To

---

[30] MFR, Gray, [1] 8 Jan 66, sub: Briefing of Gen Dunn, MACV Engr Designate, by Navy Bureau of Yards and Docks, Dunn Papers, CMH.

[31] MFR, Gray, 1 Feb 66, sub: Principal Level Conference to Resolve Funding Procedures for the South Vietnam Construction Program, Dunn Papers, CMH; Tregaskis, *Building the Bases,* pp. 204–05; Memo, McNamara for JCS and Secretaries of Military Departments, 12 Feb 66, sub: Construction Approval Procedures for South Vietnam; Memo, Department of Defense (DoD) Comptroller for JCS and Service Assistant Secretaries for Financial Management, 7 Mar 66, sub: Fiscal Procedures and Accounting for Construction in Vietnam, both in MACV Dir of Const, Plans and Opns Div, Development of the Construction Directorate, Dunn Papers, CMH.

CHART 3—MACV CONSTRUCTION DIRECTORATE, FEBRUARY 1966

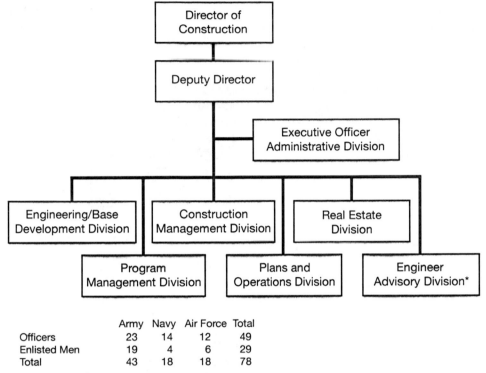

	Army	Navy	Air Force	Total
Officers	23	14	12	49
Enlisted Men	19	4	6	29
Total	43	18	18	78

*Totals do not include the Engineer Advisory Division.

Source: Raymond, Observations on the Construction Program, p. 163

reflect the joint makeup of the directorate, MACV assigned a Navy civil engineering corps captain as the deputy and an Air Force engineer lieutenant colonel as the executive officer. Army engineers headed five of the divisions while Air Force engineers headed the other two (Engineering and Base Development, and Program Management). The Engineer Advisory Division remained entirely staffed by the Army.[32] (*Chart 3*)

When Dunn reported to Westmoreland, he put any doubts about his role to rest. McNamara's memorandums in effect delegated to the director of construction the authority to decide requirements and set priorities, which bothered Westmoreland. In one of Dunn's first meetings with the MACV commander, Westmoreland alluded to the fact that the secretary of defense

---

[32] Msg, COMUSMACV MAC 05240 to CINCPAC, 18 Feb 66, sub: Construction Management Organization, MACV; MACV Directive 415–2, 15 Feb 66, *Construction: Mission and Functions of the Director of Construction*, p. 1, in MACV Dir of Const, Plans and Opns Div, Development of the Construction Directorate, Dunn Papers, CMH; Joint Tables of Distribution, Construction Directorate, MACV, 1 May 66, pp. 1–7, Historians files, CMH.

had given this authority to his construction chief. Dunn responded: "I know for whom I work, and obviously the priorities that are going to be set are those which you feel are the ones needed." After that conversation, Dunn recalled the two had a good working relationship, and "never once did he fail to back me up completely in the decisions that I reached."[33]

Dunn's first task as director of construction was to sort out the growing number of projects required by the three services and to work out a priority list. Kennedy and Raymond, the latter now Dunn's deputy, had thrashed over these same problems. But Dunn had a Defense Department charter that gave him the authority to approve what each service did with its construction funds. This authority clearly allowed him to allocate troops, contractors, or a mix of both to each major project. The breakout for the building effort remained essentially the same. Seabees and Marine Corps engineers supported all the services in I Corps. Army engineers supported the services in the other three corps. The large construction consortium of RMK-BRJ continued to work on projects throughout the country.[34]

Between 1965 and the spring of 1966, MACV's construction priorities began taking a dramatic turn. In 1965, General Westmoreland topped his list with improving airfields and related facilities, then improving roads and railroads, rehabilitating and expanding ports, and building logistics facilities. Putting logistics facilities at the bottom of the priority list, however, created another problem. Deep-draft vessels brought about 90 percent of all military supplies and equipment into the country. Of that amount, construction materials alone constituted about 40 percent. This presented an ironic situation. Increased shipping carrying the materiel for the buildup was backing up off-shore, which, in turn, delayed construction of the ports and storage facilities needed to resolve the shipping backup. The same limited berth space at the Saigon port also had to make way for ships carrying U.S. economic assistance and other commercial shipping. Secretary McNamara, during his visit to South Vietnam in November 1965, voiced his annoyance over what he considered overlapping and conflicting priorities exemplified by the glut of ocean-going ships anchored off the mouth of the Saigon River at Vung Tau waiting their turn to use the Saigon port. Colonel Raymond and the Officer in Charge of Construction came up with a quick estimate to fund the urgently needed port work and presented the defense secretary with a revised program. Following McNamara's visit, Westmoreland reevaluated the priorities and shifted more effort to port improvements and logistical facilities.[35]

For some time, Air Force officials had complained about what they termed inadequate construction support. Even when airfield work ranked at the top of the priority list, the air service criticized the lack of support

---

[33] *Engineer Memoirs, Lieutenant General Carroll H. Dunn*, p. 103 (quotation).

[34] Ibid., pp. 212–13; Tregaskis, *Building the Bases*, pp. 204–05.

[35] Dunn, *Base Development*, pp. 36, 51; Heiser, *Logistic Support*, p. 17; Tregaskis, *Building the Bases*, pp. 188–90; Msg, COMUSMACV MACJOO 42038 to CINCPAC, 28 Nov 65, sub: Priority of Construction, RVN, box 4, 69A/702, RG 334, NARA; Interv, Stuart I. Rochester with Maj Gen Daniel Raymond, 11 Jan 80, pp. 71–72, OCE Hist Ofc; MACV J–4 Briefing, 28 Nov 65, Historians files, CMH.

to complete new airstrips and improve older ones at Cam Ranh Bay, Phan Rang, Qui Nhon, and Tan Son Nhut. As far back as June 1965, the Pacific Air Forces had requested two reinforced Army engineer battalions to operate under Air Force control. Westmoreland, however, insisted on establishing an integrated priority list and keeping the Army engineer troop construction effort under Army control.[36]

When Dunn arrived in Saigon, he found his staff engaged in the planning stages reflecting the recent changes in priorities. Westmoreland had just postponed construction of a jet fighter base and supporting port complex at Tuy Hoa. The air base work, scheduled to start on 1 February, had dropped in priority following the review giving preference to ports. Although the MACV commander confirmed the need for three additional air bases, their siting would be determined by existing and programmed port projects. New air bases requiring port facilities that only supported those bases, like Tuy Hoa, dropped to a lower priority. Besides, MACV had just received $4.3 million to design and construct the first increment of a new large port complex, Project NEWPORT, to improve Saigon's port capacity. In addition, design work by the Officer in Charge of Construction continued for another high-priority job, a new MACV headquarters complex in Saigon. This project, which had Westmoreland's personal interest, required delicate, high-level negotiations for valuable real estate. Negotiations were also under way with Vietnamese officials to resolve real estate problems encountered by RMK-BRJ at its University quarry site outside Saigon.[37]

Then there were additional construction requirements driven by expected increases in U.S. and allied strength. At another joint planning conference held in Hawaii in early February 1966, Westmoreland asked Sharp for more troops. From a seventy-one battalion force arranged with McNamara the preceding summer, the MACV commander now foresaw the need to increase his combat force to 102 infantry battalions, 79 U.S. and 23 allied. The increase of American troop strength to 429,000 included many support troops, and this meant a further increase in the base construction program.[38]

By the spring of 1966, Dunn refined construction priorities using the Defense Department's fifteen functional categories. He divided these categories into five priorities. The command grouped port facilities, airfield pavements, and communications facilities under Priority I. Airfield support complexes, hospitals, maintenance support buildings, and liquid fuel

---

[36] Schlight, *Years of the Offensive*, p, 66; Msg, COMUSMACV MACJ4-EN 22789 to CINCUSARPAC, 2 Jul 65, sub: USARPAC FY 1966 Construction Requirements Priority List; Msg, COMUSMACV MACJ4-EN 27583 to CINCPAC, 6 Aug 65, sub: Engineer Troop Support for USAF in RVN, both in box 26, 77/53, RG 319, NARA.

[37] Schlight, *Years of the Offensive*, p, 66; Msg, COMUSMACV MACJ4-EN 22789 to CINCUSARPAC, 2 Jul 65, sub: USARPAC FY 1966 Construction Requirements Priority List; Msg, COMUSMACV MACJ4-EN 27583 to CINCPAC, 6 Aug 65, sub: Engineer Troop Support for USAF in RVN; Msg, COMUSMACV to CINCPAC, 23 Jan 66, sub: Airfield and Port Construction; Disposition Form (DF), MACV Dir of Const to Ch, Hist Br, MACV, 19 Feb 66, sub: Historical Summary [1–31 Jan 66]. Second and third msgs and DF in Historians files, CMH.

[38] Westmoreland, *A Soldier Reports*, pp. 192–93.

storage followed under Priority II. Priority III included ammunition storage, cold storage, warehouse storage, and administrative buildings. Projects in Priority IV encompassed lines of communications (roads and railways), shed storage, and open storage areas. Cantonments placed last in Priority V. MACV listed the NEWPORT project as the highest priority followed by more port work at Da Nang, Qui Nhon, Cam Ranh Bay, and Vung Tau. The command also listed the other Priority I projects, including the runways, aprons, and taxiways at four air bases (Da Nang, Chu Lai, Phan Rang, and Phu Cat), and several sophisticated communications systems. While outlining the priorities in answer to a query from the U.S. Embassy in Saigon, MACV also pointed out that troop self-help was to be used to the maximum to build the cantonments. Ongoing plans to move many support units out of Saigon resulted in the command giving Long Binh the highest priority in the cantonment category.[39]

A few weeks after his arrival, General Dunn presented an up-to-date status report on military construction to Admiral Sharp during his visit to Saigon. Dunn pointed out that the funding picture, including the anticipated Fiscal Year 1966 supplemental appropriation, now totaled approximately $861 million. Since troop labor involved only the cost of materials, Dunn noted that this amount would total just over $1.2 billion if the work was assigned to a contractor. He also mentioned the add-on of another $1.3 billion in contract costs for new projects outlined in the February conference in Hawaii, increasing the construction program to approximately $2.5 billion in contract values. To bring home the impact of getting the job done with his available resources, Dunn pointed to another chart depicting a month-by-month execution capability by military engineers and contractors. Together, the U.S. construction forces could complete about $601.4 million in work during 1966. This meant that, at the same rate of accomplishment, it would not be until mid-November 1967 before a start could be made on the new requirements.[40]

During the briefing, Dunn emphasized the need to have fund authorization for the added projects before beginning work. At least five to six months' lead time was necessary to obtain certain construction materials. Longer lead times were necessary to get plants and specialized equipment such as dredges. Unless advance procurement funds became available, the MACV construction chief warned that advance procurement would have to come out of the command's operating funds, thus cutting back on the rate of ongoing work.[41]

Dunn believed that by the end of 1966, minimal acceptable facilities could be completed to support tactical and logistical operations. This could only happen after giving close attention to priorities and carefully allocating resources. American forces, however, would still be left "with much less than

---

[39] Memo, Dunn for Dep Amb William Porter, 20 May 66, sub: MACV Military Construction Priorities, Historians files, CMH.

[40] MFR, MACV Dir of Const, 3 Mar 66, sub: General Dunn's Briefing for Admiral Sharp on 3 March 1966, Historians files, CMH.

[41] Ibid.

the needed facilities for the desired degree of efficiency in operations and economy of effort." Although he expected that the installation of additional DeLong piers and other port work would provide adequate port capacities by August 1966, much remained to be done in ancillary facilities—hardstands, transit storage, access roads, and other improvements—to increase the efficiency of port operations. The same applied to the need for tactical jet airfields. He cited the minimum aircraft "beddown" requirements that were developed at a Pacific Command conference in January. Dunn went on to say that the expeditionary airfield construction, and expansion at existing airfields, should meet essential operational needs. Prefabricated runways and additional aprons would be in service that year, but some effectiveness, efficiency, and safety margins would be sacrificed. The permanent airfield pavements and upgrading air base facilities, he noted, would stretch out beyond the end of the year. He warned that such projections were independent of any contingency requirements that might arise.[42]

The director of construction went on to outline starts on new projects, progress on others, and some improvements in funding and management. During McNamara's visit the previous November, MACV received authorization to start Project NEWPORT and the new MACV headquarters. Washington also made funds available for advance procurement of long-lead-time items. Work was now under way at Newport. Long Binh was being transformed into a major logistical base. Clearing a site for the new MACV headquarters building at Tan Son Nhut Air Base had begun, and prefabricated building materials were on order. Enough DeLong pier units to provide sixteen deep-draft berths were due in April, and installation was expected by September. Dunn expressed some satisfaction in the relief granted by the Defense Department in financial accounting and funding flexibility. In addition, McNamara had gone to Congress with a request for a $200 million contingency fund to meet unforeseen construction requirements worldwide.[43]

As far as construction assets were concerned, Dunn could claim some progress in the first three months of 1966. The 18th Engineer Brigade increased from nine to eleven battalions. The 30th Naval Construction Regiment would expand from four to five battalions in March. Two Air Force Red Horse squadrons had arrived at Phan Rang and Cam Ranh Bay Air Bases. RMK-BRJ's workforce now stood at 28,066, and the value of the consortium's physical plant had grown from $32.9 to $53.5 million. The Officer in Charge of Construction had brought in two dredges with seven more on the way. In appraising the construction picture to Sharp after three weeks on the job, Dunn summarized: "A lot of progress made; a long way to go; a lot of problems to solve; and, a good solid foundation has been made."[44]

The timely arrival of the right type and enough construction materials stood high among the construction chief's concerns. He communicated

[42] Ibid.
[43] Ibid.
[44] Ibid. (quotation); CHECO Rpt, PACAF, 1 Sep 69, sub: Project RED HORSE, pp. 5, 9, copy in Historians files, CMH (hereafter cited as Project RED HORSE).

directly with General Frank S. Besson Jr., commander of the Army Materiel Command, to work out a forecast of tonnages and a shipping plan. Even under ideal circumstances, Besson noted it would take between 120 to 150 days lead time to ship materials from the United States. The Materiel Command was charged with getting more DeLong piers, and Dunn prodded Besson to include these units as early as possible. To cut down on production and shipping lead times, Besson pressed forth with negotiations to have the piers built in Japan. The two generals also exchanged ideas on a proposal to take World War II T–2 (Liberty ship) tankers out of storage, convert them into electrical power-generating ships, and move them to port and base complexes. The contractor operator would then tie the power ships into a land electrical distribution system, to be built and maintained by the contractor. Dunn suggested such systems for Cam Ranh Bay, Qui Nhon, and Nha Trang, and possibly Vung Tau and Long Binh.[45]

Another method to speed up procurement of construction equipment and material entailed creative financing. Earlier, in September 1965, Secretary McNamara had authorized some advanced procurement for the services in anticipation of additional funding, but even these expedient measures fell short. In January 1966, McNamara authorized the Bureau of Yards and Docks "advance obligation authority." This temporary procedure authorized the bureau to use $200 million in the Navy Stock Fund for additional advanced procurement to support projects in Vietnam. The Navy engineering office could obligate money in contract arrangements, but no expenditures would be made until Congress appropriated the $700 million in supplemental funds requested for 1966. In late March, Dunn advised the chief of engineers that his office authorized the Officer in Charge of Construction to purchase over $40 million in equipment, over $33 million for generators, and $4 million for DeLong piers at Da Nang. He added that the advance purchase of one thousand prefabricated buildings and thirty highway bridges would follow, thus cutting down on the critical lead time to get these projects under way. When Congress appropriated the supplemental funds, the services reimbursed the Navy Stock Fund.[46]

Ironically, Dunn had to put up with some reduction in funding because of urgent requirements to expand training bases in the United States. The Army's funding requirement to improve and build facilities at training centers and schools and for new units turned out to be greater than anticipated. In early June 1966, with the Defense Department's concurrence, the Army reprogrammed some construction funds destined for Southeast Asia. This amounted to a $37.1

---

[45] Msg, Gen Frank S. Besson, CG, U.S. Army Materiel Cmd, WDC 3441 to Dunn, 19 Mar 66; Dunn SOG 0287 to Besson, 3 Apr 66; Dunn SOG 0297 to Besson, 5 Apr 66; Dunn SOG 0316 to Besson, 12 Apr 66, all in Dunn Papers, CMH.

[46] Army Buildup Progress Rpts, 6 Oct 65, p. 38, 3 Nov 65, p. 56, 2 Mar 66, p. 67, CMH; Memo, McNamara for Service Secretaries, 22 Sep 65, sub: Advance Procurement of Construction Materials and Construction Equipment for U.S. Military Construction in South Vietnam, Historians files, CMH; Tregaskis, *Building the Bases*, pp. 198–99; JLRB, Monograph 6, *Construction*, p. 104; Msg, Dunn SOG 152 to Maj Gen Thomas J. Hayes, Dep Ch of Engrs, OCE, 22 Mar 66, Dunn Papers, CMH.

million cut and caused major readjustments to Army projects in Vietnam. To make matters worse, the Officer in Charge of Construction informed Dunn of escalating contract costs. Calling upon the tenant units to help build their own cantonments reduced contract labor costs and spared engineering troops for higher-priority projects. In responding to a congressional query passed on to him by Brig. Gen. Charles C. Noble, Sheridan's chief for Southeast Asia construction, Dunn noted that many combat and support units had already made extensive use of self-help to improve their living conditions.[47]

Perhaps Dunn's severest challenge as construction chief centered on the Air Force's demand to build its own jet fighter base at Tuy Hoa. Not at all happy with Westmoreland's construction priorities, Air Force officials in Washington pressed demands for the immediate construction of more airfields. In February, the air service proposed hiring its own contractor, thus bypassing the use of Army engineers and the Navy's prime contractor. This so-called turnkey proposal invited the contractor to design and build the base in a single package and literally turn over the keys of the completed facility to the user. At first Dunn, Westmoreland, and Sharp did not agree, but the Air Force argued that its contractor would add to MACV's construction assets. Both sides continued arguing who should build the base and whether or not to shift the location to Hue. In early May, McNamara decided to let the Air Force build the base at Tuy Hoa. Before the month was out, the Air Force had selected a contractor, who moved quickly to build an interim AM2 aluminum matting runway, and then a permanent runway and all of the facilities for a complete air base. Six months later, F–100 jet fighters touched down on the interim runway.[48]

Despite the Pentagon's insistence on making Dunn the director of construction, he served in that job for only five months. By 1 July, Westmoreland had made his move to make the engineer general his J–4. Since airfields, ports, and bases remained high priorities, the MACV commander still preferred to have an engineer officer as head of his logistical staff. In early April, he discussed the matter with Secretary Vance during his visit to Vietnam. Vance, who earlier rejected Westmoreland's proposal to appoint Dunn as J–4, now agreed. "This, I hope settles the matter," Westmoreland wrote in his diary, "which was somewhat annoying to me several months ago when JCS proceeded to organize the staff and failed to support a position that I had very strongly taken." He then advised Admiral Sharp, General Wheeler, and General Johnson that Dunn would take over the MACV logistical position on 1 July and that

[47] Army Buildup Progress Rpts, 22 Jun 66, p. 45, 22 Aug 66, p. 11, CMH; Richard P. Weinert, *The Role of USCONARC in the Army Buildup, FY 1966* (Fort Monroe, Va.: U.S. Continental Army Command, 1967), pp. 172–73; Msg, Dunn SOG 0579 to Maj Gen Charles C. Noble, Chief Southeast Asia Construction, Asst Sec Def (Installations and Logistics), DoD, 21 Jun 66, Dunn Papers, CMH.

[48] Tregaskis, *Building the Bases*, pp. 219–22; Schlight, *Years of the Offensive*, pp. 87–88, 120–22, 155; Dunn, *Base Development*, pp. 62–63; Msg, COMUSMACV MACDC 06315 to CINCPAC, 27 Feb 66, sub: USAF Proposal for Additional Cost Plus Fixed Fee Construction Contractor in Vietnam, box 4, 69A/702, RG 334, NARA; Msg, Dunn SOG 152 to Hayes, 22 Mar 66, Dunn Papers, CMH.

General Dunn visits the 45th Engineer Group headquarters in Qui Nhon in late 1967 as the MACV J–4. From left to right, seated: Brig. Gen. Charles M. Duke, commanding general of 18th Engineer Brigade; General Dunn; and Col. Kenneth T. Sawyer, commanding officer of the 45th Engineer Group. Group staff in the background includes the author (second from the left).

Raymond would become the new director of construction. When Dunn took on his new job, he would be a major general and Raymond would have his star. As J–4, Dunn exercised general guidance over the Construction Directorate until his departure in September 1967. By that time, Westmoreland's top logistics priority had become supply and supply management, and he recruited a skilled manager in that field to replace Dunn.[49]

Operational Control

As more engineers arrived in Vietnam, operational control became increasingly a matter of debate and concern. Tactical commanders felt that the workload and the demands placed on them justified the need for additional combat

[49] Quote from Westmoreland Historical Notes, no. 5, 6 Apr 66; Msg, Westmoreland MAC 2751 to Sharp, Wheeler, and Gen Johnson, 7 Apr 66, Westmoreland Message files, CMH; Westmoreland, *A Soldier Reports*, p. 509; *Engineer Memoirs, Lieutenant General Carroll H. Dunn*, p. 104.

engineer units. Organic engineer units lacked the manpower and equipment to handle all the tasks involved in base development. Just to move the equipment of a division or a brigade by road to a bivouac area or base camp dictated extensive engineering effort. In such cases, nondivisional engineer units proved valuable with their greater capability to provide materials such as rock, sand, culverts, and earth fill.[50]

The operational control of units under General Ploger's command became an issue from the moment the first engineer units came ashore. The initial emphasis on base development made engineer commanders and staff officers reluctant to transfer any of the few units available to the tactical commands. At the end of 1965, the 18th Engineer Brigade had five combat battalions in country, all committed to construction, with two more expected momentarily. Under more conventional circumstances, an Army corps would have commanded at least one engineer group or perhaps one group per division, with the groups receiving directions from the corps engineer. In Vietnam, however, General Westmoreland had established the Field Force, Vietnam, at Nha Trang. Lt. Gen. Jean E. Engler, who became the deputy commanding general, U.S. Army, Vietnam, in January 1966, agreed that 18th Engineer Brigade units would not be placed under field force command. He and Ploger felt that the demands of the construction program were such that the engineer brigade had to keep control of all its construction assets, including the combat battalions. Ploger also emphasized that engineers placed under the control of tactical commanders could be misused on pet or unessential projects.[51]

When Ploger had briefed General Westmoreland the previous November, he felt then that the 18th Engineer Brigade could ill afford to attach units to the tactical commands. Besides, most of the brigade's equipment could not be easily transported aboard helicopters to support the fast-flowing combat operations. Attached units, Ploger said later, "would have spent more time traveling about the countryside than working as engineers." The 18th Engineer Brigade commander believed "every hour and every day engineer troops spent away from the work site constituted a serious waste."[52]

The command structure that Ploger and his subordinate commanders developed, both for construction and operational support, corresponded closely to the principle of general support used by artillery units. While the major construction projects influenced the location and distribution of 18th Engineer Brigade units, the centralized system allowed engineer commanders to dispatch the closest units to support tactical operations. In Vietnam, the field force commands (Field Force, Vietnam, in II Corps, later designated I Field Force, Vietnam, when II Field Force, Vietnam, was established in early 1966) had their own planning sections but no assigned troops, and had to request the use of the general support engineer units. Each supporting unit commander could disapprove the request in whole or in part. This rarely happened since Ploger made it clear to all

[50] Ploger, *Army Engineers*, pp. 140, 186–87.
[51] Interv, Bower with Ploger, 8 Aug 67, pp. 22–23.
[52] Ploger, *Army Engineers*, pp. 140–41; Interv, Bower with Ploger, 8 Aug 67, p. 17 (quoted words).

his commanders and officers that operational support missions had top priority. If they could not handle a request, they were to pass it up through the engineer chain. Although this system was designed to suit the situation in Vietnam, tactical commanders still preferred direct control over supporting engineers.[53]

The 70th Engineer Combat Battalion's general support mission at An Khe is a good example. Experience showed that it took at least one battalion-month to prepare a site for use by one division. The 70th Engineer Battalion's attempt to help set up the 1st Cavalry Division's base camp in less time pointed out the need for more engineer troops, and the battalion remained at An Khe long after the arrival of the division, improving facilities and living conditions. In time, as more engineer battalions arrived in Vietnam, the divisions found at least one of the 18th Engineer Brigade's battalions conveniently located at or near their base camps. When the 1st Cavalry Division deployed to northern Binh Dinh Province, backup support was provided by the nearby 19th and 35th Engineer Battalions. In the western highlands, the 20th and 299th Engineer Battalions at Pleiku were often summoned to help division engineers.[54]

Oddly enough, occasional circumstances caused the 18th Engineer Brigade to control divisional engineering units. An advance element of the 25th Division—Company C, 65th Engineer Battalion—served under 18th Engineer Brigade and 35th Engineer Group control until the arrival of the 3d Brigade, 25th Division. The divisional engineer company did a variety of construction work that included joining forces with elements of light equipment and dump truck companies at Dong Ba Thin airfield complex. Company C also carried out the combat engineers' secondary role as infantry by providing convoy security to the 62d Engineer Construction Battalion when it moved to Phan Rang.[55]

Ploger's insistence on a centralized system of control not only had advantages from an engineering standpoint, but also benefitted nearby tactical units. Since engineer groups and battalions operated in assigned areas of responsibility, the need to move units from other areas over long distances was greatly reduced. This also gave the enemy fewer opportunities to ambush slow-moving convoys carrying construction equipment. The assignment of a given area to an engineer unit allowed it to become familiar with the local terrain and conditions. Over a period of time, this helped the divisions and brigades when they operated in these localities. An engineer battalion based near the new area of operations of redeployed divisions and brigades could be called upon for support and local intelligence and continue to apply its experience there to the nearby combat unit's advantage.[56]

The peculiar requirements of cost accounting for construction projects also dictated the need for centralized control. The accounting procedures

---

[53] Ploger, *Army Engineers*, pp. 142, 186; Interv, Bower with Ploger, 8 Aug 67, pp. 17–18; Interv, Sowell with Ploger, 21 Nov 78, sec. 9, p. 10; Ltr, Ploger to All Officers, sub: Concept of Engineer Operations SVN, 16 Nov 66, box 31, 77/51, RG 319, NARA.

[54] Ploger, *Army Engineers*, p. 187.

[55] Ibid., p. 55; Galloway, "Essayons," p. 96; *The 25th's 25th . . . in Combat, Tropic Lightning*, pp. 187–89.

[56] Ploger, *Army Engineers*, p. 188.

laid out by the Defense Department specified that an accurate record be kept showing the dollar value of material used in and the number of man-hours expended on any given project at any point in time. The task of keeping a running account of total costs and manpower used on a project evolved into a difficult procedure even when passed up through the various headquarters. If this function had been done by divisional battalions or brigade units assigned or attached to tactical headquarters, separate reporting chains would have been necessary. Along with this responsibility would have gone the administrative burden and problems of verification and enforcement that would have complicated the already rigorous demands placed on the tactical organizations in the field. As General Ploger later pointed out, even while doing base camp self-help projects, tactical units looked to an engineer unit to do all the cost accounting. If the engineers providing the supervision and guidance for the self-help projects did not handle the flow of construction materials as well as the bookkeeping, the magnitude of the paperwork would have overwhelmed the tactical units.[57]

Toward an Engineer Command

During the second half of 1966, the 18th Engineer Brigade gained more critically needed units and group headquarters to share command and control responsibilities. The 45th Engineer Group (Construction) under Col. George M. Bush traveled from Fort Bragg, North Carolina, and arrived at Cam Ranh Bay in early June. Upon settling in at Dong Ba Thin across the bay, the 45th Group assumed control over the 20th and 39th Engineer Battalions, two light equipment companies, and a dump truck company from the 35th Group. On 31 July, Lt. Col. James L. Kelly's 577th Engineer Battalion (Construction) from Fort Benning arrived and joined the 45th Group. Also that month, the 79th Engineer Group (Construction) from Fort Lewis, Washington, set up its headquarters at Long Binh. The 79th Group, commanded by Col. David C. Clymer, took over control of the 168th and 588th Engineer Battalions and two light equipment companies from the 159th Group. Two later arrivals, the 27th Engineer Battalion (Combat) in September and the 86th Engineer Battalion (Combat) in October, expanded the group to four combat battalions and several smaller units. The 159th Group at Long Binh added the 169th Engineer Battalion (Construction) from Okinawa in May, which together with the 46th and 62d Construction Battalions comprised a construction group made up of three construction battalions. By mid-November, two more combat battalions, the 14th from Fort Bragg and the 35th from Fort Lewis, arrived, with the 14th going to the 35th Group and the 35th to the 45th Group. In addition, another divisional engineer unit served with the 18th Brigade. In October, the 15th Engineer Battalion, 9th Infantry Division, preceded the division's move from Fort Riley to III Corps and the Mekong Delta in December 1966 and January 1967. For the time being, the 15th Engineer Battalion came under the operational control of the 159th Group. In a little over one year, Army engineer

[57] Ibid., pp. 188–89.

units under the 18th Brigade had grown to 5 groups, 19 battalions, 21 companies, 2 separate platoons, and 7 detachments.[58]

Since the Army had established three principal logistical support bases at Qui Nhon, Cam Ranh Bay, and the Saigon–Long Binh area, General Ploger initially superimposed his three groups' areas of responsibility in and around these base areas. The 937th Group at Qui Nhon provided engineering construction and combat support to U.S. forces in northern II Corps. The 35th Group at Cam Ranh Bay did the same in southern II Corps, and the 159th Group at Long Binh supported both III and IV Corps. Adjustments to these geographic areas of responsibility followed the arrival of the 45th and 79th Groups. Projects and engineer units at Tuy Hoa, Nha Trang, Dong Ba Thin and the northern Cam Ranh peninsula were transferred from the 35th Group to the 45th Group. Most of III Corps and the units in the area were transferred to the 79th Group, which permitted the 159th Group to concentrate on base development projects in the Saigon–Long Binh area.[59]

A group's designation as either construction or combat had only a small effect on the nature of its assigned mission or type of units assigned to it. Although four of the group headquarters were organized to oversee construction units and one to lead combat engineers, the assignment of geographic areas and work in these areas led to the attachment of a mix of combat and construction battalions to groups regardless of their designation. By 1 November 1966, seven of the ten combat engineer battalions worked under two of the four construction groups while one of the seven construction battalions served under the direction of the brigade's only combat group, the 937th. With the arrival of the 35th Engineer Battalion in early November, another combat battalion was added to a construction group.[60]

Other measures were also taken to keep track of operations. Control of five groups put an added burden on the 18th Brigade headquarters at Tan Son Nhut. As a partial remedy, General Ploger directed that each group assign a liaison officer to the brigade's operations section. The activation of a second U.S. corps-level command, II Field Force, Vietnam, under General Seaman at Long Binh, in March 1966, to coordinate the U.S. ground war in III Corps prompted the brigade to establish a Combat Support Section. For the sake of consistency, General Larsen's Field Force, Vietnam, at Nha Trang, which filled the same function in II Corps, received a new name, becoming I Field Force, Vietnam. Serving as the link to the field forces, the Combat Support Branch coordinated operational support for the two field forces.[61]

---

[58] 45th Engr Gp Hist, 10 Sep 70, p. 1; ORLLs, 1 May–31 Jul 66, 18th Engr Bdc, 26 Aug 66, pp. 4–5, 18, 1 Aug–31 Oct 66, 18th Engr Bde, 30 Nov 66, p. 13, and 1 Aug–31 Oct 66, 577th Engr Bn, 31 Oct 66, p. 3, all in Historians files, CMH; Ploger, *Army Engineers*, pp. 131–33; Galloway, "Essayons," pp. 136, 138–39, 144, 148, 155.

[59] Ploger, *Army Engineers*, p. 132 and maps pp. 85, 132; ORLL, 1 May–31 Jul 66, 18th Engr Bde, p. 4; ORLL, 1 May–31 Jul 66, I FFV, 25 Aug 66, p. 42 and incl. 46, Historians files, CMH.

[60] Galloway, "Essayons," pp. 102–03; ORLL, 1 May–31 Jul 66, I FFV, p. 42, and incl. 46.

[61] Eckhardt, *Command and Control*, pp. 52–54; ORLLs, 1 Feb–30 Apr 66, 21 Jul 66, 18th Engr Bde, p. 1, Historians files, CMH, and 1 May–31 Jul 66, 18th Engr Bde, p. 6; Ploger, *Army Engineers*, p. 133.

Although units in the brigade submitted progress reports and communicated via radio-teletype, Ploger and subordinate commanders and staffs also kept abreast of progress by frequent visits. Helicopters and other small aircraft parceled out between the brigade and group headquarters allowed commanders and staff officers to check on even the most remote projects. Ploger made it a point to check with the supported unit commanders. In this way, he reiterated priorities and made sure 18th Brigade units complied with his policies. One policy insisted that the engineers did not enjoy any better accommodations or facilities than the other units. During his first trip to Qui Nhon not long after his arrival, Ploger found out that the engineers had built a very fine service club with materials in critically short supply. After pointing the responsible officers "in the right direction," he later reflected that the low-priority facility "could have come back to bite anybody associated with it if we hadn't changed that fairly rapidly."[62]

With five groups in country and a sixth due in, a change to the Army engineer command structure seemed imminent. Plans for a larger organization centered on an engineer command. Ploger visualized an Army command engineer assisted by a deputy overseeing two brigades. He felt that these two positions and the two brigades could be handled by four brigadier generals. Further, Ploger considered setting up an integrated staff, that is, one engineering office for both the engineer command and U.S. Army, Vietnam. General Engler, however, favored an engineer command headed by a major general with the command disassociated from the U.S. Army, Vietnam, staff as was done with the 1st Logistical Command. He preferred to have engineering matters processed through the U.S. Army, Vietnam, G–4, a common practice since most special staffs coordinated actions through one of the general staffs. Besides, Ploger on 1 November 1966 would become a major general and outrank all the general staff.[63]

Before the end of 1966, General Ploger took steps to establish the U.S. Army Engineer Command, Vietnam (Provisional). In anticipation of the reorganization and to ease coordinating with the groups in II Corps, Ploger on 18 November moved part of his staff to Dong Ba Thin under Col. Paul W. Ramee, the brigade's deputy commander. Ramee then established a provisional northern headquarters to take control of the 35th, 45th, and 937th Groups. The 921st Engineer Group (Combat), which had arrived at Dong Ba Thin between late September and early October, had remained uncommitted, and its troops and equipment joined the new northern headquarters. With the activation of the provisional engineer command on 1 December, the northern headquarters became the 18th Brigade. Later that month, as part of General Westmoreland's effort to move U.S. headquarters and troops out of Saigon, the new Engineer Command headquarters—the former 18th Brigade staff

---

[62] Interv, Sowell with Ploger, 21 Nov 78, sec. 8, pp. 18–19, 25 (quoted words, p. 19).

[63] Interv, Bower with Ploger, 8 Aug 67, pp. 7–11; Interv, Sowell with Ploger, 21 Nov 78, sec. 9, p. 6; Ltr, Maj Roy Fowler, CO, 26th Mil Hist Det, to Charles B. MacDonald, Office, Ch of Mil Hist, 18 Aug 67, Historians files, CMH.

The 18th Engineer Brigade headquarters building at Dong Ba Thin

that remained in Saigon—moved into a partially completed cantonment at Bien Hoa.[64]

Shifting of responsibilities and personnel between Engineer Command and U.S. Army, Vietnam, followed. During a transition period in December, Engineer Command took over functions such as mapping and intelligence, base development planning, engineer design, construction management, and materials. In turn, Engineer Command transferred three officers to the U.S. Army, Vietnam, G–3 to carry out staff work dealing with engineering doctrine, force development, and organization and equipment. Engineer Command also transferred nineteen officers and nineteen enlisted men to the U.S. Army, Vietnam, G–4 to man the newly organized Installations Division primarily to handle the construction program, facilities engineering, and real estate. Although he no longer had a staff at U.S. Army, Vietnam, headquarters, Ploger still served as the Engineer, U.S. Army, Vietnam. Ploger later recalled that Engler "didn't want anybody making decisions affecting engineering that didn't have the concurrence of the engineer command so, in effect, I was a staff officer for him but my staff did not function inside USARV."[65]

---

[64] Ploger, *Army Engineers*, pp. 133–35; ORLLs 1 Aug–31 Oct 66, 18th Engr Bde, p. 5, 1 Nov 66–31 Jan 67, U.S. Army Engr Cmd, Vietnam, 31 Jan 67, pp. 1, 22–24, 1 Aug–31 Oct 66, 921st Engr Gp, 15 Nov 66, p. 7, all in Historians files, CMH; General Orders 6525 and 6526, USARV, 27 Nov 66.

[65] Rpt of Visit to Various Headquarters in Vietnam, 28 Feb 67, incl. 7, p. 1, OCE Liaison Officer Trip Rpt no. 6, 6 Mar 67, OCE Hist Ofc; Interv, Bower with Ploger, 8 Aug 67, pp. 13–14

It did not take long to realize that this division of staff responsibilities would create problems. U.S. Army, Vietnam, began to refer construction matters to the Installations Division, not Engineer Command and the MACV Directorate of Construction as envisioned by Ploger. He took exception when the Installations Division, which he viewed primarily as a facilities engineering office, began to evaluate and modify construction programming and scheduling, items he previously approved. Ploger considered this a waste of scarce manpower, as evidenced by the Installations Division's assignment to three officers the nearly impossible job of keeping track of the work done by his command. Ploger tried to persuade the G–4 that no one in U.S. Army, Vietnam, headquarters had the qualifications to review construction matters. His plea met without success, and he regretted his agreement to remove Engineer Command from U.S. Army, Vietnam, headquarters.[66]

With the establishment of the provisional engineer command, Ploger's span of control improved from directly supervising five groups to supervising one brigade and two groups. The 18th Brigade retained control over the 35th, 45th, and 937th Engineer Groups and served as the major operational support link with I Field Force. Since a second brigade headquarters would not arrive for some time, Engineer Command directly controlled the other two groups and dealt directly with II Field Force. The command further adjusted geographic areas of responsibility. The 937th Group moved inland to Pleiku and assumed responsibility for the Central Highlands. The 45th Group had moved from Dong Ba Thin to Tuy Hoa and then to Qui Nhon and took over the remaining area in northern II Corps. Remaining at Cam Ranh Bay, the 35th Group extended its area north to Tuy Hoa. Reporting directly to Engineer Command, the 159th Group concentrated its base construction efforts in the Saigon–Long Binh–Vung Tau area and the new base under construction for the 9th Division at Dong Tam in the Mekong Delta. Also reporting directly to Engineer Command was the 79th Group, whose area of responsibility comprised the remainder of III and IV Corps. This arrangement would remain until March 1967, when the 34th Engineer Group (Construction) arrived and assumed area responsibility for IV Corps and the Vung Tau enclave in III Corps. The arrival of the 20th Engineer Brigade in August 1967 allowed the Engineer Command to place the groups operating in III and IV Corps under the control of the second brigade, thereby reducing to two the number of headquarters reporting directly to the command. Despite these shifts, General Ploger took some satisfaction that "there hasn't been a shade of difference in the way things get done in the field."[67]

(quotation); Ploger, *Army Engineers*, p. 135; Interv, Maj Robert H. Van Horn, 3d Mil Hist Det, with Col Fred M. Walker, Exec Officer, USARV Engr Sec, 27 Dec 66, pp. [1–2], VNIT 31, CMH.

[66] Interv, Bower with Ploger, 8 Aug 67, pp. 14–16, 18–21; ORLL, 1 May–31 Jul 67, Engr Cmd, Vietnam, p. 9; Interv, Van Horn with Walker, 27 Dec 66. Walker recalled that the Installations Division was to be responsible for the overall guidance of the Army construction program, base development, repairs and utilities, and real estate. Regardless, Ploger objected to two USARV engineer staff sections.

[67] Ploger, *Army Engineers*, pp. 136–38; Interv, Bower with Ploger, 8 Aug 67, pp. 16–18 (quoted words, p. 18).

As 1966 drew to a close, Ploger viewed the operations of Engineer Command with satisfaction. Nearly all the nondivisional engineer units in Vietnam came under his command. This command arrangement did not follow Army doctrine for a theater of operations, which typically had engineer combat groups assigned to corps and field armies and engineer commands with the construction units in the rear echelons. However, Ploger and senior engineer officers considered this the best approach under the circumstances. Engineer groups and most battalions in Vietnam were assigned geographic areas of responsibility, and they simultaneously carried out construction and operational support missions, with the latter getting priority. Engineer Command could draw upon the talents of over 16,000 soldiers in over fifty units, varying in size from detachments to battalions, to take on the full range of engineering tasks.[68]

Other Engineers

Other Army engineers and contractors also contributed to the construction effort. When not supporting operations in the field, division and brigade engineer units turned to base camp construction. The 1st Logistical Command utilities detachments, which were organized to carry out facilities engineering, had a limited construction capability. Pacific Architects and Engineers, which reported to the logistical command, continued to concentrate on construction. By the end of 1966, Pacific Architects and Engineers had increased its manpower to 16,000 people at thirty-five installations.[69]

Navy and Marine Corps engineers also expanded their construction capabilities. By the end of 1966, the 30th Naval Construction Regiment directed the efforts of eight Seabee battalions in I Corps. Four Seabee technical assistance teams, which had evenly divided their attention since late 1963 to build or improve Special Forces camps and carry out civic action projects for the U.S. Agency for International Development, which succeeded the U.S. Operations Mission, were now totally committed to support the aid mission during their six-month tours. The marines now had five engineer battalions—two supporting the 1st and 3d Marine Divisions and three heavier force battalions operating under the III Marine Amphibious Force. Like the Army engineers, the Seabees found it necessary to organize a larger command under a flag officer. On 1 June 1966, the newly established 3d Naval Construction Brigade set up shop in Saigon. R. Adm. Robert R. Wooding, who took over as the Officer in Charge of Construction in December, assumed a second hat as commanding officer of the 3d Brigade. By May 1966, RMK-BRJ's workforce expanded to more than 40,000 and 3,700 pieces of equipment. During the summer, the consortium's strength reached a peak of some 50,000 working at forty project sites. In October, the firm's monthly work in place reached $44 million and

---

[68] Rpt of Visit to the 18th Engr Bde, 29 Oct–17 Nov, incl. 7, tab A, OCE Liaison Officer Trip Rpt no. 5, 9 Dec 66, OCE Hist Ofc.

[69] PA&E History, 1963–1966, p. 42.

continued to grow at a fast pace. To supervise the consortium's efforts, the Saigon office increased its workforce in early 1967 to about one thousand.[70]

The Air Force demonstrated its air base maintenance and modest construction capabilities at the major air bases. Facility maintenance engineers at bases in the United States provided a good foundation to draw upon. When Air Force wings deployed to Vietnam, their base civil engineering maintenance forces went with them. By the end of 1966, thirty Prime Beef teams totaling nearly one thousand airmen completed projects costing nearly $5 million. In January 1966, the Air Force deployed its first Red Horse civil engineering squadron, the 555th, to Cam Ranh Bay, and before year's end five of the Air Force engineering squadrons were working at major air bases in South Vietnam. Although the squadrons had only 400 men each, they usually employed large numbers of Vietnamese workers. As noted earlier, the Air Force also received the go-ahead to administer its own construction contract to build Tuy Hoa Air Base. The contractor, Walter Kidde Constructors, Inc., of New York, had authorization to employ almost 1,700 people in its multinational workforce.[71]

Two contractors involved with construction and maintenance of facilities were added in 1966. In March, the Army Materiel Command awarded a contract to Vinnell Corporation of Los Angeles, California, to provide electrical power at major bases. The Army withdrew eleven T–2 tankers from the Maritime Reserve Fleet, and Vinnell converted them to floating electric power-generating barges for use at bases at Cam Ranh Bay, Qui Nhon, Nha Trang, and Vung Tau. The following month, Vinnell got the go-ahead to begin work on land-based electrical generation and distribution systems at these bases as well as at Long Binh. While Pacific Architects and Engineers provided the Army with base maintenance and construction support under the 1st Logistical Command in II, III, and IV Corps, the Navy in 1966 called upon Philco-Ford Corporation to provide skilled personnel to work under the Public Works Office, Naval Support Activity, Da Nang, in I Corps. The Public Works Office supervised a base maintenance force made up of Seabees, local nationals, and personnel furnished by the contractor.[72]

Engineer elements belonging to the allied forces were another construction resource. As early as March 1965, the Republic of Korea had dispatched a medical and engineer group known as the Dove Force to work on civic action projects near Saigon. The two South Korean divisions that deployed to II Corps in 1965 and 1966 had organic engineer battalions. In the summer of 1965, a company-size engineering element accompanied the Australian

---

[70] Tregaskis, *Building the Bases*, pp. 281, 287–88, 297, 300; "Helping Others Help Themselves," p. 30; Jack Shulimson, *U.S. Marines in Vietnam: An Expanding War, 1966* (Washington, D.C.: History and Museums Division, Headquarters, U.S. Marine Corps, 1982), p. 292; Army Buildup Progress Rpt, 18 May 66, p. 35, CMH; Summary Rpt of Activities of 3d Naval Const Bde, 1 Jun–30 Nov 66, in U.S. Naval Forces, Vietnam, Monthly Historical Summary, Nov 66, app. III, box 9, 69A/702, RG 334, NARA.

[71] USAF Airfield Construction in South Vietnam, pp. 4–13, 39, 70, 119–20; Project RED HORSE, pp. 5, 9, 12, 16, 26, 30; Tregaskis, *Building the Bases*, p. 158; Dunn, *Base Development*, pp. 90–91.

[72] Tregaskis, *Building the Bases*, p. 224; Dunn, *Base Development*, pp. 78–79.

Task Force. New Zealand had contributed an engineer platoon as early as mid-1964 to do civic action work. That platoon was replaced the following year by an artillery unit to support the Australian Task Force in southern III Corps. Although the Philippines did not send a combat force, it did send a 2,000-man Civic Action Group to Vietnam in late July 1966. This force, which included nearly 500 engineers, embarked on a series of civic action projects in Tay Ninh Province and the Saigon region.[73]

By the end of 1966, the engineers were well organized to carry out the construction program and support the troops in the field. The establishment of the director of construction meant that General Westmoreland exercised direct control of the construction effort in Vietnam, including direction of the Navy's Officer in Charge of Construction and its construction consortium, RMK-BRJ, in areas of project assignment, priorities of effort, and standards of construction. Through his construction chiefs, the MACV commander controlled the use and allocation of all construction resources in Vietnam. As for the Army engineers, General Ploger's creation of an engineer command eased the coordination of the increasing numbers of engineer units entering the country. The dual-role concept as commander of nearly all engineer troops and chief engineer staff officer for U.S. Army, Vietnam, was retained. Even though the field force engineers lacked their own engineer units, Ploger's principle of priority support to tactical commanders provided the engineers needed in any combat operation.

---

[73] Ploger, *Army Engineers*, pp. 155–56; Lt. Gen. Stanley R. Larsen and Brig. Gen. James L. Collins Jr., *Allied Participation in Vietnam*, Vietnam Studies (Washington, D.C.: Department of the Army, 1975), pp. 60–64, 90, 105, 123.

5

Building the Bases, 1966

Like the previous year, most engineer support in 1966 concentrated on base development. In late 1965, the 18th Engineer Brigade devoted 85 to 90 percent of its effort to construction and the balance to combat support, a ratio that remained constant until the autumn of 1966. Planning and construction focused on developing Saigon, Da Nang, Qui Nhon, and Cam Ranh Bay as the major receiving deep-draft ports—the so-called logistics islands or enclaves—and nearby roads and airfields to improve the delivery, storage, and distribution of materiel. Inland bases at An Khe, Pleiku, and the approaches to Saigon were also heavily developed. All the lodgments included supporting services: maintenance, supply, transportation, hospitalization, communications, and quarters to house troops. The development of major air bases, capable of handling large cargo and jet aircraft, in or near these complexes, served as a major link in the delivery of critically needed supplies. As the year wore on, the complexes took on an air of relative permanence as metal prefabricated and wooden buildings replaced tents and open spaces.[1]

The Cam Ranh Complex

During 1966, the sandy Cam Ranh Bay peninsula and surrounding environs grew into a vast military complex. Plans called for the site to host port facilities, airfields, and depots for all classes of supplies, hospitals, cantonments, and other facilities. Civilian contractors played a major role in the area's development. RMK-BRJ worked on the new airfield and logistical facilities, and DeLong Corporation installed portable deep-draft piers. Pacific Architects and Engineers set up facilities engineering shops. The Air Force dispatched civil engineering units to work on the air base, and the Navy sent Seabees to develop the naval facility. Cam Ranh Bay was a microcosm of the theater's construction effort.[2]

In late April 1966, two of the DeLong barges for the third general cargo pier arrived. In preparation the 497th Port Construction Company completed a sheet-pile bulkhead between the two existing piers. A routine for installing the piers had been established: the 35th Engineer Group built connecting causeways, abutments, roads, and hardstands; the DeLong Corporation

---

[1] Rpt of Visit to the 18th Engr Bde, 10 Feb–3 Mar 66, incl. 6, p. 1, OCE Liaison Officer Trip Rpt no. 1, 15 Mar 66, OCE Hist Ofc.

[2] ORLLs, 1 Feb–30 April 66, 35th Engr Gp, 15 May 66, pp. 1–2, and 1 May–31 Jul 66, 35th Engr Gp, 15 Aug 66, pp. 1–2, both in Historians files, CMH. For more on RMK-BRJ's work at Cam Ranh Bay, see Diary of a Contract, pp. 226–33.

installed the piers. When technical problems arose, the military and civilian engineers worked closely to resolve them. DeLong began installing the third pier in May and, despite some difficulty, positioned and readied the pier for caisson jacking by the end of June.[3]

Associated port work proceeded at high tempo. On the south beach, the 497th Port Construction Company completed a timber pile U-shaped marginal wharf in April and two permanent LST ramps in June. The 220-foot wharf, the first of its kind designed and built for shallow-draft vessels in the bay, could unload two barges simultaneously. The wharf was supported by 315 timber piles in seventeen feet of water and became the largest work of its type built by Army engineers since World War II. Unfortunately, it did not take long to realize that the untreated lumber used for bracing the fuel pipeline jetty and wharf had to be replaced because of the presence of marine wood borers in the bay. In late May, the 497th began to replace the bracing with treated lumber. The port construction engineers stabilized the area behind the wharf with crushed coral, which the 87th Engineer Battalion found offshore. Coral, made up of the skeletons of minute spherical animals, was chemically similar to limestone and frequently used by engineers in the Pacific during World War II. At Cam Ranh Bay, coral was used extensively for beach landing sites, roads, and hardstands. At the storage area next to the first DeLong pier, the 497th also designed and constructed a roll-on/roll-off ramp for unloading shallow-draft vessels. The company's diving section removed obstacles, including a sunken vessel that blocked a second sheet-pile bulkhead. The divers also used underwater demolitions to loosen the coral for quarrying.[4]

The 497th Port Construction Company then concentrated on placing sheet-pile bulkheads between the RMK and DeLong piers, and the 87th Construction Battalion worked on the construction of causeways to connect the new DeLong piers to the shoreline. In July, the battalion built a 600-foot causeway to the second DeLong pier. To the north, the 87th with some help from RMK-BRJ began building a 3,600-foot causeway to a new ammunition pier. The causeway alone required 27,000 cubic yards of sand, 6,000 cubic yards of blast rock, and 15,000 cubic yards of laterite on top of 11,000 cubic yards of hydraulic dredged fill. In August, DeLong completed its second cargo pier and began to place Pier Number 3 just to the south of the existing RMK pier and DeLong Numbers 1 and 2. In October, more dredging in the entrance channel got under way, and by the end of the month, the 87th began construc-

---

[3] ORLLs, 1 Feb–30 Apr 66, 35th Engr Gp, p. 3, and 1 May–31 Jul 66, 35th Engr Gp, p. 4; Yens and Clement, "Port Construction in Vietnam," pp. 21–22; Dunn, *Base Development*, p. 55; Galloway, "Essayons," p. 76; Ploger, *Army Engineers*, pp. 52–53; Quarterly Hist Rpt 1 Apr–30 Jun 66, MACV Directorate of Construction (MACDC), 11 Jul 66, pp. 14, 25–26, Historians files, CMH.

[4] ORLLs, 1 Feb–30 Apr 66, 35th Engr Gp, pp. 3–4, and 1 May–31 Jul 66, 35th Engr Gp, pp. 3–4; Yens and Clement, "Port Construction in Vietnam," p. 22; Situation Rpt on the Construction of Port Facilities in South Vietnam, incl. 12, pp. 1–2, OCE Liaison Officer Trip Rpt no. 3; Karl C. Dod, *The Corps of Engineers: The War Against Japan*, United States Army in World War II (Washington, D.C.: Office of the Chief of Military History, U.S. Army, 1966), p. 385.

The second DeLong cargo pier was completed at Cam Ranh Bay in August 1966, and work on the third DeLong pier (to the right of the first DeLong pier) had begun.

tion of an 830-foot causeway to a fourth DeLong pier. North of the main port, DeLong finished installation of the 600-foot ammunition pier in mid-October, allowing safer and more direct delivery of ammunition to its storage facility. The firm completed its third cargo pier in early December.[5]

Impressive progress was also made elsewhere on the peninsula. In the lower half, Colonel McCulloch's 87th Engineer Battalion completed storage tanks and connections for a 172,000-barrel fuel storage facility south of the air base, maintained over ten miles of road, built hardstands for ammunition and general storage areas, placed over 500 sand-cement and concrete slabs in the Army cantonment, and supervised troops building the cantonment under the self-help program. In April 1966, the 18th Brigade approved upgrading the old light aviation runway to handle C–130 transports. After a slow start, the 87th in September resumed work, and in two weeks completed the runway base and

[5] ORLLs, 1 Aug–31 Oct 66, 35th Engr Gp, 31 Oct 66, pp. 3–6, and 1 Nov 66–31 Jan 67, 35th Engr Gp, 15 Feb 67, pp. 3–4; Unit History, 35th Engr Gp (Const), n.d., pp. 12–13; Quarterly Hist Rpts, 1 Jul–30 Sep 66, MACDC, 21 Oct 66, p. 38, and 1 Oct–31 Dec 66, MACDC, 21 Jan 67, pp. 26–27, Historians files, CMH; MACV Monthly Evaluation, Aug 66, p. F-5, all in Historians files, CMH; Yens and Clement, "Port Construction in Vietnam," pp. 21–22; Galloway, "Essayons," p. 131.

surfaced it with the newly introduced M8A1 solid steel plank matting. The airfield became operational on 11 October when the Air Force landed a C–130 on the new 2,900-foot runway. Depot work by the 864th Engineer Battalion included roads, drainage, open storage areas, refrigerated storage facilities, and an automated data processing facility. By the end of June the battalion, commanded by Lt. Col. Robert A. Seelye since January when he replaced Colonel Bunch, reported that the depot as 59 percent complete. To the north at Hon Tre Island just offshore from Nha Trang, Company C and heavy equipment from Company A carved a road to the top of a mountain and cleared the crest for an air defense missile unit. Meanwhile, the 102d Construction Support Company's asphalt plant had made its first cold-mix pavement and produced its first hot mix. By summer, the roads in the depot and at the southern end of the peninsula had taken on a new look.[6]

In February 1966, Air Force Red Horse engineers arrived. The 555th Civil Engineering Squadron began to work on troop housing, roads, utilities, and other air base support facilities. Originally conceived to augment base engineers to handle heavy bomb damage or disasters, Red Horse squadrons quickly adapted to a construction role in Vietnam. In justifying an engineering force, Air Force officials emphasized that Red Horse squadrons provided an emergency capability only and did not minimize the need for Army engineer battalions to build the airfields. To General Dunn, the arrival of the Red Horse squadrons represented another construction asset, except they worked only on Air Force projects.[7]

Meanwhile, construction of the permanent runway and taxiway by RMK-BRJ lay almost dormant because of the diversion of the firm's earth-moving equipment in January to the Army ammunition and logistic support area project. Although the Air Force deployed four F–4 Phantom squadrons onto the adjacent AM2 airfield, the temporary aluminum runway and taxiways built by the contractor soon had problems. A twenty-three-inch rainfall in December 1965 raised the water table to the level of the matting, forcing workers to improvise a drainage system. Rain made the metal runway slick, and the fighters had to land with drag chutes or land at other airfields. As the dry season approached in early 1966, a constant north wind pushed the taxiway three feet south over the dry sand, while the runway edged north under the weight of the planes landing from the south. Landing the planes to the south for three weeks moved the runway back. Daily stress measurements and periodic shifts in the direction of landings kept the shifting under control. The moving sand also created bumps and dips, and Air Force work

[6] Galloway, "Essayons," pp. 79–80; Col. William L. Starnes, "Cam Ranh Army Airfield," *Military Engineer* 59 (September-October 1967): 358; Tregaskis, *Building the Bases*, pp. 275–77; Situation Rpt on the Construction of Depots and Supply Points in Vietnam, incl. 14, pp. 2–3, OCE Liaison Officer Trip Rpt no. 3; ORLLs, 1 Feb–30 Apr 66, 87th Engr Bn, 15 May 66, pp. 1–3, 1 May–31 Jul, 66, 87th Engr Bn, 15 Aug 66, pp. 2–3, and 1 May–31 Jul 66, 864th Engr Bn, 15 Aug 66, pp. 1–2, all in Historians files, CMH.
[7] Project RED HORSE, pp. 1–2, 9; USAF Airfield Construction in South Vietnam, pp. 112–13.

crews found themselves continually replacing sections of the aluminum matting and smoothing the sand.[8]

At the same time, there was an urgent need for an airfield capable of handling large transport aircraft. Although the AM2 matting had the strength to land and park large planes, the narrowness of the runway and taxiways and the limited amount of AM2 parking apron precluded the use of the Boeing 707 commercial jets or the larger C–133 cargo transports. The construction of the Army's replacement center in II Corps hinged on completing the new runway and other flight facilities. As a result, MACV decided to have the Officer in Charge of Construction redesign the permanent airfield for earlier use as an interim logistical air facility. When RMK-BRJ returned to its airfield project in June 1966, revised plans called for building the concrete runway in increments, 4,000 to 6,000 feet, with AM2 taxiways and aprons. This measure would at least allow C–130 cargo planes to use the airfield before the onset of the next monsoon season.[9]

In mid-October 1966, the firm turned over to the Air Force an 8,000-foot concrete runway. Work continued on the taxiways and high-speed turnoffs, and the hauling of fill for the remaining 2,000 feet of runway. To build up the south end of the field, three dredges excavated some 1.5 million cubic yards of hydraulic fill. By November, the parallel concrete taxiway and 90,000-square-yard AM2 parking apron were completed, permitting the Air Force's Military Airlift Command to begin operations at Cam Ranh Bay. The adjacent AM2 runway continued to be used and maintained by Air Force civil engineers. Its use, however, according to one squadron commander, continued to be "a sporty proposition." Soft shoulders and the lack of an overrun and aircraft barriers often caused planes to sink into the soft sand when they veered off the runway.[10]

In 1966, the Cam Ranh peninsula became the first site in Vietnam to receive electrical power from the T–2 power-generating tankers, now on their way from American shipyards. The first tanker, the *French Creek*, arrived in June. Once positioned and hooked up by Vinnell Corporation to a completed section of the land line distribution system, the modified tanker began to generate power to the depot on 5 September. The second tanker, the *Kennebago*, arrived shortly after the *French Creek*. Eventually Cam Ranh Bay would get five of the eleven T–2 tankers, with Nha Trang, Qui Nhon, and Vung Tau scheduled to get two each. When the ships arrived, the contractor temporarily moored them to withstand tide and weather. Vinnell was also charged with building the land-based electrical distribution systems at the four locations. By December, the firm had completed 80 percent of the forty-one miles of primary line and 5 percent of the eleven miles in the secondary line at Cam Ranh Bay.[11]

---

[8] Schlight, *Years of the Offensive*, p. 119; USAF Airfield Construction in South Vietnam, p. 10.

[9] MFR, J–4 Briefing 2 Feb 66, Base Development Cam Ranh Bay; Ltr, MACV MACJ44-USBD to CG, USARV, sub: Cam Ranh Bay USAF Base, box 11, 69A/702, RG 334, NARA.

[10] Schlight, *Years of the Offensive*, pp. 171–72 (quoted words, p. 172); Quarterly Hist Rpts, 1 Jul–30 Sep 66, p. 38, MACDC, and 1 Oct–31 Dec 66, MACDC, pp. 27–28; Diary of a Contract, pp. 227–28; USAF Airfield Construction in South Vietnam, p. 113.

[11] MACV Monthly Evaluation, Jun 66, p. A–30; "Floating and Land Based Power Plants," *Vinnell* 11 (May 1969): 6–7; Dunn, *Base Development*, pp. 78–80; Miscellaneous Detailed

Cam Ranh Bay airfield from the northeast in October 1966 after the completion of the concrete runway on the right

Although the 20th and 39th Engineer Combat Battalions encountered problems carrying out construction projects, both units quickly adapted to their new missions following their arrival on 1 January. The 20th Engineer Battalion, commanded by Lt. Col. Richard L. Harris, moved across the bay to Dong Ba Thin Army aviation base to continue the work started by Company C, 65th Engineer Battalion. The 20th had arrived in the middle of the northeast monsoon, and it concentrated on hauling fill to get the aviation units and equipment above the flood waters that inundated Dong Ba Thin from December through February. By May the battalion, with the trucks of the 513th Dump Truck Company, brought the cantonment areas to grade, built seventy-five concrete pads for UH–1 helicopters, completed a pierced steel plank parking ramp for the C–7 Caribous, and began laying planking for a taxiway parallel to the runway. Following these jobs, work crews completed landing ramps for CH–47 helicopters, the airfield taxiway, and a cantonment area. In mid-July, elements of the battalion began moving to Ninh Hoa and Nha Trang. The other combat battalion, the 39th, under Lt. Gen. Earnest E. Lane, erected buildings in the depot area and a convalescent hospital on the seaside of the peninsula. By midyear, the battalion began shifting operations

Discussion, incl. 11, pp. 4–5, OCE Liaison Officer Trip Rpt no. 4, 23 Sep 66, OCE Hist Ofc; Miscellaneous Detailed Discussion Vietnam, incl. 8. p. 5, OCE Liaison Officer Trip Rpt no. 5.

from Cam Ranh Bay to Nha Trang, Ninh Hoa, and Tuy Hoa. Both battalions were now under the control of Colonel Bush's 45th Group.[12]

During the second half of 1966, Col. William R. Starnes' 35th Group concentrated on base development projects at Cam Ranh Bay, Nha Trang, and Phan Rang. The 3,000-man group now consisted of the 62d, 87th, and 864th Construction Battalions and assorted smaller units. To keep pace with the troop buildup, the 35th accelerated work at the Cam Ranh Bay depot and other facilities at the southern end of the peninsula. Locally, the 864th Construction Battalion divided its efforts between building the depot at Cam Ranh Bay and an air-conditioned prefabricated communications center at Dong Ba Thin. Part of the battalion also worked on roads, bridges, and beach off-loading facilities at Nha Trang. The new beach facilities included four temporary LST ramps and four barge points that increased the unloading of cargo from about 9,700 short tons in January to over 21,000 short tons in December. Other work by the 87th Engineer Battalion was carried out at Phan Thiet, where a battalion task force supported an infantry battalion of the 1st Cavalry Division operating in the area. The engineer task force repaired and maintained an airfield, constructed landing craft anchorages, and improved base facilities.[13]

In July, another long-awaited construction battalion, Lt. Col. James L. Kelly's 577th Engineer Battalion (Construction) from Fort Benning, Georgia, arrived and joined the 45th Group. Not surprisingly, ships carrying the battalion's equipment did not arrive until later. The battalion assumed responsibility for projects at Dong Ba Thin. Company C, however, was detached while en route and assigned to work with the 46th Construction Battalion at Long Binh. At Dong Ba Thin Colonel Kelly initiated around-the-clock operations. Company B began building a regimental base camp for the South Korean 9th Division. In mid-August, Company D shifted to Nha Trang to build a 400-man evacuation hospital for the Koreans. Returning to Cam Ranh Bay in early October, Company D started work on facilities for an inbound replacement battalion. Once set up, the replacement battalion would handle the influx of replacements and departure of soldiers completing their one-year tours.[14]

[12] Galloway, "Essayons," pp. 80–81, 83; ORLLs, 1 Feb–30 Apr 66, 20th Engr Bn, 15 May 66, pp. 3–4, 1 May–31 Jul 66, 20th Engr Bn, 15 Aug 66, pp. 1–2, 8, 1 Feb–30 Apr 66, 39th Engr Bn, n.d., pp. 3–7, and 1 May–31 Jul 66, 39th Engr Bn, 14 Aug 66, pp. 1–4, all in Historians files, CMH; Unit History, 15 Aug 1917–1 Sep 1971, 20th Engr Bn, n.d, pp. 7–8, Historians files, CMH; Situation Rpt on the Construction of Cantonments, Administrative and Community Facilities in South Vietnam, incl. 8, p. 1, OCE Liaison Officer Trip Rpt no. 3.

[13] ORLLs, 1 Aug–31 Oct 66, 35th Engr Gp, pp. 1–2, 1 Nov 66–31 Jan 67, 35th Engr Gp, pp. 1–3, 1 Aug–31 Oct 66, 864th Engr Bn, 14 Nov 66, p. 1–3, 1 Nov 66–31 Jan 67, 864th Engr Bn, 15 Feb 67, pp. 1–4, 1 Aug–31 Oct 66, 87th Engr Bn, 14 Nov 66, pp. 1–4, and 1 Nov 66–31 Jan 67, 87th Engr Bn, 11 Feb 67, pp. 1–5, all in Historians files, CMH; Galloway, "Essayons," pp. 130–31, 133–34; 35th Engr Gp History, pp. 12–13; Ltrs, MACV MACJ322 to CG, USARV, sub: Development of Hon Tre Air Defense Complex, 11 Apr 66, and Brig Gen Carroll H. Dunn MACDC to CG, USARV, sub: Hon Tre Island Facility, 17 Jun 66, both in Historians files, CMH.

[14] ORLLs, 1 Aug–31 Oct 66, 577th Engr Bn, 31 Oct 66, pp. 1, 3–5, and 1 Nov 66–31 Jan 67, 577th Engr Bn, 31 Jan 67, p. 2; Galloway, "Essayons," pp. 127–28.

In late October, elements of the 14th Engineer Battalion (Combat) commanded by Lt. Col. William F. Brandes reached Cam Ranh Bay. Back in January, the battalion received alert orders to deploy from Fort Bragg by March, but shortages of personnel and equipment delayed its departure to September. Since the equipment did not arrive for several months, the soldiers could only use their hand tools. Companies A and C debarked at Cam Ranh Bay while the headquarters and Companies B and D continued on to Vung Ro Bay, eighteen miles south of Tuy Hoa near Highway 1. Company A helped the 577th Engineer Battalion build the replacement center, and Company C moved to Ninh Hoa to support the South Korean 9th Division. After disembarking at Vung Ro Bay, the remainder of the battalion moved to Tuy Hoa. Working with hand tools, Companies B and D maintained Highway 1 between Tuy Hoa and Vung Ro and the access road to the recently opened port at Vung Ro Bay. When the 577th Engineer Battalion received orders to move to Tuy Hoa, Colonel Brandes began to consolidate his units at Cam Ranh Bay.[15]

By the end of 1966, the combined efforts of troops and contractors had transformed Cam Ranh Bay into a major logistical base supporting over 36,000 troops. During the eighteen months, critical projects at the deep-water port, depot, and air base were completed or nearing completion. Whenever the builders finished a portion of a facility, such as new piers and warehouses, impatient operating agencies did not hesitate to put them to immediate use. By late 1966, the base boasted five deep-draft piers, four landing craft ramps, and wharves capable of handling nearly 6,000 short tons of cargo per day. More significantly, port capacity rose from approximately 73,000 short tons of cargo unloaded in January 1966 to nearly 153,000 short tons in December. A jetty with two 6-inch fuel pipelines had a rated capacity to pump ashore 30,000 barrels of product daily. Actual figures varied. Logistic planners looked forward to completing the port complex by mid-1967. Also in operation were the Air Force base with its two long runways, an Army airfield on the peninsula, and another Army airfield at Dong Ba Thin.[16] (*Map 7*)

Secondary Coastal Ports

In April 1966, Maj. Gen. Charles W. Eifler, who assumed command of the 1st Logistical Command in January 1966, began to take measures to build an all-weather port in the Tuy Hoa area. He recognized the need to do something before the approaching winter monsoons struck the central coastal lowlands in October. Before the buildup in the Tuy Hoa area, the logistical command managed to support the small number of troops by bringing in supplies over the beach during the dry season and by air during the monsoon when heavy seas halted over-the-shore operations. Vung Ro Bay appeared to offer many advantages that Tuy Hoa did not have for the development of an all-weather

[15] ORLL, 1 Nov–31 Jan 67, 14th Engr Bn, 9 Feb 67, pp. 1–5, Historians files, CMH; Galloway, "Essayons," pp. 128–29.

[16] Dunn, *Base Development*, p. 68; MACV History, 1966, p. 298; Port Development Reference Study, MACJ44, 5 Dec 66, pp. 14–15, box 8, 69A/702, RG 334, NARA; Ltr, CINCPAC to JCS, sub: Port Development Plan, RVN, 10 Jun 67, p. 3, box 3, 84/051, RG 319, NARA.

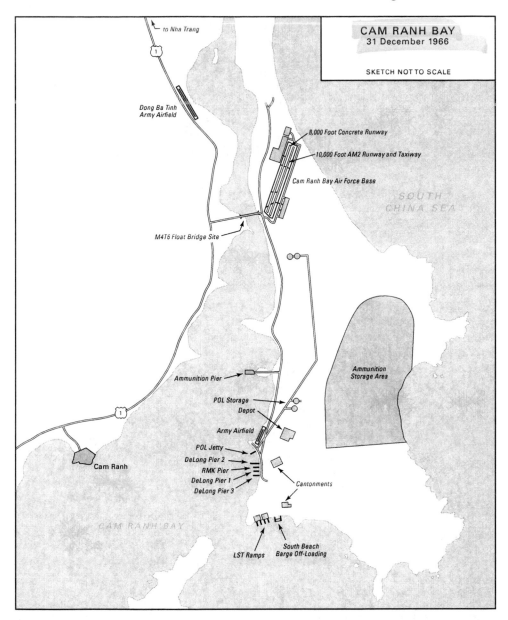

MAP 7

port, mainly a well protected, wide harbor. Only four bridges along the short distance of Highway 1 from Vung Ro to Tuy Hoa would require replacement or repair.[17]

[17] Col. John J. Sawbridge, "They Built a Port to Beat the Weather," *Army Digest* 22 (September 1967): 44; Debriefing, Maj Gen Charles W. Eifler, 13 Jun 67, p. 4, Senior Officer Debriefing Program, DA, Historians files, CMH.

On 27 April, a 39th Engineer Battalion reconnaissance party led by Colonel Lane began to survey Vung Ro Bay. While infantrymen of the 1st Brigade, 101st Airborne Division, cleared the immediate area, Lane and his men scoured the beaches, bay, road, and surrounding mountains. The engineers confirmed that the bay could fit deep-draft ships. They found ample space to build facilities, and a road could be cut through the thick jungle foliage and mountains of rock to connect the port to Highway 1. The survey completed, Colonel Lane became an enthusiastic supporter of the Vung Ro project, but he did not live to see the completed port. Several weeks later, enemy ground fire killed Lane during an aerial reconnaissance over the project area.[18]

General Larsen concurred in the need for a temporary port at Vung Ro to support combat forces around Tuy Hoa. The I Field Force commander expected the arrival of the 196th Light Infantry Brigade and a South Korean regiment. From a logistics standpoint, Larsen did not consider it feasible to supply an equivalent two-brigade force by air during the monsoon season. He also ruled out opening Highway 1 from Qui Nhon to Tuy Hoa because he lacked troops to secure the route. As for opening Highway 1 from Tuy Hoa south to Cam Ranh Bay, Larsen believed that this would require more forces and engineering effort than the Vung Ro proposal. Estimates showed that it would take only thirty days to do the road work from Tuy Hoa and improve the bay for barge traffic. Aware of the ongoing debate between the Army and Air Force over whether to construct the deep-draft port facility at Vung Ro or Tuy Hoa, Larsen prodded MACV to get started on at least a temporary port at Vung Ro. Westmoreland concurred, and on 15 July the MACV Directorate of Construction directed the 18th Engineer Brigade to do the work.[19]

On 25 July, while I Field Force initiated Operation JOHN PAUL JONES to secure the area, the development of Vung Ro Bay began in earnest. As the 1st Brigade, 101st Airborne Division, and South Korean 2d Marine Brigade kept the Viet Cong at bay, the 39th Engineer Battalion swung into action. Elements of the battalion began repairing Highway 1 and erecting several new bridges between Tuy Hoa and Vung Ro. On the twenty-seventh, Task Force SCHULTZ—consisting of Company A, elements of the Equipment Platoon, Headquarters Company, 572d Light Equipment Company, and 553d Float Bridge Company, under the command of Maj. John H. Schultz, the battalion executive officer—debarked from a landing craft and began building a port out of jungle, sand, and rock. Men and equipment cleared more than 125 acres of thick, tangled underbrush, moved some 400,000 cubic yards of earth, and used large quantities of dynamite to blast away solid rock for an access road. Within two weeks, the task force blazed a passable pioneer road two miles from the beach to Highway 1. In the port area, work crews put down a six- to eight-inch layer of laterite and rock over some 10,000

---

[18] ORLL, 1 Feb–30 Apr 66, 39th Engr Bn, p. 2; Sawbridge, "They Built a Port to Beat the Weather," pp. 44–45; Ploger, *Army Engineers*, pp. 57, 223.

[19] Msg, Lt Gen Stanley R. Larsen AVF-GD 5321 to Westmoreland, 7 Jul 66, sub: Vung Ro Bay; Msg, Westmoreland MAC 23595 to Larsen, 9 Jul 66, sub: Vung Ro Bay; Msg, COMUSMACV MACDC-BRD to CG, USARV, 15 Jul 66, sub: Vung Ro Bay. All in Historians files, CMH.

square yards of sand. They also built two concrete ramps for landing craft and installed a fuel pipeline from mooring points in the bay to a temporary storage facility inland. Elements of the 497th Port Construction Company assembled a floating pier made out of Navy cube sections for barge discharge. The steel cubes, fastened with angle irons and cable, required very little on-site construction. Cubes were joined together in connected sections or as a complete pier in the water and towed into position. The port became operational in late September, and on 11 October it opened to unrestricted traffic. The dedication of Port Lane took place five days later, and the 1st Logistical Command officially took over operation of the facility. In the first two weeks of operation, more than 565 tons of vitally needed supplies arrived daily, just before the monsoon deadline.[20]

By then MACV wanted to make Vung Ro the permanent deep-draft port for the Tuy Hoa area. While the Army proceeded to build its interim port at Vung Ro, the Air Force went ahead with its own plans for Tuy Hoa Air Base. The Air Force developed its own shallow craft harbor near Tuy Hoa to discharge materiel for the turnkey project. Planning called for the Air Force contractor, Walter Kidde Constructors, to dredge an interim port at the mouth of the Da Rang River and convert the facility into a permanent port the following year. When Pacific Command and the Joint Chiefs of Staff became aware of the development of two independent ports, they raised questions. Meetings between Army and Air Force representatives at MACV produced little more than an issue—Vung Ro versus Tuy Hoa. By mid-September, MACV predicted that the garrison strength of U.S. and South Korean troops in the area would exceed 18,000 and recommended to Pacific Command and the Joint Chiefs that Vung Ro Bay should be the permanent port.[21] (*Map 8*)

The Tuy Hoa port issue remained open until January 1967. A survey disclosed that the Tuy Hoa site involved time-consuming efforts to bring the harbor to the point of accommodating fully loaded T–1 tankers and pointed out the difficulty of tanker operations during the monsoon season, even after completion of a pier and breakwater. A tropical storm in December brought this point home. High tides and swells caused water to wash over the sandspit that separated the safe haven from the South China Sea. This water shifted the sand and silt and reduced the depth of the dredged entrance channel. MACV recommended that the port should be abandoned when the contractor no longer required it. On 13 January, the Joint Chiefs of Staff ruled in favor of Vung Ro Bay. The Air Force did not dissent. The Tuy Hoa site served its purpose by offloading construction

---

[20] ORLLs, 1 May–31 Jul 66, 39th Engr Bn, p. 1, and 1 Aug–31 Oct 66, 35th Engr Gp, p. 4; AAR, Opn JOHN PAUL JONES, 39th Engr Bn, 17 Nov 66, pp. 1–4, Historians files, CMH; Sawbridge, "They Built a Port to Beat the Weather," pp. 45–46; Galloway, "Essayons," p. 125; Rpt of Visit to 18th Engr Bde Units, incl. 7, p. 4, OCE Liaison Officer Trip Rpt no. 4; Army Construction in Vietnam, 7 Dec 66, incl. 9, p. 8, OCE Liaison Officer Trip Rpt no. 5.

[21] Msg, COMUSMACV MAC 41026 to CINCPAC, 11 Sep 66, sub: Vung Ro–Tuy Hoa Port, box 5, 69A/702, RG 334, NARA; USAF Airfield Construction in South Vietnam, pp. 75–79.

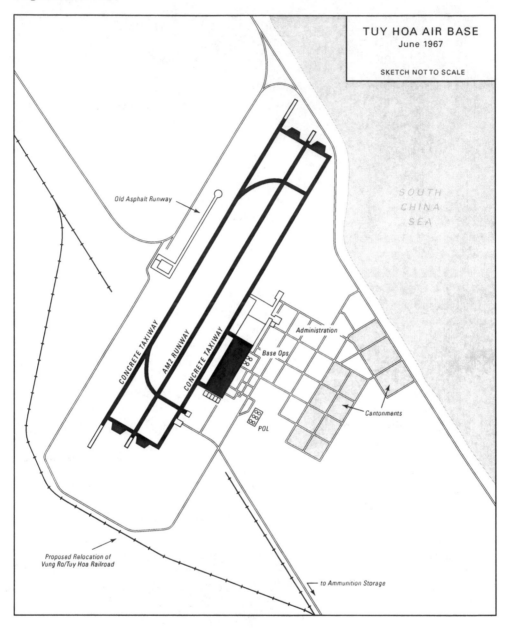

TUY HOA AIR BASE
June 1967

SKETCH NOT TO SCALE

Old Asphalt Runway

SOUTH CHINA SEA

CONCRETE TAXIWAY

AM2 RUNWAY

CONCRETE TAXIWAY

Administration

Base Ops

Cantonments

POL

Proposed Relocation of Vung Ro/Tuy Hoa Railroad

to Ammunition Storage

MAP 8

supplies, and this enabled Walter Kidde to finish the airfield project forty-five days before the 1 January 1967 deadline.[22]

[22] USAF Airfield Construction in South Vietnam, pp. 86–89; Quarterly Hist Rpt, 1 Oct–31 Dec 66, MACDC, p. 29.

Port under construction at Vung Ro Bay

The 39th Engineer Battalion, now under the command of Lt. Col. Taylor R. Fulton, divided its attention to projects in the Tuy Hoa area and improvements at Vung Ro Bay. Work included base requirements for an army logistics subarea command and cantonments for the incoming 4th Infantry Division's 1st Brigade, a regiment of the South Korean 9th Division, and an army aviation battalion. The 39th continued to upgrade the facilities at Port Lane, worked on base development projects, repaired roads, and built a heliport for aviation units that supported the 1st Brigade, 101st Airborne Brigade. The sixty-four minipad heliport consisted of 24-by-24-foot pierced steel planking placed directly over sand treated with peneprime, a dust-control material with an asphalt base. These measures provided in minimum time maximum protection for the helicopters from sand and dust. Hardstands and the connecting road to Highway 1 were paved. By the end of 1966, a DeLong pier consisting of two "A" units was installed, adding Vung Ro to the growing number of ports that could discharge cargo from oceangoing ships. Most significantly, the sheltered bay could safely harbor up to a dozen large vessels.[23]

Meanwhile, a trade off of engineer units took place in two of the coastal bases. The reason for the moves stemmed from requirements for heavy construc-

[23] Sawbridge, "They Built a Port to Beat the Weather," p. 46; Army Construction in Vietnam, 7 Dec 66, incl. 9, p. 8, OCE Liaison Officer Trip Rpt no. 5; ORLL, 1 Nov–31 Jan 67, 39th Engr Bn, 31 Jan 67, pp. 1–7, Historians files, CMH; USAF Airfield Construction in South Vietnam, pp. 88–89; Galloway, "Essayons," pp. 125, 127. For more on peneprime and dust control, see Col. Charles R. Roberts, "Trends in Engineer Support" (Student thesis, U.S. Army War College, 1969), pp. 43–46.

tion equipment at Tuy Hoa and the 35th Group's need for a combat battalion. For several days in November, the 14th Combat and the 577th Construction Battalions shuttled men and equipment between Tuy Hoa and Dong Ba Thin. The 14th took over the 577th's projects at Cam Ranh Bay, and the 577th took over the 14th's projects at Tuy Hoa. At the South Korean cantonment south of town, the 577th centered efforts on a 400-bed evacuation hospital and a CH–47 heliport. It also completed a 330-foot causeway leading to the DeLong pier at Port Lane along with other improvements. By the end of the year, the 577th and a platoon from the 643d Engineer Pipeline Company began installing an eight-inch fuel pipeline between the port and the air base with a spur leading to the allied camp. The 577th also took over rehabilitation and maintenance of Highway 1 between Tuy Hoa and Port Lane.[24]

Since Highway 1 was not opened for regular use until November, all construction supplies for the 45th Engineer Group had to be brought in by sea. Limited intercoastal shipping could only handle ammunition and rations, with construction materials put aboard on a space-available basis. This almost halted vertical construction (buildings), but the continual prodding by Colonel Bush and the opening of Port Lane brought deep-draft vessels carrying construction materials. By November, the grim logistics situation began to improve. The frequent moves of the group's battalions continued to cause replacement parts problems until the headquarters arranged for direct support maintenance on an area basis. Still, units frequently left an area just about the time previously requisitioned parts arrived, and it took several weeks for the parts to reach the unit.[25]

Some twenty-eight miles to the south at Ninh Hoa, the 45th Group found a practical way to resolve transporting large tonnages of materials from Cam Ranh Bay. Truck convoys moving north placed a heavy burden on congested Highway 1, and there was constant need for security during the trip. The group turned to the Vietnamese Railway. With the arrangement of ample government security forces, the railway started service between the two areas. The opening of the railroad reduced congestion and wear on the key coastal road.[26]

About twenty-two miles south of Ninh Hoa at Nha Trang, military engineers and RMK-BRJ improved facilities to meet the influx of U.S. Air Force and Army aviation units and support organizations. Company C, 864th Engineer Construction Battalion, completed several warehouses and a depot hardstand and started building a fuel storage area. During a two-week period, Company B, 39th Engineer Battalion, transformed a trail around the city into a single-lane bypass road. The contractor's workforce enlarged and modernized the 8th Field Hospital with twenty new buildings in a project started the previous October. In April, the firm also started to reshape the crown of the airfield's main runway and completed the paving by mid-June. Prime Beef

---

[24] ORLLs, 1 Nov 66–31 Jan 67, 18th Engr Bde, 24 Feb 67, p. 5, Historians files, CMH, 1 Nov 66–31 Jan 67, 577th Engr Bn, pp. 2–4, and 1 Nov 66–31 Jan 67, 14th Engr Bn, p. 5.

[25] Galloway, "Essayons," p. 130.

[26] ORLL, 1 Nov 66–Jan 67, 45th Engr Gp, pp. 9–10, Historians files, CMH.

construction teams built cantonment and other facilities for Air Force units at the base.[27]

Qui Nhon's Development

A corresponding surge of construction took place at the cluster of bases in and around the port city of Qui Nhon. In February 1966, the 1st Logistical Command declared that Qui Nhon would serve as one of its major Army logistics bases. The newly established Qui Nhon Support Command projected requirements for expanded storage capacity, which caused an even higher tempo of work for the 937th Engineer Group and its new commander, Col. William W. Watkin. Major improvements for the overtaxed port facilities and construction of a large maintenance facility were added to the depot construction projects. Three of the group's four battalions—19th Combat under Colonel Mathews, 299th Combat under Lt. Col. Reuben L. Anderson, and 84th Construction under Lt. Col. James D. Madison, who assumed command in January 1966—focused on Qui Nhon area projects. Operational support, for the most part, remained limited. The Navy's Officer in Charge of Construction established a construction office in Qui Nhon to oversee work done by RMK-BRJ. In May, RMK-BRJ began to move workers up Highway 1 to Phu Cat to begin preparations for constructing South Vietnam's eighth jet air base. Simultaneously, the Air Force dispatched Prime Beef teams to the Qui Nhon airfield to build cantonment facilities.[28]

In Qui Nhon, Colonel Madison's 84th Construction Battalion concentrated on depot, airfield, and port work. In the depot, the battalion built more prefabricated warehouses, sheds, and open storage areas. Many of the buildings, however, reached Vietnam without all their components, and the engineers had to improvise or locally fabricate the missing items. That expedient resolved the problem somewhat but added to the delays and costs. By April, the 84th, which adopted the motto "Never Daunted," began to construct approximately 175,000 cubic feet of refrigerated storage. At the Qui Nhon airfield, the battalion built a 1,197-foot extension to the runway and additional parking ramps. Also under way were many other projects, including the expansion of the 85th Evacuation Hospital to 480 beds.[29]

[27] Galloway, "Essayons," p. 80; Tregaskis, *Building the Bases*, pp. 149, 288; Situation Rpt on the Construction of Depots and Supply Points in South Vietnam, incl. 12, p. 2, OCE Liaison Officer Trip Rpt no. 1; Situation Rpt on Lines of Communication Construction in South Vietnam, incl. 6, p. 2, and Situation Rpt on the Construction of POL Facilities in South Vietnam, incl. 13, p. 2, OCE Liaison Officer Trip Rpt no. 3; Quarterly Hist Rpt, 1 Apr–30 Jun 66, MACDC, p. 25; USAF Airfield Construction in South Vietnam, pp. 119–20, 123–24, 126–27. For more on RMK-BRJ work at Nha Trang, see Diary of a Contract, pp. 234–35.

[28] Galloway, "Essayons," p. 73; Tregaskis, *Building the Bases*, p. 215; ORLL, 1 Feb–30 Apr 66, 937th Engr Gp, 15 May 66, pp. 3–6; USAF Airfield Construction in South Vietnam, pp. 120, 126. For more on RMK-BRJ work at Phu Cat Air Base and Qui Nhon, see Diary of a Contract, pp. 236–40, 266–70.

[29] Galloway, "Essayons," pp. 34–36, 70; ORLLs, 1 Feb–30 April 66, 937th Engr Gp, pp. 3–4, 6, 1 Feb–30 Apr 66, 84th Engr Bn, 14 May 66, pp. 1–5, and 1 May–31 Jul 66, 84th Engr Bn, 14 Aug 66, pp. 2–5, Historians files, CMH.

Further improvements in the port area were carried out by the 84th and the 1st Platoon, 497th Port Construction Company. In February, a 200-foot rock fill causeway was finished and connected to a Navy cube floating pier, 42 feet wide by 792 feet long. In another major project, the Qui Nhon peninsula was extended to receive more landing craft. When the pipeline dredge *Ann* arrived in April to work on a Navy MARKET TIME coastal surveillance base, MACV simultaneously gave its approval for the dredge to furnish fill for the extension. By the end of the month, the *Ann* completed this work and moved on to the Qui Nhon MARKET TIME site. When the engineers finished the extension and landing craft ramps, the in-transit storage area doubled in size. In early June, the port construction and construction engineers completed the installation of a four-inch submarine pipeline in the harbor. The new pipeline facilitated the transfer of fuel from tankers to a new 50,000-barrel capacity tank farm on the mainland. To stabilize the tankers while unloading, work crews installed a system of anchorages and moorings.[30]

The enlarged logistics base meant that even more work would be required to transform Qui Nhon into a major port complex. After the *Ann* completed its dredging at the MARKET TIME facility in late May, it began work on an extensive program to develop a deep-draft port. On 20 June, the hopper dredge USS *Davison* from the Corps of Engineers civil works fleet started work on the two-mile deep-draft approach channel and turning basin. Phase I planning for the expanded port included eight barge unloading points, two permanent LST ramps, and DeLong piers providing four deep-draft berths. In June, work began on a four-lane port access road one and a half miles across the bay to bypass the congested city. As part of this undertaking, RMK-BRJ began to drive sheet pile and place hydraulic fill for the bypass road's subbase. While the contractor dredged fill for the causeway leading to the pier complex, the 84th Engineer Battalion added landing craft ramps. The DeLong Corporation installed two prefabricated piers. Dredging the channel for deep-water shipping in the harbor continued. By December, the new pier complex with its four deep-draft berths boosted the cargo unloading performance from a little more than 59,000 short tons in January to over 113,000 short tons. The battalion improved the submarine fuel pipeline by more than doubling its length to 5,500 feet and began placing two additional lines. In the depot, drainage facilities were improved before the winter monsoon. To satisfy the ever-increasing demand for rock, the 84th assumed responsibility for the operation of the quarry south of Qui Nhon. The battalion pooled its equipment with the 73d Construction Support Company, and the combined capacity of the rock crushers increased to two 75-ton-per-hour units and one 225-ton-per-hour unit.[31]

[30] Quarterly Hist Rpt, 1 Apr–30 Jun 66, MACDC, 8 Aug 66, pp. 24–25; Galloway, "Essayons," p. 70; ORLLs, 1 Feb–30 April 66, 937th Engr Gp, p. 6, 1 Feb–30 Apr 66, 84th Engr Bn, p. 3, and 1 May–31 Jul 66, 84th Engr Bn, pp. 3–4; Yens and Clement, "Port Construction in Vietnam," pp. 22–23; Ploger, *Army Engineers*, p. 107.

[31] ORLLs, 1 May–31 Jul 66, 84th Engr Bn, pp. 3–5, 1 Aug–31 Oct 66, 84th Engr Bn, 14 Nov 66, pp. 2–6, and 1 Nov 66–31 Jan 67, 84th Engr Bn, 14 Feb 67, pp. 3, 5, all in Historians files, CMH; Galloway, "Essayons," pp. 108–10; MACV History, 1966, p. 299.

DeLong pier under construction at Qui Nhon in November 1966. Arrow indicates the hydraulic fill for port hardstand.

Outside Qui Nhon, the 19th Engineer Battalion carried on with an array of construction projects. Colonel Mathews departed in July and turned the reins of command over to Lt. Col. Nolan C. Rhodes. This transfer of command was literal, for it involved turning over the bit and reins of a bridle that, according to legend, came from a seahorse that rescued the drowning battalion commander during the World War II amphibious landing at Anzio, Italy. (The unit's insignia was also emblazoned with a little seahorse.) One major project involved expanding the 50,000-barrel tank farm to 112,000-barrel capacity at a second site on the outskirts of town. The fuel-thirsty 1st Cavalry Division's 450 helicopters had placed heavy demands on the logistics base and had created an urgent need for more fuel storage, including the construction of a fifty-one-mile fuel pipeline from Qui Nhon to An Khe. In late August, Company C began work on the pipeline and pumping stations. Normally, engineer pipeline companies did this type of work, and MACV made arrangements for a platoon from the 697th Pipeline Construction Company based in Thailand to help. By mid-September, work crews had placed thirty-three miles of pipeline, but a shortage of clamps forced a temporary halt. Work shifted to building pump stations and welding pipeline sections. Bad welding rods caused more delays, and a plea went out to airlift acceptable welding rods from the United States. Meanwhile, at Lane Army Airfield outside Qui Nhon, the battalion continued to make improvements. Peneprime was used extensively to cut down on the dust, and twenty-five pads were added to the heliport. In mid-February, the 19th began to build a large maintenance facility at Cha Rang near the intersection of Highways 1 and 19. The transfer of areas of responsibility between the

937th and the 45th Groups in November resulted in the transfer of the Cha Rang and the pipeline projects to the recently arrived 35th Engineer Battalion (Combat). The 19th also picked up the ammunition depot road maintenance project from the 299th Engineer Combat Battalion. Several operational missions were also assigned to the 19th later in the year.[32]

Much progress at the Qui Nhon ammunition depot had been made by the 299th Engineer Battalion. The depot site, which was perched near the base of a steep slope next to Highway 1 southwest of the city, presented a tremendous challenge in design and drainage. Company B built roads, concrete ammunition pads, earthen berms around the pads, buildings, guard towers, and fencing. However, huge runoffs of water following heavy rains repeatedly damaged roads, berms, and other structures. Except when called out on operational missions, Company C kept busy patching assigned sections of Highways 1 and 19. In late July the battalion, now commanded by Lt. Col. Richard M. Connell who replaced Colonel Anderson in April, received word to move the headquarters, Company C, and part of the 630th Light Equipment Company to Pleiku to join Company A, which had been hard at work building facilities for the 3d Brigade, 25th Division. Company B remained at the ammunition depot until relieved by the 19th.[33] (*Map 9*)

In November 1966, the 35th Engineer Battalion under the command of Lt. Col. Wesley E. Peel arrived at Qui Nhon from Fort Lewis, Washington. The battalion was unique in that it had four line companies compared to the three in the nondivisional combat battalions that arrived earlier. Newly arriving combat engineer battalions were organized under the later Table of Organization and Equipment (TOE) 5–35E series, while the battalions in country were still organized under the series 5–35D TOE. After settling in, the 35th assumed the Cha Rang depot and pipeline projects from the 19th. It also supported 1st Cavalry Division operations around Bong Son, some fifty miles north of Qui Nhon. This included keeping Highway 1 open to Bong Son during the winter monsoon already under way. Along Highway 19, the battalion and pipeline platoon completed five pump stations, two H-frame crossings, and four suspension crossings. Heavy rains and rising waters in December, however, caused the collapse of the fourth suspension bridge, and the project completion date of 31 December slipped back a week. By 20 January, with pipeline tests successfully completed, the 35th turned over the pipeline to the 1st Logistical Command. Soon fuel began to flow through the pipeline to An Khe. During this period, the battalion operated with only three lettered companies, for Company D had been ordered to Pleiku to work under the 20th Engineer Battalion's control.[34]

Before the end of the year, the 45th Group's area of responsibility and units under its control changed dramatically. By late 1966, the burgeoning Qui Nhon

[32] Galloway, "Essayons," pp. 67, 70, 71; ORLLs, 1 Feb–30 Apr 66, 19th Engr Bn, 9 May 66, pp. 3–5, 1 May–31 Jul 66, 19th Engr Bn, 13 Aug 66, pp. 1–2, 1 Nov 66–31 Jan 67, 19th Engr Bn, 31 Jan 67, pp. 1–7, and 1 Nov 66–31 Jan 67, Engr Cmd, 31 Jan 67, pp. 16–18. ORLLs in Historians files, CMH.

[33] ORLLs, 1 Feb–30 Apr 66, 299th Engr Bn, 30 Apr 66, pp. 3–4, and 1 May–31 Jul 66, 299th Engr Bn, 31 Jul 66, pp. 2–4. All in Historians files, CMH.

[34] ORLLs, 1 Nov 66–31 Jan 67, Engr Cmd, pp. 17–18, and 1 Nov 66–31 Jan 67, 45th Engr Gp, p. 4; Galloway, "Essayons," pp. 163, 173.

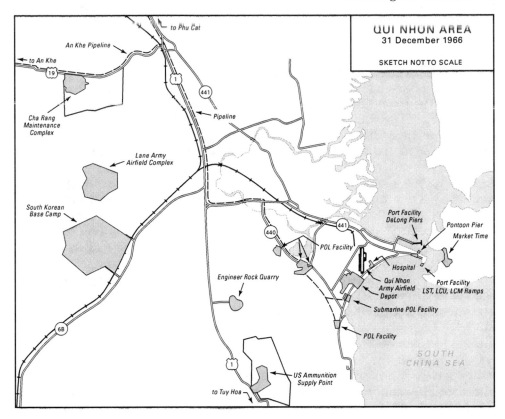

MAP 9

complex supported combat operations of over 62,000 combat and support troops. Because of the establishment of the engineer command and the shift of the 18th Engineer Brigade to Dong Ba Thin, other shifts of boundaries took place within the new brigade area. The 937th Engineer Group moved inland to Pleiku and assumed responsibility for the Central Highlands. On 10 November, Colonel Bush's 45th Engineer Group shifted its area of responsibility north to the I/II Corps boundary and moved to a more central location in Qui Nhon. In turn, the 35th Group extended its responsibility from Cam Ranh Bay to a point north of Nha Trang. The 45th Group lost the 14th Engineer Battalion to the 35th Group, and the 20th Engineer Battalion moved to the highlands to work under the 937th Group. Colonel Bush retained control of the 39th and 577th Engineer Battalions, and the 19th, 35th, and 84th Engineer Battalions in Qui Nhon along with all construction projects in the area were transferred to the 45th Group. As a result, Colonel Bush ended up commanding the largest engineer group in Vietnam—five battalions and several companies and smaller units totaling over 3,600 men.[35]

[35] Ploger, *Army Engineers*, pp. 134, 136; Dunn, *Base Development*, p. 68; ORLL, 1 Nov 66–31 Jan 67, 45th Engr Gp, p. 3; Rpt of Visit to the 18th Engr Bde, 29 Oct–17 Nov, incl. 7, tab A, OCE Liaison Officer Trip Rpt no. 5.

Highlands Camps

Work continued at a high pace at the 1st Cavalry Division's An Khe base camp. Colonel Edelstein's 70th Engineer Battalion improved access to the base by adding a one-and-a-half-mile road to Highway 19 east of An Khe City. To complete the road, the 70th spanned a river with a 260-foot double-double panel bridge. The Bailey bridge consisted of two tiers of double panels on each side of the roadway. Other projects completed were a perimeter security lighting system and a division tactical operations center. Extensive self-help construction under the supervision of the engineers continued in troop areas. In March 1966, Company B, 84th Engineer Battalion, arrived from Qui Nhon and set up a new quarry and larger rock crusher plant. The construction unit also added many wood frame buildings, one of which included a reinforced-concrete security room for use by the division's communications section, a sixty-ton-capacity ice-making plant, and a central telephone dial exchange.[36] Improving the An Khe airfield, however, turned into a major and troublesome endeavor. Through early spring, C–130s were continuing to suffer frequent tire cuts and blowouts. In late May, the rough condition of the new matting and the destructive effect of the pierced steel planking on the medium range aircraft's tires again reached an unacceptable level, forcing the closure of the airfield for repairs.[37]

To improve the airfield situation, the engineers figured three projects would have to be undertaken. First, engineer units at An Khe would build an alternate T17 membrane assault strip at the Golf Course. Second, new matting would be placed on the old airfield. Third, the engineers would construct a completely new C–130 runway. In late May, the 8th Engineer Battalion completed about one-third of the alternate airstrip's earthwork before turning that phase of construction over to Company B, 84th Engineer Battalion. On 10 June, the 70th Engineer Battalion began to lay T17 membrane while Company B finished the earthwork. Designed to serve as a rapid means to protect the runway base from the effects of moisture, the recently developed T17 membrane consisted of a neoprene-coated nylon fabric packaged in sheets measuring 303 by 60 feet. As part of the process of laying and securing the membrane

[36] Galloway, "Essayons," pp. 70, 72; ORLLs, 70th Engr Bn, 1 Feb–30 Apr 66, 7 May 66, pp. 2–8, 1 May–31 Jul 66, 70th Engr Bn, 15 Aug 66, pp. 2–8, 1 Aug–31 Oct 66, 70th Engr Bn, 31 Oct 66, pp. 1–2, 1 Nov 66–31 Jan 67, 70th Engr Bn, 31 Jan 67, pp. 2–6, 1 Feb–30 Apr 66, 84th Engr Bn, 14 May 66, pp. 3–4, 1 May–31 Jul 66, 84th Engr Bn, 14 Aug 66, pp. 3, 21, 1 Aug–31 Oct 66, 84th Engr Bn, 14 Nov 66, p. 1–2, and 1 Nov 66–31 Jan 67, 84th Engr Bn, p. 2. ORLLs in Historians files, CMH.

[37] Galloway, "Essayons," p. 30; Bowers, *Tactical Airlift*, pp. 209, 230–31; ORLLs, 1 Feb–30 Apr 66, 70th Engr Bn, pp. 2, 6, 11, and 1 May–31 Jul 66, 70th Engr Bn, pp. 2, 6; Quarterly Hist Rpt 1 Apr–30 Jun 66, MACDC, p. 23; Situation Rpt on the Construction of Airfields in South Vietnam, incl. 13, p. 1, OCE Liaison Officer Trip Rpt no. 1; Situation Rpt on Airfield Construction in SEA, incl. 13, p. 1, OCE Liaison Officer Trip Rpt no. 2, 6 May, OCE Hist Ofc; Situation Rpt on Airfield Construction in South Vietnam, incl. 7, p. 2, OCE Liaison Officer Trip Rpt no. 3; Ltr, Lt Gen Joseph H. Moore, CG, Seventh AF, to Westmoreland, 13 May 66, sub: Hazardous Airfield Conditions at An Khe, and 1st indorsement, Brig Gen Carroll H. Dunn to Cmdr, Seventh AF, 22 May 66, Historians files, CMH.

from moving during landings and takeoffs, work crews dug anchor ditches to hold the edges in place. Since the membrane provided no bearing capacity, the base required good compaction. As a finishing touch, a skid-proofing compound was applied to the membrane's surface. In a little over one week, the 70th placed T17 membrane on a 3,200-by-60-foot runway, a 900-by-180-foot parking area, and two 215-by-36-foot taxiways. The battalion also built a 100,000-square-foot compacted laterite hardstand and moved the control tower. The 70th Engineer Battalion officially completed the Golf Course project on 18 June.[38]

Air operations shifted to the T17 membrane Golf Course airstrip, and the main airfield was closed for rehabilitation. Soon the new airstrip had its own problems. A six-man maintenance crew was assigned to inspect and patch tears in the membrane after each C–130 landing. Although it did not rain very much, the impact and weight of the cargo planes repeatedly made ruts eight to twelve inches deep. This condition required frequent roller and vibratory compactor work that in itself proved damaging to the membrane. Nevertheless, the durability of the T17 membrane exceeded expectations, assuring its usefulness for forward operations, if given the proper maintenance. Unfortunately, the membrane runway at An Khe was considered less than satisfactory. On 11 July, a few days before Colonel Edelstein transferred the command of the 70th Engineer Battalion to Lt. Col. John R. Redman, the unit received a construction directive to fortify the Golf Course airstrip with AM2 matting. Since the main runway was still closed, the directive mandated that one half the strip always be open to emergency traffic. Pacific Command controlled the scarce matting, and MACV had to request the diversion of matting from the Phu Cat Air Base project. Within a week, Company A began removing and replacing the membrane after filling and compacting bad spots. AM2 matting was then placed over the runway, taxiways, and parking apron. The onset of daily rains made the job that much more difficult. On 20 August, Colonel Redman received word from the division that the strip had to be ready for use on the twenty-fifth for a tactical operation. The battalion added more men to lay matting and completed the runway and two taxiways at 0600 25 August, four hours before the first aircraft touched down.[39]

A month later, the monsoon season was in full swing, and the Golf Course airstrip again began to show signs of failure. At one location, the weight of the taxiing aircraft started to force up mud through the joints in the matting. Since the matting did not appear damaged, the airstrip remained open with

---

[38] Ploger, *Army Engineers*, pp. 111–12; *Memoirs of the First Team*, p. 181; Situation Rpt on Airfield Construction in South Vietnam, incl. 7, p. 2, OCE Liaison Officer Trip Rpt no. 3; ORLLs, 1 May–31 Jul 66, 70th Engr Bn, pp. 2, 6, and 1 May–31 Jul 66, 84th Engr Bn, pp. 1–2.

[39] AAR, 70th Engr Bn, Golf Course Airstrip, 11 Sep 66, Incl in ORLL, 1 Aug–31 Oct 66, 70th Engr Bn, pp. 12; Galloway, "Essayons," p. 70; Bowers, *Tactical Airlift*, p. 231; Quarterly Hist Rpt, 1 Apr–30 Jun 66, MACDC, pp. 23–24; Msg, CG I FFV to CG USARV, 27 May 66, sub: An Khe Airfield Requirements; Msg, COMUSMACV MACDC-CM to CINCPAC, 14 Jun 66, sub: AM2 Matting Requirements for An Khe Airfield; Msg, COMUSMACV MACDC-CM 20885 to Cmdrs Seventh AF, 1st Log Cmd, 1st Cav Div, USA Support Cmd, 18 Jun 66, sub: AM2 Matting Requirements for An Khe Airfield. Msgs in Historians files, CMH.

the suspected section closely monitored. By 30 September, the matting in this area had become deformed, but the strip remained open due to an ongoing operation. The 70th Engineer Battalion continued to monitor the airstrip, and on 7 October repair crews observed a hairline crack in the joint of the end panel of the depression. Landing and taxiing of C–130 transports caused AM2 panels in this section to become disengaged in two places. Further investigation revealed a subgrade failure that called for rehabilitation of some 300 feet of the runway. Due to the high priority given to the military operation, the 1st Cavalry Division authorized shutting down the airfield for only thirty-six hours. Taking turns in around-the-clock shifts, Companies A and C completed repairs at 0400 9 October. Another problem soon followed. Because of the lack of anchorage systems, the constant shifting of the aluminum matting continued to tear the membrane, thus allowing water to erode the subbase. The 70th again removed larger sections of matting and membrane. Work crews added and compacted crushed rock in the subgrade, applied peneprime in place of the membrane, and replaced the matting. The problems at An Khe simply reemphasized the importance of adequate drainage and the difficulties of doing earthwork during the monsoon. In December, another major change in specifications took place at the main airfield. The new asphalt runway would become the taxiway after completion of a concrete runway, the first of that type attempted by engineer troops in Vietnam.[40]

Over at the 12-million-square-foot Golf Course heliport, the 70th Engineer Battalion started building over four hundred ramps for the division's helicopter fleet. Heavy vehicle traffic had torn off much of the grass cover, and in a short time the heliport became extremely dusty. General Wright even went to the extreme of having Vietnamese laborers cut sod and transplant it to the heliport. He recognized that every little bit helped to reduce dust and helicopter maintenance problems, but even under optimum conditions putting down sod by this method was a slow process, perhaps, according to one estimate, taking up to twenty years. The engineers turned to a better method. The ramps, which varied in size according to the size of the aircraft, were leveled and topped off with an asphalt surface treatment. Workers then covered the treatment with airfield matting. Together the two surfaces reduced dust and flying particles and provided a firm foundation even when it rained.[41]

At Pleiku, more engineers arrived to help build the forward base for the 3d Brigade, 25th Division. In early January, a reinforced platoon from Company A, 299th Engineer Battalion, moved from Qui Nhon to help build facilities for the infantry brigade. A second platoon followed a week later, and by early April the entire company had moved to Pleiku. Work began on an ammuni-

[40] AAR, 70th Engr Bn, Golf Course Airstrip Repair, 9 Nov 66, Historians files, CMH; ORLLs, 1 Nov 66–31 Jan 67, 70th Engr Bn, p. 3, 1 Nov 66–31 Jan 67, 937th Engr Gp, 15 Feb 67, p. 12, 1 Aug–31 Oct 66, pp. 1–2, and 1 Nov 66–31 Jan 67, 84th Engr Bn, p. 2.

[41] Galloway, "Essayons," p. 72; Interv, Fishback with Wright, 1 Mar 83, 2:399–400; Situation Rpt on the Construction of Airfields in South Vietnam, incl. 13, p. 1, OCE Liaison Officer Trip Rpt no. 1; Situation Rpt on Airfield Construction in SEA, incl. 13, p. 1, OCE Liaison Officer Trip Rpt no. 2; ORLLs, 1 Feb–30 Apr 66, 70th Engr Bn, p. 6, and 1 May–31 Jul 66, 70th Engr Bn, p. 4.

tion depot and roads in and around the base. By June, Company A started building an aircraft maintenance facility, a cantonment area, an eighty-bed hospital, and more depot facilities. The company also ran an experiment by laying T17 membrane along a 2,000-foot section of road as a dust-control measure. To keep the surface from being punctured and torn with sharp objects, work crews closed the test section to tracked vehicles and dump trucks hauling gravel. After two weeks of use, the membrane held up well and appeared to be a good expedient road surface. On 29 July, Colonel Connell moved his battalion headquarters and most of the attached 630th Light Equipment Company to Pleiku. Company B remained at Qui Nhon working on the ammunition depot until 21 August before rejoining the battalion. In August, the battalion began making preparations to help the arriving 4th Infantry Division set up a division-size base camp outside Pleiku.[42]

The forward presence of a full division at Pleiku called for more engineers to lend a hand in base development and operational support. On 10 October, the 20th Engineer Battalion under Lt. Col. Robert L. Gilmore, who had replaced Colonel Harris in April, along with the attached 584th Light Equipment Company, moved from Ban Me Thuot to the 4th Division's new Dragon Mountain base camp. Although Colonel Gilmore took over base development responsibilities from the 299th, only a limited amount of time could be devoted to base construction. Priority went to forward airfields and roads in support of the division's operations. Whenever possible, Vietnamese laborers were hired and put to work manufacturing prefabricated buildings used in the self-help construction program. Company A initiated work on an interim water supply system by installing a pump station and pipeline from a nearby lake to a 1,000-barrel storage tank on the base. This undertaking included building a suspension bridge with a 300-foot clear span over a deep ravine and placing the pipeline underground inside the camp. Meanwhile, Company C constructed a loading ramp and crusher head wall for the 584th Light Equipment Company's quarry. The light equipment company helped build several forward airfields and upgrade Highway 19. By early December, the 20th gained a fourth line company with the attachment of Company D, 35th Engineer Battalion, to help the building effort at Pleiku. Since RMK-BRJ was already working at the air base and Camp Holloway, the firm received notice to proceed in June to build a cantonment, a hospital, and other support facilities for the Army.[43]

At nearby Pleiku Air Base, the Air Force expanded its presence from an outpost of 150 men to a complex of over 2,100 personnel. From this most inland of major air bases, the Air Force dispatched attack fighters, psychological warfare planes, gunships, forward air control observation planes, and rescue helicopters. Beginning in October 1965, a succession of Prime Beef

---

[42] ORLLs, 1 Feb–30 Apr 66, 299th Engr Bn, pp. 4–5, 1 May–31 Jul 66, 299th Engr Bn, pp. 1–2, and 1 Nov 66–31 Jan 67, 299th Engr Bn, 14 Feb 67, pp. 1–7, Historians files, CMH; AAR, Test Road T17, Co A, 299th Engr Bn, 30 Mar 66, incl. 9, tab C, OCE Liaison Officer Trip Rpt no. 3; Galloway, "Essayons," pp. 117, 119, 165.

[43] ORLLs, 1 Aug–31 Oct 66, 20th Engr Bn, 10 Nov 66, p. 3–4, 10, and 1 Nov 66–31 Jan 67, 20th Engr Bn, 12 Feb 67, p. 1–7, both in Historians files, CMH; Diary of a Contract, pp. 209–10.

teams built cantonments, aircraft ramps, operations facilities, and helped assemble fuel storage bladders and their dispensing systems. The teams also built two-story wooden troop billets and a kennel for Air Police guard dogs. By 13 September 1966, the teams had completed thirteen barracks housing over 900 men. RMK-BRJ also worked on cantonment facilities, which included an additional 1,000 kilowatts of power and a utilities system (water, sewers, sewage plant, and power distribution), and completed a 30,000-square-yard parking apron in December.[44] (*Map 10*)

On 10 November, the 937th Engineer Group, now under the command of Col. Ernest P. Braucher who replaced Colonel Watkin in August, assumed responsibility for the highlands and moved its headquarters to Pleiku. Although the group lost the 19th and 84th Engineer Battalions, it gained the 20th Engineer Battalion and Company D, 35th Engineer Battalion. The group retained the 70th and 299th Engineer Battalions and Company B, 84th Engineer Battalion. Colonel Braucher also had several smaller units to provide specialized support: the 509th and 511th Panel Bridge Companies (one platoon of the 509th remained attached to the 45th Group), the 584th and 630th Light Equipment Companies (the 630th also had a platoon attached to the 45th Group), the 585th Dump Truck Company, and two platoons and a part of a support platoon from the 554th Float Bridge Company.[45]

On the negative side, the 937th Engineer Group lost a major portion of its construction capability while taking on a larger area of responsibility. Although the 937th had given up the eastern half of its territory to the 45th Group, its area increased by 30 percent after taking over the southern highlands from the 35th Group. The recent influx of U.S. troops into the Pleiku area created requirements for construction that more than matched the resources at Colonel Braucher's disposal. This ensued despite transferring the Qui Nhon area projects and gaining the 20th Engineer Battalion. The loss of the skilled men and equipment of all but one company of the 84th, the only construction battalion in northern II Corps, could not be made up. The 84th had provided a reservoir of journeymen plumbers, fitters, electricians, and other key trades needed on the group's projects.[46]

For the most part, the 937th Group had tackled large projects typically carried out by a construction group headquarters. The small operations (S–3) and supply (S–4) sections of the group headquarters and the combat battalions required augmentation to handle the additional engineering tasks. In the end, this meant that the group and combat battalion headquarters tasked their units to locate qualified officers to assume engineering tasks at the headquarters. In turn, this situation left several companies short of officers. To some degree, the attachment of light equipment and construction support companies did bring the combat battalion on a par with the construction battalions. The 937th Group managed to get by with its two light equipment companies,

[44] Schlight, *Years of the Offensive*, pp. 172, 176; USAF Airfield Construction in South Vietnam, pp. 120, 127–28; Diary of a Contract, pp. 208–09.

[45] ORLL, 1 Nov 66–31 Jan 67, 937th Engr Gp, p. 2; Galloway, "Essayons," p. 163.

[46] ORLL, 1 Nov 66–31 Jan 67, 937th Engr Gp, pp. 6, 12.

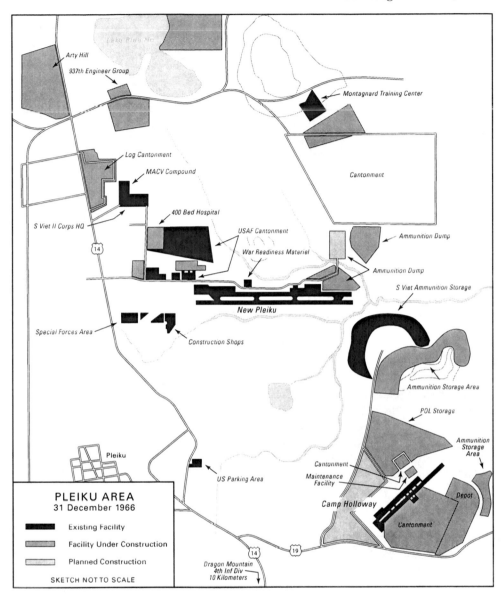

MAP 10

but the transfer of one of the two companies to the 159th Group in April severely depleted the group's construction equipment resources.[47]

In its first year in Vietnam (August 1965 to August 1966), the 937th Engineer Group had expended over 90 percent of its effort in construction.

[47] ORLLs, 1 Feb–30 Apr 66, 937th Engr Gp, p. 8, 1 May–31 Jul 66, 937th Engr Gp, 15 Aug 66, p. 10, Historians files, CMH, and 1 Nov 66–31 Jan 67, 937th Engr Gp, pp. 5–6; Galloway, "Essayons," pp. 75–76.

With such a heavy construction responsibility, the group concentrated its efforts in and around Qui Nhon and at An Khe, improving and expanding logistic and support facilities, and to a lesser degree at Pleiku. As the pace of combat operations increased in the highlands, I Field Force tasked the nondivisional engineers of the 937th Group for more operational support. With the shift to the western highlands, the proportion of operational support missions dramatically increased. Much of this effort went to building and maintaining forward airfields and tactical supply routes supporting operations. By the end of 1966, the group's construction and operational missions reached a balance. While the 70th Engineer Battalion concentrated almost entirely on construction at An Khe, the 20th Engineer Battalion devoted nearly all its effort to operational support. The 299th Engineer Battalion's effort fluctuated but broke down almost evenly between construction and operational support.[48]

More Work at Phan Rang

At Phan Rang, Colonel Triem's 62d Engineer Construction Battalion continued work at the joint Army and Air Force facility where most of the effort went into the jet fighter base. First, military engineers built an AM2 aluminum matting airfield followed by the civilian contractor, RMK-BRJ, which built a permanent 10,000-foot concrete runway, taxiways, and parking aprons. The first rains of the monsoon fell in September and October 1965, but a period of fair weather followed. Earthwork progressed satisfactorily until mid-December when heavy rains again set in. Nearly nine inches of rain fell within a ten-day period, hindering the use of heavy construction equipment, and resulting in the loss of many productive days. The 62d also suffered a high rate of equipment breakdowns and a lack of repair parts. In January 1966, the battalion and the contractor tackled the problem by pooling equipment. Military operators used the contractor's equipment at night, an arrangement that lasted until mid-February when RMK-BRJ began to work on the ammunition storage area.[49]

Although much earthwork remained to be done on the runway, the 62d began placing matting in areas as soon as the subgrade and base course were completed. All available electrical lighting was used to do earthwork at night. During the day, matting was laid even though temperatures reached 100 or more degrees. Through a trial and error method, the 62d found that small crews of soldiers proved more effective at laying the matting than Vietnamese laborers. On the other hand, the Vietnamese became proficient placing membrane sections on the subgrade and disassembling the bundles of AM2 matting. The first aircraft touched down on the interim runway on 20 February. By

---

[48] ORLL, 1 Nov 66–31 Jan 67, 937th Engr Gp, p. 6; 937th Engr Gp History, pp. 3–5; Galloway, "Essayons," pp. 108–09, 160.

[49] Quarterly Cmd Rpt 1 Jan–30 Apr 66, 62d Engr Bn, 11 May 66, pp. 1, 5–6, and ORLL, 1 May–31 Jul 66, 62d Engr Bn, 13 Aug 66, pp. 1–4, both in Historians files, CMH; Maj. Donald A. Haas, "Phan Rang Air Base," *Military Engineer* 58 (November-December 1966): 431–32; Galloway, "Essayons," pp. 45–46; Situation Rpt on the Construction of Cantonments, incl. 8, p. 2, OCE Liaison Officer Trip Rpt no. 3.

mid-March, the 10,000-by-102-foot runway and adjacent taxiways and parking aprons became operational when the first F–4 Phantoms landed, making Phan Rang the country's sixth operational jet fighter base.[50]

Work continued on the airfield, but bad weather caused more problems. After the 62d Engineer Battalion laid more matting on the north taxiway, heavy rains in May caused base course and subgrade failures. This forced the Army engineers and Air Force civil engineers of the 554th Red Horse Squadron, which arrived in January, to stop construction and make repairs. Although the AM2 matting had held up well to wear, Red Horse engineers, who were building aircraft revetments at the time, noticed depressions on the parking ramp. As the rains continued (Phan Rang had the unenviable distinction from the builder's point of view of having two rainy seasons, both winter and summer monsoons), the depressions deepened and quickly filled with water, often causing the fighter-bombers to drag their wing-hung ordnance. The ramp failures spread next to the taxiway, then the runway. With the help of RMK-BRJ, the Army engineers went to work trying to restore the taxiway and ramp. The 554th Red Horse Squadron took over runway repairs.[51]

It was determined that rain alone had not caused the base course failure. Red Horse engineers opined that Army engineers in their attempt to meet deadlines deviated from the original plans and substituted another waterproofed membrane instead of the more rugged and superior T17 membrane. Air Force engineers also attributed the failures due to hasty workmanship. In fact, the prefabricated metal surfaces could not seal water out of the joints between panels. Even the T17 and various asphalt products turned out less than satisfactory. Since each takeoff and landing produced powerful thrusts against the surface, work crews tried anchoring the matting in attempts to prevent the runway from creeping. Unfortunately, the anchorages allowed surface water access to the subgrade. By June, the 62d Engineer Battalion completed the north taxiway and turned attention to the parking ramp. Also that month, Red Horse engineers, working at night to avoid the 125-degree heat reflecting from the AM2 matting, replaced nearly 90,000 square feet of runway matting. The runway remained open for operations, though at times it was reduced to only 6,000 feet of usable length. During this time, RMK-BRJ continued work on the concrete runway.[52]

By the end of the year, the 62d Engineer Battalion's largest and most troublesome project at the air base neared completion. By then, Colonel Triem

[50] Haas, "Phan Rang Air Base," pp. 432–33; Lyons, "Aluminum Matting Runways in Vietnam," p. 246; Dunn, *Base Development*, p. 62; USAF Airfield Construction in South Vietnam, pp. 104–05. See also Quarterly Cmd Rpt, 1 Jan–30 Apr 66, 62d Engr Bn, pp. 3–4, and ORLL, 1 May–31 Jul 66, 62d Engr Bn, p. 3.

[51] ORLL, 1 May–31 Jul 66, 62d Engr Bn, pp. 2–3, 6; USAF Airfield Construction in South Vietnam, pp. 105–06; Situation Rpt on Airfield Construction in South Vietnam, incl. 7, pp. 1, 3–4, OCE Liaison Officer Trip Rpt no. 3; Msg, COMUSMACV MACDC-CM 15141 to CG, USARV, and OICC (Officer in Charge of Construction) RVN, 13 May 66, sub: Expeditionary Airfield Phan Rang, Historians files, CMH.

[52] USAF Airfield Construction in South Vietnam, pp. 107–08; Situation Rpt on Airfield Construction in South Vietnam, incl. 7, pp. 3–4, OCE Liaison Officer Trip Rpt no. 3; Ploger, *Army Engineers*, p. 114; ORLL, 1 May–31 Jul 66, 62d Engr Bn, p. 6.

had departed in late June and Lt. Col. Andrew J. Waldrop had assumed command ten days later. A notable achievement by the 62d Engineer Battalion as it entered its second year in Vietnam included the completion of the million-square-foot AM2 parking apron. Another project for the Air Force included completing a 46,000-barrel fuel storage area and connecting it to the six-inch pipeline leading to the beach. At the beach, the 62d installed two 8-inch submarine pipelines from the shore 1,200 feet out to a discharging and mooring facility. Work also began on two LST ramps and a new road to the air base. With the help of a fifty-man self-help contingent from the 1st Brigade, 101st Airborne Division, progress on the 4,700-man cantonment area began to raise the troop billets to Standard 4. The 62d also deployed forces inland to Bao Loc and to Da Lat to do airfield work.[53]

At the permanent airfield, RMK-BRJ completed the 10,000-foot concrete runway and four connecting taxiways on 12 October 1966. The new runway alleviated many operational adversities that had plagued the Army and Air Force engineers at the interim AM2 airfield. By year's end, the new runway and its supporting taxiway system were in service. All the roads were graded and most were topped with an asphalt surface. The Air Force occupied more than one-half of the sixty-seven barracks and work continued on the remainder. Work also proceeded into the final stages of installing an electrical power plant, water and sewage system, dining hall, dispensary, and other structures. To compensate for some slippage and deletions from construction schedules forced by contract costs, MACV tasked the 554th Red Horse Squadron with a six-month work program. This would provide Air Force personnel in 1967 those facilities that did not materialize in 1966.[54]

With a good share of its work completed, Colonel Waldrop received orders in late November to move the 62d Engineer Battalion to Long Binh. The second major move of the battalion while in Vietnam vividly brought out the hardships involved in relocating a 900-man heavy construction battalion. Companies continued to work on projects, phasing out construction approximately three to five days before loading on a LST for the voyage south. The move extended over a seven-week period, and as a result the battalion could barely keep up maintenance, medical, supply, engineering, and administrative support at two locations. Colonel Waldrop estimated the move had set back the battalion's maintenance effort by two months and recommended against phased movement of engineer construction battalions.[55]

[53] ORLLs, 1 Aug–31 Oct 66, 62d Engr Bn, 31 Oct 66, pp. 1–4, and 1 Nov–31 Jan 67, 62d Engr Bn, 12 Feb 67, pp. 1–4 and incl. 1, AAR, Bao Loc Airfield, both in Historians files, CMH; Galloway, "Essayons," p. 134.

[54] USAF Airfield Construction in South Vietnam, p. 110; Quarterly Hist Rpts 1 Jul–30 Sep 66, MACDC, p. 39, and 1 Oct–31 Dec 66, MACDC, p. 29. For more on RMK-BRJ work at Phan Rang, see Diary of a Contract, pp. 250–54.

[55] ORLL, 1 Nov 66–31 Jan 67, 62d Engr Bn, pp. 1–4, 11. For an account of the procedures involved in such a move, see Sgt. Maj. Edward J. Malen, "Engineer Battalion Move by Landing Craft," *Military Engineer* 60 (July-August 1968): 262–63.

Phan Rang airfield at the end of 1966 showing concrete runway and taxiways on the left and AM2 runway on the right

Long Binh and the III Corps Shield

Colonel Hottenroth's 3,000-man 159th Engineer Group pressed ahead on the Saigon base complex and tactical base camps forming a protective arc around the capital. The group's assignments included Long Binh, base camp facilities for the 1st and 25th Divisions, and the port at Vung Tau. Units carrying out this work were the 46th Construction and 168th and 588th Combat Battalions augmented by two light equipment companies and one construction support company. In addition, well-drilling detachments drilled for sources of fresh water. Besides its own troops and contractor personnel working in the area, the group also employed over 1,300 Vietnamese on a daily hire basis.[56]

At Long Binh, the 159th Group gradually transformed the former rubber plantation into a major Army base. The 46th Engineer Battalion, less Company D at Vung Tau, but reinforced with Company C, 168th Engineer Battalion, completed the 400-bed 93d Evacuation Hospital. Only shortages of electrical and plumbing supplies delayed the timely completion of the medical facility. In addition to the buildings, work at the hospital involved installation of an electrical distribution system by Pacific Architects and Engineers and construction of a minimum sewage system and access roads.

---

[56] Galloway, "Essayons," p. 83; ORLL, 1 Feb–30 Apr 66, 159th Engr Gp, 13 May 66, p. 1–7, Historians files, CMH; Rpt of Visit to the 159th Engr Gp, Long Binh, Vietnam, 15–18 Feb 66, incl. 6, tab C, p. 1, OCE Liaison Officer Trip Rpt no. 1.

Work continued at the giant ammunition supply depot, the unit's highest priority project. Seasonal rains hampered the construction of earthen berms and the extensive drainage and culvert work, but by early May approximately 20 percent of the depot, which included more than three million square feet of storage space, was done. Work on sheds and bunkers progressed at an equal pace.[57] (*Map 11*)

The 46th Engineer Battalion, commanded by Lt. Col. George Mason who replaced Colonel Hagedon on 17 January, also worked on other projects in the Long Binh area. In late January, the battalion began to clear nearby land along Highway 1A for the II Field Force headquarters. After clearing the land for the new camp, which became known as Plantation, the 46th erected seventeen vertical wall Quonset buildings. Facilities for support units followed. Pacific Architects and Engineers installed power plants and security lighting. In April, work began on a new cantonment between Long Binh and Plantation for the 90th Replacement Battalion located at the intersection of Highways 1A and 15. By the end of May, the replacement battalion moved out of Saigon to its new camp, which was capable of housing 3,000 individual replacements, soon to be expanded to a 6,000-man capacity. On the Dong Nai River, the 46th completed a barge off-loading facility known as the Cogido Dock. The new facility greatly increased the capacity for unloading barges used to transship ammunition from deep-draft ships docking downstream at Nha Be, thus avoiding hauling the dangerous cargo through populated areas to the Long Binh depot. After almost a year in country, the battalion finally took the time to complete its own maintenance shops, semipermanent headquarters buildings, and clubs.[58] (*Map 12*)

The move of Army troops from Saigon to Long Binh showed the massive scope of work still to come. The camp continued to grow in importance as Operation MOOSE (Move Out Of Saigon Expeditiously) ordered by General Westmoreland sent increasing numbers of U.S. troops to the former plantation. Long Binh had also become a major staging area for incoming men and units. From May to December, Long Binh grew from 6,000 to 20,000 men, with more troops to follow. By late 1966, the complex supported over 100,000 soldiers operating in III and IV Corps.[59]

The 18th Engineer Brigade initially handled most of the work at Long Binh. After some delay, RMK-BRJ began to mobilize its workforce and gradually assumed the bulk of construction. In late October, base development plans noted that most of the cantonment and about half the logistical facilities would be built with the $91 million earmarked from Fiscal Year 1966 supple-

---

[57] Galloway, "Essayons," p. 84; ORLLs, 1 Feb–30 Apr 66, 159th Engr Gp, pp. 3–4, 1 May–31 Jul 66, 159th Engr Gp, 12 Aug 66, p. 5, and 1 May–31 Jul 66, 46th Engr Bn, 11 Aug 66, p. 1, both in Historians files, CMH; PA&E History, 1963–1966, p. 37; Situation Rpt on Hospital Construction in South Vietnam, incl. 15, p. 1, OCE Liaison Officer Trip Rpt no. 3.

[58] Galloway, "Essayons," pp. 84–86; ORLLs, 1 Feb–30 Apr 66, 159th Engr Gp, pp. 5–6, 1 May–31 Jul 66, 159th Engr Gp, pp. 8–9, and 1 May–31 Jul 66, 46th Engr Bn, p. 2; PA&E History, 1963–1966, p. 39.

[59] Dunn, *Base Development*, p. 69; Army Construction in Vietnam, incl. 9, p. 1, OCE Liaison Officer Trip Rpt, no. 5.

mental funds. The rest of the facilities would be funded from the 1967 supplemental program at a cost of $77 million. Initial planning called for RMK–BRJ to complete the U.S. Army, Vietnam, and 1st Logistical Command permanent headquarters buildings by April 1967. Because of a scare brought on by a serious cash-flow problem that summer, the Officer in Charge of Construction slipped the projected completion date to August 1967.[60]

Throughout the summer and early fall of 1966, rains impeded progress in the Saigon area. Horizontal work (mainly earthwork) at Long Binh slowed down, but despite the mire, construction units maintained a moderately efficient rate on vertical projects. Experience gained a year earlier made the 159th Engineer Group headquarters attentive to drainage, and the commanders and staff admonished the constructing units to build drainage facilities before other work began. After almost a year commanding the group in Vietnam, Colonel Hottenroth passed the 159th Group to his successor, Col. Richard McConnell, in September.[61]

The 169th Engineer Battalion (Construction), which arrived from Okinawa in late May, became an integral part of the work under way at the logistical and headquarters complex. The battalion, commanded by Lt. Col. Marvin W. Rees, built a post headquarters and a stockade to detain U.S. soldiers charged or convicted of military offenses. The unit also cleared more storage areas, operated a prefabrication shop for the self-help construction program, began construction of the 24th Evacuation Hospital, built cantonments, and set up its asphalt plant. In August and September, the 169th built a staging area for the newly arrived 11th Armored Cavalry Regiment. Elements of the battalion helped to build the barge off-loading facility at Cogido Dock.[62]

North of Saigon, Lt. Col. Manley E. Rogers' 168th Engineer Battalion focused its efforts on the 1st Division's base camps at Di An, Phu Loi, and Lai Khe. On 28 April 1966, Lt. Col. Edwin F. Pelosky assumed command of the battalion and would hold that position for the next year. At Di An, Headquarters Company and Company A completed thirty-seven tropical buildings for the division headquarters and began building warehouses and storage areas for the Division Support Command. In January, Company B began building a C–130 airstrip for the 3d Brigade at Lai Khe. Despite the necessity of hauling all material to the base by armed convoys, work crews finished the M8 pierced steel plank strip within two months. Coinciding with the airfield work, Company B improved the base camp and completed an adjacent heliport for twenty-five helicopters. In early May, Company A took over base construction from the 588th Engineer Battalion at Phu Loi and completed a control tower, an airfield operations building, and an M8 pierced steel plank parking ramp for the 196 helicopters supporting the division.

[60] Army Construction in Vietnam, incl. 9, pp. 1–2, OCE Liaison Officer Trip Rpt, no. 5; Raymond, Observations on the Construction Program, p. 27. For more on RMK–BRJ projects at Long Binh, see Diary of a Contract, pp. 278–82.

[61] Galloway, "Essayons," pp. 138–39; ORLL, 1 Aug–31 Oct 66, 159th Engr Gp, pp. 7–8.

[62] Galloway, "Essayons," pp. 136–37; ORLLs, 1 May–31 Jul 66, pp. 2–3, 31 Jul 66, and 1 Aug–31 Oct 66, 169th Engr Bn, 31 Oct 66, pp. 4–7, Historians files, CMH.

Elements of the battalion also supported several operations and helped patrol areas around base camps.[63]

On 23 February, the 168th dispatched a platoon from Company C to Cu Chi northwest of Saigon to help build the 25th Infantry Division's base camp. General Westmoreland counted on the division to fill part of the gap in combat power and guard the approaches to Saigon from this direction. By this time, the MACV commander was convinced that major tactical headquarters and support units needed a full-time home where the individual soldier could train, take care of his equipment, and get some rest and relaxation. The 25th Division under the command of General Weyand made extensive studies before leaving Hawaii. Base development plans were put into final form following an advance party's reconnaissance. The division's 65th Engineer Battalion under Lt. Col. Carroll D. Strider assembled precut lumber kits for tents and latrines, which accompanied each unit. Upon arriving at Cu Chi, troops easily assembled the kits. Initial priority went to clearing fields of fire and constructing perimeter bunkers and wire barriers. Building semipermanent buildings followed. In early April, the 362d Light Equipment Company arrived to work on the camp's road net and drainage ditches. Before the end of the month, the division headquarters and medical personnel occupied facilities consisting of shed-type prefabricated buildings.[64]

The 159th Group's other combat engineer battalion, the 588th, commanded by Lt. Col. William T. Moore, also carried out construction work for the 1st Division. Company B worked at the Phuoc Vinh base camp northeast of Lai Khe, where it concentrated on building a cantonment, a heliport, and repairing airfield facilities. Heavy usage of the dirt airstrip (originally built by the Japanese during World War II) had caused considerable erosion, exposing large rocks that caused aircraft tire blowouts. The company remedied this problem by placing a six-inch topping of laterite on the field. Employing an expedient method to build parking ramps for helicopters, workers sprayed an asphalt cut-back treatment over the cleared area. They topped this off with pierced steel planking, thus completing a moderately dust-free heliport. On 1 June, Company B became Company C, 168th Engineer Battalion, in a transfer of companies between the two battalions. Company C, 168th Engineer Battalion, at Long Binh suddenly found itself redesignated Company B, 588th Engineer Battalion. Meanwhile, the redesignated Company C moved to the 2d Brigade's new base camp at Bearcat just off Highway 15 some five miles southeast of Long Binh. Besides the usual cantonment building tasks, the company built an eight-foot-high berm around the camp perimeter, providing a measure of security from enemy observation and direct fire.[65]

---

[63] Galloway, "Essayons," p. 87; ORLLs, 1 Feb–30 Apr 66, 159th Engr Gp, pp. 4–6, 1 Feb–30 Apr 66, 168th Engr Bn, 13 May 66, pp. 1–3, and 1 May–31 Jul 66, 168th Engr Bn, 14 Aug 66, pp. 1–3, both in Historians files, CMH.

[64] Lt. Gen. John H. Hay Jr., *Tactical and Materiel Innovations*, Vietnam Studies (Washington, D.C.: Department of Army, 1974), pp. 149–51; Galloway, "Essayons," p. 87; ORLLs, 1 Feb–30 Apr 66, 159th Engr Gp, pp. 2, 6, and 1 Feb–30 Apr 66, 168th Engr Bn, pp. 2–3.

[65] ORLLs, 1 Feb–30 Apr 66, 159th Engr Gp, p. 5, 1 May–31 Jul 66, 588th Engr Bn, 15 Aug 66, pp. 1–2, Historians files, CMH, and 1 May–31 Jul 66, 168th Engr Bn, pp. 1–2.

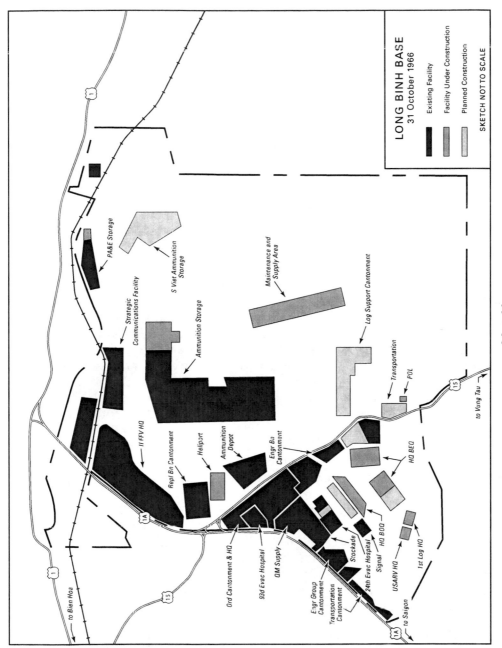

LONG BINH BASE
31 October 1966

Existing Facility
Facility Under Construction
Planned Construction

SKETCH NOT TO SCALE

to Bien Hoa
to Vung Tau
to Saigon

PA&E Storage
Strategic Communications Facility
S Viet Ammunition Storage
Ammunition Storage
Maintenance and Supply Area
Log Support Cantonment
Transportation
POL
II FFV HQ
Repl Bn Cantonment
Heliport
Ammunition Depot
Engr Bn Cantonment
HQ BEQ
Ord Cantonment & HQ
93d Evac Hospital
QM Supply
Stockade
HQ BOQ
Signal
24th Evac Hospital
USARV HQ
1st Log HQ
Engr Group Cantonment
Transportation Cantonment

MAP 11

As usual RMK-BRJ undertook the more complex construction projects. Following a terrorist bombing of the U.S. Embassy in March 1965, RMK-BRJ received a contract to build a new, more secure six-story building. In October, the consortium started work on an Army logistics depot in the eastern suburbs along the Saigon River bank. In December, workers began building a new ammunition depot, concrete parking aprons, and an Air Force cantonment at Tan Son Nhut Air Base. In April 1966, RMK-BRJ started planning for a 10,000-foot parallel runway at the air base. Lengthy negotiations with Vietnamese officials to remove graves in the path of the runway, however, delayed the start of the project to the end of the year. Also in late 1966, RMK-BRJ started building a new massive two-story prefabricated steel MACV headquarters building and complex costing $25 million adjacent to Tan Son Nhut Air Base. Projects given to the contractor at Bien Hoa Air Base included a fuel storage area, cantonment and flight line facilities, an aircraft maintenance shop, a new hangar, taxiway resurfacing, and aircraft parking aprons. Planning had also begun to add a parallel concrete runway and more aprons and taxiways. The consortium's new University Quarry—so named because of its proximity to a future Saigon University site—reached an impressive 250-ton-per-hour capacity in February 1966. (The quarry was also referred to as the SUMPCO Quarry [Saigon University Mineral Products Company] or SUMCO [Saigon University Materials Company].) In Saigon, the firm began work in April 1966 on an Armed Forces Radio and Television Service facility.[66]

Colonel Clymer's newly arrived 79th Engineer Group became operational on 20 July. With the help of the 46th Engineer Battalion, the 79th constructed cantonment facilities at nearby Plantation, where it moved the following month. The move to Plantation eased coordinating engineering support with its major customer, II Field Force. While the 159th Engineer Group concentrated on base development projects in Saigon, Long Binh, and Vung Tau, the 79th Group served a dual role as a construction and combat group. In its construction role, the 79th built bases for the 1st and 25th Divisions, and the 196th Light Infantry Brigade and the 11th Armored Cavalry Regiment following their arrival by the end of September. Simultaneously, the group provided operational support to these units. Early that month, the group lost its commander when Colonel Clymer injured his knee at a work site and had to be medically evacuated. Lt. Col. Walter C. Gelini assumed command. Gelini would become a colonel and remain in command for a full year.[67]

By the end of the year, Colonel Gelini commanded units usually found in a combat group. Starting with the two engineer combat battalions it inherited from the 159th, the 79th Group grew to four combat battalions, the 27th, 86th, 168th, and 588th, and a host of smaller units. Initially, its smaller units included the two light equipment companies, the 362d and 557th, and the three

---

[66] Tregaskis, *Building the Bases*, pp. 148, 249–52, 261, 271; Quarterly Hist Rpt, 1 Apr–30 Jun 66, MACDC, pp. 26–27; Ltr, Brig Gen Daniel A. Raymond, MACDC, to CG, Seventh AF, sub: Tan Son Nhut Parallel Runway, 14 Jun 66, Historians files, CMH. For more on RMK-BRJ work in the Saigon area, see Diary of a Contract, pp. 191–207.
[67] Galloway, "Essayons," p. 139; ORLL, 1 Aug–31 Oct 66, 79th Engr Gp, 14 Nov 66, pp. 3, 5–9, Historians files, CMH.

well-drilling detachments. In early September, the 66th Engineer Topographic Company, a corps support unit, joined the group. Although the topographic company belonged to the 79th Group, U.S. Army, Vietnam, controlled the specialized unit's mapping efforts. In early October, the 18th Engineer Brigade transferred the 617th Panel Bridge Company from the 159th to the 79th Group. In mid October, the 100th Engineer Company (Float Bridge) and in mid-December the 500th Engineer Company (Panel Bridge) joined the group. The group also increased its area of responsibility in III and IV Corps, leaving the construction at Saigon and Long Binh and the new Dong Tam base in the delta to the 159th Group.[68]

Work at the 1st Division base camps continued to preoccupy Colonel Pelosky's 168th Engineer Battalion. To keep pace with its heavy work load, the battalion operated on a seven-day, seventy-five-hour work week, easing up only Sunday mornings. Support for the battalion at Di An, Phuoc Vinh, Phu Loi, and Lai Khe included elements of two light equipment companies and three well-drilling detachments. In late October, the group transferred responsibility for Phu Loi and Lai Khe to the newly arrived 86th Engineer Battalion (Combat). The 168th shifted its main effort to Di An. This action allowed enough time for the battalion to build cantonment facilities on the northern side of the camp—known as the North Forty—to fit another move by the division's 2d Brigade in March, this time from its short stay at Bearcat. Before moving to Di An, Company A at Phu Loi and Company B at Lai Khe had spent much time maintaining troublesome airfields at the two bases. During the rainy season, the two companies fought rutting and subgrade failures on the runways with rock. To keep the clouds of dust down during the dry periods, Company A applied peneprime near helipads.[69]

The 588th Combat Engineer Battalion consolidated its efforts in the 25th Division's area of operations. In late June, Company B (formerly Company C of the 168th) at Long Binh joined the headquarters and Company A at Cu Chi. Company B then moved to Tay Ninh to begin construction of a new base camp for the 196th Infantry Brigade, then en route and expected to arrive in two weeks. On 22 August, Company C departed Bearcat and began to construct a 400-bed evacuation hospital for the 25th Division. With the hospital work moving along, the company sent a platoon to Tay Ninh in late October to build the first modern Army hospital consisting of inflatable, rubber-type buildings, formally known as MUST (medical unit, self-contained, transportable).[70]

During this period, the 588th, now under the command of Lt. Col. James F. Boylan, had grown into a considerable force. The battalion had under its operational control at Cu Chi a platoon-size force from the 100th Float Bridge Company, the 362d Light Equipment Company, and Company B, 86th

[68] Galloway, "Essayons," p. 139; ORLLs, 1 Aug–31 Oct 66, 79th Engr Gp, pp. 5–9, and 1 Nov 66–31 Jan 67, 14 Feb 66, pp. 1–5, Historians files, CMH.

[69] Galloway, "Essayons," pp. 139, 141; ORLLs, 1 May–31 Jul, pp. 1–3, and 1 Aug–31 Oct 66, 168th Engr Bn, 12 Nov 66, pp. 1–8, Historians files, CMH.

[70] Galloway, "Essayons," pp. 141–42; ORLLs, 1 May–31 Jul 66, 588th Engr Bn, pp. 1–3, 1 Aug–31 Oct 66, 588th Engr Bn, 14 Nov 66, pp. 1–3, and 1 Nov 66–31 Jan 67, 588th Engr Bn, 13 Feb 67, pp. 1–5, both in Historians files, CMH.

Engineer Battalion. Company B, which had arrived with the rest of the 86th at Vung Tau on 17 October, began building a post exchange consisting of several prefabricated metal buildings complete with electrical fixtures and concrete floors. In November, the 588th supervised a platoon from the 643d Engineer Pipeline Construction Company in the building of a 3,000-barrel fuel storage facility at Tay Ninh. After completing this project in early December, the platoon returned to Long Binh. By the end of the year, the battalion eagerly awaited the arrival of the valuable hauling assets of the incoming 67th Engineer Dump Truck Company.[71]

The 27th Engineer Battalion (Combat) commanded by Lt. Col. Charles R. Roberts supported the newly arrived 11th Armored Cavalry Regiment and the 3d Brigade, 4th Division. Arriving before the advance party and the rest of the battalion in mid-August, Company C moved to Xuan Loc, approximately twenty-five miles east of Long Binh, to build camps for the armored cavalry regiment. After completing a long sea voyage, which included a stop at Da Nang, the rest of the 800-man battalion, which had four lettered companies, arrived at Vung Tau on 30 September and moved to staging areas at Long Binh and Di An. While waiting for its equipment, the battalion headquarters worked on base development plans for the 11th Armored Cavalry Regiment's 6,000-man base camp near Xuan Loc. Meanwhile, the line companies took up their hand tools and went to work. Company B at Di An helped the 168th Engineer Battalion build the cantonment for the 2d Brigade, 1st Division. When the equipment arrived near the end of October, the headquarters and Company A moved to Xuan Loc, and joined the regiment's 919th Engineer Company on base camp construction. After completing its tasks at Di An, Company B, on 1 November, moved east of Xuan Loc to Gia Ray to open a quarry and set up a rock crusher.[72]

On 17 October, Company D, 27th Engineer Battalion, was attached to the 3d Brigade, 4th Division. When the brigade arrived that month, it did not join the rest of the division in II Corps. Instead, it was diverted to III Corps as a trade off for the 3d Brigade, 25th Division, at Pleiku. Since the 3d Brigade, 4th Division, deployed to III Corps without its usual supporting company from the 4th Engineer Battalion, Company D assumed that role and began to improve the brigade's base camp and airstrip at the vacated Bearcat camp. Within a month, the brigade joined the 25th Division northwest of Saigon at Dau Tieng. Company D continued to support the brigade by clearing South Vietnamese Army minefields around the base camp, building roads in the camp, and making improvements to the airfield. Outside the camp, Company D worked on roads, bridges, and culverts, and carried out combat engineering tasks typical for a divisional engineer company that included clearing firebases, mine clearing, and demolitions.[73]

[71] ORLLs, 1 May–31 Jul 66, 588th Engr Bn, p. 2, and 1 Nov 66–31 Jan 67, 588th Engr Bn, pp. 2–5.

[72] ORLLs, 1 Aug–31 Oct 66, 27th Engr Bn, 31 Oct 66, pp. 2, 8–9, and 1 Nov 66–31 Jan 67, 27th Engr Bn, 13 Feb 67, pp. 1–5, both in Historians files, CMH; Galloway, "Essayons," p. 144; Carland, *Stemming the Tide*, p. 166.

[73] ORLL, 1 Nov 66–31 Jan 67, 3d Bde, 4th Inf Div, p. 18, Historians files, CMH.

The other new arrival, the 86th Engineer Battalion (Combat), commanded by Lt. Col. Colin M. Carter, completed its long journey from Fort Dix, New Jersey, when it reached Vung Tau on 16 October aboard the USNS *Weigle*. After disembarking, the companies moved by air to their destinations. Headquarters and Company A deployed to Phu Loi, Company B to Cu Chi, Company C to Lai Khe, and Company D to Bien Hoa. At Phu Loi, Company A took over cantonment and airfield facility construction from Company A, 168th Engineer Battalion. Plans were under way to build seven large prefabricated maintenance hangars at the airfield, but five of the hangar sets still had not arrived. Similarly, Company C took over projects from Company B, 168th Engineer Battalion, at Lai Khe. Company B at Cu Chi worked on base development projects under the direction of the 588th Engineer Battalion. Company D at Bien Hoa completed a C–130 parking ramp at the airfield, and helipads, access roads, and cantonment projects on the Army side of the base. The battalion also received several combat support missions.[74]

At Tan Son Nhut and Bien Hoa Air Bases, RMK-BRJ made little progress on the parallel runways. By December, excavation work at Tan Son Nhut was only 3 percent complete. The fruitless negotiations to remove the two hundred or so graves came to a sudden end in early December when a Viet Cong force infiltrated the air base and had to be driven out of their hideaways, most of which were in the cemetery. Within a few days, the South Vietnamese Army appeared on the scene and without any explanation removed the tombstones. By the end of the year, excavation work had begun on the new runway. At Bien Hoa, only 50 percent of the plans were ready, and authorization for the work for the new runway, taxiways, and apron was held off until 1967.[75]

Air Force Prime Beef teams continued their work at both air bases, and advance elements of the 823d Civil Engineering Squadron had arrived at Bien Hoa Air Base in October. The 823d became the fifth Red Horse squadron deployed to Vietnam to support Air Force construction and repair requirements. By January 1967, the squadron deployed elements to Tan Son Nhut, Vung Tau, Da Nang, and Pleiku.[76]

By the autumn of 1966, Pacific Architects and Engineers began reducing its construction activities and turned attention to its facilities engineering mission. In October, Vinnell Corporation, which was also moving eleven T–2 power-generating tankers to four other locations, began construction of a central electrical power plant at Long Binh. The design called for a series of 1,500-kilowatt generators to be connected and synchronized to work on one power system. Because of the threat of night attacks, the firm's highest priority centered on installing security lighting around the ammunition depot. The contractor worked day and night to install over ten miles of security lighting around the perimeter. Power to other areas became available when workers completed portions of the pole line. Despite enemy activities and hazardous

---

[74] ORLL, 1 Nov 66–31 Jan 67, 86th Engr Bn, 14 Feb 67, pp. 9–13, Historians files, CMH; Galloway, "Essayons," p. 144.

[75] Tregaskis, *Building the Bases*, pp. 251–52; Diary of a Contract, pp. 199, 205.

[76] USAF Airfield Construction in South Vietnam, pp. 120, 126–29.

147

conditions, construction of the plant continued satisfactorily during the winter of 1966–1967.[77]

New Berths for Saigon

The backlog of ships at the commercial port in Saigon had convinced MACV of the need for more extensive port facilities. At the beginning of 1966, the port had ten deep-draft alongside berths, seven large and four small berths tied to buoys, two LST berths, and three T-shaped piers for coastal shipping. These facilities handled 90 percent of the cargo arriving in South Vietnam, but the daily average of 13,000 short tons was more than Saigon could handle. The current port, adequate by peacetime standards, just did not have enough berths, barges, landing craft ramp space, and storage space to support both the country's commercial needs and the large quantities of economic assistance and military cargo now reaching Vietnam. Also exacerbating port operations were the lack of military control and the coordination problems with Vietnamese port officials. At best, three to four berths could be reserved for military cargo, another three to four for economic assistance cargo, while the rest remained for civilian use.[78]

To resolve these problems, MACV laid plans to build a new deep-draft port for military cargo along the Saigon River. The proposed site, named Newport, was in a sparsely populated area adjacent to Highway 1A about two miles northeast of the city. To get the port operational as soon as possible, planning proceeded for seven lighterage berths and four landing craft slips and ramps. Four deep-draft berths would be built later. MACV again turned to the Officer in Charge of Construction to design the facility and select a contractor.[79]

The Navy assigned the construction portion of the Newport project to RMK-BRJ. With such high interest in the project, the Officer in Charge of Construction assigned a full-time officer as a Resident Officer in Charge of Construction to oversee work, which soon became a twenty-four-hour-a day, seven-day-a-week maximum effort. By mid-May 1966, RMK-BRJ had dredged the LST slips and drove the first piles for the barge wharves. Unfortunately, the silty material from the river could not be used to fill the former rice paddy land and bayou. Instead, local contractors furnished much of the fill, and their little sampans accounted for about 3,900 of the 9,150 cubic yards of sand hauled in from other places each day. By midyear, the director of construction reported that RMK-BRJ drove 156 piles for the barge and LST wharves. The firm's concrete precast yards also made eighteen concrete deck slabs for the deep-

[77] PA&E History, 1963–1966, pp. 42; "Vinnell Makes History at Long Binh," *Vinnell* 11 (May 1969): 8–10, copy in Historians files, CMH.

[78] MACV History, 1966, pp. 709–10.

[79] Dunn, *Base Development*, p. 59; Quarterly Hist Rpt, 1–31 Jan 66, MACDC, 19 Feb 66, Historians files, CMH; MACV History, 1966, p. 710; MACV Monthly Evaluation, May 66, p. G-5; Msg, Robert S. McNamara OSD 5096–65 to Westmoreland, 31 Dec 65, Westmoreland Message files, CMH; Msg, COMUSMACV MACJ44 to CG, USARV, 4 Jan 66, sub: Project Newport, Historians files, CMH.

Port and warehouses under construction in late 1966 at Newport

draft wharf. General Dunn informed Washington that the informal estimate to complete the demanding project now reached $30 million.[80]

Despite several problems, progress continued to be made on the docks and open-storage facilities. In August, the Joint Chiefs of Staff asked for a reevaluation of the entire project to refine the cost required to complete the port as originally designed. The new figure totaled more than $61 million, approximately double the earlier estimate. To save some money, the Army deferred some warehouses and other buildings. For a few days, a halt in the delivery of sand threatened progress. One of the Vietnamese contractors complained that "VC [Viet Cong] tax assessors" had charged the sampan operators a large fee. About ten days' supply of sand was on hand, so the Officer in Charge of Construction ended that contract and negotiated a new one with other vendors. In October, the construction office turned over four barge wharves, one LCM/LCU (landing craft, mechanized, and landing craft, utility) ramp, one LST slip, and thirty-eight acres of filled land to the 1st Logistical Command. Work then commenced on four deep-draft berths, which included four deep-draft berths, more filled land, transit warehouses, a power-generating facility, and a perimeter road.[81]

---

[80] Quarterly Hist Rpt, 1 Apr–30 Jun 66, MACDC, p. 27; Ltr, Dunn MACDC to CG, USARV, sub: Newport, Increment II, 18 May 66; DF, Raymond MACDC to CofS, MACV, sub: Status Report, 10 Jun 66, Historians files, CMH; Msg, Dunn SOG 0528 to Noble, 24 Jun 66, Dunn Papers, CMH; Tregaskis, *Building the Bases*, p. 91. Report and letter in Historians files, CMH. See also Diary of a Contract, pp. 261–63.

[81] Quarterly Hist Rpts, 1 Jul–30 Sep 66, MACDC, pp. 40–41, and 1 Oct–31 Dec 66, MACDC, p. 30; Tregaskis, *Building the Bases*, p. 246.

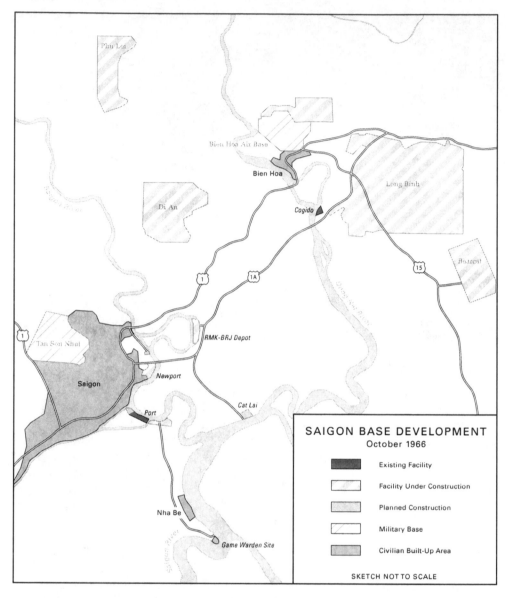

MAP 12

Other efforts were also under way to relieve the pressure on the Saigon port. By April, dredging along the Soirap River between Vung Tau and Saigon opened a second deep-draft channel, which provided an alternate route for shipping along the Saigon River. The second channel lessened the impact in case the Viet Cong did succeed in sinking a ship. Additionally, three buoy berths were installed in the Dong Nai River at Cat Lai for mooring and off-loading ammunition ships to barges.[82]

[82] Quarterly Hist Rpt, 1 Apr–30 Jun 66, MACDC, p. 27.

At Vung Tau, Company D, 46th Engineer Battalion, made headway on port facilities, the airfield, and projects for Australian units in the area. As the only company-size engineer unit at Vung Tau, Company D completed a variety of projects, including warehouses and fuel storage facilities for the 1st Logistical Command and helped the Australian task force to develop supply and troop staging areas. The fuel tank farm presented a challenge and involved removing three old storage tanks destroyed by the French when they evacuated the area. Work crews had to cut the damaged tanks with explosives, remove them from their protective shields, and replace them with new 10,000-gallon bolted steel tanks.[83]

By March, the 536th Engineer Detachment (Port Construction), the first of its kind to do this specialized work in III Corps, arrived at Vung Tau. The detachment initially concentrated on preparing designs and assembled its pile driving and diving barges before starting work on new piers and LST ramps. Plans called for the installation of a DeLong pier, but the extensive dredging requirements on the sheltered side of the peninsula prompted the Construction Directorate to change priorities. Of the seven DeLong piers intended for delivery to South Vietnam, Vung Tau dropped to the bottom of the list. The detachment also demonstrated the versatility of port construction units when it assembled and rehabilitated twenty-four barges belonging to the Army's 4th Transportation Command.[84]

Projects in the Delta

Although few U.S. troops were in the delta region, military and civilian engineers carried out extensive construction there. In early April, RMK-BRJ completed a 4,000-foot runway and other base facilities on An Thoi, a small island off the coast of the larger Phu Quoc Island near the Cambodian coast. At the Army's Vinh Long helicopter base, work that started in June 1965 continued. The contractor began further modernization at Binh Thuy Air Base. More work also awaited the firm at Can Tho, where the U.S. Navy launched patrol boat operations. In February, the Air Force began to dispatch Prime Beef teams to Binh Thuy to build billets and operational facilities for Air Force personnel.[85]

MACV had long considered the possibility of basing U.S. ground troops in the Mekong Delta. A survey team concluded that all land suitable for large tactical units was either heavily populated or occupied by the South Vietnamese armed forces. If American troops deployed to the delta, they would have to share already crowded areas or displace a portion of the population. Since the

---

[83] Galloway, "Essayons," pp. 84–86; ORLLs, 1 Feb–30 Apr 66, 159th Engr Gp, pp. 5–6, 1 May–31 Jul 66, 159th Engr Gp, pp. 8–9, and 1 May–31 Jul 66, 46th Engr Bn, p. 2.

[84] Galloway, "Essayons," pp. 84–86; ORLLs, 1 Feb–30 Apr 66, 159th Engr Gp, pp. 2, 5–6, 1 May–Jul 66, 159th Engr Gp, p. 8, and 1 May–31 Jul 66, 46th Engr Bn, p. 2; Quarterly Hist Rpt, 1 Apr–30 Jun 66, MACDC, p. 15.

[85] Tregaskis, *Building the Bases*, pp. 294–95; Quarterly Hist Rpt, 1 Apr–30 Jun 66, MACDC, p. 27; USAF Airfield Construction in South Vietnam, pp. 119, 123, 125. For more on RMK-BRJ work in the delta, see Diary of a Contract, pp. 123–26, 241–43.

command considered both courses unacceptable, the planning staff searched for other means of basing troops. The MACV Engineer proposed dredging river sand as fill material and building up an area to fit a division. Yet, the number of dredges necessary to do the work would not be available until late 1966. Although not entirely satisfied with this solution, the command believed the building of a base area by dredging operations a sound choice.[86]

Planning went forward to station a joint Army and Navy riverine force in the northern delta. Recalling that the French had dispatched small landing craft from land bases during the First Indochina War, planners envisioned river flotillas consisting of barracks ships and landing craft carrying ground troops on operations along the delta's waterways. In early 1966, General Westmoreland approved plans to develop two land bases and one mobile floating base in the delta. One land base would house an infantry division headquarters and one brigade, and a second brigade would occupy a base in the northern part of the delta, probably in Long An Province in III Corps. The third brigade would be based on the water.[87] (*Map 13*)

In May 1966, Westmoreland directed a survey to find a land base for the division headquarters and the one brigade. Tactical resupply and the transportation of materials dictated that the new delta base be accessible to a system of navigable waterways. Since the monsoon season drastically reduced the little available dry land, enough suitable fill had to be located near the site to raise an area of approximately 600 acres to a level higher than the watery countryside. The Directorate of Construction found four locations—designated Bases W, X, Y, and Z—suitable for dredging and building the land base and submitted these to Westmoreland for consideration. He picked Base W located about forty miles southwest of Saigon and about three miles west of My Tho. He believed that the site should be given a significant name in Vietnamese—something dealing with friendship and cooperation. He selected the Vietnamese term *dong tam*, which translated meant "united hearts and minds."[88]

The plan called for three major operations to transform the marshy land into a riverine base. First, the builders had to excavate a rice paddy to make a turning basin for ships. The second task consisted of dredging an entrance channel into the turning basin from the My Tho River. Third, to make the one-square-mile base, sand had to be dredged from the river and pumped into the rice paddies. For this last task alone, engineers estimated that approximately 10.5 million cubic yards of fill would be required. This figure included not only

---

[86] Maj. Gen. William B. Fulton, *Riverine Operations, 1966–1969*, Vietnam Studies (Washington, D.C.: Department of the Army, 1973), pp. 26–27. For more background on the delta and planning, see George L. MacGarrigle, *Combat Operations: Taking the Offensive, October 1966 to October 1967*, United States Army in Vietnam (Washington, D.C.: U.S. Army Center of Military History, 1998), pp. 393–97.

[87] Fulton, *Riverine Operations*, pp. 27–28.

[88] Quarterly Hist Rpts, 1 Apr–30 Jun 66, MACDC, p. 13, and 1 Jul–30 Sep 66, MACDC, p. 43; Ploger, *Army Engineers*, p. 145; Fulton, *Riverine Operations*, pp. 33, 47–48.

MAP 13

the landfill for the base but also a sufficient stockpile of sand for airfield and road projects and to make concrete.[89]

The first phase of the Dong Tam project began in July with the arrival of a clamshell dredge to prepare the work site for two larger dredges. The sixteen-inch pipeline cutterhead *Cho Gao* arrived in late July but could not start pumping until 4 August because of a mine scare. The Viet Cong, aware that the base could cripple their efforts to control the heavily populated and rice-producing delta region, soon began to make their presence known. Sappers mined the *Cho Gao*, but the Americans found and removed the mines in time. The Viet

[89] Dunn, *Base Development*, p. 53; Ploger, *Army Engineers*, p. 145; Quarterly Hist Rpt, 1 Jul–30 Sep 66, MACDC, p. 43.

Cong also mortared the dredge, causing some minor delays. Throughout July and August, the *Cho Gao* dredged the entrance channel and turning basin, pumping at an average rate of some 13,000 cubic yards of fill daily.[90]

The largest dredge in South Vietnam, the thirty-inch pipeline cutterhead *Jamaica Bay*, previously at work at Da Nang, arrived in September. Crews working among the fillpipes on shore came under sniper fire, and a plastic explosive charge damaged a dragline. Despite the harassment, the *Jamaica Bay* maintained its schedule with no breakdowns, pumping at a daily average rate of over 26,000 cubic yards. By 1 January 1967, the dredge had pumped some 2.3 million cubic yards of fill, creating 155 acres of dry land. The *Cho Gao* continued its work on the turning basin and completed 85 percent of the LST area. In October 1966, a Seabee detachment deployed from I Corps and built a concrete pontoon pier for Navy forces. Within a few weeks, the 159th Engineer Group deployed elements to start work on a 7,500-man camp and operational base for troops of the 9th Infantry Division.[91]

Logistical Troubles Continue

During 1966, Army engineers continued to have trouble obtaining building supplies and keeping equipment running. As General Ploger pointed out to General Westmoreland during his November 1965 briefing, some $25 million in building materials ordered earlier had not arrived. To carry out the construction program through June 1966, Ploger had estimated that some 325,000 tons of supplies would be needed. If the experience of previous wars continued, then about 15 percent of the total theater tonnage would consist of construction items. Translated into terms of shipping, this meant that one out of every seven ships would be loaded with lumber, airfield matting, bridge sets, generators, asphalt, cement, and prefabricated buildings. Ploger's estimate turned out to be on the modest side. By the summer of 1966, about a third of the tonnage reaching Vietnam turned out to be construction items.[92]

Even when construction supplies started reaching the depots, distribution to engineer units fell short. During the summer of 1966, two shiploads of one-inch lumber were delivered to Cam Ranh Bay where there was a shortage of two-inch lumber, and five shiploads of two-inch lumber were shipped to Qui Nhon where there was no one-inch lumber. Meanwhile, shortages of two-by-fours were exacerbated by the decision to put troops in wood frame barracks

[90] Dunn, *Base Development*, p. 53; Tregaskis, *Building the Bases*, p. 292; Quarterly Hist Rpt, 1 Jul–30 Sep 66, MACDC, p. 43.

[91] Dunn, *Base Development*, p. 53; Tregaskis, *Building the Bases*, pp. 292–93; Quarterly Hist Rpts, 1 Jul–30 Sep, 66, MACDC, pp. 43–44, and 1 Oct–31 Dec 66, MACDC, p. 33; Msg, COMUSMACV MACDC-CM to COMNAVFORV (Commander Naval Forces, Vietnam), 19 Aug 66, sub: Dredging of Deep Draft Facility at Observation Point Da Nang, box 5, 69A/702, RG 334, NARA.

[92] Ploger, *Army Engineers*, p. 202; Army Buildup Progress Rpt, 26 Oct 66, pp. 56–57, CMH; USARV Engr, Summary of Total Tonnages of Troop Construction Material for Jan 66–Jun 67, 8 Mar 66, incl. 20, tab B, OCE Liaison Officer Trip Rpt no. 2; Construction Logistics in Vietnam, incl. 9, p. 1, and 18th Engr Bde, Memo for Brigade S–4, sub: Construction Logistics, 19 Aug 66, incl. 9, tab G, p. 1, OCE Liaison Officer Trip Rpt no. 4.

The Dong Tam base camp surrounded by rice paddies began to take shape by the end of 1966.

instead of tents. Even when tents were used, two-by-fours were substituted for tent poles and used in wooden flooring. Although 1st Logistical Command increased the requisition objective for lumber from 50 to 75 million board feet as of 1 October, no one could say when the additional supplies would arrive. Another problem emerged from the reorganization of certain technical service supply units, including engineer supply units, into general depots, irrespective of the fact that a war was going on and a loss of some technical expertise was bound to result. No sooner were the new depots constituted in Vietnam than inexperienced depot troops had trouble identifying arriving inventory, especially electrical and plumbing supplies. By the summer of 1966, engineer units were regularly sending "scrounging" expeditions to the supply yards to search out items they required.[93]

The 159th Engineer Group at Long Binh continued to have the same problems from the previous year. In May, Colonel Hottenroth reported that the group continued to be plagued by shortages of construction materials and essential equipment parts and an inability to organize properly to meet the construction mission. For instance, workers had constructed troop showers up to the point of installing the shower heads only to discover that the part was not available. A 34E concrete paver remained inoperable for two months because it lacked major components, and well-drilling rigs had arrived without casings or screens. "The logistic system," Hottenroth noted, "appears to have withered due to past efforts to eliminate military spaces in supply and maintenance operations and as a result in my judgment been unable to

[93] Construction Logistics in Vietnam, incl. 9, pp. 1, 4, and 18th Engr Bde, Memo for Brigade S–4, 19 Aug 66, sub: Construction Logistics, incl. 9, tab G, p. 3, OCE Liaison Officer Trip Rpt no. 4; ORLL, 1 Aug–31 Oct 66, 35th Engr Gp, pp. 9, 16.

expand to meet the workload." He noted that seven months had gone by after he had submitted a request to modify the battalions' tables of organization and equipment. It appeared uncertain whether the Department of the Army would organize the units under the latest organizational tables. Travel to the dispersed project sites by the group's commanders and staffs posed another problem; road travel was risky and not enough aircraft were available. Finally, he concluded, the myriad of administrative and reporting tasks had become burdensome. On the plus side, the weather during the period turned out to be favorable for construction.[94]

When equipment broke down or was not available, engineers found other ways to get the job done, by hand if necessary. While visiting the 159th Group in March, the Chief of Engineers' liaison officer from Washington found that, although the group's units had their authorized complement of seven 16S concrete mixers, they did not have enough mixers to handle the large number of projects at many different sites. The group had requested twelve additional mixers the previous November, but none had arrived. In the end, the engineer company responsible for the Lai Khe base camp project resorted to mixing concrete by hand while trying to fashion a homemade mixer. A query with Colonel Lien at the 1st Logistical Command revealed an awareness of the problem, but his staff could not provide the status of the request before the liaison officer's departure.[95]

Colonel Gelini reported that logistics support for his 79th Engineer Group had been generally good, but there was a critical shortage of dump trucks. The inadequate hauling capability affected the group's construction efforts, especially the transport of rock. To make matters worse, the 79th Group had to use its trucks to pick up cargo at supply points in the Saigon–Long Binh area and haul it to its outlying units. Fortunately, the assignment of a panel bridge company with 35 five-ton dump trucks and a float bridge company with 50 five-ton bridge trucks and 39 two-and-a-half-ton cargo trucks to the 79th Group added capacity. Both bridge companies unloaded and stored their bridges at Long Binh and put their trucks on the road hauling supplies.[96]

Colonels Hart and Starnes, commanders of the 35th Group at Cam Ranh Bay, reported similar shortages. Hart complained that the regular supply system could not respond quickly enough for his units to complete their port, depot, and cantonment projects. He felt that certain special items, such as sheet piling, fuel pipeline supplies, and lumber, should be programmed in bulk instead of waiting for projects to be assigned to units, which then had to requisition the materials through normal supply channels. Hart also believed that responsibility for construction supplies should be returned to engineer commanders instead of the recent trend to general supply units. In another matter, Hart and Starnes believed that the current organization of the 497th Port

---

[94] ORLL, 1 Feb–30 Apr 66, 159th Engr Gp, pp. 7–8 (quotation, p. 7); Galloway, "Essayons," pp. 52, 89.

[95] Rpt of Visit to the 159th Engr Gp, Long Binh, Vietnam, 15–18 Feb 66, incl. 6, tab C, and Situation Rpt on Engr Equipment in South Vietnam, incl. 17, OCE Liaison Officer Trip Rpt no. 1.

[96] ORLL, 1 Aug–31 Oct 66, 79th Engr Gp, pp. 1–7; Galloway, "Essayons," p. 147.

Construction Company did not have the flexibility required to support construction work at more than one port at a time. Both men suggested changes in tables of authorization. Even so, a platoon had been detached and was hard at work on Qui Nhon port.[97]

While the 937th Engineer Group had some problems getting supplies, notably rock, electrical, and plumbing material, the main difficulty lay in transporting supplies inland to An Khe and Pleiku. Transportation units gave higher priority to ammunition, rations, and fuel, leaving little space for construction items. To support their projects, both the 299th Engineer Battalion at Pleiku and the 70th Engineer Battalion at An Khe put their trucks on Highway 19 for the long-distance haul to Qui Nhon. The 70th reported it had to use its own vehicles to haul at least 75 percent of the construction items used by the unit. Supply technicians usually accompanied the convoys to Qui Nhon to help depot troops locate needed items. Gradually, the supply situation improved, and construction supplies reached the group in increasing quantities. Repair parts became more plentiful, but Colonel Braucher, the group commander, still complained of a lack of direct-support maintenance units.[98]

The equipment situation, however, brightened. By early April, seventy-seven new HD16M bulldozers were shipped to Vietnam, forty-four to Cam Ranh Bay and thirty-three to Saigon, and more were on the way. About the same time, Caterpillar shipped 294 D7E bulldozers from Peoria, Illinois, to Army depots in the United States. Once complete packages of spare parts were available, which was expected in late April, D7Es would be shipped to Vietnam. Meanwhile, two of the bulldozers were airlifted to Saigon to acquaint operators and mechanics with the new standard bulldozer. Three technicians—one maintenance, one supply, and one factory representative—started on-the-job training courses on 18 March. At the beginning of June, the parts support buildup was approaching 90 percent completion. Shipment of the new bulldozers, however, was held up by engine failures, which had to be evaluated. On 16 June, the Army Materiel Command authorized the release of 336 D7Es from the increasing depot stocks, 268 to Vietnam and the rest for issue to units deploying to Vietnam. The bulldozers began arriving in Vietnam on 25 August. By early 1967, 196 of the D7E bulldozers that had reached Vietnam were in the hands of Engineer Command units, replacing the HD16Ms, TD20s, and TD24s.[99]

More good news concerned earthmoving equipment. By early June, 250 of the 840 Clark 290M tractors under contract had been delivered to Army depots. Some defects, however, still had to be fixed, and repair parts were not to reach the required 90 percent level before mid-July. Before being shipped to Vietnam, comparison-type testing of the new tractor with attached Euclid

[97] ORLLs, 1 Feb–30 Apr 66, 35 Engr Gp, pp. 6–9, and 1 May–31 Jul 66, 35th Engr Gp, pp. 5, 8, 10.

[98] Galloway, "Essayons," p. 119; ORLLs, 1 Feb–30 Apr 66, 937th Engr Gp, p. 8, 1 Nov 66–31 Jan 67, 937th Engr Gp, pp. 15–17, 1 Aug–31 Oct 66, 70th Engr Bn, pp. 2–3.

[99] Army Buildup Progress Rpt, 6 Apr 66, p. 47, CMH; *Arsenal for the Brave*, pp. 250–52; Rpt of Visit to the 18th Engr Bde, incl. 7, p. 5, OCE Liaison Officer Trip Rpt no. 4; ORLL, 1 Nov 66–31 Jan 67, Engr Cmd, 31 Jan 67, p. 21, Historians files, CMH.

and LeTourneau eighteen-cubic-yard scrapers was carried out. Twenty-five of the scrapers had been delivered by LeTourneau against a contract for 330, and repair parts were expected to reach the required level by early July. By January 1967, seventy-one Clark 290M wheeled tractors with LeTourneau scrapers had been received by Engineer Command units. They replaced 830Ms and other tractor-scraper combinations.[100]

Red Ball Express shipments of repair parts continued to remove equipment from deadlined status, but the rate was still high because of the variety of makes and models. Often a piece of equipment became deadlined for lack of a single part, but equipment requiring several parts usually had to wait for Red Ball items to come in piecemeal. During the period 5 December 1965 to 3 May 1966, parts received through Red Ball removed from deadline 110 of the 370 tracked and 28 of the 128 wheeled bulldozers. The 18th Engineer Brigade's overall equipment deadline rate dropped from 28.5 percent in January to 16.2 percent in early June. A vigorous maintenance program also helped. Nevertheless, by the autumn of 1966 the 18th Brigade reported that equipment productivity averaged less than 60 percent due to lack of parts.[101]

Rock continued to be a scarce commodity. An especially acute shortage of rock existed in III and IV Corps. Just two U.S. quarries, the RMK-BRJ University Quarry outside Saigon and the 159th Engineer Group quarries nearly thirty-one miles to the south at Vung Tau, were in operation in early 1966. Because of the lack of barges, only small quantities of rock could be shipped from Vung Tau to the Long Binh area. The 1st Logistical Command in late 1965 even resorted to buying rock from Singapore, which proved to be prohibitively expensive.[102]

When rock from quarries in Korea arrived in Saigon, a nightmarish situation involving customs, labor difficulties, and lighterage problems followed, producing lengthy delays and creeping costs. The engineers devised a scheme to tranship the rock on hundreds of sampans for delivery to a point near Long Binh for off-loading on trucks for the short haul to the base. At the off-loading site, it quickly became evident that the crane operator could not use the two-cubic-yard clamshell, since the weight and size of the clamshell would have easily sunk the sampans. This left the slower method of hand labor and conveyors as the only means to bring the precious rock ashore. This was the first and last shipment from Korea, and the Army ended the contract.[103]

---

[100] Army Buildup Progress Rpt, 6 Apr 66, p. 47, CMH; *Arsenal for the Brave*, pp. 250, 252; Rpt of Visit to the 1st Log Cmd, 12 and 15 Nov, incl. 10, p. 5, OCE Liaison Officer Trip, Rpt no. 5; ORLL, 1 Nov 66–31 Jan 67, Engr Cmd, 31 Jan 67, p. 21.

[101] Rpt of Visit to the 35th Engr Gp, Cam Ranh, Vietnam, 22–25 Feb 66, incl. 6, tab B, p. 1, and Situation Rpt on Engr Equipment in South Vietnam, incl. 17, OCE Liaison Officer Trip Rpt no. 1; Army Buildup Progress Rpts, 4 May 66, p. 35, 11 May 66, p. 45, and 8 Jun 66, p. 47, CMH; ORLL, 1 Aug–31 Oct 66, 18th Engr Bde, p. 8; *Arsenal for the Brave*, p. 251.

[102] Dunn, *Base Development*, p. 102; Quarterly Cmd Rpt, 1 Oct–1 Dec 65, 1st Log Cmd, 19 Feb 66, p. 14, Historians files, CMH; Situation Rpt on Construction Materials, incl. 16, OCE Liaison Officer Trip Rpt no. 1; ORLL, 1 Nov 66–31 Jan 67, 159th Engr Gp, 14 Feb 67, p. 11, Historians files, CMH.

[103] Dunn, *Base Development*, pp. 102–03.

Sampans were hired to haul rock from Vung Tau to the Long Binh area.

Additional crushing plants and associated quarry sets consisting of wagon and tracked rock drills trickled into the country. By May 1966, larger 225-ton-per-hour rock crushers reached Cam Ranh Bay and Qui Nhon. In addition, larger numbers of the smaller but versatile 75-ton-per-hour units arrived: five in the Saigon area, two at Cam Ranh Bay, and two at Qui Nhon. This equipment only put a dent in the requirements for rock. General Ploger felt larger units in the 500- and 1,000-ton-per-hour categories and larger rock-hauling trucks were needed. This situation posed another problem, for the 18th Engineer Brigade could barely provide enough men to operate the equipment already on hand. A more pragmatic approach rested with the arrival of more light equipment companies and their 75-ton-per-hour units and the purchase of rock from RMK-BRJ. By midyear things looked up for the 35th Engineer Group at Cam Ranh Bay when the 102d Construction Support Company prepared to open a new quarry site (two 10-hour shifts) consisting of a 225-ton-per-hour and three 75-ton-per-hour units.[104]

For the 79th Group, however, crushed rock needed for extensive concrete work at the base camps north of Saigon was still critically short. By way of solution, the group tried to develop quarries near construction sites. But at the Frenchman's Quarry southwest of Bien Hoa blasting work in August 1966 damaged a nearby Buddhist pagoda, and the province chief ordered a halt to operations. Blasting resumed a month later following negotiations with local officials. This time the engineers used new blasting techniques and operated only at specific times. They also repaired the pagoda.[105]

---

[104] Equipment and Equipment Maintenance, incl. 21, OCE Liaison Officer Trip Rpt no. 2, 6 May 66; Engr Equipment Situation, incl. 10, OCE Liaison Officer Trip Rpt no. 3, 28 Jun 66.
[105] Galloway, "Essayons," p. 147; ORLLs, 1 Aug–31 Oct 66, 79th Engr Gp, p. 10, and 1 Nov 66–31 Jan 67, 79th Engr Gp, 14 Feb 67, p. 9, Historians files, CMH.

159

Finance and Its Predicaments

In August 1966, the Naval Facilities Engineering Command requested a $200 million increase in construction funds to keep RMK-BRJ going in Vietnam. The deficit had developed, in part, because the contractor's mobilization costs—supplies, materials, and equipment—had been underestimated a year earlier, as had everything else related to construction, including the size of the U.S. troop commitment. In response to the request, McNamara determined that $60 million should be enough to keep the contractor in business. But he also asked Sharp and Westmoreland to furnish him with a restatement of their construction program no later than 1 October.[106]

The $200 million that the Navy asked for in August shocked many in Washington. Soon the news media began to report charges of cost overruns, waste, and mismanagement on the part of the contractor. The nationwide Huntley-Brinkley evening television news of 7 September announced: "A major scandal involving $200 million has come to light in South Vietnam." The Associated Press, in a national story, concluded that the Pentagon admitted it misled civilian contractors by overstating probable contract awards and underestimating costs. Senator John Stennis of Mississippi, chairman of the Preparedness Investigating Subcommittee of the Armed Services Committee, wrote to McNamara about a *Washington Post* article that appeared on 8 September with the caption "Builders Running $200 Million in Hole in Vietnam." McNamara tried to explain the rise in contracting costs. In response to the senator's query he wrote: "We consider the cost growth as moderate and to be expected, considering the exigencies of wartime conditions under which the initial cost estimates were made and the work performed."[107]

McNamara then went into detail. Several projects had changed, resulting in changes in design, methods, scope, and materials. For example, the initial design and construction directive issued in May 1966 for the headquarters complex at Long Binh provided for slightly more than 100,000 square feet of buildings. By late summer, several changes in scope increased the magnitude of work to 2.8 million square feet. Weather and site relocations increased costs at Phu Cat Air Base and the MACV headquarters building. Then too, recent strikes, civil unrest, wage increases, and lost time due to enemy activity had added to the higher costs. In July, 3,500 Vietnamese, 200 Koreans, and 600 Filipinos had walked off their jobs at Cam Ranh Bay because of working

---

[106] JLRB, Monograph 6, *Construction*, pp. 70, 73; Quarterly Hist Rpt, 1 Jul–30 Sep 66, MACDC, pp. 25–28; Msg, Westmoreland 25637 to Sharp, 25 Jul 66, sub: Construction Programs and Funds; Memo, Robert S. McNamara, Sec Def, for Secretary of the Navy, 7 Sep 66, sub: Inadequacy of Contract Construction Funds, RVN, both in box 44, 71A/2351, RG 330, NARA; Msg, JCS to CINCPAC, 9 Sep 66, sub: Construction Program in SVN, Historians files, CMH.

[107] Transcript of Huntley-Brinkley Evening News (first quoted words), 7 Sep 66, box 36–14, OCE History files; Ltr, McNamara to Senator John Stennis (second quoted words), 12 Sep 66, box 44, 71A/2351, RG 330, NARA; Tregaskis, *Building the Bases*, p. 333; Statement by Edward J. Sheridan, Dep Sec Def (Properties and Installations) Before the Staff of the Senate Preparedness Investigating Subcommittee, 13 Sep 66, pp. 7, 11, box 44, 71A/2351, RG 330, NARA.

conditions and wages. The defense secretary added that the transfer of the $60 million, and possibly more, would ensure continuation of the consortium's work. A reduction in RMK-BRJ's workforce from 51,000 to approximately 43,000 by the beginning of 1967 would, according to McNamara, also increase efficiency and reduce costs.[108]

Meanwhile, General Raymond, who had returned to the United States in July on a thirty-day rest and recuperation leave, found himself spending most of his leave in Washington trying to clarify the $200 million requirement. The MACV construction chief spent hours going over the funding problem with the top officials of the Naval Facilities Engineering Command and the Pentagon. He discussed the funding predicament with Assistant Secretary of Defense Ignatius and his staff. Raymond also traveled to Capitol Hill almost daily to explain the funding situation before congressional subcommittee hearings.[109]

Exercising its oversight function, Congress sent teams to examine the construction program firsthand. A team from its auditing agent, the Government Accounting Office, arrived in Vietnam in July. In reviewing the performance of RMK-BRJ, the auditors noted that during the haste of the buildup the contractor dumped equipment and materials at port staging areas, depots, and project sites, and the flood of arriving materiel inspired thievery on a grand scale. RMK-BRJ could not account for the whereabouts of approximately $120 million worth of items shipped from the United States. In October, staff members of the Armed Services Committee toured bases and received detailed briefings. In one long question-and-answer session in which many cost figures were discussed, Capt. Paul E. Seufer, the Officer in Charge of Construction in Vietnam, perhaps best summed up his organization's efforts. "If anything," he told a subcommittee staff member, "I would like you to go away from here with a feeling that you have run into an organization that is on top of the management of a rather amorphous thing and we're doing a good job." The Government Accounting Office team's report prepared in December concluded that "neither the Navy nor the contractor were adequately equipped to manage the mass buildup and neither devoted sufficient attention to maintaining a reasonable degree of management control over the fast escalating construction program."[110]

While Congress and the Pentagon continued to spar over costs, Defense Department officials were having second thoughts about the authority they

[108] Ltr, McNamara to Stennis, 12 Sep 66; Msg, Brig Gen Daniel A. Raymond SOG 882 to Lt Gen Richard D. Meyer, Director for Logistics, Joint Chiefs of Staff, J–4, 8 Oct 66, Dunn Papers, CMH; Army Buildup Progress Rpt, 21 Jul 66, p. 43, CMH; Quarterly Hist Rpt 1 Jul–30 Sep 66, MACDC, pp. 27–28.

[109] Tregaskis, *Building the Bases*, pp. 332–33; Quarterly Hist Rpt, 1 Jul–30 Sep 66, MACDC, pp. 1, 25–26.

[110] Draft, Report to the Congress of the United States, Survey of United States Construction Activities in the Republic of Vietnam, Government Accounting Office, Dec 66, pp. 2, 4, box 4, 74/167, RG 334, NARA; RVN Supplemental Data Sheet, Officer in Charge of Construction, sub: Briefing of Mr. French's Investigation for the Senate Sub-committee, p. 16 (quoted words), Incl to Ltr, OICC RVN to COMUSMACV, 1 Nov 66, sub: Senate Briefing of Preparedness Investigating Subcommittee Minority Counsel, Mr. Stuart P. French, box 16, 70/782, RG 334, NARA.

had granted Westmoreland in January to control the construction program. In May, the Pentagon asked Westmoreland to send General Dunn to Washington to draft new procedures. Following Dunn's return to Saigon, he and Raymond reviewed various drafts transmitted by message from Washington. In late June, General Noble, the Defense Department's chief for Southeast Asia construction, informed Dunn of the possibility of returning to more-detailed line-item justifications. Almost in despair, Raymond responded such a move "would be a step backward and involve time consuming detail." Dunn and Raymond managed to hold off any drastic changes for the time being. But by September, the unfavorable press coverage over contracting costs renewed Defense Department interest in tightening up construction procedures.[111]

McNamara now appeared determined to revalidate projects before approving any further work. During his visit to South Vietnam in October, he told General Raymond of his determination to reassert the authority granted to him earlier by Congress, and warned that he would require that new projects funded from the Fiscal Year 1967 Supplemental appropriation be submitted on a line-item basis instead of by categories. On 31 January 1967, McNamara made it official, forwarding a memorandum to the field rescinding the broad flexibility passed on to Westmoreland a year earlier. More changes would be forthcoming in 1967, and the Construction Directorate reluctantly began to adapt to the new details of construction management.[112]

Despite these problems and thanks to a growing U.S. military and civilian engineer workforce, the network of major and minor base complexes showed significant progress by the end of 1966. These achievements did not come about without difficulty. The engineers made mistakes. Often construction materials were not available in specified types or quantities. Equipment constantly broke down. Funding continued to plague construction managers. Meanwhile, the engineers also had to support ongoing tactical operations.

[111] Msg, Dunn MAC 3876 to Noble, 14 May 66; Msg, Dunn WDC 6375 to Raymond, 31 May 66; Msg, Noble 3557–66 to Dunn, 22 Jun 66; Msg, Dunn SOG 0538 to Noble, 28 Jun 66; Msg, Noble 3698–66 to Dunn, 29 Jun 66; Msg, Raymond SOG 0546 to Noble, 30 Jun 66; Msg, Noble WDC 13795 to Raymond, 22 Nov 66; Msg, Noble WDC 14521 to Raymond, 9 Dec 66. All in Dunn Papers, CMH.

[112] Raymond, Observations on the Construction Program, RVN, p. 52; Memo, McNamara for Secs of Svc Depts and Chairman JCS, 31 Jan 67, sub: Construction Approval Procedures for South Vietnam, in Booklet, Development of the Construction Directorate, incl. 5, tab I, Historians files, CMH; Msgs, Raymond SOG 0901 to Noble, 12 Oct 66, Raymond SOG 1085 to Noble, 25 Nov 66, Raymond SOG 1173 to Noble, 14 Dec 66, all in Dunn Papers, CMH.

PART TWO
Supporting the Offensive

6

Spoiling Attacks, January–September 1966

During 1965 and 1966, military engineers carried out construction that encompassed almost every phase of engineering endeavor. American commanders at all levels had accepted the development of the fixed, somewhat permanent bases from which to operate. The increase in the number of tactical operations in 1966 began to draw the engineers, especially those in the divisions and brigades, more deeply into their combat and operational support missions. The nature of the war focused attention on several aspects of combat engineering. Before enough forces became available to open and maintain the road system, the combat forces depended almost wholly on aerial resupply, especially when they operated any distance from the coastal bases. Engineers were needed to increase the number of airfields in and around the interior bases, and to help detect, penetrate, and destroy the enemy's well-concealed bunkers and tunnel complexes. When not occupied with high-priority base development projects, combat engineers of the 18th Engineer Brigade increasingly reinforced the division and brigade engineers.

III Corps Battles

In January 1966, allied forces in III Corps began a series of spoiling attacks, radiating out from Saigon, to clear main roads, hinder enemy operations, and improve security. MACV deployed the 1st Division and the 173d Airborne Brigade into the corps' northern and western areas. During the first week of the year, a battalion from the airborne brigade made an air assault in the delta country on the east and west sides of the Vam Co Dong River in Hau Nghia Province northwest of Saigon in Operation MARAUDER. After a brief encounter, the paratroopers pursued the Viet Cong force. Later a patrol uncovered a battalion-size base camp where the enemy left behind a sizable quantity of supplies, weapons, ammunition, and important documents. Once everything of value was removed, a demolitions team set about destroying the base.[1] (*Map 14*)

MARAUDER was followed on 7 January by Operation CRIMP in the Ho Bo Woods in southwestern Binh Duong and northeastern Hau Nghia Provinces. Both the 3d Brigade, 1st Division, and the 173d Airborne Brigade participated in the war's largest operation by U.S. forces thus far. Much of the terrain was open, except near the Saigon River, which was choked with thick jungle and overgrown rubber plantations. During the operation, General Seaman sent his 3d Brigade through the southern part of the sector. Infantrymen and engineers found and destroyed bunkers, houses, sampans, supplies, and food, and

---

[1] Carland, *Stemming the Tide*, pp. 168–69.

III CORPS TACTICAL ZONE
January–September 1966

Corps Tactical Zone Boundary

MAP 14

in one tunnel complex captured a huge collection of maps, charts, and documents. The Viet Cong did not stand and fight, using instead hit-and-run and ambush tactics to inflict casualties. In the northern part of the sector, the 173d Airborne Brigade discovered a large tunnel complex, but the enemy managed to slip away from the brigade's dragnet of one Australian and two American infantry battalions.[2]

A little over a month after CRIMP ended, General Seaman launched another spoiling attack north of Saigon—Operation MASTIFF. In a series of helicopter lifts, the 1st Division's 2d and 3d Brigades joined by the division's armored cavalry squadron deployed along three sides of a thirty-nine-square-mile area bordering the west bank of the Saigon River. Although the Americans uncov-

[2] Ibid., pp. 169–73; AAR, Opn CRIMP, 173d Abn Bde, 23 Feb 66, p. 1; AAR, Opn CRIMP, 3d Bde, 1st Inf Div, 15 Feb 66, p. 1, both in Historians files, CMH.

ered base camps, small hospitals stocked with medical supplies, training areas, rice storage caches, and small munitions factories, they met little organized resistance. Seaman terminated the operation on 25 February. Both CRIMP and MASTIFF gave U.S. commanders a frustrating reminder of the enemy's skills at avoiding a fight with large formations.[3]

In March, the 1st Division launched a series of smaller or battalion-size operations north of Lai Khe and immediately west of Highway 13 near the hamlet of Bau Bang. By reducing the size of the U.S. units sent to the field, the Americans hoped that the Viet Cong would see them as tempting targets and go after them. The tactics worked, resulting in a battle at Lo Ke Rubber Plantation just west of Bau Bang. Within a few days, all three battalions of the 3d Brigade were sweeping the battle area. Though the enemy again slipped away, he suffered heavy casualties and loss of equipment.[4]

Throughout these operations, Colonel Sargent's 1st Engineer Battalion and Captain Sowell's 173d Engineer Company supported the infantry. Combat engineers augmented every infantry company and larger unit in the field with demolition teams. They also cleared and repaired roads, built landing zones, carried out the slow and deliberate clearance of bunkers and tunnels, and set up water points. During CRIMP, Company C supported the 3d Brigade and cleared and repaired roads into the area of operations. The battalion's tankdozers and flamethrowers also supported the operation. Company B supported the 2d Brigade in Operation MASTIFF. Similarly, the 173d Engineer Company supported the 173d Airborne Brigade in Operations MARAUDER and CRIMP.[5]

To the north and east of Saigon, U.S. forces launched three major operations during the winter and early spring. In Operation MALLET, the 2d Brigade swept the area around Long Thanh southeast of Long Binh. Rice caches and base camps were uncovered with little opposition. The major impact of this operation, which ended in mid-February, was the opening of Highway 15 and the reestablishment of a government presence in the area. MALLET had not quite wound down when the 1st Division launched Operation ROLLING STONE. The purpose of the operation was to provide security for the 1st Engineer Battalion as it repaired and upgraded supply routes linking Highway 13 with Route 16 north of Saigon in Binh Duong Province, thereby improving ground traffic between the 1st Division's bases at Lai Khe and Phuoc Vinh. The operation would also allow the South Vietnamese government to extend its control over the intervening territory and cut a major enemy supply and infiltration route linking War Zones C and D. Seaman assigned the security mission to the 1st Brigade, which in turn tasked one of its three battalions to guard the engineers on a rotation basis, while the other two probed nearby to keep the enemy off balance. Two weeks into the operation, strong Viet Cong forces attempted to attack 1st

---

[3] Carland, *Stemming the Tide*, pp. 173–75.

[4] Ibid., pp. 175–78.

[5] HQ, 1st Engr Bn, 1st Engineer Battalion in Vietnam, n.d., 2 vols., vol. 1, pp. III-3 to III-4 (hereafter cited as 1st Engr Bn Hist), copy in CMH.

Brigade defensive positions outside the hamlet of Tan Binh just north of the roadwork and less than three miles west of Route 16. The attack was beat off with many enemy casualties. After the repulse at Tan Binh, the Viet Cong avoided battle, choosing instead to harass the work parties with occasional mortar and sniper fire. Although they killed three engineers and wounded twenty-nine, they were unable to stop the 1st Engineer Battalion, and by 2 March the roadwork was complete.[6]

Operation ROLLING STONE was the first major operation of the 1st Engineer Battalion as a unit. Work began on 7 February when Capt. Joseph M. Cannon, commanding officer of Company D, dispatched a platoon with a bulldozer, grader, scoop loader, and four dump trucks to open two and one-half miles of Route 2A, or Route Orange, a supply road branching off Highway 13 north of Phu Cuong and leading to Phuoc Vinh. The Viet Cong blocked the road with about fifty berms and trenches, but in a single day the platoon swept the road for mines, pushed the berms aside, and filled the ditches. On the ninth, Capt. Charles R. Kesterson, the Company A commander, deployed five mine-clearing teams to clear Highway 13 in leapfrog fashion north from Phu Loi to Ben Cat. Since the Viet Cong also ran buried wires to mines, which were command detonated from a distance, two bulldozers with rooters preceded the teams along the sides of the road to cut or expose the wires. Each team consisted of nine men: three mine detector operators with metallic detectors, and six probers. At times the mine-clearing teams were fired on, but accompanying tankdozers quickly suppressed enemy snipers with canister rounds and machine gun fire. Similarly, Company C under Capt. Robert F. Zielinski moved south from Lai Khe, clearing the road to Ben Cat. From Ben Cat attention turned to upgrading eleven miles along Routes 7B and 1A to Route 16. Engineer reconnaissance determined that a three-mile stretch of new road roughly following a cart track and ridgeline through jungle, plantations, and rice paddies just north of the existing road could be built east of Ben Cat without bridges or culvert. A bridge over the existing road had been partially destroyed, and shortly after the reconnaissance the Viet Cong, believing the bridge would be repaired, demolished it.[7]

On 10 February, the battalion less Company B moved from Di An and Lai Khe over the cleared Highway 13 to an area designated Base Camp 1 about one and one-quarter miles east of Ben Cat. While the engineers set up the field camp, equipment operators immediately went to work on the road. Throughout the operation, the weather was warm and sunny. The lack of rain made it easier for vehicles to move about but hindered compaction during the roadwork. Three cargo trucks were rigged as expedient water trucks, and a water semitrailer and a 500-gallon-per-minute water pump were borrowed from the Vietnamese Ministry of Public Works for use in compacting the base course. During the first day, the first half mile of a new pioneer road was

---

[6] Carland, *Stemming the Tide*, pp. 179–80; AAR, Opn ROLLING STONE, 1st Engr Bn, 12 Mar 66, pp. 1, 8–9, Historians files, CMH.

[7] AAR, Opn ROLLING STONE, 1st Engr Bn, pp. 2–5; 1st Engr Bn Hist, vol. 1, p. III-6; Garland, ed., *Infantry in Vietnam*, pp. 222–24.

*Engineers of the 1st Division encountered obstacles along roads such as the
berm under construction by the Viet Cong.*

cleared and a laterite pit was opened within the camp's perimeter, allowing bulldozers to clear the roadway and stockpile laterite until dusk at 1900. Pushing
east, tankdozers, bulldozers, and graders cleared vegetation to a width of sixty
feet. The fifty-five-ton tankdozer proved very effective in clearing hedgerows
and stumps when compared to the twenty-seven-ton bulldozer. Next graders,
water tankers, and pneumatic rollers went to work on a thirty-eight-foot-wide
base course. The first six-inch layer of laterite was shaped by graders and compacted by dump trucks hauling laterite. This was followed by a twelve-inch
layer of laterite compacted by steel rollers. Fortunately, the laterite used for the
wearing surface of the entire road had good moisture content and compacted
well without additional water. As the roadwork progressed, Colonel Sargent
on 16 February moved the battalion to Base Camp 2, which also contained
laterite, near the juncture of the new road and Route 7B. Similarly, on 23
February Base Camp 3 was opened farther east on Route 2A. It was on the
night of 23–24 February that the Viet Cong attacked the 1st Brigade's field
camp 1,100 yards to the northeast. On the morning of the twenty-sixth, engineer tankdozers and engineer security forces guarding minesweeping teams
found three dead Viet Cong along the road. Apparently the bodies were a
lure, and the engineers soon found themselves engaged in a heavy firefight that
lasted about thirty minutes. Work resumed, and the 1st Engineer Battalion

169

completed the thirty-foot-wide all-weather laterite road (Route Orange) from Ben Cat to Route 16 in three weeks, a half week ahead of schedule.[8]

As usual, the engineers checked for mines throughout ROLLING STONE. Thanks to the thorough sweep along Highway 13, the 1st Engineer Battalion was able to move to Base Camp 1 completely unhampered. This move contrasted sharply with previous ones in which several American soldiers had been killed in mining incidents along the same road. Venturing out from Base Camp 1 on the morning of the eleventh, mine-clearing teams discovered U.S. fragmentation bomblets emplaced by the enemy as antipersonnel mines in an area cleared just the day before. This was further evidence that the Viet Cong would take every opportunity to mine roads that were used by U.S. forces, and particularly those under construction where it was difficult to detect the presence of a concealed mine. Mine-clearing teams were dispatched every day from the field camps to clear both the area to be worked on that day and the completed portion of the road back to the starting point at Ben Cat. Despite the sweeps, some mines went undetected. On the first day out of Base Camp 2, 17 February, S. Sgt. Clyde C. Foster, while improving a secondary road to a laterite pit with his bulldozer, detonated two mines in rapid succession. Fortunately, the bulldozer absorbed much of the blast and Sergeant Foster was unharmed. On the same day, the clearing teams working on the completed portion of the road found six antipersonnel and two pressure-activated anti-vehicular mines, which had obviously been put in the preceding night after roadwork ceased. When the battalion prepared to return to Di An via Ben Cat at the end of the operation, mine-clearing teams swept the road ahead of the leading units. Company C began moving out of Base Camp 3 at 1000, 2 March, and traveled some distance without incident. Just east of old Base Camp 2, two pressure-activated mines were detonated and an armored personnel carrier and a tank were damaged. The road had been checked and cleared not more than a few hours earlier. No further incidents marred the return march, but the battalion learned that unless every inch of ground was under constant observation—a physical impossibility—the enemy would take a toll of friendly soldiers.[9]

In March, General Westmoreland launched more spoiling operations in III Corps. The 1st Division's 1st Brigade joined the 173d Airborne Brigade in Operation SILVER CITY, a sweep of the southwestern sector of War Zone D. Again the two units found and destroyed rice caches, caused numerous casualties, interdicted the enemy's lines of communications leading to Saigon, and disrupted one of the major Viet Cong sanctuaries in III Corps. Then, in anticipation of a fresh wave of enemy attacks, General Seaman, who assumed command of the newly activated II Field Force, Vietnam, sent the 1st Division on a preemptive campaign. For sixteen days, beginning on 29

---

[8] AAR, Opn ROLLING STONE, 1st Engr Bn, pp. 3, 5–8; 1st Engr Bn Hist, vol. 1, p. III-6; Garland, ed., *Infantry in Vietnam*, pp. 224–25.

[9] AAR, Opn ROLLING STONE, 1st Engr Bn, incl. 2; Garland, ed., *Infantry in Vietnam*, pp. 224–25. See also Cir 350–3–1, 1st Engr Bn, 20 May 66, sub: Mine Clearing, incl. 11, tab B, OCE Liaison Officer Trip Rpt no. 3.

A 1st Engineer Battalion mine-clearing team along Highway 13

March, the division, now under Maj. Gen. William E. DePuy, dispatched the 2d and 3d Brigades and the 1st Battalion, Royal Australian Regiment, on a sweep through a large area of Phuoc Tuy Province southeast of Saigon. The 1st Engineer Battalion provided demolition teams, water supply, clearing landing zones, roadwork, and bridge building. During the operation, called ABILENE, a U.S. infantry company encountered a Viet Cong battalion on the morning of 11 April, and the Americans beat off repeated attacks well into the night. Before reinforcements arrived the next day, thirty engineers from Company B descended through the jungle canopy on ladders hanging from the rear of hovering CH–47 helicopters. Armed with chain saws and hand tools, the combat engineers carved out a landing zone that allowed evacuation helicopters to lift out the casualties. This was the first time the 1st Engineer Battalion used this technique. Another first for the battalion took place when it used an armored vehicle launched bridge, borrowed from the 65th Engineer Battalion, 25th Division, to cross a forty-foot gap, enabling an ammunition resupply convoy to reach its destination. A culvert bypass was later built at the bridge site and the scissors bridge was withdrawn. Elsewhere, a pioneer road was blazed through dense jungle and the armored vehicle launched bridge was used for the second time, allowing artillery to move to a new location.[10]

[10] Carland, *Stemming the Tide*, pp. 181–83, 306; ORLLs, 1 Feb–30 Apr 66, II FFV, 1 Jun 66, p. 8, and 1 Feb–30 Apr 66, 1st Inf Div, n.d., p. 6, both in Historians files, CMH; 1st Engr Bn Hist, vol. 1, pp. III-4, III-6 to III-7; *Pictorial History of the 2d Brigade, 1st Infantry Division*, Company B, 1st Engr Bn sections. For more on clearing landing zones, see Cir 350–3–1, 1st Engr Bn, 17 Apr 66, sub: Construction: Helicopter Landing Zone Construction, incl. 11, tab A, OCE Liaison Officer Trip Rpt no. 3.

On 24 April, the 1st Division and its engineers entered War Zone C near the Cambodian border in Operation BIRMINGHAM—the first major drive into the enemy stronghold since 1962. During the operation, the 1st Engineer Battalion airlifted mine-detection teams to sites along Highways 1 and 22, while other elements preceded road bound convoys, replacing blown bridges and repairing roads along the way. Engineer teams also repaired Route 243 near Nui Ba Den (Black Virgin Mountain) and Route 26, bypassing Highway 22 from Tay Ninh to Go Dau Ha. Besides enemy harassment, mud and mines plagued the engineers. The monsoon season had begun in early May, and work crews had to tackle the sticky problem of rutted roads made soft by the constant pounding of tracked vehicles and heavy convoys. At this time, the 1st Engineer Battalion opened a road from Tay Ninh to the Michelin Plantation allowing supply vehicles and artillery to move by convoy. One major result of the operation was the discovery of a vast supply cache near the border. BIRMINGHAM took its toll of engineers. By the end of the operation on 16 May, the 1st Engineer Battalion lost two men killed, including the death of Captain Kesterson, Company A commander, when he stepped on a mine while rushing to the aid of his minesweeping team struck down by a blast from a claymore mine. The battalion also suffered sixteen wounded.[11]

During June and July, the 1st Division and the South Vietnamese 5th Division mounted a series of operations along the eastern flank of War Zone C. In Operation EL PASO II, the aim was to open Highway 13 from Saigon to rubber plantations in Binh Long Province and to seek out and destroy a Viet Cong division, then in the process of preparing to seize and hold the province capital of An Loc and several district capitals. By the end of the operation, highly effective counter-ambush tactics based on the armored cavalry's firepower and the rapid reaction of helicopter-borne infantrymen uncovered supply caches and forced the Viet Cong to withdraw into sanctuaries along the border. Again the 1st Engineer Battalion supported the division by minesweeping and opening supply roads, clearing landing zones, and serving as infantry.[12]

Keeping the roads open for resupply convoys remained one of the 1st Engineer Battalion's major responsibilities. In late August, the battalion supported the 1st Brigade to clear and secure Routes 16 and 1A for a resupply convoy from Di An to Phuoc Vinh. Code-named AMARILLO, the operation started out as a routine road-clearing job. The road was in poor condition and two culverts had been blown. While Company C worked north, Company A started at Phuoc Vinh and worked south. When the two infantry battalions guarding the road were committed to action against a reinforced Viet Cong battalion about five miles west of Route 16, the 1st Engineer Battalion was tasked to assume the security mission formerly assigned to the infantry battalions. By then, a north-bound convoy was about halfway to Phuoc Vinh.

[11] Sharp and Westmoreland, *Report*, p. 125; MACV History, 1966, pp. 383–84; ORLL, 1 Feb–30 Apr 66, II FFV, pp. 8–9; 1st Engr Bn Hist, vol. 1, pp. III-7 to III-8; AAR, Opn BIRMINGHAM, 1st Inf Div, n.d., pp. 23–24, Historians files, CMH; *Always First, 1965–1967,* "Mud and Mines."

[12] Sharp and Westmoreland, *Report*, p. 126; MACV History, 1966, pp. 385–86; Galloway, "Essayons," p. 150.

Lt. Col. Joseph M. Kiernan, who had recently assumed command from Colonel Sargent, called out the remaining elements of the battalion from Di An. The road was secured, and the convoy passed without incident. At nightfall, Company A formed a defensive perimeter for the brigade command post and artillery base at Ap Bo La near the intersection of Routes 16 and 1A. A few miles to the south, the rest of the battalion set up a defensive perimeter. During the night, both perimeters were probed by Viet Cong, and Company A killed two of the enemy with small-arms fire. For three critical days, the 1st Engineer Battalion acted as infantry while the infantry battalions engaged the enemy in bitter fighting. On 26 August, the rest of the battalion moved north to protect the brigade command post. During the move, a 2½-ton truck from Headquarters Company was destroyed by a command-detonated claymore mine. The blast was so strong that pellets penetrated the engine and cracked the block. Six of the eight engineers in the truck were wounded. On the night of the twenty-seventh, the Viet Cong again made probing attacks on the 1st Brigade's perimeter. Company E, on the west side came under attack and before the night was over six Viet Cong were killed. There were no American casualties. On 28 August, the battalion resumed its mission as engineers. They destroyed tunnel complexes discovered by the infantry, cleared jungle and landing zones, and continued road and bridge work.[13]

In September, the battalion returned to Route 16 in Operation LONGVIEW. Just south of Tan Uyen, work crews replaced a failing bridge, weakened by heavily laden convoys, with two 90-foot triple-single Bailey bridges supported at midpoint on a pier. Plans called for the Ministry of Public Works to build the pier, but work progressed too slowly and the piles were too short to carry the design load. The 1st Engineer Battalion took over and assembled a reinforced raft from its float bridge set for use as a pile-driving platform. After splicing additional piles to the Vietnamese piles, work crews, using a pile driver and operator from the 169th Engineer Battalion, drove the fourteen piles. The bridge opened to convoy traffic at 1230, 27 September. Meanwhile, Companies A and D cleared Route 16 of mines and obstacles. At one location, known as Ambush Corner, Company D bulldozed the jungle on both sides of the road to hamper future ambushes.[14]

The following month, the 1st Engineer Battalion directed its attention to Highway 13, by now dubbed Thunder Road, in Operation TULSA. Again, obstacles had to be removed, and monsoon rains inundated the road in many places. Potholes and mine craters turned into bottomless pits. Working continuously to keep the road open to convoys and commercial traffic, the battalion turned to expedients such as placing timbers and perforated steel planks to surface the wet, sagging artery. Other improvisations included replacing a damaged bridge within two hours with a rock-filled french drain. When a mine ripped apart a bridge north of Phu Cuong, the battalion called on CH–47 Chinook helicopters to lift sections of the dry span bridge to the site. To the

[13] 1st Engr Bn Hist, vol. 1, pp. III-8 to III-9; *Always First, 1965–1967,* "Committed as Infantry." For more on AMARILLO, see Carland, *Stemming the Tide,* pp. 325–33.

[14] 1st Engr Bn Hist, vol. 1, pp. III-9 to III-10; *Always First, 1965–1967,* "Back to Route 16."

dismay of the engineers, only one helicopter showed up, and the engineers had to place an armored vehicle launched bridge over the almost completed dry span bridge in time for a scheduled convoy.[15]

Two other engineer mainstays in III Corps were the 168th and 588th Combat Battalions. Both concentrated on building the arc of forward bases protecting Saigon as well as carrying out operational support missions. Headquarters Company and Company A, 588th Engineer Battalion, worked on 1st Division projects at Phu Loi until late April when the 168th Engineer Battalion took over. In early May, the two companies of the 588th moved to the 25th Division's Cu Chi base camp, relieved the 168th platoon there, and supported the 65th Engineer Battalion's base development projects. A month later, Company B at Long Binh joined the battalion. In addition, the 588th Engineer Battalion maintained the airfield at Dau Tieng, improved the Loc Ninh Special Forces camp, and kept roads open and cleared jungle from a potential Viet Cong assembly area near Cu Chi. Company C at Bearcat joined the battalion in late August. When the 1st Division entered War Zone C near the Cambodian border in late April in Operation BIRMINGHAM, elements of the 588th Engineer Battalion and the 617th Panel Bridge Company helped the 1st Engineer Battalion keep Highways 1 and 22 open between Cu Chi and Tay Ninh. In mid-May, the 588th sent men to Di An, where they joined troops of the 1st Engineer Battalion to train on tunnel destruction techniques. Following the training, the 588th became the 18th Engineer Brigade's tunnel destruction force.[16]

At the beginning of 1966, the 168th Engineer Battalion had units at Di An, Lai Khe, and Long Binh carrying out construction and security tasks. Headquarters and Headquarters Company and Company A were at Di An, the 1st Division's main base, where Company A was busy building tropical wooden buildings, drainage ditches, roads, hardstands, latrines, and showers. First priority was for kitchens and dining halls, and a total of sixteen of the former and thirty-two of the latter were completed before the end of April. A carpenter shop was set up and began fabricating latrines and roof trusses. Most wood frame buildings included three-foot-high masonry walls. At Di An, Company A also ran combat patrols and night ambushes. In late April, most of Company A moved to Phu Loi to take over projects from the 588th. In addition to an array of tropical wooden buildings, metal prefabricated warehouses, sheds, and hangars, the company kept busy improving the Phu Loi airfield and the camp's drainage system. Company B did work both on the cantonment and airfield at Lai Khe. In early June, when the 1st Division launched Operation EL PASO II, Company B took over perimeter security. The battalion's third lettered company, Company C, augmented the 46th Engineer Construction Battalion's many projects at Long Binh. These included the 93d Evacuation Hospital and the head-

---

[15] 1st Engr Bn Hist, vol. 1, pp. III-10 to III-11; AAR, Opn TULSA, 1st Engr Bn, 22 Nov 66, OCE Hist Ofc; *Always First, 1965–1967,* "Miracle of Thunder Road."

[16] ORLLs, 1 May–31 Jul 66, 588th Engr Bn, pp. 1–3, 1 Aug–31 Oct 66, 588th Engr Bn, pp. 1–3, and 1 Nov 66–31 Jan 67, 588th Engr Bn, pp. 1–5; Galloway, "Essayons," pp. 141–42.

quarters complex for II Field Force at nearby Plantation. To help carry out these labor-intensive projects, Company C employed approximately 140 Vietnamese workers as masons, carpenters, and general laborers.[17]

The II Corps War

In II Corps, U.S., South Korean, and South Vietnamese forces launched a series of operations in the highlands and the key coastal province, Binh Dinh. Until late January 1966, existing airfields adequately supported operations of the 1st Cavalry Division, and forward airfield work mainly consisted of repairs. Other than this, the division's 8th Engineer Battalion limited its combat support to landing zones and road repairs. When airmobile operations shifted farther afield at the end of January, the division increased its dependence on cargo planes for logistical support, and airmobile engineers stepped up efforts to clear landing zones, build forward airfields, and open supply roads. Between 26 January and 5 June, Colonel Malley's 8th Engineer Battalion carried through an extensive airfield building program resulting in seven new airfields, lengthening two fields to twice their previous size, and repairing and maintaining many others. Other airfields in II Corps were being improved by the 18th Engineer Brigade.[18] (*Map 15*)

From 4 to 18 January, the 1st Cavalry Division and the 3d Brigade, 25th Division, moved into the border area north of the Chu Pong Mountain to locate infiltration routes from Cambodia. Following a first phase, which involved clearing Highway 19 between An Khe and Pleiku, the infantry moved by helicopters into newly cleared landing zones in the heavily forested areas. During Operation MATADOR, the 8th Engineer Battalion descended down troop ladders from hovering CH–47 helicopters for the first time in combat. Once on the ground, work crews used power saws to clear areas to fit one or two Huey helicopters. The next wave of helicopters delivered light engineer equipment to expand the landing zone to fit Chinook helicopters. In other cases, the Air Force dropped napalm and fragmentation bombs, ranging from 250 to 750 pounds, to cut holes in the forests. During the operation, troops uncovered several base camps and bunkers, and the enemy offered little resistance.[19]

In late January, the 1st Cavalry Division shifted operations from Pleiku and Kontum Provinces to the valleys along the coastal plains of northeastern Binh

---

[17] ORLLs, 1 Feb–30 Apr 66, 168th Engr Bn, pp. 1–3, 1 May–31 Jul, 168th Engr Bn, pp. 1–3, and 1 Aug–31 Oct 66, 168th Engr Bn, pp. 3–5; Galloway, "Essayons," pp. 141–42.

[18] Rpt, 14th Mil Hist Det, 1st Cav Div, 9 Jun 67, sub: Seven Month History and Briefing Data (September 1965–March 1966), p. 20; Lt. Col. Robert J. Malley, "Forward Airfield Construction in Vietnam," *Military Engineer* 59 (September-October 1967): 318–22. See also Lt. Col. Robert J. Malley, "Engineer Support of Airmobile Operations" (Student essay, U.S. Army War College, 1967), pp. 11–15.

[19] MACV History, 1966, p. 372; ORLL, 1 Feb–30 Apr 66, I FFV, 15 May 66, p. 7, Historians files, CMH; Rpt, 14th Mil Hist Det, 1st Cav Div, 9 Jun 67, sub: Seven Month History and Briefing Data (September 1965–March 1966), pp. 57–60; Tolson, *Airmobility*, p. 92.

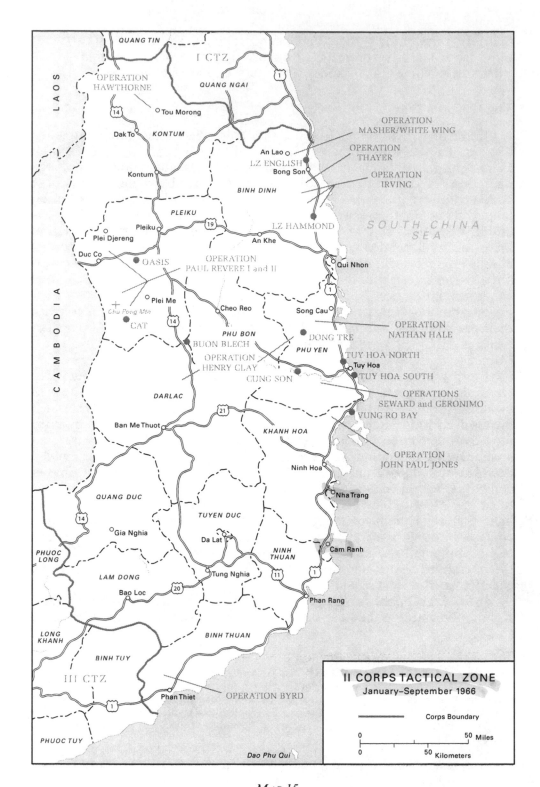

QUANG TIN

I CTZ

OPERATION
HAWTHORNE

QUANG NGAI

LAOS

Tou Morong

14

Dak To

KONTUM

Kontum

An Lao

LZ ENGLISH
Bong Son

OPERATION
MASHER/WHITE WING

OPERATION
THAYER

OPERATION
IRVING

BINH DINH

PLEIKU

Pleiku

19

SOUTH CHINA
SEA

Plei Djereng

An Khe

LZ HAMMOND

Duc Co

OASIS

OPERATION
PAUL REVERE I and II

Qui Nhon

1

Plei Me

Chu Pong Mtn

CAT

14

Cheo Reo

Song Cau

OPERATION
NATHAN HALE

PHU BON

DONG TRE

CAMBODIA

BUON BLECH

OPERATION
HENRY CLAY

PHU YEN

CUNG SON

TUY HOA NORTH
Tuy Hoa
TUY HOA SOUTH

DARLAC

21

OPERATIONS
SEWARD and GERONIMO
VUNG RO BAY

Ban Me Thuot

KHANH HOA

OPERATION
JOHN PAUL JONES

Ninh Hoa

QUANG DUC

TUYEN DUC

Nha Trang

PHUOC
LONG

14

Gia Nghia

Da Lat

NINH
THUAN

Cam Ranh

LAM DONG

20

Bao Loc

Tung Nghia

11

1

Phan Rang

LONG
KHANH

BINH THUAN

BINH TUY

III CTZ

1

Phan Thiet

OPERATION BYRD

PHUOC TUY

Dao Phu Qui

II CORPS TACTICAL ZONE
January–September 1966

━━━━━━━━━━ Corps Boundary

0   50 Miles

0   50 Kilometers

MAP 15

A CH–47 Chinook helicopter hovers while troops practice climbing down a ladder.

Dinh Province. Reinforced by South Vietnamese and South Korean troops, the 3d Brigade began Operation MASHER. Washington had problems with such a provocative name like MASHER, and MACV renamed the operation WHITE WING. When the remaining two brigades joined the hunt, the operation phased into a massive search-and-destroy effort called WHITE WING. By the time the operation ended on 6 March, the division's troop carrying helicopters had completed a full circle of airmobile sweeps around the Bong Son Plain. The operation, also known as the Bong Son Campaign, temporarily cleared the plain of

177

strong enemy forces. Several enemy battalions were destroyed and the division captured a large number of arms, but Binh Dinh was not yet secure.[20]

During MASHER/WHITE WING, the 8th Engineer Battalion teamed up with South Vietnamese, South Korean, and 937th Group engineers to maintain air and ground lines of communication from Qui Nhon to the combat area. While most of the 8th Engineer Battalion supported the forces in the field, elements of the battalion built two forward airfields. Beginning on 28 January and working around the clock for seventy hours, Company A and the 2d Equipment Platoon, Headquarters Company, opened the airstrip at Landing Zone DOG (later renamed ENGLISH) to Caribou aircraft, a few miles north of the town of Bong Son. Continuing on their own for another day, they improved the dirt airstrip to C–123 standards. In early March, Headquarters Company completed a 3,200-foot airfield for C–130s at a landing zone four miles north of Phu Cat (later named Landing Zone HAMMOND).[21]

South of HAMMOND another hasty airstrip took shape. A small force led by 1st Lt. Thomas O. Boucher and S. Sgt. Sherman L. Kohlway from Company A, 19th Engineer Battalion, completed a strip for the South Vietnamese 22d Infantry Division at An Nhon about eleven miles north of Qui Nhon just off Highway 1. Elements of the 84th Engineer Battalion and the 630th Engineer Light Equipment Company provided equipment support. The completed 1,800-foot-long, 80-foot-wide airstrip with 10-foot shoulders and a 35-foot cleared area on both sides was able to accommodate Caribou aircraft. The engineers hauled 40,000 cubic yards of fill during the round-the-clock operation. Although built primarily for use by the South Vietnamese division, the airstrip could also be used to relieve the traffic flow to the Qui Nhon airfield. The South Vietnamese division commander dedicated the airstrip to the memory of Maj. Stuart M. Andrews, U.S. Air Force, who was missing in action since being downed in March.[22]

In late March, the 1st Cavalry Division returned to the highlands to begin Operation LINCOLN. Teaming up with the 3d Brigade, 25th Division, the airmobile division reentered the Ia Drang area in another search-and-destroy mission, which resulted in moderate contact with the enemy. To support this and future operations, the 8th Engineer Battalion moved a platoon by CH–47 helicopters to Landing Zone OASIS thirty-one miles southwest of Pleiku. Within four hours, the platoon cleared a new dirt airstrip for Caribou transports. Next, Company A arrived to build a separate runway for C–130s. Using the battalion's lightweight bulldozers, graders, scrapers, and pneumatic rollers, the reinforced company carved out a new 3,500-by-60-foot runway. Instead of covering the runway with metal matting, work crews used T17 membrane for the first time in Vietnam. On 2 April, the first aircraft landed. During the

[20] Rpt, 14th Mil Hist Det, 1st Cav Div, 7 Jun 67, sub: Seven Month History and Briefing Data (September 1965–March 1966), pp. 61–67; Carland, *Stemming the Tide*, pp. 202–15; Tolson, *Airmobility*, pp. 92–93.

[21] Rpt, 14th Mil Hist Det, 1st Cav Div, 7 Jun 67, sub: Seven Month History and Briefing Data (September 1965–March 1966), pp. 20–21; Malley, "Forward Airfield Construction in Vietnam," p. 319; Galloway, "Essayons," p. 66.

[22] "19th Engineers Extend Airstrip," *Castle Courier*, 23 May 1966.

operation, the runway supported over five hundred landings and take-offs without significant damage.[23]

While the infantry swept over the western highlands, the 8th Engineer Battalion went to work on more airfields. Since vehicles could not get to all of the sites, helicopters had to move men and equipment. Some of the equipment could not be moved in one lift and had to be broken down into sections. In addition to its own CH–47 Chinooks, the division used CH–54 Tarhes or Flying Cranes, large cargo-carrying helicopters, to lift sections of D6B bulldozers, Caterpillar Model 112 graders, and MRS 100 tractor-scrapers. The bulldozer and grader required two lifts by the Flying Cranes, and sections of the tractor-scraper were delivered in four loads. Although the CH–47 could lift the bulldozer's track and blade, CH–54s usually carried out the initial sorties. After the tracks were laid out, a CH–54 brought in the bulldozer's body to be set on the tracks.[24]

On 3 April, two platoons and equipment arrived at the first site, Landing Zone CAT, in the trackless jungle some 43.5 miles southwest of OASIS, and close to the Cambodian border. An advance work party cleared a landing zone using hand tools and chain saws. Next, helicopters brought the equipment: two D6B bulldozers, two motorized graders (one with a scraper), a self-propelled pneumatic roller, and vibratory compactor. More helicopters delivered a light dump truck, which also served as a makeshift water distributor carrying a 500-gallon fuel bladder. By 5 April, the first Caribou landed on the hastily built strip, and the next day C–123 transports landed and took off from the completed 2,300-foot airstrip. This mix of equipment became the standard package when future airfield work required helicopters to transport all construction equipment.[25]

In twenty days, the 8th Engineer Battalion built four forward airfields and extended one more. Equipment moved by truck, helicopter, and fixed-wing aircraft. One D6B bulldozer, used in all but two of the projects, had moved 92 miles by truck, over 170 miles in four moves by helicopter, and 50 miles by C–130. During the twenty-day period, this bulldozer was in operation sixteen days.[26]

On 5 June, the 8th Engineer Battalion added another airfield, this one at Buon Blech Special Forces camp in northern Darlac Province. About four weeks earlier, Company B, augmented by two equipment platoons, moved to the camp under the protection of a 1st Cavalry Division task force. While an infantry company, artillery, and helicopter gunships of the task force secured the area, the engineer company fought the effects of heavy rains, cleared the site, and moved large quantities of earth. On 25 May, work crews began to

---

[23] MACV History, 1966, p. 374; ORLL, 1 Feb–30 Apr 66, I FFV, p. 8; AAR, Opn LINCOLN, 8th Engr Bn, 13 Apr 66, p. 2, Historians files, CMH; Rpt, 14th Mil Hist Det, 1st Cav Div, 7 Jun 67, sub: Seven Month History and Briefing Data (September 1965–March 1966), p. 21; Malley, "Forward Airfield Construction in Vietnam," pp. 319–20; Galloway, "Essayons," p. 92; *Memoirs of the First Team*, p. 181.

[24] Malley, "Forward Airfield Construction in Vietnam," p. 320; Rpt, 14th Mil Hist Det, 1st Cav Div (Ambl), 7 Mar 67, sub: Seven Month History and Briefing Data (April–October 1966), pp. 48–51, Historians files, CMH; Tolson, *Airmobility*, pp. 269, 274.

[25] Malley, "Forward Airfield Construction in Vietnam," p. 320; Galloway, "Essayons," p. 92.

[26] Malley, "Forward Airfield Construction in Vietnam," p. 320; Galloway, "Essayons," p. 95.

place T17 membrane on the 3,200-foot runway and parking apron that could accommodate five C–130s.[27]

Other airfields supporting operations in the highlands were assigned to the 18th Engineer Brigade. General Westmoreland expressed concern about the airstrip at Cheo Reo, about halfway between Pleiku and Tuy Hoa near Route 7B. On 27 February, he directed that the airstrip be upgraded to C–130 standards. The Field Force, Vietnam, engineer tasked the 937th Group to carry out the improvements. In turn, the group passed the mission to Colonel Anderson's 299th Engineer Combat Battalion. On 26 March, Company C set out from Qui Nhon on a 160-mile road march along Highway 19 to Pleiku then south a few miles on Highway 14 and then southeast along Route 7B to Cheo Reo. Due to security concerns along Route 7B, the unit took all necessary supplies and materials and completed the grueling trip in one day. After reaching Cheo Reo, the company repaired soft spots on the runway, added 700 feet to the runway, and topped it off with M8 matting. A T17 membrane-covered parking ramp was added. Completing its work on 23 April, the company returned to Qui Nhon on the twenty-fourth, repeating the long trip in a single day.[28]

Although busy with various construction projects in the Qui Nhon and Pleiku areas, the 299th Engineer Battalion received a steady flow of operational missions elsewhere in II Corps. In mid-May, Company C ventured out again with the 630th Light Equipment Company for three weeks to Plei Kly, twenty-five miles south of Pleiku, to build an airfield to support the 3d Brigade, 25th Division, in Operation PAUL REVERE. During PAUL REVERE, U.S. troops reentered the Chu Pong–Ia Drang area for the first time since the late 1965 campaign. The 299th task force built a 3,100-by-60-foot runway, a turnaround, and a parking apron, all covered with T17 membrane. On 8 June, two platoons from Company A left Pleiku to provide additional support to the 1st Brigade, 101st Airborne Division, in the Dak To area. The paratroopers, who had moved into the area following search-and-destroy operations in coastal Phu Yen Province, had just begun HAWTHORNE, an eighteen-day search-and-destroy operation in Kontum Province. One of the engineers' major efforts was to keep the asphalt-surfaced Dak To airfield in operation.[29]

At midyear, U.S. forces continued spoiling attacks in II Corps. In June, the 1st Brigade, 101st Airborne Division, returned to Phu Yen Province to carry out Operation NATHAN HALE, a twelve-day search-and-destroy mission. The 1st Cavalry Division returned to Binh Dinh Province in September and launched Operations THAYER I, IRVING, and THAYER II. These and other combined operations with South Korean and South Vietnamese forces resulted in large North Vietnamese Army and Viet Cong losses and the capture of large quantities of supplies. By October, elements of the 4th

[27] Galloway, "Essayons," p. 95; Situation Rpt on Airfield Construction, incl. 7, p. 3, OCE Liaison Officer Trip Rpt no. 3.

[28] Msg, CG, FFV to COMUSMACV, 2 Mar 66, sub: Repairs to Cheo Reo Airfield, Historians files, CMH; ORLLs, 1 Feb–30 Apr 66, I FFV, p. 26, and 1 Feb–30 Apr 66, 299th Engr Bn, 30 Apr 66, p. 3, Historians files, CMH.

[29] MACV History, 1966, pp. 372, 374, 377; ORLL, 1 May–31 Jul 66, 299th Engr Bn, 31 Jul 66, pp. 1–3, Historians files, CMH; Galloway, "Essayons," p. 73.

Infantry Division and the 3d Brigade, 25th Division, followed later by elements of the 1st Cavalry Division, moved to the western highlands near the Cambodian border to continue Operation PAUL REVERE. Again, the sweeps disrupted enemy plans and raised havoc to his base areas. Operations in Phu Yen Province late in the year included SEWARD, which found the 1st Brigade, 101st Airborne Division, protecting the rice harvest, and GERONIMO I, a combined operation that included the 1st Brigade, 4th Division, and 1st Brigade, 101st Division. Long-running campaigns such as Operation BYRD in Binh Thuan Province demonstrated that an airmobile infantry battalion task force from the 1st Cavalry Division, supported by a platoon from the 8th Engineer Battalion, could operate alone over an extended period of time against small enemy forces.[30]

In July, two brigades of the 1st Cavalry Division swept west through Phu Yen, Phu Bon, and Darlac Provinces to the Cambodian border in Operation HENRY CLAY. Engineering tasks for the 8th Engineer Battalion teams attached to the assaulting infantry battalions included sweeping roads for mines and repairing forward airstrips. Repairing the airstrips posed some problems because much of the battalion's equipment remained committed to higher-priority projects such as the An Khe airfield. As usual, ingenuity was brought into play. Lt. Col. Charles G. Olentine, who replaced Colonel Malley on 19 June, noted that at Buon Blech one of his Company A platoons used a quartermaster laundry dryer as an expedient to dry the ground under the T17 membrane, allowing the glue to hold satisfactorily.[31]

During the series of PAUL REVERE operations near the Cambodian border, the 3d Brigade, 25th Division, saw heavy combat. In August, the 1st Cavalry Division diverted forces from Binh Dinh Province to reinforce the 3d Brigade. Operation PAUL REVERE II brought forth the heaviest commitment by the 8th Engineer Battalion in any single operation to date. Colonel Olentine moved his headquarters to OASIS, marking the first time that the headquarters operated outside Camp Radcliff since arriving in Vietnam. Monsoon rains had returned, and Olentine committed all of his resources during the month-long operation to keep the C–130 OASIS airstrip open and to carry out other missions to support the operation. Landing zones were cleared, enemy bunkers destroyed, and bridges and culverts built to cross the rising rivers and streams. Additional help came from 18th Engineer Brigade units of the 937th Engineer Group. The 299th Engineer Battalion, augmented with trucks and equipment from the 19th and 70th Engineer Battalions, the 630th Light Equipment Company, and the 509th Panel Bridge Company units hauled rock, put in culverts, and helped maintain roads. When the 1st Cavalry Division returned to Binh Dinh

---

[30] MACV History, 1966, pp. 378–79; ORLLs, 1 May–31 Jul 66, I FFV, 25 Aug 66, pp. 12–13, 1 Aug–31 Oct 66, I FFV, 30 Nov 66, pp. 10–12, 17–22, 23–28, and 1 Nov 66–31 Jan 67, I FFV, 6 Mar 67, pp. 9–18, all in Historians files, CMH; *Memoirs of the First Team*, pp. 182–83.

[31] ORLL, 1 May–31 Jul 66, I FFV, pp. 12–13; Rpt, 14th Mil Hist Det, 1st Cav Div, 7 Mar 67, sub: Seven Month History and Briefing Data (April–October 1966), pp. 100–103; AAR, Opn HENRY CLAY, 8th Engr Bn, 7 Aug 66, pp. 1–3, OCE Hist Ofc.

in mid-September, the 3d Brigade remained to carry out Operations PAUL REVERE III and IV.[32]

In southern Phu Yen Province, the 39th Engineer Battalion under Lt. Col. Taylor R. Fulton moved from Cam Ranh Bay and Nha Trang to Tuy Hoa to concentrate on the Vung Ro Bay port complex and carry out operational support for nearby combat units. Between 10 and 25 July, the battalion headquarters, Companies A and B, and the 572d Light Equipment Company moved by sea aboard landing craft to Tuy Hoa where they joined Company C. During Operation JOHN PAUL JONES, a three-phase series of sweeps running from 21 July to 5 September intended to secure the Vung Ro area and ensure that farmers around Tuy Hoa could harvest their rice crop without interference, elements of the battalion supported the 1st Brigade, 101st Airborne Division. Once the paratroopers secured Vung Ro Pass and Highway 1, Company C reopened and maintained Highway 1 from Tuy Hoa to the port project area. At Tuy Hoa South airfield, Company B built sixty-four 12-by-12-foot helicopter landing pads and two refueling pads, all on peneprime-treated sand and topped off with pierced steel planking, for aviation units supporting the airborne brigade. During Operation SEWARD, another sweep around Tuy Hoa lasting from 5 September to 26 October, the battalion's major tasks supporting the 1st Brigade, 101st Airborne Division, were building an airstrip capable of handling C–130s at Dong Tre Special Forces camp southwest of the town of Song Cau and repairing the gravel runway at Cung Son Special Forces camp airstrip twenty-two miles west of Tuy Hoa. Tasks also included minesweeping and maintaining Routes 6B and 7B to the two camps. Later, Company B began extending and placing T17 membrane surfacing at the Cung Son airstrip.[33]

Prior to Operation JOHN PAUL JONES, the valley area south of Tuy Hoa had been under Viet Cong control for several years, and Highway 1 was blocked by numerous ditches cut across the pavement and blown bridges and culverts. The key to the success of the Vung Ro Bay port project was the reopening of the road from Ban Nham on the Song Ban Thach, about ten miles south of Tuy Hoa, to Vung Ro. Four major bridges had been destroyed in this stretch of highway, and engineer reconnaissance teams accompanying the infantry in their advance relayed information back to the 39th Engineer Battalion concerning the bridges and the condition or existence of bypasses. Although a partly destroyed 850-foot concrete T-beam bridge was still standing at Ban Nham, the usable concrete spans, piers, and superstructure were old and damaged. The destroyed spans had been replaced by French Eiffel bridging; at best, the bridge might take ten-ton (Class 12) vehicles. No bypass was possible.

---

[32] ORLL, 1 Aug–31 Oct 66, I FFV, pp. 17–19; Rpt, 14th Mil Hist Det, 1st Cav Div, 7 Mar 67, sub: Seven Month History and Briefing Data (April–October 1966), pp. 104–10; ORLL, 1 Aug–31 Oct 66, 8th Engr Bn, 31 Oct 66, pp. 1–2; AAR, Opn PAUL REVERE II, 8th Engr Bn, 8 Sep 66, pp. 1–5, both in OCE Hist Ofc; *Memoirs of the First Team*, p. 182.

[33] AAR, Opn JOHN PAUL JONES, 39th Engr Bn, 17 Nov 66, pp. 1, 4; AAR Opn SEWARD, 39th Engr Bn, 17 Nov 66, pp. 1–4, Historians files, CMH; ORLL, 1 May–31 Jul 66, 39th Engr Bn, pp. 1–2.

A decision was made, therefore, to construct a Class 60 float bridge just downstream from the existing bridge.[34]

Colonel Fulton assigned the job to Company C and the 553d Engineer Company (Float Bridge). The bridging was transported on one ship and on 21 July was offloaded by transportation units on a beach just south of Tuy Hoa. When the bridge company arrived the following day, the bridging was loaded on bridge trucks in an around-the-clock operation for transport to Company C at the bridge site. Construction of the bridge began at dawn on the twenty-fourth. The bridge center line was located adjacent to and twenty feet downstream from the old bridge, but it was necessary to place the three work sites upstream because of shallow water and lack of space for work sites downstream. One work site was devoted to building the end sections. At this time it was discovered that sections of Class 60 steel treadway end ramps were missing, and plans were promptly made for expedient construction of the end ramps from M4T6 dry spans. The M4T6 bridge decking, known as balk (hollowed aluminum beams), was designed to be assembled by hand. At the other two work sites, work crews began assembling the successive bays. First, air compressors inflated two floats for each bridge section or bay. Only two cranes were available to lift and emplace the heavy Class 60 decking on the floats, so construction of each bay was slow. Upon completion, each bay was floated downstream under the damaged bridge and joined to the floating span. Anchor cables were secured to the existing bridge piers and bridle lines were run to each bay as it was joined. By early the next morning, the first 375-foot span was completed to a sandbar island at midstream and the makeshift end sections were placed. A hasty roadway was bulldozed across the island and topped off with pierced steel planking. New work sites were prepared on the island for construction of the span to the far shore. Since the channel on this side of the island was too shallow for a 27-foot power boat and construction of floats, a crane with a clamshell attachment was rushed to the site to dredge and deepen the river. As work continued into the second night, successive bays and the end sections were assembled and pulled into place by hand. By 1100, the bridge was open to traffic.

Although the project took a day longer to complete than anticipated, the rest of Highway 1 to Vung Ro was opened on the morning of 27 July. Along the way, Company C repaired a damaged concrete-decked bridge and built five bypasses, which were improved by the construction of a 110-foot double-single Bailey bridge and four 23-foot M4T6 dry spans. Meanwhile, improvements were made at the float bridge site. Due to the approaching monsoon, the sandbar probably would be flooded with six to eight feet of water, making it unusable as a ford. This predicament was remedied by constructing a Class 60 treadway fixed span on Class 50 trestles across the island. M4T6 dry spans were again used as ramps to connect the fixed spans to the floating Class 60 bridges on both sides of the island. This remedy eliminated the need for the pierced steel plank roadway and made it possible to adjust the height of the overland fixed span

---

[34] Lt. Col. Taylor R. Fulton, "Conglomerate Tactical Bridging," *Military Engineer* 59 (September-October 1967): 323; AAR, Opn JOHN PAUL JONES, 39th Engr Bn, p. 1.

During Operation John Paul Jones, *men of the 39th Engineer Battalion built a conglomerate bridge made up of several different tactical bridges.*

to the changing flood depths. Since the 800-foot span was composed of seven parts—four M4T6 dry span end sections, two Class 60 float bridges, and a Class 60 dry span trestle—it was called the conglomerate bridge.[35]

By midyear, more combat engineers arrived in the western highlands of II Corps. The 299th Combat Engineer Battalion in shifting operations to Pleiku helped build the 4th Division's base camp, maintained roads in the area, and carried out various operational support missions. Heavy wheeled and tracked vehicle traffic combined with monsoon rains had turned a section Highway 19 west of Pleiku into a slippery sea of mud. The battalion determined that only rock would save the road. Throughout August, the 299th supported by other 937th Engineer Group units hauled some 8,000 cubic yards of rock from quarries as far away as An Khe. Elements of the battalion also completed the improvement of defenses at the Duc Co Special Forces camp near the Cambodian border. During the period, another company of the battalion ventured into the highlands to improve and build C–130 airstrips south and southeast of Pleiku.[36]

---

[35] Fulton, "Conglomerate Tactical Bridging," pp. 323–25; AAR, Opn John Paul Jones, 39th Engr Bn, pp. 1, 4.

[36] ORLLs, 1 May–31 Jul 66, 299th Engr Bn, pp. 1–2, and 1 Nov 66–31 Jan 67, 299th Engr Bn, 14 Feb 67, pp. 1–7, Historians files, CMH; Galloway, "Essayons," pp. 117, 119, 165.

The 20th Engineer Combat Battalion was also shifting operations to the western highlands. In mid-August, Company B moved by convoy to Ban Me Thuot to build a brigade-size bivouac area for the 4th Division. When this work neared completion, the company moved a month later to the remote Phu Tuc Special Forces camp southeast of Pleiku to extend an existing Caribou airstrip for C–130s. By 10 October 1966, the battalion and the attached 584th Light Equipment Company completed the move to the 4th Division's Dragon Mountain Base Camp outside Pleiku. Within a short time, the battalion was carrying out a large share of the 937th Engineer Group's operational support missions in the highlands, devoting most of its efforts to working on forward airfields and roads in support of tactical operations.[37]

Manpower

In April and May 1966, many Army engineers neared the end of their twelve-month tour of duty. Since nearly all had arrived with their units on the same ship, the rotation dates would be the same. The prospect of losing so many seasoned men at the same time threatened the operational expertise of many units. Seabee battalions faced a different circumstance. After completing a tour of approximately eight months, an entire battalion returned to a home base, and another battalion arrived in time to take over projects, equipment, and base camp.[38]

While the 18th Engineer Brigade worked on getting replacements, group commanders initiated programs to reduce the so-called rotational hump. The approach usually taken involved adjusting rotation dates and exchanging men between units with different deployment dates. For instance, Colonel Bush's 45th Engineer Group had two battalions that departed the United States the same day in December 1965. Excluding earlier departures and extensions, approximately 717 enlisted men and 33 officers were expected to depart in December 1966. To offset this excessive turnover, the group set a goal restricting battalion losses to no more than 25 percent of their soldiers in a one-month period. To achieve this goal, Bush promulgated a four-step program. First, battalions were authorized 10 percent overstrength. Second, some individual tours were shortened by as much as one month, allowing the administrative load to be spread over at least two months. Third, the tours of 10 percent of the men eligible for rotation were extended by one month or longer if necessary, using a combination of voluntary or involuntary methods. Last, the group exchanged soldiers who had less Vietnam service with men from other battalions, including divisional engineer battalions, to lessen the impact of the loss on the unit.[39]

Other manpower factors affected productivity. In April 1966, one battalion commander reported that only 89.7 percent of the battalion's authorized strength

---

[37] ORLLs, 1 May–31 Jul 66, 20th Engr Bn, pp. 1–2, 8, 1 Aug–31 Oct 66, 20th Engr Bn, pp. 1–4, 6–8, and 1 Nov 66–31 Jan 67, 20th Engr Bn, pp. 1–3.

[38] Ploger, *Army Engineers*, pp. 185–86.

[39] Ibid., p. 130; ORLLs, 1 May–31 Jul 66, 18th Engr Bde, 26 Aug 66, p. 4, Historians files, CMH, and 1 Aug–31 Oct 66, 45th Engr Gp, p. 5; Galloway, "Essayons," p. 108.

of 525 enlisted men was on hand, and only 40 percent worked on projects related to the construction program. To support the continuity of operations, Bush kept company and battalion mess, maintenance, supply and administrative troops, and equipment operations at or slightly above authorized levels. Guard duty took away more men. As a result, shortages turned up mostly in the line squads, about three men in each squad, leaving platoons with only about sixteen to eighteen men on the job each day. Hiring Vietnamese laborers helped somewhat, but the commander warned that the effects of the manpower shortages would be most severely felt when the battalion started on projects requiring technical skills and carried out operational support missions. To make matters worse, about 270 of the battalion's enlisted men were scheduled to depart in July and August 1966.[40]

General Ploger attempted to bring relief to the manpower problem. In March 1966, he asked U.S. Army, Vietnam, for additional troops to augment 18th Brigade units. He contended that the brigade needed an enlisted over-strength of between 10 to 15 percent to take full advantage of the dry season to complete as much work as possible before monsoon weather curtailed activity. The brigade had already fallen 700 men below authorized strength and 1,160 by the end of the month. Since engineer fillers were beginning to arrive in the number required, Ploger indicated that he could get by with filling one-half to two-thirds of this shortage with soldiers from other branches. Some of these men without engineering skills had relieved engineer soldiers previously tied down with security duties, which at Long Binh alone amounted to 2 officers and 77 enlisted men daily. Between late August and early September, the 18th Engineer Brigade, with a strength now approaching 13,000 soldiers, had received 1,100 men from other branches to learn engineering skills or assume other duties. This measure helped ease the manpower problem.[41]

Hiring Vietnamese civilian laborers helped the 18th Engineer Brigade take up some of the manpower slack. By September 1966, the brigade supplemented its strength with approximately 5,400, mostly unskilled, workers. The contractors had exploited the skilled labor market to such a degree that those hired by the brigade worked on jobs requiring few or no technical skills. Engineer units began training programs to instruct Vietnamese workers in such skills as carpentry, masonry, mechanics, and vehicle operation. By late 1966, the 159th Engineer Group at Long Binh began to graduate a class of heavy equipment operators every two weeks, thus alleviating a serious shortage of military operators.[42]

Efforts were also under way to ease a growing shortage of engineer officers. One way to relieve the problem at brigade headquarters was to designate

[40] Extract of a Letter from a Battalion Commander to a Group Commander, incl. 6, tab G, OCE Liaison Officer Trip Rpt, no. 2.

[41] Ploger, *Army Engineers*, pp. 130–31; Rpt of Visit to the 18th Engr Bde and Subordinate Units, 9–18 Apr 66, p. 1, incl. 6, OCE Liaison Officer Trip Rpt no. 2; Rpt of Visit to the 18th Engr Bde and Subordinate Units, 28 May–11 Jun 66, p. 1, incl. 5, OCE Liaison Officer Trip Rpt no. 3; Rpt of Visit to the 18th Engr Bde, p. 1, incl. 7, OCE Liaison Officer Trip Rpt no. 4; Army Buildup Progress Rpt, 10 Aug 66, p. 63.

[42] Ploger, *Army Engineers*, p. 131; Rpt of Visit to the 18th Engr Bde, p. 5, incl. 7, OCE Liaison Officer Trip Rpt, no. 4.

positions that could be filled by officers from other branches. Engineer group and battalion headquarters did the same. Signal Corps officers took over communications positions, and Adjutant General Corps officers replaced engineers in administrative jobs. Ploger also asked U.S. Army, Vietnam, to locate soldiers not assigned engineering duties but who had engineering backgrounds for reassignment to the brigade. In Washington, an Army study found that to fill officer positions in the new engineer units being raised, especially captains and lieutenants, an increasing number of engineer officers had to be diverted from other assignments. The solution pointed to temporarily assigning officers from other branches to the Corps of Engineers for eighteen months. Newly commissioned officers in other branches were also diverted to the Corps of Engineers. Officers with engineer and engineer-related degrees but assigned to other branches such as Armor or Field Artillery suddenly received orders to report to engineering jobs. This was the case in the 84th Engineer and the 589th Engineer Battalions (Construction) outside Qui Nhon. A field artillery officer with an engineering degree served as the 84th's operations officer. Similarly, an armor officer was assigned as operations officer in the 589th when it arrived in early 1967. Reserve officers volunteering to come on active duty were eagerly accepted.[43]

The long-term solution to the shortfall in officers, however, lay in the United States, and programs were soon in train to raise Reserve Officers Training Corps enrollments at universities and colleges, increase the size of classes at the U.S. Military Academy, and turning to the time-honored expedient for quick expansion by increasing the output of the Officer Candidate Schools. With the demands of Vietnam, the Army also stepped up its efforts to develop promising enlisted men, particularly college graduates and those with some college training, as junior officers. The Engineer Officers Candidate School at Fort Belvoir was reactivated in the fall of 1965. By June 1966, 1,132 junior engineer officer graduates had been commissioned. The number would climb steadily, reaching some 4,000 in 1968. When the school closed in January 1971, 5,859 new second lieutenants had been assigned to the Corps of Engineers. Altogether, the number of graduates reached 10,380, the balance being assigned to other branches.[44]

Forward Airfields

One task many engineer soldiers faced in South Vietnam was building forward airfields for cargo aircraft. Throughout South Vietnam, airfields gained importance as the final, vital link sustaining military operations. The land lines of communication—roads and rail—presented major security and

---

[43] ORLL, 1 May–31 Jul 66, 18th Engr Bde, p. 10; Army Buildup Progress Rpts, 6 Apr 66, p. 3, and 4 May 66, p. 6, CMH.

[44] Ploger, *Army Engineers*, pp. 183–86; Interv, Bower with Ploger, 8 Aug 67, pp. 22–23; Ronald H. Spector, "The Vietnam War and the Army's Self-Image," in *Second Indochina War Symposium*, ed. John Schlight (Washington, D.C.: U.S. Army Center of Military History, 1986), pp. 176–77; Ronald H. Spector, *After Tet: The Bloodiest Year in Vietnam* (New York: Free Press, 1993), pp. 32–34.

maintenance problems, and did not exist in many areas. The fluid nature of the war and the absence of battle lines prompted American forces to respond to tactical requirements in as mobile a fashion as possible. When compared to the massive effort required to build and maintain the land lines, it became apparent that an expanding network of airfields would better serve tactical mobility.[45]

Although General Westmoreland shifted some construction units to do port construction, he also pushed building and upgrading airfields for transport aircraft up to and including the C–130 Hercules. Existing and new major air bases under construction could easily accommodate the medium-range Hercules. While each of combat base camps occupied by U.S. divisions and brigades had an airstrip suitable for C–123 landings, the capabilities and locations of forward airstrips still fell far short of U.S. Air Force criteria for the more desirable, larger-capacity C–130. Coupled with the need to upgrade existing airstrips, Westmoreland pursued the goal of establishing a checkerboard pattern of forward airfields covering the entire country.[46]

The growing importance of forward airfields in tactical planning led Westmoreland to push for a coordinated upgrading effort. In May, a coordinating group representing MACV, U.S. Army, Vietnam, and Seventh Air Force (which succeeded the 2d Air Division) reviewed the airfield construction priority list and identified the urgent need for C–130 airfields. The airfields had to have an all-weather capability to handle sixty sorties a day to support division-size operations. In early June, Westmoreland approved a master plan and priority listing of forty airfields submitted by the group, now known as the Joint Airfield Evaluation Committee. Top priority for upgrading between July and September went to ten sites in Pleiku, Kontum, and Darlac Provinces in II Corps, and Binh Long and Phuoc Long Provinces in III Corps. Airfields in the coastal regions of I and II Corps had second priority, and airfields not covered in priority one in III Corps and all of IV Corps became third priority. Because of a shortage of airfield matting, the committee decided to limit the upgrade to the first priority fields at Kontum, Plei Me East, and Loc Ninh.[47]

The MACV Directorate of Construction designated Army funds for airfield construction. In late June, General Dunn informed U.S. Army, Vietnam, that the Military Construction Program covered funding requirements for airfields supporting logistical bases and division and brigade base camps. Top priority airfields by corps areas in this group included the division bases at

---

[45] MACV History, 1966, p. 300; Bowers, *Tactical Airlift*, p. 185.

[46] MACV History, 1966, pp. 300–301; Ltr, Westmoreland 001093 to CG, FFV, 10 Dec 65, sub: Tactical Employment of US Forces and Defensive Action; Interv, Rochester with Raymond, 11 Jan 80, p. 86; Bowers, *Tactical Airlift*, pp. 224–25; Situation Rpt on Airfield Construction, incl. 7, p. 1, OCE Liaison Officer Trip Rpt no. 3; Ploger, *Army Engineers*, p. 108. For a list of airfields and C–130 airfield criteria, see MACV Concept of Operations, 30 Aug 65.

[47] MACV History, 1966, pp. 301–02; Quarterly Hist Rpts, 1 Apr–30 Jun 66, MACDC, pp. 7–9, and 1 Jul–30 Sep 66, MACDC, p. 15; Bowers, *Tactical Airlift*, pp. 186, 230; Msg, COMUSMACV 18054 to CGs, III MAF, I FFV, and II FFV, 26 May 66, sub: Master Plan for Upgrading Airfields in RVN; Msg, COMUSMACV 20225 to CG, USARV, et al., 13 Jun 66, sub: Airfield Evaluation, both in box 5, 69A/702, RG 334, NARA.

An Khe in II Corps and Cu Chi in III Corps and a brigade base in IV Corps. Materials used by troops to build the forward deployment airfields required for tactical operations fell within the other funding categories, Operations and Maintenance, Army (O&MA), and Procurement of Equipment and Missiles for the Army (PEMA). These included fields at Kontum and Cheo Reo in II Corps and Song Be and Loc Ninh in III Corps.[48]

By midyear, the upgrading program had achieved a degree of coordination, but more mundane troubles plagued the construction efforts. The weight of the C–130s, with high landing impact and maximum braking action, frequently caused runway surface problems. The M8A1 matting proved especially troublesome. The newly developed T17 membrane surfacing also presented high maintenance requirements on soils with low-bearing ratios. Under such conditions, AM2 and MX19 matting became the choice for surfacing runways, the M8A1 matting for taxiways and parking, and the T17 membrane principally for C–123 airfields.[49]

The need for continuing maintenance and shortages of matting presented difficulties for the engineers, who were still spread thin. There were more than two hundred airfields, including simple dirt strips, that were candidates for maintenance and upgrading. Airfield matting came under tight control, and in several cases MACV diverted matting from other projects—for example, shifting the AM2 originally programmed for Phu Cat Air Base to An Khe's Golf Course airstrip. By September, airfields supporting Marine Corps operations in I Corps received special attention. The sharp burred edge of the mat adjacent to the connector hooks in newly installed M8A1 matting at several airstrips near the Demilitarized Zone caused blowouts to C–130 tires. More AM2 matting was diverted to Dong Ha and an alternate airfield at Khe Sanh farther inland, and the Seabees completed both C–130 airfields in mid-October before the northeast monsoon came in full force, much to Westmoreland's relief.[50]

By the end of 1966, the number of airfields and airstrips had grown to 282, which represented an increase from 243 listed airfields in July. In I Corps, the Seabees also improved An Hoi, Duc Pho, and Quang Ngai to C–130 capability, and fifteen other airfields throughout South Vietnam received similar priority listing. In November, MACV listed over sixty airfields requiring upgrading to C–130 capability, and the command expected to upgrade twenty-four of these by mid-1967.[51]

[48] Ltr, Dunn to CG, USARV, 25 Jun 66, sub: C–130 Airfield Requirements, box 5, 69A/702, RG 334, NARA.

[49] MACV History, 1966, pp. 302–03; Msg, CG, I FFV, to CG, USARV, 18 Aug 66, sub: Landing Mat, box 5, 69A/702, RG 334, NARA.

[50] MACV History, 1966, pp. 303–05; Quarterly Hist Rpts, 1 Apr–30 Jun 66, MACDC, pp. 20, 23–24, 1 Jul–30 Sep 66, MACDC, pp. 15–16, and 1 Oct–31 Dec 66, MACDC, pp. 8–9, 24; Tregaskis, *Building the Bases*, pp. 304–06.

[51] MACV History, 1966, pp. 300, 305.

Tunnels

During search-and-destroy operations, U.S. forces often came across enemy base areas containing vast networks of bunkers and tunnels. These networks had evolved over time as a natural response by the guerrillas to the advantages in aircraft and artillery held by the French, then the South Vietnamese, and finally the Americans. While U.S. units could move or destroy caches of supplies and munitions with relative ease, they had to develop individual and unit skills in the techniques of detecting, penetrating, and destroying the well-concealed bunker and tunnel complexes. Specially trained teams were established to search the tunnel complexes for prisoners, equipment, and documents. The soldiers in these teams, usually made up of men of small physical stature because the Vietnamese built entrances no larger than eighteen inches, proudly called themselves "tunnel rats." Combat engineers contributed men, equipment, and the demolitions expertise to destroy the enemy tunnels.[52]

As allied operations increased during 1966, the troops discovered even more extensive tunnel complexes. The local Viet Cong used hastily built simple, shallow structures, while larger forces established well-constructed excellent camouflaged systems usually found in uninhabited areas. No two tunnel complexes were exactly alike, and lengths varied from about 6 to over 4,300 yards. Passageways leading from the small entrances from well-concealed, hidden trap doors expanded to about 2 feet wide and 2½ to 3 feet high and extended in straight sections to about 50 feet long. The sections joined at various angles to one another, thus forming a zigzag pattern. This pattern served to protect the Viet Cong from observation. It also protected them from shock waves caused by explosives dropped into the tunnel, from shrapnel of detonating grenades, from direct-fire weapons, and made it easier to ambush tunnel rats during their search. The larger and more permanent tunnel complexes, especially those in the war zones, were elaborate, having as many as four distinct levels that held compartments usable as quarters, hospitals, and mess facilities. Rooms discovered during Operation CRIMP measured approximately 4 feet by 6 feet by 3 feet high and were found about every 100 yards. Shelves were provided along one side wall together with various types of seats. In some cases, air shafts were dug inside by rodents held against the tunnel roof in cages. Some tunnel complexes contained facilities for manufacturing and storing war materials. Booby traps were used extensively, both inside and outside the entrance and exit trap doors. Often, grenades placed in trees near the exit trap doors could be activated by pulling a wire from under the trap door or by the trap door itself.[53]

Moving from ad hoc methods practiced during CRIMP, allied units developed techniques and teams to deal with tunnels. Two-man teams usually explored the

---

[52] Ploger, *Army Engineers*, p. 92; Sharp and Westmoreland, *Report*, p. 122; Maj. Glen H. Lehrer, "Viet Cong Tunnels," *Military Engineer* 60 (July–August 1968): 244.

[53] Rpt, MACV, 18 Apr 66, sub: Lessons Learned no. 56: Operations Against Tunnel Complexes, pp. 1–8, Historians files, CMH; Ploger, *Army Engineers*, p. 94; Lehrer, "Viet Cong Tunnels," pp. 244–45; Capt. Francis E. Trainor, "Tunnel Destruction in Vietnam," *Military Engineer* 60 (September–October 1968): 341; Diary of an Infiltrator, Extracts from Captured Diaries, Dec 66, U.S. Mission in Vietnam, box 1, 70A/782, RG 334, NARA.

A 1st Division soldier enters a tunnel during Operation CRIMP.

tunnels, one member staying at the entrance and the other descending into the tunnel. Equipped with a telephone, communications wire, compass, bayonet, flashlight, and pistol, the man in the tunnel explored the network, keeping in communication with his partner at the entrance. Riot-control chemicals and acetylene gas—along with explosives—appeared to be the best way to deny use of the tunnels to the enemy. In the 1st Division, tunnel destruction was assigned to the division's chemical officer, but the division engineer soon assumed most of this responsibility. It also became obvious that such an unnatural and stressful task should be given to volunteers. In April 1966, MACV incorporated lessons learned during CRIMP in a document that spelled out recommended techniques to detect and destroy tunnel complexes, especially those found in the war zones and Viet Cong base areas.[54]

Combat engineers on the scene came forth with proposals to neutralize the tunnels. Colonel Sargent, commander of the 1st Engineer Battalion, believed the best approach lay in denying the enemy use of a tunnel instead of trying to collapse it with explosives. It took large quantities of explosives to destroy tunnels, and Sargent suggested using Mighty Mite smoke blowers to find the

---

[54] ORLL, 1 Feb–30 Apr 66, 1st Inf Div, p. 20; Operations Against Tunnel Complexes, pp. 9–13; Ploger, *Army Engineers*, p. 94; Tom Mangold and John Penycate, *The Tunnels of Cu Chi* (New York: Random House, 1985), pp. 101–03.

191

exits, then place CS gas sacks inside the tunnel and a small explosive charge at each exit. Detonating explosives and gas simultaneously, he believed, would seal the tunnel and trap the riot-control tear gas. This method worked well in collapsing tunnels with less than eight to ten feet of overburden.[55]

The acetylene gas tunnel demolition kit looked promising. The kit, which originated in a MACV request for help in November 1964 and was developed by the U.S. Army Munitions Command's Picatinny Arsenal in Pennsylvania, worked on the principle of mixing calcium carbide and water to produce a highly explosive acetylene gas. Mighty Mite or other blowers pushed the gas into the tunnel. A small explosive charge detonated the combination of the gas and oxygen.[56]

A Department of the Army team, consisting of Maj. Jack E. Mowery from the Chief of Engineers and two civilians from Picatinny Arsenal, arrived in Vietnam in May 1966 to test the system and train combat engineers in its use. The team gave a three-day training course at Di An for selected troops from the 1st, 65th, 168th, and 588th Engineer Combat Battalions. A similar course at An Khe followed for members of the 8th and 70th Engineer Combat Battalions and Company A, 326th Airborne Engineer Battalion. In turn, these teams were expected to train other tunnel destruction teams.[57]

The Department of the Army team also tested the kit in the field. Near the village of Ba Ria, some thirty-one miles north of Vung Tau, the 173d Airborne Brigade found a tunnel complex, heavily fortified trenches, and a machine gun bunker. The 173d Engineer Company and tunnel demolition troops from the 588th Engineer Battalion spent two days exploring the tunnels, and after removing enemy material attempted to destroy the complex, with varying degrees of success. Destruction of an enemy-occupied tunnel complex found by the 25th Division in the Ho Bo Woods north of Cu Chi met with more success. While the assistant division commander and staff members assembled to observe the demonstration, Company C, 65th Engineer Battalion, sealed the entrances. The combination of three acetylene generators and a Mighty Mite blower pushed the gas through a hose into the tunnel. The explosion set off by electric blasting caps produced good results. About eight feet of overburden dropped into the tunnel, and loose earth sealed the entrances.[58]

Satisfied with the results, the U.S. Army, Vietnam, adopted the system for immediate use. The equipment brought by the Army team remained in Vietnam, with more sets to follow. Acetylene gas tunnel demolition kits were

---

[55] Ltr, Lt Col Howard L. Sargent to Lt Gen William F. Cassidy, OCE, 13 Jun 66; Ltr, Sargent to Cassidy, 16 Jul 66, both in OCE Hist Ofc.

[56] Rpt, DA Tunnel Destruction Team, sub: The Employment of Acetylene in the Destruction of Viet Cong Tunnels, n.d., pp. 1–2, Historians files, CMH; Tech Info Rpt 33.8.7.5 HQ, Army Materiel Command, Oct 66, sub: Interim Report, Tunnel Destruction Demolition Set, XM69, pp. 1–8, box 8, 84/051, RG 319, NARA.

[57] Rpt, DA Tunnel Destruction Team, pp. 3–5; Ltr, HQ USARV, AVEN-O, to USARV G–2, G–3, CGs, I FFV, II FFV, and 1st Div, and CO, 3d Bde, 25th Div, 2 Jul 66, sub: Tunnel Destruction, incl. 7, tab F, OCE Liaison Officer Trip Rpt no. 4, 23 Sep 66.

[58] Rpt, DA Tunnel Destruction Team, pp. 6–17; Rpt, 65th Engr Bn, n.d., sub: Destruction of Tunnels and Bunkers, p. 12, OCE Hist Ofc.

issued based on one tunnel destruction team for each division and four to the 18th Engineer Brigade. The command also charged the engineer brigade to provide teams and equipment to support the separate brigades and to develop a light acetylene system.[59]

But demolition teams continued to experience mixed results in their attempts to destroy the complexes. Returning to the Ho Bo Woods in support of a 25th Division operation in late June, the team from the 65th Engineer Battalion encountered sniper fire. The battalion reported that the bulky acetylene kit should not be used in assault demolition work, but only in more deliberate denial operations. The 588th Engineer Battalion had more success in supporting the Australians and other units. In July, a team helped the 45th Engineer Group to destroy six bunkers near Tuy Hoa "with 100% effectiveness." In November, the 588th trained more four-man teams from the 1st, 15th, 27th, 86th, and 168th Engineer Battalions and the 919th Engineer Company, 11th Armored Cavalry.[60]

Overall, results were frequently less than desired because of technical and tactical factors. General Ploger noted that operators had difficulty obtaining a complete mix of the explosive gases. Using oxygen when the acetylene was generated produced better results but added to logistical requirements. U.S. commanders preferred to carry regular explosives, which were no heavier than the generators and oxygen bottles and could stand rougher treatment. The 65th Engineer Battalion had some success with bangalore torpedoes, five-foot-length pipe-type explosives normally joined together to breach barbed-wire obstacles, by placing them throughout the tunnels. Because tunnel rats usually searched the tunnels before destruction, some commanders felt this obviated the need for acetylene generators. Deliberate tunnel destruction with the generators also took time. This conflicted with the preference of most tactical commanders to move on rather than to keep their troops in a recently won area. As a result, the short duration of U.S. operations allowed the Viet Cong to reoccupy and rebuild their tunnels. The Viet Cong later boasted of the invincibility of their complexes. Meanwhile, the engineers, whether using different types of explosives or acetylene generators or attempts to flood the tunnels, continued to seek faster and better means of destruction.[61]

[59] Rpt, DA Tunnel Destruction Team, p. 17; Ltr, HQ USARV, 22 Jul 66, sub: Tunnel Destruction.

[60] Rpt, 65th Engr Bn, sub: Destruction of Tunnels and Bunkers, pp. 12–14; ORLLs, 1 May–31 Jul 66, 588th Engr Bn, pp. 1–2, 1 Aug–31 Oct 66, 588th Engr Bn, p. 2, and 1 Nov 66–31 Jan 67, 588th Engr Bn, pp. 1–2; Combat Activities and Operational Support, incl. 11, pp. 3–4, OCE Liaison Officer Trip Rpt no. 5.

[61] Ploger, *Army Engineers*, p. 94; Combat Activities and Operational Support, incl. 11, pp. 3–4, OCE Liaison Officer Trip Rpt no. 5; Rpt, 65th Engr Bn, sub: Destruction of Tunnels and Bunkers, pp. 7, 10; ORLL 8–66, DA, Engineer Notes no. 1, 13 Oct 66, p. 26; USARV Combat Lessons Bull, no. 5, 8 Feb 67, pp. 3–4; Sedgwick D. Tourison, *Talking with Victor Charlie: An Interrogator's Story* (New York: Ivy Books, 1991), pp. 282, 287.

193

The Land-Clearing Weapon

Since the heavily forested areas of Vietnam also concealed Viet Cong base areas, MACV planners agreed something had to be done to deny the sanctuaries, either by chemical defoliants, burning, or simply cutting and clearing the forest growth. The use of herbicides began with the early American buildup in 1961. In considering various techniques to turn the tide against the insurgents, the Kennedy administration authorized the South Vietnamese to run tests of a chemical plant killer. Helicopters and fixed-wing aircraft sprayed the herbicide Dinoxol along a section of road north of Kontum and a section of Highway 13 north of Saigon. Although the results disappointed the Americans, President Diem became a staunch supporter of the defoliation program. Despite concern that the United States could be charged with employing chemical or biological warfare, President Kennedy approved the use of defoliants selectively and under careful control. In January 1962, the Air Force deployed spray-equipped C–123s to South Vietnam under the code name Operation RANCH HAND. Similarly, the South Vietnamese deployed helicopters on defoliation missions. The controversial program gained momentum in 1966, improving observation from both the air and ground and the defense of fixed bases. Agent Orange proved the most powerful of the herbicides, killing foliage and producing drying in four to six weeks. Agent Blue, a drying agent on contact, resulted in leaves drying up and dropping off, but the leaves could grow back. Still, most of Vietnam remained a verdant land, and even in the defoliated areas the enemy had some tree canopy and reoccupied his bases. Accordingly, MACV formulated plans to deny the Viet Cong his forest sanctuaries permanently by cutting and clearing with specially equipped bulldozers operated by engineer units.[62]

General Westmoreland's interest in land clearing went back to 1965 while attending one of the Hawaiian deployment conferences. By a chance encounter, he met a B. K. Johnson, who claimed to have cleared rain forests

[62] William A. Buckingham Jr., *Operation RANCH HAND: The Air Force and Herbicides in Southeast Asia, 1961–1971* (Washington, D.C.: Office of Air Force History, United States Air Force, 1982), pp. 11–12, 16–17, 20–22, 30–31, 121–29, 185; Bowers, *Tactical Airlift*, pp. 89–90, 392–93; Schlight, *Years of the Offensive*, pp. 42, 91–92; F. Clifton Berry Jr., *Gadget Warfare*, The Illustrated History of the Vietnam War (New York: Bantam Books, 1988), pp. 48–57; MACV History, 1966, pp. 774–75; Westmoreland, *A Soldier Reports*, p. 280. Spraying the herbicides from aircraft reached its height in 1967 and 1968. Spraying continued at a reduced rate until April 1970 when use of Agent Orange was halted. By May 1971 all spraying stopped. Meanwhile, in October 1970 Congress directed the Defense Department to commission the National Academy of Science to review the herbicide program. In addition, the Engineer Strategic Studies Group (later the Engineer Studies Center) was requested to study the military effects of herbicides. The Studies Group was picked because of the large background of information on hand of the geology, climatology, flora of the area, other surveys of the country, and the capability to do war-gaming. This completed study sparked more controversy when it concluded herbicides could be useful in future conflicts depending on a variety of factors. For more on Air Force and the Engineer Studies Center participation, see Bernard C. Nalty, *Air War Over South Vietnam, 1968–1975*, United States Air Force in Southeast Asia (Washington, D.C.: Air Force History and Museums Program, United States Air Force, 2000), and Baldwin, *Engineer Studies Center*, pp. 172–78.

in Australia by shattering big trees with a crane mounting a 5,000-pound steel ball on a chain. This intrigued the MACV commander. Returning to Saigon he dispatched Capt. Robert L. Stuart, MACV Engineer Section, to observe the clearing operations on the King Ranch property in Australia and to assess its possible application to military operations. While visiting the ranch, Stuart observed large D8 and D9 bulldozers, two or four in a tandem hook-up dragging a chain, or a large hollow steel ball and ship anchor chain, or a chain and a wire rope. The nine-foot-diameter ball kept the middle of the chain up off the ground. Its weight also prevented the chain from sliding over the top of bent trees. The bulldozers were also modified with a reinforced canopy to protect the operator, reinforced push bars and undercarriages, and an extension on the blade. The simple extension on the blade, locally called a tree horn, allowed the operator more latitude in pushing down a tree. By positioning the blade high enough so that the bulldozer did not sit on the tree's root system, the full force of the tractor could knock down a large tree in less than one minute. In late September 1965, Stuart returned and reported to Westmoreland that the procedures used in Australia could be effectively used in Vietnam. Westmoreland agreed and tasked U.S. Army, Vietnam, to obtain similar equipment from the United States for tests.[63]

In February 1966, the Department of the Army offered to obtain six sets of a special land-clearing blade, cab guard and screen, instead of the equipment requested by MACV. The Rome Plow Company of Cedartown, Georgia, had developed a special cutting blade for mounting on a standard military bulldozer. The sharp blade cut medium trees and brush at or near ground level, and a "stinger," which protruded from one corner of the blade, split longer trees. MACV still preferred the King Ranch idea and heavier bulldozers because the equipment could clear seven acres an hour compared with two for the Rome plow. The Department of the Army, however, suggested testing the Rome Plow Company blades first.[64]

In late June, a company representative arrived to brief the engineers. On 1 August, four Rome plow blades mounted on Allis Chalmers HD16M tractors reached the port of Saigon. The 169th Construction Battalion ran controlled tests in the Long Thanh area from 9 through 15 August and reported favorable results. The 588th Engineer Combat Battalion ran the second test from 17 to 27 August at the abandoned Filhol Rubber Plantation near the 25th Division's Cu Chi base camp. On the tenth day, three of the bulldozers struck mines, causing extensive damage but no casualties. The test ended the next day. Results revealed that the bulldozers equipped with Rome plows could clear

---

[63] Quarterly Hist Rpt, 1 Jul–30 Sep 66, MACDC, p. 18; Ltr and incls, MACV to CofS Army, no. 0172, 5 Feb 66, sub: Clearing of Rain Forests in Republic of Vietnam, CMH; Historical Notes, no. 1, 24 Oct 65, p. 11, Westmoreland Papers, CMH; Interv, U.S. Army Center of Military History with Gen William C. Westmoreland, 6 Dec 89, p. 102, CMH; Interv, Bower with Ploger, 8 Aug 67, p. 53; Interv, Sowell with Ploger, 21 Nov 78, sec. 9, p. 7; Westmoreland, *A Soldier Reports*, p. 280; Ploger, *Army Engineers*, pp. 97–98.

[64] Ploger, *Army Engineers*, p. 98; Quarterly Hist Rpt, 1 Jul–30 Sep 66, MACDC, p. 19; Ltr, Raymond MACDC-AD to CG, USARV, sub: Clearing Rain Forest in RVN, 11 Jun 66, box 11, 69A/702, RG 334, NARA.

Rome plows clear vegetation during Operation PAUL BUNYON.

trees at approximately twice the rate of a standard bulldozer. Since the equipment worked best as a team, the 79th Engineer Group assigned all four plows and operators to the 557th Light Equipment Company. Soon the plows went to work clearing 500-yard-wide strips around base camps and 100-yard-wide strips on both sides of key roads. With their value now proven and capabilities made known, the team's Rome plows could not keep up with the requests.[65]

In September 1966, all activities pertaining to the Rome plows took on the code name PAUL BUNYON. Westmoreland appointed General Raymond to chair a task force to develop a land-clearing plan. Raymond decided that the best approach was the formation of special teams made up of approximately one hundred men and thirty D7E tractors and plows. Estimates based on experience projected that such a team could clear over 250 acres a day of heavy brush and scattered trees in the Viet Cong sanctuaries. In turn, the 18th Engineer Brigade recommended fifty-six additional Rome plow kits and disk harrows to keep areas free from regrowth. In early October, Westmoreland reluctantly agreed that the only way he could get the scarce bulldozers as soon as possible would be by holding up the deployment of three engineer battal-

---

[65] Quarterly Hist Rpt, 1 Jul–30 Sep 66, MACDC, p. 19; Miscellaneous Detailed Discussion, incl. 11, pp. 3–4, OCE Liaison Officer Trip Rpt no. 4; ORLLs, 1 Aug–31 Oct 66, 79th Engr Gp, 14 Nov 66, p. 10, 1 Aug–31 Oct 66, 169th Engr Bn, 1 Nov 66, p. 4, both in Historians files, CMH; ORLL, 1 Aug–31 Oct 66, 588th Engr Bn, p. 6; Msg, Raymond SOG 0798 to Cassidy, 15 Sep 66, Dunn Papers, CMH.

ions. The completed PAUL BUNYON plan called for the 18th Engineer Brigade to undertake land-clearing operations in coordination with the field forces, with the initial phase taking place in III Corps. Control of the land-clearing teams and their priorities remained with the U.S. Army, Vietnam, commander. Since large-scale land-clearing operations could literally change the face of the country, Westmoreland vetted the plan with Ambassador Henry Cabot Lodge during a Mission Council meeting.[66]

Westmoreland, emphasizing the need to "think big," still saw merit in the King Ranch ball-and-chain technique to level Viet Cong base areas. MACV requested the heavy chain from the Navy, but the Pacific Command could only provide a lighter one and one-half-inch commercial chain, which MACV claimed would not work. Although the towed ball worked well in the Australian jungle, difficulties in fabricating sets of balls and chains, the large size of the ball, troubles in transporting it, and inadequate equipment stood in the way of early success. The maintenance requirements of the heavier D9 bulldozer were also a concern. Instead, the 18th Engineer Brigade tested the Australian method of land clearing with the more common D7E bulldozers using smaller, locally fabricated steel balls, and a Navy chain.[67]

The tests of the Rome plows continued with good results. By November, three plows remained at Long Binh, clearing fields of fire and perimeter strips around the ammunition storage area. Three other Rome plows moved to Pleiku to help clear the 4th Division's new base camp. In December, military airlift delivered six more kits, four going to Long Binh and two to Pleiku, bringing the total in country to twelve. Clearing requirements for the 9th Division camp at Bearcat resulted in the move of two Rome plows from Long Binh. By the end of the year, U.S. Army, Vietnam, had placed the order for the remaining fifty Rome plows.[68]

By the autumn of 1966, operational support tasks had picked up as the tempo of tactical operations increased. By then, five engineer battalions and four companies organic to a like number of divisions, separate infantry and airborne brigades, and an armored cavalry regiment, backed by eleven engineer combat battalions and an assortment of separate companies of the 18th Engineer Brigade, kept resupply roads open, built and reopened forward airfields, and explored and destroyed enemy tunnels and bunkers. The combat

[66] Quarterly Hist Rpts, 1 Jul–30 Sep 66, MACDC, pp. 20–21, and 1 Oct–31 Dec 66, MACDC, p. 6; Msg, Ploger MAC 7326 to Besson, 23 Aug 66, sub: Information on Rome Plow; Msg, Raymond SOG 0798 to Cassidy, 15 Sep 66, both in Dunn Papers, CMH; MFR, MACJ02, 23 Aug 66, sub: Forest Clearing; DF, Asst CofS, J–4 to Dir of Cons, sub: MACV Task Force on Land Clearing Operations, 25 Aug 66, both box 12, 69A/702, RG 334, NARA; MACV Planning Directive 10–66, Opn PAUL BUNYON, 23 Sep 66, pp. 1–2; Memo, MACDC for Dep COMUSMACV, 3 Oct 66, sub: Paul Bunyon Program, both in 69A/702, RG 334, NARA; Msg, COMUSMACV 44465 to CGs, II FFV and USARV, 6 Oct 66, sub: Control and Use of Rome Plows, box 5, 69A/702, RG 334, NARA.

[67] Ploger, *Army Engineers*, p. 98; Quarterly Hist Rpts, 1 Jul–30 Sep 66, MACDC, p. 20, and 1 Oct–31 Dec, 66, MACDC, pp. 6; Interv, Sowell with Ploger, 21 Nov 78, sec. 9, pp. 7–9; Interv, Bower with Ploger, 8 Aug 67, pp. 53–54.

[68] Quarterly Hist Rpt, 1 Oct–31 Dec 66, MACDC, pp. 6–7; Sharp and Westmoreland, *Report*, p. 122.

engineers also applied new developments in accomplishing their missions. These included the T17 membrane matting for forward airfields, various tunnel exploration kits, and specially designed Rome plow bulldozer kits to clear the land. The ensuing months would see an even greater role played by the engineers in combat operations.

7

The Campaign Widens, October 1966–June 1967

The American ground offensive in Vietnam dates from the late weeks of summer 1966, when General Westmoreland finally received the troops he needed to challenge the in-country enemy sanctuaries and inflict heavy casualties. This influx of troops had come after a lengthy pause in deployments, when Westmoreland had been reduced to carefully husbanding his resources for about six months and fighting a transitional campaign with limited forces. Starting in July, however, the flow resumed of infantry, armor, artillery, and their supporting forces, adding importantly to the pressure that could be brought to bear in the Central Highlands, on the II Corps coast, and in the corridors leading out from Saigon. By January, the Army's infantry, tank, and armored cavalry strength had nearly doubled from summer levels, and engineer strength had risen by almost a third to twenty-three battalions. With these reinforcements, the thrust of U.S. strategy shifted, becoming one of operating at a rapidly rising tempo and giving no respite to the Viet Cong and the North Vietnamese—in short, of expanding the war on a major scale with the expectation of significant results. The plan was to launch a heavy series of operations just after the start of the new year, but the offensive began sooner than Westmoreland expected.[1]

ATTLEBORO

In October 1966, the discovery in Tay Ninh Province of a Viet Cong division and a North Vietnamese regiment led to the largest U.S. operation of the war to that point. The trigger was an American element new to Vietnam, the 196th Light Infantry Brigade, which, while searching for rice and other Communist supplies on the southern fringes of War Zone C, became dispersed in inhospitable terrain. Seeing a tantalizing opportunity to destroy an American unit, the enemy turned on the 196th. The battle that followed under the code name Operation ATTLEBORO involved over 22,000 U.S. and allied troops, including the 1st Infantry Division, two brigades from the 4th and 25th Infantry Divisions, the 173d Airborne Brigade, the new 11th Armored Cavalry Regiment, and the 196th. Control of the operation passed from the 196th to the 1st Division and finally to General Seaman's II Field Force, making it the first Army operation in South Vietnam controlled by a corps-size headquarters. The series of engagements extended into late November. The enemy took serious casualties and lost huge quantities of weapons, ammunition, and supplies. In one battle, the 1st Division's 1st Battalion, 28th Infantry, repulsed an

---

[1] MacGarrigle, *Taking the Offensive*, pp. 13–15; Sharp and Westmoreland, *Report*, p. 131.

assault by two North Vietnamese battalions, inflicting heavy casualties on the attackers and discovering a large cache of enemy ordnance nearby. The Viet Cong division was so badly mauled in the fighting that it would not reappear in combat until the following spring.[2]

ATTLEBORO brought forth heavy demands for engineers to destroy enemy tunnels and bunkers, clear jungle, repair roads and airfields, and build bridges. Directly involved were the 1st Engineer Battalion, 1st Division; the 65th Engineer Battalion, 25th Division; the 173d Engineer Company, 173d Airborne Brigade; and the 175th Engineer Company, 196th Infantry Brigade. The 3d Brigade, 4th Division, which had just arrived in Vietnam and participated later in the operation, was supported by Company D, 27th Engineer Combat Battalion, 79th Engineer Group.[3]

Missions performed by Colonel Kiernan's 1st Engineer Battalion during ATTLEBORO reflected the support given by other engineer units. Initial assignments included upgrading nine miles of road between Tay Ninh and Suoi Da and the construction of a ford and later a bridge over a stream north of Suoi Da. Since the Communists usually tried to destroy all bridges that they did not find useful, Company D built a combination timber trestle and M4T6 dry span bridge to cross the sixty-three-foot gap. The design allowed for the forty-five-foot dry span to be removed and replaced on short notice. The engineers left enough lumber at the site so that villagers could build a light bridge to replace the dry span for use by bicycles and motor scooters. The 1st Engineer Battalion also removed a low-capacity Eiffel bridge and replaced it with a two hundred-foot panel bridge across the Saigon River at Dau Tieng and opened twelve and a half miles of road to the town, allowing movement of artillery and troops to the Michelin Plantation. While Company A concentrated on the bridge, Company C bulldozed a wide swath of rubber trees to reduce the possibility of ambushes. On 20 November, Company C moved by helicopters to the top of Nui Ong Mountain. There in a little over one day, the engineer troops cut vegetation and blasted and leveled the peak into a firing position for two artillery batteries.[4]

The bridge across the Saigon River presented a major engineering challenge. To make room for the higher-capacity panel bridge (a Military Class 60 bridge to support heavy division traffic), Company A had to remove the old bridge without damaging its two piers. By carefully using acetylene torches and steel-cutting charges, welders successfully dropped all three spans. In the meantime,

[2] Sharp and Westmoreland, *Report*, p. 29; MACV History, 1966, pp. 386–87; ORLL, 1 Nov 66–31 Jan 67, II FFV, 25 Apr 67, p. 19, Historians files, CMH; Lt. Gen. Bernard W. Rogers, *Cedar Falls–Junction City: A Turning Point*, Vietnam Studies (Washington, D.C.: Department of the Army, 1974), pp. 8–12; and especially MacGarrigle, *Taking the Offensive*, pp. 31–59.

[3] ORLLs, 1 Aug–31 Oct 66, 27th Engr Bn, 31 Oct 66, p. 2, 1 Aug–31 Oct 66, 4th Inf Div, 22 Dec 66, p. 46, and 1 Nov 66–31 Jan 67, 3d Bde, 4th Inf Div, 23 Feb 67, p. 18. All in Historians files, CMH.

[4] This operation was code-named BATTLE CREEK for the 1st Division. Unless otherwise noted, this section is based on AAR, Opn BATTLE CREEK, 1st Engr Bn, 23 Dec 66; Ltr, Lt Col Joseph M. Kiernan to All Men 1st Engineer Bn, Thanksgiving Day 1966, sub: 1st Engineer Actions in Operation Battle Creek, both in OCE Hist Ofc; *Always First, 1965–1967*, "Prepared for the Unexpected" and "The Bulldozer in the Attack."

the troops rebuilt the abutments and used CH–47 Chinook helicopters to position I-beam caps on the piers. The 617th Panel Bridge Company readied the bridge for assembly, and work crews, now working around the clock, completed the bridge on 21 November.

During ATTLEBORO, elements of the 1st Engineer Battalion accompanied the combat forces and more than once served as infantry. When the 1st of the 28th Infantry uncovered the large enemy cache, a platoon from Company C descended seventy-foot ladders from Chinook helicopters and hacked out a clearing in the jungle. Helicopters shuttled in and out of the landing zone to remove the captured material. An engineer squad used satchel charges to destroy four enemy bunkers, a machine shop, and a hospital. When the 1st Division's infantry battalions became heavily committed, Colonel Kiernan assigned company-size engineer task forces in the role as infantry to protect artillery firebases.

Colonel Kiernan led the 1st Engineer Battalion, 1st Infantry Division, from 16 August 1966 to 3 June 1967 when he was killed in a helicopter accident.

By Thanksgiving, ATTLEBORO had drawn to a close, and the units returned to their combat bases. Although frequently under fire, the 1st Engineer Battalion did not suffer casualties and claimed to have killed nineteen Viet Cong. As the year ended, the battalion busied itself with preparations to support a major clearing operation in the Iron Triangle. ATTLEBORO proved the time had come to use larger forces and the added firepower now available to shift the search-and-destroy operations to the corps level.[5]

The Western Highlands

In the western highlands, General Larsen's I Field Force continued its series of border surveillance operations in Pleiku and Kontum Provinces. In mid-October, elements of the 4th Infantry Division under the command of Maj. Gen. Arthur S. Collins and the attached 3d Brigade, 25th Infantry Division, launched PAUL REVERE IV in the hope of finding the enemy and his major enemy supply base in the Plei Trap Valley, a jungle-covered river basin some twelve and a half miles wide, marked on the west by the Nam Sathay River and

[5] Rogers, *Cedar Falls–Junction City*, p. 12.

Cambodia, and on the east by the Se San River. The 2d Brigade, 1st Cavalry Division, screened the attacking force's southern flank in the Duc Co area. Heavy resistance was encountered, and in early December General Larsen airlifted the 1st Brigade, 101st Airborne Division, from Phu Yen Province into the northern Plei Trap. Assaults on well-fortified and dug-in enemy bases typified the action to the end of the year. PAUL REVERE IV disrupted a planned North Vietnamese offensive, inflicting over 1,200 enemy killed and captured, but the sharp engagements and enemy sniping caused substantial American losses. The remote and rough terrain and restrictions placed on operating near the border made the operation that much more difficult. By the second week in December, the North Vietnamese had vanished into Cambodia.[6] (*Map 16*)

On New Year's Day 1967, the 4th Division resumed the border campaign with SAM HOUSTON, but the campaign did not get going until the division crossed the Nam Sathay River in mid-February. By then, enemy activity in the area increased sharply, and several battalion-size engagements took place before the operation ended in early April. Maj. Gen. William R. Peers, who took over command of the 4th Division in early January, claimed the enemy was hurt severely, but because of the dense, rugged terrain, which limited the full use of American firepower, the area favored the enemy during the vicious close fighting. PAUL REVERE IV and SAM HOUSTON featured the building of landing zones and firebases and the opening of a supply route to the Se San, bridging the river, and extending the road west to a newly opened firebase. Because the 4th Division did not enjoy the vast airmobile resources of the 1st Cavalry Division, it relied heavily on roads for resupply. During the second phase of SAM HOUSTON, enemy mines made it increasingly difficult to keep the roads open west of Pleiku City. Although combat engineers swept the roads daily, fifty-three vehicles suffered mine damage, with over 90 percent of the incidents taking place between 14 February and 5 April. The 4th Division then moved into another frontier campaign in Operation FRANCIS MARION, which would run until October 1967, with the aim of destroying enemy forces and tracking down food and ammunition caches, rest areas, and infiltration routes.[7]

Directly supporting the attack into the Plei Trap Valley were Company B from Lt. Col. Norman G. Delbridge's 4th Engineer Battalion; Company D, 65th Engineer Battalion; and Company C, 20th Engineer Combat Battalion. Company B carried out close combat support missions, building and improving supply roads and clearing helicopter landing zones as the division's 2d Brigade advanced into the Plei Trap. In early December, the company built a pioneer road known as Route 509B from Plei Djereng to the Se San River, a

---

[6] MACV History, 1966, pp. 378–79; ORLLs, 1 Aug–31 Oct 66, I FFV, pp. 10–12, 17–22, 23–28, and 1 Nov 66–31 Jan 67, I FFV, pp. 9–18. For more on operations in the highlands, see MacGarrigle, *Taking the Offensive*, pp. 61–76, 166–77.

[7] MACV History, 1967, pp. 375–78; Interv, George L. MacGarrigle, CMH, with Lt Gen William R. Peers, CG, 4th Inf Div, 21 Oct 75, pp. 1–2, Historians files, CMH; ORLLs, 1 Nov 66–31 Jan 67, I FFV, pp. 9–18, and 1 Feb–30 Apr 67, I FFV, 30 May 67, pp. 19–20, Historians files, CMH; AAR, Opn SAM HOUSTON, 4th Inf Div, 16 May 67, with Cover Ltr, Maj Gen William R. Peers, 16 May 67, Historians files, CMH.

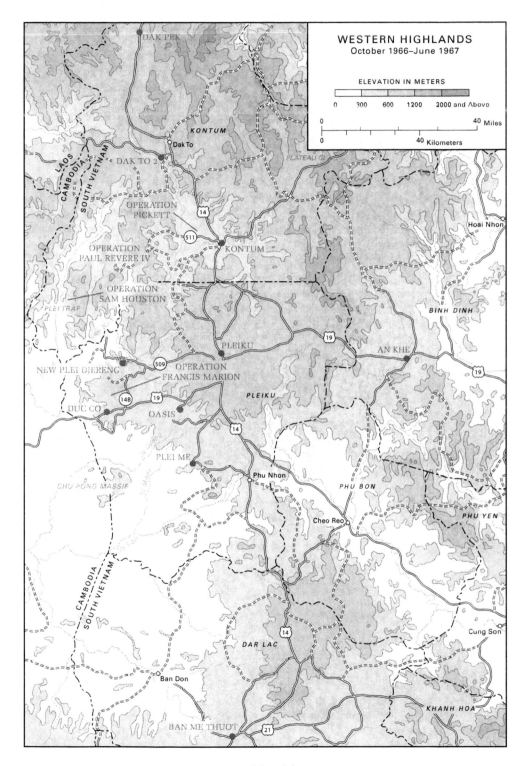

WESTERN HIGHLANDS
October 1966–June 1967

ELEVATION IN METERS

0 300 600 1200 2000 and Above

0 40 Miles

0 40 Kilometers

DAK PEK

KONTUM

Dak To

DAK TO 2

LAOS

CAMBODIA

SOUTH VIETNAM

OPERATION PICKETT

14

511

OPERATION PAUL REVERE IV

KONTUM

OPERATION SAM HOUSTON

PLEI TRAP

PLATEAU GI

Hoai Nhon

BINH DINH

19

PLEIKU

AN KHE

NEW PLEI DJERENG

509

OPERATION FRANCIS MARION

19

PLEIKU

DUC CO

14B

19

OASIS

14

PLEI ME

Phu Nhon

CHU PONG MASSIF

PHU BON

PHU YEN

Cheo Reo

CAMBODIA

SOUTH VIETNAM

DAR LAC

14

Cung Son

Ban Don

KHANH HOA

BAN ME THUOT

21

MAP 16

Aerial view the Se San River bordering the rugged Plei Trap Valley, the area where the North Vietnamese Army had a major supply base

distance of fourteen miles. When the 4th Division pushed across the Se San to the west, an armored vehicle launched bridge and an M4T6 dry span were used to span the river. From mid-November to early December, elements of Company C, 20th Engineer Battalion, cut a three-mile extension of the road through thick jungle west of the Se San. A firebase for 4th Division artillery was built and some clearing and surveying were done for a C–130 airfield. Two bulldozers equipped with Rome plow blades were used extensively for the first time in this area to level the dense vegetation. The Rome plows made the initial cut and a bulldozer followed doing the rough road construction. Demolitions were taken along to fell trees considered too big for the Rome plows, but it soon became apparent the plows could easily knock down any sized tree much faster than the time required to do the demolition work. In early January, part of Company C returned to the Plei Trap to extend Route 509B another four miles to the northwest and built several firebases. This time there were two Rome plows and two bulldozers to do the clearing. The road was improved from pioneer status into a well-shaped, adequately drained one-way road. Altogether, the engineers placed six armored vehicle launched bridges and three M4T6 dry span bridges along Route 509B. Likewise, Company D, 65th Engineer Battalion, supported the advance of the 25th Division's 3d Brigade as it advanced into the Plei Trap. Platoons were attached to each of the infantry battalions, and tasks included clearing landing zones, destroying enemy bunkers, and building defensive perimeters and helipads.[8]

Close support from the 4th Division's engineers intensified during Operation SAM HOUSTON. Engineer squads from Company A, 4th Engineer Battalion, were attached to infantry battalions on a mission basis. The rest of the company, using demolitions and chain saws, cleared and improved firebases. This was a major effort, and D4B bulldozers were airlifted into the landing zones by 1st Cavalry Division CH–54 Flying Cranes. By 17 March, the company began widespread minesweep operations along major supply routes. On 4 April, elements of Company A accompanied by an infantry company and armor platoon providing security began improving Highway 19 from Duc Co to the Cambodian border. The job involved widening the existing gravel road, which was overgrown with vegetation and damaged from craters caused by artillery rounds. Two bulldozers and one Rome plow (the battalion had four Rome plows with protective cab assemblies considered sufficient to support most combat operations) cleared brush and widened and rough graded the road. Since there was some rock on the road, limited traffic during wet weather could be supported. Little contact was made with the enemy, but his presence in the area was evident when mines and trip flares were discovered. Undetected mines, however, damaged one truck and one bulldozer. Meanwhile, Company B provided one squad to each of the 2d Brigade's three infantry battalions to help clear landing zones and firebases. At the brigade's field command post, the 1st Platoon built defensive bunkers, an interior road net, and helicopter pads. Peneprime was generously applied to reduce dust at helicopter pads, refueling points, and roads. When the brigade moved west of the Se San, it became necessary to commit one engineer platoon to each infantry battalion to clear landing zones and firebases. From 14 to 22 March, the 1st Platoon replaced an armored vehicle launched bridge with a one-way timber pile bent bridge. Similarly, the 2d Platoon extracted two of the launched bridges and a dry span and then moved to Landing Zone OASIS to work on interior roads and helicopter pads. Company D, which was usually held in general support, on 6 February dispatched a squad to accompany the 2d Battalion, 35th

---

[8] ORLLs, 1 Aug–31 Oct 66, 4th Inf Div, 22 Dec 66, p. 47, 1 Nov 66–31 Jan 67, 4th Inf Div. 20 Mar 67, p. 30, 1 Nov 66–31 Jan 67, 3d Bde, 25th Inf Div, 14 Feb 67, pp. 29–30, 1 Nov 66–31 Jan 67, 20th Engr Bn, 12 Feb 67, pp. 3–4, 9, 17–18, all in Historians files, CMH, and 1 Nov 66–31 Jan 67, 937th Engr Gp, p. 9; Operational Support, USAECV [U.S. Army Engineer Command, Vietnam] (Prov), p. 1, incl. 7, tab R, OCE Liaison Officer Trip Rpt no. 6.

Infantry, 25th Division, when it was placed under the operational control of the 1st Brigade. Over the next seven weeks, the squad cleared thirteen landing zones. On 6 March, two D4Bs were airlifted into one landing zone to help. When the bulldozers were first placed in the landing zone there was room for only one helicopter, but after a day's work there was enough cleared area to land six helicopters. The 4th Battalion's bridge company, Company E, lent its bridging expertise in erecting, repairing, and removing M4T6 dry span and float bridges. During SAM HOUSTON, the battalion did not go unscathed. Two engineers were killed and seven wounded. Equipment losses were six 5-ton dump trucks, one bulldozer, two water supply points, one tankdozer, a trailer, and a fuel pod.[9]

During FRANCIS MARION, Companies A and B of the 4th Engineer Battalion continued sweeping for mines, clearing landing zones, firebases, and helipads, carrying out road maintenance and improvements, and working on base development at Dragon Mountain. Both companies built two new east–west roads running from Highway 14B to the west providing access to firebases. Company A with the 1st Brigade swept Highway 19 east to Duc Co, north on Highway 14B to Route 509, then west on Route 509 to New Plei Djereng Special Forces camp. The minesweeps, which included infantry and armored cavalry escorts, emanated from several locations to allow rapid and simultaneous movement by the mine-clearing teams. On 12 April, Company A augmented with two Rome plows blazed a ten-mile pioneer road (Route 4A) from the junction with Highway 14B west to a firebase. The company then helped further develop the firebase by working on the helipad and defenses and supporting infantry battalions in clearing vegetation at landing zones and other firebases. Company B supported the 2d Brigade at OASIS by improving interior roads to all-weather use, built a 250-by-750-foot helicopter loading pad, and prepared protective berms in ammunition storage areas. On 9 April, the company, augmented by two Rome plows, cleared 800 acres for the resettlement of 10,000 Montagnard villagers. The Rome plows cleared an average of twenty-five acres per day in light to medium vegetation. Some delays were caused by small leaves from bamboo thickets clogging the bulldozers' radiator grills, which resulted in overheating and frequent stops to cool the engines. Using air-compressor hoses to blow the leaves out of the radiators helped but did not solve this problem. FRANCIS MARION would continue for another six months.[10]

Supporting the 1st Cavalry Division's 2d Brigade during its screening operation in the Duc Co area was Company B, of Colonel Olentine's 8th Engineer Battalion. The company moved with the brigade and set up operations at OASIS, where the cavalrymen encountered clouds of dust stirred by scores of helicopters landing and taking off every day. The dry season from

[9] ORLL, 1 Feb–30 Apr 67, 4th Inf Div, 15 Jun 67, p. 29, Historians files, CMH; AARs, Opn SAM HOUSTON, 4th Engr Bn, 23 Apr 67, pp. 1–9, 12, Historians files, CMH, and Opn SAM HOUSTON, 4th Inf Div, p. 7.

[10] ORLLs, 1 Feb–30 Apr 67, 4th Inf Div, pp. 29–30, 1 May–31 Jul 67, 4th Inf Div, pp. 37–38; Engineer Support of Combat Operations, 26 May 67, incl. 8, p. 4, OCE Liaison Officer Trip Rpt no. 7.

206

September to April was now in effect. To overcome the discomfort, flight hazards, and increased wear on engines caused by the dust, Company B answered with peneprime, generously spraying the helipads. A modification to the sticky petroleum mixture not only significantly reduced the dust but also prevented vehicle traffic from ripping up the hardened peneprime adhering to the laterite soil. Company B also developed mobile minefield kits to hasten the placement and retrieval of antipersonnel mines. The enemy's ability to move back and forth across the Cambodian border had given him a tactical advantage. As a countermeasure, U.S. troops looked more to using mines to block or canalize enemy movements, protect artillery firing positions, and strengthen ambush sites. Conventional methods, however, required large amounts of material and time, considerable drawbacks in quick-moving airmobile operations. With the mobile minefield kits—consisting of 150 small M14 antipersonnel blast mines, 6 claymore antipersonnel mines, 6 trip flares, a spool of communications wire, 18 ammunition box rods, 10 minefield markers, and 3 sledge hammers—designed for quick installation and removal, engineer squads were able to place a fifty-five-yard-wide minefield in forty-five minutes and pick it up in about the same time.[11]

Another novel weapon used by combat engineers in the western highlands were armored personnel carriers mounted with flamethrowers from the Flame Platoon, Headquarters Company, 4th Engineer Battalion. During SAM HOUSTON, three of the platoon's four flamethrower tracks and a fuel carrier were attached for nine days (24 February to 4 March) to Company A to clear ambush sites along Route 509B on both sides of the Se San. It took fifty-seven loads of napalm to burn off 55- to 110-yard strips of vegetation on each side of approximately ten miles of the road. Progress was limited to shooting only nine loads a day due to 1½- to 2-hour curing times needed for the napalm to thicken, and the engineers looked into ways for a faster incendiary mix. On 30 April, the flame platoon's four flamethrower tracks were combined with the four tankdozers from the four line companies and a tank retriever into an engineer armored task force, which initially provided security for convoys hauling sand between Dragon Mountain and Kontum City. This naturally led to guarding work crews on the roads and at bridge sites and dump trucks hauling fill, reducing the dependence on infantry and armor for security.[12]

Though the 4th Engineer Battalion did a commendable job of carrying out assignments, some equipment shortages affected operations. At the beginning of 1967, the battalion did not have any of its authorized four launchers for the rapidly erected armored vehicle launched bridges. One launcher arrived for use in SAM HOUSTON, but the shortage of three launchers seriously limited the reaction time in relocating the launched bridges. The battalion expected three new M60A1 launchers by May. During SAM HOUSTON, the shortage of 600-gallon-per-hour water purification units meant that the 1,500-gallon-per-hour unit had

[11] AARs, Opn PAUL REVERE IV, 8th Engr Bn, 26 Jan 67, p. 7, an. A, and Opn THAYER II, 8th Engr Bn, 28 Feb 67, p. 26; ORLL, 1 Nov 66–31 Jan 67, 8th Engr Bn, 31 Jan 67, pp. 4, 7–8. All in Historians files, CMH.
[12] ORLLs, 1 Feb–30 Apr 67, 4th Inf Div, p. 30, 1 May–31 Jul 67, 4th Inf Div, p. 38.

to be used on a part-time basis at the most critical location, which reduced potable water at other sites. Still, the battalion was able to supply more than 1.2 million gallons to using units during the operation. Two 600-gallon-per-hour units
eventually arrived, but they were minus their components. By the end of April,
the battalion was able to support the 1st and 2d Brigades with one water point
each with two others at Dragon Mountain. There was still a shortage of four
units. Fortunately, water points run by the 20th and 299th Engineer Battalions in
the field, at project sites, and at the division base camp helped supply the needed
water. On the plus side, the battalion received all of the replacement D7E bulldozers and all eight Clark 290M tractors with LeTourneau-Westinghouse scrapers authorized in the unit's modified table of organization and equipment.[13]

General engineering support to the 4th Division and other forces in
the western highlands fell to Colonel Braucher's 937th Engineer Group. In
November 1966, the group headquarters and elements of the 509th Engineer
Company (Panel Bridge), 585th Engineer Company (Dump Truck), and 554th
Engineer Company (Float Bridge) moved from Qui Nhon to Pleiku City. In a
switch of areas of responsibility, the 45th Group moved to Qui Nhon and took
over the eastern half of northern II Corps while the 937th Group assumed
the western half of the 45th Group's area and some of the 35th Group's area,
resulting in an increase of the 937th Group's area of responsibility by approximately 30 percent. With this realignment of territory, the 937th Group lost
two battalions (the 19th Combat and 84th Construction), gained one (the 20th
now located at the 4th Division's Dragon Mountain base camp), and retained
two (the 299th Engineer Combat Battalion also near Pleiku City, and the 70th
Engineer Combat Battalion at An Khe). Although down to three engineer
combat battalions, the group's engineer combat companies were expected to
increase by three in 1967 when the three battalions were to expand to four
letter companies under the E-series of organization and equipment. In addition, the group also had several company-size and smaller units, including
Company B, 84th Engineer Battalion, with the 70th Engineer Battalion at An
Khe; Company D, 35th Engineer Battalion, with the 20th Engineer Battalion;
and either all or parts of two light equipment companies, two panel bridge
companies, one dump truck company, and one float bridge company. The
attachment of the 102d Construction Support Company in March and the
arrival of the 815th Engineer Construction Battalion in April increased the
group's strength by 24 percent and released the combat battalions in the Pleiku
City area of most of their base camp construction duties, freeing them to concentrate on tactical operations and roadwork.[14]

As the tactical situation gradually improved and U.S. and allied forces
moved from reaction missions to carefully planned operations, the 937th
Group followed suit with more orchestrated planning to meet the requirements
of the combat forces. One of these requirements took place in the latter part of

---

[13] ORLLs, 1 Feb–30 Apr 67, 4th Inf Div, pp. 32–33, 1 May–31 Jul 67, 4th Inf Div, p. 39;
AAR, Opn SAM HOUSTON, 4th Engr Bn, pp. 9–10, 12.
[14] ORLLs, 1 Nov 66–31 Jan 67, 937th Engr Gp, pp. 2, 4–5, 1 Feb–30 Apr 67, 937th Engr
Gp, pp. 2–3.

PAUL REVERE IV in support of the 1st Brigade, 101st Airborne Division, when it deployed into the northern Plei Trap. On 3 December 1966, group headquarters tasked the 299th Engineer Battalion under Lt. Col. Walter G. Wolfe to support the airborne brigade's search-and-destroy Operation PICKETT in Kontum Province. Company A, elements of the 554th Float Bridge Company, and Company E, 4th Engineer Battalion, were charged with building a float bridge across the Krong Poko River leading to Polei Kleng Special Forces camp west of Kontum City. Company C was to provide backup support on Route 511 west of Kontum City and Highway 14 between Pleiku and Kontum City. On the fourth, the 2d Platoon, Company A, left Pleiku and arrived at Kontum City, dropping off a grader and bulldozer at the airfield to ready it for the C–130 airlift of the airborne brigade. The next day, the platoon moved out on Route 511 to begin improving crossing sites for the trucks carrying the float bridging. The six-mile stretch was little more than a trail, which crossed many small streams and swamps. Four crossings were upgraded by using 240 feet of 36-inch culvert. One of the crossings, however, washed out after the platoon made it across. When the platoon reached the Krong Poko River, it began laying out the bridge site. The rest of the Company A task force, which departed Pleiku on 5 December, installed culvert at the washed-out site to allow passage before it washed out again. Upon arrival at the Krong Poko, the task force began constructing the float bridge. But the poor near-shore landing and lack of space there to inflate the floats made it necessary to pull the operation about a mile back from the river. Inflated floats atop bridge trucks were shuttled to the bridge site where a crane lifted the floats into the water for assembly. On completion of the 420-foot bridge in the early morning of 7 December, two platoons of Company A crossed over and began working on the short distance to the Special Forces camp while the 2d Platoon continued improving the road to the east.[15]

The 299th Battalion's support of Operation PICKETT continued at a high pitch until its termination on 21 January 1967. Company A continued to improve the Kontum airfield's taxiways, ramps, bivouac areas, and roads. On 26 December, an engineer squad was committed to the airborne brigade to carry out minesweeps and hasty road repairs northwest of Kontum City. At Polei Kleng, the company applied peneprime to helicopter landing areas, improved the airstrip to handle C–123 transports, built a refueling point, and supported the 101st Airborne Division elements at the Special Forces camp. Since there was a forward supply element at Polei Kleng, the battalion S–4 section set up a water supply point, which provided approximately 2,000 gallons of potable water a day. On 6 January, T17 membrane was placed on the airstrip, and the job was finished by the eleventh. As traffic increased on Highway 14, the wooden bridges rapidly deteriorated, and it became necessary to put Company C on repairing bridge damage as well as filling potholes on bridge approaches. As the operation began to wind down, Company C dispatched a

---

[15] AAR, Opn PICKETT, 299th Engr Bn, 8 May 67, pp. 1–3, 5, incl. 3 to ORLL, 1 Feb–31 [*sic*] Apr 67, 299th Engr Bn, 8 May 67, Historians files, CMH; Operational Support, USAECV (Prov), p. 1, incl. 7, tab R, OCE Liaison Officer Trip Rpt no. 6.

platoon and a bulldozer to upgrade seven miles of road so that an artillery battery that had been airlifted to a firing position could depart overland. On 20
January, Company A returned to Polei Kleng to remove the float bridge from
the Krong Poko. The float bridge and the dry span were removed the next day,
and the engineers returned to Pleiku.[16]

Keeping supply routes in the western highlands open and building new tactical roads were among the 937th Group's major duties. Of note was Highway
19, the eighty-one-mile supply route between Qui Nhon and Pleiku City, where
heavy convoy traffic became routine. The key road, which had virtually disin-
tegrated because of heavy traffic and weather, underwent upgrades, including
new bridges, base course work, widening, and preparation for paving. Roads
supporting operations west of Pleiku City to the Cambodian border were
repaired to carry two-lane all-weather traffic. Increased emphasis was placed
on Highway 19W from Dragon Mountain to Duc Co and Highway 14B and
Route 509 between the intersection of Highway 19W to Plei Djereng, altogether forty-seven miles, the only roads leading to the western sector of the 4th
Division's area of responsibility. Route 509B extended twenty-eight miles from
Plei Djereng Special Forces camp into the Plei Trap Valley, and a new six-mile
stretch of Route 511, formerly a trail, extended to Polei Kleng Special Forces
camp. Maintaining these mostly unpaved roads became a never-ending task,
especially during the rainy season.[17]

Not all bridges were erected to stay in place. The Kon Bring Bridge is an
example of the detailed planning arrangements and last-minute adjustments
even for a supposedly simple bridge-building task. In early November 1966,
the 299th Engineer Battalion was given the task to build and remove a M4T6
dry span bridge at the hamlet of Kon Bring on Highway 14 about two miles
northwest of the town of Dak To. The existing bridge, an Eiffel structure, had
been destroyed in October, and the dry span would enable a team of the 539th
Engineer Control and Advisory Detachment to withdraw its equipment from
Dak Seang Special Forces camp, some fifteen and a half miles up the road. The
engineer team had just completed building the camp and CV–2 Caribou airstrip
and was ready to move to its next assignment. On 10 November, Colonel Wolfe
dispatched his operations officer to coordinate the move and bridge site security
with the Special Forces staff at Kontum City, where he learned that the engineer team also needed low-bed trailers to carry its equipment. Eventually three
low-beds would be needed to carry a combination front loader and backhoe, a
small bulldozer, a cement mixer, and a 2½-ton truck that could not be towed. A
grader could make it out on its own. The 2d Platoon, Company C, was tasked to
build the bridge. On 9 November, the platoon picked up forty-five feet of M4T6
dry span from the 554th Float Bridge Company, and practiced erecting the
bridge on dry land at its camp outside Pleiku City. At 0630, 12 November, the
2d Platoon departed Pleiku City with the bridge and two low-beds. On reaching

[16] AAR, Opn PICKETT, 299th Engr Bn, pp. 3–5; Operational Support, USAECV (Prov), p. 1,
incl. 7, tab R, OCE Liaison Officer Trip Rpt no. 6.

[17] ORLLs, 1 Nov 66–31 Jan 67, 937th Engr Gp, pp. 7–11, 13–14, and 1 Feb–30 Apr 67, 937th
Engr Gp, pp. 7–12.

Tan Canh, a village just south of Dak To and east of the Dak To Special Forces camp, the platoon turned over the low-beds to Company A, which had been working at the airfield. In turn, the 2d Platoon was to borrow a bulldozer and front loader from Company A and continue on to the bridge site. Company A, however, needed the bulldozer to build a bypass around a bridge incapable of supporting heavy loads. This caused a delay of two and one-half hours for the Company B platoon, and it did not reach the bridge site until 1330. The bridge was installed by 1730, and the platoon dug in with the South Vietnamese escort for the night. The next morning, the low-beds loaded with supplies for Dak Seang and a Civilian Irregular Defense Group escort left Dak To, crossed the bridge, picked up the equipment at Dak Seang, and made the return trip to Dak To by noon. Company C then removed and reloaded the bridge and joined Company A at the Dak To airfield. The next day, Company C made the return trip to Pleiku City.[18]

Between October 1966 and June 1967, I Field Force had tasked the 937th Engineer Group to build and upgrade several C–130 airstrips near the Cambodian border to all-weather capability to support combat operations. Nine airfields in the western highlands were upgraded to all-weather capability (eight to C–130 and one to C–123) to be used as jumping-off points for operations. Like the roads, these airfields routinely required maintenance and repairs. Typically, the airfield projects were company-size and spread over a wide geographic area. Taking advantage of the dry weather, the 20th Engineer Battalion built new airfields at New Plei Djereng and Duc Lap Special Forces camps and rehabilitated and extended the T17 covered fields at Landing Zone OASIS, Phu Nhon southeast of OASIS, Phu Tuc Special Forces camp farther to the southeast, and Buon Blech Special Forces camp southwest of Phu Tuc. At New Plei Djereng five miles south of the existing dirt airstrip at Plei Djereng, five work crews of Company D, 35th Engineer Battalion, surfaced the 3,500-by-60-foot runway with the new MX19 matting, four-foot square aluminum sheets encasing an aluminum honeycomb. This airstrip, built at a better site just south of the old strip, marked the first use of this new matting in Vietnam. After completing the work at New Plei Djereng in time to support SAM HOUSTON, Company D moved to OASIS, where it removed the T17 membrane and replaced it with an asphalt-treated vinylon cloth—a strong, burlap-type, synthetic fiber material—on the repaired subgrade. The runway was topped off with MX19 matting. The problem with all prefabricated metal surfacing in Vietnam was the inability to seal water out of the joints between panels. This technique offered the subgrade some protection.[19]

[18] ORLL, 1 Nov 66–31 Jan 67, 299th Engr Bn, 14 Feb 67, p. 2, and incl. 2, AAR, Kon Bring Bridge, 299th Engr Bn, 27 Nov 66; Unit History, 1966, 539th Engr Det, n.d., pp. 15–16, both in Historians files, CMH.

[19] ORLLs, 1 Nov 66–31 Jan 67, I FFV, pp. 58–60, 1 Feb–30 Apr 67, I FFV, pp. 65–67, 69, 1 Nov 66–31 Jan 67, 937th Engr Gp, pp. 8–11, and 1 Feb–30 Apr 67, 937th Engr Gp, pp. 7–8; Operational Support, USAECV (Prov), p. 1, incl. 7, tab R, OCE Liaison Officer Trip Rpt no. 6; AAR, Oasis Airfield, 20th Engr Bn, 10 Jun 67, pp. 1–7; Construction Performance Rpt [Plei Djereng], 20th Engr Bn, 5 Feb 67, pp. 1–7; Completion Rpt, Project 20–6 [Phu Tuc], 20th Engr Bn, 22 Mar 67, incls. 1–7. All 20th Battalion reports in box 4, 72A/2315, RG 319, NARA.

*A soldier of the 20th Engineer Battalion places MX19 matting at New Plei
Djereng Special Forces camp.*

The T17 airfield project at Duc Lap, code-named Operation DUCHESS,
typified some of the challenges—long distances, security, and logistics—
facing the engineers in the highlands. Upon completing work at Buon
Blech airstrip, Company C, 20th Engineer Battalion, was ordered to move
to Duc Lap Special Forces camp some 150 miles south of Pleiku City to
build the new C–130 airstrip. A convoy carrying construction materials
from the battalion at Dragon Mountain would make the fifty-three-mile
trip down Highway 14, rendezvous with Company C at Buon Blech Special
Forces camp, and proceed with Company C to Duc Lap. Meanwhile, a
reconnaissance along the convoy route found the three-span Eiffel bridge
over the Ea Krong River south of Ban Me Thuot could not support the
convoy's heavy loads. The bridge could not be bypassed, and reinforcing
it was not considered feasible. Battalion headquarters determined the best
way to cross the river in insecure territory was by raft. On 27 January, sections of a M4T6 raft were assembled and airlifted by six C–130s to Ban Me
Thuot for pickup and installation by the South Vietnamese 23d Division's
engineers. The four-float raft was quickly assembled at the bridge site
and secured. On 1 February, ninety-five trucks loaded with construction
materials and a security force consisting of a South Vietnamese Special
Forces company and their advisers from Buon Blech set out from Pleiku.
The convoy reached Buon Blech in less than eight hours without difficulty.

An M4T6 raft carries a 20th Engineer Battalion vehicle across the Ea Krong River on the way to build a new C–130 airstrip at Duc Lap Special Forces camp.

Company C and the convoy carrying construction materials—altogether 110 vehicles—proceeded to Ban Me Thuot and remained overnight. The next morning, the engineer column left Ban Me Thuot, crossed over the Ea Krong River by raft, and reached Duc Lap in the early afternoon, completing the trip without a single breakdown or major incident. Clearing and grubbing of the area began on 3 February, and the following morning all equipment and troops were hard at work. Twelve days later, six C–7A Caribous (on 1 January 1967 all Army Caribous were transferred to the Air Force and their designation changed from CV–2 to C–7A in the process) landed on the partly completed airstrip. A C–123 airstrip was completed in eighteen days, and a hasty C–130 dirt airstrip was ready in twenty days. Special emphasis was placed on proper drainage, including the airstrip's crown, ditches, and culverts. As the airstrip work neared completion, construction began on the parking apron and its two 40-by-80-foot access ramps and a two-mile access road to Highway 14. A six-inch layer of high-quality laterite was spread and shaped, wetted down with a water distributor, and compacted with a steel-wheeled roller. Laying of T17 membrane on the airstrip, parking ramp, and access ramps began on 15 February and lasted for twenty-five days. This work was done at night to avoid excessive wrinkles caused by heat expansion during the day. As workers placed the membrane, another work party followed applying a nonskid compound.

Surveying and site preparation at Duc Lap

By 7 April, the C–130 airstrip was complete, and Company C was ready to return home. On the first leg of the return, a mine caused some damage to a tractor-trailer, and the brakes failed on a grader causing it to overturn down an embankment and seriously injuring two men. On the way back, the raft at the Ea Krong was retrieved. The following afternoon, 8 April, the convoy reached Dragon Mountain without any further incidents, bringing Operation DUCHESS to an end.[20]

The 299th Engineer Battalion was also assigned several airfield projects of note. Besides the work at Kontum City and Polei Kleng, the Dak To airfield project was a major undertaking involving considerable planning, roadwork, and logistical support. The undertaking was generated by I Field Force's requirement to support a brigade-size force out of the Dak To forward supply area, on this occasion the 1st Brigade, 101st Airborne Division, during its foray into the northern Plei Trap. The existing 4,200-foot bituminous surface treatment airfield had sustained severe damage during Operation HAWTHORNE several months earlier from the combined effects of heavy use and seasonal rains. To get to Dak To, however, required improving Highway 14 from Kontum City to Dak To. Company A was given the job of improving the airfield and the initial opening of the road to Class 31 traffic and upgrading four of the northernmost water crossings while Company B was tasked to

---

[20] For detailed accounts of Duc Lap, see Maj. Darryle L. Kouns, "Combat Engineers in Operation DUCHESS," *Military Engineer* 62 (May-June 1967): 173–76; and Completion Rpt, Project 01–937/OS–66 [Duc Lap], 20th Engr Bn, 21 Apr 67, box 4, 72A/2315, RG 319, NARA.

Work crews apply a nonskid compound to Duc Lap airstrip.

improve the three southernmost crossings from Kontum City. Repairs to the Dak To airfield began on 30 October. As much as possible, work was done at night and during periods of slack air traffic, thereby keeping the airfield open throughout the repair period. Numerous soft spots and rutted areas were refilled with crushed rock, compacted with a steel-wheeled roller, and seal-coated. The patches and seal coat, despite some bleeding of asphalt, held out well, and C–130 aircraft regularly used the airfield with no detrimental effects. On 20 November, its work done, Company A departed Dak To. After all the heavy equipment passed over the dry span and M4T6 float bridge, work crews dismantled the bridges and loaded them on bridge trucks for the trip home.[21]

The Dak To project illustrated that administration, logistics, and communications were needed for engineer projects at remote locations. Some 210 cubic yards of 3/4- and 3/8-inch aggregate were needed for the airfield patching, but neither the aggregate nor the transport was available to move that amount to Dak To. The 937th Group came to the rescue arranging for trucks from other units and enough aggregate from Qui Nhon, some 100 cubic yards, to get started. Another 40 cubic yards came later, and the 299th produced the remaining 70 cubic yards of 3/4-inch rock, which was carried in battalion dump trucks. Asphalt required for patch work and seal coat was moved by transportation units and the battalion's vehicles.

[21] ORLL, 1 Nov 66–31 Jan 67, 299th Engr Bn, p. 2, and incl. 3, AAR, Dak To Airfield and Route 14B (Kontum to Dak To Airfield), 299th Engr Bn, 18 Dec 66, pp. 1–6, 9; ORLL, 1 Nov 66–31 Jan 67, 937th Engr Gp, pp. 7–8.

One of the most critical resupply problems was getting repair parts and third-echelon mechanics. The group again helped by using its helicopters, but still this was not good enough to prevent lost time because of deadlined equipment. The repair problem was alleviated by the tenth day when an ordnance contact team and truck arrived at Dak To. As for food and potable water, the battalion attached a water supply team for the duration of the operation. The company also carried a fifteen-day supply of B-rations with fresh A-rations occasionally flown in by group helicopters or more frequently carried in resupply convoys.[22]

As a result of the engineers' efforts, the new and improved roads and airfields made possible sustained operations in the highlands and on the border. These operations gave rise to some cautious optimism. For although the enemy controlled the tempo of fighting near the border, making it a costly undertaking to send U.S. troops there, those troops would nonetheless return whenever sizeable enemy forces sallied forth over the border from Cambodia.[23]

On the Coast

In Binh Dinh and Phu Yen Provinces, U.S., South Korean, and South Vietnamese forces continued their attempts to weaken the enemy's hold on the agriculturally rich coastal region. The 1st Cavalry Division's return to Binh Dinh Province in September 1966 marked the beginning of a series of battles that kept the division in action for seventeen months. Operation THAYER I began with a five-battalion air assault into the mountains of the Kim Son Valley, popularly known as the Eagle's Claw or Crow's Foot because of the shape of the tributary valleys. The cavalrymen did not meet much resistance, but they did uncover significant supply caches. Operation IRVING unfolded on 2 October when elements of the 1st Cavalry Division moved east from the valleys and maneuvered into position around Landing Zone HAMMOND, the division's main forward supply base at the entrance to the Suoi Ca Valley. There they joined South Korean and South Vietnamese troops in an effort to trap the enemy in a coastal pocket. In October 1966 and January 1967, THAYER II and PERSHING extended the preceding operations as part of an all-out effort to pacify eastern Binh Dinh Province. The size of the forces in the operations ranged from a brigade with two battalions to all three brigades. Besides inflicting casualties on the enemy, the operations greatly reduced his dominance in the critical province.[24] (*Map 17*)

---

[22] AAR, Dak To Airfield and Route 14B (Kontum to Dak To Airfield), 299th Engr Bn, pp. 6–8.

[23] MacGarrigle, *Taking the Offensive*, pp. 76, 177; ORLLs, 1 Nov 66–31 Jan 67, 937th Engr Gp, pp. 7–11, 13–14, and 1 Feb–30 Apr 67, 937th Engr Gp, pp. 7–12.

[24] ORLLs, 1 Aug–31 Oct 66, I FFV, pp. 21–27, and 1 Nov 66–31 Jan 67, I FFV, p. 15; Tolson, *Airmobility*, pp. 117–18, 124; Rpt, 14th Mil Hist Det, 1st Cav Div, 7 Mar 67, sub: Seven Month History and Briefing Data (April–October 1966), pp. 111–32; *Memoirs of the 1st Air Cavalry Division*, p. 33–34. See also Carland, *Stemming the Tide*, pp. 256–74, and MacGarrigle, *Taking the Offensive*, pp. 85–89, 160–84.

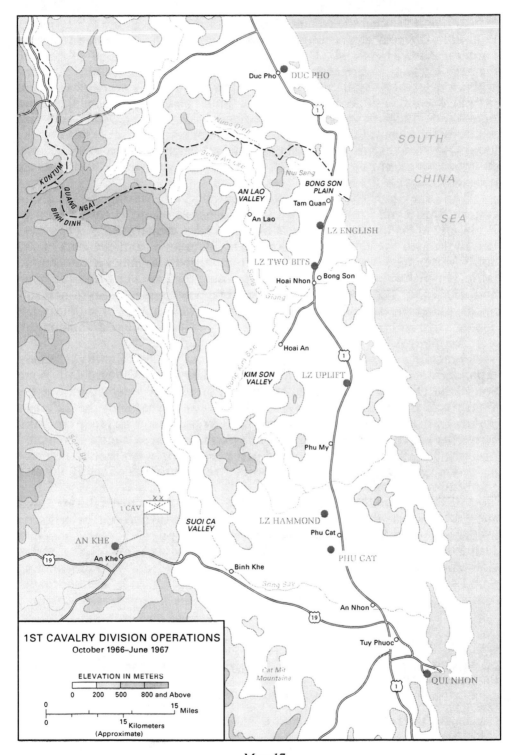

1ST CAVALRY DIVISION OPERATIONS
October 1966–June 1967

ELEVATION IN METERS

0   200   500   800 and Above

0   15 Miles

0   15 Kilometers
(Approximate)

MAP 17

Colonel Olentine, the commanding officer of the division's 8th Engineer Battalion, directed engineering support from a forward command post at Landing Zone HAMMOND. Companies A, B, and C continued supporting the 1st, 2d, and 3d Brigades, respectively. In turn, the companies attached a platoon to each assaulting battalion. Typically, the engineers cleared and expanded landing zones and artillery positions and destroyed bunkers and tunnel systems. When the division found several caves in the coastal mountains in use by the enemy, the demolitions men used shaped charges instead of the usual C4 plastic explosives. The shaped charges, normally used to make large craters in roads, proved highly effective in breaking up the rock, thus making it easier to destroy the caverns.[25]

While the 1st Cavalry Division concentrated on clearing the enemy in eastern Binh Dinh Province, 937th Group engineers kept Highway 1 open. Beginning in September with Operation THAYER, Company B, 19th Engineer Battalion, with the help of a platoon from Company C and elements of the South Vietnamese Army 6th Engineer Group, coped with mines and torrential rains. The highway and railroad bridges crossing the Lai Giang River at Bong Son still stood. The 1,600-foot Eiffel truss highway bridge, however, had suffered damage to several spans, reducing loads to twelve tons. Miraculously, the unused railroad bridge a few hundred yards to the east remained undamaged, and Company A laid a plank decking for vehicular use.[26]

Coinciding with the beginning of THAYER, the 1st Cavalry Division expanded operations north of the Lai Giang to the An Lao Valley and the Bong Son Plain. Division plans during the autumn of 1966 included upgrading the airstrip at Landing Zone ENGLISH to C–130 traffic. Because of the difficulty in moving heavy earthmoving and construction equipment up Highway 1 and across the Lai Giang, the 937th Engineer Group decided to transport men and equipment by sea to a beach site used earlier by the 1st Cavalry Division east of the village of Tam Quan. The engineers then were to proceed to Highway 1 and head south to the airstrip.[27]

The 937th Group's plan to rebuild the access route to Highway 1 and airfield work was called Operation DUKE and was carried out by elements of the 19th and 84th Engineer Battalions in two phases. Colonel Rhodes' 19th Engineer Battalion kicked off the Phase I roadwork. On 13 September, Company A and the 509th Panel Bridge Company loaded aboard the Landing Ship Dock *Gunston Hall* at Qui Nhon and headed north. After some difficulty coming ashore in a high surf the following day, the task force rendezvoused

---

[25] AARs, Opn THAYER, 8th Engr Bn, 29 Oct 66, pp. 1–3, 7–8, Opn IRVING, 8th Engr Bn, 7 Nov 66, pp. 1–3, 9, and Opn THAYER II, 8th Engr Bn, 28 Feb 67, pp. 1–2, 4–5; ORLLs, 1 Aug–31 Oct 66, 8th Engr Bn, pp. 2–6, and 1 Nov 66–31 Jan 67, 8th Engr Bn, 31 Jan 67, pp. 1–4. All in Historians files, CMH.

[26] AARs, Opn THAYER, 8th Engr Bn, pp. 1–2, Opn IRVING, 8th Engr Bn, pp. 1–2, and Opn THAYER II, 8th Engr Bn, pp. 2, 5; ORLL, 1 Nov 66–31 Jan 67, 19th Engr Bn, pp. 7–9; Galloway, "Essayons," p. 113; Lt. Col. Nolan C. Rhodes, "Operation DUKE," *Military Engineer* 61 (September-October 1969): 332–33.

[27] Unless otherwise noted, the narrative covering Operation DUKE is based on Rhodes, "Operation DUKE," pp. 330–33, and Galloway, "Essayons," pp. 111, 113.

with South Vietnamese forces and set up a defensive perimeter. Work crews quickly repaired a causeway that had been partly washed out; built a 250-foot Bailey bridge across the tidal inlet (an earlier bridge had been removed); replaced a damaged seventy-foot Bailey bridge leading to Highway 1; and proceeded inland to Highway 1 and turned south, clearing mines and making road repairs along the way. Company B, 84th Engineer Battalion, which had landed on the sixteenth to do the airfield work, followed. When Company B reached the airstrip at ENGLISH the following day, the 19th Engineer Battalion task force proceeded down Highway 1 to Qui Nhon, completing its portion of Phase I.

In early October, the 937th Engineer Group began Phase II of Operation DUKE: the return of Company B, 84th Engineer Battalion, from ENGLISH to the beach site for the return trip to Qui Nhon. By then the company had extended the runway at ENGLISH to over 3,300 feet and built an aircraft parking area. On 4 October, elements of the 19th Battalion started back up Highway 1 from Qui Nhon to secure the beach site and remove the two Bailey bridges from the beach access road. Included in the convoy were two hundred tons of steel and lumber for use on the railroad bridge over the Lai Giang at Bong Son. After unloading the materials and leaving two float bridge platoons to help work on the railroad bridge, the task force crossed the highway bridge without difficulty. South Vietnamese security forces escorted the convoy past ENGLISH to the Tam Quan beach site where the engineers set up a defensive perimeter that evening. When Company B, 84th Engineer Battalion, joined the task force on the beach, work crews removed the two Bailey bridges and replaced them with a steel stringer bridge and a suspension footbridge. An enemy probe on the night of 9 October and heavy mortar fire the next evening, which wounded eighteen men, coaxed the engineers to finish the job and get off the beach. When the LST that had delivered the stringers and taken aboard the bulldozers, graders, dump trucks, and other equipment could not free itself from the beach, the Navy brought in smaller LCUs. As a result, it took several more days to return the troops and equipment to Qui Nhon.

The airfield at Landing Zone ENGLISH got more attention the following month when the 19th Engineer Battalion returned on 19 November to upgrade it to an all-weather capability. Company C took six days to place M8A1 matting on the runway, but heavy rains forced the engineers to switch from T17 membrane to the metal matting on the taxiway and parking area. To keep on schedule, Colonel Rhodes sent two platoons from Company B to assist. Still, the heavy rainfall (over thirty inches fell in November and December) and a lack of matting caused delays. The two companies completed the surfacing, including a peneprime treatment of the runway's shoulders, on 20 December. By this time, Landing Zone ENGLISH had been transformed into a large logistical base and a key airfield to move 1st Cavalry Division troops long distances. The combination of heavy use and rains at ENGLISH made maintenance an endless task for the engineers of several units.[28]

---

[28] ORLL, 1 Nov 66–31 Jan 67, 19th Engr Bn, pp. 7–8, 20–22.

Meanwhile, efforts were under way with South Korean forces south of Qui Nhon to reopen gaps along Highway 1 in Phu Yen Province. In December and January, elements of the 19th Engineer Battalion, now under the command of Colonel Bush's 45th Engineer Group, pushed south with roadwork while the 39th Engineer Battalion at Tuy Hoa pressed northward. As the South Korean Capital Division swept the area southward, Company A, 19th Engineer Battalion, reopened the road south of Qui Nhon from the Cu Mong Pass to Song Cau, filling sixty-five cuts in the road and building three panel bridges. As the South Korean 9th Division pushed northward from Tuy Hoa, Colonel Fulton pressed his 39th Engineer Battalion to work on the road and bridges. The reopening of the road for the first time in years was cause for some symbolic gesture. To emphasize the importance of this occasion, the 1st Brigade, 101st Airborne Division, dispatched a thirty-four-vehicle convoy (dubbed Operation ROADRUNNER) from Kontum City by way of Highway 19 to Qui Nhon, then down Highway 1 to the brigade's home base at Phan Rang. To help speed up the crossing over the Song Cau at Ha Yen, the 39th Engineer Battalion dispatched two platoons from Company A and part of the 553d Float Bridge Company to assemble a light tactical raft to supplement the South Vietnamese M4T6 raft at the ferry site. The restored road's original asphalt surface was in fair condition and held up under local traffic. The 45th Engineer Group remained responsible for keeping the north-south highway open and assigned maintenance to the battalions in the area.[29]

By mid-1967, operations launched by allied forces in the coastal lowlands took away the previous dominance enjoyed by the Viet Cong and the North Vietnamese. The enemy was forced out of populated areas, and his large-scale operations became extremely risky. In Binh Dinh Province, the 1st Cavalry Division inflicted heavy losses on the enemy, but he still contested the province. The engineers contributed by reopening roads and making airfield improvements. Coastal Highway 1, albeit in poor condition and dangerous in many places, was passable during daylight traffic almost its entire length. Newly opened stretches included thirty-one miles from Landing Zone ENGLISH north to Duc Pho and another thirty-one miles from the Cu Mong Pass to Tuy Hoa. Interior roads were declared safe up to twelve and a half miles into the mountains. Meanwhile, completion of jet-capable runways at Phu Cat by RMK-BRJ and Tuy Hoa by Walter Kidde and the addition of second permanent runways at Cam Ranh Bay and Phan Rang by RMK-BRJ allowed for the stationing of more Air Force tactical fighter-bombers to provide support to the ground forces in II Corps. Work also progressed on the new permanent concrete runway at An Khe.[30]

[29] ORLLs, 1 Nov 66–31 Jan 67, 19th Engr Bn, p. 9, and 1 Nov 66–31 Jan 67, 39th Engr Bn, p. 3.

[30] MacGarrigle, *Taking the Offensive*, pp. 83, 92, 322–24; ORLL, 1 Feb–30 Apr 67, I FFV, pp. 51, 66–69.

The Delta

By the spring of 1967, the Mobile Riverine Force, the joint U.S. Army–Navy afloat element, was in place at Dong Tam and was beginning a long series of operations under the code name CORONADO. The 9th Division's 2d Brigade provided the ground forces from three infantry battalions and other attached units. The Navy provided some one hundred boats and crews. From two barracks ships, the troops went on operations in armored troop carriers, preceded by minesweeping craft and escorted by armored boats called monitors. When the flotilla reached the objective, the men debarked under the protective fire of weapons mounted on the river craft, including 105-mm. howitzers on Ammi pontoon barges. When operations began, the Viet Cong countered with ambushes from the shore and in battalion strength. As time passed, the Mobile Riverine Force pressed forward into enemy-held areas, with results that lent encouragement that riverine operations might succeed.[31]

Army engineers organized for operations in this environment. The 1,015-man 15th Engineer Battalion, the 9th Division's organic engineer battalion, under Lt. Col. William E. Read continued to divide its efforts among the division's three brigades. Battalion headquarters, Company A, and Company E, the bridge company, were located near the division's and its 1st Brigade base camp at Bearcat east of Saigon. Company C was located at the 3d Brigade's base camp at Tan An southwest of Saigon, and Companies B and D supported the 2d Brigade at Dong Tam also southwest of Saigon. The 15th Engineer Battalion used a broad array of equipment, including Rome plows, tankdozers, and bridging to support the division's widely scattered operations in III and IV Corps. Meanwhile, general support for the 9th Division's operations in the delta fell to the recently arrived 34th Engineer Group.[32]

Riverine operations by their nature reduced the requirements for combat engineer support and eliminated many demolitions and minesweeping tasks. Consequently, combat engineers attached to the riverine force frequently filled out infantry units that were short of riflemen. The usual allocation of engineers from the divisional engineer battalion was one engineer platoon to support each infantry battalion in the field. For example, Company D at Dong Tam typically supported the same two infantry battalions of the 2d Brigade on operations, providing a platoon to each battalion, while the remaining platoon and equipment worked at the Dong Tam camp. Lt. Col. Thomas C. Loper, who assumed command of the 15th Engineer Battalion in September 1967, reported that habitually fragmenting the two platoons into two- or three-man demolition teams to support infantry companies of the two battalions of the 2d Brigade in the field did not seem economical. Consequently, the battalion began field tests by setting aside one of the platoons at a central location aboard a barracks ship to respond to calls while the other platoon dispatched

---

[31] Westmoreland, *A Soldier Reports*, p. 209; MacGarrigle, *Taking the Offensive*, pp. 402, 411–30; Fulton, *Riverine Operations*, pp. 68, 73–74, 84–85.

[32] ORLL, 1 Nov 67–31 Jan 68, 15th Engr Bn, 31 Jan 68, pp. 1, 4, Historians files, CMH; Galloway, "Essayons," pp. 244–45.

Men of the 15th Engineer Battalion use an airboat in the delta.

demolition teams to the infantry companies. Loper also put two of his M113 armored personnel carriers mounting flamethrowers aboard armored troop carriers for use against enemy positions along the waterways. The remaining two M113s with flamethrowers were used to clear heavily booby trapped areas. Until then, the battalion's flamethrower platoon had been used for job site security and supporting the division's three brigades and the Thai regiment.[33]

Among the 15th Engineer Battalion's unusual elements were the 1st and 2d Airboat Platoons, activated and attached to the battalion in December 1967. These unique Army units were used to support the 3d Brigade in Long An Province and Thai forces in the Nhon Trach District of Bien Hoa Province. This coincided with the Navy dispatching three Patrol Air-Cushion Vehicles, or PACVs, to the Mobile Riverine Force. PACVs, also known as hovercraft that could skim right over river banks and dikes, had been demonstrated earlier in the Plain of Reeds, southwest of Saigon. The Army's smaller and considerably less expensive airboats (each hovercraft cost close to $1 million) with mounted machine guns had been successfully used by the Special Forces. Because the 15th Engineer Battalion had powered bridge erection boats, reconnaissance and pneumatic assault boats, responsibility for river reconnaissance, and some maintenance capability, assigning the twenty-eight airboats to the battalion was a logical choice. A short training program was conducted at Long Binh and concluded in a joint operation with the 720th Military Police Battalion.

---

[33] ORLLs, 1 Feb–30 Apr 67, 34th Engr Gp, pp. 5–9, and 1 May–31 Jul 67, 34th Engr Gp, pp. 1–3, 5–11, 1 Nov 67–31 Jan 68, 15th Engr Bn, pp. 4–6, 9–10; Fulton, *Riverine Operations*, pp. 139, 141; Galloway, "Essayons," pp. 244–45.

Within a few days, the airboats began supporting tactical operations. On 16 December, the 1st Airboat Platoon departed Long Binh with sixteen boats and infantry machine gunners to support 3d Brigade operations in Long An Province. From the eighteenth through the twenty-ninth, the boats carried out screening and reconnaissance missions, moved troops into battle, and operated under day and nighttime conditions. On 5 January, five airboats began working with the Thai regiment. The boats were used almost daily, and by the eighteenth, eleven boats were working with the Thais. Some success was reported by the battalion with reconnaissance and screening missions, but there was less success in transporting troops. Although the airboats were rated to carry seven men, their speed and maneuverability decreased when exceeding the optimum load of three (operator, gunner, and one passenger). The dry season precluded the entry of airboats into the rice paddies, the primary intended use, thus confining them to rivers and streams. Their loud noise gave the enemy enough time to set up ambushes along the channelized waterways. Maintenance and supply problems included a lack of special tools and spare parts. Supplying the specified aviation fuel to remote areas was another problem for a while. Refueling took a long time because the size of the fuel intake hole was too small. A special nozzle was made, but it still took up to three hours to refuel one platoon. The battalion also reported the need for better communications, since only five radios were authorized for each platoon. The battalion felt that each boat needed a radio, while the command and control boat needed two, one for communicating within the platoon and the other for communicating with the supported unit and requesting artillery fire.[34]

III Corps Again

As the threat along the border in III Corps abated following Operation Attleboro, General Westmoreland in early 1967 launched the first of several large clearing operations north of Saigon. Operation Cedar Falls began on 8 January 1967 with the objective of destroying the Viet Cong *Military Region 4* headquarters harbored in the Iron Triangle. Under the direction of General Seaman's II Field Force, the corps-size sweep involved the 1st and 25th Divisions, the 173d Airborne Brigade, the 196th Infantry Brigade, and the 11th Armored Cavalry Regiment. After maneuvering units into position, Seaman launched airmobile assaults to seal the enemy bastion, exploiting the natural barriers of the Saigon and Thi Tinh Rivers that formed two of its boundaries. A series of sweeps followed to push the enemy toward the blocking forces. During the nineteen-day campaign, the enemy preferred to hide or infiltrate through allied lines than fight. Still, some seven hundred Viet Cong were killed, and approximately the same number were taken prisoner. When Cedar Falls ended on 26 January, the combined American and South Vietnamese

---

[34] ORLL, 1 Nov 67–31 Jan 68, 15th Engr Bn, pp. 12–14, 20–22; Berry, *Gadget Warfare*, pp. 35–38; Edward J. Marolda, *By Sea, Air, and Land: An Illustrated History of the U.S. Navy and the War in Southeast Asia* (Washington, D.C.: Naval Historical Center, Department of the Navy, 1994), pp. 167–68.

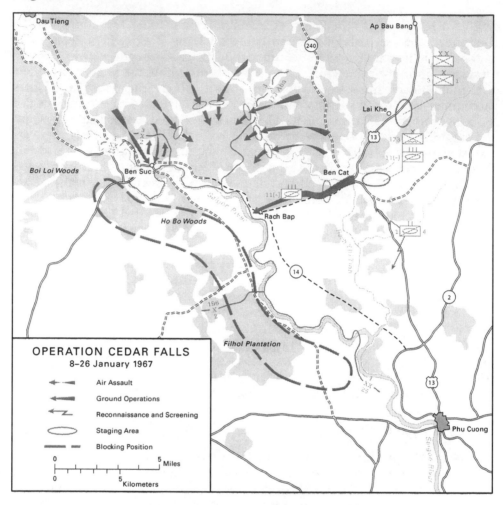

OPERATION CEDAR FALLS
8–26 January 1967

- ◄━━ Air Assault
- ◄━━ Ground Operations
- ◄⌁ Reconnaissance and Screening
- ⬭ Staging Area
- ▬▬ ━ ━ Blocking Position

0 — 5 Miles
0 — 5 Kilometers

MAP 18

force had found and destroyed major installations, including over 1,100 bunkers, 400 tunnels, and 500 other structures, including fortifications. They had also captured large quantities of food and supplies, confiscated many enemy documents, and cleared land for future operations.[35] (*Map 18*)

Before and during CEDAR FALLS, the Engineer Command and the 1st Division's 1st Engineer Battalion formed a task force to support the combat forces gathering around the Iron Triangle. Colonel Kiernan, the division engineer and commander of the 1st Engineer Battalion, coordinated these efforts. Units in the task force included elements of his battalion; flame-throwing platoons from the division's 1st Squadron, 4th Cavalry; and tunnel rats from the

---

[35] Westmoreland, *A Soldier Reports*, p. 249; Sharp and Westmoreland, *Report*, pp. 133, 137, 152; ORLL, 1 Nov 66–31 Jan 67, II FFV, pp. 7–8; MacGarrigle, *Taking the Offensive*, pp. 96–112.

242d Chemical Detachment. Most of the rest of the engineers came from the 79th Engineer Group (elements of the 27th, 86th, 168th, and 588th Combat Engineer Battalions, and the 557th Light Equipment Company) and the 159th Engineer Group (elements of the 169th Engineer Construction Battalion). In addition, the 1st Logistical Command's Saigon Support Command provided a detachment from the 188th Maintenance Battalion to render backup maintenance support to the bulldozer fleet being formed for large jungle-clearing operations. The bulldozers were consolidated under the headquarters of one engineer battalion, Colonel Pelosky's 168th. Pelosky organized four teams, each consisting of six to twelve bulldozers. The engineer task force also set up a composite maintenance team that could fabricate parts.[36]

In a deception operation called NIAGARA FALLS (5 to 7 January), the 1st Division moved into attack positions. The 600-man engineer task force helped lead the way. The task force, supported by fifty-nine bulldozers, four Rome plows, and tunnel demolition teams from the 79th and 159th Engineer Groups in a short shakedown operation, cleared one-half square miles of jungle bordering the Thi Tinh River southeast of the Iron Triangle. On 5 January, a 1st Engineer Battalion bridge platoon with a pile-driving crane from the 169th Engineer Construction Battalion completed a 160-foot panel bridge across the Thi Tinh River west of Ben Cat. This route provided a rapid crossing for the 11th Armored Cavalry Regiment. Shortly after the start of CEDAR FALLS, a span of the new bridge collapsed under the combined load of a recovery vehicle towing a disabled M48 tank. Work crews from the 1st Engineer Battalion worked throughout the night removing the damaged section and installing an armored vehicle launched bridge to cross the gap. An engineer task force followed the 11th Armored Cavalry, clearing jungle and building landing zones, roads, and support areas. While the majority of engineers provided the customary support for a combat operation, they also took the opportunity to make CEDAR FALLS one of the most unique operations in the war up to that point. For the first time, engineers deployed bulldozers and Rome plows on a massive scale to open large jungle areas during a combat operation. Until then, the Rome plows had worked in safe or partially secure areas. Now they became part of an assault team or dozer-infantry team to destroy enemy fortifications. The dozer part of the team included standard D7E bulldozers, tankdozers,

---

[36] Unless otherwise noted, engineer roles in CEDAR FALLS and NIAGARA FALLS are based on Rogers, *Cedar Falls–Junction City*, pp. 29, 31–41, 60–79; Ploger, *Army Engineers*, pp. 139–40; *Always First, 1965–1967*, "The Bulldozer in the Attack"; Combat Lessons Bulletin, HQ USARV, no. 5, 8 Feb 67, sub: Engineer Operations in the Iron Triangle; Lt. Col. Joseph M. Kiernan, "Combat Engineers in the Iron Triangle," *Army* 17 (June 1967): 42–45; AAR, Opn CEDAR FALLS, 1st Engr Bn, 2 Mar 67, Historians files, CMH; Ltr, Kiernan to All Men 1st Engr Bn and Dozer and Demolition Teams, 79th Gp, 29 Jan 67, sub: Engineer Actions on Operations NIAGARA FALLS and CEDAR FALLS, both in OCE Hist Ofc; Interv, Capt George E. Creighton Jr., 17th Mil Hist Det, with Lt Col Joseph M. Kiernan, CO, 1st Engr Bn, 20 Apr 67, Historians files, CMH; Self Interv, Maj John D. Simpson, S–3, 168th Engr Bn, 8 Jul 67, pp. 1–77, VNIT 48, CMH; ORLL, 1 Nov 66–31 Jan 67, 168th Engr Bn, 14 Feb 67, pp. 4, 8, 10–12, Historians files, CMH; Rpt of Visit to Various Headquarters in Vietnam, incl. 7, tab R, p. 2, OCE Liaison Officer Trip Rpt no. 6.

and Rome plows. The 1st Engineer Battalion also created an instant navy with its fleet of bridge erection boats to patrol the rivers.

Dozer-infantry teams moved into the jungle together, simultaneously clearing vegetation, searching for the enemy, and destroying his fortifications. One technique involved placing two tankdozers at the point to clear vegetation and detonating booby traps, followed by four bulldozers abreast, some equipped with Rome plow blades, with two more bulldozers to clean up the windrows of cut vegetation. Infantry worked alongside the dozers, at the same time carrying out search-and-destroy missions. A second technique employed a Rome plow in the lead followed by troops in armored vehicles. When the plow uncovered an enemy position, the vehicles provided fire support for the attacking infantry. Experience during NIAGARA FALLS revealed the continuous need for on-the-spot servicing and small repairs by the bulldozer operators and maintenance teams.

The duration of CEDAR FALLS did not allow for jungle clearing in the entire Iron Triangle. Instead, as Kiernan recommended, the engineers cleared only strategic areas, 7 or 8 percent of the sixty-three square miles of jungle. By the end of clearing operations on 22 January, the 1st Division engineer task force had cleared a total of 2,711 acres or 4 square miles. This included 55- to 110-yard strips of jungle on each side of major roads and thirty-four landing zones spaced throughout the bastion for helicopter resupply and medical evacuation. Engineers lifted into the jungle by helicopters cleared three of the landing zones, and the dozer-infantry teams cleared the remainder. At the northwest corner of the triangle, the village of Ben Suc, a suspected enemy fortified supply and political center, was sealed off and villagers evacuated. Bulldozers, tankdozers, and demolition teams then leveled buildings, collapsed tunnels, and obliterated bunkers and underground storage rooms. The teams also cut numerous swaths in the Iron Triangle, usually in widths of 33 to 55 yards, to permit rapid movement of mechanized and airmobile units in future operations. The effect of this clearing lasted for a time, but vegetation always returned.

While the engineers cleared the jungle above ground, the tunnel rats and demolition teams explored and destroyed Viet Cong complexes found in villages and logistical base areas. The 1st and 168th Engineer Battalions discovered an effective method to destroy deeper tunnels by using conventional demolitions with acetylene gas. Demolition satchel charges, which were thirty-pound charges of TNT, and forty-pound cratering charges placed at critical locations (rooms, tunnel junctions, exits, and entrances) or spaced at about 27-yard intervals, provided a powerful boost for the acetylene. The combined effect helped to collapse tunnels as deep as 15 to 20 feet below the surface.

Basically, however, the 1st Engineer Battalion's tunnel rats had the main job, going where a flashlight guided the way and a pistol was the primary weapon. "Charlie is in there. All we have to do is dig him out," explained Pfc. Michael R. Tingley, who served in a team of seven tunnel rats working in conjunction with the division's chemical detachment. Other team members carried gas masks, nauseating gas, grenades, smoke grenades, a telephone, and fifty-foot lengths of reinforced rope. One team member, Pfc. Roger L. Cornett, noted, "Handling

Detonating cord used to initiate an explosive charge is lowered into a tunnel by troops of the 173d Engineer Company, 173d Airborne Brigade, during Operation CEDAR FALLS.

explosives is real touchy at times. You can't really worry about it but you often wonder." Another tunnel rat with six months' experience with explosives, Pfc. Stephen E. Sikorski, added, "I'm most concerned about booby traps," and regarded his work as a "specialty that can either build or destroy."[37]

[37] "Engineers Turn Tunnel Rats," *Pacific Stars and Stripes*, 11 June 1967, extracted from *Always First: 1st Engineer Battalion, 1967-1968* (1st Engr Bn, n.d.), copy in CMH.

Men of the 1st Engineer Battalion use an M4T6 raft as a patrol boat.

During NIAGARA FALLS and CEDAR FALLS, 1st Division engineers attempted to seal off enemy escape routes. General DePuy, the division commander, suggested mounting quad .50-caliber machine guns (four heavy machine guns mounted on a single pedestal and fired by one gunner) on platforms placed at the confluence of the Saigon and Thi Tinh Rivers. Kiernan turned the task over to his bridge company. Using components of a M4T6 floating bridge, Company E assembled two rafts at the Vietnamese Engineer School at Phu Cuong, some seven and a half miles downstream. Each raft contained two bridge floats connected with aluminum decking, one quad .50- and six .30-caliber machine guns for additional firepower. Sandbags were placed on the floats and decking to protect the riflemen and grenadiers. A twenty-seven-foot utility boat propelled each raft. The makeshift armed rafts were given the imposing name of monitors. In addition, the engineer navy included two armed utility boats, dubbed U-boats, and several pneumatic assault boats, each manned by two engineers and thirteen infantrymen for river patrols. Early on 5 January, the engineer flotilla left Phu Cuong and moved up the Saigon River. Nearing their destination, the lead elements came under sniper fire. The rafts and an armed helicopter overhead returned the fire, killing four of the enemy. By nightfall, the monitors took up position at the juncture of the two rivers, one raft tied to the east bank and the other anchored in midstream.

While the monitors remained in the same general area during CEDAR FALLS, the engineer utility and assault boats of the 1st and 25th Divisions and 173d Airborne Brigade patrolled the rivers. The patrols checked all Vietnamese boats and took several Viet Cong suspects to interrogation points. In addition, the boats ferried supplies to outposts along the rivers, and returned with large quantities of rice and other enemy supplies extracted from nearby inlets.

Encouraged by the results of CEDAR FALLS, allied forces in February intensified their pressure on the enemy in the largest operation of the war to that time, JUNCTION CITY, an attack on War Zone C. While South Vietnamese forces remained near the populated areas, elements of the U.S. 1st and 25th Divisions, the 196th Infantry Brigade—altogether six brigades—and the 11th Armored Cavalry maneuvered into position to establish a horseshoe-shaped cordon around the war zone. On 22 February, a battalion of the 173d Airborne Brigade, in the only combat paratroop jump of the war, dropped into a blocking position near the Cambodian border. Combined armored and mechanized elements of the 2d Brigade, 25th Division, and the 11th Armored Cavalry thrust northward through the open end of the horseshoe. Meeting little resistance at the outset, advancing troops found numerous enemy base camps, which they destroyed. Before the end of Phase I on 17 March, several sizable engagements took place, with two major battles occurring at or near Prek Klok, approximately halfway between Tay Ninh City and the Cambodian border, along Route 4.[38] (*Map 19*)

As JUNCTION CITY entered its second phase, American forces concentrated their efforts in the eastern part of War Zone C, close to Highway 13. A brigade from the 9th Infantry Division replaced the 173d Airborne Brigade, and units began search-and-destroy operations in their assigned sectors. Enemy troops struck sharply against night defensive positions and firebases in attempts to isolate and defeat individual units. The combined firepower of the U.S. units, supporting artillery, and close air support, however, forced the attackers to break contact after suffering large losses.

By mid-April, the scale of JUNCTION CITY had tapered off. The third phase saw most Army units withdrawn, either to return to their bases or to take part in other operations. For the next month, a joint U.S.–South Vietnamese Army task force roamed throughout the war zone, but the temporarily shattered enemy had retreated into Cambodia. The operation officially came to an end at midnight on 14 May.

Engineer tasks in JUNCTION CITY paralleled those in CEDAR FALLS. Although the region was too large to level with bulldozers, tankdozers, and Rome plows, the engineers managed to destroy some of the major hideouts. In the western sector of operations, the 65th Engineer Battalion, 25th Division, supported by elements of the 27th, 86th, and 588th Engineer Combat Battalions; Company C, 4th Engineer Battalion, 4th Division; the 175th Engineer Company, 196th Infantry Brigade; and the 500th Panel Bridge Company cleared land along the sides of roads and built landing zones and bridges along the way. The engineers also provided minesweeping and demolition teams and maintained forty-one miles of roadway. In the eastern half of War Zone C, the 1st Engineer Battalion—backed by the 168th Engineer

---

[38] MacGarrigle, *Taking the Offensive*, pp. 113–43; Westmoreland, *A Soldier Reports*, p. 249; Sharp and Westmoreland, *Report*, pp. 133–34, 137, 152–53; ORLL, 1 Feb–30 Apr 67, II FFV, 15 May 67, pp. 21–23, Historians files, CMH.

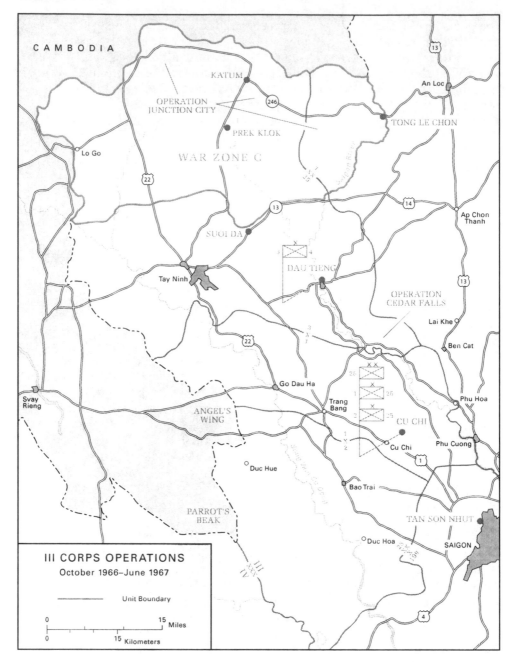

Map 19

Battalion during Phase I, the 27th Engineer Battalion during Phase II, and the 173d Engineer Company—opened, cleared, improved, and maintained over forty-six miles of roads and opened and repaired several airstrips. At the start of the operation, Company D and Headquarters Company moved from the just-completed C–130 airstrip at Suoi Da to begin work on an airstrip capable of handling C–123s at Katum, completing the laterite-topped airstrip in six days. The 168th Engineer Battalion built a new Special Forces camp and a 3,100-foot M8A1 airstrip that was capable of handling C–130s at Prek Klok, helped complete the airstrip at Katum, and repaired the T17-covered airstrip at Suoi Da. Similarly, the 27th Engineer Battalion constructed a Special Forces camp at Tong Le Chon.[39]

The 168th Engineer Battalion played a major part in the second battle at Prek Klok. The battalion shared defense responsibilities inside the circular wagon train–type perimeter with 1st Division units—a mechanized infantry battalion and an artillery battalion. On the night of 10 March, the Viet Cong launched a heavy mortar barrage followed by the main attack on the perimeter's eastern sector. During the attack, the engineers fought as infantry, filling in the gaps from their foxholes in the front lines. Air strikes and fire from other firebases helped drive away the attackers with heavy losses. U.S. losses were three killed and thirty-eight wounded, including two engineers killed (one from the 168th and one from attached elements of the 27th Engineer Battalion) and seven wounded.[40]

After JUNCTION CITY, Colonel Kiernan reviewed the accomplishments of his engineers. He told his troops that the battalion had built "more than one-half of all Bailey Bridging employed to date in Vietnam . . . and constructed half of all C–130 airfields built by U.S. Engineers in the II Field Force area." While Company D worked on the airfields during JUNCTION CITY, the other companies worked on roads and bridges.[41]

The battalion completed two panel bridges: the first on Route 246 over the Saigon River and the second on Route 1A over the Song Be River. To meet a short deadline to allow passage of the 11th Armored Cavalry, Companies A and B helped by Company E's bridge experts worked day and night to complete a two-span 220-foot bridge over the Saigon River at Tong Le Chon. Company A then deployed to the Song Be site south of Phuoc Vinh to rebuild a three-span bridge, claimed by Kiernan as one of the longest and most sophisticated Bailey bridges built by an engineering unit since World War II. Since the Vietnamese government would not allow removal of the concrete truss approach spans, the battalion had to modify panel bridge parts in order to fit the new center span between the old French trusses. Kiernan reported the new span was partially built on the existing center span, then jacked up from the old steel truss, and returned to a final position after the older truss dropped into the river one hundred feet below "with a beautifully executed demolition charge." The battalion completed the 430-foot cantilever bridge just as

[39] Ploger, *Army Engineers*, p. 140; Galloway, "Essayons," p. 193; Rogers, *Cedar Falls–Junction City*, pp. 122–23, 152–53, 155, 159; AAR, Opn JUNCTION CITY, 1st Engr Bn, 14 Apr 67, pp. 2, 4–5; Ltr, Kiernan to All Men 1st Engr Bn, 14 Apr 67, sub: 1st Engineer Actions on Operation JUNCTION CITY, both in OCE Hist Ofc; *Always First, 1965–1967*, "The Big Push into War Zone C"; Engineer Support of Combat Operations, 26 May 67, incl. 8, pp. 5–6, OCE Liaison Officer Trip Rpt no. 7, OCE Hist Ofc.

[40] Rogers, *Cedar Falls–Junction City*, pp. 117–21; Interv, Simpson, 8 Jul 67, pp. 79–115.

[41] Ltr, Kiernan to 1st Engr Bn, 14 Apr 67.

JUNCTION CITY came to a close, and the previously rated Class 23 (thirty-ton capacity) bridge could now carry tanks and fully loaded supply vehicles in the next operation by the division.[42]

In the wake of JUNCTION CITY, allied forces resumed clearing operations closer to Saigon and ventured into War Zone D. Just north of the Iron Triangle, the 1st and 25th Divisions began Operation MANHATTAN. The 1st and 65th Engineer Battalions continued extensive land-clearing projects and the destruction of hundreds of bunkers, tunnels, and installations. Monsoon rains began to fall, and keeping the unpaved routes open became a major challenge. In May, the 1st Division started Operation DALLAS by entering the rolling, jungle-covered area of War Zone D to the east of Route 1A. The following month, the division moved north of Phuoc Vinh in Operation BILLINGS. Notable 1st Engineer Battalion achievements during BILLINGS, which ended on 26 June, included clearing landing zones, destroying bunkers, and reopening the twenty-mile stretch of Route 1A from Phuoc Vinh to Dong Xoai. Carving infantry landing zones in the heart of the jungle was dangerous. During one landing zone clearing mission, the 1st Platoon, Company C, had to stop work when the infantry security force tried to beat off an attack. During the three hours of fierce fighting against an estimated North Vietnamese battalion, the engineers doubled as infantrymen and self-appointed medics.[43]

Combat engineers suffered their share of casualties during these operations, and just as often became victims of accidents. The 1st Engineer Battalion suffered one man killed in action and seven wounded during CEDAR FALLS; one killed and forty-two wounded during JUNCTION CITY; and sixteen wounded in Operations MANHATTAN, DALLAS, and BILLINGS. By its very nature, military engineering is dangerous. Working with construction equipment, handling explosives, and traveling by vehicles and aircraft, coupled with long hours and fatigue, meant that engineers in even so-called secure areas could be killed or injured. For instance, during JUNCTION CITY, two tankdozers of the 1st Engineer Battalion with leading elements of the 3d Brigade moving north along the road between Suoi Da and Katum were lost to mines on the first morning of the operation. A party looking for laterite was attacked and suffered three casualties within six hundred feet of the battalion command post. At the Katum airstrip, surveying was often interrupted by sniper fire. Shortly after Operation DALLAS, the 1st Engineer Battalion suffered a deadly blow when a helicopter carrying the battalion's command group struck a high-tension wire while flying low under the flight pattern at Bien Hoa Air Base and crashed to the ground. Killed in the incident were Kiernan, who only had seven days to go before returning home; Lt. Col. Rodney H. Smith, executive officer and commander-designate; Maj. Millard L. Treadwell, operations officer; and Sgt. Maj. Terry M. Rimes, who was about to assume the post of battalion sergeant

---

[42] Ibid.

[43] ORLL, 1 May–31 Jul 67, II FFV, 18 Sep 67, pp. 19–20, Historians files, CMH; 1st Engr Bn Hist, vol. 1, pp. III-21 to III-22; Engineer Support of Combat Operations, 26 May 67, incl. 8, pp. 6–7, OCE Liaison Officer Trip Rpt no. 7. See also MacGarrigle, *Taking the Offensive*, pp. 145–55, 339–43.

major Maj Edwin C. Keiser, assistant division engineer, assumed temporary command of the battalion.[44]

By mid-1967, Army engineers had proven their ability to support fast-moving combat operations in II and III Corps and more recently in IV Corps. In III Corps, they kept resupply roads open; built and reopened forward airfields; cleared the land for landing zones and mutually supporting firebases, cut swaths through dense vegetation that cloaked the enemy; and destroyed enemy base areas, tunnels, and bunkers. Similarly, in much larger II Corps, key roads were being opened and more airfields sprinkled the hinterlands. Most impressive was the gradual return of coastal Highway 1 to service with accompanying benefit to the local economy. Much work, however, remained in pushing the air and road supply lines into the inhospitable interior. As for the engineers' prodigious efforts, local commanders expressed their appreciation, but some still wanted them under their direct control. Senior engineers continued to emphasize priority to support combat operations, but at the same time striving to complete base development projects and turning attention to the growing highway restoration program.

New techniques and equipment and reliance on technology did not always result in complete success. Although a great boon as an expedient runway surface, the T17 membrane, like the new metal mattings, required a good base, and frequent failures caused by heavy rains required extensive airfield repairs. The tunnel destruction teams in III Corps using the acetylene and explosives together achieved good results, but the Viet Cong had built too many burrows for the few trained tunnel rats and engineers to explore and map. Similarly, the impressive results of the bulldozers and Rome plows in land-clearing operations in III Corps were tempered by the large areas of dense vegetation and heavy maintenance requirements. In the course of a day's operations, it was common for most of the Rome plows to sustain disabling damage of some sort. Troops in land-clearing units suffered higher casualty rates both from enemy action and natural hazards. Aside from falling trees, the operators often faced swarms of bees. In the end, elusive enemy troops still had abundant jungle cover to move about and build concealed supply caches. The jungle also quickly reclaimed the cleared areas. Although the engineers reopened many roads, the enemy came back at night to bury mines and destroy bridges. Sweeping for mines became a daily morning task. Heavy rains and traffic required continuous maintenance on the unsurfaced roads. Much work remained to improve the coastal road network and to extend road and air supply lines to the contested interior. Meanwhile, the military and civilian engineers remained heavily committed to base development, for the logistics bases and ports played a crucial role in supporting the combat forces.

The allied offensives had set the enemy back, but he remained a major threat in the countryside. A discouraging feature of the war was the fact that the allies could not muster enough troops to occupy the enemy strongholds and thereby prevent him from returning. As soon as American troops left areas they

---

[44] 1st Engr Bn Hist, vol. 1, pp. III-19 to III-22; Rogers, *Cedar Falls–Junction City*, p. 155; *Always First, 1965–1967*, "The Big Push into War Zone C."

had cleared such as the Iron Triangle, the enemy returned. Meanwhile, trouble was brewing in I Corps. In 1967, the threatening situation against the marines along the Demilitarized Zone intensified. General Westmoreland, concerned that the Marine forces there were insufficient, decided to send Army troops to the northern provinces.

8

To I Corps

In early 1967, a growing North Vietnamese threat below the Demilitarized Zone between the two Vietnams brought infantry units of the U.S. Army to I Corps for the very first time. General Westmoreland had planned to put the Army temporarily in southern I Corps in connection with his dry-season campaign beginning in the spring. But during the winter, a quickening of enemy activity in the A Shau Valley and Quang Ngai Province, and his conviction that the marines in I Corps were stretched to the limit, forced him to postpone his campaign indefinitely and to rush in reinforcements on a schedule not quite his own. On 28 February, he completed an initial planning action preparatory to the deployment of units northward, approving a new makeshift division to be created from brigades in II and III Corps, but deferred actually selecting the units until the division headquarters could be formally activated. Five weeks later, on 6 April, with eight North Vietnamese regiments counted in the vicinity of the Demilitarized Zone, and enemy artillery hammering Marine bases along the border, Westmoreland ordered I Field Force to enter southern I Corps by the next evening so that the marines there could be sped at once to the threatened north. With this order, which implemented the provisional division called Task Force OREGON, the brigades fell into place. But because the brigades were still scattered through the interior, with the nearest one too far south to meet the schedule, I Field Force called on its most agile unit to serve as the tip of the knife, and within hours elements of the 1st Cavalry Division were air-assaulting north into Duc Pho to relieve the marines.[1]

Duc Pho and Chu Lai

What the 1st Cavalry Division found in southern I Corps were facilities scarcely adequate for the marines, let alone the Army. For two years, the marines had treated southern I Corps as a backwater holding action while looking to the north. Da Nang had become the command and support hub of the corps-level III Marine Amphibious Force, nourishing satellite bases at Chu Lai and Phu Bai in consonance with the strategy, which was to hold the line as far north as possible and to pacify the rural population around the cities and towns. Well to the south lay Duc Pho, a district headquarters just above

---

[1] MacGarrigle, *Taking the Offensive*, pp. 203–10; Lt. Gen. Willard Pearson, *The War in the Northern Provinces, 1966–1968*, Vietnam Studies (Washington, D.C.: Department of the Army, 1975), pp. 12–13; Maj. Gary L. Telfer, Lt. Col. Lane Rogers, and V. Keith Fleming Jr., *U.S. Marines in Vietnam: Fighting the North Vietnamese, 1967* (Washington, D.C.: History and Museums Division, Headquarters, U.S. Marine Corps, 1984), pp. 75–78.

the I/II Corps line, a town that had only become important to the Americans in January when the marines dispatched a battalion to cope with a rise in Viet Cong activity and to cut an enemy supply line to the coast. Despite two months of heavy fighting, the marines were barely dug in, still without beach and pier facilities or an airfield, or a secure lifeline down Highway 1 from the port at Chu Lai, and therefore dependent on helicopter resupply from ships offshore. Westmoreland's orders to I Field Force were to the point: turn the town of Duc Pho into a robust support base—and do so quickly—and a staging area as large as necessary for the Army commitment.[2]

To spearhead that commitment, Maj. Gen. John J. Tolson, the 1st Cavalry Division's commander, chose his reserve unit, the 2d Battalion, 5th Cavalry. At first light on 7 April, the cavalrymen boarded Caribous and Chinooks at An Khe for the flight to Landing Zone ENGLISH where they transferred to Huey helicopters for the journey to Duc Pho. By nightfall, the 2d Battalion relieved the marines outside the town. The 1st Cavalry Division's 2d Brigade headquarters flew in the following morning and assumed control of all Army and Marine forces in the area, now designated LE JEUNE. A second cavalry battalion followed the headquarters. Because the marines at the time had too few helicopters to effect their redeployment north, they remained under the control of the 2d Brigade until 21 April.[3] (*Map 20*)

With follow-on forces on their way to Duc Pho, one of the first priorities was construction of a rough but ready airfield. The assignment fell to the division's 8th Engineer Battalion. Building on a reconnaissance of the site the previous month, advance elements of the battalion, including Company B, arrived on 7 April to start work east of the town in an open area called Landing Zone MONTEZUMA. Over a period of forty-eight hours, helicopters delivered thirty-one pieces of equipment weighing over two hundred tons— making this the largest movement for the airmobile engineer battalion up to that point. The movement of the equipment, much of it partially disassembled, required twenty-four CH–54 Flying Crane sorties and fifteen CH–47 Chinook sorties. Work proceeded at a rapid pace around the clock, with operators running their equipment at night under the lights of vehicles and a floodlight set. This caused some concern to the marines who had endured nightly sniper and mortar fire over the past three months. The rumble of the bulldozers, graders, and other earthmoving equipment turned out to be the only activity to disturb their sleep that night. By dawn of the second day, the engineers completed one-half of the runway, and the work continued without interference.[4]

After twenty-four hours of continuous work, the 8th Engineer Battalion completed a 1,400-foot Caribou strip and proceeded into the next night to

---

[2] MacGarrigle, *Taking the Offensive*, pp. 205, 211; Pearson, *War in the Northern Provinces*, p. 15; Tolson, *Airmobility*, p. 131; Telfer, Rogers, and Fleming, *Fighting the North Vietnamese*, pp. 52–57, 63.

[3] MacGarrigle, *Taking the Offensive*, pp. 210–11, 227–43; Tolson, *Airmobility*, pp. 130–33; Telfer, Rogers, and Fleming, *Fighting the North Vietnamese,* pp. 78–79.

[4] Maj. Gene A. Schneebeck and Capt. Richard E. Wolfgram, "Airmobile Engineer Support for Combat," *Military Engineer* 59 (November-December 1967): 397–98; AAR, Opn LE JEUNE, 8th Engr Bn, 4 May 67, pp. 1–2, Historians files, CMH; Tolson, *Airmobility*, pp. 132–33.

MAP 20

expand the runway for C–123 use. On 9 April, the first Caribous began delivering sections of culvert to be used to carry a drainage channel under the expanded runway. Air traffic was extremely heavy on that first day with over a thousand landings by helicopters and fixed-wing aircraft. Work crews tried to reduce the immense dust clouds of the light sandy soil stirred by the heavy helicopter traffic by spreading peneprime on the helipads and refueling areas. On the evening of the tenth, work began on the installation of the culvert. Two scrapers cut a trench while a squad completed the culvert assembly for the drainage channel. Two bulldozers slowly moved the culvert into position and pushed it into the trench. The heavy equipment continued the cut, fill, and compaction work. By dawn the next day, the compacted earth C–123 runway, 2,500 feet long and 50 feet wide, was done. A few days later, work began on a second Caribou airstrip placed 145 feet west of the C–123 runway's centerline.

237

The second strip, completed on the evening of 15 April, became necessary to allow further expansion of the original airstrip to handle C–130s. Upon its completion, the Caribou runway then reverted to a taxiway.[5]

On 10 April, an advance party of the 39th Engineer Combat Battalion arrived from Tuy Hoa to coordinate support requirements with the 8th Engineer Battalion and to prepare for the 39th's main mission—engineering support for Task Force OREGON. The main body of the battalion traveled by ship. On 12 April, Company D came ashore two and a half miles east of Duc Pho at Landing Zone GUADALCANAL, also called RAZORBACK BEACH. A few days later, Company A and parts of the Headquarters Company and the 4th Platoon, 554th Float Bridge Company, came ashore. Two companies, Company B at Tuy Hoa and Company C at Vung Ro Bay, remained behind for the time being. Other elements of the 554th Float Bridge Company and a bridge platoon of the 509th Panel Bridge Company were also scheduled to join the 39th Engineer Battalion.[6]

Since the 1st Cavalry Division could not completely resupply the 2d Brigade by airlift and roads, the 39th Battalion's first job was to build a road from the beach to Duc Pho as a ground link for supplies delivered by landing craft. Despite heavy rains and enemy mines, the 39th Battalion task force blazed a trail to Landing Zone MONTEZUMA on the twenty-second. Within three days, the trail was upgraded to a pioneer road. While Company D continued to improve the road, Company A moved men and equipment to Duc Pho to take over the C–130 airfield work. Company A, which endured some sniper fire, completed topping off the runway with MX19 aluminum matting on 15 May. The first C–130 landed on the new 3,500-by-60-foot strip with 300-foot overruns that evening. The following morning, Company A joined the expanded all-weather C–130 runway to the temporary Caribou airstrip. The Duc Pho airfield, renamed BRONCO, was officially completed on 31 May. The 39th Engineer Battalion also built helicopter pads and an ammunition supply point, and connected an access road from the airfield to Highway 1. On 30 June, the battalion began building revetments for all aircraft parked at BRONCO.[7]

Direct support to the 2d Brigade's two infantry battalions during operations in the LE JEUNE area came from the 8th Engineer Battalion's Company B, which carried out minesweeps, tunnel and bunker destruction, and general support. The latter chores included building helicopter revetments and pioneer roads, installing 6,000 yards of concertina barbed wire around the Landing Zone MONTEZUMA perimeter, and spreading peneprime to reduce dust. One

[5] Schneebeck and Wolfgram, "Airmobile Engineer Support for Combat," pp. 398–99; AAR, Opn LE JEUNE, 8th Engr Bn, pp. 3, 6–7; Tolson, *Airmobility*, p. 133.

[6] ORLL, 1 May–31 Jul 67, 39th Engr Bn, 15 Aug 67, pp. 2–3, 5; AAR, Opn BAKER/MALHEUR I, 39th Engr Bn, 16 Sep 67, pp. 1–2, both in Historians files, CMH.

[7] AAR, Opn BAKER/MALHEUR I, 39th Engr Bn, pp. 2–6; AAR, Opn LE JEUNE, 8th Engr Bn, pp. 1–2, 5–6; ORLL, 1 May–31 Jul 67, 39th Engr Bn, pp. 1–2, 5; Schneebeck and Wolfgram, "Airmobile Engineer Support for Combat," pp. 398–99; Quarterly Hist Rpt, 1 Apr–30 Jun 67, MACDC, 11 Jul 67, p. IV-8, Historians files, CMH; ORLL, 1 May–31 Jul 67, Task Force OREGON, 5 Nov 67, p. 35, Historians files, CMH.

Men of the 39th Engineer Battalion lay MX19 matting at Duc Pho.

of the battalion's water points was also airlifted from Landing Zone ENGLISH to Duc Pho on the eighth. Because the only source of water lay some 550 yards outside the perimeter, troops every morning had to sweep the road to the water point for mines.[8] (*Map 21*)

On 12 April, Task Force OREGON, under the command of Maj. Gen. William B. Rosson, was officially activated and eight days later opened its headquarters at Chu Lai some thirty-seven miles north of Duc Pho. The task force relieved the 1st Marine Division on the twenty-sixth and formally took over responsibility for the Chu Lai and Duc Pho sectors. By then the 196th Light Infantry Brigade, which had been withdrawn a few days earlier from the final phase of JUNCTION CITY, completed its move to Chu Lai. Reinforcing the brigade were a battalion from the 3d Brigade, 25th Infantry Division, and a squadron from the 11th Armored Cavalry Regiment. When, on 22 April, two battalions from the 3d Brigade, 25th Division, replaced the 2d Brigade, 1st Cavalry Division, at Duc Pho, the LE JEUNE operation ended. During the first week in May, the 1st Brigade, 101st Airborne Division, put ashore at RAZORBACK BEACH and joined Task Force OREGON, making the task force equivalent to a three-brigade division.[9]

[8] Schneebeck and Wolfgram, "Airmobile Engineer Support for Combat," p. 399; AAR, Opn LE JEUNE, 8th Engr Bn, 4 May 67, pp. 4–7.

[9] MacGarrigle, *Taking the Offensive*, pp. 213–14, 229, 232, 236; Sharp and Westmoreland, *Report*, p. 153.

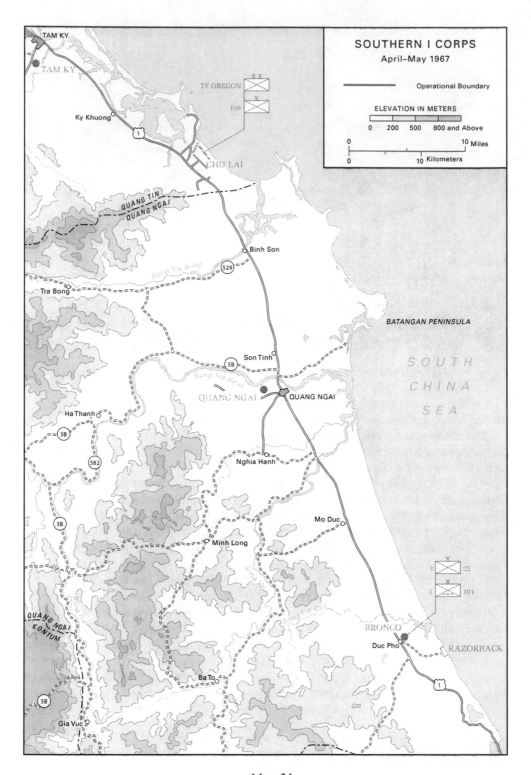

SOUTHERN I CORPS
April–May 1967

Operational Boundary

ELEVATION IN METERS

0   200   500   800 and Above

0   10 Miles

0   10 Kilometers

TAM KY

TAM KY

TF OREGON

196

Ky Khuong

1

CHU LAI

QUANG TIN
QUANG NGAI

Binh Son

529

Song Tra Bong

Tra Bong

BATANGAN PENINSULA

SOUTH
CHINA
SEA

Son Tinh

5B

Song Tra Khuc

QUANG NGAI   QUANG NGAI

Ha Thanh

5B

582

Nghia Hanh

Mo Duc

5B

3   25

Minh Long

1   101

QUANG NGAI
KONTUM

BRONCO

Ba To

Duc Pho   RAZORBACK

1

5B

Gia Vuc

MAP 21

On 11 May, Task Force OREGON launched its first major operation, MALHEUR, with the airborne brigade making a sweep inland from Duc Pho. The heliborne search-and-destroy operation lasted until the end of May, followed by MALHEUR II, which ended on 2 August. Operations HOOD RIVER and BENTON followed. Although the operations resulted in large enemy losses and the capture of food, weapons, and ammunition caches, they did not draw enemy forces into large-scale fighting.[10]

Army engineers allocated to Task Force OREGON, in addition to the 39th Battalion, were drawn from units that supported the three brigades. Company A, 326th Engineer Battalion, deployed with the 1st Brigade, 101st Airborne Division; the 175th Engineer Company with the 196th Infantry Brigade; and Company C, 65th Engineer Battalion, with the 3d Brigade, 25th Division. Technically the separate companies of these brigades were considered part of the 39th Engineer Battalion, but they remained attached to their respective brigades.[11]

As soon as the 39th Engineer Battalion completed the Duc Pho airfield, attention turned to reopening Highway 1 to traffic between Chu Lai and Duc Pho and south to the I/II Corps boundary. Logistics planners emphasized that this stretch would be the only dependable means of supply during the forthcoming monsoon. The road was secure between Chu Lai and Quang Ngai City, but between Quang Ngai and Duc Pho and south to the corps border it was controlled by the enemy and needed extensive repairs. General Westmoreland reemphasized the importance of the highway when he learned of plans to build a Caribou-capable airstrip at Mo Duc, about half-way between Duc Pho and Quang Ngai City. The airfield would support a nearby firebase and the district advisory team, but Westmoreland announced during a visit to Chu Lai that airfield work could be delayed until 1 September.[12]

With elements of Task Force OREGON providing security, the 39th Engineer Battalion, and the 9th Marine Engineer Battalion, which had stayed behind in southern I Corps to support the Army, were ordered to bring the road to pioneer status, followed by more improvements. A new 75-ton-per-hour rock crusher was shipped to the 39th Engineer Battalion so it could set up a quarry approximately three miles south of Duc Pho. Limited rock production started on 10 September.[13]

Company A opened the road to traffic northward from Duc Pho, filling craters, grading, and building bypasses around destroyed bridges, followed

[10] MacGarrigle, *Taking the Offensive*, pp. 229–43; Sharp and Westmoreland, *Report*, p. 154; Telfer, Rogers, and Fleming, *Fighting the North Vietnamese*, p. 119.

[11] Engineer Support of Combat Operations, 26 May 67, incl. 8, p. 2, OCE Liaison Officer Trip Rpt no. 7.

[12] Msgs, COMUSMACV 18616 to COMNAFORV, 8 Jun 67, sub: Support of Project OREGON, and Task Force OREGON to III MAF, 31 May 67, sub: Visit of COMUSMACV, 30 May 67; Memo, 1st Log Cmd, 18 Jun 67, sub: Trip Report, Visit to Chu Lai, 5 Jun 67, pp. 1, 3; Memo, Lt Gen Bruce Palmer Jr., Dep CG, USARV, for COMUSMACV, 5 Nov 67, sub: Engineer Requirements in American Division TAOR [Tactical Area of Responsibility], p. 2. Messages and memos in Historians files, CMH.

[13] ORLLs, 1 May–31 Jul 67, Task Force OREGON, p. 23, 1 May–31 Jul 67, 39th Engr Bn, pp. 3, 9, and 1 May–31 Jul 67, 45th Engr Gp, p. 4.

Once the airfield work at Duc Pho neared completion, the engineers turned their attention to reopening Highway 1 to Chu Lai.

by repairing or rebuilding the bridges. In many cases culverts were installed instead of bridges. Along the way, Company A provided its own security and deployed demolition teams and a bulldozer to destroy nearby tunnels and bunkers. When the Viet Cong retaliated by destroying a new forty-five-foot dry span on the evening of 1 June, work crews returned the next day and replaced the damaged span with two culverts. Company C, which had now rejoined the battalion, was given the job to improve the road southward from Quang Ngai City. On the fourteenth, the two companies met, and the fifty-mile stretch of Highway 1 between Chu Lai and Duc Pho was now open to convoy traffic. South of Duc Pho, Company D opened the road to the corps border on 8 July, linking up with the 8th Engineer Battalion working north from Bong Son. The next day the first convoy traversed this stretch.[14]

Although now open between Chu Lai and the border during daylight hours, Highway 1 still resembled a trail with many bypasses, culverts instead of bridges, and temporary bridges. Much work still needed to be done. The danger of mines and ambushes remained foremost in the minds of engineers working on the road. During the three-month period ending on 31 July, the 39th Engineer Battalion suffered three killed and twenty-six wounded, mostly along Highway 1. In July, the southernmost one-third of the 39th Engineer

[14] ORLLs, 1 May–31 Jul 67, 39th Engr Bn, pp. 2–5, and 1 May–31 Jul 67, 45th Engr Gp, p. 4; Quarterly Hist Rpt, 1 Jul–30 Sep 67, MACDC, 21 Oct 67, p. IV-8, Historians files, CMH; MacGarrigle, *Taking the Offensive*, pp. 242–43.

Battalion's Highway 1 responsibility was transferred to the 19th Engineer Battalion, which began to deploy north from Qui Nhon to work on the road from Bong Son to Duc Pho. Meanwhile, Seabees based in Chu Lai upgraded the highway from Quang Ngai City to Chu Lai, and Marine engineers worked northward from Chu Lai. Improvements to Highway 1 gradually advanced from a pioneer dry-weather roadbed and bypasses to a road with semipermanent bridges and a road surface which, if properly maintained, could bear divisional loads year-round.[15]

Yet for all of the good work along Highway 1 as well as at Duc Pho airfield, the 39th Engineer Battalion was not trouble-free. Logistical problems abounded. Sustaining U.S. forces around Duc Pho was the job of Task Force GALLAGHER, a special group of support units organized to manage supply for all Army units in the area. For the engineers, this included Class IV construction and barrier materials. Equipment and supplies not available from Task Force GALLAGHER came from the Qui Nhon Support Command with help from the 45th Group headquarters, which was conveniently located next door to the support command. Construction work, however, was slowed by Task Force GALLAGHER's lack of forklifts to load heavy materials. Work was also slowed because the MX19 airfield matting kits did not have turndown adapters, and the engineers had to improvise time-consuming anchorage systems to hold the Duc Pho runway in place. The 39th Engineer Battalion's shortage of seven scoop loaders, one twenty-ton crane-shovel, a welding set, an air compressor, two 10-kilowatt generators, and several vehicles and trailers further affected the unit's capability. Efforts were made through the 45th Group supply officer to speed the delivery of these items.[16]

Another problem facing the 39th Engineer Battalion was maintenance support. Initially, there were no direct or general support maintenance units in the area, and replacement parts for engineer equipment were almost nonexistent. An eleven-man contact team from a maintenance company based at Tuy Hoa provided some repair capability to the 39th Engineer Battalion but no repair parts. By mid-June, the lack of spare parts for engineer equipment had become critical. When the problem came to the attention of General Westmoreland during a visit to Chu Lai, Task Force OREGON, U.S. Army, Vietnam, and the Qui Nhon Support Command reacted by checking all requisitions and initiating a crash program to improve supply flow. By 5 July, the engineer repair parts situation at Duc Pho had become "marginally adequate."[17]

In stark contrast to Duc Pho, Army troops at Chu Lai found a well-developed complex supporting some 19,000 marines and sailors on a base still undergoing construction. Following the construction of an 8,000-foot AM2 airstrip there in mid-1965, the Seabees added a 4,000-foot AM2 crosswind runway and separate helicopter facilities the following year. In late 1966, RMK-BRJ

---

[15] ORLLs, 1 May–31 Jul 67, 39th Engr Bn, pp. 6–9, and 1 May–31 Jul 67, 45th Engr Gp, p. 3.

[16] ORLL, 1 May–31 Jul 67, 39th Engr Bn, pp. 6–8; AAR, Opn BAKER/MALHEUR I, 39th Engr Bn, p. 6.

[17] ORLLs, 1 May–31 Jul 67, Task Force OREGON, p. 36 (quoted words), 1 May–31 Jul 67, 39th Engr Bn, pp. 8–9; AAR, Opn BAKER/MALHEUR I, 39th Engr Bn, p. 6; Msg, Task Force OREGON to III MAF, 31 May 67, sub: Visit of COMUSMACV, 30 May 67.

completed a 10,000-foot concrete runway. The shallow-draft port had several LST and LCU ramps, a barge discharge pier, and pipelines for offshore fuel discharging. LSTs, however, could enter the port area only at high tide, and monsoonal storms and typhoons threatened the facilities. The contractor was nearing completion of a fifth hangar and was working on a large power plant when it received word to phase down. On 1 May 1967, RMK-BRJ turned over uncompleted projects, concrete and asphalt plants, shops, and work camps to the Navy, which in turn divided the facilities and projects among Army, Navy, and Marine Corps engineers on the scene. Total contract cost for work done during the twenty-two months RMK-BRJ had been at Chu Lai came to some $37 million. The Seabees, in addition to completing work for the 6,500 Marine aviation, communications, and naval support troops staying behind, had already begun planning for Task Force OREGON. The former Marine Corps commander in southern I Corps, a brigadier general, remained as the installation coordinator.[18]

Marine ground units began leaving Chu Lai in April, and they took with them their generators and water purification sets. As a result, water rationing became necessary for several weeks. Existing wells could not support the troops housed by Task Force OREGON, and Army engineers had to drill a new well. After repeated efforts, they received enough pumps and filtration units, and on 20 May water rationing was no longer required. Additional wells were dug in August. As for generators, the 1st Logistical Command temporarily assumed facilities engineering responsibility for the task force's headquarters, and generators were furnished to the headquarters repair and utilities section. On the other hand, the marines left behind some refrigeration equipment and a limited ice-making capability, supplemented by small amounts of ice purchased on the economy. In order to provide more ice, the Seabees began building a fifteen-ton-capacity ice-making plant in May and completed it in mid-July.[19]

In mid-June, General Westmoreland decided to leave Task Force OREGON in southern I Corps throughout the autumn monsoon. As a result, base construction requirements at Chu Lai and Duc Pho took on a new perspective. Initially, the 39th Engineer Battalion, the 9th Marine Engineer Battalion, and Naval Mobile Construction Battalions 71 and 8 (the latter replaced by Naval Mobile Construction Battalion 6 in early August), concentrated on completing the most urgently needed facilities to support operations. Between 1 and 16 May, two large complexes for the Army's 14th Aviation Battalion at Chu Lai were completed. The aviation facilities consisted of M8A1 landing mat aprons, helicopter pads, taxiways, and refueling pads, plus liberal applications of peneprime for dust control. With the approaching northeast monsoon in mind, attention turned to upgrading Highway 1 to an all-weather supply route, improving ammunition supply points, forward supply areas, and helipads,

[18] Telfer, Rogers, and Fleming, *Fighting the North Vietnamese*, pp. 320–23; MacGarrigle, *Taking the Offensive*, p. 299; MACV Complex Review, 1 Dec 66, pp. 40–47; Diary of a Contract, pp. 244–46; Memo, 1st Log Cmd, 18 Jun 67, sub: Trip Report, Visit to Chu Lai, 5 Jun 67, p. 1.

[19] ORLL, 1 May–31 Jul 67, Task Force OREGON, pp. 34–35; Quarterly Hist Rpt, 1 Jul–30 Sep 67, MACDC, p. IV-4; Msg, Maj Gen Charles W. Eifler, CG, 1st Log Cmd, to Gen Dwight E. Beach, CINCUSARPAC, 31 May 67, sub: Leventhal Report, Historians files, CMH.

While the base at Duc Pho had very little work done, the base at Chu Lai was consistently improved by the Seabees and RMK-BRJ.

upgrading living quarters to Standard 2 tentage with raised wooden floors and tent frames at Chu Lai and Duc Pho, and providing revetments for each aircraft at the two bases. Other work slated for the 39th Engineer Battalion was the new airfield at Mo Duc.[20]

[20] ORLL, 1 May–31 Jul 67, Task Force OREGON, pp. 28, 35; Quarterly Hist Rpt, 1 Apr–30 Jun 67, MACDC, p. IV-5.

With the added workload, the 39th Engineer Battalion pressed to reclaim its remaining company left behind in II Corps. On 19 May, Company B had moved sixty-two miles north over Highway 1 to Qui Nhon where it stayed for one month with the 19th Engineer Battalion at Long My. During its stay at Long My, the company prepared to rejoin its parent unit and helped the 19th with its depot projects. On 21 June, Company B departed Qui Nhon aboard an LST and reached Chu Lai the same day. Soon the company was at work building roads in support of tactical operations, an ammunition supply point at Quang Ngai City, and taking on several base camp projects at Chu Lai.[21]

Meanwhile, Lt. Col. Joseph F. Castro, who replaced Colonel Fulton in July as commanding officer of the 39th Engineer Battalion, learned that Engineer Command did not have jurisdiction to issue construction directives for construction in I Corps, still a Navy and Marine Corps preserve. As a result, units could not draw materials to make tent frames and latrines for self-help base development. The problem was passed to the MACV Directorate of Construction. In turn, the directorate prepared a list of additional and unprogrammed projects, including $5.2 million for Task Force OREGON and passed these requirements on to Washington for approval and funding. With this action and approval, steps were taken to obtain materials and get tentage construction under way to keep the troops out of the mud before the rains arrived in a few months. In addition, the 1st Logistical Command began preparations to have Pacific Architects and Engineers take over facilities engineering responsibilities at Chu Lai and Duc Pho.[22] (*Map 22*)

Progress at Chu Lai continued at an impressive rate through the remainder of 1967. Improvements continued on the offshore fuel pipelines by extending them out another 1,500 feet. By July, the lines were ready for T–2 tanker operation. In June, Naval Mobile Construction Battalion 8 began expanding fuel storage tank capacity on the beach. Five 10,000-barrel tanks were completed by Naval Mobile Construction 6 in August and September, but the tanks soon developed leaks. The Seabees applied an epoxy seal, and the tanks were in service by November. Additional ammunition storage pads for Marine Corps jet fighters based at Chu Lai were started in July and finished in October. An approach and center-line runway lighting system for the 10,000-foot runway was installed in September. Two of the hangars under construction for the marines were ready for use in late October. Naval Mobile Construction Battalion 6 completed an engine maintenance shop in October and neared completion of an avionics repair facility in December. Quarry operations run by Naval Mobile Construction Battalion 71 picked up momentum in June with a two-shift operation. In May and June, the Seabees were even able to slip in the construction of a new chapel. As evidence of permanency, the Seabee

[21] ORLLs, 1 May–31 Jul 67, Task Force OREGON, pp. 23, 35, 1 May–31 Jul 67, 39th Engr Bn, pp. 2–3; Msg, CG, USARV, AVDF-GC 623 to Engr Cmd, 30 May 67, sub: Request for Additional Engineer Company, Historians files, CMH.

[22] Memo, 1st Log Cmd, 18 Jun 67, sub: Trip Report, Visit to Chu Lai, 5 Jun 67, p. 3; Msg, Brig Gen Mahlon E. Gates, MACV Dir of Const, SOG 1343, to Brig Gen Daniel A. Raymond, Ch of Southeast Asia Const, Dept of Def, 1 Jul 67, Historians files, CMH.

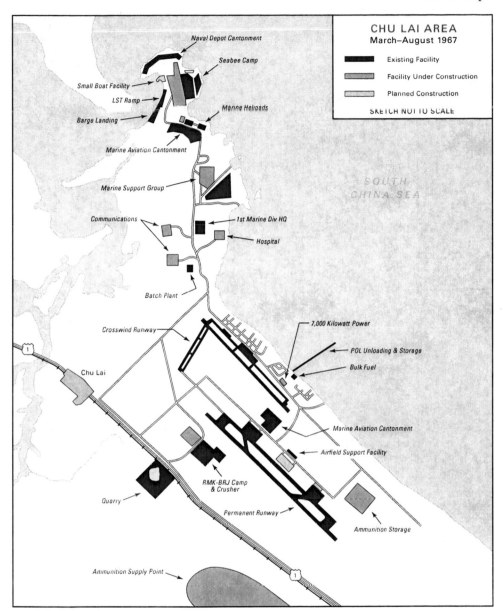

MAP 22

battalion began construction of a dairy plant in early June. The plant was completed in August, and Foremost Dairy, one of two firms contracted to produce milk and dairy products in Vietnam, began production in August. Also in August, MACV directed U.S. Army, Vietnam, to start shipping 40-by-90-foot prefabricated buildings from Army stocks in Qui Nhon to Chu Lai. Four of the eighteen buildings in stock were to be erected by the

Chu Lai base hosted a jet airfield and logistic facilities.

Seabees for use as warehouses for Class II and IV storage, a post office, and a post-exchange addition.[23]

At Duc Pho, extensive development was also under way at the combat base. Plans were developed to provide an all-weather capability for the C–130 airfield and additional crushed rock for roadwork. Procurement action on a runway lighting system was initiated in August. In late August, Naval Mobile Construction Battalion 71 was charged with providing technical help on self-help construction projects in the cantonment area to replace tents with wooden buildings. During October and November, the Seabee battalion completed a refueling facility that included a 16,000-barrel tank farm. By the end of the year, the brigade-size camp was well on its way to become a semipermanent installation, including facilities engineering support by Pacific Architects and Engineers.[24]

As 1967 drew to a close, Task Force OREGON transformed itself into a legitimate division. General Westmoreland resigned himself to the fact that Army troops would have to remain in southern I Corps indefinitely. On 25 September, Task Force OREGON officially became the 23d Infantry Division (Americal).

[23] Quarterly Hist Rpts, 1 Apr–30 Jun 67, MACDC, pp. IV-4 to IV-5, 1 Jul–30 Sep 67, MACDC, pp. IV-3 to IV-4, V-3, and 1 Oct–31 Dec 67, MACDC, 21 Jan 68, pp. IV-3 to IV-4, last in Historians files, CMH. For more on dairy products, see U.S. Army Procurement Agency, Vietnam, *Procurement Support in Vietnam, 1966–1968* (Japan: Toshio Printing Co., n.d.) pp. 142–44.

[24] Quarterly Hist Rpts, 1 Jul–30 Sep 67, MACDC, p. IV-7, and 1 Oct–31 Dec 67, MACDC, p. IV-6.

The American designation came about because of comparable circumstances in the Southwest Pacific Theater during World War II. U.S. Army units on New Caledonia were transformed into the American Division to support the Marine Corps on Guadalcanal. In Vietnam, the title seemed appropriate as a means to promote good working relationships between the Marine Corps and the Army. Similarly, the American remained the only division to be formed outside the United States in the two wars. But rather than completely restructure the task force, the MACV commander allowed the three constituent brigades to retain their identities. The new division consisted of the 196th Light Infantry Brigade at Chu Lai; the 3d Brigade, 4th Infantry Division at Duc Pho; and the 1st Brigade, 101st Airborne Division, now at Tam Ky, the capital of Quang Tin Province about eighteen and one-half miles northwest of Chu Lai.[25] A fourth brigade, from the 1st Cavalry Division, was put under the operational control of the new division and relieved the marines in the Que Son Valley, a populated, fertile area extending inland from Highway 1 along the Quang Nam–Quang Tin provincial border some thirty-one miles northwest of Chu Lai. In October, the 198th Light Infantry Brigade arrived at Hoi An in Quang Nam Province, replacing the 1st Brigade, 101st Airborne Division, and in December the 11th Light Infantry Brigade reached Duc Pho and replaced the 3d Brigade, 4th Division. The two new brigades had their own engineer companies, the 55th with the 198th Brigade and the 6th assigned to the 11th Brigade.[26]

For the most part, existing engineer units were merged into the division's new 26th Engineer Battalion. Senior engineer officials, including General Cassidy, the chief of engineers; General Ploger at Engineer Command; and Brig. Gen. Andrew P. Rollins Jr., who assumed command of the 18th Engineer Brigade in September, had supported the idea to leave the infantry brigades' organic engineer companies with the brigades and to activate a battalion at division level. This option, however, did not prevail, with the final decision by the Department of the Army in early November. Companies A, B, and C of the newly activated 26th Engineer Battalion were formed by assigning new designations to the three companies assigned to the 11th, 196th, and 198th Infantry Brigades. To flesh out the rest of the new battalion, Ploger and Rollins transferred men and equipment from the 18th Engineer Brigade, mostly from the 39th Engineer Battalion, to form the headquarters and Company D. The command also gave the 554th Float Bridge Company a new designation: Company E, the 26th Battalion's bridge company. Elements lost by the 39th Engineer

[25] On 1 August 1967, the 3d Brigade of the 25th Infantry Division was redesignated the 3d Brigade of the 4th Infantry Division. The 4th Division's original 3d Brigade, which had been operating under the 25th Division in III Corps since entering Vietnam, was simultaneously redesignated the 3d Brigade of the 25th Division. In addition, both engineer companies attached to the two brigades exchanged designations. This change of title resolved some administrative problems of two brigades located so far away from their divisions. It also brought the 3d Brigade, 4th Division, in southern I Corps a step closer to rejoining its parent division in the highlands. MacGarrigle, *Taking the Offensive*, pp. 245, 267.

[26] ORLL, 1 Aug–31 Oct 67, American Div, 26 Nov 67, pp. 1–2, Historians files, CMH; Memo, Palmer for COMUSMACV, 5 Nov 67, sub: Engineer Requirements in American Division TAOR, p. 1; Sharp and Westmoreland, *Report*, p. 138.

Battalion were reconstituted from Engineer Command assets. Colonel Castro moved over to the 26th Engineer Battalion in mid-December as its first commander. On 20 January 1968, the 39th Engineer Battalion was released from its attachment to the American Division and returned to 45th Group control. The battalion, now under the command of Lt. Col. James M. Miller, continued to provide general support to the American Division.[27]

While the introduction of an Army division threw the enemy off balance in southern I Corps, the North Vietnamese kept relentless pressure on the marines all along the Demilitarized Zone. By September, the increased shelling endangered the construction of strongpoints and combat bases parallel to the Demilitarized Zone. The construction by Marine engineers, begun in April, was part of the long-debated electronic anti-infiltration barrier, the "McNamara Line," ordered by Secretary McNamara to be built from the coast to the Laotian border. Mounting casualties prompted Westmoreland and the Marine commander, who both had doubts whether the system would work, temporarily to halt construction and concentrate on improving existing strongpoints and combat bases. As 1967 came to an end, Marine units constantly pounded by North Vietnamese artillery continued to be pinned down inside their fortified bases, hindering their operations.[28]

Meanwhile, the American Division continued the slow process in a seemingly never-ending search for elusive enemy units. In September, Westmoreland assigned all of Quang Ngai and Quang Tin Provinces and part of Quang Nam Province to the American Division, allowing the marines to shift more forces north. Operation WHEELER, which ran from 11 September to 15 October, was an attempt to clear the area in the interior northwest of Chu Lai. In October, the deadly game of cat and mouse continued north of WHEELER in the Que Son Valley as Operation WALLOWA. This time the 1st Cavalry Division's 3rd Brigade, operating under the control of the American Division, was brought north from Binh Dinh. After relieving elements of the 1st Marine Division, the cavalry brigade launched a series of air assaults in the valley. This operation, combined with the 1st Brigade, 101st Airborne Division, and later the 196th Infantry Brigade, extended into the following year. To the south the 3d Brigade, 4th Division, in the Duc Pho and Mo Duc districts faced the same situation. While these operations took a toll against the enemy, the North Vietnamese in the area remained an effective threat. In the end, the dry-season

[27] Rpt of Visit to Various Headquarters in Vietnam, 22 Nov–13 Dec 67, incl. 7, p. 1, OCE Liaison Officer Trip Rpt no. 10, 19 Jan 68, OCE Hist Ofc; ORLLs, 1 Nov 67–31 Jan 68, Engr Cmd, Jan 68, p. 15, 1 Nov 67–31 Jan 68, 39th Engr Bn, 9 Feb 68, pp. 2, 8, both in Historians files, CMH; Msg, Maj Gen Robert R. Ploger, CG, Engr Cmd, GVP 921 to Brig Gen Andrew P. Rollins Jr., CG, 18th Engr Bde, 3 Nov 67, sub: Engineer Battalion of 23d Inf Div (American); Msg, Rollins ARV 1978 to Ploger, 9 Nov 67, sub: American Engineer Battalion, both in Robert R. Ploger Papers, CMH; Memo, Palmer for COMUSMACV, 5 Nov 67, sub: Engineer Requirements in American Division TAOR, pp. 1–2.

[28] MacGarrigle, *Taking the Offensive*, pp. 255–63. For more on the Demilitarized Zone barrier, see Pearson, *War in the Northern Provinces*, pp. 21–24; Shulimson, *An Expanding War*, pp. 414–19; and Telfer, Rogers, and Fleming, *Fighting the North Vietnamese*, pp. 86–94.

operations in southern I Corps were not the ambitious sweeps into the interior that Westmoreland had anticipated.[29]

Highway 1 and Other Assignments

During most of 1967, the 1st Cavalry Division continued to put pressure on the enemy in neighboring Binh Dinh Province. Operation PERSHING began on 11 February with the purpose of eliminating Viet Cong and North Vietnamese forces from that rice-rich coastal province. A continuation of the cavalry division's campaigns in Binh Dinh Province since early 1966, PERSHING continued into early 1968.[30]

While operating on the coastal plains of Binh Dinh, the division combined the surprise of cavalry with the shock of armor to overwhelm North Vietnamese Army and Viet Cong defenses. Enemy fortifications were well organized and typically prepared in a series of hedgerows quickly thrown up in an elaborate perimeter that gave excellent cover and concealment around an entire village. During an attack, enemy troops remained ensconced in their bunkers, and 1st Cavalry Division forces could only kill or capture them by destroying the bunkers. This led to the attachment of a tank company from the 4th Division. Both the M48A3 tank's 90-mm. gun and heavy armor proved the effectiveness of using armor to deal with the fortifications. The tank's heavy weight alone, or in concert with the engineers' bulldozers, was able to crush trench lines and bunkers. In September, the division received another armor capability with the attachment of the 1st Battalion (Mechanized), 50th Infantry, a recent arrival in Vietnam. General Tolson did not confine this newest unit to mechanized infantry operations, directing the battalion to train for airmobile warfare. With the armored personnel carriers at a central position near Landing Zone UPLIFT, troops could be flown from any landing zone when a mission needed a mechanized unit. Engineer minesweeping teams and bulldozers normally accompanied the armored columns. The onset of the monsoon limited the movement of armor cross-country, and engineering equipment, especially bulldozers and portable bridging, helped overcome obstacles and gaps along the way.[31]

The 1st Cavalry Division deployed troops to I Corps more than once. By the spring of 1967, the enemy had suffered severe troop and supply losses. Relentlessly pursued in every direction of Binh Dinh Province, the remaining elements of the North Vietnamese *3d People's Army of Vietnam (PAVN) Division* sought refuge in adjacent Quang Ngai Province. These frequent retreats to the north, to rest and regroup, contributed to the 1st Cavalry Division's participation in the LE JEUNE operation in April. In May, elements

[29] MacGarrigle, *Taking the Offensive*, pp. 264–83; Telfer, Rogers, and Fleming, *Fighting the North Vietnamese*, pp. 119–20.

[30] Tolson, *Airmobility*, pp. 129–30; MacGarrigle, *Taking the Offensive*, pp. 180–85.

[31] Tolson, *Airmobility*, pp. 130, 139–41; MacGarrigle, *Taking the Offensive*, pp. 186–89, 321; ORLLs, 1 Feb–30 Apr 67, 1st Cav Div, 23 May 67, p. 29, 1 May–31 Jul 67, 8th Engr Bn, 31 Jul 67, pp. 2–3, 12–13, 1 Aug–31 Oct 67, 8th Engr Bn, 31 Oct 67, p. 4; AAR, Opn PERSHING, 1st Cav Div, 29 Jun 68, pp. 13–15. ORLLs and AAR in Historians files, CMH.

of the 3d Brigade based in the An Lao Valley started a series of battalion-size air assaults in southern Quang Ngai Province. In August, the brigade made a major reconnaissance in force into Song Re Valley, for years a sacrosanct Viet Cong stronghold northwest of Duc Pho. By late September, the steady enemy buildup in I Corps generated another call for reinforcements. This time the 3d Brigade moved to Chu Lai and, operating under the control of the Americal Division, began Operation WALLOWA. After relieving elements of the 1st Marine Division, the cavalry brigade launched a series of air assaults in the Que Son Valley. This operation, combined with the 1st Brigade, 101st Airborne Division, and later the 196th Infantry Brigade (WHEELER/WALLOWA), extended into the following year.[32]

In late 1967, the division also fought a major battle in northern Binh Dinh Province during Operation PERSHING. The Battle of Tam Quan took place late in the campaign between 6 and 20 December. By chance a scout team discovered an enemy radio antenna near the town of Tam Quan. The 1st Cavalry Division used "piling on" tactics that had proved successful in earlier operations. An infantry battalion and elements of the mechanized infantry battalion deployed first. The 1st Brigade added forces and teamed up with South Vietnamese troops in a battle characterized by massive use of artillery, tactical air support, and air assaults. The mechanized force, which included flame-throwing armored personnel carriers, closed in to destroy and crush enemy trenches. Combat engineer bulldozers from the 19th Engineer Battalion supported the attackers by building a causeway over the spongy ground, burying trench lines, and clearing areas for helicopter medical evacuation. Enemy forces, consisting primarily of a North Vietnamese Army regiment, lost 650 men during this fierce engagement.[33]

During these operations, the 8th Engineer Battalion provided the full range of support in the field and at the combat bases. In northeastern Binh Dinh, the division dotted the area of operations with many landing zones of various types, which enabled air cavalry forces to move and fight anywhere quickly. The 8th Engineer Battalion helped build or improve existing landing zones at mountaintop firebases (HUMP and LARAMIE), Caribou airstrips (LITTS and MAHONEY), and forward operating bases (the division's tactical command post at TWO BITS near Bong Son, and GERONIMO, ENGLISH, PONY, UPLIFT, and HAMMOND). Landing Zone TWO BITS also included a C–123 runway, and Landing Zone ENGLISH, thanks to continuous improvements made by 45th Engineer Group units, could handle C–130 transports. The engineers also helped pacification efforts to return territory to government control. This activity centered on reopening and improving Highway 1 and local roads.[34]

Within the framework of pacification, the 8th Engineer Battalion carried out contrasting missions of destruction and building. While supporting the 2d

[32] Tolson, *Airmobility*, pp. 142–44, 147–49; MacGarrigle, *Taking the Offensive*, pp. 186–92, 263, 275–81, 315–19.

[33] Tolson, *Airmobility*, pp. 149–50; MacGarrigle, *Taking the Offensive*, p. 322; AAR, Opn PERSHING, 1st Cav Div, p. 16.

[34] Engineer Support of Combat Operations, 26 May 67, incl. 8, pp. 2–3, OCE Liaison Officer Trip Rpt no. 7; ORLL, 1 May–31 Jul 67, 8th Engr Bn, pp. 1–7.

Brigade's operations on the coast some twenty-five miles north of Qui Nhon, Company B destroyed several well-prepared bunkers in and around the villages. During the first week in June, Company A working with the 1st Brigade north of Landing Zone ENGLISH reopened a five-mile section of Highway 1 to the I Corps boundary. After airlifting earthmoving equipment to the work site, the engineers opened a pioneer road in four days. When the 1st Brigade pushed into the An Lao Valley northwest of Bong Son, Company A reopened a section of Route 514, erecting several timber bridges along the way. Company A also answered the brigade's call to destroy fortified villages in the valley.[35]

Repairing damage caused by enemy attacks on the forward bases and countering his increased use of mines and booby traps drew heavily on the 1st Cavalry Division's engineers. In one seventy-day period during Operation PERSHING, the 8th Engineer Battalion reported 268 discoveries of mines or booby traps. On 6 June, the ammunition dump at Landing Zone ENGLISH was blown up, and the battalion straightaway committed troops and equipment to build a new seven-bay ammunition storage area.[36]

Meanwhile, the 45th Engineer Group began a northward shift in northern II Corps and southern I Corps to reopen and upgrade Highway 1. In May, the 18th Engineer Brigade moved the group's southern boundary northward from Ninh Hoa to Tuy Hoa. Concurrently, the 35th Group took control of the 577th Engineer Construction Battalion and responsibility for projects at Tuy Hoa and Port Lane. The arrival of the 589th Engineer Battalion (Construction) from Fort Hood, Texas, in late April restored the 45th Group's construction capability, allowing Col. Kenneth T. Sawyer, who replaced Colonel Bush in May, some flexibility to refocus the efforts of his three combat battalions from base construction to tactical operations and roadwork. By this time, the 35th and 39th Engineer Battalions were supporting the 1st Cavalry Division and Task Force OREGON on a regular basis. Both battalions were also hard at work reopening and upgrading Highway 1 in their areas of responsibility. In addition, Colonel Sawyer planned to move the 19th Engineer Battalion northward and position it between the 35th and 39th Engineer Battalions to work on the road between Bong Son and Duc Pho.[37]

The 35th Engineer Battalion had been providing the 45th Group's principal backup engineering support to the 1st Cavalry Division since late 1966. At the height of the northeast monsoon, Colonel Peel had moved his Headquarters and Company A from Cha Rang Depot to Landing Zone HAMMOND to support Operation THAYER. This mission continued during PERSHING, and by late April the battalion centered its efforts on supporting the division and improving Highway 1 between Qui Nhon and Landing Zone ENGLISH. Company B

[35] Engineer Support of Combat Operations, 1 May–1 Jul 67, incl. 7, p. 29, OCE Liaison Officer Trip Rpt no. 8, 18 Aug 67, OCE Hist Ofc.

[36] Engineer Support of Combat Operations, 26 May 67, incl. 8, p. 3, OCE Liaison Officer Trip Rpt no. 7; Engineer Support of Combat Operations, 1 May–31 Jul 67, incl. 7, p. 2, OCE Liaison Officer Trip Rpt no. 8; ORLLs, 1 Feb–30 Apr 67, 1st Cav Div, p. 29, 1 May–31 Jul 67, 8th Engr Bn, p. 2.

[37] ORLLs, 1 Feb–30 Apr 67, 45th Engr Gp, 11 May 67, pp. 1–5, and 1 May–31 Jul 67, 45th Engr Gp, 14 Aug 67, pp. 2–6, both in Historians files, CMH.

moved by air and road to ENGLISH to maintain a section of Highway 1 and the airstrip. Despite the earlier work by the 19th Engineer Battalion, the runway at ENGLISH started to fail. When the heavy rains slackened in mid-February, Company B removed the matting, repaired the subgrade, and laid new M8A1 matting.[38]

Colonel Peel immediately assigned the attached Company D, 20th Engineer Battalion, the job of building all semipermanent bridges along Highway 1 between Phu Cat and Bong Son, a mission that soon extended to the battalion's entire length of the road. By the end of July, Company D, which on 1 May had been redesignated Company D, 35th Engineer Battalion, averaging two to three new bridges each week, completed thirty timber trestle bridges. The new bridges consisted of timber pile abutments, bents, and piers, and wood and steel stringers depending on the lengths of spans. This arrangement lasted until the autumn, when the impending monsoon dictated a change in the battalion's construction techniques, and each company took charge for all work along its assigned section of the road.[39]

When Lt. Col. David N. Hutchison took command of the 35th Battalion on 1 July, he inherited a battalion now committing most of its resources to improving Highway 1. The emphasis on the roadwork reflected the high priority that MACV and the South Vietnamese government gave to reopening major roads, especially Highway 1 running from the Demilitarized Zone to Saigon and on to the Cambodian border. Each morning mine detection teams checked the road between Landing Zones HAMMOND and ENGLISH, later extending north to the I Corps border and south to the Phu Tai ammunition depot south of Qui Nhon. Work parties followed to work on sections of the road assigned to each company. Using specifications set by the MACV Construction Directorate, the troops widened and elevated the road, placed culverts, and repaired and replaced bridges.[40]

Replacing some old narrow French-built concrete spans along Highway 1 required their destruction. Tight working space in populated areas prevented the construction of the new spans parallel to the old bridge, which was done when possible. After improving a bypass around the bridge to carry traffic, the engineers sometimes had to destroy the old bridge with explosives. On the outskirts of one town north of Qui Nhon, however, the concussion of the blast dislodged some roof tiles of several nearby buildings. As the dust and smoke cleared, the onlookers could hear the sound of crashing tiles as they slid off the roofs of nearby buildings and struck the ground.[41]

Specialized engineer units augmented the 35th Engineer Battalion's road-building capability. Attachments from two separate companies that arrived

---

[38] ORLL, 1 Feb–30 Apr 67, 35th Engr Bn, 9 May 67, pp. 1–5, Historians files, CMH.

[39] Road Construction by Army Engineer Troops, incl. 9, tab F20, p. 1, OCE Liaison Officer Trip Rpt no. 7; ORLLs, 1 Feb–30 Apr 67, 35th Engr Bn, p. 5, 1 May–31 Jul 67, 35th Engr Bn, 10 Aug 67, pp. 2, 4–5, and 1 Aug–31 Oct 67, 35th Engr Bn, 8 Nov 67, p. 6, last two in Historians files, CMH.

[40] ORLLs, 1 May–31 Jul 67, 35th Engr Bn, p. 2, and 1 Aug–31 Oct 67, 35th Engr Bn, p. 2.

[41] ORLL, 1 Aug–31 Oct 67, 35th Engr Bn, p. 7; MFR, 21 Feb 92, sub: Author's Recollections, 45th Engr Gp, p. 1, Historians files, CMH.

Mines were a constant hazard as the 35th Engineer Battalion worked along Highway 1 between Qui Nhon and Bong Son.

in June included the 2d Platoon, 137th Light Equipment Company, and the 1st Platoon, 70th Dump Truck Company. The haul capability of the dump truck platoon's 24 five-ton dump trucks greatly increased the amount of fill, subcourse, and base course materials used by the 35th Engineer Battalion. The arrival of the 517th Light Equipment Company in July further increased the battalion's equipment capabilities. Once settled in, the 517th replaced the 137th's platoon and began operations on 1 August. The small rock crusher, earthmovers, bucket loaders, graders, cranes, and bulldozers of the light equipment company made possible the amount of work done by the battalion before the monsoon rains began that autumn. The cranes also increased pile-driving operations, thus allowing the simultaneous construction of several bridges. At Bong Son, the 553d and 554th Float Bridge Companies helped Company B span the Lai Giang with a 224-foot M4T6 floating bridge strung between two causeways. South Vietnamese Army engineers also did some bridge work on Highway 1 with materials provided by the 35th Engineer Battalion.[42]

The road improvement effort in Binh Dinh Province received extensive support from RMK-BRJ's Phu Cat industrial site. Initially established to support the construction of Phu Cat Air Base, the quarry and asphalt plant set aside an increasing amount of production for Army projects. Sometimes this placed

[42] Engineer Support of Combat Operations, 26 May 67, incl. 8, p. 3, OCE Liaison Officer Trip Rpt no. 7; ORLLs, 1 May–31 Jul 67, 35th Engr Bn, p. 2, 5, and 1 Aug–31 Oct 67, 35th Engr Bn, pp. 2, 5.

RMK-BRJ paves a portion of Highway 1 north of Qui Nhon.

the two services in competition for the same products, especially crushed rock. The 45th Group needed rock for the road, but the air base commander insisted on getting two-inch clean rock to fill aircraft revetments. Fortunately, the efficient plant served both needs. By mid-1967, the consortium's asphalt plant was in full operation, as shown by newly paved streets in Qui Nhon and twenty-one miles of new pavement along Highway 1 from the port city to the air base. Large belly dump trucks delivered hot asphalt to the firm's paving machine. Once the asphalt was spread by the paving machine, rollers compacted the two-inch surface. Operations then shifted to Highway 19. Following behind the 589th Engineer Battalion's base course work, the contractor hauled and placed the asphalt surface to An Khe. The forty-one-mile stretch between the intersection with Highway 1 and the city of An Khe and the gate of Camp Radcliff was finished before the end of the year.[43]

Convoy traffic on both Highways 1 and 19 greatly increased but moved along briskly thanks to the newly surfaced roads. An asphalt surface provided a swifter all-weather supply route link and made it difficult for the enemy to mine. Since the trucks and semitrailer moving inland on Highway 19 had to negotiate sharp curves along the An Khe Pass, the contractor paved the shoulders. The Viet Cong could still place command-detonated mines by tunneling

---

[43] Engineer Support of Combat Operations, 26 May 67, incl. 8, p. 4, OCE Liaison Officer Trip Rpt no. 7; MFR, 21 Feb 92, sub: Author's Recollections, 45th Engr Gp, p. 1.

a hole from the embankment, but at least the drivers could see any tampering with the road's surface.[44]

To get a jump on paving Highway 1, Colonel Sawyer arranged with local RMK-BRJ officials to work on Sundays. The foremen agreed to operate the asphalt plant on their off day and to lend the paving equipment to the 35th Engineer Battalion. During the week, Company A prepared the road with a six-inch base course, which included a three-inch layer of crushed stones obtained from the Phu Cat quarry. Using its dump trucks to haul the asphalt, the battalion paved almost one mile each Sunday. By the end of October, the total reached six miles. Commenting on the team work between the group and the contractor, Colonel Sawyer explained that the RMK-BRJ industrial site at Phu Cat had given his group the capability of an additional engineer construction support company.[45]

In mid-July, the 19th Engineer Battalion started its move north. Colonel Rhodes and his staff coordinated the deployment with the two major tactical units on both sides of the corps boundary, the 1st Cavalry Division and the 3d Brigade, 25th Division. The 19th Battalion reconnoitered several potential base camp sites. One hilltop site near Tam Quan had a commanding view of the highway and the South China Sea, but it had to be rejected. An access road would have to be built, and considerable earthwork would be required to build a base camp atop the hill. Also looking for a suitable site for the battalion's main camp, Colonel Sawyer spotted from his helicopter a slight rise of land amid rice paddies just off Highway 1, some six miles north of Landing Zone ENGLISH. The group commander selected this location, known by the 1st Cavalry Division as Sniper's Island. Before the battalion's move north, Rhodes departed Vietnam, leaving the executive officer, Maj. Richard W. Stevens, temporarily in charge. Following the dispatch of an advance party to Sniper's Island, a task force from battalion headquarters, Companies A and B, and elements of the 137th Light Equipment Company departed the Long My base camp before dawn on Sunday, 16 July. The task force, led by the operations officer, Maj. Adrian G. Traas, rendezvoused with minesweep teams from the 35th Engineer Battalion north of Qui Nhon. Traveling some sixty-two miles up Highway 1, the convoy arrived at its destination without incident. That afternoon an infantry company from the 1st Cavalry Division arrived at Sniper's Island to help guard the hastily drawn perimeter for the first night. A few days later, Lt. Col. Andrew C. Remson Jr. flew in by helicopter to take command of the battalion.[46]

At Sniper's Island, the 19th Engineer Battalion set forth to build a defensible base camp and to start work on Highway 1. Scrapers and bulldozers moved earth to form a protective berm around the perimeter. Using lumber

[44] MFR, 21 Feb 92, sub: Author's Recollections, 45th Engr Gp, p. 1.

[45] ORLLs, 1 Aug–31 Oct 67, 35th Engr Bn, 8 Nov 67, p. 3, Historians files, CMH, and 1 Aug–31 Oct 67, 35th Engr Bn, p. 3; MFR, 21 Feb 92, sub: Author's Recollections, 45th Engr Gp, p. 1.

[46] MFR, 21 Feb 92, sub: Author's Recollections, 19th Engr Bn, pp. 1–2; Interv, Maj Paul D. Webber, CO, 26th Mil Hist Det, with Lt Col Andrew C. Remson, CO, 19th Engr Bn, 22 Apr 68, incl. 2, p. 1, VNIT 80, CMH; ORLL, 1 May–31 Jul 67, 19th Engr Bn, 31 Jul 67, p. 2. Both in Historians files, CMH.

hauled from Qui Nhon, troops built a tactical operations center and mortar-proof live-in bunkers along the berm and inside the perimeter. An artillery forward observer stayed with the battalion for several days to register artillery fire around the perimeter. In addition, an artillery unit took up position next to the camp for a short time. Meanwhile, reconnaissance parties set out to examine the road and bridge sites between Bong Son and Duc Pho. Soon work was under way. Abutments on several destroyed bridges were removed in preparation for the construction of timber trestle bridges, and concrete was used to patch up holes along the road.[47]

By late September, the 19th Battalion had completed its deployment. Company D, which was newly formed in June, remained at Long My until late August when it moved to a new bivouac site called Landing Zone THUNDER approximately five miles south of Duc Pho. In late September, Company C moved six miles north of the battalion headquarters, now known as Landing Zone ENGLISH NORTH. Like the 35th Engineer Battalion, the 19th Engineer Battalion with its attached units grew into a large reinforced battalion with over one thousand men. The 1st Cavalry Division considered a force this size straddling Highway 1 beneficial because the engineers' presence helped block the free movement of the enemy in several places. The 137th Light Equipment Company opened quarries at Tam Quan and Duc Pho, and the 513th Dump Truck Company added a section of eight 5-ton dump trucks. In mid-December, the 73d Construction Support Company, which had provided the 45th Group with rock crushing and paving in and around Qui Nhon, moved from its Phu Tai complex to join Company C at Landing Zone LOWBOY and set up operations at Tam Quan and Duc Pho. At the same time, the battalion's responsibility was extended ten miles north of Duc Pho to Mo Duc, and Company B moved to Landing Zone MAX six miles north of Duc Pho.[48]

At first the battalion's roadwork centered on clearing and upgrading Highway 1 to the I/II Corps border and improving on the 8th Engineer Battalion's pioneer work to upgrade the road to all-weather traffic. This involved filling all large holes and building hasty bypasses, covering the road with a sand-asphalt mix seal coat, using culverts where necessary, and spanning sixteen gaps where the enemy had destroyed the bridges. New spans included seven panel bridges (totaling 690 feet long), three decked railroad bridges used as substitute highway bridges, and six timber trestle bridges. Work was completed ahead of schedule on 8 September. The battalion moved to the next stage of work to upgrade the road, dubbed Seahorse Highway in tribute to the unit's insignia depicting a seahorse, all the way to Duc Pho to MACV standards.[49]

---

[47] MFR, 21 Feb 92, sub: Author's Recollections, 19th Engr Bn, pp. 1–2; ORLL, 1 May–31 Jul 67, 19th Engr Bn, p. 5.

[48] MFR, 21 Feb 92, sub: Author's Recollections, 19th Engr Bn, p. 2; ORLLs, 1 Aug–31 Oct 67, 19th Engr Bn, 31 Oct 67, pp. 1–3, and 1 Nov–31 Jan 68, 19th Engr Bn, 31 Jan 68, pp. 1, 3–4, both in Historians files, CMH.

[49] ORLLs, 1 Aug–31 Oct 67, 19th Engr Bn, pp. 2–4, and 1 Nov–31 Jan 68, 19th Engr Bn, 31 Jan 68, p. 3.

After moving north of Bong Son, the 19th Engineer Battalion set up a main base camp on the rise of land overlooking the surrounding rice paddies.

By the end of the year, six U.S. engineer battalions of three services shared minesweeping and roadwork along Highway 1 in northern Binh Dinh and Quang Ngai and Quang Tin Provinces. The 35th Engineer Battalion continued its work from Qui Nhon to Bong Son; the 19th Engineer filled the gap between Bong Son and Duc Pho and later to Mo Duc; and the 39th Engineer Battalion from Mo Duc to Chu Lai. The newly activated 26th Engineer Battalion of the Americal Division took on a portion of the minesweeping duties along this stretch of road. The 9th Marine Engineer Battalion and Naval Mobile Construction Battalion 6 carried out minesweeps and roadwork north of Chu Lai.[50]

The improved road helped the flow of combat troops and cargo and civilian traffic. In late December, Highway 1 served as the most practical means to move the arriving 11th Light Infantry Brigade from Qui Nhon to Duc Pho, where it joined the Americal Division. Pacification efforts in the area also made civilian travel less dangerous, at least during daylight. This measure of security and further improvements, as shown by the construction of new bridges and paving, prompted a resurgence of commercial traffic and a growing confidence on the part of the South Vietnamese government, which tried to reclaim control over the region. A determined enemy, however, struck back. With renewed vigor, the Viet Cong waged a relentless campaign of mining the roads, ambushing vehicles, and damaging or destroying recently built bridges.

---

[50] ORLLs, 1 Aug–31 Oct 67, 45th Engr Gp, pp. 4–5, 1 Nov 67–31 Jan 68, 45th Engr Gp, pp. 4–5, 1 Nov 67–31 Jan 68, Americal Div, 8 Feb 68, p. 30, last in Historians files, CMH.

Even though enemy activity along the Americal Division's section of Highway 1 was extremely high, the road was never closed for an entire day. Destroyed or damaged bridges were usually discovered during early morning reconnaissance overflights or by minesweep teams. Normally, the road was open to traffic by noon or at least open to traffic before nightfall.[51]

As usual, the Viet Cong displayed ingenuity when using mines and explosives to disrupt traffic. The 35th Engineer Battalion reported that the enemy had taken advantage of the American soldier's reluctance to get dirty if he did not have to, by burying mines deep under mud puddles or covering them with a layer of water buffalo dung. Scoop loader operators working in borrow and fill pits also had to be wary of mines buried deep in the sand and rock. In October 1966, Viet Cong sappers had expertly destroyed one of the central concrete piers of the 1,352-foot highway bridge over the Lai Giang at Bong Son. The explosion dropped two spans into the river and left a 194-foot gap. Fortunately, traffic could still cross over the railroad bridge. Later, the engineers responded with several projects that improved crossings over the river. Starting in late May 1967, Company B of the 35th Engineer Battalion built two new timber pile piers and raised and repaired the double Eiffel span, closing the gap on 29 July. Also, with the end of the monsoon several months earlier, the battalion returned the M4T6 floating bridge to service. In July, Company B replaced the timber decking on the railroad bridge. Three spans were now available to cross traffic over Lai Giang at Bong Son.[52]

At other crossings, the Viet Cong resorted to a simple method of damaging the new timber trestle bridges by setting the decks on fire. The engineers countered by placing sheet metal under the treadway and dirt, gravel, and asphalt on the decking. The battle of wits between the adversaries continued. The enemy next started to burn the creosoted piles and sun-dried timbers of the substructures and the undersides of the superstructures. One bridge near Tam Quan became a frequent target. During the initial attack, only the high level of the water under the bridge prevented the piles from being burned completely to the ground. The 35th Engineer Battalion rebuilt the center span of the 221-foot bridge with steel I-beams and stringers on top of two hollow-steel pile bents, each made up of four piles filled with concrete. The Viet Cong's success in destroying bridges this way also depended on the steadfastness of the security forces. North of Qui Nhon, soldiers of the South Korean Capital Division zealously defended their positions. Farther north, the Viet Cong usually brushed past the South Vietnamese bridge guards without much resistance.[53]

Although travel along Highway 1 between Qui Nhon and Bong Son became fairly secure during the day, the enemy still ambushed vehicles and work parties. One of these ambushes took place at twilight on 9 October when a small party from the 35th Engineer Battalion en route to its base camp at Landing

[51] ORLL, 1 Nov 67–31 Jan 68, Americal Div, pp. 23, 31, 33.

[52] ORLLs, 1 Feb–30 Apr 67, 35th Engr Bn, p. 4, 1 May–31 Jul 67, pp. 4, 6–7. For more on the Bong Son Bridge repair, see Capt. Gerald W. Kamicka, "Rebuilding the Bong Son Highway Bridge," *Military Engineer* 60 (March-April 1968): 112–13.

[53] ORLLs, 1 Aug–31 Oct 67, 45th Engr Gp, pp. 8–9, 1 Aug–31 Oct 67, 35th Engr Bn, p. 9.

Zone HAMMOND came under fire from both sides of the road. The barrage wounded the officer in charge, 2d Lt. Robert E. Knadle. He directed his men to defensive positions while remaining in the open near his disabled jeep to radio for reinforcements and medical evacuation. Knadle operated the radio until wounded again, this time by an enemy grenade. Although mortally wounded, the lieutenant remained in charge, leading his men against the attackers until reinforcements arrived. The U.S. Army posthumously awarded Lieutenant Knadle the nation's second highest award, the Distinguished Service Cross.[54]

Farther north the 19th and 39th Engineer Battalions, working in the contested territory between Bong Son and Chu Lai, suffered even more casualties. Soon after arriving at Landing Zone ENGLISH NORTH, the 19th Battalion experienced its first casualty. During a minesweeping operation along Highway 1 north of the base, a jeep detonated a mine, which killed one soldier. Another ambush on the same road struck vehicles of the 137th Light Equipment Company, resulting in several casualties. From November to the end of January, a period that included the battalion's participation in the Battle of Tam Quan, the 19th Engineer Battalion and its attached units lost three killed and five wounded. Along the 39th Engineer Battalion's section of the highway, Company C was frequently fired upon by snipers and lost men and vehicles to mine explosions. From May through July, Companies C and D suffered three killed and fifteen wounded along the road. Elements of Company B sweeping a local road for mines south of Chu Lai discovered two Viet Cong placing a mine. The team's point man fired his grenade launcher and killed one enemy soldier. In late July, the enemy ambushed a platoon along the same road twice, resulting in two dead and three wounded Americans.[55]

Mines were especially insidious. From May through July, the 39th Battalion discovered sixty mines and thirty-one booby traps. Twelve mines and eight booby traps were detonated accidently. Most of the mines consisted of bamboo firing devices with an electric blasting cap and about twenty pounds of explosive. Seldom were the mines marked in any way. Most of the booby traps were hand grenades with trip wires connected to the pin. The 35th Engineer Battalion reported the enemy's clever use of empty chambers so that minesweep teams could pass by. After the minesweep team and first convoy passed by, the Viet Cong returned, removed the board over the top of the chamber, quickly placed a mine, and topped off the chamber with fill to make the road appear undisturbed. Several vehicles returning as little as three hours later were destroyed.[56]

October brought forth the annual northeast monsoon. In the 19th Engineer Battalion's sector, the first heavy rains, twelve inches in one twenty-four-hour period, caused washouts along Highway 1. The battalion rallied by building

---

[54] General Order 6340, USARV, 10 Dec 67, copy in CMH; Robert G. McClintic, "Clearing the Way," *Engineer* 1 (Fall 1971): 9–10; "35th Lt. Receives DSC Presented Posthumously," *Colt 45 News* 2 (2 April 1968): 1, Historians files, CMH.

[55] MFR, 21 Feb 92, sub: Author's Recollections, 19th Engr Bn, pp. 2–3; ORLLs, 1 May–31 Jul 67, 39th Engr Bn, pp. 3–5, 1 Aug–31 Oct 67, 19th Engr Bn, p. 2, 1 Nov 67–31 Jan 68, 19th Engr Bn, p. 3.

[56] ORLLs, 1 May–31 Jul 67, 39th Engr Bn, pp. 9–10, 1 Feb–30 Apr 67, 35th Engr Bn, pp. 5–6.

two panel bridges, one dry span, and one float bridge in less than seventy hours. The 35th Battalion faced similar problems in several areas where new bridges remained incomplete and the bypasses exposed to washouts. One evening, the rising waters threatened a washout of a bypass north of Qui Nhon despite the efforts of the 35th Engineer Battalion, which had used all the available crushed rock from the RMK-BRJ Phu Cat quarry. Although the quarry did have some rock destined for aircraft revetments at the air base, employees on the scene hesitated to release the more expensive two-inch washed rock normally used in concrete work. The 35th Engineer Battalion appealed to the group commander, Colonel Sawyer. Sawyer dispatched Major Traas, now the group contract liaison and installation master planning officer, to persuade local RMK-BRJ officials, who lived across the street from the headquarters in Qui Nhon, to issue the rock. By using the substitute rock as fill, the 35th Engineer Battalion managed to keep the crucial road to Bong Son open the next morning.[57]

In southern I Corps, a tropical storm dumped seventeen inches within twenty-four hours. Flooding washed out sections of Highway 1, and it took the 39th Engineer Battalion one week to repair the damage. The rains also closed RAZORBACK BEACH at Duc Pho, and supplies could only be delivered by air. Looking for an alternate over-the-beach site, the American Division ordered engineers to build a limited facility at the small port town of Sa Huynh located just north of the I/II Corps border. The 19th Engineer Battalion built a one-half-mile access road through a marshy area from Highway 1 and a 165-by-250-foot off-loading area. With this work completed on 15 October, the odds to continue over-the-beach supply operations during the monsoon for units based at Duc Pho were improved. Meanwhile, the engineers continued roadwork, improved the Duc Pho airstrip, and expanded ammunition and fuel storage sites.[58]

Besides roadwork and daily minesweeps, the 45th Group's three combat battalions carried out other tasks. The 35th Engineer Battalion continued to maintain the ENGLISH airstrip, built artillery-firing platforms at Landing Zones UPLIFT and PONY, rehabilitated the Phu Cat Regional Forces Training Center, and cleared fields of fire at Landing Zone PONY and Phu Cat Air Base. The 19th Engineer Battalion built bunkers and helicopter revetments at Landing Zone TWO BITS for the 1st Cavalry Division. During the Battle of Tam Quan, elements of Company A and the 137th Light Equipment Company helped destroy bunkers and tunnel complexes, killing ten enemy soldiers in the process. The 39th Engineer Battalion supported the American Division with jungle clearing, bunker construction, and destroying enemy tunnel and bunker complexes and captured munitions.[59]

[57] ORLL, 1 Aug–31 Oct 67, 19th Engr Bn, p. 4; MFR, 21 Feb 92, sub: Author's Recollections, 45th Engr Gp, p. 1.

[58] MacGarrigle, *Taking the Offensive*, p. 269; ORLLs, 1 Aug–31 Oct 67, American Div, p. 35, 1 Aug–31 Oct 67, 45th Engr Gp, p. 4, 1 Aug–31 Oct 67, 19th Engr Bn, p. 3.

[59] ORLLs, 1 Aug–31 Oct 67, 45th Engr Gp, pp. 4–5, and 1 Nov 67–31 Jan 68, 45th Engr Gp, p. 4.

By the end of 1967, it seemed certain that the U.S. Army's role in I Corps would grow. Earlier that year, other than advisers and Special Forces troops, the Army only had several batteries of 175-mm. guns and 105-mm. self-propelled howitzers in northern I Corps—the 175-mm. guns delivering long-range fire support into and across the Demilitarized Zone and the howitzers giving direct support to the marines. By year's end with the Americal Division and supporting troops on the scene, Army troops in I Corps numbered over twenty thousand with more to come. Together, Army, Navy, and Marine Corps engineers developed the combat bases at Chu Lai and Duc Pho and upgraded roads, especially Highway 1. With the anticipated arrival of even more troops in I Corps, additional Army engineers would be drawn northward.[60]

[60] Pearson, *War in the Northern Provinces*, p. 12; Tolson, *Airmobility*, p. 150.

9

Completing the Bases

The momentum of combat that carried over into 1967 did not diminish the construction effort. Military engineers and contractors continued an all-out effort that had started in 1965 to finish the theater's bases, a task that had been given urgency with the resumption of the troop buildup in the summer of 1966. At the start of 1967, roughly half of the envisioned construction was done, but new requirements were still being added. Cost containment was also a growing issue, leading Engineer Command to pass on more work to engineer troops, to approve essential construction only, and to step up coordination between the contractor and troop efforts to save money and materiel. Meanwhile, the completion of projects did not end engineer responsibilities. There were increased demands on the facilities engineers to maintain, repair, and upgrade the expanding bases, which, in turn, meant budgeting more operations and maintenance funds and enlarging the utilities workforce.[1]

Southern and Central II Corps

In southern II Corps, the 35th Engineer Group, Air Force engineers, and the contractors continued to push ahead on base development. And because it constituted the crown jewel of the widening system of II Corps bases, Cam Ranh Bay received the lion's share of engineer attention. As 1967 dawned, the peninsula continued to be dominated by scrub and sand. But all along the fringes, and into the interior as well, lay evidence of a maturing infrastructure of an expanding army. Off the north, near the neck, stood the Air Force base, and farther north still the Army replacement center and a hospital. To the west facing the bay clustered the piers, pipelines, and electrical power ships, while behind the port lay the Army depot, cantonments, and ammunition stores, with Army and Navy maintenance sheds clinging to the southern shore.[2]

The 35th Engineer Group committed the equivalent of two to three battalions to projects in the bay area. Most work went to the 1st Logistical Command's Cam Ranh Bay Support Command, whose 15,000 troops were stationed on every part of the peninsula. In addition to paving roads and building barracks, warehouses, ammunition pads, and tank farms, the engineers constructed an air-conditioned automatic data processing facility for

[1] Dunn, *Base Development*, pp. 71, 74–75; Quarterly Hist Rpts, 1 Oct–31 Dec 66, MACDC, tab A, 1 Oct–31 Dec 67, MACDC, 21 Jan 68, tab A, Historians files, CMH; Letter of Instruction, 18th Engr Bde, 24 Nov 66, tab J, incl. 7, OCE Liaison Officer Trip Rpt no. 6, p. 4.

[2] Dunn, *Base Development*, pp. 68–69; Ploger, *Army Engineers*, pp. 59–60; Quarterly Hist Rpts, 1 Oct–31 Dec 66, MACDC, p. 27, 1 Oct–31 Dec 67, MACDC, p. IV-10; MACV History, 1966, p. 298; ibid., 1967, vol. 2, pp. 831–32.

During 1968, wooden tropical barracks and administrative buildings, designed and built by the 35th Engineer Group such as these buildings in the replacement center, had replaced tents at Cam Ranh Bay.

the logistical command's 14th Inventory Control Center, and also a strategic communications facility on the other side of the bay at Dong Ba Thin. For the inventory control center, Company D, 864th Battalion, had to lay a dust-proof floor and create an extensive power distribution system in the 2,000-square-foot masonry building. At Dong Ba Thin, the 864th's Company B cleared ten acres of heavily vegetated land, constructed access roads, protective berms, generator pads, and administrative buildings, and prepared concrete bases for eleven large antennas.[3]

For its part, RMK-BRJ worked on as many as twenty projects at once at Cam Ranh Bay. In April, the contractor completed a 2,000-foot concrete runway extension thereby giving the air base two 10,000-foot runways. The extended runway was now open to all aircraft, including the ever-increasing number of commercial jets transporting troops to and from Vietnam. The depot area neared completion with the addition of a 200,000-barrel fuel tank farm and more storage facilities. RMK-BRJ made many of the port improvements, which included dredging the ammunition pier's turning basin. The dredged material was put to use as filling behind the bulkheads between the four DeLong piers. And not least of all, RMK-BRJ supplied a steady flow of

---

[3] Galloway, "Essayons," pp. 176, 178, 180–81, 217, 219–20; ORLLs, 1 Nov 66–31 Jan 67, 864th Engr Bn, pp. 3–4, 1 Feb–30 Apr 67, 864th Engr Bn, 13 May 67, pp. 2–5, 1 May–31 Jul 67, 864th Engr Bn, 8 Aug 67, pp. 2–4, all in Historians files, CMH.

ready-mix concrete, crushed rock and asphalt, and building supplies from its depot to Army and Air Force engineers.[4]

Linking the peninsula to the mainland became a reality when RMK-BRJ began driving piles for a new bridge in late June. The combination railroad and highway bridge, located just south of the My Ca ferry and float bridge site, required eight pile bents, fourteen spans (seven each for the railroad and the highway), approaches, abutments, and decking. In September the bridge was well along, the east and west approaches being filled with dredged material, with an estimated date of completion the spring of 1968.[5]

Bringing electricity to the entire peninsula became possible when the Army and Air Force worked out a distribution arrangement. Part of the Army complex, including the convalescent and replacement centers, was located a long way from the planned T–2 tanker power plant and was separated from the main Army installation to the south by the concrete runway. Meanwhile, the Air Force had to plan for electrical power to serve its facilities on either side of the runway. To avoid building two power plants, the Air Force agreed to construct a power plant north of the runway and provide 5,000 kilowatts to the Army complex, while the Army reciprocated by supplying a like amount of power to Air Force units south of the runway. Early in the year, Vinnell connected two of the five T–2 tanker power plants to the electrical distribution system, with a third coming online in late February. The land-based diesel electrical power plant of eight 1,100-kilowatt units, a switching station to protect the primary line, and the last two tanker connections were completed in May. All told, the tankers, in conjunction with the land-based plant, generated 34,200 kilowatts over 121 miles of pole lines (60 miles primary line, 27 miles secondary line, and 34 miles street lighting).[6]

The Nha Trang base complex also saw significant progress. In the late spring of 1967, the Officer in Charge of Construction, to avoid more cost overruns, began phasing down the RMK-BRJ contract at Nha Trang and some other locations and transferring remaining work to engineer troops. Altogether, the firm completed over $31 million in base construction in and around the former sleepy resort. Top effort went to paving the beach road, completing a warehouse and maintenance complex and 2,500-man cantonment for the 1st Logistical Command, and work along Highway 1. The logistical complex and cantonment project were challenging because of the technical work required. These projects featured hot and cold potable water systems and a waterborne sewage system. Meanwhile, Vinnell moved

---

[4] Diary of a Contract, pp. 226–33; Quarterly Hist Rpts, 1 Jan–31 Mar 67, MACDC, pp. 34–35, 1 Apr–30 Jun 67, MACDC, pp. IV-1 to IV-11, and 1 Jul–30 Sep 67, MACDC, pp. IV-12 to IV-13.

[5] Quarterly Hist Rpts, 1 Jan–31 Mar 67, MACDC, pp. 34–35, 1 Apr–30 Jun 67, MACDC, pp. IV-1 to IV-11, and 1 Jul–30 Sep 67, MACDC, pp. IV-12 to IV-13.

[6] "Floating and Land Based Power Plants," pp. 6–7; Development of Cam Ranh Bay, 30 Oct 67, 1st Log Cmd, pp. 6–1 to 6–3, box 13, 68A/4975, RG 319, NARA; Dunn, *Base Development*, p. 80; Fact Sheet, 6 Feb 67, tab J, incl. 8, OCE Liaison Officer Trip Rpt no. 6.

By May 1967, five T–2 power tankers and a land-based power plant began producing electricity at Cam Ranh Bay.

two T–2 power tankers into place, and by July the electrical contractor was capable of distributing 15,000 kilowatts online.[7]

In the Tuy Hoa area, the 577th Construction Battalion split the workload with the 39th Engineer Combat Battalion. The 577th mostly carried out base development, and the 39th handled combat support missions with the exception of Company B, which continued to work at Port Lane, the subsidiary port at Vung Ro Bay. The 577th's major projects included the expansion of the South Korean cantonment at Tuy Hoa and the upgrading of Highway 1 between Vung Ro and the allied encampment just south of Tuy Hoa. Companies B and D maintained and patched the road with cold-mix asphalt, a forerunner to the envisioned widening to twenty-four feet, hot-mix asphalt surface, six-foot shoulders, and Class 35 two-way bridges.[8]

One of the projects undertaken by Company D, 577th Engineer Battalion, was a 400-bed hospital at Phu Hiep near Tuy Hoa, which officially opened on 17 June. No comparable medical facility existed for a radius of eighty miles. From 7 December 1966 to 15 March 1967, the men worked, often in high winds and driving rain, to ready the facility for its first patients. In mid-March, a heli-

[7] Diary of a Contract, pp. 234–35; "Floating and Land Based Power Plants," pp. 6–7; ORLL, 1 Aug–31 Oct 67, 864th Engr Bn, 7 Nov 67, pp. 2–5, Historians files, CMH; Dunn, *Base Development,* p. 80; Fact Sheet, 6 Feb 67, tab J, incl. 8, OCE Liaison Officer Trip Rpt no. 6; Quarterly Hist Rpt, 1 Apr–30 Jun 67, MACDC, p. IV-11.

[8] Galloway, "Essayons," pp. 174–75, 220; ORLLs, 1 Feb–30 Apr 67, 577th Engr Bn, n.d., pp. 2–4, 1 May–31 Jul Oct 67, 577th Engr Bn, n.d., pp. 2–4, and 1 Aug–31 Oct 67, 577th Engr Bn, 11 Nov 67, pp. 2–4, ORLLs in Historians files, CMH.

copter arrived carrying several wounded infantrymen. Although not officially open, the hospital treated those soldiers and subsequently cared for several hundred allied casualties. Almost all of the hospital's plumbing and fixtures and major parts of the generators, maintenance buildings, latrines, and water towers were improvised by the engineers. Many Vietnamese children were also treated there. Capt. Richard F. Hill, the Company D commander, said: "The men don't say much about their work, but you can see the enthusiasm and pride in their eyes when an injured or sick child is treated and released from the hospital."[9]

Air Force engineers complemented the construction program in southern II Corps. In the spring of 1967, the 554th Civil Engineering Squadron at Phan Rang became the first Air Force engineering unit to set up and operate a concrete batch plant. Most of this production was used for large aircraft dispersal hardstands. The employment of Vietnamese workers increased from 11 in April 1966 to 685 by September 1967. At first, most were hired as unskilled laborers or as painters, carpenters, and masons and had to be closely supervised. Eventually, a semiskilled labor pool was built up and included vehicle mechanics. At Tuy Hoa, Walter Kidde, working under the turnkey concept, finished the airfield lighting in February, and the entire $52 million permanent facility—a 9,000-foot interim AM2-surfaced runway, a 10,000-foot concrete primary runway and taxiway with parking aprons, and a logistic complex that included an extensive fuel storage facility and two maintenance hangars and cantonment—by the end of June 1967, one year after the initial contract award. Though the turnkey idea, where the contractor built the base in a single package, was a success at Tuy Hoa, no similar construction project was planned elsewhere because the Air Force did not need another jet-capable base in South Vietnam.[10]

During 1967, the equipment problems plaguing the Army engineers in the 35th Group subsided. In the spring, the status of earthmoving equipment improved considerably with the arrival of the Clark 290M tractors and scrapers, which replaced the worn-out 830M tractors and scrapers. Likewise, relief was brought to the 864th Battalion's badly worn bulldozers, as HD16 models were gradually replaced by the D7Es. At Cam Ranh Bay's quarry, one of the largest troop-operated quarries in Vietnam set up by the 864th Engineer Battalion and currently operated by the 87th Engineer Battalion, equipment and manpower shortages eased when the 610th Engineer Company (Construction Support) arrived in August.[11]

[9] "Engineers Build Jungle Hospital," *Castle Courier*, 30 June 1967.

[10] Schlight, *Years of the Offensive*, pp. 224–25; Tregaskis, *Building the Bases*, p. 223; USAF Airfield Construction in South Vietnam, p. 95; Directorate of Construction, MACV, Construction Program South Vietnam (Complex Review), 15 Jan 68, p. 11–1, Historians files, CMH (hereafter cited as MACV Complex Review, date).

[11] ORLLs, 1 Nov 67–31 Jan 68, 35th Engr Gp, p. 9; 1 Nov 66–31 Jan 67, 864th Engr Bn, pp. 1–2, 1 Feb–30 Apr 67, 87th Engr Bn, p. 3, 1 May–31 Jul 67, 864th Engr Bn, p. 3. For more on the Clark 290M tractor and scraper, see Pfc. Randy Hunter, "The 290M: Where There's a Man . . . There's a 290," *Kysu'* 1 (Summer 1969): 22–25.

The Clark 290M tractor and scraper required skilled operators and mechanics. For example, Pfc. Gerald Schreiner of the 557th Engineer Company (Light Equipment) was one of the few men qualified to operate a 70,000-pound machine capable of moving 20 cubic yards of earth while leveling hills and filling low spots at the Phuoc Vinh base camp. On a typical day, he moved 400 to 600 yards of laterite to a new revetment area for an Army assault helicopter unit. By 0800, the thermometer already registered 105°F. "I guess I make 20 or 30 trips a day between the job and laterite pit," he yelled over the noise made by the 400-horsepower engine to a reporter from the Engineer Command's public affairs office. "The trick is to roll with the ship," he grinned, while shifting the heavy-duty transmission into second gear for a particularly steep hill. Schreiner controlled the 290's main components by three levers to the right of the operator's seat to fill the scraper with dirt and empty it at the job site. Seven hours later, his face resembled a topographical map with rivulets of sweat racing through reddish dust-caked plains. At night he tried to get six hours of sleep from the mosquitoes, muggy heat, and occasional rocket attacks.[12]

A great deal of maintenance had to be done to keep a machine like Schreiner's 290M working. "The heat really fouls up the turbochargers," declared Sfc. Henry Flagg, the light equipment company's maintenance chief. The turbochargers could double the engine's horsepower, but the combination of too much heat and too little oil could cause the engine to freeze up and reduce its performance. The heat also took its toll on the four 24-volt batteries often cracking them. Flagg's mechanics could replace two engines a day, but because engine exchange was considered third-echelon work, this practice had to cease. Still, there was enough maintenance work like checking the oil level in each of the five large sumps and tightening connections to keep the 290M running. "You know," Flagg told the reporter, "we whip 90 percent of the problems we have with these babies by simply practicing a little preventive maintenance." In closing, he remarked: "The men are pretty good about it too; they know that when you got a piece of machinery like this you got to take care of it."[13]

By the end of the year, Cam Ranh Bay, Tuy Hoa, Nha Trang, and Phan Rang were large operating installations with the full range of appurtenant facilities, including administrative buildings, hospitals, warehouses, barracks, repair shops, mess halls, and a deep-draft port at Cam Ranh Bay. The construction effort at Cam Ranh Bay, eventually costing over $145 million, had transformed the peninsula into a major logistical center. As of 31 December 1967, the port offered deep- and shallow-draft berthing facilities with one permanent and four DeLong piers, six LST ramps, an ammunition pier, a fuel discharging jetty, 1,700 feet of sheet-pile wharf, and a barge wharf. The port handled an average of 6,800 short tons of dry cargo and 30,000 barrels of petroleum products daily. The peninsula abounded with depots and maintenance shops, which when completed would total over 1.4 million square feet of covered storage, 1.2 million square feet of open ammunition storage, and

---

[12] Hunter, "The 290M: Where There's a Man," pp. 22–25.
[13] Ibid., p. 25.

bulk storage for over 775,000 barrels of fuel. Supplementing the storage facilities were standard military containers and trailer van containers transportable by specially designed ships put into use in 1967 by Sea-Land Services, Inc. Electrical power came from central systems relieving the insufficient tactical generators. Intensive well drilling improved the water production and quality. Some 25,000 troops from all the services worked and lived on the base, and they enjoyed access to many of the comforts offered on bases in the United States. On the negative side, central waterborne sewage systems were originally provided at few locations. Burn-out latrines, locally manufactured from fifty-five-gallon drums cut in half and partially filled with diesel fuel, were widely used. This primitive system proved practical but distasteful because of the inevitable odors and the dense, foul, black smoke generated during burning. Troops were particularly unhappy when assigned this duty. On one side of the peninsula the Army had burn-out latrines, while on the other side the Navy and Air Force enjoyed a central sewage system. In any case, more construction was programmed to improve operations and living conditions, including a complete water and sewage system.[14]

As combat bases, Cam Ranh Bay, Tuy Hoa, and Phan Rang hosted large jet-capable airfields, each with permanent 10,000-foot runways and necessary facilities supporting large numbers of fighter-bomber and cargo aircraft. With air base construction complete, the operations tempo increased dramatically. At Cam Ranh Bay, the average number of Air Force aircraft on hand increased from 87 to 128 assigned to four F–4 Phantom tactical fighter-bomber squadrons, two C–130 Hercules squadrons, and two C–7 Caribou squadrons. The number of sorties flown jumped from 8,349 during the last quarter of 1966 to 33,342 during the last three months of 1967. In addition, the Navy and the Army housed another forty aircraft on the air base. Army airfields at Cam Ranh Bay and Dong Ba Thin housed over seventy fixed-wing aircraft and helicopters. In terms of landings and takeoffs, Cam Ranh Bay topped 43,000 in October, ranking it just behind the monthly averages at the top three airports in the United States. Phan Rang increased from 86 to 111 Air Force planes assigned to three F–100 Super Sabre fighter-bomber squadrons, one B–57 Canberra medium-bomber squadron, and three C–123 Provider transport squadrons. Sorties increased from 8,709 during the last three months of 1966 to 23,950 during the same period in 1967. During the last quarter of 1966, the new Tuy Hoa Air Base bedded down an average of seventeen F–100s. A total of 778 sorties were flown during this period. By the end of 1967, the average number of aircraft increased to sixty-two (three F–100 squadrons), and the sorties flown in the last three months of the year leaped to 9,211.[15]

[14] MACV Complex Review, 15 Jan 68, pp. 8-1 to 8-2; Dunn, *Base Development*, p. 84; MACV History, 1967, vol. 2, p. 833; Development of Cam Ranh Bay, pp. 2-3, 3-1; Heiser, *Logistic Support*, pp. 26–27, 171–73.

[15] MACV Complex Review, 1 Apr 67, pp. 138, 192; ibid., 15 Jan 68, pp. 8-3, 8-5 to 8-6, app. V; Development of Cam Ranh Bay, pp. 2-1 to 2-3; Schlight, *Years of the Offensive*, app. 1; U.S. Air Force Statistical Digest, FY 1967, Headquarters, USAF, pp. 50, 58, 64, and U.S. Air Force Statistical Digest, FY 1968, Headquarters, USAF, pp. 52, 62, 68, both at Office of Air Force History, Bolling Air Force Base, Washington, D.C.

Even though the main war effort did not take place along the envisioned Cam Ranh Bay–Ban Me Thuot–Pleiku axis, Cam Ranh Bay continued to play an important role in the logistics picture, serving as a sustaining base for the entire Army theater of war. It served as a major juncture for transshipping supplies from oceangoing vessels to coastal-type shipping. The ammunition storage areas allowed the Army to keep large stocks of ammunition in country relatively safe from enemy attack, and the cold-storage facilities allowed fresh vegetables to be brought down from Da Lat and stored prior to distribution.[16]

Farther North

Impressive progress was also made to bases at Qui Nhon, An Khe, and Pleiku. Expanding logistical facilities supporting the 1st Cavalry and 4th Divisions and South Korean forces in northern II Corps remained a high priority. Operations by the 1st Cavalry Division and the South Korean Capital Division coupled with a severe northeast monsoon in the coastal region placed a heavy dependence on Highway 1 from Bong Son south to Ninh Hoa. Responsibility for engineering support in the eastern half of northern II Corps now fell to the 45th Engineer Group following the departure of the 937th Group to Pleiku. To improve control of its five battalions and retain a major engineer headquarters in Qui Nhon, Colonel Bush moved the 45th Group headquarters from Tuy Hoa. After Bush's departure in May, Colonel Sawyer concentrated much of the group's efforts in the Qui Nhon area, employing three to four battalions. As work on the Qui Nhon depot complexes proceeded, enlargement of logistical facilities in the hinterlands gained momentum. This task fell to Colonel Braucher's 937th Group which, following its move from Qui Nhon, could now concentrate on base construction and operational support in the western half of northern II Corps. Upgrading Highway 19 from Qui Nhon to Pleiku was shared by the two engineer groups.[17]

The two groups took different approaches in implementing an Engineer Command directive in November 1966 to establish Contract Liaison and Installation Master Planning offices to serve as go-betweens with civilian contractors and Army users. Duties assigned to the contract liaison offices included offering technical advice to installation commanders to help carry out their master planning responsibilities, monitoring troop self-help construction, providing a contracting officer's representative on specified contracts, and participating in final inspections of completed projects. In the 45th Group, a separate office was established consisting of a captain (later a major), a clerk and jeep driver, and a Vietnamese Army sergeant who served as an interpreter. Coordination with the RMK-BRJ field office located at Phu Cat Air Base mainly concerned rock and asphalt production and paving. The office also maintained liaison with Vinnell and Pacific Architects and Engineers to check the status of their projects. In addition, the 45th Group's contract liaison officer handled civil affairs matters and

---

[16] Heiser, *Logistic Support*, p. 27.
[17] Galloway, "Essayons," p. 169.

served as point of contact with the Vietnamese Railroad and U.S. Agency for International Development office in Qui Nhon. The 937th Engineer Group also set up a separate office but assigned the duties to the S–3. He was assisted by one officer and one noncommissioned officer. Regional Contract Liaison and Installation Master Planning officers were assigned at installations in the group's area of responsibility to keep watch over RMK-BRJ's projects.[18]

A major development for the 45th and 937th Groups was the conversion of more combat battalions to the new TOE 5–35E series, which called for four lettered companies. In February, four companies were split from the new 31st Engineer Battalion (Combat) at Fort Hood and deployed to Vietnam. These companies became Company D of the 20th and 299th Battalions of the 937th Group and the 39th Battalion of the 45th Group (the fourth company was assigned to the 588th Engineer Battalion in III Corps), and gave the former D-series battalions a 25 percent increase in capability to carry out operational support and construction missions. Conversion of the other combat battalions—the 19th and 70th—came over the following months. The stripping of the four companies from the 31st Engineer Battalion, however, meant that the battalion would have to be rebuilt. Its deployment from Fort Hood to Vietnam would be delayed by more than a year.[19]

In Qui Nhon, Colonels Bush and Sawyer found functioning port, depot, hospital, and cantonment facilities well on their way to completion. Base development plans for the 45th Engineer Group's chief customer, the Qui Nhon Support Command, called for more port and depot facilities, maintenance shops, and administrative buildings. For almost a year now, Qui Nhon was coequal with Cam Ranh Bay, and requirements for storage facilities greatly increased. The port served as the entry point for the material needs of the two U.S. and one South Korean divisions and Phu Cat Air Base. Envisioned was a logistical complex supporting over 100,000 troops in northern II Corps.[20]

During 1967, the 45th Group completed major projects for the Qui Nhon Support Command. Men of the 84th Engineer Battalion and the 643d Engineer Pipeline Company completed a refrigerated warehouse, a causeway in the port area, and more than four miles of eight-inch fuel line within the city and twenty miles of six-inch line from a tank farm outside Qui Nhon to Phu Cat Air Base. Dredging in the harbor was done by two dredges, the *Ann* operated by RMK-BRJ and the USS *Davison* manned by the Corps of Engineers. A new turning basin was completed in mid-March, and the dredged material was put to good use as fill for an access road to the docks and behind the bulkhead.[21]

[18] ORLLs, 1 Nov 66–31 Jan 67, 45th Engr Gp, pp. 7–8, 1 Nov 66–31 Jan 67, 937th Engr Gp, pp. 14–15.

[19] Galloway, "Essayons," p. 158.

[20] Dunn, *Base Development*, p. 71; MACV Complex Review, 15 Jan 68, p. 4-1.

[21] Galloway, "Essayons," pp. 170, 213; ORLLs, 1 Nov 66–31 Jan 67, 84th Engr Bn, pp. 2–7, 1 Feb–30 Apr 67, 84th Engr Bn, 14 May 67, pp. 2–5, Historians files, CMH, 1 May–31 Jul 67, 45th Engr Gp, p. 5, 1 Aug–31 Oct 67, 45th Engr Gp, p. 5, and 1 Nov 67–31 Jan 68, 45th Engr Gp, p. 5.

RMK-BRJ retained a significant presence in Qui Nhon. MARKET TIME facilities for the Navy's coastal patrol operations, under construction at Qui Nhon since February 1966, were completed in May 1967. At the Qui Nhon airfield, the firm completed an airfield lighting system, a 3,400-foot taxiway extension, most of the work on an operations building, two maintenance hangars, and two large warehouses before turning the project over to Army engineers in May. Atop Vung Chua Mountain overlooking the city, the firm helped Page Communications Engineers, the Army Signal Corps contractor charged with building fixed communications systems, by laying foundations for buildings and antennas, and building fuel storage, water, and sewage systems, access roads, bunkers, security lighting, and security fencing.[22]

North of Qui Nhon, RMK-BRJ's projects at Phu Cat Air Base neared completion. The concrete runway, 10,000 feet long by 125 feet wide, was completed on 14 March, and all work on the taxiways and warm-up apron was completed on 25 March. In May and June, two squadrons of F–100s were transferred from Bien Hoa and Phan Rang and began operations. During the last three months of 1967, the average number of aircraft stood at sixty-seven (mostly F–100s and C–7As). A total of 15,057 sorties were flown between 1 October and 31 December. RMK-BRJ and Air Force engineers pressed on with more projects, adding a control tower, runway lighting, paved roads, aircraft revetments, and improvements in living and working conditions.[23]

In early 1967, Vinnell Corporation mobilized its workforce to carry out its electric power-generation project for bases in the Qui Nhon area. Two T–2 tankers had arrived but not positioned because dredging was in progress. Upon completion of the dredging, the tankers were moved to mooring dolphins built by the 45th Group. Simultaneously, Vinnell erected pole lines to military bases in the city, and the first increment of power came online in April. By summer, the tankers were providing up to 15,000 kilowatts of electricity. In November, Vinnell's contract was modified to add power plants and distribution systems to bases outside Qui Nhon. Altogether, plans called for the installation of six 1,500-kilowatt generators in the Phu Tai Valley southwest of the city and three 1,500-kilowatt generators for the South Korean encampments in ROK Valley just west of Qui Nhon. A power plant for Cha Rang depot and maintenance facility was also in the works.[24]

South of Qui Nhon, work progressed on depot facilities. The 19th Engineer Battalion continued improving the Phu Tai ammunition depot and began to build new depots for the Qui Nhon Support Command and the 34th Aviation Group at Long My, eight miles south of the city. To speed up the tempo of construction, Colonel Bush ordered the battalion to move to the

[22] Diary of a Contract, pp. 236–40; Tregaskis, *Building the Bases*, p. 287.

[23] Diary of a Contract, pp. 266–70; Tregaskis, *Building the Bases*, pp. 286–87; Schlight, *Years of the Offensive*, pp. 155, 158, 173, app. 1; Quarterly Hist Rpts, 1 Jan–31 Mar 67, MACDC, p. 36, 1 Apr–30 Jun 67, MACDC, p. IV-12, 1 Jul–30 Sep 67, MACDC, p. IV-10, 1 Oct–31 Dec 67, MACDC, p. IV-8; U.S. Air Force Statistical Digest, FY 1968, p. 62.

[24] Quarterly Hist Rpt, 1 Jul–30 Sep 67, MACDC, p. IV-13; Fact Sheet, 6 Feb 67, tab J, incl. 8, OCE Liaison Officer Trip Rpt no. 6; "Floating and Land Based Power Plants," p. 10; MACV Complex Review, 15 Jan 68, p. 4–13.

In 1967, construction of depot facilities outside Qui Nhon were under way such as this Butler warehouse building being built by the 19th Engineer Battalion at Long My.

work site. Extensive earthwork and vertical construction were required, and the group attached several 830M tractors and scrapers, and operators from the 84th Engineer Battalion to move fill down an adjacent mountain to the depot site. Operating heavy construction equipment did have its hazards, and on one occasion one of the scrapers tipped over coming down the slope. Fortunately, the operator was not seriously injured. By April, level plateaus had been shaped for the depots and the battalion's base camp, later to become the depot's cantonment. A concrete plant borrowed from a paving detachment at An Khe was pressed into service to make concrete floors for the 120-by-200-foot Butler warehouse and 40-by-200-foot storage sheds. The Butler buildings were especially challenging because few of the combat engineers had experience erecting huge prefabricated metal buildings. In July, the 19th received long-awaited orders to move north of Bong Son to begin Highway 1 improvements, and the 84th Battalion took over the three depot projects.[25]

West of Qui Nhon, the newly arrived 589th Engineer Battalion (Construction) from Fort Hood took over the jobs at Cha Rang from the 35th Battalion in April. The battalion set up camp along Highway 19 near the intersection with Highway 1, putting it in a position to take on projects at the South

[25] Galloway, "Essayons," pp. 170, 214; ORLLs, 1 Feb–30 Apr 67, 19th Engr Bn, 30 Apr 67, pp. 2–6, 1 May–31 Jul 67, 19th Engr Bn, 31 Jul 67, pp. 2, 4, both in Historians files, CMH.

Korean Capital Division's base camp in ROK Valley, expanding on the work previously done by the 19th and 35th Engineer Battalions.[26]

At Camp Radcliff just outside An Khe, construction of the 1st Cavalry Division's base camp also neared completion. The division's base development planners predicated their requirements on supporting a troop strength expected to increase to 31,800 by the end of 1967. The 70th Engineer Battalion and Company B, 84th Engineer Battalion, completed a 65,000-barrel fuel tank farm, upgraded tactical roads east of An Khe, prepared Highway 19 from An Khe to the An Khe Pass for paving, supported self-help construction, and finished the 4,365-by-72-foot concrete runway. Heavy rains during the southwest monsoon, however, penetrated the AM2 matting at the Golf Course airstrip, and breaks in the T17 membrane resulted in saturation and breakdown of the subgrade. Repairs had to be repeated to what was believed to be an overly safe drainage system. With most of the base camp work completed, the 70th Engineer Battalion moved on to the Engineer Hill encampment outside Pleiku. By October, the battalion, in a period of twenty-six months, had played a major role in turning scrub land into a major camp.[27]

In October 1966, RMK-BRJ received notice to build an electrical power plant and distribution system at Camp Radcliff. Since the firm had no projects at An Khe, it first had to set up camp on the base. The setting of the poles began in January, and soon, thanks to help from Army Signal Corps units, 230 poles were in place. Work on concrete slabs for the generators, transformers, and powerhouse was completed in April, and the site was ready for the six 1,500-kilowatt Army-furnished generators still en route. By June, the locomotive-size generators, hauled by Army tractor-trailers from Qui Nhon, finally arrived. On 1 August, the electrical distribution line work was turned over to the 589th Engineer Battalion.[28]

At Pleiku, the pace of base development continued to grow following the arrival of the 4th Division at the Dragon Mountain Base Camp, six miles south of the city. But the added workload imposed a strain on the 937th Engineer Group, which also had commitments to support the division and other combat units in the western half of II Corps. The group had become the center of base development activity, particularly after 1 April 1967 when RMK-BRJ ended operations in Pleiku. By then the balanced distribution of effort between combat support and construction shifted to a 40:60 ratio. Heavy combat loads west and north of Pleiku kept a steady demand on engineer support in building and maintaining forward airstrips and supply roads. In mid-April, help arrived when the 815th Engineer Battalion (Construction)

[26] Galloway, "Essayons," pp. 213–14; ORLLs, 1 May–31 Jul 67, 45th Engr Gp, pp. 4–5, 1 Aug–31 Oct 67, 45th Engr Gp, pp. 5–6, and 1 Nov 67–31 Jan 68, 45th Engr Gp, p. 5.

[27] Galloway, "Essayons," pp. 167–68, 209–10; MACV Complex Review, 15 Jan 68, pp. 5-1, 5-8; ORLLs, 1 May–31 Jul 67, 45th Engr Gp, pp. 4–5, 1 Aug–31 Oct 67, 45th Engr Gp, pp. 5–6, 1 Nov 67–31 Jan 68, 45th Engr Gp, p. 5, 1 Nov 66–31 Jan 67, 70th Engr Bn, pp. 2–5, 1 Feb–30 Apr 67, 70th Engr Bn, 10 May 67, pp. 6–8, 1 May–31 Jul 67, 70th Engr Bn, 10 Aug 67, pp. 4–9, 1 Aug–31 Oct 67, 70th Engr Bn, 14 Nov 67, pp. 1, 3–6, 1 Nov 67–31 Jan 68, 70th Engr Bn, 31 Jan 68, pp. 6–7, last four in Historians files, CMH.

[28] Diary of a Contract, pp. 275–76.

As a new transformer station went online, RMK-BRJ continued to work on an electrical power plant and distribution system at Camp Radcliff.

deployed from Fort Belvoir and soon became the group's main construction force. The 815th took over running the contractor's two 225-ton-per-hour rock crushers, construction materials storage area, and the bulk of equipment left behind by the contractor. RMK-BRJ did finish an asphalt parking apron and cantonment work at Pleiku Air Base, and Army cantonment construction was nearly completed. The remaining work was transferred to the 937th Engineer Group. Like An Khe, the string of camps around Pleiku was transformed into a livable semipermanent base complex for a division and supporting units. The move of the 70th Engineer Battalion from An Khe also helped balance the workload between combat support and construction in the area.[29]

During the course of the year, the 45th and 937th Engineer Groups noted a gradual reduction in equipment and supply deficiencies. In early 1967, getting supplies to the job sites in the 937th Engineer Group's area of responsibility improved to the point that Colonel Braucher reported the increased availability of Class IV construction materials was the most noteworthy development in the three-month period since 1 November. In An Khe, requisitions for building supplies were submitted first through the group and routed to the 70th

---

[29] Ibid., pp. 208–10; Galloway, "Essayons," pp. 163, 206; Quarterly Hist Rpts, 1 Jan–31 Mar 67, MACDC, p. 36, 1 Apr–30 Jun 67, MACDC, pp. IV-12 to IV-13, 1 Jul–30 Sep 67, MACDC, pp. IV-11, IV-12, 1 Oct–31 Dec 67, MACDC, p. IV-8.

Engineer Battalion S–4 for submission to the Qui Nhon Support Command. Since there was no Class IV yard in the An Khe logistical support activity, the 70th Battalion was chosen to run the building materials storage facility. In Pleiku, requisitions were submitted through group headquarters to the Pleiku logistical support activity, which did operate a Class IV yard. Still, shortages of electrical and plumbing supplies and lumber and cement delayed several projects at An Khe and Pleiku, as did shortages of equipment. Of forty-eight authorized graders, the 937th Group was short ten. Only twenty-seven scoop loaders of sixty-one authorized and three motorized rollers of seven authorized were on hand. Repair parts were more plentiful. The chief maintenance concern was the lack of direct-support maintenance units. As for the 45th Group, Colonel Bush complained that the lack of compaction equipment needed in earthmoving work—thirty-five-ton compactors, sheepsfoot rollers, wobbly-wheel rollers, ten-ton steel-wheel rollers, asphalt finishing rollers, and water distributors—hampered road, airfield, and depot expansion projects.[30]

Some heavy construction equipment transferred from RMK-BRJ helped the equipment situation. Running the contractor's former rock crushers, however, became a challenge for the 815th Battalion at Pleiku. One was in a state of disrepair and turned out to be a maintenance nightmare. The second crusher was new and had not been assembled. There were no maintenance catalogs or plans, but the battalion's maintenance team persisted and brought the crusher, like the first crusher, a Pettibone, online in August.[31]

By the end of 1967, the 45th and 937th Engineer Groups and the contractors had accomplished most of their base construction projects. Qui Nhon, An Khe, and Pleiku became major staging areas for operations in northern II Corps. Qui Nhon developed into a major port and logistical complex supporting American and Korean troops in the region. Like other deep-draft ports under development, Qui Nhon's added port capability reduced the pressure on LSTs, permitting their greater use for deliveries to shallow-draft ports instead of LCUs and LCMs. By the end of December, the daily port discharge capacities increased to 6,275 short tons of cargo, not much below Cam Ranh Bay, and 59,000 barrels of petroleum products, almost double the capacity at Cam Ranh Bay, up from 4,450 short tons and 14,000 barrels in January 1967. Among the improvements in logistics was the new depot complex under construction at Long My. The demand for logistical facilities was so great that the Qui Nhon Support Command took over each building and storage area the moment they were ready for use and began relocating supplies stacked up on the beach. Qui Nhon also boasted three 400-bed hospitals, two American and one South Korean. Improved roads emanating from Qui Nhon, Highway 1 north to Phu Cat and beyond and Highway 19 to An Khe, were surfaced with asphalt. A new jet-capable airfield at Phu Cat was in operation, and

[30] Galloway, "Essayons," pp. 168, 175, 210; ORLLs, 1 Nov 66–31 Jan 67, 937th Engr Gp, pp. 15–17, 1 Feb–30 Apr 67, 937th Engr Gp, pp. 13–14, 1 Aug–31 Oct 67, 937th Engr Gp, 31 Oct 67, p. 13, 1 Feb–30 Apr 67, 45th Engr Gp, 11 May 67, pp. 5–6, last two in Historians files, CMH.

[31] Unit History, 815th Engr Bn, n.d., p. 9, Historians files, CMH; ORLLs, 1 May–31 Jul 67, 815th Engr Bn, 9 Aug 67, pp. 2, 4–8, 1 Nov 67–31 Jan 68, 815th Engr Bn, 31 Jan 68, pp. 2–7, both in Historians files, CMH; Galloway, "Essayons," p. 209.

improvements were under way to upgrade inland forward airstrips to C–130 capability. Inland camps at An Khe and Pleiku were taking the form of World War II stateside camps. Work would continue at a slightly lower pace to complete ongoing projects, enhance existing facilities, and improve working and living conditions. To ensure that the bases received constant maintenance and utilities services and some new construction, Pacific Architects and Engineers routinely hired more workers and set up shops at many locations.[32]

III and IV Corps

Base construction in III and IV Corps made great strides in 1967 thanks to the arrival of a second engineer brigade headquarters, a sixth group headquarters, and six newly raised construction battalions from the United States. On 3 August, the 20th Engineer Brigade headquarters, commanded by Brig. Gen. Curtis W. Chapman Jr., arrived at Bien Hoa Air Base from Fort Bragg. Two days later, Engineer Command placed the 34th, 79th, and 159th Groups under the 20th Brigade. The new brigade's mission encompassed operational support, troop construction, and civic action in III and IV Corps. This command arrangement simplified Engineer Command's span of control to two brigades with each brigade commanding three groups. Each brigade also supported the field force in its area of responsibility.[33]

Headquarters, 20th Engineer Brigade, was a new unit activated only three months before deploying to Vietnam. Despite several problems, the brigade headquarters company was able to meet the requirements for overseas deployment. Based on the experiences of the 18th Engineer Brigade, changes were made to the table of organization and equipment to fit the situation in Vietnam. An intelligence section was established, and the operations branch was expanded to four sections dealing with plans, construction operations, operational support, and liaison. General Chapman decided that most design work was to be done at group, battalion, and port construction company level. Standard drawings, such as barracks, mess halls, and showers, that had been developed by the three groups were catalogued for use by all the groups. A maintenance section was organized within the S–4. In so doing, the brigade put itself in a position to give technical help and speed up delivery of spare parts.[34]

Noteworthy among the many bases under development in III and IV Corps, Long Binh continued to grow into a major logistical and headquarters complex. By mid-1967, General Westmoreland's campaign (Operation Moose) to reduce the U.S. troop presence in Saigon resulted in relocating approximately half of the soldiers to Long Binh. On 15 July, U.S. Army, Vietnam, headquarters announced its move from Saigon to Long Binh even as construction

---

[32] Heiser, *Logistic Support,* p. 171; MACV Complex Review, 15 Jan 68, pp. 4–5, 4–6, 4–12; MACV History, 1967, vol. 2, p. 833; CINCPAC Port Development Plan, 12 Jan 67, p. 17, Historians files, CMH.

[33] Galloway, "Essayons," pp. 156–60, 221; ORLL, 1 Aug–31 Oct 67, 20th Engr Bde, 31 Oct 67, p. 2, Historians files, CMH; Ploger, *Army Engineers,* p. 138.

[34] ORLL, 1 Aug–31 Oct 67, 20th Engr Bde, pp. 2–6, 12–13.

workers rushed to finish the new facilities. Other major headquarters, including U.S. Army Engineer Command and the 1st Logistical Command, had also taken up residence on the base. During 1967, more requirements for storage, airfield, maintenance, and community facilities were added to accommodate some 35,000 troops. This number would eventually increase to over 40,000. There was enough room, for the base covered over twenty-six square miles, nearly ten undeveloped. As a result, there was a large backlog of work facing the 5,000-man 159th Engineer Group.[35]

To carry out this work, Colonel McConnell, the group commander, relied on four construction battalions. By midyear, the 62d Engineer Battalion, which had completed its move from Phan Rang in January, finished expanding the II Field Force tactical operations center at Plantation and built the Long Binh central dial telephone facility. One of the challenging projects assigned to the 169th Battalion entailed shoring up the failing upper structure of one of the U.S. Army, Vietnam, headquarters buildings erected by RMK-BRJ. The 46th Engineer Battalion devoted much of its effort to the Long Binh ammunition depot. Considered to be one of the largest ammunition storage areas in the world at the time, the depot contained 225 storage pads offering 250,000 square yards of storage and thirty-nine miles of roads. Following a spring Viet Cong attack on the depot, the battalion's Company B repaired damaged pads and improved the protective berms around the pads. Work assigned to the 92d Engineer Battalion included landscaping around the new U.S. Army, Vietnam, headquarters and building bunkers at the MACV headquarters at Tan Son Nhut.[36]

The timely flow of the right type of construction supplies to the 159th Group's project sites still had hurdles to overcome. The combined efforts of engineer units and self-help programs exhausted stocks of electrical and plumbing supplies, water tanks for mess halls and showers, and two-inch lumber as soon as they arrived. This shortfall took place despite the large quantities of supplies reaching the depots. Asphalt and peneprime needed at Long Binh, Bien Hoa, and Bearcat for dust control on roads and helipads were in short supply. Getting rock and sand remained a problem because of the lack of convenient sources and transportation. A contract with a Korean company to deliver 100,000 cubic yards of rock from Korea was plagued with problems of gradation control, shipping schedules, port facilities, equipment availability, and Vietnamese bureaucracy. By April, only 23,000 cubic yards had been delivered, and the 159th Engineer Group cancelled the contract. Hiring a Vietnamese contractor to haul rock from the Vung Tau quarry by

[35] MACV Complex Review, 1 Dec 66, p. 197; ibid., 15 Jan 68, pp. 13-1 to 13-2; "Long Binh Post," *Army* 23 (April 1968): 48; Dunn, *Base Development*, p. 145.

[36] Galloway, "Essayons," pp. 182–83, 224–25; ORLLs, 1 Nov 66–31 Jan 67, 159th Engr Gp, pp. 7–10, 1 Nov 66–31 Jan 67, 46th Engr Bn, pp. 1–5, 1 Nov 66–31 Jan 67, 62d Engr Bn, pp. 1–2, 1 Nov 66–31 Jan 67, 169th Engr Bn, pp. 2, 5–8, 1 May–31 Jul 67, 62d Engr Bn, 31 Jul 67, pp. 1–6, 1 Aug–31 Oct 67, 62d Engr Bn, 31 Oct 67, pp. 4–6, 1 Nov 67–31 Jan 68, 62d Engr Bn, 31 Jan 68, pp. 2, 5–8, 1 Aug–31 Oct 67, 169th Engr Bn, 11 Nov 67, pp. 2–9, 1 Feb–30 Apr 67, 46th Engr Bn, 15 May 67, pp. 1–5, 1 May–31 Jul 67, 46th Engr Bn, 14 Aug 67, pp. 1–6, 1 Aug–31 Oct 67, 46th Engr Bn, 1 Nov 67, pp. 1–8, last seven in Historians files, CMH.

truck helped somewhat, but that contract expired without renewal on 30 June. Using barges to haul rock from Vung Tau to Dong Tam and Long Binh was only partially successful at first because tugs were not readily available to move loaded barges. Overcoming this problem and the Vietnamese bureaucracy resulted in gradually improved barge traffic. While a search continued for suitable quarry sites, the group was able to procure much of the rock for Long Binh from RMK-BRJ's University Quarry outside Saigon. Then there was the constant challenge of keeping up with rising standards of living, especially at the large Long Binh base. Standards were elevated nearly to those used in the United States.[37]

The contractors working at Long Binh took much of the strain off the 159th Group. During 1967, RMK-BRJ carried out approximately $25 million of work on the base. Most visible were the headquarters and support facilities for U.S. Army, Vietnam, and 1st Logistical Command. Building the headquarters complexes involved erecting six 2-story buildings forming an "H" for the Army headquarters and four 2-story buildings for the logistical command; adding asphalt roads and parking; and installing water, storm, and sewer lines. The buildings featured amenities such as air conditioning, indoor plumbing, and vinyl tiled floors, things that customers now expected and incorporated into designs. Such items were not readily available, and U.S. Army, Vietnam, and U.S. Army Engineer Command, Vietnam, recognized this problem and focused on getting the materials rather than lowering standards. Construction on this priority project for the U.S. Army, Vietnam, headquarters complex started in December 1966. After working day and night, the contractor had two of the buildings ready for occupancy in early July 1967. All six buildings were ready by the end of the month. Three of the logistical command's buildings were completed in late October and the fourth the following month. Since the windows of these buildings were designed not to open, occupancy was held up until air conditioning was installed in December. Other projects related to the arrival of the two headquarters were quarters for officers and enlisted men, dispensaries, chapels, support buildings, and utilities, including water and sewer distribution systems. Another high-priority project was constructing two buildings for the 14th Inventory Control Center, which managed the Army's depot assets. The hope was that the third-generation computers expected in late 1967 would bring some order out of the tremendous influx of supplies coming over the beaches and arriving at the ports. One building housed the computers, and the other contained the administrative center. Using Army-furnished materials, RMK-BRJ met the beneficial occupancy date of 28 June.[38]

---

[37] Galloway, "Essayons," pp. 184, 225, 227; ORLLs, 1 Nov 66–31 Jan 67, 159th Engr Gp, p. 11, 1 Feb–30 Apr 67, 159th Engr Gp, 14 May 67, pp. 18–19, 1 May–31 Jul 67, 159th Engr Gp, 14 Aug 67, pp. 14–15, last two in Historians files, CMH.

[38] Quarterly Hist Rpts, 1 Jul–30 Sep 67, MACDC, p. IV-15, 1 Oct–31 Dec 67, MACDC, 21 Jan 68, p. IV-12, Historians files, CMH; Diary of a Contract, pp. 278–82; Army Construction in Vietnam, 11 Aug 67, incl. 8, p. 5, OCE Liaison Officer Trip Rpt no. 8; Galloway, "Essayons," p. 184; Heiser, *Logistic Support*, p. 23; Dunn, *Base Development*, pp. 114–15.

At Long Binh, RMK-BRJ built several headquarters complexes, including U.S. Army, Vietnam, which were ready for occupancy by the end of June 1967.

The power plant at Long Binh begun the previous October by Vinnell started coming online in 1967. The firm's responsibility in this project included not only the design, which Vinnell subcontracted, but also the construction, operation, and maintenance of the plant claimed to be the only one of its kind in the world. For the first time, twenty-one 1,500-kilowatt generators were connected and synchronized to work simultaneously on one power system, or in electrical terms on a single loop. As soon as pole lines were strung, power was made available to users. Because of the strong possibility of enemy attacks, extra measures and precautions were designed into the system. Ten-foot-high revetments were built around generator complexes. If enemy fire hit one complex, the damage would be confined to that one complex. Besides physical protection, the plans called for fail-safe devices to ensure that power would continue to flow to the major headquarters, the hospital, and the computer center. In case any one part of the plant or distribution system was damaged, a series of reclosers would lock that part out of the remaining system. Work continued, and by November 1968, 19,000 kilowatts of power were being delivered. Vinnell also built power plants and distribution systems at Long Thanh, Bearcat, Di An, Xuan Loc, Bien Hoa, Phu Loi, and Cu Chi.[39]

Fifteen miles northeast of downtown Saigon, RMK-BRJ carried out projects at Bien Hoa Air Base for Army and Air Force tenants, and the 159th

[39] "Vinnell Makes History at Long Binh," pp. 8–10.

Engineer Group committed forces to improve the Army side of the base. A long-delayed authorization to begin a second 10,000-foot concrete runway (parallel to the one built by the contractor in 1962 and 1963) was finally made in early April 1967. This work was justified, for Bien Hoa had become the busiest air base in South Vietnam. Excavation soon began, and a night shift was added. Concrete work started in the second week of June, but by then annual rains held up progress. Work on the new runway, taxiway, warm-up aprons, and perimeter roads extended to April 1968.[40] The contractor also ran into delays in the ammunition storage area originally authorized to be built in 1965. Work begun in April 1966 lasted to mid-1967. Other facilities built for the Air Force included an engine test stand; blast pads; a fire station; a 10,000-foot-8-inch welded-steel fuel pipeline from the Dong Nai River to the air base; water; utilities systems; airmen's dormitories; and runway shoulder rehabilitation and replacement airfield lighting cables. RMK-BRJ also built a telephone exchange building and drilled nine water wells on the Army side of the base. Projects carried out by the 159th Engineer Group included building a communications center with air conditioning, general officers quarters, and a road that would provide a direct link between Bien Hoa and Long Binh.[41]

Also greatly improved was the string of bases for the 1st and 25th Divisions defending Saigon from the north. The 79th Engineer Group at Plantation under Colonel Gelini, who was replaced by Col. Joseph A. Jansen in July, continued to carry out base camp construction simultaneously with its operational support missions and roadwork. Work at the divisions' base camps progressed from minimum essential facilities to more sophisticated and more permanent support facilities. Making this possible was the arrival in April and May of the 554th Engineer Battalion (Construction) from Fort Knox and the 34th Engineer Battalion (Construction) from Fort Stewart. The two construction battalions increased the group's construction capability and allowed the group's combat engineer battalions to boost operational support to the 1st and 25th Divisions during a period of increased combat activity.[42]

East of Saigon, the 27th Engineer Combat Battalion took up the development of the 11th Armored Cavalry Regiment's Blackhorse base camp at Xuan Loc. Along with its attached units—the 591st Light Equipment Company at Blackhorse, the 94th Quarry Detachment at Gia Ray, and two sections of the 2d Platoon, 67th Dump Truck Company, at Blackhorse—the battalion's strength was almost 1,100, with about 870 men in the battalion and over 200 in

---

[40] Bien Hoa was not only the busiest airport in Vietnam in terms of landings and take-offs (over 70,000 in January and over 83,000 in September 1967), but Bien Hoa and Tan Son Nhut even exceeded Chicago's O'Hare, the busiest airport in the United States, which averaged slightly over 47,000 landings and takeoffs per month in 1966. MACV Complex Review, 15 Jan 68, app. V.

[41] Tregaskis, *Building the Bases*, p. 252; Quarterly Hist Rpts, 1 Oct–31 Dec 67, MACDC, 21 Jan 68, p. IV-11, 1 Apr–30 Jun 68, MACDC, 24 Jul 68; ORLL, 1 Aug–31 Oct 67, 1st Log Cmd, 11 Nov 67, p. 114, Rpts and ORLL in Historians files, CMH; Diary of a Contract, pp. 203–07; Galloway, "Essayons," pp. 224; ORLL, 1 May–31 Jul 67, 159th Engr Gp, pp. 9, 11–12.

[42] Galloway, "Essayons," pp. 184–85, 187–89, 233, 235, 239; ORLLs, 1 Nov 66–31 Jan 67, 79th Engr Gp, pp. 2–5, 1 Feb–30 Apr 67, 79th Engr Gp, 13 May 67, pp. 2–5, 1 May–31 Jul 67, 79th Engr Gp, 8 Aug 67, p. 2, last two in Historians files, CMH.

the attached units. Company A concentrated on building a 1,500-foot airstrip and a sixty-bed surgical hospital, while Company C helped build cantonment areas. East of Xuan Loc at Gia Ray, Company B did cantonment construction and ran a quarry from the base of Nui Chua Chan.[43]

When the 34th Engineer Group (Construction) from Fort Lewis under Col. Joseph M. Palmer arrived at Vung Tau in late March, it assumed a wide area of responsibility for southern III Corps (less the Long Binh–Saigon area) and all IV Corps. During the late spring and the summer of 1967, the group expanded to five battalions with the May, June, and September arrivals of the 69th, 93d, and 36th Construction Battalions. At this time, the group's construction commitments increased with the assumption of responsibility from RMK-BRJ for work at Vung Tau and Can To on top of expanding the base camps at Dong Tam, Long Thanh, and Xuan Loc. Despite this and the addition of the construction units, the group's ratio of operational support to base construction showed a significant increase, from 16 to 33 percent by the end of July.[44]

At Vung Tau, RMK-BRJ, Vinnell, and DeLong transformed the once small port into a key base. Several miles north of the city at Cat Lo, which served as a hub for GAME WARDEN naval patrol activities, RMK-BRJ built a base camp and storage and repair facilities for U.S. and South Vietnamese naval forces. At the port, the firm dredged a turning basin for deep-water ships and a 12,700-foot-long channel. Other jobs included facilities for Page Communications and the Signal Corps and included antenna and microwave antenna footings; slabs for communications trailers and buildings; water and sewage lines; and roads, drainage, and walkways. These jobs were finished in March 1967. DeLong completed the pier facility on 22 April. In late 1967, Vinnell finished installing both primary and secondary electrical distribution systems, and two power ships were online providing a generating capacity of 11,000 kilowatts. Simultaneously, Pacific Architects and Engineers hooked up buildings to the system.[45]

Vung Tau was one of the places selected to phase down RMK-BRJ operations. The 69th Engineer Battalion from Fort Hood arrived at Vung Tau on 1 May and took over the contractor's projects at the ammunition depot and the fuel storage area, which involved building a 50,000-barrel addition to the existing tank farm. The battalion also built a barge-loading facility to support the movement of supplies to the Mekong Delta. On 20 September, the 36th Engineer Battalion from Fort Irwin, California, reached Vung Tau, marking the sixth and final construction battalion scheduled for Vietnam in 1967. On

[43] Galloway, "Essayons," pp. 185, 187, 227, 229; ORLLs, 1 Nov 66–31 Jan 67, 27th Engr Bn, 13 Feb 67, pp. 3–5, 1 Aug–31 Oct 67, 27th Engr Bn, 8 Nov 67, pp. 3–4, 10–14, 1 Nov 67–31 Jan 68, 27th Engr Bn, 8 Feb 68, pp. 1–2, 8–13. ORLLs in Historians files, CMH.

[44] ORLLs, 1 Feb–30 Apr 67, 34th Engr Gp, 18 May 67, pp. 5–9, and 1 May–31 Jul 67, 34th Engr Gp, 10 Aug 67, pp. 1–3, 5–11, both in Historians files, CMH; Galloway, "Essayons," pp. 227, 244–45.

[45] Quarterly Hist Rpts, 1 Jan–31 Mar 67, MACDC, p. 41, 1 Apr–30 Jun 67, MACDC, pp. IV-17, IV-19, 1 Jul–30 Sep 67, MACDC, pp. IV-17, IV-23, 1 Oct–31 Dec 67, MACDC, p. IV-14; Diary of a Contract, pp. 222–24.

1 October, the battalion took over the projects at Vung Tau, and the 69th Battalion moved to Dong Tam and Can To. The 36th Battalion continued work on port facilities, a 5,000-man cantonment, rehabilitated the airstrip, and designed and built three 50,000-barrel fuel storage tanks and their concrete pads, a manifold system, and a jetty for off-loading T–2 tankers.[46]

Meanwhile, the 34th Group began moving more units into the delta region to take over projects started by RMK-BRJ and to build the bases for the 9th Division and support units. In May, company-size task forces from the 27th and 69th Engineer Battalions were on the job in Vinh Long and Can To doing airfield and cantonment construction. By midyear, one company from the Long Thanh–based 86th Engineer Combat Battalion moved to Dong Tam to help the building effort, and the rest of the battalion deployed to delta bases to work on base construction and support the 9th Division in the field. When the 69th Engineer Battalion moved to Dong Tam in October, it assumed construction responsibility for the base.[47]

The Viet Cong made concerted efforts to hinder construction at Dong Tam. The insurgents knew the base held a commanding position in the delta and could hamper their control of the region. By the end of 1966, the dredges had made substantial progress in filling in the base area and clearing approach channels. On 9 January 1967, Viet Cong sappers succeeded in sinking the *Jamaica Bay*, the largest dredge in Vietnam. This attack prompted General Westmoreland to send elements of the arriving 9th Division to the partially filled Dong Tam site to take over security from the South Vietnamese and begin work on the base. Two companies of the division's 15th Engineer Battalion accompanied the advance elements. Within a few weeks, Company C, 577th Engineer Battalion, moved from Long Binh to build facilities for the 7,500-man camp and operational base. Company C also replaced a weak Eiffel bridge with two panel bridges connecting the camp with the city of My Tho to the east.[48]

Meanwhile, the Navy's Officer in Charge of Construction diverted another dredge from Cam Ranh Bay, and in March the *Jamaica Bay*'s sister dredge, the *New Jersey*, arrived. On 7 May, the first LST entered the turning basin, relieving the smaller LCMs supplying the base for other use. By then the engineers had completed 20 percent of the planned construction, including a runway for light Army planes and helicopters, a sixty-bed MUST (medical unit, self-contained, transportable) hospital consisting of inflatable buildings, and numerous cantonment buildings.[49]

---

[46] Galloway, "Essayons," pp. 230, 232–33; ORLLs, 1 May–31 Jul 67, 69th Engr Bn, 31 Jul 67, pp. 2, 6–8, 1 Nov 67–31 Jan 68, 36th Engr Bn, 13 Feb 68, pp. 2, 5–7, both in Historians files, CMH.

[47] Galloway, "Essayons," pp. 185, 229–30, 232; ORLL, 1 May–31 Jul 67, 69th Engr Bn, pp. 2, 6–8.

[48] MacGarrigle, *Taking the Offensive*, p. 402; Tregaskis, *Building the Bases*, pp. 292–93; Dunn, *Base Development*, p. 53; Ploger, *Army Engineers*, p. 145; Galloway, "Essayons," pp. 183–84; Quarterly Hist Rpt, 1 Jan–31 Mar 67, MACDC, p. 41.

[49] Tregaskis, *Building the Bases*, p. 293; Ploger, *Army Engineers*, pp. 145–46; Quarterly Hist Rpt, 1 Apr–30 Jun 67, MACDC, pp. IV-17, IV-18.

After overcoming the challenging soil conditions, engineers completed the Dong Tam base in late 1968.

The builders of Dong Tam faced challenges found nowhere else in Vietnam. At Dong Tam, the rain was severe enough that it almost halted construction for two months, and all efforts went into saving what had been built. The instability of the ground in the delta and the high water table caused special problems. Large buildings required supporting piles. When underground storage tanks popped out of the ground after removal of their contents, workers had to place concrete collars around the tanks. Holes dug for communications and power poles were shored up with fifty-five-gallon drums. Scarce rock moved by barge from the Vung Tau quarry, a five- to ten-day trip, and was used only on the most important concrete structures. Sand-cement sufficed for other work. By June 1967, the growing base occupied 4.6 square miles. In addition to the cantonment and storage facilities for the 9th Division's 2d Brigade, the installation had a 1,640-foot airstrip and a 0.8-square-mile turning basin for shipping, with boat and barge landing sites. Summer saw the completion of more permanent buildings, water and electrical distribution systems, a water-borne sewage system, more ramps for landing craft, warehouses, hardstands, and maintenance shops. The Dong Tam project would extend to late 1968 when it was officially declared complete.[50]

---

[50] MacGarrigle, *Taking the Offensive*, p. 411; Ploger, *Army Engineers*, pp. 146–49; ORLL, 1 Feb–30 Apr 67, 159th Engr Gp, p. 19. For more on Dong Tam, see Maj. Bryon G. Walker, "Construction of a Delta Base," *Military Engineer* 60 (September-October 1968): 333–35.

Though the 34th Group had many of the same problems as the other engineer groups, it had to put up with the challenges of working in the delta. Rain, while not a major problem at the better-developed Vung Tau complex, played havoc in the delta, especially at Dong Tam and the newer sections of Long Thanh, areas that had been built since the 1966 monsoon. After assuming responsibility for transporting construction supplies and rock to the delta from the 20th Brigade, the group continued to work closely with Transportation Corps officials in scheduling the tows of 300- and 700-ton barges to off-loading sites. From May to the end of the year, twenty-four barges were hauling approximately 7,000 tons of bulk construction material each month to the delta. Rock requirements set up by MACV, however, were much larger, some 38,000 tons per month from Vung Tau to the delta. During an eight-month period, the barges only moved 50,000 tons. Since the 1st Logistical Command controlled the majority of barges, the group recommended that the logistical command take over responsibility for moving the rock. On 15 January 1968, the 1st Logistical Command took over the operation while the group retained responsibility for the loading and unloading of rock, a task also under consideration for transfer to the logistical command.[51]

The Saigon Projects

In Saigon, RMK-BRJ completed several eye-catching projects. One of these was the new U.S. Embassy complex located about four blocks from Independence Palace, the Vietnamese White House. The old leased building downtown had been bombed several times and was susceptible to attack, and work began on the new site in September 1965. Delays in plans and funding, however, delayed occupancy until September 1967. Surrounded by a concrete wall and encased in a masonry rocket shield, the peculiar structure resembled a giant pillbox rising above the surrounding neighborhood. At first plans called for a four-story building, but this was changed in February 1967 to six floors topped off with a penthouse helicopter landing pad.[52]

Next in line was the $25 million MACV headquarters complex. The project had started on a site in the Chinese quarter of Cholon in west Saigon in February 1966, but strong protests by the inhabitants halted work in May. As a result, the project shifted to the northern edge of Tan Son Nhut Air Base. Surveying the new site began on 1 July, but stringent security requirements for construction workers resulted in slow hiring. It was not until September that construction got into high gear. At its peak, the project employed over 100 Americans, 300 Koreans and Filipinos, and over 1,200 Vietnamese. By the end of June 1967, less than twelve months from the time excavation started for the security fence and less than eleven months from the time the first concrete slab

---

[51] Galloway, "Essayons," pp. 229, 233; ORLLs, 1 Feb–30 Apr 67, 34th Engr Gp, p. 7, 1 May–31 Jul 67, 34th Engr Gp, pp. 15–16, 1 Nov 67–31 Jan 68, 34th Engr Gp, 1 Feb 68, p. 14, last in Historians files, CMH.

[52] Tregaskis, *Building the Bases*, p. 250; Quarterly Hist Rpt, 1 Jul–30 Sep 67, MACDC, 21 Oct 67, p. IV-22, Historians files, CMH; Diary of a Contract, p. 191.

was poured, the project was 96 percent complete. In August, MACV head-quarters began moving in. The two-story building contained a foundation of concrete caissons, 310,000 square feet of concrete floors with vinyl tile cover-ing, an insulated metal roof, insulated double-wall metal siding, suspended acoustical ceilings, plastered fixed partitions, and plywood movable partitions. Also part of the new MACV complex were a three-story metal prefabricated barracks and mess hall, a one-story metal prefabricated building containing headquarters commandant offices and a warehouse, a power plant, a chiller plant, and a telephone exchange building. A chapel and automated data pro-cessing buildings were added and completed in the autumn. The complex had the largest air-conditioning plant in Southeast Asia, its own fresh water system with wells and treatment plant, and almost 1¼ mile of water and sewage lines. Surrounding the buildings were paved roads and four acres of parking lots, and 6,000 feet of 12-foot-high wire mesh security fence with lights every 90 feet. It was no wonder that the MACV headquarters became known as the Pentagon East, or the Little Pentagon, for it was more than one-third (twelve acres of enclosed space) the size of the Pentagon in Washington (thirty-four acres). MACV headquarters now had enough space for 4,000 troops and civil-ian workers in one complex.[53]

Next door at Tan Son Nhut, RMK-BRJ worked on airfield expansion and base development. All official objections to building the parallel runway through an old Vietnamese cemetery vanished overnight on 4 December 1966 when the Viet Cong infiltrated the base and had to be routed out of their hiding places, including the cemetery. Within a few days, the South Vietnamese Army appeared on the scene with bulldozers and leveled the obstructing tombstones. Progress was fast thereafter, and the new runway was in use on 15 April 1967 and dedicated on 3 June. The entire parallel runway project consisted of a 150-by-10,000-foot main runway of 10 inches of concrete (11 inches on the 1,000-foot touchdown section of either end); runway lighting; 930,000 square feet of taxiway and apron of 12 inches and 11 inches of concrete, respectively; 1 million square feet of compacted fill for the shoulders; and 400,000 square feet of soil-cement overruns, of which the 150 linear feet next to the runway on both ends was overlaid with 3 inches of asphalt. More 10-inch-thick concrete parking aprons were also built. Among the various air base facilities added by the contractor at Tan Son Nhut was a hardened concrete command post for the Air Force. During 1967, Tan Son Nhut bedded down an average of 142 Air Force aircraft. Over 126,000 sorties were flown in 1967.[54]

RMK-BRJ also wrapped up other projects in Saigon. The firm built the facilities for the country's first television broadcasting stations for the Armed Forces Radio and Television Service and the Vietnamese government. About four blocks from Tan Son Nhut, RMK-BRJ undertook the rehabilitation of

---

[53] Tregaskis, *Building the Bases*, pp. 250–51; Quarterly Hist Rpt, 1 Jul–30 Sep 67, MACDC, p. IV-21; Diary of a Contract, pp. 194–96; Dunn, *Base Development*, p. 145.

[54] Tregaskis, *Building the Bases*, pp. 251–52; Quarterly Hist Rpt, 1 Jul–30 Sep 67, MACDC, p. IV-21; Diary of a Contract, pp. 197–202; U.S. Air Force Statistical Digest, FY 1967, pp. 60–61, and U.S. Air Force Statistical Digest, FY 1968, p. 66.

*The MACV headquarters complex at Tan Son Nhut (the airfield is in the back-
ground) was completed by RMK-BRJ in the summer of 1967.*

the 3d Field Hospital. The improved complex was completed in May 1967.
On Saigon's southern outskirts at the Phu Lam signal site, the contractor had
been at work since July 1966. By May 1967, the project neared 50 percent
completion, and the automatic switching center was activated in 1968. Some
twenty other projects in Saigon were accomplished ranging from repairing of
the fender system on a pier at the port, a penthouse addition to the MACV
communications center, and a number of rehabilitation jobs on military offices
and billets.[55]

East of Saigon, RMK-BRJ and Army engineers completed Newport.
After finishing landing craft slips and ramps and barge off-loading wharfs
in December 1966, the contractor began work on the deep-water berths
and port buildings. When completed, the complex—consisting of a 2,400-
foot wharf, four deep-draft berths, and two transit sheds, each containing
four interconnected buildings, 120 by 200 feet—totaled 290,000 square feet.
The inner 140 feet of the pier structure involved building a wharf of free-
standing pipe or H-pile with steel caps, stringers, and a concrete deck. Two
piers made up the outer 140 feet supported by prefabricated steel jackets and
concrete deck sections. The jackets and deck trusses were made at the firm's
fabrication yard at Poro Point in the Philippines. The piers were the idea of
H. William Reeves, who adapted a design for oil platforms he built earlier

[55] Tregaskis, *Building the Bases*, p. 251; Diary of a Contract, pp. 191–93; MACV History,
1967, vol. 2, p. 793; Bergen, *A Test for Technology*, p. 302.

A ribbon-cutting ceremony marked the opening of the Newport deep-draft port facility in Saigon when Brig. Gen. Shelton E. Lollis, commanding general of the 1st Logistical Command, officiated by cutting the ribbon as (left to right) *Generals Ploger, Dunn, and Raymond and Admirals Husband and Seufer observe.*

in the Gulf of Mexico, where piles were driven through steel tubes designed as part of the pier sections. Reeves, who had been sent to Vietnam earlier as a consulting engineer, returned to work for Brown and Root as part of the combine. Like the DeLong piers, the Reeves piers were prefabricated but with smaller sections and permanent. In late January 1967, the first berth was declared open and commemorated in a ribbon-cutting ceremony as the first ship came alongside. The second berth was completed in March, and all four were finished in July. Construction of a power plant and a maintenance shop proceeded concurrently with the pier work. At the peak of the port project, RMK-BRJ employed some 1,800 Vietnamese workers and 210 Americans and third-country nationals. The work schedule consisted of two 10-hour shifts per day, seven days a week. Meanwhile, fill operations for the port area continued. Fill material was hauled by the firm's trucks from the University Quarry and subcontractors' sampans. Newport could now accommodate simultaneously four oceangoing vessels, four shallow-draft landing craft, and seven barges. As with most projects, more improvements would be made. The cost up to that point was nearly $25 million.[56]

---

[56] Quarterly Hist Rpts, 1 Jul–30 Sep 67, MACDC, p. IV-21, 1 Oct–31 Dec 67, MACDC, pp. IV-2 to IV-21; Diary of a Contract, pp. 262–63; Tregaskis, *Building the Bases*, pp. 237, 246; Dunn, *Base Development*, p. 143.

Funding, Overbuilding, and Logistics

During 1967, construction funding was on the downswing. By midyear, the total authorization since 1965 for military construction in Vietnam was $1.482 billion, including the $397 million Fiscal Year 1967 supplemental. The funding for 1967 as compared to $850 million in 1966 indicated that the greatest part of the construction effort was financed and well on its way to completion. Nevertheless, Pacific Command and MACV believed the downward trend for Fiscal Year 1968 did not provide for all the required facilities and highway upgrading. But when Pacific Command requested additional supplemental funds, the Johnson administration demurred. By September and October 1967, added construction requirements totaled $341 million, primarily to support the increased troop levels. In the end, MACV received $164.5 million supplemental funds for Fiscal Year 1968, which was much better than the $77 million Fiscal Year 1968 regular appropriation. Only $69.1 million would be forthcoming for Fiscal Year 1969, bringing the total funds for Vietnam military construction to $1.71 billion.[57]

Meanwhile, General Raymond and his Construction Directorate faced a second funding overrun that came to light in January 1967. The Navy's Officer in Charge of Construction in Saigon continued to underestimate the full degree of RMK-BRJ's mobilization costs. Costs also swelled when the scope of projects expanded, and the Navy's Officer in Charge of Construction in Saigon faced cash-flow problems. Up to this point, planning had been based on about $600 million equally divided between the contractor and troop effort. This figure did not include the $200 million to complete the contractor's portion of the construction program, and now the Officer in Charge of Construction disclosed that the contractor required more than the $200 million estimated the previous July. In light of a reduced Fiscal Year 1967 supplemental program, Raymond could not expect relief in the planned Fiscal Year 1968 regular program submission, and he had to wait until about 1 November before these funds would become available. Raymond had no choice but to take steps to align future work within the new funding level. Since most of operational and logistical facilities were done, he looked to the increasing troop capability to take over a major share of the remaining work.[58]

The Naval Facilities Engineering Command also took steps to fix the overrun problem and dispatched its comptroller, Capt. Donald G. Iselin, to Saigon in February. One option that seemed to have the most merit allowed the consortium to complete current jobs along with some new work and the highway restoration program, turning the smaller and remote sites such as Pleiku over to the troop units. Besides easing the transition, Iselin believed that an infusion of cash for new work in the program would go a long way.

[57] CINCPAC History, 1967, vol. 2, pp. 897–99; MACV History, 1967, vol. 2, pp. 843–44; MACV History, 1968, vol. 2, p. 670; Quarterly Hist Rpt, 1 Jan–31 Mar 67, MACDC, pp. 11–12; Quarterly Hist Sum, 1 Jan–31 Mar 69, MACDC, 19 Apr 69, app. 1, Historians files, CMH.

[58] Raymond, Observations on the Construction Program, pp. 35–36; Quarterly Hist Rpt, 1 Jan–31 Mar 67, MACDC, pp. 11–16; Msg, Raymond SOG 0138 to Noble, 23 Jan 67, Dunn Papers, CMH.

The Officer in Charge of Construction would only need part of that money to do the new jobs, enough to pay the labor since equipment and materials had already been purchased. The rest of the money could then be used to pay for the labor on current jobs. Iselin and Raymond traveled to Washington to recommend this approach to Defense Department officials. During the briefing, Assistant Secretary of Defense Ignatius went along with Iselin's proposal, agreeing that the only way to handle the problem was to establish some level of effort. When Iselin drew up his plan he called it, for lack of anything better, the Level of Effort. Following this approach, the contractor was allocated a specific amount of funds at each of its remaining locations. Productive use of equipment, manpower, and materials was the major consideration at each site. The plan matched the firm's given capability to a known workload, purposely established at such a level that work would always be ready, waiting for the capability rather than the consortium's original posture of having more than enough capability and frequently waiting for work. Armed with Pentagon concurrence of the Level of Effort concept, Iselin returned to Saigon in March. Raymond agreed to put the Level of Effort plan into action. Iselin held a series of meetings to work out the details with the Construction Directorate, Officer in Charge of Construction, RMK-BRJ, and the three services. On the twenty-fifth, Raymond directed the implementation of the system by 1 April 1967.[59]

By June, the Construction Directorate and the Officer in Charge of Construction had taken steps to absorb the overrun and phase down the contract with RMK-BRJ. The Navy's Saigon office came up with a final estimate of $300 million to finish the contractor's projects. Other measures included transferring some of the contractor's projects, materials, and equipment to troop units; consolidating, closing out, or transferring the firm's operating sites; and where possible, reducing the scope of projects. The Officer in Charge of Construction hastened plans to reduce the number of employees from 48,800 to 17,000 by August and to phase out the contract by 1 July 1968. Since the reduction released workers with construction skills, the planning assumed that most of them would be able to find employment with other construction organizations. Losing all the firm's construction capability, however, did not make much sense. In July 1967 MACV, after consulting with Honolulu and Washington, decided to retain the contractor and a workforce of at least 15,000 to finish some projects and to carry out highway restoration.[60]

[59] Tregaskis, *Building the Bases*, pp. 336–40, 342–45; Intervs, Richard Tregaskis with Capt Donald G. Iselin, Comptroller, Naval Facilities Engr Cmd, 29 Jan 71, pp. 6–21, and Brig Gen Daniel A. Raymond, Former Dir of Const, MACV, 20 Jul 70, pp. 19–21, both in Richard Tregaskis Papers, Naval Facilities Engr Cmd Hist Ofc, Port Hueneme, Calif.; Quarterly Hist Rpt, 1 Jan–31 Mar 67, MACDC, pp. 16–17, 19–20; MACDC Paper, n.d., sub: Level of Effort Concept, Dunn Papers, CMH.

[60] MACV History, 1967, vol. 2, pp. 840–42; Raymond, Observations on the Construction Program, pp. 36–37, 60–61; Quarterly Hist Rpts, 1 Jan–31 Mar 67, MACDC, pp. 16–20, 1 Apr–30 Jun 67, MACDC, 11 Jul 67, pp. III-4, IV-2, IV-3, 1 Jul–30 Sep 67, MACDC, 21 Oct 67, p. III-3; Msg, COMUSMACV 09024 to CINCPAC, 17 Mar 67, sub: Reduction in RMK-BRJ Work Force; MACDC Fact Sheet, 5 Apr 67, sub: Construction Contractor Demobilization, RVN; Msg, CINCPAC to COMUSMACV, 20 May 67, sub: Construction Contractor Capability; Msg, COMUSMACV 17260 to CINCPAC, 26 May 67, sub: Construction Contractor

One way to cut down the number of projects and save money and engineering resources was to make sure facilities, especially troop billets, were not overbuilt. Since tents could be expected to survive little more than one year in such a climate, General Westmoreland had allowed construction of wooden tropical buildings. It was also recognized that roads, mess halls, latrines, showers, dispensaries, water towers, chapels, post offices, and other facilities would have to be built sooner or later. After reviewing the cantonment standards in 1966, U.S. Army, Vietnam, continued to use the field, intermediate, and temporary cantonment standards based on expected tenure of occupancy. In October 1966, MACV revised Directive 415–1 prescribing the three cantonment standards, which with minor modifications remained in effect for the remainder of the conflict. General Westmoreland concurred in using troops to build cantonments, but he considered it inappropriate to provide base camps beyond the mere necessities for units that operated and lived in the field most of the time. He was surprised, therefore, during a visit to Phan Rang in September 1967 to learn that new facilities were being built for the 1st Brigade, 101st Airborne Division, particularly since the unit spent most of its time away from camp and did not make full use of the buildings already erected. On returning to Saigon he told General Palmer, the deputy commanding general of U.S. Army, Vietnam, to stop the work at Phan Rang and to look into the status of base camp construction.[61]

General Palmer, who in March 1967 had left the Army staff as chief of operations to command II Field Force, and on 1 July succeeded General Engler as deputy commander of U.S. Army, Vietnam, took immediate steps. No friend of the base camp concept, Palmer considered the policy "pernicious," and the "manpower it soaked up was appalling, not to mention the waste of material resources and the handicap of having to defend and take care of these albatrosses." It was also obvious to him that the camps were being overbuilt, and construction was too sophisticated. He himself became aware of a situation involving the 196th Infantry Brigade at Tay Ninh. And at Pleiku, Palmer found the 4th Engineer Battalion building concrete sidewalks for the 4th Division while the ammunition and fuel dumps remained impassable mud holes in the rainy season. His personal view was that all units should live in the field, with minimum facilities serving all the forces in a large area. He set up a base development study group under the chairmanship of the G–4, comprising staff officers of U.S. Army, Vietnam, and 1st Logistical Command, to evaluate all base camp construction requirements. The twelve-man study group reviewed base development plans for tactical and logistical bases and the reasons for their locations, degree of development, and their support to the tactical forces. Altogether, thirty-two Army bases and their satellites were visited and evaluated. At Qui Nhon, the study

Demobilization; Msg, COMUSMACV 17572 to CINCPAC, 29 May 67, sub: Construction Contractor Demobilization. Two reports, messages, and fact sheet in Historians files, CMH.

[61] Dunn, *Base Development*, pp. 45, 73–74; Ltr, Westmoreland to Dep CG, USARV, 23 Jan 67, sub: Cantonment Construction Policy; Westmoreland Historical Notes, no. 21, 6 Sep 67, both in Westmoreland History files, CMH. MACV Directive, 415–1, 20 Oct 66, *Construction: Cantonments, Standards of Construction To Be Used by US and FWMA Forces in RVN*, an. A.

group recommended that the port and logistical complex continue its base development program at the construction pace then under way, but that Tay Ninh, with the departure of the 196th Infantry Brigade to I Corps and a now-outdated base development plan, be considered for possible closing. At Bien Hoa, the study group found that the 173d Airborne Brigade only used 22 percent of its base camp capacity at any given time and recommended any additional construction for the brigade should be deferred. With reference to Bien Hoa, the team also noted that the 20th Engineer Brigade and newly arrived 34th Engineer Battalion were living and working in tents considered below acceptable standards and recommended that the construction battalion be housed in a portion of the airborne brigade's cantonment. At Phan Rang, the study group verified Westmoreland's concern that the brigade of the 101st Airborne Division had not occupied its cantonment for more than a few days. The study group's report suggested several reprogramming actions to assure the continuation of only essential base construction. General Palmer approved these recommendations and a proposed "hotel" concept. This meant that in any given base camp, the Army would not try to bed down every man since a large part of maneuver units would always be in the field.[62]

Despite MACV's efforts to tighten the construction program, the Defense Department expanded upon the already constricting procedures of the construction authority. On 3 April, Secretary McNamara issued another procedural memorandum that did away with the concept that treated the entire country as one installation and instead divided the country into nineteen base complexes along with their satellite sites. Although the fifteen broad functional facility category groups had worked well for MACV program managers, they apparently did not provide the detail needed by the Defense Department. The Defense Department now preferred more detailed facility category groups, which nearly doubled the number of the superseded categories, but they did provide more detail and gave a better picture of what was being built. For example, the MACV list depicted hospitals as one category group, and dental clinics and dispensaries were lumped under cantonments. The Defense Department listed hospitals, dental clinics, and dispensaries as separate groups. MACV planners were now faced with managing over five hundred entities compared to the previous fifteen.[63]

The command again faced what amounted to a system that one could expect under normal peacetime procedures. Programming, reprogramming, reevaluation, rejustification, and resubmittals, with all of the attendant

[62] Dunn, *Base Development*, pp. 74–75; General Bruce Palmer Jr., *The 25-Year War: America's Military Role in Vietnam* (Lexington: University Press of Kentucky, 1984), pp. 69–71 (quoted words, p. 69); Report of HQ USARV Ad Hoc Group to Study Base Development, 18 Sep 67, pp. 2, 6, 8, 14–15, an. C, apps. II, XXII, XXIV, Historians files, CMH; Army Construction in Vietnam, incl. 8, p. 1, OCE Liaison Officer Trip Rpt no. 9, 24 Oct 67, OCE Hist Ofc.

[63] JLRB, Monograph 6, *Construction*, pp. 74–76; Raymond, Observations on the Construction Program, pp. 51–52, 211–13 (tab V, Office of the Secretary of Defense Memorandum, 3 Apr 67, sub: Construction Approval Procedures for SVN); Quarterly Hist Rpt, 1 Jan–31 Mar 67, MACDC, pp. 20–21.

administrative burdens became a way of life for the construction managers. Since the Defense Department required copies of all construction directives, the Construction Directorate faced a possible increase of approximately 1,500 entities to manage for each of the annual and supplemental appropriations. General Raymond later reported that even though the MACV commander retained some flexibility, "the new system has imposed a monumental paperwork task since, one there are so many separate projects and two copies of all amendments must be forwarded to DOD." He noted that one reprogramming action alone could include five to ten amendments to the construction directives. As the Defense Department's logistics study in 1969 pointed out, the modified system "provided maximum visibility and minimum flexibility." The engineers had to abide by these restrictions and assigned more troops and civilians to do the bookkeeping.[64]

Bookkeeping also applied to construction supplies, especially those used by the contractors. In late 1966, RMK-BRJ had instructions to start construction on projects estimated to cost $823 million, some two-thirds of the entire construction program. Once the Seabees and the contractor had drawn large amounts of equipment and materials from Navy stockpiles, it became apparent that RMK-BRJ still needed equipment and materials in quantities far greater than could be replenished by Navy depots. This would bring the total equipment requirements to 5,000 items, an increase of 4,000, plus another 1,000 pieces of rental equipment like barges, tugs, dredges, and pile drivers. The firm scoured the Far East for dredges and cement. When Westmoreland decided to house troops in tents with wood frames and wood frame buildings, RMK-BRJ added to its growing lists a huge demand for lumber. The firm's Saigon office placed an order for 78.4 million board feet of lumber, and the home office at San Bruno, California, almost cornered West Coast lumber sources. This not only increased the price of lumber but also made it difficult for Defense procurement officials to satisfy the needs of the services within a specified time limit and at reasonable prices.[65]

Sorting and inventorying these vast stores of building materials and equipment during the hectic buildup posed major challenges. At first RMK-BRJ dumped equipment and materials at port staging areas, depots, and project sites. The flood of arriving materiel inspired thievery on a grand scale. Changes in project priorities and criteria also left the contractor with excesses of some materials. It became apparent that the consortium had to build its own depots and work camps in Vietnam to secure and house the growing stocks, which by May 1966 had grown in value to $162 million.[66]

It did not take long for Congress to become concerned. A team from its auditing agent, the General Accounting Office, arrived in Vietnam in July 1966

[64] Raymond, Observations on the Construction Program, p. 52 (quoted words); JLRB, Monograph 6, *Construction*, pp. 74–77, 80 (quoted words).

[65] Tregaskis, *Building the Bases*, pp. 199–201, 225, 281; JLRB, Monograph 6, *Construction*, p. 174; Report to the Congress of the United States, Survey of United States Construction Activities in the Republic of Vietnam, General Accounting Office, Dec 66, pp. 1–2, 6, 31, box 4, 74/167, RG 334, NARA (hereafter cited as GAO Rpt).

[66] Tregaskis, *Building the Bases*, pp. 201, 212, 239–40; GAO Rpt, pp. 52–70, 77–81.

to survey construction activities. In December, the team's report concluded that problems it uncovered came about because of the speed in which RMK-BRJ had to mobilize. In noting the contractor's loss of control over materials and equipment, the team pointed out the magnitude of the problem. RMK-BRJ could not account for the whereabouts of approximately $120 million worth of supplies shipped from the United States. The report prompted a lively exchange of views between the auditors and the commands. In response to the findings concerning inadequate depots, MACV and Pacific Command noted that the contractor's facilities were at least equal to those provided to the military forces.[67]

The Navy and RMK-BRJ made a concerted effort to fix the problem. In July 1966, the Officer in Charge of Construction in Saigon established a Material Department to monitor the contractor's procurement, shipping, inventory, property, and equipment programs. RMK-BRJ built major depot and repair complexes near Saigon, at Cam Ranh Bay and Da Nang, with subsidiary storage areas at other locations. Near Saigon, RMK-BRJ began to capitalize on plans made earlier in the year to transform a marshy cigar-shaped island at Thu Duc into a vast storage and maintenance area. A large amount of fill was needed for the firm's depot, but the Newport project had a higher priority and received the lion's share of crushed rock from the firm's University Quarry. This slowed progress somewhat. Nevertheless, some 1.2 million cubic yards of sand and 240,000 cubic yards of rock were used to stabilize an area of over 4 million square feet at Thu Duc. The firm completed most of the vertical construction (administration buildings, warehouses, and maintenance shops), utilities systems, surface stabilization, barge anchorages, and floating piers in early 1967. At the end of May, the inventory value of material (excluding spare parts) at Thu Duc totaled approximately $30 million. Altogether, the firm's depot facilities in South Vietnam grew to twenty sites with ninety-seven warehouses providing 800,000 square feet of floor space. The firm also improved its security force by hiring Koreans and ethnic Chinese Nung mercenaries. While doing all this, the firm finally got the go-ahead to improve its camp sites for its American and third-country employees.[68]

Indicative of the development of the bases was the corresponding expansion of the facilities engineering contractor, Pacific Architects and Engineers. During 1967, 5,000 more employees were hired, increasing the firm to over 20,000 workers, including 1,500 Americans, 3,000 third-country nationals, and 16,000 Vietnamese. At the start of the year, Pacific Architects and Engineers had responsibility for forty-nine permanent camps. These included major installations at Qui Nhon, Da Nang, Cu Chi, Long Binh, Vung Tau, Tuy Hoa, Nha Trang, Cam Ranh Bay, and Phan Rang. By the end of the year, the firm had facilities engineering responsibility for 39 primary sites (classified as A installations), 57 smaller permanent offices (B and C installations), and 200

---

[67] GAO Rpt, pp. 2, 16–17, 52; Msg, CINCPAC to OSD, 6 Jan 67, sub: GAO Draft Report, Survey of U.S. Construction Activities in the Republic of Vietnam, 16 Dec 1966. Messages and correspondence concerning the report are in the same file with the report.

[68] Tregaskis, *Building the Bases*, pp. 201, 248–49, 281–84; Diary of a Contract, pp. 262–63.

isolated Special Forces and adviser sites maintained by roving maintenance teams. The Long Binh complex alone required a facilities engineering staff of 2,153 (80 U.S., 334 third-country nationals, and 1,739 Vietnamese). Taking care of the U.S. Army, Vietnam, headquarters complex required the dedicated services of 2 Americans, 6 third-country nationals, and 200 Vietnamese workers. The firm was still doing some $1 million in new construction each month, including jobs like building kennels for sentry dogs guarding a communications site at Long Binh. Though not in its facilities engineering charter, the firm pitched in to help hook up 2,200 telephones and run 200,000 feet of cable at MACV headquarters and 1,095 telephones and 25,000 feet of cable at Long Binh. The contractor also built engineer construction materiel yards for the logistical command in Saigon, Long Binh, and Vung Tau, stocking items from nails to prefabricated buildings.[69]

Because the Army supply system fell short of the contractual requirements with Pacific Architects and Engineers, the firm had unique supply problems of its own. Unlike the Navy's contract with RMK-BRJ and the Air Force's contract with Walter Kidde at Tuy Hoa Air Base, which allowed the two firms to procure the bulk of their equipment and supplies, the contract with Pacific Architects and Engineers called for the Army to provide these items. Despite the efforts of the 1st Logistical Command engineer and the U.S. Army Procurement Agency, Vietnam, Pacific Architects and Engineers received little of the equipment it was authorized. More than once the contractor resorted to purchasing materials and equipment from civilian sources. When the Army renewed the contract in mid-1966, the firm only possessed $3 million of the approximately $22 million of equipment and tools specified in the contract. The contractor's difficulties varied from cancellations of requests for the missing equipment to critical shortage of space aboard ships and aircraft bound for Vietnam. In one instance, more than $1 million of equipment, vehicles, and prefabricated buildings purchased in Singapore reached Saigon almost six months later. The ever-expanding bases exacerbated the facilities engineering contractor's materiel shortages, particularly refuse trucks, fire trucks, prefabricated buildings, and hand-operated insecticide foggers and rodenticide. Because of the lack of refuse trucks, Pacific Architects and Engineers had to hire local contractors, who demanded increasing prices to haul away the military's trash. In late 1966, the contractor had on hand only 78 of the authorized 626 pieces of "rolling stock," but by mid-1967 the number on hand soared to 674. By then, however, the shortfall in equipment was great because requirements had quadrupled over the authorized levels.[70]

Nevertheless, Pacific Architects and Engineers reported some improvement as equipment began to trickle in. Cement mixers, road graders, jeeps, assorted trucks, generators, and small shipments of water treatment plants, requested

---

[69] *History, Pacific Architects and Engineers, Incorporated: Repairs and Utilities Operations for U.S. and Free World Military Forces, Republic of Vietnam, Calendar Year 1967* [Saigon, n.d.], pp. 3, 6–7, 9, 14, 20, 23, copy in CMH.

[70] JLRB, Monograph 6, *Construction*, pp. 173, 175; ORLL, 1 Nov 66–31 Jan 67, 1st Log Cmd, n.d., p. 75, Historians files, CMH; PA&E History, 1963–1966, pp. 47–50.

when the troop population began to rise, started arriving in September 1966. When thirty pickup trucks arrived in Saigon that month, Vietnamese customs red tape held up distribution for a month. Another contractor accidentally picked up and stored some transformers in Phan Rang, and it took the firm one month to locate them.[71]

The rapid growth of the facilities engineering contract made it evident that the Army needed better control. Although the 1st Logistical Command's engineer monitored the contractor's work, the command's procurement agency managed the contract. To keep track of the firm's performance, the procurement agency relied on contracting officer's representatives, a title and responsibility typically given to staff officers as an additional duty. On 1 December 1966, Col. Robert W. Fritz, who replaced Colonel Lien as the 1st Logistical Command engineer, established a Contract Management Branch at the firm's Contract Management Office in Saigon. The branch, which later expanded to a division of three branches, supervised and controlled all technical aspects of the contractor's operations. Fritz assigned duties as contracting officer's representatives to the staff engineers of the Saigon, Qui Nhon, and Cam Ranh Bay support commands. In turn, the three commands appointed contracting officer's representatives among the staff engineers of the installations within the support command areas. These assignments corresponded with Pacific Architects and Engineers' three area and forty-five installation offices.[72]

Summing Up

By the end of 1967, the bases, all in full operation, were nearing completion. The builders had made tremendous headway to reduce the backlog, especially at the major complexes in and around Long Binh, Cam Ranh Bay, Da Nang, and Qui Nhon. Since 1965, a vast base development program provided the operational, logistic, and support facilities that enabled the United States to deploy and operate a force now approaching 500,000. The construction of deep-draft ports with satellite shallow-draft ports, major air bases, depots and logistical complexes, and troop bases proceeded concurrently with the arrival of troops and their supplies. The anticipated long duration of operations made economical sense to upgrade troop housing from tents to tropical wood-framed buildings and install better utilities beyond the designs called for in technical manuals. Statistics of completed Army facilities alone showed the immense effort put into the construction program. Over 4.4 million square yards of airfields were cleared, graded, and surfaced; 160 miles of pipeline and storage tanks capable of storing more than 37 million gallons of fuel were built; and 11.6 million square feet of covered and open storage for dry cargo were in place at the supply depots. Hospitals were capable of handling 2,800

[71] PA&E History, 1963–1966, pp. 48–49.

[72] Dunn, *Base Development*, p. 92; Rpt of Visit to the 1st Log Cmd, 12 and 15 Nov 66, p. 2, incl. 10, 7 Dec 66, OCE Liaison Office Trip Rpt, no. 5; ORLL, 1 Nov 66–1 Jan 67, 1st Log Cmd, pp. 73–74; U.S. Army Procurement Agency, Vietnam, *Procurement Support in Vietnam, 1966–1968*, n.d., pp. 30, 36, Historians files, CMH.

Among the major accomplishments in building the bases in South Vietnam were the deep-draft ports such as Newport shown here in October 1967.

patients. Engineer troop strength increased as more construction battalions activated in 1966 crossed the Pacific, and command and control improved with the arrival of a second engineer brigade and the sixth engineer group. The long-awaited arrival of new equipment greatly helped, but getting construction supplies continued to be a challenge. With the construction program over the hump, planning was set in motion in 1967 to make sure there was no overbuilding and to cut back the expensive contract with RMK-BRJ. Washington also reduced funding and imposed more restrictions. Congress allocated over $1.5 billion in construction funds to the Defense Department for Fiscal Year 1968 starting 1 July 1967, but only $300 million was slated for Southeast Asia, as compared to $1.1 billion the previous year. Still, the expansion of combat operations generated new requirements for roads, airfields, cantonments, and advanced logistical bases, an almost never-ending process as combat forces deployed to new areas. Priority now shifted to highway improvements. As 1968 dawned, the engineers looked forward to finishing the bases, making headway on the road improvement program, and supporting tactical operations.[73]

---

[73] Dunn, *Base Development*, pp. 133–34; Annual Report of the Secretary of the Army in Department of Defense Annual Report for Fiscal Year 1967 (Washington, D.C.: Government Printing Office, 1969), p. 218; Annual Report of the Department of Defense for Fiscal Year 1968 (Washington, D.C.: Government Printing Office, 1971), p. 90.

10

Inland and to the Borders, June 1967–January 1968

During his visit to the United States in April 1967, General Westmoreland spoke of the importance of persevering on all fighting fronts simultaneously. He was now at the midpoint of his three-phase strategy that began with protecting his logistical bases with occasional forays. Currently, the allies were trying to gain the initiative and eliminate the enemy's base areas and possibly his main forces. Westmoreland's third phase envisioned moving into sustained ground combat to mop up the last of the main forces, or at least push them across the borders and try to contain them. "At one and the same time," he told a gathering of the Associated Press managing editors in New York City, "we must fight the enemy, protect the people, and help them build a nation in the pattern of their choice." A few days later in an address before a joint session of Congress, Westmoreland expanded on his war-winning strategy. He explained that the enemy, despite suffering large losses on the battlefield, still clung to the belief that he could defeat the allies through a clever combination of psychological and political warfare. "The only strategy which can defeat such an organization," he stated, "is one of unrelenting but discriminating military, political, and psychological pressure on his whole structure—at all levels."[1]

Strategy in Mid-Passage

This strategy in the mid-passage of the war was incorporated in the Joint Vietnamese-U.S. Combined Campaign Plan for 1967. With enlarged forces, added firepower, and improved mobility, Westmoreland carried the battle to the enemy regularly throughout the year. The plan provided for highly mobile American forces to carry the bulk of the offensive effort against Viet Cong and North Vietnamese Army main force units. South Vietnamese forces took on the primary role in pacification. Westmoreland viewed the U.S. effort as complementary—to drive the enemy main forces away from the priority pacification areas. He estimated that over half the U.S. combat forces would operate close to the heavily populated areas against guerrillas and local forces. These units would reinforce Vietnamese units in pacification, just as designated Vietnamese units would help carry the attack against the main forces.[2]

The Combined Campaign Plan also contained a blueprint for the security and restoration of the roads. As the priority construction of ports, airfields, and

---

[1] Quoted in Sharp and Westmoreland, *Report*, pp. 131–32; Westmoreland, *A Soldier Reports*, pp. 224–29.
[2] Sharp and Westmoreland, *Report*, pp. 131–36; MACV History, 1967, vol. 1, pp. 317–22.

bases neared completion, more effort combining U.S. and South Vietnamese military and civilian resources could go to the highway restoration program. Planners viewed a good highway system as a high-priority requirement supporting military operations and nation-building. Early in 1967, MACV's Directorate of Construction developed plans to upgrade about two thousand miles of roads. Dependable roads would also reduce dependency on the expensive air and slow water transportation systems. Sabotage, monsoon rains, and increasingly heavy use by allied military forces together had intensified the poor condition of the roads. Repairs were far beyond the capabilities of the South Vietnamese, and military engineers usually made minimum repairs and improvements to support military operations and the vital flow of commerce. The engineers would often repeat this process pending a fully funded restoration program. To carry out the program, the Directorate of Construction estimated an annual cost of about $130 million over the course of three years employing troop and contractor capability.[3]

While he prepared to upgrade the highways, additional troops and weaponry permitted Westmoreland to increase the scope and pace of offensive operations. In March, Westmoreland regarded the Army's force ceiling under Program 4 of the buildup to be inadequate for the assigned missions, and he asked Washington for more troops. Washington reluctantly relented, and in October approved a force ceiling (Program 5) of 525,000 for 1968. U.S. strength increased significantly during 1967, from 385,000 to 486,000. By the end of the year, the number of maneuver battalions available to allied forces rose from 256 to 278. U.S. Air Force aircraft in South Vietnam increased from 834 to 1,085, and twenty-eight tactical fighter squadrons could be called upon to provide close air support and assist in the interdiction campaign. The number of B–52 sorties rose sharply. Army and Marine Corps helicopter units could muster over 3,000 helicopters of all types.[4]

By the autumn of 1967, Westmoreland had refined his earlier concept of a three-phase war to one of four phases. He saw 1968 as the year of the third phase, in which the Americans would continue to help strengthen the South Vietnamese military. The United States would also turn over more of the war effort to the South Vietnamese. In what he termed the fourth—and decisive—phase, Westmoreland could see the "U.S. presence becoming superfluous as infiltration slowed, the Communist infrastructure was cut up, and a stable government and capable Vietnamese Armed Forces carried their own war to a successful conclusion." He also expected that the enemy would try to do something to change this trend.[5]

Although events later in 1967 pointed to some kind of change in Hanoi's plans, probably something big, General Westmoreland had no intention of sit-

[3] MACV History, 1967, vol. 1, pp. 326–30, vol. 2, pp. 762–63; MACDC-EBD Fact Sheet, sub: Highway Restoration Program in Vietnam, n.d., box 3, 70A/782, RG 334, NARA; Quarterly Hist Rpts, 1 Jan–31 Mar 67, MACDC, p. 1, and 1 Oct–31 Dec 67, MACDC, p. II-1.

[4] Sharp and Westmoreland, *Report*, p. 131; MACV History, 1967, vol. 1, pp. 135–36, 147–49; Schlight, *Years of the Offensive*, p. 159. For more on the troop ceiling requests, see MACV History, 1967, vol. 1, pp. 135–66 and Cosmas, *Years of Escalation, 1962–1967*, pp. 451–66.

[5] Quoted in Sharp and Westmoreland, *Report*, p. 136.

ting back to await the enemy's next move. The promulgation of the Combined Campaign Plan for 1968 in November 1967 did not foresee any major changes in allied plans. U.S. ground forces would carry on with the destruction of the enemy's main forces and base areas, contain the border areas and his use of infiltration and invasion routes, and help South Vietnamese forces to open and secure roads and the countryside. Westmoreland had concluded in mid-1967 that the enemy was in a position of weakness. Hanoi had nothing tangible to show but increasing losses in personnel and materiel after a little more than a year of fighting the growing number of American troops. He regarded Hanoi's choices as few, and intended to keep them that way. Indeed, as additional U.S. ground forces arrived, he saw an opportunity to employ the 1st Cavalry Division as a theater exploitation force in areas where good weather prevailed. In a series of operations code-named YORK, the division would deploy to the III Corps border during the December to April dry season in the south. During the May to September dry season farther north, the cavalry division would then move to I Corps to sweep the four provinces along the Laotian border. Reestablishing control in the A Shau Valley could set the stage for an invasion of Laos, which, along with a possible amphibious assault around the Demilitarized Zone, could, Westmoreland believed, end the war.[6]

The Saigon Arc

Integral to these hopes for the future was progress in III Corps, and by mid-1967 II Field Force had completed a series of large-scale operations following JUNCTION CITY. With the onset of the monsoon season, U.S. and allied forces had settled in assigned areas meant to block the approaches to Saigon. The command's maneuver forces included the 25th Infantry Division guarding the northwest approaches, the 1st Infantry Division to the north and northeast of the capital, the 9th Infantry Division east of Long Binh and to the southwest in the delta, the 11th Armored Cavalry Regiment farther to the east, and the 199th Light Infantry Brigade in Gia Dinh Province surrounding Saigon. II Field Force focused on strengthening security in the countryside, holding off enemy attempts to sabotage the national elections, and continuing to strike enemy base areas and defeat and destroy his forces. Other forces in the area included South Vietnamese regular and territorial units, the 1st Australian Task Force southeast of Saigon, and the Royal Thai Volunteer Regiment.[7]

By this time, the deployment of Rome plows and bulldozers had become a routine part of operations in the III Corps Tactical Zone. Their effectiveness during CEDAR FALLS had bolstered MACV's earlier requests for land-clearing teams. Although standard bulldozers had helped to prove the effectiveness of the land-clearing concept, their use in these operations had affected construction projects. MACV warned that similar use in the future would slow work at Long Binh and elsewhere until the land-clearing capability increased. As a

---

[6] Ibid., p. 157. For details on the Combined Campaign Plan and the YORK series of operations, see MACV History, 1967, vol. 1, pp. 338–45, ibid., 1968, vol. 1, pp. 13–24.

[7] ORLL, 1 Aug–31 Oct 67, II FFV, 15 Nov 67, pp. 21–22, 57, 61, Historians files, CMH.

result, II Field Force deferred large forest-clearing operations (PAUL BUNYON) pending the arrival of land-clearing teams being formed at Fort Lewis, Washington.[8]

Rome plow kits reaching combat engineering units gradually increased. In December 1966, the Rome Plow Company rushed six kits to South Vietnam by air shipment. By early February, the Georgia-based company began weekly shipments of nine blades that continued over the next several weeks. As of late March, thirty medium and six light blades were in South Vietnam, but only twelve were mounted. The remainder still required attachment to the bulldozers. Totals in III Corps included four kits to the 1st Division, four to the 25th Division, two to the 11th Armored Cavalry Regiment, eight to the 79th Engineer Group, and four to the 159th Engineer Group. At midyear, plans called for a total of 174 blades, including 90 for three new land-clearing teams that began to arrive in May.[9]

The 27th, 86th, and 35th Land Clearing Teams reported for duty in May, June, and July, respectively. Originally proposed as self-contained teams, each consisting of 112 men and thirty Rome plows, the teams were reduced in scope by the Army to 64 men. The teams now depended on engineer battalions for maintenance, logistical, and administrative support. After undergoing several weeks of intensive training, the teams became operational approximately one month after their arrival. Two teams were attached to combat engineer battalions in III Corps, the 27th to the 168th Engineer Battalion at Di An and the 86th to the 86th Engineer Battalion at Bearcat. The third team, the 35th, deployed to II Corps, where it joined the 35th Engineer Battalion north of Qui Nhon.[10]

Battalion commanders used different approaches to assimilate the teams. Lt. Col. John R. Manning, who assumed command of the 168th Engineer Battalion in May, organized a land-clearing task force to support the 27th Land Clearing Team. Besides creating a small task force headquarters, the battalion formed a maintenance section consisting of a warrant officer, six mechanics, plus a contact team from a direct-support maintenance company. Another augmentation included a small communications section to provide a radio link between the task force and security forces. Lt. Col. James F. Miley of the 86th Engineer Battalion placed his land-clearing team in one of his line companies, which, in turn, designated two platoons to supervise and support the two sections. A mechanized infantry battalion usually provided security,

[8] Quarterly Hist Rpt, 1 Jan–31 Mar 67, MACDC, pp. 6–7; Msg, COMUSMACV 03824 to CINCPAC, 1 Feb 67, sub: Engineer Land Clearing Teams, box 18, 84/051, RG 319, NARA; Msg, COMUSMACV 09572 to CGs, USARV and II FFV, 22 Mar 67, sub: Forest Clearing Operations, box 5, 70A/782, RG 334, NARA. For more background on the land-clearing teams, see Army Buildup Progress Rpts, 23 Feb 67, p. 14, and 8 Mar 67, Supp, p. 1, CMH.

[9] Ltr, J. T. Soules, Vice President, Rome Plow Co, to Ploger, 9 Feb 67, Historians files, CMH; Quarterly Hist Rpts, 1 Oct–31 Dec 66, MACDC p. 7, 1 Jan–31 Mar 67, MACDC, p. 6, and 1 Apr–30 Jun 67, MACDC, pp. II-6 to II-7.

[10] Quarterly Hist Rpts, 1 Apr–30 Jun 67, MACDC, p. II-7, and 1 Jul–30 Sep 67, MACDC, pp. II-5 to II-6; ORLLs, 1 Aug–31 Oct 67, Engr Cmd, p. 27, 1 May–31 Jul 67, 79th Engr Gp, p. 8, 1 May–31 Jul 67, 34th Engr Gp, p. 2, and 1 May–31 Jul 67, 45th Engr Gp, pp. 2–3. For more on the deployment of the teams, see Army Buildup Progress Rpts, 19 Apr 67, p. 1, and 12 Jul 67, Supp, p. 1, CMH.

A modified D7E bulldozer equipped with a Rome plow and an armored cab

and the armored personnel carriers also carried an engineer demolition squad, welder, and mechanics to support each land-clearing section. The extensive supervision and maintenance support required by the teams in the field forced engineer commanders to look for a more permanent solution. Col. Joseph A. Jansen, the commanding officer of the 79th Engineer Group, suggested organizing a separate land-clearing company. The U.S. Army, Vietnam, engineer staff agreed and began to prepare a table of organization for such a unit.[11]

The arrival of the land-clearing teams started a new phase of exposing enemy sanctuaries, base camps, and infiltration routes. During combat operations, the teams leveled large areas of jungle, forests, and rubber plantations. Bulldozers equipped with the plows cleared areas on both sides of supply routes and reduced the threat of ambushes and increased security, at least during the daylight hours. During August, search-and-destroy operations, which included extensive road and jungle clearing, penetrated known and suspected base camps and staging areas and prevented the Viet Cong's infiltration into populated areas. By September, the land clearers opened large areas previously inaccessible to friendly forces. Operations in the Ong Dong jungle by the 1st Division, and the Filhol Plantation and the Ho Bo Woods by the 25th Division did much to deny the enemy previously secure base camps and forced him to abandon established lines of communications for less desirable routes. Simultaneously, the jungle-clearing effort improved allied base camp security by expanding fields of fire and clearing concealed avenues of

---

[11] Engineer Support of Combat Operations, 1 Jul–1 Sep 67, incl. 9, pp. 1–2, OCE Liaison Officer Trip Rpt no. 9, 24 Oct 67, OCE Hist Ofc.

approach leading to the bases. Not only did cleared areas make it less difficult to detect enemy movements, but they also complemented efforts to improve base camp defenses from rocket and mortar attack. Land-clearing rates averaged 1.2 acres per hour when cutting and piling vegetation up to six inches in diameter, more than double the standard D7E bulldozer's 0.5 acres. The ratio for cutting dense vegetation, including larger trees, remained the same: 1.04 acres with the Rome plows and 0.4 acres with the standard bulldozer. By the end of the summer, the teams had leveled over 30,000 acres of vegetation.[12]

The 1st Infantry Division centered its efforts north and northeast of Saigon. Two operations specifically designed to deny the enemy his secure base areas, LAM SON 67 and PAUL BUNYON, epitomized the contribution made by the land-clearing teams. During LAM SON 67, a pacification operation that began in February, the 27th Land Clearing Team cleared a jungle area often used by the Viet Cong for mortar attacks on Bien Hoa Air Base. Between 3 and 16 July, the team cleared 2,610 acres, averaging 186 acres per day, and discovered and destroyed eight tunnels, nine base camps, and several mined booby traps. Three days later, the team joined Company A, 168th Engineer Battalion, and the 1st Engineer Battalion in the first phase of Operation PAUL BUNYON in the Ong Dong Jungle, some ten miles north of Di An. Between 19 July and 13 August, they cleared 7,740 acres at the rate of 309 acres per day and destroyed five base camps and more booby traps. On 16 August, the effort shifted to Highway 13, dubbed Thunder Road. During Phase II of PAUL BUNYON, the land clearers cut 5,200 acres along the road averaging 325 acres per day. In addition, the 1st Engineer Battalion's Rome plow section, imitating an earlier experiment by the 9th Division, connected 180 feet of heavy anchor chain from a Navy cruiser between two bulldozers and cleared some 1,500 nearby rubber trees. The clearing operation, which continued into September, also resulted in finding and destroying thirteen base camps, fourteen tunnel complexes, and many mines and booby traps. By the end of the month, the 1st Engineer Battalion with help from the 79th Engineer Group had cleared selected sectors of jungle along Highway 13 and Route 1A toward An Loc and Phuoc Vinh, respectively. During Operation BLUEFIELD II the same month, the 1st Engineer Battalion reopened Highway 13 and cleared vegetation a little over one hundred yards on each side of Route 301. This effort opened a second land route to Phuoc Vinh from the west.[13]

Similar operations took place to the east and southeast of Saigon. On 10 July, the allies initiated Operation PADDINGTON to open a jungle area used by the Viet

---

[12] ORLL, 1 Jul–31 Oct 67, II FFV, pp. 21, 28; Miscellaneous Detailed Discussion, incl. 9, p. 3, OCE Liaison Officer Trip Rpt no. 8, 11 Aug 67; Ltr, HQ USARV, AVHEN-MO, to MACV, sub: Technical Report, Paul Bunyon Land Clearing Teams, n.d., incl. 9, tab B, p. 3, OCE Liaison Officer Trip Rpt no. 9; Galloway, "Essayons," p. 201. See also ORLL, 1 Jul–31 Oct 67, Engr Cmd, incls. 3 and 4, pp. 1–6; DA Pam 525–6, *Military Operations, Lessons Learned: Land Clearing*, 16 Jun 70.

[13] ORLL, 1 Jul–31 Oct 67, II FFV, p. 28; 1st Engr Bn Hist, vol. 1, pp. III-22 to III-26; Engineer Support of Combat Operations, 1 Jul–1 Sep 67, incl. 9, p. 3, OCE Liaison Officer Trip Rpt no. 9; Interv, Maj Paul B. Webber, 26th Mil Hist Det, with Lt Col John R. Manning, CO, 168th Engr Bn, 10 Apr 68, VNIT 106, pp. 1–2, CMH. For more on PAUL BUNYON, see Technical Report, Paul Bunyon Land Clearing Teams, incl. 9, tab B, OCE Liaison Officer Trip Rpt no. 9.

An aerial view of an anti-ambush clearing on each side of a road by Rome plows

Cong as an operations and logistics base in Phuoc Tuy Province. The combined operation by the 1st Brigade, 9th Division; the 1st Squadron, 11th Armored Cavalry Regiment; the 1st Australian Task Force; and South Vietnamese army and marine units cleared strips along Route 23 and other local roads. As part of an engineer task force deployed by the 9th Division, the 15th Engineer Battalion's Company B used fourteen Rome plows and three bulldozers to clear the jungle back a little over two hundred yards. Also in July, the 9th Division attached elements of the 11th Armored Cavalry Regiment to protect the 86th Engineer Battalion's land-clearing task force in Operation EMPORIA, a continuation of jungle-cutting along roads. Between 21 July and 15 August, Company B and the 86th Land Clearing Team in its first operation cleared 1,586 acres on both sides of Highway 20, the key land route from the vegetable-producing area around Da Lat. The task force next cleared another 1,005 acres along Highway 1 between Xuan Loc and Gia Ray. In September, Rome plows moved south on Route 2 cutting vegetation from the 11th Armored Cavalry's camp at Blackhorse to Ba Ria. In another jungle-clearing operation under the 9th Division's 1st Brigade (AKRON III), the 86th team's Rome plows, while cutting the thick underbrush and towering trees near the borders of Phuoc Tuy, Bien Hoa, and Long Khanh Provinces, uncovered a fresh trail on 8 October. This led to the discovery of the largest weapons cache to date in III Corps.[14]

---

[14] ORLLs, 1 May–31 Jul 67, II FFV, pp. 23–24, 1 Aug–31 Oct 67, II FFV, p. 33, 1 May–31 Jul 67, 34th Engr Gp, p. 8, and 1 Aug–31 Oct 67, 34th Engr Gp, pp. 5–6; Engineer Support of Combat Operations, 1 Jul–1 Sep 67, incl. 9, p. 4, OCE Liaison Officer Trip Rpt no. 9; AAR, Opn PADDINGTON, 15th Engr Bn, 23 Sep 67, p. 3, OCE Hist Ofc.

307

Another land-clearing device, the so-called transphibian tactical tree crusher, made its appearance at this time. Like the towed ball and chain experiments, the massive ninety-seven-ton machine proved only marginally effective. General Ploger pointed out the difficulty in deploying such a massive piece of equipment. He noted that for the price of one tree crusher, approximately $300,000, he could buy seven of the proven Rome plows with the tractors. He did, however, accede to the 1st Logistical Command's proposal to evaluate the machine. In March 1967, the logistical command contracted with LeTourneau Corporation of Longview, Texas, for a six-month rental of two tree crushers. The tree crushers arrived at Saigon's Newport in mid-July, and by the end of the month the logistical command's provisional detachment began to clear land around Long Binh Post. During the test period, which lasted to the middle of November, the two crushers used their pusher bars against large trees and used their cleated drums to chew up felled trees and vegetation. Approximately 2,000 acres in the depot and ammunition storage areas were cleared. The crushers did clear all kinds of vegetation averaging 2.1 acres per hour, but both machines spent about half the time in the repair shop.[15]

Following several modifications to alleviate the maintenance problems, the detachment and tree crushers were transferred to Ploger's command to support tactical operations. The detachment, operating under the control of the 93d Construction Battalion, leveled 1,300 acres supporting the 9th Division near the Binh Son Rubber Plantation, thirty-five miles southeast of Saigon. Again the equipment spent too much time in the shop. Fully satisfied that the machine's limitations outweighed the benefits—its downtime, vulnerability to enemy action, and difficulty working in shallow water with mud bottoms of inadequate bearing and where insufficient water depth could not float the crusher free from the suction of the mud—Ploger had the equipment returned to the United States after the lease expired.[16]

Land clearing had proved to be an effective weapon against the enemy, and the planners made preparations for even larger operations in 1968. Open areas along the cleared roads lessened his capability to mount ambushes and reduced the effectiveness of those that did take place. During the month following the clearing operations along Highways 1 and 20 and Route 2, the 11th Armored Cavalry Regiment reported no ambushes where previously many had occurred. Captured documents revealed that land clearing had a major

[15] Quarterly Hist Rpts, 1 Apr–30 Jun 67, MACDC, p. II-6, and 1 Jul–30 Sep 67, MACDC, p. II-5; Ploger, *Army Engineers*, pp. 98–99; Hay, *Tactical and Materiel Innovations*, p. 87; ORLLs, 1 May–31 Jul 67, 1st Log Cmd, 15 Aug 67, p. 112, 1 Aug–31 Oct 67, 1st Log Cmd, 11 Nov 67, pp. 113–14, and 1 Nov 67–31 Jan 68, 1st Log Cmd, 14 Feb 68, pp. 89–90, all in Historians files, CMH; Ltr, Ploger to CG, USARV, 15 Mar 67, sub: LeTourneau Transphibian Crusher, Ploger Papers, CMH; Interv, Bower with Ploger, 8 Aug 67, pp. 51–53; Interv, Sowell with Ploger, 16 Nov 78, sec. 9, pp. 11–12.

[16] Quarterly Hist Rpt, 1 Oct–31 Dec 67, MACDC, pp. II-5 to II-6; Hay, *Tactical and Materiel Innovations*, p. 87; ORLL, 1 Nov 67–31 Jan 68, 93d Engr Bn, 14 Feb 68, p. 10, Historians files, CMH; "Tree Crusher Tested in Vietnam," *The Army Engineer in Vietnam*, U.S. Army Engineer School Pamphlet (Fort Belvoir, Va.: n.d.), pp. 3–5; Interv, Bower with Ploger, 8 Aug 67, pp. 52–53.

The limitations of the transphibian tactical tree crusher outweighed its benefits, and the equipment was returned to the United States after the lease expired.

influence on the enemy, who regarded these operations as a threat to his plans. Newly cleared strips along roads reduced the need for convoy escorts and also became an effective barrier countering the enemy's free movement of troops and supplies. In the wake of these successful operations, II Field Force headquarters shifted priority in III Corps to area clearing (some 400,000 acres), whereas I Field Force continued to use its land-clearing team in II Corps mainly for road clearing.[17]

In late September, General Westmoreland ordered III Marine Amphibious Force to prepare for large-scale land-clearing operations in I Corps. The Directorate of Construction helped the marines with planning details, and by the end of the year ten Rome plow blades for the marines' bulldozers arrived in Da Nang. After considering operational requirements, III Marine Amphibious Force asked for two Army land-clearing teams to support future operations, but MACV suggested that the best source of land-clearing capability would be for that command to develop its own organic teams. The Construction Directorate also studied possible use of Rome plows by the South Vietnamese to support pacification in selected provinces. It soon became apparent, however, that Saigon's military engineers had limited operating and maintenance capabilities and could not be expected to join the land-clearing effort.[18]

The arrival of the southwest monsoon in late spring placed heavy demands on the combat engineers to keep the deteriorating ground supply

[17] Technical Report, Paul Bunyon Land Clearing Teams, incl. 9, tab B, p. 4, OCE Liaison Officer Trip Rpt no. 9.

[18] Ibid., pp. 4–5; Quarterly Hist Rpt, 1 Oct–31 Dec 67, MACDC, pp. II-6 to II-7.

routes open, while simultaneously clearing back the encroaching vegetation. As the 1st Division moved operations northward to the Cambodian border, the 1st Engineer Battalion used hand labor and unsophisticated expedients to rebuild and maintain Highway 13 and Rolling Stone Road between Ben Cat and Phuoc Vinh. Sometimes the troops had to blow stopped-up culverts and build short fixed span bridges over the water-filled gaps and timber roadways along muddy sections of the road. Several fixed spans were lifted intact to work sites by CH–47 helicopters, thus saving days of working time. During BLUEFIELD II, Company A also built the superstructure of a twenty-foot timber trestle bridge airlifted to a prepared site by a CH–47. Northwest of Saigon, the 65th Engineer Battalion and the 79th Engineer Group's 588th Engineer Combat Battalion supported the 25th Division by keeping supply routes open between Tay Ninh and Dau Tieng, Tri Bi, Prek Klok, and the Cu Chi area. The 9th Division's 15th Engineer Battalion and other engineer units continued to work on the roads to the east, southeast, and southwest of Saigon. Work proceeded along Highways 1 and 20 as far as the II Corps boundary, Highway 15 and Route 2 toward the coast and the port of Vung Tau, and Highway 4 into the delta region. The roads were open, but no one could say for how long.[19]

The heavy rains also played havoc at the austere, hastily built forward airstrips, especially at Special Forces camps near the Cambodian border. In May, the 79th Group's 168th Engineer Battalion airlifted troops to the Tong Le Chon airstrip west of An Loc to make repairs to the runway, turnaround, and parking apron. In June, the battalion deployed a task force to Chi Linh northeast of Saigon to repair the airstrip and enlarge the parking apron. By August, the 168th also extended the existing C–130 runway and added a turnaround and parking apron at Dong Xoai and began to upgrade the airfield at Loc Ninh to C–130 standards. Next, men and equipment boarded CH–54 and CH–47 helicopters and flew to the Bu Dop Special Forces camp to begin upgrading the runway for C–123 aircraft.[20]

A variety of work continued in support of pacification in the arc of provinces around Saigon. Operation ENTERPRISE, similar in scope to LAM SON 67, began in February in Long An Province southwest of the city and lasted throughout the year. While the 9th Division's 3d Brigade and South Vietnamese territorial units kept up the pressure on the Viet Cong, the 86th Engineer Battalion built several operating bases and artillery firing pads on rice paddies. What set this work apart was the need to import large quantities of rock from the Vung Tau quarry. The 34th Group arranged with transportation units for twenty-one barges, which carried the rock to the base camp sites in the delta. At Tan An, the engineers pushed up the rice paddy mud and

[19] ORLL, 1 Aug–31 Oct 67, II FFV, p. 28; 1st Engr Bn Hist, vol. 1, pp. III-22 to III-26; ORLL, 1 Aug–31 Oct 67, 34th Engr Gp, pp. 8–9; Engineer Support of Combat Operations, 1 Jul–1 Sep 67, incl. 9, p. 4, OCE Liaison Officer Trip Rpt no. 9; Galloway, "Essayons," pp. 238, 243–44.

[20] Engineer Support of Combat Operations, 1 May–1 Jul 67, incl. 7, pp. 5–6, OCE Liaison Officer Trip Rpt no. 8; Engineer Support of Combat Operations, 1 Jul–1 Sep 67, incl. 9, pp. 4–5, OCE Liaison Officer Trip Rpt no. 9; Interv, Webber with Manning, 18 Mar 68, p. 2.

topped it off with fill and rock to provide the necessary areas for cantonments and hardstands for a brigade base. This work also included the building of a C–123 airfield and heliport. Similar work took place at Ben Phuoc, a forward operating base for a mechanized infantry battalion, and Ben Luc, a forward support base for a composite artillery battalion. Near Saigon, the 199th Light Infantry Brigade and South Vietnamese units continued Operation FAIRFAX, a long-term pacification effort to keep the Viet Cong from reestablishing influence in the area. The 199th's 87th Engineer Company split its efforts between supporting operations in the field and improving the brigade's base camp at Long Binh.[21]

Operational Control Revisited

Increased emphasis on land clearing and road work illustrated the steady shift from base construction to combat support missions in 1967. Brig. Gen. Curtis W. Chapman's 20th Engineer Brigade, which arrived in August to take control of the three engineer groups in III and IV Corps, reported a twofold increase in tactical missions from August through September. By 20 September, 24 percent of the brigade's total productive effort was being devoted to combat support. In II Corps, a shift of combat operations toward the border brought forth similar patterns of engineer support.[22]

With tactical missions on the rise, the old operational control question resurfaced with a vengeance in early 1967. It was raised by General Palmer, who had taken over as commander of II Field Force in March and who wasted no time in requesting a change in who would command the nondivisional engineers. General Engler turned him down, noting that the field force had never used more than one and a half battalions at any time in tactical operations. General Westmoreland agreed.[23]

In late spring, General Engler returned to the United States and General Palmer took over as U.S. Army, Vietnam, deputy commander. Palmer soon took steps to formalize a more precise means of identifying engineer support for the field forces. He asked that the field force commanders and the Engineer Command draw up agreements identifying certain engineer units to be on call for duty in tactical operations. On 17 May, Ploger signed a memorandum of understanding with Maj. Gen. Frederick C. Weyand, who moved up from the 25th Division to command II Field Force. Brig. Gen. Charles M. Duke, the 18th Brigade commander, prepared a similar agreement with General Larsen of I Field Force.[24]

---

[21] Engineer Support of Combat Operations, 1 May–1 Jul 67, incl. 7, p. 4, and Miscellaneous Detailed Discussion, incl. 9, p. 6, OCE Liaison Officer Trip Rpt no. 8; ORLLs, 1 May–31 Jul 67, 34th Engr Gp, p. 7, and 1 Aug–31 Oct 67, 34th Engr Gp, p. 5; Galloway, "Essayons," p. 245.

[22] Engineer Support of Combat Operations, 1 Jul–1 Sep 67, incl. 9, p. 1, OCE Liaison Officer Trip Rpt no. 9; Galloway, "Essayons," pp. 221, 243.

[23] Ploger, *Army Engineers*, p. 141; Msg, Westmoreland MAC 3653 to Palmer, 17 Apr 67, Westmoreland Message files, CMH.

[24] Ploger, *Army Engineers*, p. 141.

Because Ploger foresaw few changes caused by the agreements to the operations of his command, he did not resist entering such arrangements. The agreement with Weyand put two battalions of Colonel Gelini's 79th Engineer Group, the 168th and the 588th, under II Field Force's operational control effective 6 June. The Engineer Command also earmarked three separate companies (the 100th Float Bridge, the 500th Panel Bridge, and the 362d Light Equipment) for the field force. Although II Field Force now had formal authority during combat operations, the 79th Group retained responsibility for all formal construction projects assigned these units. Ploger could not help but note that by August the units reported no significant change in their day-to-day operations.[25]

The memorandum of understanding drawn up between the 18th Engineer Brigade and I Field Force did not go into effect. The memorandum called for General Duke's brigade to provide one engineer battalion or equivalent force from each group and such float and panel bridge and light equipment company forces as required. Larsen, however, expressed his satisfaction with the support given by the 18th Brigade and preferred to keep the same kind of relationship that existed before General Palmer's request. He saw little advantage in the new arrangements because designating only certain engineer units could reduce the flexibility previously held by his commanders. Considering the larger geographic area of II Corps, it seemed unwise to have to move designated units a considerable distance, while bypassing units located in the area of an operation. Area support continued, with operational support missions retaining precedence over all other engineer support requirements. When General Larsen prepared to turn over his command in July, he wrote to General Duke commending the cooperation and support given by the 18th Engineer Brigade. He said the brigade's efforts "surpassed all expectations, even to the point that operational control of engineer elements by I Field Force was never required."[26]

Soon after succeeding General Ploger in August, General Duke, now with two stars, became aware that General Weyand still had some reservations concerning the responsiveness of support given by Engineer Command. On 16 September, he wrote to Weyand reaffirming that the 20th Engineer Brigade's units would respond immediately to operational support requirements. In urgent cases, Duke recommended that the field force commander have his staff engineer bypass the brigade headquarters and "transmit the requirement directly to the engineer group—or even the battalion—headquarters immediately concerned." He stressed that all routine

---

[25] Ibid.; Interv, Bower with Ploger, 8 Aug 67, p. 24; Rpt of Visit to Various Headquarters in Vietnam, 9–23 Jul 67, incl. 6, pp. 4–5, OCE Liaison Officer Trip Rpt no. 8. The memorandum of understanding is at tab L.

[26] Ploger, *Army Engineers*, p. 141; Interv, Bower with Ploger, 8 Aug 67, p. 24; Rpt of Visit to Various Headquarters in Vietnam, 9–23 Jul 67, incl. 6, p. 4, OCE Liaison Officer Trip Rpt no. 8. The memorandum of understanding is at tab K. Also Rpt of Visit to Various Headquarters in Vietnam, 15–30 Sep 67, incl. 7, p. 2, OCE Liaison Officer Trip Rpt no. 9. General Larsen's letter is at tab A.

or planned requirements would continue to be processed through the 20th Engineer Brigade.[27]

Ten days later, Duke met with Maj. Gen. George S. Eckhardt, the deputy commanding general of II Field Force, to discuss again the matter of operational control. According to Eckhardt, the operational support requirements had increased to such a degree, including responsibilities for rural development and the training of Vietnamese Army troops, that an entire group could be kept fully occupied. He then asked for operational control of a group headquarters plus all the combat battalions, light equipment companies, and bridge companies in the 20th Brigade. Since the field force had other units placed under its operational control to carry out its requirements, it seemed only natural to place the engineers in the same status. This new and enlarged request surprised Duke considering the evident success of past operations. He could not agree to such a request unless II Field Force accepted responsibility for the construction projects assigned to these units. Eckhardt stated that these conditions were unacceptable, suggesting that field force responsibility for combat operations would be incompatible with taking on construction projects outside the tactical realm.[28]

Since the two generals could not reach a new agreement, the procedure outlined in Duke's letter to General Weyand remained in effect. General Ploger later wrote that engineer units carried out operational support missions "with the ease and timeliness that made any change in the tactical arrangements seem unnecessary." He maintained that neither the support for tactical operations nor the progress on the base development program suffered appreciably. If more engineer units were available, he might have seen his way to place more units under the operational control of the field forces.[29]

The lack of further agreements did not deter engineer support of combat operations. In fact, this support increased. The Engineer Command reported that between 1 September and 15 November approximately 20 percent of its productive effort was devoted to operational support. The two brigades and all six groups reported increases. For example, the 45th Group in II Corps, with three combat and two construction battalions and several separate companies, calculated that 70 percent of its effort included land clearing, road and airfield maintenance and upgrading, minesweeping, firebase construction, and river crossings.[30] (*Chart 4*)

On the III Corps Border

On 29 September, the 1st Infantry Division launched Operation SHENANDOAH II. The operation had a two-pronged purpose, combining

[27] Ploger, *Army Engineers*, pp. 142–43; Rpt of Visit to Various Headquarters in Vietnam, 15–30 Sep 67, incl. 7, p. 2, OCE Liaison Officer Trip Rpt no. 9. General Duke's letter is at tab B.

[28] Ploger, *Army Engineers*, p. 143; Rpt of Visit to Various Headquarters in Vietnam, 15–30 Sep 67, incl. 7, p. 2, OCE Liaison Officer Trip Rpt no. 9.

[29] Ibid.

[30] Rpt of Visit to Various Headquarters in Vietnam, 22 Nov–13 Dec 67, incl. 9, p. 1, OCE Liaison Officer Trip Rpt no. 10, 19 Jan 68, p. 7, OCE Hist Ofc.

CHART 4—ORGANIZATION OF U.S. ARMY ENGINEER COMMAND, VIETNAM,
NOVEMBER 1967

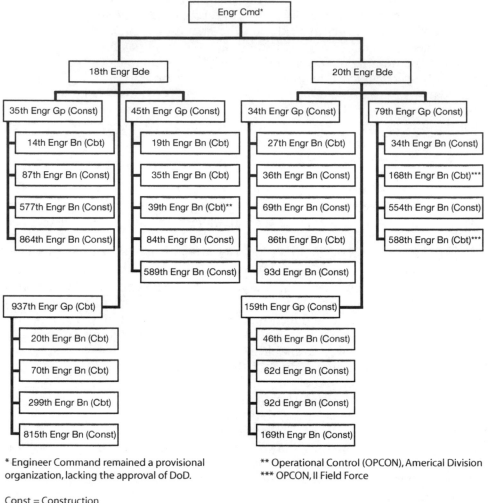

* Engineer Command remained a provisional
organization, lacking the approval of DoD.

** Operational Control (OPCON), Americal Division
*** OPCON, II Field Force

Const = Construction
Cbt = Combat

Source: Ploger, *Army Engineers*, p.135

search-and-destroy missions to inflict maximum casualties upon the newly arrived North Vietnamese forces in III Corps and roadwork, which included maintenance and upgrading and land clearing. The area of operation reached out from the division's base camp at Lai Khe approximately nineteen miles to the north, and west to the edge of the Michelin Rubber Plantation. While a mechanized infantry battalion from the 1st Brigade swept an area north of Lai Khe, the 3d Brigade began securing Route 240 between Ben Cat to the edge of the rubber plantation. Next, Rome plows began to push back the jungle about a little over one hundred yards on each side of the road. Simultaneously, the

During Operation SHENANDOAH, *the 1st Engineer Battalion supported the movement of the division's armored personnel carriers by erecting tactical bridging such as this armored vehicle launched bridge.*

1st Engineer Battalion started to reopen the road, by then reduced to little more than a footpath because of enemy obstacles and jungle overgrowth.[31]

Most of the 1st Engineer Battalion moved to the area of operations. Companies B and C worked along Route 240, upgrading the road and cutting down 100-foot-tall trees with explosives, and encountering resistance along the way. On 13 October, North Vietnamese troops ambushed a Company C column by setting off a command-detonated mine and firing small arms. The engineers returned fire and continued on their way after the enemy faded away. Four days later, Company C gained some vengeance by uncovering and destroying a ten-ton rice cache. Meanwhile, Companies A and D airlifted precut timber trestle bridges and men to worksites, installing the twenty-foot spans across gaps in the road. Together, the 1st Engineer Battalion with the help of the 168th Engineer Battalion's 27th Land Clearing Team cleared nearly 1,000 acres of dense jungle. They also cleared areas for five firebases, allowing 105-mm. howitzers to cover the division's western tactical area of responsibility. Since the rainy season continued throughout the campaign, the 1st Engineer Battalion again used timbers to build a corduroy base along the impassible muddy sections of the road. The battalion then topped off the road

---

[31] ORLL, 1 Aug–31 Oct 67, II FFV, p. 29. For more on the SHENANDOAH operation, see AAR, Opn SHENANDOAH II, 1st Inf Div, 12 Apr 68, Historians files, CMH; MacGarrigle, *Taking the Offensive,* pp. 347–49.

315

with over 700 five-ton dump truckloads of laterite and rock. All this work was accomplished in two weeks and enabled the 1st Division to penetrate one more enemy sanctuary. Although the division soon left the area, the message was clear. Allied forces could return at a moment's notice.[32]

Enemy forces gave little resistance during the initial stage of SHENANDOAH, but in October they turned on their tormentors and launched a series of attacks on South Vietnamese bases near the Cambodian border. It became obvious to General Westmoreland that the North Vietnamese had shifted efforts to the border areas of II and III Corps and were caught while building staging areas and base camps. He claimed that the enemy hoped to achieve important psychological victories to divert allied attention from the urban areas and I Corps and to offset the inauguration of the newly elected Thieu and Ky government. On 4 October, the 1st Division's mechanized infantry battalion came under attack six miles southwest of Chon Thanh. Two days later, an infantry battalion that moved into the area also came under attack. On the seventeenth, elements of another U.S. infantry battalion were ambushed eight and a half miles southwest of Chon Thanh. Friendly artillery and air strikes helped to beat off a Viet Cong regiment, but the enemy's heavy use of small arms and automatic weapons took a heavy toll on the American battalion.[33]

The first border battle began early in the morning of 27 October near the village of Song Be. North Vietnamese forces assaulted a South Vietnamese infantry battalion, but the defenders repulsed every assault and inflicted heavy losses on the attackers. Much the same happened two days later at Loc Ninh, where a Viet Cong regiment struck the district town and a nearby rubber plantation eight miles south of the border. Phase II of SHENANDOAH started when U.S. air power and infantry units went to the aid of the South Vietnamese. Battalions of the 1st Division launched airmobile assaults, swept adjacent rubber plantations, and repelled attacks on night defensive positions. By the time the enemy broke off the battle on 8 November, he had suffered nearly 1,000 killed. Operation SHENANDOAH II ended on 19 November.[34]

The 1st Engineer Battalion's major effort during the operation took place on supply routes north of Lai Khe. Following the national election, enemy forces had sabotaged Highway 13 to the point it could no longer support the division's supply convoys to its Quan Loi base just to the east of An Loc. On 17 October, Company A moved by fixed-wing aircraft from Phuoc Vinh to Quan Loi where it began to build sections of culvert for the road work to start on 1 November. The day before the enemy, in a carefully planned act of sabotage, blew three large craters along a two-mile section of the road,

---

[32] 1st Engr Bn Hist, vol. 1, p. III-26; *Always First, 1967–1968*, "Opn SHENANDOAH."

[33] ORLL, 1 Aug–31 Oct 67, II FFV, pp. 29–30; Westmoreland, *A Soldier Reports*, p. 288; CHECO Rpt, PACAF, 15 May 68, sub: VC Offensives in III Corps, Oct–Dec 67, p. 3, copy in Historians files, CMH. For more on these actions, see MacGarrigle, *Taking the Offensive*, pp. 349–61.

[34] ORLLs, 1 Aug–31 Oct 67, II FFV, app. 30–33, and 1 Nov 67–31 Jan 68, II FFV, pp. 24–25; Westmoreland, *A Soldier Reports*, pp. 287–88; Interv, Webber with Manning, 10 Apr 68, p. 2.

approximately halfway between sources of rock needed for the repairs. On 1 November, two platoons of Company B moved by CH–47 Chinooks from Di An to a site midway between Chon Thanh and An Loc to build a seventy-foot timber trestle bridge. The platoons, using the proven techniques of preparing abutments with demolitions and precutting and airlifting timber to the bridge site, completed the planned two-day project on the first day. With the early completion of the bridge, the battalion headquarters changed Company C's mission to load rock at Lai Khe that night and to move at first light. On the second day, Company A continued to move south, filling craters and repairing damaged culverts. Company C collected additional dump trucks at Chon Thanh, crossed over the new bridge, and pushed north filling craters and potholes along the way. Upon linking, both units joined forces to fill the remaining crater by 1400. A half hour later, the main supply column moved north, one day ahead of schedule. Work along Highway 13—filling potholes, installing new culverts, grading, and clearing trees along the road—continued throughout the operation.[35]

Keeping the heavily worn and damaged roads open continued after SHENANDOAH II. The 1st Division carried out operations near Highway 13 and built hasty night defensive positions and firebases along the road. Convoys moved more freely, and civilian traffic and economic activity increased. The Viet Cong and North Vietnamese, who once moved men and supplies along the road and crossed it at will, launched a series of night assaults against the American bases. During one attack on 10 December, elements of a North Vietnamese regiment struck a battalion-size task force five miles southwest of An Loc. U.S. ground fire and tactical air support drove the attackers back. Meanwhile, work along the key north–south road included a new bypass around the town of Ben Cat by Company C, 1st Engineer Battalion.[36]

In late November, the enemy again switched targets in a pattern of off-and-on offensives and attacked two South Vietnamese outposts in Phuoc Long Province not far from the Cambodian border. Just after midnight on 29 November, a reinforced Viet Cong battalion attacked the district headquarters compound at Bo Duc. Mortar rounds also struck the Bu Dop Civilian Irregular Defense Group–Special Forces camp a little more than a mile to the north. Friendly air and artillery strikes helped to oust the enemy from the district compound, and South Vietnamese and U.S. reinforcements moved by helicopters to the area. After setting up its night defensive position, a 1st Division infantry battalion soon came under heavy attack. Again firepower beat off the attackers. The 1st Division deployed more infantry battalions to probe the area. Several light encounters followed, and on the night of 8 December a battalion beat off another battalion-size attack. For the remainder of the month, one U.S. infantry battalion remained around Bu Dop while another moved to Song Be, where it carried out reconnaissance in force operations.[37]

[35] AAR, Opn SHENANDOAH II, 1st Engr Bn, 10 Dec 67, pp. 2–7, Historians files, CMH.
[36] ORLL, 1 Nov 67–31 Jan 68, II FFV, p. 37; *Always First, 1967–1968*, "Opn SHENANDOAH."
[37] ORLL, 1 Nov 67–31 Jan 68, II FFV, pp. 36–37.

With Bu Dop under threat, the engineers found themselves again repairing bridges, this time along Highway 14A. The enemy had severed the only road link between the Bu Dop Special Forces camp and the MACV advisory compound in Bu Duc. Company E, 1st Engineer Battalion, received a mission to replace a damaged Eiffel bridge that cluttered the stream crossing. The bridge company's engineers literally threaded the stiffeners of a helicopter-borne fifty-four-foot fixed span bridge, the longest bridge of this type installed by the unit, between the trusses of the older French bridge. Within a week, saboteurs destroyed the new bridge, but the determined engineers again replaced it.[38]

Summer Battles in the Highlands

In the western highlands of II Corps, General Westmoreland planned to continue screening the Cambodian border with light forces, introducing reinforcements only when large North Vietnamese units entered South Vietnam to mount offensive operations. Throughout the southwest or summer monsoon from April to October 1967, General Peers' 4th Infantry Division carried out this mission in an operation called FRANCIS MARION. Moving back to the flat rolling hills of western Pleiku Province south of the Se San River, the 1st and 2d Brigades tried to stifle large enemy movements through the area. This resulted in a series of border brushes and heavy contacts, and by late May MACV dispatched one of its strategic reserve forces, the 173d Airborne Brigade, to reinforce the hard-pressed division.[39]

Then in June, a growing threat developed farther north in Kontum Province. As a countermeasure, General Peers deployed two of the 173d's battalions to the Dak To 2 airstrip and Special Forces camp, a few miles southwest of the town of Dak To. This operation, initially dubbed GREELEY, which began on 17 June and lasted through 11 October, ranged over a rugged wilderness covered by thick double- and triple-canopy jungle. Thick bamboo fifteen to twenty feet high impeded movement in the only open areas of the mountainous rain forest. Within a few days, an airborne company clashed with a North Vietnamese Army infantry battalion and suffered heavy casualties in a violent battle. A week later, the rest of the 173d Airborne Brigade, a South Vietnamese airborne task force, and the 3d Brigade, 1st Cavalry Division, moved to Kontum City. The next few months involved grueling marches in western Kontum Province, where U.S. forces suffered mounting losses against firmly entrenched North Vietnamese units. Throughout July, the cavalry brigade systematically searched the area north of Kontum City. In August, a South Vietnamese infantry battalion moved to the area northwest of the province capital to stave off an impending attack at Dak Seang Special Forces camp north of Dak To. The battalion got hung up trying to assault a fortified hilltop, and two South Vietnamese airborne battalions moved to the field of battle. A few days later, a battalion threw back five mass attacks on its night defensive perimeter.

---

[38] *Always First, 1967–1968*, "Opn SHENANDOAH" and "Company E."
[39] Sharp and Westmoreland, *Report*, pp. 133, 153; ORLLs, 1 May–31 Jul 67, I FFV, pp. 9–11, and 1 Aug–31 Oct 67, I FFV, p. 10.

When the enemy withdrew to the west after suffering heavy losses, the South Vietnamese found three hilltops forming a regimental-size base area, many bunkers, and an elaborate mockup of the Dak Seang camp.[40]

The fighting at Dak Seang and mortar attacks on Dak Pek Special Forces camp to the north gave credence to a possible attack on Dak Pek. A sweep of the area by the 173d Airborne Brigade and Civilian Irregular Defense Group elements located a few bunkers but not the enemy. This operation may have thwarted an attack on Dak Pek. On 20 and 21 August, the two airborne battalions were moved by helicopters to the Dak To area. Further sweeps by U.S. and South Vietnamese forces failed to find the enemy. Meanwhile, most of the reinforcements had deployed elsewhere: the 1st Cavalry Division's brigade to the coast, the South Vietnamese airborne units to their home base near Saigon, and the bulk of the 173d Airborne Brigade to coastal Tuy Hoa. The remaining battalion of the 173d continued to sweep the area northeast of Dak To until 11 October when Operation GREELEY ended. The 4th Division's border operations resumed the following day under the name of Operation MACARTHUR.[41]

During Operation FRANCIS MARION, the 4th Division's 4th Engineer Battalion sent Companies A and B to the field to support the 1st and 2d Brigades, respectively. In turn, the companies usually doled out platoons to the infantry battalions and squads to infantry companies. The engineers cleared landing zones, swept roads for mines, improved forward base camps, destroyed enemy bunkers, and built firebases and their access roads. Company A, augmented with two Rome plows, built over ten miles of fair-weather tactical road to firebases near the Cambodian border. The plows also cleared the jungle for fields of fire around the bases. CH–54 helicopters airlifted smaller D6B bulldozers to hilltop outposts for clearing tasks. Two of the bulldozers under the direction of Company B also cleared eight hundred acres for the resettlement of ten thousand Montagnards. The engineer battalion's tankdozers and armored personnel carrier flamethrowers provided security.[42]

Other elements of the 4th Engineer Battalion moved forward to support the operation. Company D, which continued to do some base development work, built a road linking the Plei Me Special Forces camp to Highway 14. Company E stood by, ready to haul bridging and cargo. Headquarters Company augmented the companies in the field with equipment and security elements from the flame-throwing platoon. In addition, the battalion commander, Lt. Col. Norman G. Delbridge, organized an armored task force. This ad hoc organization used the tankdozer from each of the four line companies and armored personnel carriers from the flame-throwing platoon to accompany work parties and convoys. Initially the task force protected convoys hauling sand between Dragon Mountain base camp and Kontum City. Delbridge envisioned deploying the

[40] ORLLs, 1 May–31 Jul 67, I FFV, pp. 10–11, and 1 Aug–31 Oct 67, I FFV, p. 10–11. For more on GREELEY, see AAR, Opn GREELEY, 16 Dec 67, 4th Inf Div, Historians files, CMH; and MacGarrigle, *Taking the Offensive*, pp. 299–304, 309.

[41] ORLL, 1 Aug–31 Oct 67, I FFV, p. 10–11; AAR, Opn GREELEY, 4th Inf Div, pp. 10–11.

[42] ORLLs, 1 Feb–30 Apr 67, 4th Inf Div, 15 Jun 67, pp. 29–30, and 1 May–31 Jul 67, 4th Inf Div, 20 Aug 67, pp. 37–38, both in Historians files, CMH; Engineer Support of Combat Operations, incl. 8, p. 4, OCE Liaison Officer Trip Rpt no. 7, 2 Jun 67.

armored force to guard engineers working on roads and bridge sites, such as the road to Plei Me.[43]

During GREELEY, the 4th Engineer Battalion built an M4T6 floating bridge across the Dak Ta Tan River on Route 512 just west of the town of Tan Canh. Route 512 led to the two airstrips serving Dak To: Dak To 1 (or Old Dak To) and Dak To 2. Companies D and E worked together on the night of 17 June to build a 120-foot span across the river. Pre-inflated floats were carried forward by Company E's bridge trucks to a point near the bridge site. The floats were then lifted by CH–47 helicopters to the bridge site and placed in the river ready for assembly. Despite working at night and occasional mortar attacks, the engineers completed the bridge in twelve hours and it was ready for traffic by morning.[44]

Additional engineer support for the two operations came from the 937th Engineer Group. The 20th Engineer Battalion dedicated most of its efforts to operational support missions in the area west of Pleiku. Major tasks included upgrading forward airfields at Duc Co and Buon Blech and improving Highway 19 and other supply routes before the monsoon arrived. Then the battalion shifted the road work to Route 509 and a connecting road to Highway 19. In August, the 18th Engineer Brigade deployed the 35th Land Clearing Team to Pleiku. Under the direction of the 20th and 299th Engineer Battalions, detachments of Rome plows started clearing vegetation along the road between Highway 19 and Route 509 and Highway 19 between Mang Yang Pass and Pleiku. The 299th Engineer Battalion divided work between base facilities at Pleiku and Highway 14 from Pleiku to Dak To. The battalion also maintained nearby roads and made emergency repairs to the rain-damaged T17 airstrip at Polei Kleng. Companies B and D also supported Operation GREELEY at Dak To and Kontum, respectively, building helicopter revetments and temporary ammunition storage dumps.[45]

Ambushes remained a constant risk to the engineers while working and traveling these roads. The 299th Engineer Battalion was ambushed four times between November 1967 and January 1968. One of these ambushes that took place on 11 November 1967 along Highway 14 north of Kontum was especially brutal. A platoon from Company B, 299th Engineer Battalion, was en route to continue repairs to a previously damaged timber trestle bridge when it was ambushed by some fifty to eighty Viet Cong, who attacked with grenades, rockets, automatic weapons, and mortars. Sfc. John K. McDermott was in the lead vehicle when a recoilless rifle round exploded against his truck, but he managed to keep control of the vehicle and drove it off the road to allow the rest of the convoy to get by. McDermott then jumped from the truck and directed his men to safety. He killed two Viet Cong attackers before two more leaped at him bringing him to the ground. After struggling free, McDermott

---

[43] ORLLs, 1 Feb–31 Apr 67, 4th Inf Div, p. 30, and 1 May–31 Jul 67, 4th Inf Div, p. 38.

[44] ORLL, 1 May–31 Jul 67, 4th Inf Div, p. 38.

[45] Galloway, "Essayons," pp. 206, 208–09; Engineer Support of Combat Operations, 1 May–1 Jul 67, incl. 7, pp. 2–3, OCE Liaison Officer Trip Rpt no. 8; Engineer Support of Combat Operations, 1 Jul–1 Sep 67, incl. 9, p. 2, OCE Liaison Officer Trip Rpt no. 9; Engineer Support of Combat Operations, 1 Sep–15 Nov 67, incl. 9, p. 1, OCE Liaison Officer Trip Rpt no. 10.

wounded two of the attackers and proceeded under heavy fire to drive several trucks out of the ambush site. Despite intensive enemy rifle fire and exploding grenades, he rallied his men in a counterattack. Meanwhile, S. Sgt. Frank J. Walker, a squad leader who was riding behind McDermott when the ambush occurred, leapt from his truck, which was also hit, and charged into enemy machine gun fire, knocking out several Viet Cong positions. These heroic actions allowed the convoy to move out of danger. Within a half hour, reinforcements arrived and drove off the enemy. Six engineers died and four were wounded in the encounter. Nine Viet Cong were confirmed dead. Extensive damage was made to one 5-ton dump truck, one 3/4-ton truck, and one jeep. General Westmoreland later awarded the Distinguished Service Cross for valor to both sergeants in a ceremony at Pleiku.[46]

The 937th Group's other two battalions, the 70th Combat and the 815th Construction, worked in and around the Pleiku bases. The 815th relieved the 299th of many base development projects in the area. By midsummer, the construction battalion was hard at work on a heliport, a petroleum products storage area, the road network for the Pleiku evacuation hospital, and many smaller projects. In October, Col. Robert C. Marshall moved his fourth battalion, the 70th Engineer Battalion, from An Khe to Pleiku. Until then, the battalion had spent twenty-six months transforming the An Khe base camp into a major city and completing the only Army-built concrete runway in country. The 70th took over the 299th's responsibility for Highway 14 from Pleiku to Kontum and a share of construction projects in the area.[47]

Marshall had to contend with several shortcomings. All three of his combat battalions were charged with at least some base development work. Some of these projects, which required more engineering and vertical construction talent not found in the tables of organization for a combat group, were highly sophisticated. Shortages of electrical and plumbing supplies held up projects, and shortages of repair parts for equipment inherited from RMK-BRJ resulted in excessive downtime. Although the group took steps to cope with expected problems brought on by the monsoon, some projects had to be discontinued or delayed. The rains delayed paving at Camp Enari and completing site preparations at the Duc Co airfield. The improved roads stood up well during the monsoon season, but those that were not all weather turned into quagmires. The 937's engineer battalions struggled to keep Route 512, an alternate six-mile fair-weather route to Dak To, open to convoys. Since the 299th Engineer Battalion lacked sufficient rock for the road, wheeled vehicles endured considerable wear and tear trying to move through the muck. Ground mobility in the area was lost. For approximately twenty-eight days from September through October, the only movement to Dak To was by air. The battalion went so far as to recommend that tactical commanders conduct

[46] ORLL, 1 Nov 67–31 Jan 68, 299th Engr Bn, 31 Jan 68, pp. 7–8, 12, Historians files, CMH; "Two 299th Sergeants Awarded DSCs," *Castle Courier*, 10 February 1968.

[47] Galloway, "Essayons," pp. 209–10; ORLL, 1 Aug–31 Oct 67, 937th Engr Gp, pp. 11–12.

operations in the highlands during the monsoon season only when there were all-weather roads in the area.[48]

Dak To

In November, a few days after the battles at Loc Ninh and Song Be in III Corps, a third border battle erupted near Dak To. General Westmoreland later reported that the fierce fighting along the northwestern corner of II Corps was the pivotal battle of the last quarter of 1967. Here, too, North Vietnamese and Viet Cong forces took advantage of nearby sanctuaries in Cambodia and Laos and short supply lines. Just a few miles from the borders, Dak To stood astride a natural infiltration route into Kontum and Pleiku Provinces. Generally, Dak To consisted of the town and district head-quarters on Highway 14, the short Dak To 1 airstrip a few miles to the south-west along Route 512, and the C–130 capable Dak To 2 airstrip and Special Forces camp farther west on Route 512. Dak To 2 lay on a valley floor close to a river, surrounded on all sides by peaks and ridges varying between 2,600 to 4,300 feet high. Tall, thick trees, up to a hundred feet tall and topped with double and triple canopies, covered the steep slopes.[49]

When the 4th Division began Operation MACARTHUR in mid-October, the sweeps throughout the region produced only sporadic contact with the enemy, mostly in Pleiku Province. This soon changed. Near the end of the month, allied intelligence revealed that a North Vietnamese division controlling five regiments was moving northeastward into adjacent Kontum Province. The 4th Division deployed the 1st Brigade headquarters and an infantry battalion by air to Dak To. There they reinforced a battalion in the area guarding a work site at Ben Het, where construction had begun on an airstrip and preliminary clearing for a proposed Civilian Irregular Defense Group camp. Also at Ben Het, Westmoreland planned to place 175-mm. guns that could reach the ene-my's base camps inside Cambodia and southern Laos. In early November, a defecting enemy soldier divulged that the North Vietnamese division intended to capture the Dak To camp and Ben Het. Sharp contacts with enemy patrols confirmed the earlier intelligence. General Rosson, who had replaced General Larsen at I Field Force in August, began to airlift more U.S. and South Vietnamese units to the airstrip at the Dak To Special Forces camp. Before the enemy could launch his attack, the 4th Division had deployed a brigade of three battalions, including a battalion from the 173d Airborne Brigade. The division then set up firebases and began to assault the enemy's forward bases.[50]

The twenty-two or so days of fighting around Dak To became one of the largest battles of the war to date and the largest in the highlands. Seldom

[48] Galloway, "Essayons," p. 211; ORLLs, 1 Aug–31 Oct 67, 937th Engr Gp, pp. 7, 13–14, 1 Aug–31 Oct 67, 20th Engr Bn, pp. 1–2, and 1 Aug–31 Oct 67, 299th Engr Bn, p. 12.

[49] Sharp and Westmoreland, *Report*, pp. 138–39; Westmoreland, *A Soldier Reports*, p. 289; MACV History, 1967, vol. 1, p. 378. For more on enemy activity and terrain, see AAR, Battle for Dak To, 4th Inf Div, 3 Jan 68, pp. 2–5, Historians files, CMH.

[50] Sharp and Westmoreland, *Report*, p. 139; Westmoreland, *A Soldier Reports*, p. 288; AAR, Battle for Dak To, 4th Inf Div, Peers cover ltr, pp. 1–2, main rpt, pp. 1–2, 6–8.

during the war did enemy units stand their ground, but here they stubbornly defended heavily fortified trenches along the ridges and peaks of the mountains surrounding Dak To. The enemy had carefully selected this region as the site for his campaign. Hill 875 stood as a dominating terrain feature in relative isolation and beyond the range of fixed artillery positions. It appeared the enemy intended to draw the guns to temporary positions and also lure allied troops deep into the battlefield, pinning and wearing them down, and eventually annihilating them in a major battle. Initially, the 4th Division's 1st Brigade sent one infantry battalion to occupy the ridgeline running east to west south of the airstrip. A second battalion deployed southwest of Dak To, and the airborne battalion moved west to Ben Het to set up a firebase. During the first week of November, the brigade encountered enemy units moving to preselected and sometimes previously prepared positions. These fierce encounters prompted General Creighton W. Abrams—Westmoreland was in Washington at the time—to let General Rosson deploy the 173d Airborne Brigade with two more battalions to the Ben Het area. In a series of helicopter assault landings, the paratroopers and infantrymen moved from ridge to ridge, where they faced strong resistance from the entrenched enemy. Meanwhile, South Vietnamese troops took up blocking positions east of Dak To. They were joined by additional infantry and airborne troops to the north and northeast initially to block and then attack a North Vietnamese regiment moving down a valley from the northeast. To complete the ring around Dak To, the 1st Brigade, 1st Cavalry Division, deployed south and southeast to block possible attacks and intercept any withdrawing enemy. To the west of Ben Het, a North Vietnamese regiment began to cover the withdrawal to the southeast of two hard-hit regiments. This resulted in a violent five-day struggle beginning on 17 November for Hill 875, which involved two battalions of the 173d Airborne Brigade and a battalion from the 4th Division. When a battalion of the 173d found a large enemy force entrenched on the slopes and suffered moderately heavy casualties trying to reach the peak, the brigade commander quickly replaced it with another battalion. The hill was taken following the heaviest concentration of tactical air and artillery on any single terrain feature in the II Corps area. Northeast of Dak To, South Vietnamese forces inflicted heavy losses on the enemy in a fierce two-day battle.[51]

By the beginning of December, it became obvious that the U.S. and the South Vietnamese forces had the upper hand in the battle for the highlands.

---

[51] AAR, Battle for Dak To, 4th Inf Div, Peers cover ltr, pp. 1–2, main rpt, pp. 8–15. After action reports prepared by the 1st Brigade, 4th Infantry Division; 1st Brigade, 1st Cavalry Division; and the 173d Airborne Brigade are also in this report as inclosures 6, 7, and 8. Published accounts of the hill battles collectively known as Dak To are contained in Berry, *Sky Soldiers*, pp. 89–125; Edward F. Murphy, *Dak To* (Novato, Calif.: Presidio Press, 1993); and Shelby L. Stanton, *The Rise and Fall of an American Army: U.S. Ground Forces in Vietnam, 1965–1973* (Novato, Calif.: Presidio Press, 1985), pp. 168–78. For the opponent's view of the battle, see *Luc Luong Vu Trang Nhan Dan Tay Nguyen Trong Khang Chien Chong My Cuu Nuoc* [*The People's Armed Forces of the Western Highlands During the War of National Salvation Against the Americans*] (Hanoi: Nha Xuat Ban Quan Doi Nhan Dan [People's Army Publishing House]), 1980, pp. 56–69, copy in CMH (hereafter cited as *Western Highlands*).

During the fighting at Dak To, strong enemy defenses on Hill 875 included trenches.

Before the battle ended, the allies had temporarily reinforced to a strength of three U.S. brigades that included nine battalions and six South Vietnamese battalions. Massive B–52 and tactical air strikes, using targets located by long-range reconnaissance patrols, supported the friendly forces. In the heavy fighting throughout November, four enemy regiments lost 1,600 men and were virtually destroyed. General Westmoreland described the Dak To fight as "an engagement exceeding in numbers, enemy losses, and ferocity even the Ia Drang Valley Campaign of 1965." The allied attacks drove the enemy back into Laos and Cambodia and nullified his ability to stage major operations in the Central Highlands, at least for the time being. The fighting also caused heavy losses to the 4th Division and the 173d Airborne Brigade. The Americans lost nearly 300 killed and over 1,000 wounded. South Vietnamese casualties included nearly 100 killed and over 200 wounded.[52]

Throughout Operation MACARTHUR and the fighting around Dak To, engineer troops kept up a high pitch of supporting missions. Elements of the 4th Engineer Battalion, mainly Company A, supported the 1st Brigade, 4th Division; most of Company A, 8th Engineer Battalion, accompanied the 1st Cavalry Division's brigade; and the 173d Engineer Company supported the 173d Airborne Brigade headquarters and three of its battalions in the area. (The brigade had increased to four airborne infantry battalions with the arrival of the 3d Battalion, 503d Infantry, in October.) The 937th Engineer Group supported the campaign by building bridges and maintaining roads and airfields while simultaneously carrying out other tasks in the western highlands. The group's 299th Engineer Battalion bolstered tactical units in the Dak To and Kontum areas. Elements of the 20th and 70th Engineer Battalions, taking advantage of the approaching dry season, helped by working on roads leading into the area and nearby airfields.[53]

Again the organic engineers supporting the three combined arms brigades were parceled out to the infantry battalions and companies. Company A, 4th Engineer Battalion, provided a platoon to each of the three battalions of the 1st Brigade, 4th Division. The platoon headquarters remained in the battalion firebases, with the squads often accompanying the infantry companies. Work varied from clearing landing zones and firebases to helping the infantry set up defensive positions and laying barbed wire. Small D6B bulldozers airlifted by CH–54s sped up the clearing of six firebases. Demolitions, including the preferred C4 plastic type, TNT, and the XM37 Demolition Kit, helped speed up the clearing of heavy timber, bamboo, and brush. When the infantry companies went on operations, the engineer squads did demolition tasks and used mine detectors to help locate buried weapons and ammunition. Enemy soldiers, however, took steps to bury their caches deeper than nine to twelve inches, too deep for effective use by the detectors. The 173d Engineer Company followed different procedures. Two-man demolition teams were attached to each infantry

---

[52] Quoted in Sharp and Westmoreland, *Report*, p. 139. See also Westmoreland, *A Soldier Reports*, pp. 289–90; AAR, Battle for Dak To, 4th Inf Div, main rpt, pp. 16, 20–21, 32–35.

[53] AAR, Battle for Dak To, 4th Inf Div, main rpt, pp. 25–30; ORLL, 1 Nov 67–31 Jan 68, 937th Engr Gp, pp. 7–8.

company, with a noncommissioned officer remaining at the infantry battalion headquarters to control the teams and advise the staff. The rest of the company remained with the brigade, working in the base camp areas at Dak To 1 and Dak To 2 and dispatching troops on specific missions. Company A, 8th Engineer Battalion, deployed two platoons with the 1st Cavalry Division's 1st Brigade, with each platoon directly supporting an infantry battalion.[54]

Lt. Col. Domingo I. Aguilar, who had taken command of the 299th Engineer Battalion in mid-October, moved his forward command post to the airfield at Dak To 1. There the bulk of the intelligence, operations, supply, and maintenance and equipment elements coordinated and carried out their own supporting missions. The S–4 section's water purification teams provided the bulk of water for units in the Dak To area. These were supplemented by smaller water purification units of the 4th Engineer Battalion at Dak To 2 and the 173d Engineer Company at Ben Het. Aguilar's S–4 also arranged for the transportation of the heavy volume of construction materials required for the battalion's projects. Fleets of tractors and lowboy trailers from the battalion and from units of the 937th and 45th Engineer Groups hauled construction materials, particularly airfield matting. The maintenance section modified two D6B bulldozers allowing them to be broken down into two pieces, one with the tracks, chassis, and blade, the other with the engine and body. These were lifted by CH–54s to clear the jungle for artillery emplacements, ammunition storage areas, helicopter landing zones, and fields of fire.[55]

During the battle of Dak To, the 299th and its attached 15th Light Equipment Company provided general engineering support, and at various times took on direct-support missions for the three combat brigades. Company A continued to upgrade Route 512 between Dak To and the Highway 14 intersection in the town of Tan Canh. At the Dak To 1 airstrip, then only capable of handling C–7A Caribous on its 1,500-foot dirt runway, troops began to extend the runway to 2,300 feet. Support for the 173d Airborne Brigade at its Dak To forward bases included excavating holes for three tactical operations centers, building revetments for fuel bladders, and clearing fields of fire. Company B, located near the Kontum airfield, directed its efforts to maintaining Highway 14 from Kontum to Dak To, improving Route 511 to the west, and improving the airfield. With two sections of the 35th Land Clearing Team under its operational control, the company cleared about one hundred yards of jungle back on both sides of the road. Company B also built helicopter revetments and expanded ammunition storage for elements of the 1st Cavalry Division and the 173d Airborne Brigade. Earlier, Company C had moved to Ben Het on 22 October to build a C–7A airfield with provision for later expansion for C–130 transports. The company also began preparations for a Special Forces camp and stood ready to support 4th Division and 173d Airborne Brigade units in the area. The 937th Group furnished a section of the Rome plow platoon to clear vegetation around the Dak To 2 airfield and Special Forces camp and

---

[54] AAR, Battle for Dak To, 4th Inf Div, main rpt, pp. 29–30, and incl. 6, 7, and 8.

[55] AAR, Battle for Dak To, 4th Inf Div, main rpt, p. 25; AAR, Battle of Dak To/Opn MacArthur, 299th Engr Bn, 10 Dec 67, pp. 2–3, 10, Historians files, CMH.

east along Route 512 toward the intersection with Highway 14 at Tan Canh. While the Rome plows cleared both sides of the road back about one hundred yards, Company C graded the road and ditches and placed culvert. At Dak To 2, Company D supported combat units, worked on its section of Route 512 to Ben Het, and maintained the C–130 airfield.[56]

Enemy attacks on the forward bases and supply routes also kept the 299th busy. On the morning of 15 November, a mortar attack at Dak To 2 destroyed two of the three C–130s on the parking apron. In the late afternoon, another mortar attack destroyed the ammunition storage area. Two days later, Company D started hauling fill to build an expedient taxiway that served as a bypass to the damaged aircraft. A team of engineers topped off the earthen ramp with a dust-proofing treatment and peneprime, completing the job on the nineteenth. Meanwhile, other troops from Companies A and D built a new five-cell ammunition storage area, finishing this task on 28 November. To keep the supply roads open, the 299th also installed culvert bypasses next to the new bridges. The enemy tried to disrupt traffic by repeatedly attacking the spans. For example, on the night of 7 November sappers blew a bridge and its bypass along Highway 14. Company B immediately dispatched a platoon from Kontum and reopened the bypass by 1330 and within three days replaced the decking on the bridge.[57]

The 4th Engineer Battalion and the 937th Engineer Group weathered the demands of the operation with some diversions from regular tasks. Some confusion did arise concerning priority of work at the growing Dak To base camp. The 4th Division's after action report noted that equipment time was lost because of inadequate or incorrect work. An appointment of a base commander and a board composed of representatives of major units, the report stated, could have resolved this problem. When the 4th Engineer Battalion needed help, Lt. Col. Emmett C. Lee Jr., who replaced Delbridge in July, could call on Colonel Marshall's 937th Engineer Group. Marshall also reinforced his battalions. On 17 November, he placed one company from the 70th Engineer Battalion under the operational control of the 299th. Company C moved to Dak To and started to build a panel bridge over the Dak Poko River, replacing the 4th Engineer Battalion's M4T6 float bridge. The new Bailey bridge (a 240-foot double-single) was opened to traffic on 15 December. During its stay at Dak To, Company C also furnished dump trucks and front loaders, repaired two aircraft revetments, and took on some combat support missions for infantry and artillery units in the area. Meanwhile, the 70th's Company A continued to upgrade Route 509 west of Pleiku to a one-way fair-weather road. To the east, Company D kept open Highway 19 to the 937th Group's eastern boundary at Mang Yang Pass. The 20th Engineer Battalion kept two companies, A and D, and elements of the 584th Light Equipment Company working along Highways 14 and 19 west of Pleiku. In late October, despite the many base development projects at Pleiku, the 20th also shifted a platoon

---

[56] AAR, Battle for Dak To, 4th Inf Div, main rpt, pp. 25–29; AAR, Battle of Dak To/Opn MacArthur, 299th Engr Bn, pp. 2–8.
[57] AAR, Battle for Dak To, 4th Inf Div, main rpt, pp. 26–28; AAR, Battle of Dak To/Opn MacArthur, 299th Engr Bn, pp. 2, 4–8.

from Company C south to Buon Blech. The forty-man unit supported the 4th Division's 2d Brigade by upgrading the T17 airstrip to M8A1 matting and improving the firebase. On 2 November, another platoon from Company C interrupted base development work to move south to Phu Nhon airstrip to repair its damaged T17 membrane surface and subgrade.[58]

Although battles raged along the border and intelligence showed massive enemy troop movements still taking place, General Westmoreland in December 1967 believed the military situation in South Vietnam was turning in the allied favor. The South Vietnamese government succeeded in holding national elections. A new coordinated pacification program was under way. The protective arc around major cities expanded, and more U.S. troops arrived, allowing operations to shift to the remote border regions. When summoned to Washington by President Johnson in November, Westmoreland effused his most optimistic assessment of the war. "It is significant," he told the National Press Club, "that the enemy has not won a major battle in more than a year. In general, he can fight his large forces only at the edges of his sanctuaries. . . . His guerrilla force is declining at a steady rate. Morale problems are developing within his ranks."[59]

Though remaining heavily committed to completing the base infrastructure, the Army engineers' support to the combat forces increased with military operations. To support the flow of troops and supplies to these more remote areas, combat engineers pushed ahead on road and airfield work and cleared landing zones. Planners now considered the Rome plows used in clearing areas and along supply routes as a tactical weapon. Although the demand for combat engineers had increased, Generals Ploger and Duke managed to keep control of the bulk of engineer troops, parceling out units as needs arose.

General Westmoreland and his intelligence chief, Brig. Gen. Phillip B. Davidson, concluded that the border battles were North Vietnamese failures. Davidson later wrote that General Vo Nguyen Giap paid a bloody price for the tactical lessons that he and his staff learned to avoid direct attacks on American positions. Giap apparently learned this lesson well. During the attacks of the Tet offensive in late January 1968, the North Vietnamese and their Viet Cong allies carefully avoided American combat units, concentrating instead on South Vietnamese forces and some U.S. military headquarters. If Giap had hoped to draw U.S. units and command attention to the peripheries of South Vietnam in a series of diversionary battles, he failed. The U.S. forces' strategic mobility permitted them to move to the borders, smash Giap's attacks, and redeploy back to the interior in a mobile reserve posture. The enemy must have realized the futility of these attacks. One Viet Cong colonel who defected in 1968 characterized the "border battles" as "useless and bloody."[60]

[58] ORLLs, 1 Nov 67–31 Jan 68, 937th Engr Gp, pp. 7–8, 1 Nov 67–31 Jan 68, 70th Engr Bn, p. 4, and 1 Nov 67–31 Jan 68, 20th Engr Bn, pp. 1–4, 9–10; AAR, Battle for Dak To, main rpt, 4th Inf Div, pp. 29, 55–56; AAR, Battle of Dak To/Opn MACARTHUR, 299th Engr Bn, p. 8.

[59] Quoted in William M. Hammond, *Public Affairs: The Military and the Media, 1962–1968*, United States Army in Vietnam (Washington, D.C.: U.S. Army Center of Military History, 1988), p. 334.

[60] Quoted in Phillip B. Davidson, *Vietnam at War: The History, 1946–1975* (Novato, Calif.: Presidio Press, 1988), p. 469.

PART THREE

Changing Course

11

Tet, January–March 1968

In early 1968, the war reached a critical stage when Hanoi launched its Tet offensive throughout South Vietnam. The allies had suspected that heavy attacks were in the offing, but they only began to realize the scope and schedule of the offensive when the assaults began just after midnight on 30 January, coinciding with the start of the Vietnamese Tet celebrations ushering in the lunar New Year. By the next day, fighting erupted almost everywhere, as the Viet Cong and North Vietnamese attacked 36 of 44 provincial capitals and 64 of 242 district towns, as well as 5 of 6 of South Vietnam's autonomous cities, among them Hue and Saigon. Local force units and sappers spearheaded the assaults and attempted to hold their designated objectives until reinforcements could arrive in strength from outside the cities. Political cadres accompanied the assault groups with the intent, in vain as it turned out, to coax the local population into civil rebellion. The fighting at Khe Sanh, Hue, and Saigon was especially vicious, but throughout the country heavy rocket and mortar fire and demolition charges caused damage to airfields, logistical facilities, and supply routes. And while the South Vietnamese government and its forces constituted the prime enemy target, American units were swept into the turmoil.[1]

Defending the Coast

The earliest attacks took place in the coastal cities of II Corps. At 0035 on 30 January, only a half hour into the Vietnamese New Year, mortar rounds landed near the South Vietnamese Navy Training Center in Nha Trang. At 0410, two Viet Cong battalions attacked Qui Nhon. Cam Ranh Bay experienced its first major threat when sappers swam into the bay and set off a demolition charge on a tanker, leaving it with a diamond-shaped hole above the water line. Within two days, the enemy attacked major population centers in nine of the corps' twelve provinces. Highway 19 was cut just west of Qui Nhon, as were roads from Cam Ranh Bay to Da Lat and Ban Me Thuot. Heavy fighting at Da Lat, Ban Me Thuot, and Kontum shut down the airfields in all three towns.[2] (*Map 23*)

[1] For background on the Tet offensive, see Sharp and Westmoreland, *Report*, pp. 157–88; Westmoreland, *A Soldier Reports*, pp. 310–49; Davidson, *Vietnam at War*, pp. 473–74; and, for the best general account, Don Oberdorfer, *Tet!* (Garden City, N.Y.: Doubleday, 1971).

[2] Westmoreland, *A Soldier Reports*, p. 322; Bowers, *Tactical Airlift*, p. 322; ORLL, 1 Feb–30 Apr 68, U.S. Army Support Cmd, Cam Ranh Bay, n.d., p. 6, Historians files, CMH. For more on the attacks in II Corps, see ORLL, 1 Feb–30 Apr 68, I FFV, 15 May 68, pp. 11–14, Historians files, CMH; MACV History, 1968, vol. 1, p. 386, vol. 2, pp. 889–92; Col. Hoang Ngoc Lung, *The*

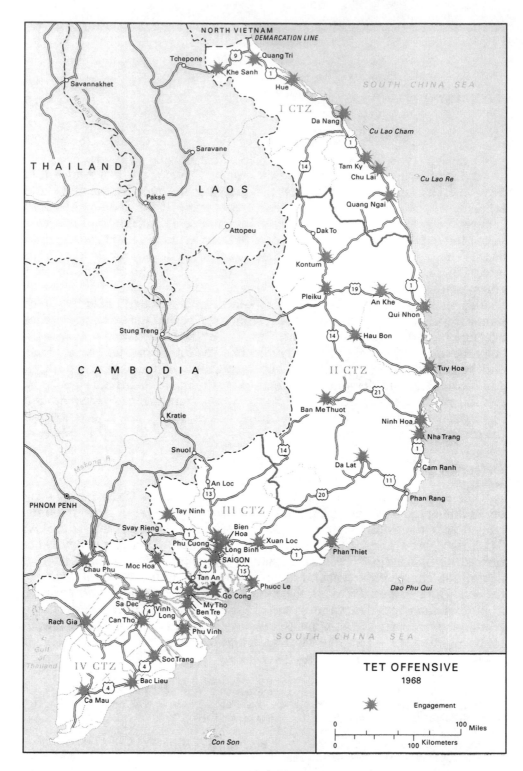

NORTH VIETNAM

DEMARCATION LINE

THAILAND

LAOS

CAMBODIA

SOUTH CHINA SEA

Tchepone

Savannakhet

Saravane

Paksé

Attopeu

Stung Treng

Kratie

Snuol

PHNOM PENH

Svay Rieng

Quang Tri

Khe Sanh

Hue

Da Nang

Cu Lao Cham

I CTZ

Tam Ky

Chu Lai

Cu Lao Re

Quang Ngai

Dak To

Kontum

Pleiku

An Khe

Qui Nhon

Hau Bon

Tuy Hoa

II CTZ

Ban Me Thuot

Ninh Hoa

Nha Trang

Cam Ranh

Da Lat

Phan Rang

An Loc

Tay Ninh

III CTZ

Bien Hoa

Phu Cuong

Long Binh

SAIGON

Xuan Loc

Phan Thiet

Dao Phu Qui

Moc Hoa

Chau Phu

Tan An

Go Cong

Phuoc Le

Sa Dec

My Tho

Ben Tre

Rach Gia

Vinh Long

Can Tho

Phu Vinh

SOUTH CHINA SEA

Gulf of Thailand

IV CTZ

Soc Trang

Bac Lieu

Ca Mau

Con Son

Mekong R.

TET OFFENSIVE
1968

Engagement

0   100 Miles

0   100 Kilometers

MAP 23

The fighting in the Qui Nhon area lasted from 30 January well into February and was typical of the allied experience in coastal II Corps. Before the attack on Qui Nhon, South Vietnamese forces had uncovered local Viet Cong hideouts and captured several cadre and prerecorded tapes to be used following the seizure of the radio station. Expecting an attack, the province chief announced a ban on fireworks, which was ignored by some people, and a curfew. Despite these precautions, a Viet Cong sapper battalion and a local force battalion slipped into the city, some of the insurgents disguised as South Vietnamese soldiers. They seized and briefly controlled the radio station, freed the Viet Cong captives, and occupied the railway station. The South Vietnamese regrouped and with the help of troops from the South Korean Capital Division ousted the attackers. The counterattack, however, destroyed the radio station and damaged the train station's workshop, including a new General Electric locomotive, and other buildings in town.[3]

Since the Viet Cong's main objectives in Qui Nhon were South Vietnamese facilities, the attackers initially bypassed the Americans unless they happened to be in the way. At daybreak on 30 January, three employees of Pacific Architects and Engineers traveling to work in a company jeep passed near the radio station. The Viet Cong opened fire killing all three. In addition, seven U.S. soldiers died and several were wounded in Qui Nhon, including three soldiers who also drove by the radio station during the fighting.[4]

Army engineer units in the Qui Nhon area immediately took up arms to help defend U.S. installations. Just before the offensive, the 45th Engineer Group deployed a provisional infantry platoon from the 35th Engineer Battalion to guard RMK-BRJ's quarry and asphalt plant near Phu Cat. When the 35th received orders to head for I Corps, the security mission transferred to a platoon from the 523d Port Construction Company. The engineers assumed additional infantry duties when the U.S. Army Support Command, Qui Nhon, set in motion a rear area protection plan, which included mobile reaction forces drawn from the 45th Group. Within days of the Tet attacks, the group commander, Col. George W. Fink, ordered one company from the 84th Engineer Construction Battalion to hold the high ground west of the city and one from the 589th Engineer Construction Battalion to move into Qui Nhon proper and secure the airfield. Even so, the Viet Cong continued to attack. On 3 and 13 February, rockets struck fuel storage tanks near the camp of the 84th Battalion west of town, leaving holes but no other damage. On 3 February, a squad of sappers infiltrated the camp itself, and the engineers lost two killed and two wounded. A second attack occurred on the twenty-sixth, along with an assault on a nearby South Vietnamese ammunition depot. This time the

General Offensives of 1968–69, Indochina Monographs (Washington, D.C.: U.S. Army Center of Military History, 1981), pp. 49–51; and Lt. Col. Phan Van Son and Maj. Le Van Duong, eds., *The Viet Cong Tet Offensive, 1968*, trans. J5/Joint General Staff Translation Board (Saigon: Printing and Publications Center, Republic of Vietnam Armed Forces [RVNAF], 1969), pp. 359–424, copy in CMH.

[3] Lung, *General Offensives of 1968–69*, pp. 50–51; Son and Duong, eds., *Viet Cong Tet Offensive, 1968*, pp. 381–92.

[4] PA&E History, Jan–Jun 1968, p. 30; Son and Duong, eds., *Viet Cong Tet Offensive, 1968*, p. 388.

engineers killed two sappers without suffering any casualties. Meanwhile, at the 84th Battalion's Phu Tai quarry, a mortar attack on 11 February resulted in the construction battalion's third combat death.[5]

Viet Cong attacks on the roads leading out of Qui Nhon caused additional hazards for the engineers. Several convoys to Pleiku and Landing Zone ENGLISH were ambushed. With the departure of the 35th Engineer Battalion to I Corps, the 84th Engineer Battalion assumed responsibility for all of Highway 1 between Cu Mong Pass south of the city and Bong Son to the north. Five bridges required immediate repair, and sniper fire wounded sixteen engineers working along the way. Farther north along the dangerous stretch to Duc Pho, the 19th Engineer Battalion had to repair five bridges destroyed by fire, two destroyed by demolitions, and three more damaged by sapper attacks. The sappers also destroyed thirty-seven culverts using 100 to 200 pounds of explosives or artillery rounds. The battalion reported 255 contacts with enemy forces, who fired on the engineers in ninety-four separate incidents. Minesweep teams and work parties were caught in thirteen well-planned ambushes. In all, the 19th Engineer Battalion and attached units suffered seven killed and eighty-eight wounded. The 589th Engineer Battalion faced similar problems along Highway 19 to An Khe. When sappers destroyed two spans over the Phu Phong River, a little over four miles east of Binh Khe on the night of 27 February, Company C built a 160-foot panel bridge over a wrecked span and reopened the road to traffic by 1600 the following day. The next night, the company reinforced the span to carry heavier traffic. Another panel bridge was built downstream to allow two-way traffic along the busy road. This second crossing later served as a bypass when the engineers began to reconstruct the permanent bridge. The repaired bridge, consisting of a new timber pile pier, steel beams, and timber decking, opened to traffic on 10 April. Company D cleared mines and improved the road from the base of Mang Yang Pass eastward to An Khe. When the company finished repairing a large hole blown in one of the bridge's concrete decks, it began fixing three spans of another badly damaged bridge. Despite the enemy's efforts, the engineers quickly reopened the main roads and kept them open.[6]

By mid-March, Col. John A. Hughes Jr. had moved his 35th Engineer Group from Cam Ranh Bay to the 45th Group's small cantonment at Qui Nhon. This relocation was prompted by the impending move of the 45th Group to I Corps to take charge of engineer units slated to support Army troops deploying there. The move to Qui Nhon allowed Hughes and his 35th Group headquarters better control of operations in an area of responsibility that had more than doubled in size. Hughes also inherited three battalions

---

[5] MFR, 21 Feb 92, sub: Author's Recollections, 45th Engr Gp, p. 2; ORLLs, 1 Feb–30 Apr 68, 45th Engr Gp, 30 Apr 68, pp. 4, 8–9, 1 Feb–30 Apr 68, 84th Engr Bn, 14 May 68, pp. 3–4, and 1 Feb–30 Apr 68, 589th Engr Bn, 15 May 68, p. 5, all in Historians files, CMH; Interv, Maj Paul B. Webber, 26th Mil Hist Det, with Lt Col James F. Fraser, CO, 84th Engr Bn, 22 Feb 68, pp. 1, 8, VNIT 108, CMH.

[6] ORLLs, 1 Feb–30 Apr 68, 84th Engr Bn, p. 4, 1 Feb–30 Apr 68, 19th Engr Bn, 30 Apr 68, pp. 2–4, Historians files, CMH; 1 Feb–30 Apr 68, 589th Engr Bn, pp. 7–8, 10; and 1 Feb–30 Apr 68, 1st Log Cmd, p. 70.

*Before and during the Tet offensive, the Viet Cong consistently attacked bridges
like this one burnt along Highway 1 north of Qui Nhon.*

(the 19th, 84th, and 589th) and several separate companies, making his the
largest engineer group in Vietnam. It now consisted of 6 battalions, 9 separate
companies, and 7 detachments, nearly 7,000 troops. Since the 14th and 35th
Engineer Battalions were slated to move north with the 45th Group, Hughes
faced the task of dividing the vast area among the remaining battalions to
carry out tasks formerly assigned to eight battalions.[7]

Fortunately, most of the vast logistical complex and highway improvements in the Qui Nhon area were nearing completion. The 35th Group could
concentrate on extending road improvements and operational support during
the coastal dry season. The 19th Engineer Battalion, using crushed rock from
the Tam Quan quarry that began operations on 11 January, steadily added a
base course along Highway 1 between Bong Son and Mo Duc in southern I
Corps. Simultaneously, the battalion carried out operational support missions
for the 1st Cavalry Division and its successors, the 3d Brigade, 4th Division,
and, in turn, the 173d Airborne Brigade and the Americal Division. The 19th
Engineer Battalion made airfield repairs at Landing Zone ENGLISH, completed
a small port at Sa Huynh, and built an access road to an artillery unit near Duc
Pho. Minesweep teams carried out the repetitive but dangerous job of daily
checks along Highway 1. Roadwork along Highway 1 by the 84th Engineer
Battalion consisted of readying the base course to Bong Son for an asphalt

[7] ORLL, 1 Feb–30 Apr 68, 35th Engr Gp, 11 May 68, pp. 2–4, Historians files, CMH.

surface and widening the Cu Mong Pass south of Qui Nhon. The attached 523d Port Construction Company helped the 84th and other engineer units in several projects. Chief among the versatile company's accomplishments in Qui Nhon were the completion of a third fuel storage tank farm (three 50,000-barrel and four 10,000-barrel welded-steel tanks) and a second causeway bridge in the port. The 589th Engineer Battalion devoted most of its efforts to Highway 19 and An Khe's logistical and airfield facilities. With the attached 51st Asphalt Platoon's asphalt plant at An Khe in full production shortly after 15 March, the 589th Battalion began paving west of the city.[8]

Roads and Bases in the Highlands

The first attacks in the highlands took place around 0130 on 30 January at Ban Me Thuot, about an hour following the opening strike at Nha Trang. As celebrants set off a string of fireworks in the streets, a barrage of mortar and rocket rounds struck the town, followed by a ground assault by two Viet Cong battalions. Simultaneously, a battalion-size attack took place against the small district capital of Tan Canh near Dak To. Thirty minutes later, a North Vietnamese regiment and two separate battalions, almost 3,000 men, moved against Kontum City, and a half hour later the long-expected attack on Pleiku began.[9]

The 4th Infantry Division's preparations before Tet included assigning an infantry role to its 4th Engineer Battalion. Two brigades (the 3d Brigade remained with the Americal Division until the end of March) continued to carry on Operation MacArthur, guarding the border regions and searching for the enemy in an immense area of operations. On the morning of 25 January, Colonel Lee, the 4th Engineer Battalion's commander, received orders to organize two companies as infantry and prepare to deploy to one of three possible blocking positions outside Pleiku. Since Lee could not call back Companies A, B, and C from their commitment to the three brigades, he assembled the remaining units at Camp Enari, the battalion's base camp just west of Pleiku, and organized the task force around Company D, most of Company E (bridge company), a platoon from Company A, and a makeshift armor platoon of tankdozers from the land-clearing team. The division also attached an armor company (Company A, 69th Armor) to the task force and provided an artillery liaison officer and a forward observer. The engineers gained additional fire power by borrowing twelve .50-caliber machine guns from other units and issuing two hand grenades per man. Material for overhead protection and bunkers was loaded on trucks and included concertina wire, heavy timbers, salvaged M8A1 airfield matting, and sandbags. The task force was declared ready for deployment by 1630 and spent the evening and the next day reviewing artillery adjustment and medical evacuation procedures

---

[8] ORLLs, ibid., p. 2, 1 Feb–30 Apr 68, 45th Engr Gp, pp. 7–9, 1 Feb–30 Apr 68, 19th Engr Bn, pp. 1–6, 1 Feb–30 Apr 68, 84th Engr Bn, pp. 2–5, and 1 Feb–30 Apr 68, 589th Engr Bn, pp. 2–9.

[9] Westmoreland, *A Soldier Reports*, p. 322.

and firing weapons. The infantry mission, however, was canceled, and division headquarters returned the engineers to normal duties at 1530, 26 January.[10]

At 0930 on 30 January, a Viet Cong local force battalion stormed into the provincial capital of Pleiku, and the division ordered the 4th Engineer Battalion to help retake the city. Luckily for the defenders, a North Vietnamese regiment failed to rendezvous with the local forces, which then carried out the attack alone. The Viet Cong penetrated Pleiku's defenses only after crossing a large open field at great cost. At 1500, the engineers helicoptered to a landing zone outside town and set up a defensive perimeter. The armor company's tanks joined the task force that afternoon, followed that evening by the 4th Engineer Battalion's own armor force. The next morning, the task force and South Vietnamese troops advanced into town, took enemy fire, and continued the sweep until dusk. Company E took three prisoners, two of whom were members of sapper teams on suicide missions. The next morning, the engineers formed two skirmish lines and resumed the sweep, this time with the tank company. This drive resulted in the killing of twenty-one Viet Cong and the capture of several weapons. As the advance proceeded northward, resistance ceased. By nightfall, the engineer task force's command post moved to a South Vietnamese artillery compound. One platoon reinforced the Artillery Hill Camp west of town, and another platoon with a platoon of tanks guarded the power plant. The engineers also manned a nearby observation post equipped with two starlight scopes (optical instruments that intensified low light allowing its users to see at night). By 1930, the observation post reported enemy troops moving toward town. Heavy concentrations of firepower delivered by artillery, helicopter gunships, and Air Force AC–47 Spooky gunships halted the advance. South Vietnamese troops moving through the area the next day found some blood-soaked clothing, bandages, and equipment but no bodies. Several sweeps of the area followed without any traces of the enemy. Its mission as infantry completed, the task force returned to Camp Enari at 1800, 1 February, and its troops resumed normal chores as engineers.[11]

At Kontum City, elements of the 4th Engineer Battalion helped allied forces repel the enemy attack. Heavy fighting developed around government facilities and the airfield. By noon on 30 January, it became apparent that the South Vietnamese regular and territorial defenders needed help. The South Vietnamese rushed in a reinforced battalion, and the 4th Division deployed a task force centered on the 1st Battalion, 22d Infantry. In a campaign that lasted through 12 February, the 4th Division's task force (divided into teams and supported by two batteries of artillery; a company of armor; a troop of air cavalry; and two Dusters, or twin 40-mm. self-propelled guns) set up blocking positions, helped retake the city in house-to-house fighting, and pursued the enemy through nearby hamlets and heavily defended hills to the north. Supporting the 4th Division task force, the 3d Platoon, Company B, 4th Engineer Battalion, placed a squad with

[10] ORLL, 1 Nov 67–31 Jan 68, 4th Inf Div, 7 Mar 68, pp. 39–40, 50–51, Historians files, CMH.

[11] Ibid., pp. 40–41, 50–51.

each infantry company. Tasks included clearing mines, disposing of munitions, dispensing potable water at the task force's firebase, and destroying bunker complexes. Additional help and equipment came from the rest of the division engineer battalion or nondivisional engineers nearby. An armored vehicle launched bridge supporting the armor team came from the battalion's bridge company, and the 299th Engineer Battalion sent two D7 bulldozers to clear a patrol base for an infantry team. The engineers also made abundant use of their chain saws. One squad sent to support the armor team cleared a shroud of dense foliage that obstructed the tanks' fields of fire. Another squad cleared a landing zone, permitting helicopters to extract an infantry company.[12]

The 937th Engineer Group's construction program suffered because of the initial and follow-up attacks on roads and convoys. Mines and destroyed or damaged bridges blocked the main supply routes, and the nearly completed pipeline running parallel to Highway 19 was sabotaged. Rocket and mortar attacks caused some damage to buildings and airfield surfaces, which were quickly repaired. Quarries previously operating on a twenty-four-hour schedule were limited to daylight work. In March, Col. William J. Talbot assumed command of the group from Colonel Marshall. His main concern shifted to repairing the destruction caused by the enemy's attacks and carrying on the mix of operational support and construction tasks. One chore included sending the 299th Engineer Battalion's Company B to Kontum City, where it cleared some 1,500 cubic yards of debris. As for roadwork, the 20th, 70th, and 299th Engineer Combat Battalions, each backed by a light equipment company, labored to keep the critical supply routes open with daily mine sweeps and repairing road surfaces, culverts, and bridges. The 815th Engineer Construction Battalion, augmented by a construction support company and part of a pipeline company, continued base development tasks and paving Highway 19 east of Pleiku. Two dump truck companies and a land-clearing platoon also supported the group.[13]

Repairs and improvements at forward airstrips and helicopter facilities also resumed at full force. On 19 February, the 20th Engineer Battalion finished renovations at Buon Blech, and the runway and parking apron topped with new M8A1 matting could now handle C–130 aircraft. On the same day, the battalion started comparable upgrading at the Ban Don airstrip. The 20th Battalion also completed a C–7A airstrip (Operation FLORIDA) and various facilities at the Tieu Atar Special Forces camp and began to extract its lightweight airmobile equipment on 1 April. At Polei and Old Cheo Reo, the 70th Engineer Battalion labored to upgrade the airstrips to C–130 and C–123 traffic, respectively. Both Dak To airfields had more work done by the 299th

---

[12] ORLL, 1 Feb–30 Apr 68, I FFV, p. 13; AAR, Battle of Kontum/Tet Counteroffensive, 1st Bn, 22d Inf, n.d., pp. 1, 3–6, and an. D, pp. 2–3, Historians files, CMH.

[13] Engineer Support of Combat Operations, incl. 6, p. 1, OCE Liaison Officer Trip Rpt no. 11, 6 May 68, OCE Hist Ofc; ORLLs, 1 Feb–30 Apr 68, 937th Engr Gp, 30 Apr 68, pp. 3–4, 6–8, 1 Nov 67–31 Jan 68, 299th Engr Bn, 31 Jan 68, pp. 2–3, and 1 Feb–30 Apr 68, 299th Engr Bn, 30 Apr 68, p. 3, all in Historians files, CMH.

Engineer Battalion, which prepared Dak To 1 for C–123 traffic and Dak To 2 to withstand the effects of the southwest monsoon.[14]

After Tet, convoys carrying men and construction materials moved at great risk, and ambushes took an ever-increasing toll in casualties and damage to equipment. Dump trucks continued to haul sand on Highway 14 from Kontum, the only source of the material in the area, twice daily to the asphalt plant in Pleiku. The enemy, believing that U.S. and South Vietnamese forces would be dispersed while reacting to the attacks on the cities, made it a point to attack isolated vehicles and large convoys alike. Since Highway 14 served as the main supply road between the two province capitals, the engineers cleared a little over one hundred yards of vegetation from each side to reduce ambushes, but this did not discourage the attackers. On 6 March and again on 13 March, the 937th Group took heavy losses. In the second attack, the engineer vehicles at the rear of a 150-vehicle convoy were struck from both sides of the road. The enemy, estimated to range from two reinforced companies to two battalions, had taken positions behind berms of foliage and earth formed by the land clearing. After the initial fusillade of automatic weapons and rocket fire, sappers rushed toward the vehicles, throwing satchel charges at the trucks. The South Vietnamese armored cavalry squadron providing security entered the battle, and air strikes broke the attack. A few vehicles managed to run the ambush and clear the killing zone, but one hit a truck and blocked the road. A relief convoy consisting of wreckers and a company from the 70th Engineer Battalion moved to the ambush site to remove the disabled vehicles. Casualties from the three engineer companies (the 70th and 586th Dump Truck and the 509th Panel Bridge) totaled twelve dead and nineteen wounded. Equipment losses consisted of five trucks destroyed and another sixteen damaged.[15]

An investigation revealed the need for better coordination between the engineers and security forces and changes in convoy procedures. When the two officers in the serial containing the engineer vehicles were killed, communications with the convoy commander and the armored cavalry escort ceased. This left the engineers virtually helpless until the armored cavalry arrived. To counter future ambushes, the group assigned platoons from two combat engineer battalions as infantry to provide added security for engineer vehicles on Highway 14. Modifications to vehicles included armored plating welded to the doors of trucks or sand and gravel filled in door cavities. Makeshift armored trucks, usually a five-ton dump truck with the bed and sides sandbagged, were intermingled with every six dump trucks. Each armored truck carried a noncommissioned

---

[14] ORLLs, 1 Feb–30 Apr 68, 937th Engr Gp, pp. 4–6, 1 Feb–30 Apr 68, 20th Engr Bn, 30 Apr 68, pp. 2–3, 9–10, Historians files, CMH, 1 Feb–30 Apr 68, 70th Engr Bn, 30 Apr 68, pp. 6, 8, Historians files, CMH, and 1 Feb–30 Apr 68, 299th Engr Bn, pp. 1–8. For more on Operation FLORIDA, see Intervs, Maj Paul B. Webber, 26th Mil Hist Det, with Lt Col James H. Phillips, CO, 20th Engr Bn, 21 Mar 68, and Capt Leroy T. Cool, CO, Co B, 20th Engr Bn, 14 Mar 68, VNIT 175, CMH.

[15] Engineer Support of Combat Operations, incl. 6, p. 1, OCE Liaison Officer Trip Rpt no. 11; ORLLs, 1 Feb–30 Apr 68, 937th Engr Gp, p. 6, 1 Feb–30 Apr 68, 1st Log Cmd, p. 70; *Western Highlands*, pp. 85–86. For a full account of the 13 March ambush, see AAR/Intervs, 937th Engr Gp, 10 Jul 68, 26th Mil Hist Det, VNIT 176, CMH.

officer with a radio and three to five well-armed troops. These measures must have helped, for on 30 April the 937th Group reported no more ambushes of engineer convoys along Highway 14.[16]

The engineers tried to reduce ambushes by doing a more thorough job of land clearing. Although trees and vegetation were cut back about one hundred yards along many stretches of roads, the windrows of fallen timber left behind provided good cover and concealment for enemy troops. Company A, 70th Engineer Battalion, took care of this problem by using the "haystack" method. Since burning the fallen timber in the windrows was time consuming and inefficient, the battalion bulldozed the debris into piles, which were left to dry out for a few days, and then burned and the ashes scattered. The company, one of the engineer units assigned to protect convoys along Highway 14, also used two Rome plows and two bulldozers to clear an additional one hundred yards on each side of the road in areas most frequently used by the enemy. Rome plows of the 35th Land Clearing Team, attached to the 20th Engineer Battalion, were also busy at work on several supply routes. On 19 February, two sections of the team completed work along Route 7B west to the intersection of Highway 14. From there, the twenty Rome plows moved onto Highway 14 and by 9 March cleared 1,180 acres south of Phu Nhon toward Ban Me Thuot. Operations then moved to Highway 19 to widen the cleared areas from about one hundred to about three hundred yards between An Khe Pass and Mang Yang Pass. The team's third section of ten Rome plows, previously attached to the 35th Group at Phan Rang, moved to Ban Me Thuot and 70th Engineer Battalion control in late February and began clearing east along Highway 21. Hanoi's account of the war in the highlands acknowledged these prodigious efforts to keep the land lines open.[17]

The Fight for Saigon and the Delta

Although the allies in III Corps had expected trouble at Tet, the number of Viet Cong who slipped through the defenses around Saigon and the extent and fury of the offensive came as a shock. South Vietnamese and U.S. troops reacted quickly and turned back the enemy within a week, save for some holdouts in Cholon, the Chinese quarter of Saigon. The attack in Saigon began with the raid on the American embassy, a rectangular six-story fortress completed by RMK-BRJ the previous September. The embassy was a dubious military objective but of significant psychological value. After breaching the wall and entering the grounds and chancery building, but not the embassy building, all the sappers were killed. Simultaneous assaults at Tan Son Nhut Air Base and the nearby Joint General Staff compound, the Presidential Palace, and other installations in Saigon also failed, although some enemy soldiers did briefly penetrate the back side of the air base and the South Vietnamese headquarters

---

[16] ORLLs, 1 Feb–30 Apr 68, 937th Engr Gp, p. 6, and 1 Feb–30 Apr 68, 70th Engr Bn, pp. 3, 5–6; AAR/Intervs, 937th Engr Gp, 10 Jul 68, p. 6.

[17] ORLLs, 1 Feb–30 Apr 68, 937th Engr Gp, p. 4, 1 Feb–30 Apr 68, 20th Engr Bn, pp. 2–4, and 1 Feb–30 Apr 68, 70th Engr Bn, pp. 3, 7, 14; *Western Highlands*, p. 85.

complex. The assaults on Bien Hoa, Long Binh, the surrounding provinces, and in IV Corps were also short lived, although a second series of attacks in and around Saigon took place on 17 February, beginning with rockets striking Tan Son Nhut and MACV headquarters. Their intensity, however, was much less than the initial assaults. These attacks lasted intermittently until early March, highlighted by several fire fights inside Saigon and Cholon.[18]

The enemy interdicted all major roads from Saigon, delaying convoys on Highways 1 and 22 to Tay Ninh and Route 1A to Phuoc Vinh. Highway 13 to Lai Khe was cut repeatedly by sabotage and roadblocks. Convoys caught in outlying areas often waited for days for the engineers to clear the way. Ground transportation in the delta was also at a standstill since the Vietnamese drivers, fearing reprisals, stayed away from work. Most highway improvements in III and IV Corps ceased, including RMK-BRJ's project to upgrade the narrow and deteriorated Highway 4 in the delta. River barges that normally plied the waterways in the region were held up at Vung Tau. Rail movements stopped entirely. For several days, only priority airlifts could reach some of the inland bases.[19]

When Long Binh, Bien Hoa, and nearby posts came under fire on the morning of 31 January, Army engineers were already manning defensive positions. At the Long Binh post, parts of the perimeter were held by the 159th Engineer Group. The 79th Group headquarters and the attached 66th Engineer Topographic Company based at II Field Force headquarters at Plantation put about one-half their troops in perimeter bunkers. Just up the road at the 199th Light Infantry Brigade's Camp Frenzell-Jones, the 79th Group's 100th Float Bridge Company helped man perimeter positions. At 0300 on 31 January, the Viet Cong launched an intense rocket and mortar barrage at Plantation and Camp Frenzell-Jones. About the same time, rockets struck Bien Hoa Air Base and the adjacent III Corps headquarters compound, followed by ground assaults. When the barrage lifted at Plantation, the rest of the 79th Group's soldiers—clerk typists, carpenters, draftsmen, surveyors, and mapmakers—joined the troops on the line. Viet Cong soldiers, in position in Ho Nai Village on the other side of Highway 1A, opened up with a torrent of rifle and machine-gun fire and rocket-propelled grenades against the camp, succeeded by a ground attack aimed at the II Field Force headquarters. The defenders responded with automatic weapons fire and grenades, which stopped the assault and pinned down the attackers. Military police and Army helicopter gunships joined the fighting. Soon elements of the 199th Infantry Brigade, 9th Infantry Division, and 11th Armored Cavalry Regiment swept the area, leaving sixty enemy dead in Ho Nai. Some Viet Cong were caught milling around in the village (nicknamed "Widows' Village" because of the widows and orphans of South Vietnamese soldiers

---

[18] Sharp and Westmoreland, *Report,* p. 159; MACV History, 1968, vol. 1, pp. 390, 397; ORLL, 1 Feb–30 Apr 68, II FFV, 20 May 68, pp. 38–42, Historians files, CMH. For more on the fighting in III and IV Corps, see MACV History, 1968, vol. 1, pp. 390, 397–99, vol. 2, pp. 894–906; AAR, Tet Offensive, II FFV, 5 Aug 68, Historians files, CMH; Lung, *General Offensives of 1968–69,* pp. 51–75; General Donn A. Starry, *Mounted Combat in Vietnam,* Vietnam Studies (Washington, D.C.: Department of the Army, 1979), pp. 118–29.

[19] Bowers, *Tactical Airlift,* p. 322; MACV History, 1968, vol. 2, p. 653.

Bunkers like this were the first line of defense during the enemy attack on Plantation, the headquarters of II Field Force.

living there). Determined to liberate a prisoner of war camp just west of the village, the attackers had gotten lost, and were quickly killed or captured.[20]

The intensity of the opening attacks immediately shifted the engineers in the Saigon area from base construction to improving base defenses and operational support. At Plantation, the 79th Group's headquarters was heavily damaged, due largely to the concussion from an explosion set off by sappers who infiltrated through a thick bamboo grove between Plantation and the Long Binh ammunition dump. This gave Col. John H. Elder Jr., the group commander, good reason to rebuild an improved headquarters and a new tactical operations center protected in a reinforced-concrete block structure covered with sandbags. He also borrowed the 34th Group's 86th Engineer Battalion Land Clearing Team to eliminate the bamboo grove. Some nine hundred acres were cleared in a week. At Long Binh, Col. Harvey C. Jones, the 159th Engineer Group commander, quickly committed his four construction battalions (46th, 62d, 92d, and 169th) to improving the base perimeter, which involved adding more barbed wire fencing, reaction force bunkers, and access roads. Similar work went on at the U.S. Army camp at Bien Hoa and nearby firebases. Also under way at Long Binh was construction of a temporary tactical operations

---

[20] ORLLs, 1 Feb–30 Apr 68, 79th Engr Gp, 14 May 68, pp. 3, 9, 11, and 1 Feb–30 Apr 68, 159th Engr Gp, n.d., p. 5, both in Historians files, CMH; "Tet Truce Offensive, 1968," and "In Defense of Plantation, 1968," *Hurricane,* II FFV Magazine (April–June 1971): 2–4; Lung, *General Offensives of 1968–69,* p. 69; Engineer Support of Combat Operations, incl. 6, p. 1, OCE Liaison Officer Trip Rpt no. 11; Neil Sheehan, *A Bright Shining Lie: John Paul Vann and America in Vietnam* (New York: Random House, 1988), pp. 707, 714.

center for the 1st Logistical Command. This entailed excavating a 21-by-51-by-15-foot-deep pit, erecting heavy timbers with a blast cover of a single layer of 12-by-12-foot timbers, and a 1-foot reinforced-concrete floor on top of two inches of compacted soil. All exposed areas were covered with roofing felt and sprayed with peneprime. Altogether, the bunker had 604 square feet of working area, plywood paneled walls, fluorescent lighting, and electrical outlets. With so much construction, and reconstruction, under way, the group's lumber requirements between February and April rose sharply to more than 1 million board feet, up from 700,000 board feet in the three months before Tet. By March, the Long Binh depot had run out of 80-foot timber piles. Although 250 piles were found at the Qui Nhon depot, movement of the timber to Long Binh was delayed for a month, and construction there on the ammunition off-loading pier came to a halt. Transportation of engineer construction materials again took a lower priority to ammunition, fuel, and rations.[21]

There were other examples of engineers under attack in III and IV Corps. At Can Tho Airfield, Headquarters Company of the 69th Engineer Construction Battalion lost five men killed and several wounded while stopping an attack of over one hundred Viet Cong. North of Saigon at Phu Cuong, a reinforced Viet Cong battalion on the morning of 1 February struck the district headquarters and the South Vietnamese Army Engineer School just west of town. Because of the Tet holidays, the school was defended by only eighty-seven Vietnamese Army soldiers and four American advisers. By 0700, the enemy had overrun the northern half of the school. If the school had fallen, the Viet Cong would have dominated the high ground, which also overlooked the bridge site where men of the 41st Port Construction Company and Company B, 92d Engineer Construction Battalion, were building the steel and concrete bypass bridge over the Saigon River. The school troops were outnumbered and low on ammunition, and casualties were mounting. A request for help from the American senior adviser went to the engineers at the bridge site, and twenty-five men in a makeshift platoon were sent to the school. Guided by the advisers and aided by an air strike, the engineers counterattacked and drove the enemy back. Soon afterward, a South Vietnamese tank-infantry force arrived, and the Viet Cong were evicted from the school grounds. The U.S. engineers lost two killed and several wounded. Less than forty-eight hours later, while fighting was still going on, the engineers were back at work on the bridge.[22]

In the meantime, engineers assigned to the divisions and brigades were heavily engaged. Ongoing operations such as the 25th Division's YELLOWSTONE and SARATOGA; the 9th Division's RILEY, ENTERPRISE, and CORONADO; and the 1st Division's LAM SON 68, which succeeded LAM SON 67 on 1 February, continued.

---

[21] ORLLs, 1 Feb–30 Apr 68, 79th Engr Gp, pp. 3, 11, 1 Aug–31 Oct 67, 159th Engr Gp, p. 15, 1 Feb–30 Apr 68, 159th Engr Gp, pp. 5–10, 19–21, 1 Feb–1 Apr 68, 46th Engr Bn, 14 May 68, p. 7, Historians files, CMH, and 1 Feb–30 Apr 68, 34th Engr Gp, 1 May 68, p. 6, Historians files, CMH; Engineer Support of Combat Operations, incl. 6, p. 1, OCE Liaison Officer Trip Rpt no. 11.

[22] Msg, Maj Gen Charles M. Duke, CG, Engr Cmd and USARV Engr, ARV 263 to Cassidy, 5 Feb 68, Charles M. Duke Papers, CMH; Lt. Col. Walter R. Hylander Jr., "Port Construction Engineers in Combat," *Military Engineer* 60 (November-December 1968): 420–21.

The 1st Engineer Battalion's support of 1st Division operations typified the role of a divisional engineer battalion during February. On 4 February, Company B entered the village of An My with the 2d Brigade. With the help of the battalion's tunnel rats, the company found a large North Vietnamese Army staging base. Rome plows from the 27th Land Clearing Team moved in and razed the facility, deterring an expected attack on the Phu Loi base camp. During rocket and mortar attacks on the Lai Khe base camp, Company C put out fires that endangered the aviation refueling point and the base's ammunition supply point. With Highway 13 cut by roadblocks, craters, and mines, Company C just south of Ben Cat filled the largest mine crater encountered by the battalion to date, requiring some thirty-six truckloads of laterite. On the eighteenth, while working toward Phu Loi, a Company C platoon-size task force and its mechanized infantry escort came under heavy antitank and automatic weapons fire. A fierce battle followed for several hours with the engineers covering the maneuvering infantry with fire until the attackers withdrew, leaving behind fifty-five dead. The engineers suffered four wounded and damage to several vehicles.[23]

Because of the extensive damage to the cities and towns, and the flow of refugees, the Saigon government, with American urging, launched a massive recovery program. When the immediate threat subsided and the most important tasks had been completed, engineer troops turned their attention to helping Vietnamese civilians. General Westmoreland made American engineers available and authorized the drawing of large stocks of roofing, concrete, and other building materials, and providing emergency food supplies, drinking water, and medical help. South Vietnamese and allied engineers joined the effort. At Hoc Mon north of Tan Son Nhut Air Base, the 159th Engineer Group provided materials and foodstuffs for the South Vietnamese 5th Engineer Group's damaged dependent housing area. Carpenter shops and block shops were set up there, and work began on homes for some 540 people. Another effort, with commerce in mind, was Operation PEOPLE'S ROAD, aimed at reestablishing security and improving Highway 4 from the III Corps border to Can Tho before the onset of the rainy season. RMK-BRJ had started the work in January, but because of the threat to unarmed work parties along the road, MACV decided to have the 20th Brigade work on the My Tho to My Thuan section. While the 1st Brigade, 9th Division, provided security, the 34th Group's 86th Engineer Battalion started work on 2 March. By the end of the month, the battalion had completed the base course and grading and had paved two miles with a double bituminous surface treatment. The contractor did not do much between Saigon and My Tho in February but did resume work in early March. The South Vietnamese 40th Engineer Group, helped by the 69th Engineer Battalion, worked along the remaining sections to Can Tho.[24]

---

[23] *Always First, 1967–1968*, "Operation LAM SON" and "Company B and C"; Intervs, Maj William E. Daniel Jr., 17th Mil Hist Det, with 1st Lt William P. Francisco, Plt Ldr, 2d Plt, Co C, 2d Bn, 2d Inf, 1st Inf Div, pp. 1–8, and 1st Lt Terrence L. Hueser, Plt Ldr, 1st Plt, Co C, 1st Engr Bn, 1st Inf Div, 18 Jan 68, VNI 161, pp. 1–4, CMH.

[24] Westmoreland, *A Soldier Reports*, pp. 404–05; Sharp and Westmoreland, *Report*, p. 166; MACV History, 1968, vol. 1, p. 536, vol. 2, p. 653; ORLLs, 1 Feb–30 Apr 68, II FFV, p. 49,

Restoring Logistics in I Corps

In northern I Corps, North Vietnamese troops attacked Hue and Quang Tri City and cut off Tan My, one of the three ports above Da Nang. The other two, Dong Ha and Cua Viet, fell within artillery range from the Demilitarized Zone, making their use tenuous. The enemy also cut the Hai Van Pass along Highway 1 linking Da Nang to Hue. Supplies could be delivered by air, but the monsoon weather during February made flying and landing over-the-beach uncertain at best. Meanwhile, the North Vietnamese had increased their pressure on Khe Sanh. Earlier General Westmoreland had decided to hold the base, which he believed could tie down large North Vietnamese forces that otherwise could have moved unhindered around allied positions into the populated areas. On 5 February, heavy ground attacks followed by intense artillery fire the next day struck Khe Sanh and nearby Lang Vei Special Forces camp. Although the enemy took Lang Vei, the defenders at Khe Sanh and the surrounding hilltop outposts held out against heavy bombardments and ground assaults.[25]

On the eve of Tet, Brig. Gen. Willard Roper's 18th Engineer Brigade and Colonel Fink's 45th Engineer Group were preparing to send units to northern I Corps to support the 1st Cavalry Division, which had just deployed there to reinforce the marines. Fink and members of his staff made several trips to Phu Bai to coordinate the engineer deployment with the MACV Forward Command Post, an advanced headquarters that General Westmoreland established on 25 January to observe, direct, and, if necessary, control operations in the threatened northern provinces. On the twenty-ninth, General Roper, the commander of the 18th Engineer Brigade, and Fink traveled north to reconnoiter the group's new area of responsibility. The following day—the same day the attacks began in the northern half of the country—the group dispatched a reconnaissance party to the Gia Le combat base, just south of Hue. One of the party's helicopters was shot down with heavy casualties, including the death of the group's operations officer.[26]

The first unit of the 45th Group to start north was the 35th Engineer Battalion from Landing Zone HAMMOND in northern II Corps. On 2 February, Lt. Col. John V. Parish Jr., who had assumed command of the battalion two

1 Feb–30 Apr 68, 159th Engr Gp, pp. 5, 24, and 1 Feb–30 Apr 68, 34th Engr Gp, pp. 9–10; Quarterly Hist Rpt, 1 Jan–31 Mar 68, MACDC, 20 Apr 68, pp. IV-17 to IV-18, Historians files, CMH.

[25] Westmoreland, *A Soldier Reports*, pp. 336–41, 345–47; Sharp and Westmoreland, *Report*, pp. 162–64; Tolson, *Airmobility*, pp. 165–68; Pearson, *War in the Northern Provinces*, pp. 73–78. For more on the Tet offensive in I Corps, see MACV History, 1968, vol. 1, pp. 376–79, vol. 2, pp. 883–89; Jack Shulimson, Lt. Col. Leonard A. Blasiol, Charles R. Smith, and Capt. David A. Dawson, *U.S. Marines in Vietnam: The Defining Year, 1968* (Washington, D.C.: History and Museums Division, Headquarters, U.S. Marine Corps, 1997), pp. 113–223; Bowers, *Tactical Airlift*, pp. 295–316.

[26] Col. George B. Fink, "Engineers Move to I Corps," *Military Engineer* 60 (September-October 1968): 358; MFR, 21, Feb 92, sub: Author's Recollections, 45th Engr Gp, p. 2; Engineer Support of Combat Operations, incl. 6, pp. 1–2, OCE Liaison Officer Trip Rpt no. 11; ORLL, 1 Feb–30 Apr 68, 45th Engr Gp, p. 7; Ploger, *Army Engineers*, p. 150.

The Hai Van Pass was often obscured by clouds and fog.

weeks earlier, was alerted to get ready for the move. The battalion began its preparations. Initially, Parish hoped to make an amphibious landing off to the north of Hue, build an access road to Highway 1, and work south to Camp Evans, the 1st Cavalry Division's forward base camp. This approach seemed quicker and preferable to disembarking at Da Nang, the scene of a hectic buildup of Army support units. Colonel Parish was also concerned that the battalion's arrival in Da Nang could result in changing the unit's mission to do local projects.[27]

The Parish plan never got off the ground. No sooner had he received his alert to move than the whole tactical situation turned dire along the coastal plain from Quang Tri south to the Hai Van Pass, making foolhardy any attempt to insert the engineers into the teeth of North Vietnamese resistance. Pending stabilization of the battle for Hue and improvement in allied fortunes in the countryside nearby, which did not seem likely for some time, the engineers would start from the safe haven of Da Nang, working overland with infantry to reopen Highway 1 as far as the Hai Van Pass. And there was no room for delay—the fighting north of the pass was consuming supplies at an alarming

[27] ORLL, 1 Feb–30 Apr 68, 45th Engr Gp, p. 8; AAR/Intervs, 35th Engr Bn, 29 Mar 68, 26th Mil Hist Det, p. 2, and incl. 8, Interv, Maj Paul B. Webber, 26th Mil Hist Det, with Lt Col John V. Parish Jr., CO, 35th Engr Bn, 29 Mar 68, VNI 10, p. 1, CMH. VNI 10 also contains maps, photographs, and other documents.

North of the Hai Van Pass, the enemy destroyed bridges like this French-type concrete bridge.

rate, threatening a deficit in logistical support that a land line alone had the earliest chance to avert.[28]

With time of the essence, the battalion made the trip to Da Nang in less than a week. Parish began by moving men and equipment down Highway 1 to the port at Qui Nhon, some thirty miles to the south. Initial plans had Navy LSTs transporting the unit over a two-week period. The sudden availability of a much larger capacity contract freighter, the *Maryland* of Seatrain Lines adapted to haul containers besides vehicles and general cargo, accelerated the deployment. The first two LSTs with Companies B and C less some equipment departed Qui Nhon on the tenth. Two more LSTs arrived the following day to load the two companies' remaining equipment and part of the battalion's largest unit, the attached 517th Light Equipment Company. When the *Maryland* showed up on the twelfth, a day ahead of schedule, the rest of the battalion, with some difficulty, loaded the freighter in two days. Seven 290M tractors and scrapers could not fit through the cargo hole, so three more LSTs carried these items, a second attached company (the 511th Panel Bridge), and the balance of the battalion. The brisk pace of loading resulted in the abandonment of tent frames and floors, items that the battalion had planned to take.

---

[28] Msg, Westmoreland MAC 01858 to Wheeler and Sharp, 9 Feb 68, Westmoreland Message files, CMH; Interv, Webber with Parish, 29 Mar 68, p. 1; Pearson, *War in the Northern Provinces*, pp. 58–59.

Remaining elements reached the new camp, which was at the Nam O Beach site seven and a half miles north of Da Nang just off Highway 1, by midnight 17–18 February.[29]

The enemy had done a thorough job in cutting Highway 1 between Da Nang and Hue. Sappers closed the narrow winding Hai Van Pass with three major breaks and many obstacles. Along a straight stretch farther north, dubbed the Bowling Alley, they destroyed every bridge and culvert and excavated large trenches along the way. During the first two weeks of February, attempts by the 7th Marine Engineer Battalion and Naval Mobile Construction Battalion 62 to clear the road resulted in heavy casualties and loss of equipment. When the 35th Engineer Battalion arrived, it too could not proceed due to the strong presence of enemy forces. On 14 February, the battalion's first reconnaissance party, which included Colonel Parish, the operations officer, and battalion sergeant major, ran into a command-detonated mine and did not reach Hai Van Pass. Repeated efforts over the next few days resulted in enemy fire and several casualties. Only the arrival of the infantry, which swept the area west of the pass and similar efforts by the marines to the north, made it possible for the 35th Battalion to reopen the road.[30]

Real progress began on the morning of 21 February. Minesweep teams reached the pass, and the following day work parties and equipment began filling cuts and replacing culverts along the roadway. Although Army and Marine infantry reduced direct attacks by the enemy, he still harassed the engineers. Mines also posed a threat. The minesweep teams could not find all the mines, particularly homemade nonmetallic types that did not show readings on the mine detectors. Many vehicles passed over deeply buried mines without setting them off, but heavier loads such as tractor-trailers would usually cause the pressure-type mines to detonate. This happened on the twenty-second, resulting in heavy damage to a tractor-trailer and seriously wounding the driver. By the twenty-fourth, Companies C and D worked their way to Hai Van Pass. On that day, a D7 bulldozer working on the road hit a mine, damaging the bulldozer and wounding several engineers. The next day, Companies A and B moved farther north to a bivouac on the sea. From there, the units began reopening the Bowling Alley to traffic. Work included the around-the-clock building of a 105-foot, single-lane M4T6 trestle bridge. By 29 February, Company B met elements of the 32d Naval Construction Regiment working south. In the early morning hours of 1 March, a mine-clearing team swept the road and reported it clear. Engineers working on the road briefly took a break and watched a 100-vehicle convoy from Da Nang file past. This marked the first time since the start of the Tet attacks a month earlier that vehicles could make the trip to Hue. In all, the 35th Battalion in the period from 21 to 29

---

[29] AAR/Interv, 35th Engr Bn, 29 Mar 68, p. 2, and incl. 10, Interv, Maj Paul B. Webber, 26th Mil Hist Det, with Maj Wilbur C. Buckheit, XO [Executive Officer], 35th Engr Bn, 29 Mar 68, VNI 10, pp. 1–2, CMH.

[30] ORLL, 1 Feb–30 Apr 68, 35th Engr Bn, p. 3; Fink, "Engineers Move to I Corps," p. 358; Commander in Chief Pacific Fleet, Pacific Area Naval Operations Review, Feb 68, pp. 85–87, copy in Historians files, CMH; Interv, Webber with Parish, 29 Mar 68, pp. 1–2.

At the top of Hai Van Pass, the 35th Engineer Battalion deployed equipment like this 290M scraper to widen the road and reduce sharp curves.

February suffered damage to one 5-ton tractor trailer, two D7 bulldozers (one a total loss), and three 5-ton dump trucks. Ten men were wounded.[31]

The reopening of Highway 1 eased the logistical crisis around Hue, but officials still faced the task of supporting the action farther north and reconstituting the forward stockpiles, especially the fuel and ammunition, that would be needed for the relief of Khe Sanh. By one estimate, another 1,000 tons a day would be needed, nearly four times the rated capacity of Highway 1, if the counteroffensive was to kick off as planned on 1 April. With no other practical alternative but supply by sea, the surest way to move materiel in bulk, the U.S. command now gambled that the battle for the northern coast was about to be won. It immediately laid plans for a logistical-over-the-shore, or LOTS, site east of Quang Tri City, confident that the receipts would eliminate the deficit looming over future operations. The facility was to open in mid-March.[32]

With a LOTS site now on the drawing board, logistics figured heavily in Colonel Fink's decision to locate the headquarters of his 45th Group in Da Nang. Initially he considered placing the headquarters farther north at Phu Bai near the MACV Forward Command Post and the 1st Cavalry and

[31] AAR/Interv, 35th Engr Bn, 29 Mar 68, pp. 3–5; ORLL, 1 Feb–30 Apr 68, 35th Engr Bn, pp. 3–4; Fink, "Engineers Move to I Corps," pp. 358–59; HQ, 35th Engr Bn, Summary of Operations in I Corps, incl. 6, tab F, OCE Liaison Officer Trip Rpt no. 11.

[32] MACV History, 1968, vol. 1, pp. 477, 479, vol. 2, p. 619; Msgs, COMUSMACV to CO, 3d Nav Const Bde, 16 Feb 68, sub: Hai Lang Beach Road Construction, and CG III MAF to COMUSMACV, 18 Feb 68, sub: LOTS Operation Northern I CTZ, both in box 1, 71A/354, RG 334, NARA.

101st Airborne Divisions. Setting up operations temporarily in Da Nang, however, made it easier to coordinate operational and logistical matters for the group's projected four combat engineer battalions and attached units. These units initially included the 35th Engineer Battalion working on Highway 1, the 39th Engineer Battalion supporting the Americal Division, and the 14th Engineer Battalion, preparing to move to its new assignment, the over-the-shore site near Quang Tri. The group's fourth battalion, the 27th, was alerted in late March to move from its base sixty-two miles east of Saigon to support the 101st Airborne Division at Camp Eagle, south of Hue. An advance party from the group headquarters had arrived in Da Nang on 15 February. In mid-March, the group headquarters completed its relocation from Qui Nhon and moved into buildings provided by the 7th Marine Engineer Battalion next to the 1st Marine Division's headquarters just west of Da Nang Air Base. There, the 45th Group had easy access to the III Marine Amphibious Force headquarters in Da Nang, the 3d Naval Construction Brigade across the road at Red Beach, and the logistical complexes throughout the metropolitan area.[33]

Some modifications to command and control and logistical support for the engineers in northern I Corps now became necessary. The 45th Group remained an assigned unit under the 18th Engineer Brigade at Dong Ba Thin, but III Marine Amphibious Force exercised operational control and the U.S. Navy provided most of the logistical support. On 25 February, the Marine command assigned the Army group the mission to minesweep, maintain, and upgrade Highway 1 between the bridge at Nam O and Phu Loc. The command also directed the 3d Naval Construction Brigade and Marine Corps engineer units in the area to help the Army engineers with supplies. Also made available were materials from the Seabee brigade's pre-positioned and packaged war reserve stocks and heavy bridge timbers and large culverts programmed by the 3d Naval Construction Brigade for future major highway improvements. Save for the extreme southern part of I Corps, the 45th Group's units relied on supplies provided by the Navy.[34]

Meanwhile, after considerable delay, the 14th Engineer Battalion began arriving at the over-the-shore site on 20 March. On 6 February, Lt. Col.

---

[33] ORLLs, 1 Feb–30 Apr 68, 45th Engr Gp, pp. 1–10, and 1 Feb–30 Apr 68, 27th Engr Bn, 7 May 68, p. 9, Historians files, CMH; Engineer Support of Combat Operations, incl. 6, p. 2, OCE Liaison Officer Trip Rpt no. 11; Msg, COMUSMACV to CG, USARV and CG, III MAF, 2 Feb 68, sub: Engineer Support for Northern I CTZ, box 5, 71A/354, RG 334, NARA; Msg, CG, III MAF to COMUSMACV, 18 Feb 68, sub: LOTS Operation Northern I CTZ; Quarterly Hist Rpt, 1 Jan–31 Mar 68, MACDC, p. II-3; MFR, 21 Feb 92, Author's Recollections, 45th Engr Gp, pp. 2–3.

[34] ORLLs, 1 Feb–30 Apr 68, 45th Engr Gp, pp. 1, 5–6, 11–12, and 1 Feb–30 Apr 68, 1st Log Cmd, p. 4; Ltrs, CG, III MAF to CO, 45th Engr Gp, sub: Assignment of Responsibility for Land Lines of Communication, 25 Feb 68, tab D, and CG, III MAF to CO 45th Engr Gp, sub: Interim Guidance on Tactical Support Functional Components [TSFC], Program and Reporting Procedures for Lines of Communications, 26 Feb 68, tab E, both in incl. 6, OCE Liaison Officer Trip Rpt no. 11. For more on the two services' logistical support arrangements, see MACV History, 1968, vol. 2, pp. 616–18; Heiser, *Logistic Support*, pp. 216–17; and Merdinger, "Civil Engineers, Seabees, and Bases in Vietnam," pp. 244–46.

Bennett L. Lewis received the warning order to prepare for immediate deployment from Cam Ranh Bay to I Corps. A Navy survey revealed, however, that LSTs could not beach but would require a 900-foot amphibious causeway to discharge cargo. The pontoon causeway, which had to be shipped from Japan and assembled by the Seabees, was not expected to be completed until 15 March. Lewis had to postpone the deployment, and planning continued. When the 35th Engineer Battalion reopened Highway 1 north of Da Nang on 1 March, the 14th Engineer Battalion put the move in motion. The heavy equipment and cargo went directly by sea to the beach site, now called Utah Beach. On 14 March, the battalion, less Company C at Phan Thiet, departed for Da Nang aboard the SS *Carolina*. Following delays at Da Nang's congested harbor, the troops and vehicles moved by land to Utah Beach. Two LSTs transported Company C and the remaining cargo at Cam Ranh Bay directly to Utah Beach.[35]

After arriving at the beach site, the 14th Engineer Battalion took over construction from Naval Mobile Construction Battalion 10. So far the Seabees had reopened an old narrow French road (Route 602) to Hai Lang where it connected to Highway 1, laid out sand roads, emplaced matting in open storage areas, and dug pits for fuel storage bladders. Company A started to improve the access road to two lanes topped with laterite and to construct two timber pile bridges, 120 and 140 feet long. It also built facilities and defense positions for the 159th Transportation Battalion task force, which ran the over-the-shore operation. Company B began building a sand-cement road system connecting sand-cement access ramps and storage pads built by the other two companies at the beach. The Earthmoving Platoon, Company C, 589th Engineer Battalion, joined the 14th Battalion on the beach. The platoon's 290M scrapers, bulldozers, graders, and water distributors increased the 14th's capabilities significantly. Company C of the 14th Battalion swept the access road for mines. Its greatest challenge was an antivehicular mine similar to the Soviet TMB2 mine with a pressure device that contained only a small amount of metal. The mine proved extremely difficult to detect.[36]

The over-the-shore facility was a major accomplishment. A joint effort by Army engineers, Navy Seabees, Army transportation units, Navy landing ships, and a Marine Corps fuel detachment transformed the beach into a key logistical facility. LSTs and other landing craft discharged cargo over the pontoon causeway while amphibious resupply cargo barges and lighters brought

---

[35] MACV History, 1968, vol. 2, p. 619; Msgs, CG, III MAF to CG, 18th Engr Bde, 15 Feb 68, sub: Hai Lang Beach Road Construction, CG, III MAF to Commander in Chief Naval Forces, Vietnam, 15 Feb 68, sub: Pontoon Causeways at Hai Lang Beach, and CG, 18th Engr Bde to CG, III MAF, 16 Feb 68, sub: Unit Deployment, all in box 1, 71A/354, RG 334, NARA; Ploger, *Army Engineers*, p. 150; AAR, 14th Engr Bn, 7 May 68, sub: Relocation of 14th Engineer Battalion to I Corps Tactical Zone, pp. 16–23, incl. 1, ORLL, 1 Feb–30 Apr 68, 14th Engr Bn, 30 Apr 68, Historians files, CMH; ORLLs, 1 Feb–30 Apr 68, 14th Engr Bn, pp. 2–6, and 1 Feb–30 Apr 68, 45th Engr Gp, pp. 9–11.

[36] ORLL, 1 Feb–30 Apr 68, 14th Engr Bn, pp. 4–6; Interv, 1st Lt Raymond F. Bullock, 26th Mil Hist Det, with Lt Col Bennett L. Lewis, CO, 14th Engr Bn, 20 Aug 68, VNIT 241, p. 2, CMH; Interv, Maj Alexander E. Charleston, 30th Mil Hist Det, with Lt Col Charles H. Sunder, CO, 159th Trans Bn, 7 Jul 68, VNI 63, pp. 1–2, CMH.

supplies of all types ashore from deep-draft ships. Over time, Utah Beach contained extensive ammunition and fuel storage areas, a helicopter refueling point, a road net with a two-lane connection to Highway 1, and fuel pipelines laid offshore and from the beach to Highway 1 and then north to Dong Ha. Each day truck convoys moved cargo inland from the beach storage area to the forward support bases of the combat divisions. Although daily deliveries of under 350 tons were originally predicted, the beach facility usually averaged over 1,000 tons. (Because of its achievements, the facility shortly took on the name of Wunder Beach, a play on words by the personnel on the beach of the 159th Transportation Battalion's commanding officer, Lt. Col. Charles H. Sunder.) As an additional benefit, the new inland road also crossed and effectively cut a main North Vietnamese supply route leading into Hue.[37]

The fighting to retake Hue lasted three weeks. Viet Cong and North Vietnamese troops, moving under the concealment of low fog, infiltrated the old imperial capital with the help of accomplices inside the city. They quickly captured most of the city on the south bank of the Perfume River and later seized the bulk of the northern half, including the Imperial Citadel. U.S. Marines drove the enemy from most of the south bank in a few days. The fierce battle for the Citadel, in which U.S. and South Vietnamese units suffered heavy casualties, raged until the end of February.[38]

To the north and northeast, the 1st Cavalry Division began striking back at the enemy. The 1st Brigade helped South Vietnamese troops clear Quang Tri City. The 3d Brigade, which had been conducting operations between Chu Lai and Da Nang, was ordered to reinforce weakened local forces and block the enemy's approaches to Hue. Eventually four battalions were involved at Hue in some of the heaviest fighting of the war. By the end of the operation, the brigade had killed over four hundred enemy, nearly all North Vietnamese.

Supporting the 1st Cavalry Division were Army, Navy, and Marine Corps engineers. A platoon from Company A of the division's 8th Engineer Battalion accompanied the first cavalry battalion alighting at Landing Zone EL PASO near Hue on 17 January. Joined two days later by a second platoon, the divisional engineers, using their own and borrowed heavy engineer equipment, concentrated on opening Camp Evans. As the rest of the division arrived, the 8th Engineer Battalion committed its resources to daily minesweeps along Highway 1 between Quang Tri and Hue, developing new landing zones and firebases, and supporting units closing into the area. Additional engineering support came from Seabee units—mainly Naval Mobile Construction Battalions 5 and 10—before the arrival of the 14th and 35th Engineer Battalions. At Camp Evans, Naval Mobile Construction Battalion 10 helped the 8th Engineer Battalion build a C–7 Caribou airstrip, which officially opened on 13 March. Work progressed to expand the airfield to C–130 capability. On 11 March,

---

[37] Pearson, *War in the Northern Provinces*, pp. 59–62; Interv, Charleston with Sunder, 7 Jul 68, pp. 1–3.

[38] Westmoreland, *A Soldier Reports*, pp. 329–31; Sharp and Westmoreland, *Report*, p. 160; Pearson, *War in the Northern Provinces*, pp. 39–48; Tolson, *Airmobility*, pp. 158–64; Lung, *General Offensives 1968–69*, pp. 75–85. For more on the battle for Hue, see Keith William Nolan, *Battle for Hue: Tet, 1968* (Novato, Calif.: Presidio Press, 1983).

part of the battalion operations section moved to Ca Lu to survey the area and coordinate the construction of a new C–123 airstrip and base facilities (Landing Zone STUD) by Marine Corps engineers and Seabees. These and other projects typified the Seabees' responsiveness. When the division's chief of staff had trouble with the III Marine Amphibious Force staff concerning the space requirements for the division headquarters, the Seabees' commander simply asked what the division wanted done and in what order to have it done. General Tolson later wrote, "This was just the first example of the magnificent support we received from the Seabees."[39]

The 1st Cavalry Division's shift northward also brought home the 8th Engineer Battalion's need for a fourth line company and additional equipment. Although the Seabees and Marine Corps engineers provided invaluable construction support, the 8th Engineer Battalion still lacked the general support capability found in the other divisions' engineer battalions. This meant that Col. Edwin S. Townsley, the battalion commander, had to reduce the amount of direct support to the brigades in the field and take troops from the line companies to help the headquarters set up camp and do other tasks. The daily minesweeps along Highway 1 took up about 40 percent of the battalion's effort, with most of the troops drawn from the line companies. Since the 3d Brigade was collocated with the division headquarters at Camp Evans, Company C, which provided direct support to the brigade, saw 50 percent of its capability drawn off to work on bunkers, tactical operations centers, and perimeter defenses. Townsley also pointed out the need for more backhoes and airmobile bulldozers, which were considered vital to firebase construction. Since the brigades usually opened at least two firebases during operations, he recommended increasing the number of authorized backhoes from three to six and boosting the number of bulldozers from the two on hand to the six currently authorized. This would enable each line company to build two firebases concurrently.[40]

Return to Normal

Operational support missions continued at a high pitch through March, absorbing two-thirds of Engineer Command's capacity, then began to subside as the Tet offensive petered out. By April, most of the roads had been reopened and the bridges repaired, allowing most engineer units, except those in I Corps, to return to a more typical allocation of effort between ordinary construction missions and providing for the divisions and brigades. One pacing factor in this return to normality was the extent to which the Vietnamese labor force felt safe to go back to work.

III Corps returned to normal fairly quickly. By the end of March, the 79th Engineer Group, operating north of Saigon, had put the Tet cleanup behind

---

[39] Quoted words from Tolson, *Airmobility*, p. 160; ORLL, 1–31 Mar 68, 8th Engr Bn, 1 Apr 68, pp. 1–2, Historians files, CMH; *Memoirs of the First Team*, p. 184.

[40] ORLLs, 1–31 Mar 68, 8th Engr Bn, pp. 4–5, and 1–30 Apr 68, 8th Engr Bn, 4 May 68, pp. 6–7, Historians files, CMH.

it and had started to stockpile construction material in preparation for the approaching rainy season when the weather would focus the expenditure of effort on base camp development projects. Three-fourths of its Vietnamese workers were back on the job. Closer to the capital, the 159th Engineer Group turned its attention to long-term defense, constructing combat-essential roads and hardening such facilities as airfields, generators, and fuel storage sites. Cantonments, laboratories, and storage yards were now also a construction priority. In II Corps, the picture was mixed. At An Khe and Cam Ranh Bay, little touched by the Tet offensive, base construction projects continued as before. Elsewhere, worker absenteeism and the closing of several quarries slowed the postattack cleanup, and as March ended the 35th Engineer Group had yet to complete its repairs to facilities at Phan Thiet, Nha Trang, and Tuy Hoa. Ordinary construction was picking up, but so far at reduced priority.[41]

The contractors faced some of the same problems as the troop units because of the absenteeism of their Vietnamese employees. The fighting near the bases forced many workers to stay home or seek safety elsewhere. Those who did report for work or were at work during the attacks faced the same risks as the military. Pacific Architects and Engineers, the Army's facilities engineering contractor, had several facilities outside the base perimeters with security guards neither armed nor trained to defend against attacks. Civilians working side-by-side with soldiers often lacked the minimum protection of helmets and flak jackets. Sometimes the civilian engineers and technicians were barred from entering the bases. Fifteen employees of Pacific Architects and Engineers were killed and several others wounded during the Tet attacks. Eight Americans were missing, and seven were reported taken prisoner. Despite the intensity of the attacks countrywide, the firm evacuated only one site—at Vinh Long in the delta. The displaced Americans and third-country nationals returned to Vinh Long on 4 February and began repairs on electrical distribution lines, perimeter lighting, and power system for the water pumping station. By 12 February, 120 of the 177 Vietnamese had reported to their jobs, with conditions returning to near normal. At many locations, the contractors continued to carry out critical operations using American supervisors, Filipino and Korean workers, and the Vietnamese who reported to work. Essential services as mundane as trash collection resumed with military escorts, often with American, Filipino, and Korean supervisors and technicians doing this necessary chore. By early March, the facilities engineering contractor reported it had returned to 80 percent of its normal level of work. Construction work by RMK-BRJ slipped by one month, but by early March the firm reported that 82 percent of its workforce was on the job.[42]

[41] Engineer Support of Combat Operations, incl. 6, p. 1, OCE Liaison Officer Trip Rpt no. 11; ORLLs, 1 Feb–30 Apr 68, 18th Engr Bde, 30 Apr 68, pp. 1–2, Historians files, CMH, 1 Feb–30 Apr 68, 79th Engr Gp, pp. 2, 4, 1 Feb–30 Apr 68, 159th Engr Gp, pp. 5–19, 1 Feb–30 Apr 68, 34th Engr Gp, pp. 2, 5–8, 1 Feb–30 Apr 68, 35th Engr Gp, pp. 5–6, 1 Feb–30 Apr 68, 937th Engr Gp, pp. 3–5.

[42] PA&E History, Jan–Jun 1968, pp. 28, 33–34, 38–39, 44–45; Quarterly Hist Rpt, 1 Jan–31 Mar 68, MACDC, pp. III-2 to III-4; MACV History, 1968, vol. 2, p. 671.

Army engineers were now carrying out tasks throughout South Vietnam. From the Mekong Delta to the northernmost provinces of I Corps, engineer troops were hard at work in support of combat operations, road and airfield improvements, and base construction. Much of this work was carried out under hazardous conditions. More severe tests would come following the enemy's Tet offensive as the allies readied themselves for counteroffensive operations.

12

The Aftermath, April–December 1968

By the end of March, the storm of the Tet offensive had abated. Far from collapsing, the South Vietnamese government had rallied, and although fighting appeared touch-and-go in many places, no South Vietnamese units were destroyed, and their casualties were surprisingly low. More than simply rallying, however, the government also mobilized, approving a MACV plan to increase the armed forces to over 800,000 men in the next two years. To establish a self-sufficient logistical system, the expansion plan included four construction battalions. Senior U.S. Army engineers in Washington and Saigon also made preparations to organize and train South Vietnamese Army land-clearing companies.[1]

Americans at home saw a different picture, especially when word got out that even more U.S. troops would be needed in Vietnam. Dramatic images showing the sapper attack on the American embassy in the heart of Saigon, the besieged marines at Khe Sanh, and the bitter fight for Hue dimmed General Westmoreland's contention of an allied victory. President Johnson's claims of progress in the war, already doubted by many, lost more credibility, and he showed signs of frustration with the war. Defense Secretary McNamara, also disenchanted, left office on 1 March, replaced by Clark M. Clifford, an adviser and close friend of the president. Public skepticism over the war swelled after the media revealed Westmoreland's request for an additional 206,000 troops. At the time he had about 500,000 of the 525,000 American troops promised, and he needed all of them. Though most of the new troops were intended for the strategic reserves, the size of the request cast further doubts on the conduct of the war. Westmoreland did get some modest reinforcements in February. The 27th Marine Regimental Landing Team took up positions south of Da Nang and freed other Marine Corps units to move north. Meanwhile, the 3d Brigade, 82d Airborne Division, which included Company C from the division's 307th Engineer Battalion, joined the 101st Airborne Division at Phu

---

[1] MACV History, 1968, vol. 1, pp. 26 27; Sharp and Westmoreland, *Report*, pp. 161–62, 164; Jeffrey J. Clarke, *Advice and Support: The Final Years, 1965 1973*, United States Army in Vietnam (Washington, D.C.: U.S. Army Center of Military History, 1988), pp. 293–94, 313–14; Msgs, Maj Gen Robert R. Ploger, Dir of Topography and Mil Engineering, OCE, GVP 297 to Maj Gen Charles M. Duke, CG, Engr Cmd and USARV Engr, 4 Apr 68, sub: Land Clearing Companies; Duke ARV 989 to Ploger, 16 Apr 68, sub: Army of the Republic of Vietnam (ARVN) Land Clearing Companies; Brig Gen Andrew P. Rollins, MACV Dir of Const, MAC 05396 to Ploger, 23 Apr 68, sub: ARVN Land Clearing Companies, all in Duke Papers, CMH.

Bai as its third brigade. This measure allowed the 101st Airborne Division's 3d Brigade to remain in the Saigon area.[2]

The small increases in U.S. force levels came after much debate in Washington. Without mobilization, the United States had become overcommitted. The strategic reserves had dwindled during a period of international tension. North Korea captured an American naval vessel, the USS *Pueblo*, a week before the Tet offensive. Chronic problems in the Middle East persisted. Civil unrest in the United States between 1965 and 1968 required the use of Army troops. After extensive deliberation, Johnson chose the modest force increase that would meet Westmoreland's first troop request. A mechanized infantry brigade, including an attached engineer company, from the 5th Infantry Division would replace the 27th Marines in July. President Johnson also overcame his reluctance to mobilize the reserves. There were two reserve call-ups to meet and sustain these deployments, one in March and one in May. The call-ups totaled 62,000 men, nearly 54,000 of them Army, and included a National Guard engineer combat battalion and a light equipment company as part of the 13,500-man support package for Vietnam. This raised the manpower authorization to 549,500, the peak U.S. strength in Vietnam. For a disappointed Westmoreland, this meant that he would have to make the best possible use of American troops, and the South Vietnamese would have to take on a larger share of the war effort. As a signal to foster negotiations, Johnson also curtailed air strikes against North Vietnam. When the president outlined his decision in a nationwide address on 31 March, he surprised the nation and the world by announcing that he would not seek reelection and would devote his full attention to resolving the conflict. Though Hanoi had suffered a military defeat, Johnson's actions implied that the Communists had achieved a political and diplomatic victory.[3]

To keep the number of mobilized support troops low, Washington authorized civilian substitutes. In addition to the 13,500 support troops, MACV received authority to hire 13,035 additional local civilians to augment logistic and construction units. MACV already had initiated a civilianization program in January, but the Tet offensive forced the command to defer the program for six months. Civilianization would have converted some military spaces, but using local nationals required a degree of stability in Vietnam, and fixed units and bases. During Tet, many Vietnamese failed to show up for work. If the U.S. Army, Vietnam, had completed the program, many logistical and engineering units would have lacked the flexibility to carry out their

---

[2] Westmoreland, *A Soldier Reports*, pp. 350–59; Sharp and Westmoreland, *Report*, pp. 170, 184; MACV History, 1968, vol. 1, pp. 225, 244–45; Hammond, *Military and the Media, 1962–1968*, pp. 344–45, 357–66, 371–73, 375–82. For more on the request for additional forces, see *The Pentagon Papers: The Defense Department History of United States Decisionmaking on Vietnam*, Senator Gravel ed., 4 vols. (Boston: Beacon Press, 1971), 4:539–604; Davidson, *Vietnam at War*, pp. 492–520.

[3] MACV History, 1968, vol. 1, pp. 227–28. For the text of Johnson's 31 March speech, see *Pentagon Papers* (Gravel) 4:596–602. See also Herbert Y. Shandler, *The Unmaking of a President: Lyndon Johnson and Vietnam* (Princeton: Princeton University Press, 1977), pp. 194–217, 229–36.

missions. For Engineer Command, the conversion would have amounted to almost two thousand spaces, about 20 percent of the total planned for U.S. Army, Vietnam. Each of the five construction battalions involved in the program would have lost ninety-nine military positions. Similar substitutions of Vietnamese workers would have occurred in smaller units such as construction support companies. MACV also considered transferring construction projects from engineer units to RMK-BRJ and raising the consortium's manning level, enough to replace a construction battalion. Ironically, Saigon's mobilization was expected to hinder MACV's civilianization program and the contractor's ability to hire more workers. Viet Cong reprisals against Vietnamese employees also provoked fear and reluctance on the part of potential workers. On the other hand, as the government inducted more men in the armed forces, women entered the workforce in larger numbers. That spring, Pacific Architects and Engineers hired about 200 women trainees. At the end of June, 161 women successfully completed training, and Pacific Architects and Engineers hired them as electricians, welders, drivers, mechanics, and plumbers.[4]

On 1 April, Westmoreland believed the allied forces were ready to counterattack. He told his senior commanders that forces in I Corps were "now in position to seek decisive battles in Quang Tri and Thua Thien." Along the coastal regions and in the western highlands "our troops are orienting their operations on the enemy to keep him away from population centers and resources." Moving south, he noted, "Allied campaigns throughout III and IV Corps are diligently securing and restoring lines of communications and installations (economic lifelines) that are vital to these rich and populated areas." Westmoreland concluded: "We must go after the enemy throughout the country; we must hound him and hurt him. We can achieve a decisive victory and we must do so at once, to restore the perspective with which the world sees this war. We must demonstrate by our actions that we are, in fact, winning the war."[5]

Westmoreland's admonition to increase the tempo of pressure against the enemy came soon after the announcement that he would be leaving Vietnam. On 23 March, he was informed of his appointment as Army chief of staff. His deputy, General Abrams, would replace him that summer. In the few remaining months as MACV commander, Westmoreland directed counteroffensive operations, including the relief of Khe Sanh and a drive into the A Shau Valley. The war had reached a peak and, as usual, Army engineers, now spread from the Demilitarized Zone to the delta, were heavily involved in those operations.[6]

---

[4] Sharp and Westmoreland, *Report*, p. 165; Westmoreland, *A Soldier Reports*, p. 359; MACV History, 1968, vol. 2, p. 678; Quarterly Hist Rpt, 1 Jan–31 Mar 68, MACDC, pp. II-10 to II-11; PA&E History, Jan–Jun 1968, p. 17. For more on the military force levels and civilianization program, see MACV History, 1968, vol. 1, pp. 225–33; Msg, Palmer ARV 1569 to Gen Creighton W. Abrams, Dep COMUSMACV, 9 Jun 68, sub: Program 6 Civilianization; Msg, Gen Johnson WDC 8527 to Palmer, 11 Jun 68, sub: Civilianization Program—Program 6; Msg, Palmer ARV 1622 to Gen Johnson, 14 Jun 68, sub: Civilianization Program, all in Creighton W. Abrams Papers, CMH.

[5] Quoted in MACV History, 1968, vol. 1, pp. 26–27.

[6] Westmoreland, *A Soldier Reports*, p. 439.

Relieving Khe Sanh

In mid-March, the North Vietnamese apparently had given up at Khe Sanh and began to pull back into Laos just as Westmoreland decided to reestablish a land link with the Marine base. He did not consider the base in great peril but became anxious to reestablish contact "if for no other reason than to silence dolorous critics and allay President Johnson's concern." MACV Forward Command Post under Westmoreland's deputy, General Abrams, had developed a plan, code-named Operation PEGASUS, before the end of February, but weather conditions and logistic preparations held up launching the campaign. Planning continued while allied forces retook Hue, more Army reinforcements reached the area, engineers reopened the lines of communication, and logisticians reconstituted stockpiles. When the logistics-over-the-shore supply facility at Wunder Beach began to discharge over one thousand tons a day, MACV set 1 April as the D-day for the relief of Khe Sanh.[7]

Meanwhile, five widely scattered U.S. divisions, three Army and two Marine, had taken up positions in I Corps, and Westmoreland considered the force too large to be controlled from a single headquarters. The recent arrivals included the 101st Airborne Division, which deployed from III Corps to Phu Bai on 19 February. When the 101st took over responsibility for the 1st Cavalry Division's old operating area, the cavalrymen moved to a new base and airstrip called Landing Zone STUD at Ca Lu in central Quang Tri Province to prepare for PEGASUS. At that point, Westmoreland decided to upgrade the III Marine Amphibious Force to the level of a field army headquarters with a subordinate Army corps. On 10 March, accordingly, MACV Forward at Phu Bai officially became Provisional Corps, Vietnam, and General Abrams returned to Saigon. Lt. Gen. William B. Rosson, who had earlier commanded Task Force OREGON and then I Field Force, and was currently serving as Abrams' deputy at MACV Forward, assumed command of the new Army corps and operational control of the two Army divisions, the 3d Marine Division, and other supporting forces in northern I Corps.[8]

Operation PEGASUS began at 0700 on 1 April with American and Vietnamese forces moving out from Landing Zone STUD along Highway 9 toward the Khe Sanh base. This followed an allied deception operation initiated a few days earlier northeast of Dong Ha. The PEGASUS force, under the control of the 1st Cavalry Division, included the division's three brigades, a Marine Corps regiment, a South Vietnamese airborne task force, and the garrison at Khe Sanh, over 30,000 troops. The 1st Marines led the ground attack, securing and repairing the road as it advanced westward. Poor weather delayed the airmobile part of the operation to 1300, when the 1st Cavalry Division's 3d Brigade began seizing high ground in a series of airmobile assaults. On 3 April, the 2d Brigade landed three battalions southeast of Khe Sanh and attacked to the northwest. During the fourth day, the marines continued their push while the

---

[7] Quoted in ibid., p. 347; Pearson, *War in the Northern Provinces*, p. 68.

[8] Pearson, *War in the Northern Provinces*, pp. 68–69; Shulimson, et al., *U.S. Marines in Vietnam, 1968*, p. 284.

cavalrymen applied pressure throughout the battle area. The following day, elements of the 26th Marines sallied forth from Khe Sanh and seized Hill 471 southeast of the base. On 5 April, North Vietnamese troops tried in vain to retake the hill, but the marines, sustained by artillery and close air support, cut down the attackers in one of the major fights of the campaign. That same day, units of the 1st Cavalry Division's 1st Brigade air-assaulted into positions overlooking Highway 9 south of the base. More air cavalry troops landed on Hill 471 to relieve the marines in the first relief of Khe Sanh's defenders. The official linkup came the following afternoon when the 1st Cavalry Division airlifted South Vietnamese airborne troops to the base. On the morning of the eighth, the 3d Brigade cleared the remaining section of road and linked up with the 26th Marines. Two days later, the 1st Brigade seized the old Lang Vei Special Forces camp. Provisional Corps, Vietnam, officially ended PEGASUS on 15 April.[9] (*Map 24*)

PEGASUS depended much on the logistical portion of the plan, which in turn depended on the engineers. Carrying out the logistic operation hinged on the construction at Landing Zone STUD. Before this work could start, the 3d Marine Division had to clear Highway 9 westward from the Marine artillery base at the Rockpile. Supplies, fuel, ammunition, and construction material were then stockpiled at STUD. On 14 March, a joint task force of engineers under the operational control of the 1st Cavalry Division—consisting of elements of the 11th Marine Engineer Battalion, Naval Mobile Construction Battalion 5, and the 8th Engineer Battalion—reached STUD and began work. Within twelve days, the tri-service engineers completed a C–7A Caribou airstrip 1,500 feet long by 150 feet wide. The Seabees, who were augmented with heavy equipment from the 32d Naval Construction Regiment, concentrated on the airfield. When PEGASUS jumped off on 1 April, Landing Zone STUD had reached about 83 percent completion. By the time the operation ended, the base was in full operation and included an extended C–123 airstrip with parking ramps, ammunition storage areas, aircraft and refueling facilities, road nets, and strengthened defenses. Many vital supplies for the operation were brought up from Wunder Beach, where the 14th Engineer Battalion had arrived to join the Seabees working at the over-the-beach logistics facility.[10]

When the 1st Marines began its attack, two companies of Lt. Col. Victor A. Perry's 11th Marine Engineer Battalion and an attached bridge platoon from the 7th Marine Engineer Battalion followed, repairing Highway 9 as the advance continued. On the first day, the 11th Engineer Battalion cleared one mile of

[9] Sharp and Westmoreland, *Report*, pp. 164, 186; Tolson, *Airmobility*, pp. 169–78; Pearson, *War in the Northern Provinces*, pp. 81–89. See also Shulimson, et al., *U.S. Marines in Vietnam, 1968*, pp. 283–90; Operation PEGASUS, Historical Study 3–68, 31st Mil Hist Det, HQ Prov Corps Vietnam, May 1968 (hereafter cited as Operation PEGASUS), copy in Historians files, CMH.

[10] Westmoreland, *A Soldier Reports*, p. 347; Tolson, *Airmobility*, pp. 170–71; Capt. Moyers S. Shore II, USMC, *The Battle for Khe Sanh* (Washington, D.C.: History and Museums Division, Headquarters U.S. Marine Corps, 1969), pp. 133, 150; ORLL, 1 Feb–30 Apr 68, 8th Engr Bn, p. 2; Operation PEGASUS, pp. 6–7, 12; Fink, "Engineers Move to I Corps," p. 359; Shulimson, et al., *U.S. Marines in Vietnam, 1968*, p. 284.

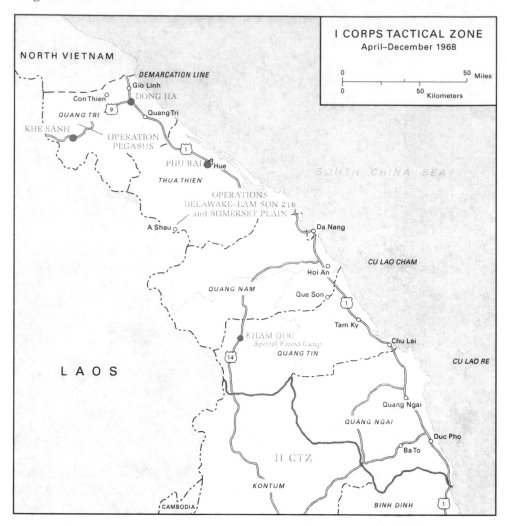

I CORPS TACTICAL ZONE
April–December 1968

NORTH VIETNAM

DEMARCATION LINE
Gio Linh
Con Thien
DONG HA
QUANG TRI
9
Quang Tri
KHE SANH
OPERATION
PEGASUS
1
PHU BAI
Hue
THUA THIEN
OPERATIONS
DELAWARE–LAM SON 216
and SOMERSET PLAIN
A Shau
Da Nang
SOUTH CHINA SEA
CU LAO CHAM
Hoi An
QUANG NAM
Que Son
1
Tam Ky
KHAM DUC
Special Forces Camp
Chu Lai
QUANG TIN
CU LAO RE
14
LAOS
Quang Ngai
QUANG NGAI
Duc Pho
II CTZ
Ba To
KONTUM
CAMBODIA
BINH DINH
1

MAP 24

road and built four bypasses. The following day the engineers, protected on the flanks by Marine infantry, cleared almost two miles of road and built two more bridges and two bypasses. By the seventh, they had gone over five miles and replaced or rebuilt five of the nine bridges to Khe Sanh. On 8 April, helicopters lifted a third company to another bridge site. Meanwhile, back at Dong Ha, the 11th's fourth line company and a bridge company assembled prefabricated bridge sections and bridge parts. The 1st Cavalry Division's CH–54s lifted the finished bridge sections and M4T6 floating bridges to three sites. The last bridge went in on 11 April, and the Class 50 road, capable of supporting the standard divisional loads of approximately fifty tons, officially opened at 1600. In eleven days, the 11th Engineer Battalion and attached units reconstructed over eight miles of road, filled craters, cleared landslides, repaired or replaced bridges, and constructed bypasses. It seemed fitting that the first vehicle to make the trip from

Dong Ha to Khe Sanh since the summer of 1967 should be the jeep of the 11th Battalion's command group.[11]

As the 11th Marine Engineer Battalion worked overland, elements of the Colonel Townsley's 8th Engineer Battalion accompanied the leapfrogging 1st Cavalry Division units to the besieged base. Company C provided direct support to the 3d Brigade's battalions air-assaulting into Landing Zones CATES, MIKE, and THOR. This support included hasty engineering tasks such as demolitions, field fortifications and wire, tactical operations center construction, and artillery positions. At Landing Zone CATES, the company established and maintained the brigade's water supply point. As the division approached Khe Sanh, Company C opened several landing zones, and furnished light bulldozers to speed up reopening the road link to the Marine base. Company B accompanied the 2d Brigade's assaults into Landing Zones TOM and WHARTON. Company A already had a platoon at Landing Zone STUD, where it built the division's tactical operations center and remained in general support. The remainder of the company opened Landing Zone SNAPPER for the 1st Brigade, supported the South Vietnamese 3d Airborne Task Force on Landing Zone SNAKE, and sent out a demolition squad and minesweep team to support the assault on Lang Vei.[12]

Although PEGASUS was a joint engineering effort, the need for more Army engineers in northern I Corps had become obvious. The commitment of the two Army divisions to the area had placed heavy demands on the available engineers to build combat and logistical bases and carry out road improvements. Both the 1st Cavalry Division's 8th Engineer and the 101st Airborne Division's 326th Engineer Battalions were small organizations consisting of only three instead of four letter companies. They not only lacked a bridge company authorized in infantry and armor divisions but also lacked the equipment to build new camps. For the time being Company D, 14th Engineer Battalion, at Camp Evans served as the 1st Cavalry Division's fourth engineer company, a temporary measure at best. On 15 March, Provisional Corps, Vietnam, informed the III Marine Amphibious Force that it urgently needed additional Army engineering support for the two divisions and requested another engineer battalion.[13]

The assignment fell to the 27th Engineer Combat Battalion, located at the time about sixty-two miles east of Saigon. The 27th Engineer Battalion had been carrying out base development work and operational support for the 11th Armored Cavalry Regiment at Blackhorse base camp and other bases nearby. During the Tet offensive, the battalion reopened roads cut by the Viet Cong and designated two line companies as a ready reaction force, freeing armored cavalry units for operations outside the base camp. On 24 March, a warning

---

[11] Tolson, *Airmobility*, pp. 173, 178; Pearson, *War in the Northern Provinces*, pp. 83–84, 88; Shulimson, et al., *U.S. Marines in Vietnam, 1968*, pp. 284, 589; Operation PEGASUS, pp. 29, 40, 46, 56.

[12] ORLL, 1 Feb–30 Apr 68, 8th Engr Bn, pp. 2–3.

[13] Operation PEGASUS, pp. 8–9; Fink, "Engineers Move to I Corps," p. 359; ORLLs, 1 Feb–30 Apr 68, 14th Engr Bn, p. 6, and 1 Aug–31 Oct 68, 14th Engr Bn, 31 Oct 68, p. 2, Historians files, CMH.

An Army CH–54 Flying Crane lowers a dry span bridge over a gap as the Marine 11th Engineer Battalion worked to reopen Highway 9 to Khe Sanh during Operation PEGASUS.

order dispatched through the 20th Engineer Brigade and 34th Engineer Group to Lt. Col. Kent C. Kelly, the battalion commander, ordered him to prepare the unit for the move north. The group quickly transferred projects to the 36th Engineer Construction Battalion. Colonel Kelly established a staging area at Bien Hoa, and on 4 April an advance party flew to Phu Bai and drove a short distance north to Gia Le to establish a base camp near Camp Eagle. In the next month, most of the 27th Battalion and the attached 591st Light Equipment Company moved by land, sea, and air, making this the longest of the engineer moves in Vietnam to date. Wheeled vehicles and cargo packed in Conex containers moved to Saigon for loading aboard a ship. A few days later, 325 troops flew aboard six C–130s to Da Nang to meet the ship and begin off-loading. After moving to a staging area at the 35th Engineer Battalion's camp north of Da Nang, the vehicles convoyed over the Hai Van Pass to Gia Le. The rest of the troops flew directly to Phu Bai airfield. Engineer equipment that would have been difficult to move over the road was placed aboard LSTs at Newport for shipment beyond the Hai Van Pass to the port at Tan My. This movement required six LST loads spread over a twenty-day period, with the last LST reaching Tan My on 30 April. Two of the 27th Battalion's units stayed behind. Company D remained at Phan Thiet and was placed under the 87th Engineer Construction Battalion's control. The 27th Land Clearing Team continued its mission in III Corps under the 168th Engineer Combat Battalion.[14]

The A Shau

The 27th Engineer Battalion arrived just in time to support a major operation, an airmobile assault against the North Vietnamese redoubt in the A Shau Valley. Located less than seven miles from the Laotian border and running northwest to southeast, the valley extends between two high and steeply sloped mountain ranges that climb to over three thousand feet. Among the few man-made features were three abandoned airfields spread along the valley floor. North Vietnamese troops had controlled the A Shau since March 1966 when they overran the Special Forces camp in the southern end and established a major staging area to infiltrate troops and supplies through Laos into Thua Thien Province. After reviewing detailed intelligence of the North Vietnamese bastion, General Tolson exclaimed: "This was his [the enemy's] Cam Ranh Bay, so to speak." The valley's peculiar location and topography expose it to both the northeast and southwest monsoons, and the brief interval between monsoons—mid-April to mid-May—prompted Westmoreland's planners to propose a quick strike. During the last days of Operation PEGASUS, Provisional Corps, Vietnam, completed final preparations for the assault, called Operation DELAWARE–LAM SON 216, committing elements of Tolson's 1st Cavalry Division and the 101st Airborne Division along with South Vietnamese forces.

[14] ORLL, 1 Feb–30 Apr 68, 27th Engr Bn, pp. 1–2, 8–10, 14–16; Ploger, *Army Engineers*, p. 151; Fink, "Engineers Move to I Corps," p. 359.

A corps reserve was established by moving the 196th Light Infantry Brigade to Camp Evans.[15]

The operation began on the morning of 19 April with preplanned air and artillery strikes. But almost at once, heavy antiaircraft fire met the 1st Cavalry Division's 3d Brigade, and poor weather played havoc with helicopter assaults and Air Force resupply missions. Just getting the desired firebases in place to support an assault on the old French airstrip at A Luoi took until the twenty-fourth. Meanwhile, to the east, the 1st Brigade, 101st Airborne Division, drove westward along Route 547, a narrow, winding, overgrown, unimproved road. On the second day, the cavalry division's 3d Brigade expanded operations with one battalion pushing southeast and another blocking Route 548 that entered the valley from Laos. The following day, 21 April, fighting increased as the allied units moved deeper into the sanctuary. As the weather temporarily improved, the 1st Brigade, 1st Cavalry Division, inserted a battalion into the central part of the A Shau, about a mile south of the A Luoi airstrip. Two more battalions followed, sweeping the surrounding area. During the rest of April, the buildup of forces and supplies continued around the airstrip, and by early May North Vietnamese resistance lessened. On 12 May, the 1st Cavalry Division and South Vietnamese airborne troops pushing westward along Routes 547 and 547A met some six miles east of A Luoi. As the operation progressed, American and South Vietnamese forces uncovered vast enemy caches. One find of interest to engineers was a large maintenance area and two Soviet-manufactured bulldozers used in building the logistical base and supply routes. By mid-May the rains of the southwest monsoon intensified, and the allies began to pull out. Operation DELAWARE–LAM SON 216 officially ended on 17 May.[16]

During the A Shau operation, the engineers played a vital role. Company C, 8th Engineer Battalion, moved with the cavalry division's 3d Brigade into landing zones and concentrated on setting up and improving firebases. Most of Company A accompanied the 1st Brigade's assault battalions into Landing Zones CECILE and STALLION (A Luoi airfield). The unit also made minesweeps, constructed defensive positions and tactical operations centers, cleared fields of fire, and prepared artillery positions. While the 1st Brigade continued its buildup, CH–54 helicopters brought in construction equipment broken down into sections to work on the airstrip. On 1 May, a team from Headquarters Company, augmented with equipment from the 326th Engineer Battalion, opened the 2,900-foot runway to C–7A Caribous and four days later to C–130s. Company A of the 326th supported the 1st Brigade, 101st Airborne

---

[15] Interv, Capt Joseph W. A. Whitehorne, 14th Mil Hist Det, with Maj Gen John J. Tolson, CG, 1st Cav Div, 27 May 68, p. 5 (quoted words), Historians files, CMH; Tolson, *Airmobility*, p. 182; Pearson, *War in the Northern Provinces*, pp. 3–4, 89; Spector, *After Tet*, p. 138. For more on the planning of the operation, see MFR, Brig Gen John R. Chaisson, Director, MACV Combat Operations Center, 17 Apr 68, sub: Report of Visit by COMUSMACV to HQ, PCV [Provisional Corps, Vietnam] 14 Apr 68, Westmoreland History files, CMH.

[16] Sharp and Westmoreland, *Report*, pp. 165, 187; Tolson, *Airmobility*, pp. 184–92; Pearson, *War in the Northern Provinces*, pp. 89–92; Bowers, *Tactical Airlift*, p. 333; Spector, *After Tet*, pp. 139–40.

Division, and concentrated on clearing landing zones and constructing fire-bases. Prior to the operation elements from the 27th Engineer Battalion, still in the process of moving from III Corps, cleared and opened Route 547 from Hue to the valley. On 14 April, Company A, 27th Engineer Battalion, started roadwork to Firebase BIRMINGHAM, approximately twenty miles east of the valley. Within a few days, a platoon from the 27th Battalion's Company B reached Firebase BASTOGNE to work on the defenses and the road to Firebase BIRMINGHAM. Another platoon moved farther west to the 101st Airborne Division's Firebase VEGHEL to support the South Vietnamese airborne task force. During the operation, the 591st Light Equipment Company organized an ad hoc land-clearing platoon of ten D7E bulldozers mounting Rome plows to clear vegetation out to three hundred yards along Route 547. Seabee units in I Corps provided additional equipment such as bulldozers and maintenance troops.[17]

During the operation, Army engineers noted some disturbing trends. In several cases, the 8th Engineer Battalion's minesweep teams were blown up by command-detonated mines. Security elements were alerted to get at least one hundred yards to either side and forward of the minesweep team and to use grappling hooks along the ditches to uncover the connecting wires. To make detection more difficult and dangerous, the enemy was also burying mines deeper than normal while placing scrap metal just above the mine. The 27th Engineer Battalion warned its mine-detector operators to keep checking suspected areas for signals that might reveal real mines.[18]

During DELAWARE–LAM SON 216, the abysmal weather affected air operations, especially the efforts to keep A Luoi airstrip open. The operation did not call for major units to remain in the area after the operation, and the provisional corps overruled opening Route 547 all the way into the A Shau Valley. Instead, troops and supplies arrived by air. The 8th Engineer Battalion should have reached A Luoi on the second day to begin rebuilding the airstrip, but the poor weather and intense antiaircraft fire delayed its arrival ten days to 29 April. Meanwhile, supplies were delivered by helicopters and C–130 airdrops until the engineers opened the airstrip to the first C–130 on 4 May. Heavy rainstorms turned the red dust of A Luoi into mud, and the airstrip remained closed from midday 7 May until the morning of the ninth. Withdrawal of troops began on 10 May, but heavy rains the following day again closed the strip to fixed-wing landings. Since the airstrip did not have ample time to dry out, helicopters had to be used to lift out the remaining supplies and troops as the operation ended.[19]

---

[17] ORLLs, 1 Feb–30 Apr 68, 8th Engr Bn, 4 May 68, pp. 2 4, and 1 May–31 Jul 68, 8th Engr Bn, 1 Aug 68, pp. 1–4, Historians files, CMH; Bowers, *Tactical Airlift*, p. 340; ORLL, 1 Feb–30 Apr 68, 27th Engr Bn, pp. 11–12, and incl. 2, ORLL, 1 Feb–30 Apr 68, 591st Engr Light Equip Co, pp. 2, 7–9; Ploger, *Army Engineers*, pp. 151–52; Fink, "Engineers Move to I Corps," p. 359; Tolson, *Airmobility*, pp. 187, 189–90.

[18] ORLL, 1 Feb–30 Apr 68, 8th Engr Bn, pp. 4–5.

[19] Interv, Whitehorne with Tolson, 27 May 68, p. 6; Bowers, *Tactical Airlift*, pp. 332–33, 339–41; Tolson, *Airmobility*, p. 190; Stanton, *Rise and Fall of an American Army*, pp. 262–63.

Allied officers declared PEGASUS and DELAWARE–LAM SON 216 tactical and strategic successes, but it did not take long for North Vietnamese forces to return to Khe Sanh and the A Shau. Though no longer cut off or threatened with imminent attack, Khe Sanh remained under siege. North Vietnamese artillery, dug deep into the mountains in nearby Laos, continued to bombard the base. Badly mauled North Vietnamese units withdrew across the border to regroup and refit, and by early May enemy troops returned in large numbers. Operation DELAWARE–LAM SON 216 initially seemed worth the cost. The operation resulted in an impressive list of captured ammunition, weapons, and equipment; disrupted future enemy operations; and effectively showed that he had no sanctuaries. Still, truck sightings in the valley less than a week after the allied withdrawal confirmed that the interdiction would not be long-lived. After PEGASUS, senior U.S. Army and Marine Corps commanders in I Corps continued to declare the Khe Sanh base a liability. They recommended that the base be abandoned and destroyed and the area defended with airmobile troops from Landing Zone STUD. The marines had long advocated quitting Khe Sanh and employing a mobile defense along the Demilitarized Zone. Westmoreland agreed in principle, but he disagreed with the recommended timing and deferred the decision to his successor, General Abrams. Preparations already under way to close the base were canceled, and the marines dug in again. On 15 April, fresh Marine battalions and two battalions from the 1st Cavalry Division launched a new operation, SCOTLAND II. In the ten weeks that followed PEGASUS, American troops in or near Khe Sanh suffered more than twice the casualties reported during the siege. In June, Abrams became theater commander, and he quickly decided to close the combat base. With shells continuing to rain down on the base, the marines completed the evacuation on the evening of 5 July. The decision to leave Khe Sanh caused some bewilderment in the United States. Abrams justified this move by pointing out that not tying down troops to specific terrain would better the chances to counter future threats.[20]

Two More Enemy Highpoints

Hanoi's actions suggested a renewed offensive, and it began on 5 May. Its timing coincided with the opening of the long-heralded Paris peace talks after President Johnson ordered a halt to bombing North Vietnam above the 20th Parallel. This offensive, known by American troops as the "Mini-Tet," turned out to be a reduced version of the Tet offensive. At the same time, it showed the conferees in Paris that Hanoi would continue to fight. Except at Saigon and a few other places, the May offensive featured rocket and mortar attacks against towns, cities, and U.S. installations. Attacks on bridges and airfields were usually repulsed. In northern I Corps, the allied forces anticipated the attacks and blocked enemy advances toward Dong Ha. After a series of bloody engagements in the eastern Demilitarized Zone,

[20] Interv, Whitehorne with Tolson, 27 May 68, p. 5; Bowers, *Tactical Airlift*, pp. 341–42; Tolson, *Airmobility*, p. 192; Spector, *After Tet*, pp. 128–29, 140, 228–31.

the planned attack on Hue never got under way. Strong North Vietnamese forces did succeed in taking the Special Forces camp at Kham Duc, located about ninety miles southwest of Da Nang near the Laotian border. In the area around Pleiku and Kontum, preemptive attacks by the 4th Division, elements of the 173d Airborne Brigade, and the 3d Brigade, 101st Airborne Division, plus heavy B 52 strikes, forced the enemy to cancel attacks and withdraw into Cambodia. The main attack took place in Saigon. American and South Vietnamese troops on the approaches to the capital intercepted and destroyed many small guerrilla bands. Those that got into the city holed up in the Cholon district and fought stubbornly, often setting fires in the highly flammable shanty neighborhoods. This gave the impression of a city under siege and was designed to embarrass the government and influence the talks in Paris. Allied efforts to rout the enemy created more destruction. By mid-May, the allied forces had rooted the Communists out of most of their strongholds, ending the worst of the street fighting. Rocket and mortar attacks continued, however, and renewed attacks in Cholon and near the Phu Tho racetrack took place later that month.[21]

Army engineers witnessed the only real North Vietnamese battlefield success—Kham Duc. Kham Duc and its satellite camp at Ngoc Tavik, five miles to the south, served as the last remaining Special Forces camps along the Laotian border in I Corps after the fall of Lang Vei. The camp sat astride Highway 14, which also served as an avenue for the North Vietnamese coming down the roads and tracks of the Ho Chi Minh Trail just across the border. Although the two outposts could not completely block the infiltration into South Vietnam, they kept the enemy's activities under observation and frequently hindered his movements. Placed in one of the most rugged border regions in Vietnam only ten miles from the Laotian border, the Kham Duc camp, village, and airfield sat in a milewide bowl surrounded by hills rising abruptly to heights of two thousand feet. Earlier in January when General Westmoreland canceled the first of his projected YORK operations, he had still wanted the airfield upgraded to handle sustained C–130 traffic and pads built for a radio navigation system. These tasks were passed down through command and engineer channels to Company A, 70th Engineer Battalion, 937th Engineer Group, at Pleiku. To meet the 1 April completion date, the 70th Engineer Battalion made plans to move the company by air. Delivering the unit's equipment and construction material, mostly M8A1 matting, would have required some forty C–130 sorties. Due to the scarcity of airlift at the time, the Air Force balked at handling the requirement until early April. Concerned that increasing rains during April would create construction problems, group headquarters tried to delay the project until the next dry season. The request was denied, and on 9 April the first platoon flew to Kham Duc and moved to a nearby abandoned

[21] Sharp and Westmoreland, *Report*, pp. 166–68; Westmoreland, *A Soldier Reports*, pp. 359–61; MACV History, 1968, vol. 1, pp. 132–33, 393. For more on the fighting during this period, see ORLLs, 1 May–31 Jul 68, II FFV, pp. 22–35, and 1 May–31 Jul 68, I FFV, 15 Aug 68, pp. 13–19; Starry, *Mounted Combat*, pp. 129–31; Lung, *General Offensives of 1968–69*, pp. 91–106; Spector, *After Tet*, pp. 142–83; Keith William Nolan, *The Magnificent Bastards: The Joint Army-Marine Defense of Dong Ha, 1968* (Novato, Calif.: Presidio Press, 1994).

ENGINEER
CAMP

An aerial view of the Kham Duc airfield with engineer camp to the left of the airfield

camp located on higher ground adjacent to the airfield. Work was well under way by the time the last element arrived on 15 April. Still, it would be several days before the company's two D7E bulldozers, each too large for a C–130, arrived aboard C–124 Globemasters.[22]

Company A made good progress until 10 May, when the North Vietnamese launched their attack. First Lt. Daniel W. Waldo Jr., former executive officer and now company commander, assigned tasks to each of the three line platoons. The 2d Platoon developed a sand pit and access road, and the 1st Platoon started on the pads, which were leveled and waterproofed with a coating of RC–3, an asphaltic cutback material. This was covered by a one-inch layer of sand and topped off with M8A1 matting. With the arrival of the first bulldozer on 18 April, the 1st Platoon began clearing the new parking apron area. Meanwhile, the 3d Platoon worked on the runway and existing parking apron, patching and sealing the many holes and improving drainage. After the

[22] Bowers, *Tactical Airlift*, p. 343; Spector, *After Tet*, p. 166; AAR/Interv, Upgrading Kham Duc Airfield (Opn SANTA BARBARA), 26th Mil Hist Det, 24 Jul 68, pp. 1–4, and incl. 2, Interv, Maj Paul B. Webber, 26th Mil Hist Det, with Capt Daniel W. Waldo Jr., CO, Company A, 70th Engr Bn, 24 Jul 68, VNIT 214, pp. 1–2, 4, CMH.

first mortar rounds struck the base at 0630, the engineers resumed work on the airfield. Another attack at 1000 brought all work to a halt.[23]

Within a few days, Kham Duc was hastily evacuated. Until then, Company A helped defend the base and tried to keep the runway open. During the second mortar attack, one soldier was wounded while repairing a telephone line between the engineer and the Special Forces camps. Following this attack, the 3d Platoon spotted North Vietnamese soldiers on the southeast side of the airfield and took them under fire. An intense firefight followed, and three more engineers were wounded. After one and one-half hours, the platoon withdrew to its camp. A major ground attack now appeared imminent, and later that morning an infantry battalion task force from the Americal Division began arriving by air and assumed command at Kham Duc. Company A spent the rest of the day clearing fields of fire and improving defensive positions. Earlier in the day, Ngoc Tavak also came under heavy attack, and the defenders were forced to abandon the camp and withdraw to Kham Duc before darkness. Scattered mortar fire rained down on Kham Duc on 11 May as the last of the reinforcements arrived. By then U.S. commanders had second thoughts about another protracted battle, and they recommended vacating the camp. After reviewing the situation that evening, General Westmoreland approved the withdrawal. Kham Duc, Westmoreland later wrote, "had none of the importance of or defensive potential of a Khe Sanh, and I ordered evacuation." Just before midnight, the 70th Engineer Battalion ordered Lieutenant Waldo to get out with as much equipment as possible, but because of intense enemy fire early the next day this was easier said than done. Army and Marine helicopters managed a few pickups during the morning, but a downed CH–47 blocked the runway. Since the two bulldozers had been disassembled for evacuation, the engineers used a bucket loader to move the damaged Chinook. Fuel leaking from the burning helicopter, however, set the bucket loader ablaze, and the engineers had to abandon it. Frantically, they reassembled a bulldozer and finally moved the helicopter, permitting the landing of C–123s and C–130s. When one C–130 took off down the cratered and shrapnel-littered runway, mortars burst on all sides flattening a tire. The crew aborted the takeoff, off-loaded the passengers, and with the help of engineer maintenance troops cut away the ruined tire. With fuel streaming from holes in the wings, the damaged and nearly empty aircraft managed to take off. At 1100, a C–123 took off with forty-six engineers, and CH–47s carried out the remaining sixty-four between 1100 and 1500. That afternoon, two C–130s were shot down within minutes, but helicopters and transports, protected by close-in air strikes, completed the evacuation by 1600. Following the evacuation, air strikes demolished the remains of the camp, including Company A's equipment and construction materials.[24]

[23] AAR/Interv, Upgrading Kham Duc Airfield, pp. 4–6, and incl. 2, Interv, Webber with Waldo, 24 Jul 68, pp. 1–2, 6; Bowers, *Tactical Airlift*, p. 343.

[24] Westmoreland, *A Soldier Reports*, p. 360 (quotation); Bowers, *Tactical Airlift*, pp. 343–47; Spector, *After Tet*, pp. 166–75; AAR/Interv, Upgrading Kham Duc Airfield, pp. 4–11, and incl. 2, Interv, Webber with Waldo, 24 Jul 68, pp. 5–10.

Enemy mortar attack at Kham Duc as shown from the engineer camp

The forced evacuation of the engineers at Kham Duc greatly affected Company A. Those men aboard the C–123 were flown to Da Nang, and those carried out by helicopters were dropped off at Landing Zone Ross where other helicopters transported them to Landing Zone Baldy. Within a few days, the engineers were flown to their home base at Pleiku. Thirty-one men were wounded, almost a third of the engineers at Kham Duc. None died. Company A lost almost all its equipment and some weapons and communications gear, including both D7E bulldozers, eight 5-ton dump trucks, two bucket loaders, one grader, other assorted equipment, and all squad and platoon tool kits. Construction material left behind and assumed destroyed by the Air Force included 432 bundles of M8A1 airfield matting, 360 drums of asphalt, and 1,200 bags of cement. These were approved as combat losses, but by the end of June the unit still lacked its replacement bulldozers, dump trucks, and grader. As a result, the unit restricted its work to construction tasks in the Pleiku area.[25]

[25] Quarterly Hist Rpt, 1 Apr–30 Jun 68, MACDC, p. II-9, Historians files, CMH; AAR/ Interv, Upgrading Kham Duc Airfield, pp. 9–11, and incl. 2, Interv, Webber with Waldo, 24 Jul 68, pp. 11–12.

The May attacks in the Saigon area proved more destructive than Tet. Some 15,000 homes were destroyed creating nearly 104,000 new refugees. With the monsoon on the horizon, the South Vietnamese government began a crash reconstruction program. Fortunately, the government's Central Recovery Committee established during the Tet offensive was still carrying out Project RECOVERY, and was able to move quickly to erect temporary housing. Media claims that allied firepower had caused much of the damage prompted General Westmoreland to lend assistance. On 13 May, he ordered General Rollins, the director of construction, to organize a joint engineer task force of U.S. Army, Navy, and Air Force engineers to help build some 1,500 temporary family units. Westmoreland also endorsed naming the cooperative effort Operation DONG TAM, the same Vietnamese term used for the delta base, literally meaning "united hearts and minds." Under Rollins' direction, the participating organizations—the Directorate of Construction, the Navy's Officer in Charge of Construction, and the three services' engineer units—formed Task Force DONG TAM. The South Vietnamese chief of engineers, Col. Nguyen Chan, dispatched two liaison officers to the task force's headquarters, and the Navy's Officer in Charge of Construction provided a civilian draftsman and used an architectural engineering contractor to design the housing areas. The United States Agency for International Development provided cement and roofing from its stocks, and the U.S. Army furnished the rest of the materials, which the agency later replenished. All work came under the general control of the Central Recovery Committee.[26]

Within a few days Task Force DONG TAM began operations, which lasted through the summer. Since the 46th Engineer Construction Battalion at nearby Long Binh ran a prefabrication lumber yard, it became the U.S. Army's most logical representative unit. On 19 May, Lt. Col. Pendleton A. Jordan, who just assumed command of the battalion, deployed Capt. Theodore B. McCulloch's Company B and other elements to the Phu Tho racetrack in Saigon. McCulloch's task force, which eventually numbered 217 officers and men plus Vietnamese laborers, quickly established a bivouac and set up a prefabrication yard. Simultaneously, the South Vietnamese Army's 301st Engineer Combat Battalion based in the Saigon area and the Ministry of Public Works began clearing rubble and debris. Soon a team from the Air Force's 823d Civil Engineering Squadron, Red Horse, from Tan Son Nhut Air Base arrived to help the 301st Engineer Battalion construct housing. On 23 May, elements of the 46th Engineer Battalion began site work in another part of the city. A week later, Naval Mobile Construction Battalion 58 in Da Nang deployed a detachment of fifty-two Seabees to a third site in metropolitan Saigon. Enemy attacks in late May and coordination problems caused by the hasty bringing together of the diverse workforce initially hampered the operation.

[26] MACV History, 1968, vol. 1, pp. 540, 542, vol. 2, pp. 675–76; Quarterly Hist Rpt, 1 Apr–30 Jun 68, MACDC, p. V-1; MFR, Brig Gen Andrew P. Rollins, 17 May 68, sub: Saigon Emergency Housing, Westmoreland History files, CMH; AAR/Interv, Opn DONG TAM, 26th Mil Hist Det, 12 Sep 68, with enclosed AAR, Task Force HELPER under Opn DONG TAM, 26th Mil Hist Det, 10 Sep 68, p. 1, and incl. 1, Interv, 1st Lt Raymond F. Bullock, 26th Mil Hist Det, with Maj Gen Andrew P. Rollins, MACV Dir of Const, 31 Aug 68, VNIT 253, pp. 1–2, CMH.

After two weeks on the job, the original completion date was moved back two weeks to the end of July. Plans called for 20-by-60-foot standard wooden barracks shells divided into five family units. Each unit consisted of 240 square feet of living space under corrugated metal roofing, lightly topped mortar or wooden floors, and one ceiling light fixture and outlet in areas provided with electrical power. Shared toilet facilities were built at most of the sites. By late June, the task force had completed over 300 housing units, with the 46th Engineer Battalion completing 120. As work progressed, additional tasks included road networks, drainage systems, running water, and electricity for each dwelling, which extended the project another four weeks. In early August, the Construction Directorate decided to keep the 46th Engineer Battalion's prefabrication yard open for another five weeks so that the Vietnamese could build 1,200 more family units. By early September, the project finally neared completion. The 46th Engineer Battalion alone had cleared 141,000 square feet of area; removed over 300 truckloads of rubble; and used more than 4 million board feet of lumber, 10,000 bags of cement, and 3,500 truckloads of gravel, rock, sand, and laterite fill. A week later, the battalion began restoration work at the Phu Tho racetrack.[27]

Though it appeared Operation DONG TAM achieved its intended purpose, the engineers had some misgivings about how they had gotten the job done. General Rollins noted that projects of this type were "not a very efficient way of using engineer troops." He believed that the operation would have been better organized under one service. Since Westmoreland considered DONG TAM a one-time venture, Rollins concluded the political goal was accomplished, and he hoped the South Vietnamese engineers would do more of this kind of work. Colonel Jordan agreed, noting that his battalion happened to be the most accessible Army unit in the area. He reckoned a single reinforced engineer battalion "would have done the job much more efficiently." There were other problems. The high level interest in the success of the housing project prompted excessive supervision by higher headquarters. Company B's junior officers, under pressure to get the job done, initially overlooked the importance of equipment maintenance. The company's deadline rate soon far exceeded the rates in the rest of the battalion. To make matters worse, the arrival of the rainy season caused flooding in the lower bivouac area, and many trucks and even bulldozers got stuck in the mud. Jordan quickly ordered improvements to the maintenance area and more mechanics to fix the equipment. Still, construction deadlines were not met because of poor estimates, and unstable soil conditions were made worse by the rains. Army engineers also faced a hostile populace during the rubble cleanup. The people apparently

[27] MACV History, 1968, vol. 1, p. 676; Quarterly Hist Rpt, 1 Apr–30 Jun 68, MACDC, pp. V-1 to V-2; AAR/Interv, Opn DONG TAM, 26th Mil Hist Det, with enclosed AAR, Task Force HELPER under Opn DONG TAM, 26th Mil Hist Det, pp. 2–4, and incl. 1, Interv, Bullock with Rollins, 31 Aug 68, pp. 2–4, incl. 2, Interv, 1st Lt Raymond F. Bullock, 26th Mil Hist Det, with Lt Col Pendleton A. Jordan, CO, 46th Engr Bn, 27 Aug 68, pp. 1–2, and incl. 5, Briefing on Opn DONG TAM for VIPs, n.d., p. 1, all in VNIT 253, CMH; Msg, COMUSMACV MACDC 22618 to CG, USARV, 27 Aug 68, sub: Opn DONG TAM, box 1, 71A/0354, RG 334, NARA; MACDC Fact Sheet, 6 Sep 68, sub: Opn DONG TAM, box 1, 71A/0354, RG 334, NARA.

The DONG TAM *project provided emergency housing for Vietnamese civilians such as this unit under construction in Saigon.*

did not expect any help and were initially suspicious. As work neared completion, the mood changed and many inhabitants welcomed the Americans with smiles, hand waving, and military salutes.[28]

In mid-August, Hanoi launched the last and weakest of its three offensives in 1968. Mortar and rocket attacks on populated areas and attacks on remote outposts were typical. In I Corps, the Americal Division's armored cavalry squadron blocked North Vietnamese forces heading toward Tam Ky City. In II Corps, the enemy avoided contact with two exceptions. One attack took place against Duc Lap district headquarters near the Cambodian border. Another struck the Dak Seang Special Forces camp farther north along the Laotian border. In III Corps, the U.S. 25th Division fought off enemy attacks at Firebase BUELL, two miles northeast of Tay Ninh City, and a night defensive position west of Dau Tieng. Other isolated attacks included an unsuccessful assault on

[28] AAR/Interv, Opn DONG TAM, 26th Mil Hist Det, with enclosed AAR, Task Force HELPER under Opn DONG TAM, 26th Mil Hist Det, pp. 4–5, and incl. 1, Interv, Bullock with Rollins, 31 Aug 68, pp. 2–4, incl. 2, Interv, Bullock with Jordan, 27 Aug 68, pp. 1–4, incl. 3, Interv 1st Lt Raymond F. Bullock, 26th Mil Hist Det, with Capt Theodore B. McCulloch, CO, Company B, 46th Engr Bn, 25 Aug 68, VNIT 253, pp. 1–4, CMH. VNIT 253 also contains photographs, site drawings, and a map.

the Duc Lap Special Forces camp and a damaging ambush on a supply convoy traveling between Long Binh and Tay Ninh. During the attack on the convoy, an engineer squad from the 65th Engineer Battalion and escorting mechanized infantry returning from a minesweeping operation fought off the attackers until relieved by an armored cavalry troop. The battered convoy reached Tay Ninh, but some two dozen ruined vehicles littered the road.[29]

August also marked the arrival of the 1st Brigade, 5th Infantry Division (Mechanized), the last major U.S. tactical unit sent to Vietnam. The mechanized brigade consisted of one tank battalion, which came with two armored vehicle launched bridges; two infantry battalions (one of them mounted on armored personnel carriers); a self-propelled artillery battalion; an armored cavalry troop; and an engineer company (Company A, 7th Engineer Battalion). In a short time, the brigade, which deployed along the Demilitarized Zone under 3d Marine Division control, surprised the North Vietnamese who had never battled true armor formations in this area.[30]

Finally, two U.S. Army reorganizations in midsummer created a new corps headquarters in I Corps and a second airmobile division. On 15 August, the Provisional Corps, Vietnam, became the XXIV Corps, a headquarters that previously saw combat service in the Pacific during World War II. Lt. Gen. Richard G. Stilwell, who had replaced General Rosson in July, assumed command of the corps and operational control over the 1st Cavalry, 101st Airborne, and 3d Marine Divisions and the 1st Brigade, 5th Division. Like its predecessor, XXIV Corps remained under the operational control of the III Marine Amphibious Force. The Department of the Army had considered converting the 101st Airborne Division to an airmobile configuration before the division deployed. Conversion began on 1 July. A year would pass before the full complement of helicopters was available and support facilities at Camp Eagle were expanded.[31]

Hanoi's offensives did not deter the allies from trying to regain the initiative through a series of operations aimed at supply trails, caches, and way stations. Along the Demilitarized Zone, U.S. Marines, no longer tied down to fixed bases, expanded airmobile operations. They followed the Army's lead in setting up mutually supporting firebases atop jungle peaks as the infantry fanned out below searching for trails and supplies. In early August, allied forces returned to the A Shau Valley. A brigade-size task force from the 101st Airborne Division and elements of the South Vietnamese 1st Infantry Division were airlifted into the valley in Operation SOMERSET PLAIN. Along with finding and destroying caches and base camps, the invading troops this time left the area littered with minefields. These obstacles hardly daunted the North Vietnamese, however, and they quickly moved back from neighboring Laos. After withdrawing from the A Shau, the 101st carried out an integrated airmobile, naval, and ground cordon

---

[29] Starry, *Mounted Combat*, pp. 131–36; Lung, *General Offensives of 1968–69*, pp. 109–12; Spector, *After Tet*, pp. 235–40; Stanton, *Rise and Fall of an American Army*, p. 276.

[30] Starry, *Mounted Combat*, pp. 139–41.

[31] MACV History, 1968, vol. 1, p. 219, 239–42; Tolson, *Airmobility*, pp. 195–97; Msg, CG, 101st Abn Div, to CG, USARV, 12 Aug 68, sub: Construction Facilities at Camp Eagle, box 1, 71A/0354, RG 334, NARA.

Mountaintop firebase of the 101st Airborne Division in the A Shau Valley during Operation SOMERSET PLAIN

operation on the enemy sanctuary at Vinh Loc Island, east of Hue. II Corps saw a continuation of reconnaissance and reconnaissance in force operations. Around Saigon, a large-scale campaign, Operation TOAN THANG ("Complete Victory"), started in April aimed at enemy units in the area. In the delta, the U.S. 9th Division fought several battles, destroying a main force battalion southwest of Can Duoc near the III/IV Corps border. To support the effort in the delta, a study done by Army and Navy headquarters in Vietnam concluded that additional units of the 9th Division could be stationed at Dong Tam without further dredging. With a modest addition of naval craft to the river assault squadrons, an expanded force could support two brigades afloat and one at Dong Tam. By midyear, planning was under way to reorganize the 9th Division to a mobile river configuration and move it to IV Corps. On 25 July, the division headquarters moved from Bearcat to a new base camp at Dong Tam.[32]

During Operation SOMERSET PLAIN, which took place from 3 to 19 August, elements of the 101st Airborne Division's 326th Engineer Battalion and the 45th Group's 27th Engineer Battalion furnished engineering support. Expansion of Firebase BIRMINGHAM as a forward supply point included helicopter rearming and refueling points and more pads for troop and cargo helicopters. Again, the 27th Engineer Battalion kept the main supply route, Route 547, to BIRMINGHAM open and supported the divisional engineers. Four

[32] MACV History, 1968, vol. 1, pp. 242–43, 392–94; ORLLs, 1 Aug–31 Oct 68, XXIV Corps, 15 Nov 68, p. 15, Historians files, CMH, and 1 Aug–31 Oct 68, I FFV, pp. 11–15; Spector, *After Tet*, pp. 231–35; Fulton, *Riverine Operations*, pp. 168–69.

new firebases covering the five infantry battalions in their sweeps were built. While paratroopers provided security, work crews from Company A of the 326th alighted from helicopters and cleared the landing zones with chain saws and demolitions. When there was enough room, CH–47 helicopters brought in small bulldozers to level the sites. Usually within two days, the engineers had enlarged the area enough to allow a CH–54 to deliver a larger D5A bulldozer. The altitude at Firebase EAGLE'S NEST, however, was too great to airlift that bulldozer, and the engineers borrowed a smaller T6 bulldozer from the Seabees. The new firebases included artillery positions, bunkers for command posts, fire direction centers, communications equipment, ammunition storage, and passenger and cargo helipads. Steep terrain in several instances required the fastening of handholds so that the engineers could pull themselves up the slopes, proof of their determination to support the operation. Before departing the A Shau, the 326th Engineer Battalion placed minefields at three defiles along Route 548. Chemical delay fuzes were adapted for use with the M16 antipersonnel and M18 antitank mines and set to self-destruct in forty-five days. Electronic sensors were also placed in the minefields to detect any attempts to remove the mines.[33]

The move of additional 9th Division units and headquarters to Dong Tam caused a major shift of the 34th Engineer Group's units to the delta base. Previously the group had carried out operational support and construction tasks in all of IV Corps and southern III Corps, which included Long An Province, part of Gia Dinh Province, Phuoc Tuy Province, the Rung Sat Special Zone, and Vung Tau. The number of units attached to the Vung Tau–based headquarters usually stood at three construction battalions, one to two combat battalions, and an assortment of smaller units, altogether some 3,900 to 4,200 men. Of these units, elements of the 69th and 93d Construction Battalions were completing a 7,500-man cantonment at Dong Tam for the division's riverine brigade. In mid-June, the rest of the 93d Construction Battalion and the attached 67th Dump Truck Company and 702d Power Distribution Detachment moved from Long Thanh North. Companies B and D, 86th Combat Battalion, which had been working along Highway 4, reinforced the 93d's efforts to meet the added base camp requirements. In late August, more help arrived when the 36th Construction Battalion dispatched two construction platoons from Company B and one construction platoon from Company D. In September, the 113th Concrete and Paving Detachment arrived at Dong Tam to help the 93d Construction Battalion's concrete production.[34]

---

[33] Maj. Emmett Kelly, "Into the A Shau," in *A Distant Challenge: The US Infantryman in Vietnam, 1967–1972* (New York: Berkley Publishing, 1985), pp. 115–21; 45th Engr Gp (Const) Operational Support, 1 Jul–1 Nov 68, incl. 5, tab A–2, pp. 1–2, OCE Liaison Officer Trip Rpt no. 13; ORLL, 1 Aug–31 Oct 68, XXIV Corps, p. 15.

[34] 34th Engr Gp (Const) Status of Principal Construction Projects, 15 Nov 67–15 Feb 68, incl. 5, tab B–1, pp. 2–3, OCE Liaison Officer Trip Rpt no. 11, 6 May 68; 34th Engr Gp (Const) Troop Construction, 1 Feb–1 Jul 68, incl. 2, tab E–1, pp. 3–4, OCE Liaison Officer Trip Rpt no. 12, 23 Sep 68, OCE Hist Ofc; ORLLs, 1 May–31 Jul 68, 34th Engr Gp, 1 Aug 68, pp. 1–2, 6, 13–14, 1 May–31 Jul 68, 93d Engr Bn, 14 Aug 68, pp. 2, 5, and 1 Aug–31 Oct 68, 93d Engr Bn, 15 Nov 68, pp. 1, 9, 11, all in Historians files, CMH.

Assessing Construction

Hanoi's offenses had caused some disruption to military construction, but the program steadily neared completion. American military and civilian engineers had in less than three years built an impressive array of operational, logistic, and support facilities needed during the buildup. Under the overall direction of MACV's Directorate of Construction, a carefully managed program combined the needs of the three services and allied forces. Between 1 January 1965 and 1 January 1968, the United States had provided $1.48 billion in Military Construction funds to build Army, Navy, and Air Force facilities in Vietnam. Comparing the funds available to the work done to this point showed the Military Construction Program to be 65 percent complete. Funding would reach $1.71 billion by year's end.[35]

As a rule, the Construction Directorate ensured that only minimum essential construction using austere standards consistent with operational and tactical requirements was accomplished. This included the use of DeLong piers and prefabricated buildings that could be salvaged when U.S. forces left the area. The military now had four major ports along with additional smaller ports and over-the-shore beach facilities, relieving congestion and ending the temporary disruption to civilian ports. There were now depots with over 11 million square feet of covered storage, over 5 million yards of open storage, 2.5 million cubic feet of cold storage, and hospitals providing 8,250 beds. Five major jet air bases had been built since 1965, increasing the total to eight. Over one hundred smaller fields, many with newly developed aluminum and steel matting, allowed transport aircraft to airlift material to dispersed forces. Many key roads had been reopened and restored, with increased efforts devoted to expanding road networks and improving secondary roads. As the construction program progressed, the task of maintaining, repairing, and operating the facilities also grew. Facilities engineering contractors—the Army's Pacific Architects and Engineers and the Navy's Philco-Ford—took on most of the repairs and utilities operations at the expanding bases.[36]

Meanwhile, despite General Palmer's admonition to tighten up the base construction program, overbuilding remained a concern. In May 1968, General Duke, shortly before turning over command of engineer troops in Vietnam, stated in his debriefing report that the number and size of base camps continued to be a problem and bore watching. Nevertheless, unauthorized self-help construction by troop units continued as commanders arrived and departed. Commanders and troops characteristically leaned toward improving living and working conditions, and they found ways such as bartering to obtain building materials.[37]

In May, Pacific Architects and Engineers marked its fifth anniversary as the U.S. Army's facilities engineering contractor in the Republic of Vietnam.

---

[35] MACV History, 1967, vol. 2, p. 843; ibid., 1968, vol. 2, p. 670.

[36] Sharp and Westmoreland, *Report*, pp. 257–64.

[37]Debriefing, Maj Gen Charles M. Duke, 14 May 68, pp. 5–6, Senior Officer Debriefing Program, DA, Historians files, CMH.

Manpower ceilings limited direct hiring, and the Army called upon Pacific Architects and Engineers to expand its organization as the buildup and pace of facilities construction increased. Approximately 24,750 employees (1,750 Americans, 4,000 third-country nationals, and 19,000 Vietnamese) manned seventy-two major company installations and an assortment of supply and equipment maintenance facilities. When Army units moved into I Corps in 1967, the firm expanded operations in areas not supported by the Navy. The sheer magnitude of its buildings and grounds gives an idea of the size of the facilities placed in its care. In mid-1968, the company maintained over 43 million square feet of building space and 11.5 million square yards of roads. It ran 1,128 generating plants ranging from small single units to a cluster of seven 1,000-kilowatt machines at MACV headquarters. Total monthly output neared 30 million kilowatt-hours and electrical distribution lines extended 650 miles. That summer, Pacific Architects and Engineers took over responsibility for the new central power plant at An Khe consisting of six 1,500-kilowatt generators, the first of several large complexes to increase and stabilize electricity at major bases. The company operated 137 water treatment facilities purifying some 300 million gallons a month. Among its six sewage treatment plants, the Long Binh facility included a huge sewage lagoon that processed 70,000 gallons a day. Other tasks included entomology services, refrigeration and air-conditioning maintenance, and trash collection. A fleet of 123 compactor trucks that arrived in late 1967 hauled away about 350,000 cubic yards of refuse every month and allowed the facilities engineering contractor to do away with most of its refuse collection subcontractors. With military units and subcontractors, the company disposed of 500,000 cubic yards of trash each month in sanitary fills. Forty-three ice-making plants produced 300 tons of potable ice each day such as the small but vital plant at Da Lat, which provided ice to preserve fresh vegetables delivered to troops throughout the country. By mid-1968, the company's fire department had grown to forty-one stations with a roster of nearly one thousand people, over a hundred pumpers and tankers, and a newly opened school in Vung Tau.[38]

The Army's power plant arrangements with Vinnell Corporation also expanded in the three years following the first contract. These contracts, awarded by Army Materiel Command, involved construction of power plants and distribution systems under the Military Construction, Army, program, plus operation and maintenance at seven locations. Cam Ranh Bay now had five T–2 power ships in operation, each with an output of 3,100 kilowatts. Power plants at Qui Nhon, Nha Trang, and Vung Tau contained two ships, each with an output of 4,300 kilowatts. Vinnell brought the last ship, which arrived March 1967, online on 6 September 1967. At Long Binh, work progressed on a large land-based power plant, which would include twenty 1,500-kilowatt diesel-driven generators when completed in late 1968. Land-based plants were also at Cam Ranh Bay, Phu Loi, and Bien Hoa, with more

---

[38] PA&E History, Jan–Jun 1968, pp. 1–6; Dunn, *Base Development*, p. 93. For more on Pacific Architects and Engineers' five years in Vietnam, see *PA&E News Vietnam*, 1 and 15 May 68, copy in Historians files, CMH.

Pacific Architects and Engineers' headquarters in Saigon

programmed. In November 1967, Army Materiel Command modified the contract to build twelve more land-based plants and distribution systems. Planning for Fiscal Year 1969 envisioned land-based power plants replacing the power ships. Vinnell also had contracts to run a field maintenance shop, a Class IV Engineer Material Yard, and stevedoring operations for the 1st Logistical Command at Cam Ranh Bay.[39]

With cutbacks in construction funds already happening, planners in Washington and Saigon examined the roles of engineering troops and contractors. The contracts awarded to RMK-BRJ, Pacific Architects and Engineers, and Vinnell had resulted in tailor-made forces to do a job. Although the contractors required considerable time to mobilize, they were, because of their tailored makeup, more efficient than troop units. On the other hand, only the cost of construction materials was charged to the troop projects. Decreases in construction funds in 1967 reduced RMK-BRJ's workforce to 18,000, and MACV planned to end the civilian construction effort in 1968. As the contractor phased out, troops would assume more of the construction workload. Fortunately, most of the major work that best suited the contractors' talents, such as technical expertise, special equipment, and organizing for specific tasks would be completed.[40]

---

[39] Dunn, *Base Development*, pp. 78–79; Debriefing, Duke, 14 May 68, pp. C-6 to C-7, Senior Officer Debriefing Program, DA, Historians files, CMH; "Floating and Land Based Power Plants," "Vinnell Makes History in Long Binh," "12 Additional Power Plants," *Vinnell* 11 (Spring 1969): 6–11; Procurement Support in Vietnam, 1966–1968, pp. 50–51. See also Thomas E. Spicknall, "Civilian Repairs and Utilities in the Combat Zone," *Military Engineer* 61 (March–April 1969): 77–80.

[40] JLRB, Monograph 6, *Construction*, p. 126; MACV History, 1967, vol. 2, pp. 841–42; ibid., 1968, vol. 2, p. 678.

Vinnell Corporation's central control room at the Long Binh power plant

These plans were modified in 1968, when MACV transferred projects assigned to military engineer units in the major enclaves to RMK-BRJ, freeing the troops to do projects in more remote areas. Troop to contractor transfers took place in areas where RMK-BRK had a mobilization capability. These included work sites in Saigon, Da Nang, and Cam Ranh Bay, and satellites in Chu Lai, Can Tho, Dong Tam, and Vung Tau. After further study that summer, MACV decided to obligate available funds before the end of the fiscal year and raise the contractor's workforce to 25,000, decreasing the backlog of construction from eleven to eight months. As of 7 December, RMK-BRJ had 23,804 people on its payroll (2,256 Americans, 1,766 third-country nationals, and 19,782 Vietnamese).[41]

During 1968, the buildup of engineer units reached its peak. Engineer troops comprised approximately 11 percent of all U.S. forces in Vietnam, and Army engineers exceeded all other branches except the infantry. Between July and December 1967, military engineer construction battalions had grown from fourteen to fifteen Army and from ten to eleven Seabee. This did not include a variety of smaller units and repairs and utilities specialists. Adding the eleven Army nondivisional combat battalions, three Fleet Marine Force battalions, and five Air Force Red Horse squadrons (but excluding the Army's eight divisional battalions and the companies with the separate brigades and the Marine Corps' two divisional battalions) the construction force had risen to forty-five battalions. Army engineers gained two more nondivisional combat battalions in

---

[41] MACV History, 1968, vol. 2, pp. 678–79; Quarterly Hist Rpts, 1 Jan–31 Mar 68, MACDC, p. V-18, 1 Apr–30 Jun 68, MACDC, p. V-19, and 1 Jul–30 Sep 68, MACDC, p. V-13, Historians files, CMH.

1968. Naval engineering forces, which reached twelve Seabee construction battalions that summer, would level off to ten battalions by the end of the year.[42]

At the same time, in order for U.S. Army, Vietnam, to stay under the current troop ceiling, Engineer Command had to share in troop cutbacks. Plans at midyear called for the fifteen construction battalions and six separate companies (one float bridge and five construction support) to lose 4,561 spaces. Each construction battalion would lose 204 military spaces and convert to a modified Type B organization, which could be augmented by local civilian employees. Several units were further reduced, brought to zero strength, or scheduled for inactivation. The 62d Construction Battalion at Long Binh lost its three line companies and became the controlling headquarters for three land-clearing companies. At Cam Ranh Bay, the 87th Engineer Construction Battalion was scheduled for inactivation on 31 March 1969. Also reduced to zero strength were a float bridge company, a pipeline company, a panel bridge company, a dump truck company, and two detachments. Although assigning infantrymen as engineer replacements made up some of the shortages, there were soon critical deficiencies in construction foremen, equipment operators, mechanics, electricians, and plumbers.[43]

Streamlining the Effort

Since the early days of the buildup, U.S. Army Engineer Command, Vietnam, had directed troop construction, and 1st Logistical Command facilities engineering. In turn, logistical command had charged its engineering staff and U.S. Army Procurement Agency, Vietnam, to carry out the facilities engineering mission and to administer the Pacific Architects and Engineers and Vinnell contracts. By early 1967, it had become evident that Pacific Architects and Engineers' rapid growth demanded better control. As a result, 1st Logistical Command established a Contract Operations Branch under its engineer at the firm's Contract Management Office in Saigon and increased the number of contracting officers' representatives within the staff engineers at the Saigon, Qui Nhon, and Cam Ranh Bay Support Commands and large installations. The Contract Management Office helped identify and resolve many problems while directing and analyzing Pacific Architects and Engineers' activities and operations. By that time, however, the contract had risen to $100 million a year. Army planners, accordingly, sought ever-greater control over soaring costs and the potential for overbuilding.[44]

[42] MACV History, 1967, vol. 2, pp. 842–43; ibid., 1968, vol. 2, p. 679; Sharp and Westmoreland, *Report*, p. 263; Tregaskis, *Building the Bases*, pp. 300–301.

[43] Rpt of Visit to Various Headquarters in South Vietnam, 11–31 Jul 1968, incl. 1, pp. 2–4, OCE Liaison Officer Trip Rpt no. 12; Rpt of Visit to Various Headquarters in South Vietnam, 1–21 Nov 68, incl. 3, pp. 7–9, OCE Liaison Officer Trip Rpt no. 13; Quarterly Hist Rpt, 1 Oct–31 Dec 68, MACDC, p. II-3, Historians files, CMH; ORLLs, 1 Nov 68–31 Jan 69, 18th Engr Bde, pp. 2–3, 1 Aug–31 Oct 68, 20th Engr Bde, pp. 3–4, 1 Nov 68–31 Jan 69, 20th Engr Bde, pp. 2–3, and 1 Nov 68–31 Jan 69, 35th Engr Gp, p. 5; FM 5–1, *Engineer Troop Organizations and Operations*, September 1965, pp. A-B-2, A-B-41.

[44] Dunn, *Base Development*, pp. 92; Procurement Support in Vietnam, 1966–1968, pp. 36, 38.

It seemed logical that one organization, such as Engineer Command, should combine those responsibilities, but the Army still had difficulty justifying an engineer command. After succeeding General Ploger in August 1967, General Duke continued to use the integrated staff that combined those of the command and the engineer section of headquarters U.S. Army, Vietnam. In practice, Duke found that he spent almost 80 percent of his time as USARV Engineer and 20 percent on Engineer Command matters.[45]

In September 1967, Duke took steps to develop a new organization. He seemed satisfied that the two engineer brigades had done an impressive job in accomplishing their combat support, road improvements, and construction missions in that order of priority. Still, he wanted to address Washington's concerns about his ability to account for Military Construction, Army, funds, and he wanted to do so with an organization that would not downgrade combat and operational support, require an additional base structure, or add troop spaces. After an exchange of ideas with Maj. Gen. Frederick J. Clarke, the deputy chief of engineers, Duke gained General Palmer's approval for what he believed to be the most workable solution. Thus, construction and facilities engineering functions were merged under a new organization, the U.S. Army Engineer Construction Agency, Vietnam (Provisional). U.S. Army, Vietnam, activated the new command on 15 March 1968, and on the twenty-ninth Brig. Gen. William T. Bradley took command. Basically, the new command served as a management organization to coordinate the construction and facilities engineering efforts carried out by engineer units and contractors. The Construction Agency assumed responsibilities for construction, base development planning, and the Army's part of the highway, railroad, bridge restoration, and revolutionary development support programs. The new command also served as a link with MACV's Directorate of Construction on matters dealing with the Army portion of the Military Construction Program. Troop projects were directed to the two brigades, and other construction projects were assigned to the Navy's Officer in Charge of Construction for RMK-BRJ. On 1 April, the 1st Logistical Command transferred its real estate functions to the Construction Agency. Three months later on 1 July, the agency took over the facilities engineering responsibilities, which included the management of the Pacific Architects and Engineers and Vinnell contracts and the supervision of the utilities detachments.[46]

[45] Rpt of Visit to Various Headquarters in Vietnam, 15–30 Sep 67, incl. 7, p. 1, 4, OCE Liaison Officer Trip Rpt no. 9.

[46] Ltr, Palmer to Gen Johnson, 14 Nov 67, Harold K. Johnson Papers, MHI; Ploger, *Army Engineers*, pp. 143–45; Dunn, *Base Development*, pp. 92–93; Debriefing, Duke, 14 May 68, p. 5, Senior Officer Debriefing Program, DA, Historians files, CMH; Interv, Maj Paul B. Webber, 26th Mil Hist Det, with Maj Gen Charles M. Duke, CG, Engr Cmd and USARV Engr, 14 May 68, VNIT 164, pp. 20–27, CMH; Quarterly Hist Rpt, 1 Apr–30 Jun 68, MACDC, p. II-4; Rpt of Visit to Various Headquarters in Vietnam, 22 Nov–13 Dec 67, incl. 7, p. 1, OCE Liaison Officer Trip Rpt no. 10; Rpt of Visit to Various Headquarters in Vietnam, 11–31 Mar 68, incl. 4, pp. 4–5, tabs L–N, OCE Liaison Officer Trip Rpt no. 11; Msg, Duke ARV 2130 to Maj Gen Frederick J. Clarke, Dep Ch Engr, 3 Dec 67, sub: Engineer Organization; Msg, Duke ARV 016 to Clarke, 3 Jan 68; Msg, Clarke GVP 015 to Duke, 9 Jan 68; Msg Duke ARV 061 to

The Construction Agency's organization closely paralleled Corps of Engineers districts and facilities engineering offices in the United States. Offices and divisions in the headquarters dealt with comptroller, administration, automatic data processing, engineering, real estate, supply, safety, public information, and real property functions. When the agency took over the facilities engineering function, it set up district engineer offices at Saigon (Southern), Cam Ranh Bay (Central), and Qui Nhon (Northern). In turn, the district engineers oversaw the installation engineers. For example, the Southern District Engineer's area of responsibility included the Cu Chi, Long Binh, Tan Son Nhut, Lai Khe, Vung Tau, Dong Tam, Can Tho, and Saigon port installations. Large installations like Long Binh controlled the nearby Bien Hoa and Long Thanh Army bases. This structure provided a vertical command channel from the agency through the district engineers to the installation engineers, who could now operate free from the whims of local commanders. As more Army troops moved into northern I Corps, the Northern District added Phu Bai, Da Nang, Camp Evans, and Camp Eagle to its list of bases. Eventually each installation engineer managed all Army-funded construction programs, facilities engineering contracts, and real estate matters in his area of responsibility. He also commanded the engineer detachments, mostly utilities, at a base or bases. Facilities engineering support for forty-five Integrated Wideband Communications Systems sites built and maintained by Page Communications Engineers would be added to the facilities engineering contract in Fiscal Year 1969. The Construction Agency also provided troops to operate and maintain large-capacity air conditioners and generators at each site. In its aim to control and improve technical performance, particularly Pacific Architects and Engineers' operations, the agency nearly tripled the number of supervisors from 73 to 212.[47]

Despite these developments, 1st Logistical Command's procurement agency retained administrative control of the facilities contracts. General Duke proposed giving the Construction Agency contracting officer authority for the contracts, but 1st Logistical Command insisted on keeping the old arrangement, and Department of the Army agreed. To make matters more complex, the U.S. Army Procurement Agency, Vietnam, not only came under the control of 1st Logistical Command, but its procurement chain went to U.S. Army, Japan. The construction and procurement agencies, accordingly, worked out a memorandum of understanding in which installation engineers

Clarke, 10 Jan 68; Msg, Clarke GVP 031 to Duke, 11 Jan 68; Msg, Duke ARV 126 to Clarke, 18 Jan 68, all messages in Duke Papers, CMH.

[47] Dunn, *Base Development*, p. 92; Interv, Webber with Duke, 14 May 68, pp. 35–38; Debriefing, Duke, 14 May 68, pp. 6, C-7, Senior Officer Debriefing Program, DA, Historians files, CMH; JLRB, Monograph 1, *Advanced Base Facilities Maintenance*, pp. 27–30; Rpt of Visit to Various Headquarters in South Vietnam, 11–31 Jul 68, incl. 1, p. 6, OCE Liaison Officer Trip Rpt no. 12. Trip report no. 12 also contains U.S. Army Engineer Construction Agency, Vietnam (USAECAV), organization charts, a chart depicting the flow of base development plans and construction requests, and Letter of Instruction from CG, Engr Troops, Vietnam, to CG, USAECAV, 31 Jun 68. See also USAECAV Organization and Functions Manual, 27 May 68, OCE Hist Ofc.

served as the key link to the two organizations. While the installation engineers now reported through command channels to the Construction Agency, they continued as the Procurement Agency's contracting officers' representatives evaluating the performance of Pacific Architects and Engineers.[48]

General Duke and the Army Procurement Agency seemed satisfied with this arrangement. After only two months, Duke claimed the new command had already proved its worth. Construction programming became more effective within a single command. "You have one agency," Duke stated in an interview, "that develops the requirements, that designs the facilities, that directs its construction and inspects it during the construction. . . . As far as I know, this has never been done before. And frankly, I think it's superb."[49]

The establishment of the Army Engineer Construction Agency, Vietnam, also resulted in a name change for Engineer Command and a broadening of its functions. Since Defense Department approval of an engineer command or an organization with command in its name seemed so doubtful, U.S. Army, Vietnam, inactivated the command and on 29 March activated Engineer Troops, Vietnam. Duke carried on as a commander and staff engineer. In the first role, he commanded three major subordinate organizations, the 18th and 20th Engineer Brigades and the Army Engineer Construction Agency, Vietnam. As the U.S. Army, Vietnam, engineer, he directed a staff charged with planning, developing policies, and setting construction priorities. His staff also continued to carry out mapping and engineer intelligence functions. The old Engineer Command had handled all personnel matters, including assigning all field grade officers by name. Through an agreement with the U.S. Army, Vietnam, G–1, Duke was still able to select his battalion and group commanders and play a role in the assignment of Corps of Engineers colonels.[50]

In May 1968, as General Duke prepared to leave Vietnam, he outlined major problems to be solved. Not only did Army engineers fail to gain contracting authority, but also they still had to contend with the existing cost-plus-fixed-fee contracts. He believed that control over the facilities engineering contract would have allowed the Army Engineer Construction Agency, Vietnam, to take over contracting authority for Military Construction, Army, construction when the large tri-service contract with RMK-BRJ no longer proved practical. Each service would then directly contract its own construction, but this did not happen. When comparing contractors, Duke rated Pacific Architects and Engineers' performance below that of RMK-BRJ and Vinnell. The facilities engineering contract, originally negotiated in 1963 and renewed annually, remained little changed, thus reducing incentives that would come from a cost-plus-award-fee contract. Although Pacific

[48] Dunn, *Base Development*, pp. 92–93; Interv, Webber with Duke, 14 May 68, pp. 17–20; Debriefing, Duke, 14 May 68, p. 6, Senior Officer Debriefing Program, DA, Historians files, CMH.

[49] Interv, Webber with Duke, 14 May 68, pp. 37–46 (quotation, pp. 37–38); Debriefing, Duke, 14 May 68, p. 5, Senior Officer Debriefing Program, DA, Historians files, CMH.

[50] Rpt of Visit to Various Headquarters in South Vietnam, 11–31 Mar 68, incl. 4, p. 4, and HQ USARV, General Order 1381, Engr Troops, Vietnam (Prov), 28 Mar 68, incl. 4, tab L, OCE Liaison Officer Trip Rpt no. 11; Interv, Webber with Duke, 14 May 68, pp. 28–32.

Architects and Engineers had to compete for the Qui Nhon area contract in 1968, the cost-plus-award-fee contract would not go into effect until 1970. Pacific Architects and Engineers' familiarity with operations in Vietnam gave the firm a distinct advantage, and it won the contract. Duke believed the competition resulted in better performance, especially in road maintenance.[51]

In a different area, although he viewed his relations with the Directorate of Construction and the Navy's Officer in Charge of Construction as good, some of their practices bothered him. He preferred that the Construction Directorate confine its activities to programming and reprogramming projects and funds and setting construction

General Parker

standards. Duke became concerned, however, when Construction Directorate's staff officers visiting projects and work sites returned to their offices and wrote disparaging inspection reports. The Navy construction office's handling of funds also came in for criticism. He was disturbed by the Navy's operation "in letting us see the behind the scenes manipulations of theirs insofar as cost is concerned." The Army's inspection system of cost-plus contracts, he claimed, was much more elaborate.[52]

General Duke's departure in May preceded the changeover of all senior Army engineers in Vietnam. On 21 July, Maj. Gen. David S. Parker took over as U.S. Army, Vietnam, Engineer and Commanding General, Engineer Troops, Vietnam (Provisional), formerly known as Engineer Command, Vietnam. In October, General Bradley, who temporarily replaced Duke for three months, replaced General Rollins as the MACV director of construction. Col. John H. Elder Jr. replaced General Roper at 18th Engineer Brigade in September. Roper, who had commanded the brigade since November 1967, then assumed command of the U.S. Army Engineer Construction Agency, Vietnam, during his remaining months in Vietnam. Col. Harold R. Parfitt took command of the 20th Engineer Brigade from General Chapman in November. (Both Elder and Parfitt were on the brigadier generals promotion list.) Parker concurred in the integrated engineer headquarters and the two brigades but had doubts about a separate construction agency. He saw some duplication and overlap

---

[51] Debriefing, Duke, 14 May 68, p. 6, Senior Officer Debriefing Program, DA, Historians files, CMH; Dunn, *Base Development*, p. 93.

[52] Interv, Webber with Duke, 14 May 68, pp. 38–48.

between the agency and his engineer staff sections and would have preferred to combine the two organizations. Since the construction agency had greatly improved the control and quality of the construction program and got better performance from the facilities engineering contractor, Parker decided that such a quick change would have generated too much turbulence. At some point in time, after American forces started to phase out, Parker visualized consolidating the two organizations.[53]

New Approaches

In the aftermath of Hanoi's offensives, both sides changed approaches and consolidated forces. After heavy troop, equipment, and supply losses, the North Vietnamese and Viet Cong retired to their sanctuaries. On the allied side, American troop strength continued to rise, and in-country moves took place. In September, the 101st Airborne Division's 3d Brigade, which had stayed behind in III Corps, rejoined the division in exchange for the 3d Brigade, 82d Airborne Division. In late October, the 1st Cavalry Division moved by sea and air to III Corps to respond to a growing threat in the south. The move, code-named Operation LIBERTY CANYON, involved redeploying the division nearly four hundred miles between the two fronts. By mid-November, the division had set up its headquarters at Phuoc Vinh and its support at Bien Hoa, and had begun operations along the Cambodian border. As for IV Corps, the 9th Division, which had moved its headquarters and a second brigade to Dong Tam, readied itself for the dry season. The campaign would be known as SPEEDY EXPRESS.[54]

The latest deployments caused some changes in the allied order of battle. In I Corps, III Marine Amphibious Force continued to direct U.S. ground operations of one Army corps and two divisions. North of the Hai Van Pass, the XXIV Corps had operational control of the 101st Airborne Division, 3d Marine Division, and the 1st Brigade, 5th Infantry Division. South of the pass to the I/II Corps border, III Marine Amphibious Force directed operations of the 1st Marine and Americal Divisions. Other forces in I Corps included two South Vietnamese infantry divisions and infantry regiment and a South Korean Marine brigade. I Field Force now had only two major tactical units, an infantry division and an airborne brigade, in II Corps. The 4th Infantry Division continued operations in the western highlands around the border regions. Most of the 173d Airborne Brigade remained in Binh Dinh Province with a battalion-size task force in the south. Two South Vietnamese and two South Korean Infantry divisions covered the rest of II Corps. In III Corps, II

[53] Debriefing, Maj Gen David S. Parker, USARV Engr and CG, Engr Troops, Vietnam, 14 Oct 69, pp. 7–8, Senior Officer Debriefing Program, DA, Historians files, CMH; Parker's Comments on OCE Manuscript, 14 Dec 77, pp. 1–6, Historians files, CMH; Debriefing, Brig Gen Willard Roper, CG, 18th Engr Bde, 7 Oct 68, p. 5, Senior Officer Debriefing Program, DA, Historians files, CMH; Debriefing, Brig Gen Curtis W. Chapman, CG, 20th Engr Bde, 30 Oct 68, p. 2, Senior Officer Debriefing Program, DA, Historians files, CMH.

[54] MACV History, 1968, vol. 1, pp. 401; Tolson, *Airmobility*, pp. 209–13; *Memoirs of the First Team*, p. 42.

Field Force controlled the U.S. 1st and 25th Infantry Divisions; 1st Cavalry Division; 3d Brigade, 9th Infantry Division; 3d Brigade, 82d Airborne Division; 199 Light Infantry Brigade; and the 11th Armored Cavalry Regiment. Allied forces included most of a Thai infantry division and the Australian and New Zealand task force. Three South Vietnamese infantry divisions and airborne and Marine contingency forces also ringed the area around Saigon. In IV Corps, the U.S. corps advisory group had operational control over the rest of the 9th Infantry Division. South Vietnamese forces in IV Corps consisted of three infantry divisions.[55]

Amid these adjustments, engineer dispositions also changed. After the 45th Group and the 14th, 27th, and 35th Engineer Battalions moved to I Corps, the two engineer brigades shifted areas of responsibility several times to fit the tactical moves and engineering priorities. On 1 November, the 18th Brigade's last adjustment of the year took place when the 937th Group at Pleiku took over responsibility for the northern half of II Corps and the 35th Group the southern half. As a result, the 937th Group gained the 19th and 84th Engineer Battalions, formerly 35th Group units in Binh Dinh Province. In turn, the 35th took control of the 70th Engineer Battalion at Ban Me Thuot. Since the 937th now had responsibility for the Qui Nhon area, the 35th Group's headquarters returned to a more central location at Cam Ranh Bay. In III and IV Corps, the arrival of the 1st Cavalry Division and emphasis given to the highway restoration program prompted realignments in the 20th Brigade as well. The 79th Group at Plantation near Long Binh kept northern III Corps; the 159th Group at Long Binh took over more of the southern half, including Saigon, Long Binh, Bien Hoa, Phu Loi, Lai Khe, Xuan Loc, and Blackhorse; and the 34th Group, which moved to Dong Tam in December, held on to parts of southern III Corps and all of IV Corps. Units affected by boundary changes and moves included the transfer of the 34th Construction and 168th Combat Battalions from the 79th to the 159th Group. In return, the 159th transferred the 62d Engineer Battalion to the 79th.[56] (*Map 25*)

Both LIBERTY CANYON and SPEEDY EXPRESS actively involved the engineers. During LIBERTY CANYON, which lasted until 30 December, squads from the 8th Engineer Battalion cleared fields of fire by cutting down trees with chain saws and demolitions and helped fortify battalion bases. The three line companies focused efforts at forging firebases and landing zones out of the dense jungle near the Cambodian border. Within sixty days, Company C built ten firebases and more than thirty landing zones. Most of the 79th Engineer Group bolstered the 1st Cavalry Division's engineers at Phuoc Vinh, Quan Loi, Song Be Tay Ninh, and other locations. One combat battalion devoted all

[55] MACV History 1968, vol. 1, pp. 218, 244, 372, 401, 646.

[56] Rpt of Visit to Various Headquarters in the Republic of Vietnam, 1–21 November 1968, incl. 3, p. 7 and tab O, OCE Liaison Officer Trip Rpt no. 13; Quarterly Hist Rpt, 1 Jul–30 Sep 68, MACDC, p. II-3; ORLLs, 1 Nov 68–31 Jan 69, 20th Engr Bde, 10 Feb 69, pp. 2–3, 1 Aug–31 Oct 68, 35th Engr Gp, 31 Oct 68, pp. 2–3, 1 Nov 68–31 Jan 69, 35th Engr Gp, 31 Jan 69, p. 4, 1 Nov 68–31 Jan 69, 937th Engr Gp, 31 Jan 69, p. 2, 1 Nov 68–31 Jan 69, 79th Engr Gp, 14 Feb 69, p. 6, incl. 6, 1 May–31 Jul 68, 34th Engr Gp, pp. 1–2, 1 Aug–31 Oct 68, 34th Engr Gp, 1 Nov 68, pp. 1–2, and 1 Nov 68–31 Jan 69, 159th Engr Gp, 14 Feb 69, p. 2, all in Historians files, CMH.

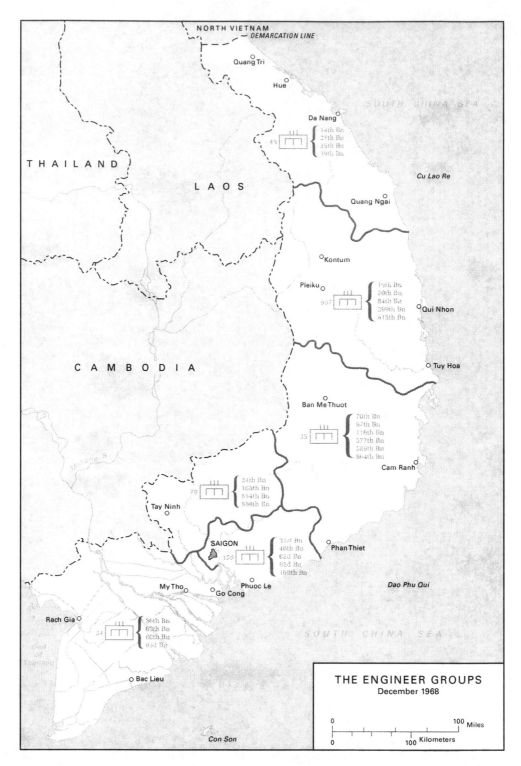

Quang Tri

Hue

Da Nang

45 { 14th Bn
27th Bn
35th Bn
39th Bn

Quang Ngai

Kontum

Pleiku

937 { 19th Bn
20th Bn
84th Bn
299th Bn
815th Bn

Qui Nhon

Tuy Hoa

Ban Me Thuot

35 { 70th Bn
87th Bn
116th Bn
577th Bn
589th Bn
864th Bn

Cam Ranh

Tay Ninh

79 { 34th Bn
168th Bn
554th Bn
588th Bn

SAIGON

159 { 31st Bn
46th Bn
62d Bn
92d Bn
169th Bn

Phan Thiet

My Tho

Go Cong

Phuoc Le

Rach Gia

34 { 36th Bn
69th Bn
86th Bn
93d Bn

Bac Lieu

THE ENGINEER GROUPS
December 1968

0 — 100 Miles
0 — 100 Kilometers

MAP 25

its efforts to the airmobile divisions' deployment while the group's other two combat battalions and one construction battalion spent about 50 percent of their resources supporting the operation. Much work went into readying the base camps for the cavalry division's fleet of aircraft. The group erected protective revetments around helicopters and fixed-wing aircraft and expanded airfield facilities. Other tasks included upgrading base defenses, and logistical, medical, and communications facilities. This work caused a general slowdown of other construction projects and continued into 1969.[57]

Operation SPEEDY EXPRESS, due to begin on 1 December, diverted much of the 34th Group's highway improvement efforts for months. Col. Earnest Graves Jr., who assumed command of the group in September, recalled that SPEEDY EXPRESS curtailed plans to upgrade Highway 4, the major road south of Saigon. The engineers hastily built forward bases to be used as jumping-off points to seal the Cambodian border, but frequent changes in engineer priorities in the delta caused turmoil. After one month's work, the engineers had to change plans to include the addition of a 1st Cavalry Division brigade, which called for a different distribution of troops and materials. The airmobile brigade, however, was not used in the operation. The operation slipped into the following year, and the initial number of forward bases was scaled back from eleven to four. Finding land in the densely populated delta became a major problem. Formal real estate arrangements involved a tedious process through U.S. and South Vietnamese channels. Some land was reclaimed by dredging or by drying out paddy mud.[58]

With the tactical emphasis shifting to III and IV Corps, the II Corps command worried that it would lose engineers. In November, the I Field Force commander, General Peers, met with Colonel Elder and discussed the effects of the 18th Brigade's troop reductions and possible deployment of units to support SPEEDY EXPRESS. Elder reported the impending loss of nearly two thousand military spaces, and he considered inactivation of whole units as less objectionable than civilianization. The brigade considered zeroing out two battalions and transferring a third battalion to III Corps, actions that Peers viewed as "totally unsatisfactory." He informed General Abrams, citing road and airfield maintenance problems, which, added to other requirements in a vast area of operations, were "such that we will have difficult time maintaining the status quo to keep our head above water." Peers concluded that replacing soldiers with the local national civilians was "infeasible, undesirable and in combat areas may be impossible." Abrams told Peers that the civilianization program must be completed. He added, however, that a recent lifting of Department of the Army restrictions on the replacement pipeline had given U.S. Army, Vietnam, a chance to consider increasing engineer unit manning

---

[57] Hist Rpt, 14th Mil Hist Det, 1st Cav Div, Opn LIBERTY CANYON 1–69, p. 5; 8th Engr Bn Hist, 1968, pp. 14–15; ORLL, 1 Nov 68–31 Jan 69, 79th Engr Gp, incls. 4, 6; Quarterly Hist Rpt, 1 Oct–31 Dec 68, MACDC, p. II-4.

[58] Quarterly Hist Rpt, 1 Oct–31 Dec 58, MACDC, p. II-4; Interv, Capt Raymond F. Bullock, 26th Mil Hist Det, with Col Earnest Graves, CO, 34th Engr Gp, 11 Jul 69, VNIT 445, pp. 3–4, CMH; AAR, Opn SPEEDY EXPRESS, 16 Mil Hist Det, 14 Jun 69, pp. 29–30, Historians files, CMH.

*Despite delays caused by Operation SPEEDY EXPRESS, the 34th Engineer Group
continued to make progress along Highway 4.*

levels. Abrams also noted that a U.S. Army, Vietnam, engineer study would
review the future disposition of engineer units in II Corps. Generally, the
number of U.S. Army engineer battalions in II Corps remained about the
same: ten following the Tet deployments to I Corps, eleven between September
1968 to March 1969, and ten again when the 87th Engineer Battalion was
zeroed out.[59]

Facing the possible loss of the 18th and 20th Engineer Brigade units, U.S.
commanders turned to greater use of allied engineers. While expressing his
concerns about the civilization program, General Peers proposed using
Vietnamese soldiers just completing basic engineer training to flesh out the
U.S. engineer units. He noted that General Parker already had initiated an
affiliation program with nearby Vietnamese engineer units. The Vietnamese
soldiers would not have to be assigned to the American units. Instead, Peers
suggested the soldiers could remain assigned to the associated unit but work
with the U.S. unit. Abrams responded that an on-the-job training program
and increased participation by Vietnamese engineer units in the highway restoration program were under consideration. He also advised Peers to see if the
South Korean engineers supporting their two divisions in II Corps could take

---

[59] Msg, Lt Gen William R. Peers, CG, I FFV, NHT 1900 to Abrams, 13 Nov 68; Msg,
Abrams MAC 16033 to Peers, 20 Nov 68. Both in Abrams Papers, CMH.

over some road and airfield responsibilities. Peers also cited an urgent need to improve South Vietnamese engineer units in II Corps. To this Abrams replied that ultimately more Vietnamese Army engineer units would be activated and given the same equipment as U.S. units. Meanwhile, an affiliation program for logistical units called Project BUDDY had been advocated by Maj. Gen. Joseph M. Heiser Jr., commanding general of the 1st Logistical Command. He felt that firsthand observation and experience on the job would be the best way to improve the skills of the South Vietnamese Army. Abrams approved the BUDDY proposal in January 1969, and the engineers enthusiastically adapted General Heiser's concept to their needs.[60]

As 1968 came to a close, the U.S. Army engineers reluctantly carried out the civilianization program. In late October and early November, the 18th and 20th Engineer Brigades reported hiring 6,600 Vietnamese workers. By the end of the year, the 20th Brigade was authorized a local national workforce of 4,358 employees. Authority for an additional 73,200 man-days for the quarter also allowed the brigade to hire daily laborers with Assistance-in-Kind funds. Senior engineer officers, including General Parker, still declared their doubts about the program. On 16 November, Parker and his staff briefed Lt. Gen. Frank T. Mildren, General Palmer's successor as Deputy Commanding General, U.S. Army, Vietnam. During the briefing, the engineers pointed out several problems with the civilianization program. Fewer troops in the construction units meant fewer supervisors. Local nationals could neither be employed in hostile areas nor stand guard or defend base camps. Their holidays conflicted with American work schedules. The engineers considered some jobs too complex for the civilians' limited experience, and cited their poor vehicle and equipment maintenance habits, and possible labor problems. Parker insisted that the construction battalions could not absorb any more local nationals. Engineer units at the time the program started, he noted, were already employing as many civilians as they could find, train, and supervise. Eventually, the two brigades had difficulty in recruiting skilled workers from the reduced manpower pool, which the contractors also depended on for workers. One group, the 35th, reported that moving its construction units to rural areas had caused further problems in recruiting skilled employees. Parker succinctly concluded, "We cannot operate on a mobile basis in a counterinsurgency environment at reduced TOE strength." He bluntly declared the civilianization program a complete failure. Generals Roper and Chapman echoed these concerns in their debriefing reports near the end of 1968. Roper preferred to use the civilians above the authorized military strengths, as was true earlier. Chapman suggested restricting civilianization to one company so that the other companies in a battalion could retain their "complete integrity and usefulness."[61]

[60] Msg, Peers NHT 1999 to Abrams, 1 Dec 68, sub: Upgrading ARVN Engineer Capabilities; Msg, Abrams MAC 16667 to Peers, 5 Dec 68, sub: Upgrading ARVN Engineer Capabilities, both in Abrams Papers, CMH; Debriefing, Parker, 14 Oct 69, p. 4, Senior Officer Debriefing Program, DA, Historians files, CMH; Ploger, *Army Engineers*, pp. 169–70; Heiser, *Logistic Support*, pp. 240–41.

[61] Rpt of Visit to Various Headquarters in South Vietnam, 1–21 Nov 68, incl. 3, pp. 7–9, OCE Liaison Officer Trip Rpt no. 13; ORLLs, 1 Nov 68–31 Jan 69, 20th Engr Bde, p. 4, and

Fortunately, several new units joined the two brigades that year, easing troop losses. During August and September, the 116th Engineer Combat Battalion, a National Guard unit from Utah called to active duty in May following Tet, deployed from Fort Lewis, Washington, to Phan Rang Air Base to begin a ten-month tour of duty. The last element of the 800-man battalion, the largest single National Guard unit deployed to Vietnam, arrived at the air base in mid-September. After additional training and attachment to the 35th Engineer Group, the 116th deployed to three locations in southern II Corps: battalion headquarters and Company D inland to Bao Loc, Companies A and C inland to Di Linh, and Company B south to coastal Phan Thiet. At Bao Loc and Di Linh, the battalion's efforts centered on maintaining and upgrading Highway 20. It offered general engineering support at Phan Thiet. A second National Guard engineering unit sent to Vietnam, the 131st Light Equipment Company from Vermont, joined the 35th Group in September. After a short stay with the 577th Engineer Battalion at Tuy Hoa, the company moved inland in mid-November, joining the 70th Engineer Battalion at Ban Me Thuot. There it undertook the rebuilding of Highway 21, airfield work, and civil affairs projects that included moving Montagnards to more secure locations. Also in 1968, the last Regular Army battalion arrived in Vietnam. On 24 April, the 31st Engineer Combat Battalion reached Vung Tau aboard the USNS *Barrett*. Back in November 1967, General Duke had asked the chief of engineers to deploy the battalion, then based at Fort Bliss, Texas, two months ahead of schedule. General Ploger, assigned to the Office of the Chief of Engineers after leaving Vietnam, had cited the unit's shortage of almost four hundred enlisted men and experienced officers and noncommissioned officers. He pointed out that an early deployment required strong justification and an "acceptance of serious reduction in training." When the battalion did arrive, it moved to Blackhorse and accomplished a mix of combat support and construction projects, initially under the control of the 34th Group, and from July to November, under the 159th Group. In November, the battalion moved to Phuoc Vinh where it was committed to Operation LIBERTY CANYON and the 1st Cavalry Division. In December, the 31st Battalion was transferred to the 79th Group.[62]

1 Nov 68–31 Jan 69, 35th Engr Gp, p. 6; Ltr, 1st Lt Raymond F. Bullock, CO, 26th Mil Hist Det, to Cmd Historian, HQ USARV, 19 Nov 68, sub: Civilianization of Engineer Construction Battalions (Briefing of Lt Gen Frank T. Mildren, DCG, USARV), 16 Nov 68, Historians files, CMH; Debriefing, Parker, 14 Oct 69, pp. 8–9 (quoted words, p. 9), Senior Officer Debriefing Program, DA; Debriefing, Roper, 7 Oct 68, pp. 5–6, Senior Officer Debriefing Program, DA; Debriefing, Chapman, 30 Oct 68, pp. 4–5 (quoted words, p. 5), Senior Officer Debriefing Program, DA. Debriefings in Historians files, CMH.

[62] Ltr, HQ 18th Engr Bde to CG, Engr Troops, Vietnam, 18 Sep 69, sub: Evaluation of National Guard Units, incl. 1, 116th Engr Bn (Cbt), p. 6, incl. 2, 131st Engr Co (Light Equip), p. 37; Msgs, Rollins ARV 1928 to Ploger, 1 Nov 67, sub: Acceleration of Engr Bn, and Ploger GVP 935 to Rollins, 9 Nov 67, sub: Acceleration of Engineer Battalion, both in Duke Papers, CMH; ORLLs, 1 Feb–30 Apr 68, 34th Engr Gp, p. 2, 1 Aug–31 Oct 68, 159th Engr Gp, p. 2, and 1 Nov 68–31 Jan 69, 79th Engr Gp, p. 6; Annual Hist Sum, Aug 69 to Sep 70, 31st Engr Bn, n.d., pp. 5–6, Historians files, CMH.

A crane belonging to the 116th Engineer Battalion from the Idaho National Guard, which carried out engineer support in the highlands, including highway work

The disposition of Army engineer forces at the end of 1968 reflected the emergence of the highway improvement program, which came up during the meeting between Generals Parker and Mildren. The discussion turned to engineer support in the four corps. In I Corps, seventeen Army, Navy and Marine nondivisional engineer battalions, four of these Army combat battalions, supported the equivalent of four and one-third U.S. divisions. II Corps had eleven Army engineer battalions (five combat and six construction) backing one division and a separate brigade. III Corps had ten Army battalions (four combat and six construction) supporting the equivalent of four divisions. And IV Corps had three construction battalions supporting two-thirds of an American division. Mildren criticized this breakdown, noting that the seventeen battalions in I Corps accomplished the same work as ten battalions in II Corps. Most of the battalions in I Corps, however, were not Army but Navy and Marine engineers under III Marine Amphibious Force control. The deputy U.S. Army commander pointed out that the bulk of engineer work would be concentrated in III and IV Corps. This meant moving the 19th Engineer Battalion from II to III Corps and the 35th Engineer Battalion from I Corps to IV Corps. In the end, the 19th remained in II Corps and continued working on Highway 1 between Phu Cat and Mo Duc. As the year ended, the 35th and its attached 517th Light Equipment Company began its second major move in one year. This time the battalion moved almost the length of the country from Da Nang to Binh Thuy to bolster the 34th Groups' work along Highway 4 in the delta. Army engineers were now working on major sections of the nation's highways: 35 miles in I Corps, 290 miles in II Corps, 253 miles in III Corps, and 45 miles in IV Corps. The coming year would see even more military engineers and civilian workers committed to renovating the nation's highways.[63]

---

[63] Ltr, Mildren Briefing, 16 Nov 68; ORLLs, 1 Nov 68–31 Jan 69, 45th Engr Gp, p. 41, and 1 Nov 68–31 Jan 69, 19th Engr Bn, 31 Jan 69, p. 1, Historians files, CMH.

13

The Land Lines of Communication

Before the arrival of U.S. forces, South Vietnam's land lines of communication consisted of the system of highways and railroads built by the French. Though the French administrators of Indochina had constructed an effective if modest transportation system over the years, the destructive effects of World War II, the First Indochina War, and Viet Cong interdiction later had left the network in ruins. By 1965, the severing of these transportation links had disrupted economic development, restricted the government's military movements, and raised serious questions about the government's claim that it still controlled a fair portion of the countryside. As U.S. planners began considering intervention with ground combat troops to stave off a Communist victory, it was plainly evident that the system of roads and railroads would not support a modern expeditionary force. Other lines of communication, the ports and airfields, would have to be relied on first. Once the situation improved, ground transportation would be reclaimed.[1]

Early Roads and Rails

The most direct and preferred link between major cities during the French colonial period was rail. Before World War II, Indochina possessed some 1,800 miles of narrow one-meter gauge track built to high standards. It was a single track line with double track or sidings at stations holding trains of varying lengths. Roadbeds consisted of steel ties (treated wooden ties near the coast) and deep rock ballast. The stretch running from the Demilitarized Zone to Saigon had been built so that trains did not have to climb up grades steeper than 1 percent. Designs also included gradual curves with few having a radius of less than 1,300 feet. French and Vietnamese engineers made deep cuts and fills well above the waterways, built tunnels, and built bridges to cross the many rivers—shallow most of the year but susceptible to major floods during the monsoon—descending from the highlands. The main line running from Saigon to Hue to Hanoi was completed in 1936, or some forty years after it was begun. Considered one of the finest railroad systems in prewar Southeast Asia, service operated daily between Hanoi and Saigon at average speeds of

---

[1] For more political and economic background of the Vietnamese ground transportation system during the colonial period, see Joseph Buttinger, *Vietnam: A Dragon Embattled*, 2 vols. (New York: Praeger, 1967), 1:7–8, 21–22, 26–29, 32–33, 2:970–71; Stanley Karnow, *Vietnam: A History* (New York: Viking Press, 1983), pp. 115–18.

twenty-seven miles per hour. A trip between the two cities usually took about forty-two hours.[2]

The colonial administrations simultaneously developed the highway system. By the beginning of World War II, Indochina could claim one of the finer road systems in the Far East with a network of approximately 22,000 miles, approximately two-thirds in the Vietnamese colonies. The French improved the indigenous "Mandarin Road," originally built by the Annamite emperors to connect Hue with both extremities of their domain. Redesignated Colonial Route 1 (later Highway 1), the stretch between Saigon and Hanoi became a modern all-weather road. Although, in contrast to the rail lines, the main highways were marginally designed, narrow, partially surfaced asphalt or macadam roads; they were still considered quite an accomplishment and appropriate for the light and intermittent traffic of the period. Road widths varied from sixteen to twenty feet and narrowed in mountainous regions and at bridges, typically at one-half of the road widths allowing only one vehicle to cross at a time. Steep grades and rugged terrain in the highlands, rice paddies in the lowlands, and seasonal monsoons presented formidable obstacles to the road builders. The French improvised construction techniques and relied heavily on Vietnamese workers doing the manual labor and using local materials to place the base course. Workers usually topped off this "Telford base," which consisted of large uniformly graded rock placed in a single layer over a compacted subgrade, with an asphalt or a light penetration macadam. Although adequate for prewar World War II traffic loads, the Telford base became unsuitable for heavier loads since wheel loads bearing down directly on to a single rock and the weaker soil beneath punched the rock into the subgrade. This defect was made worse during wet weather by the impact and vibration of fast, heavy moving loads. Maintenance became a process of annual reconstruction. The only way to resolve this built-in structural deficiency for heavier traffic was the complete removal of the Telford base.[3]

During the First Indochina War, fighting practically destroyed the highways and railroads of South Vietnam. More than 60 percent of the 11,800-mile road system received damage in varying degrees, with most of the major bridges destroyed or damaged. The encroaching jungle reclaimed many sections of road, and only ruts and mud remained in sections still claimed to be open to traffic. Approximately two-thirds of Highway 1 between Saigon and Hue was impassable to motor traffic. Highway 19 from Qui Nhon to the western border had 310 permanent bridges; 240 of them had to be reconstructed. Only some 560 miles of the 1,400-mile railway system remained open, with enormous gaps separating the former link between Saigon and the new border at the Demilitarized Zone. The war left nearly all the stations, depots, and other rail facilities in a damaged state and little in the way of rolling stock and

[2] Dunn, *Base Development*, p. 108; *Indo-China*, Geographical Handbook Series, Great Britain (Cambridge, England: Naval Intelligence Division, December 1943), pp. 424, 429–30; Jerry A. Pinkepank, "Rails Through Vietnam—1," *Trains* 29 (March 1969): 27–28.

[3] Buttinger, *Vietnam: A Dragon Embattled*, 1:33; *Indo-China*, pp. 409–10; Vietnam Transportation Study, Based on Field Studies Conducted April 1965 through May 1966 for the Government of Vietnam and the U.S. Agency for International Development (USAID) (Washington, D.C.: Transportation Consultants, Inc., June 1966), pp. 101–02.

*The Mandarin Road (shown here in the early 1930s) ran adjacent to the South
China Sea from the Chinese border to Saigon.*

personnel to run the system. In 1955, the United States began to fund road
and railway upgrading under the Foreign Aid Program.[4]

Much of the railroad was temporarily restored by the Vietnam Railway
System, a semiautonomous railroad agency formed after the 1954 Geneva
Agreement. Upgrading included modernization of shop facilities, mechaniz-
ing track maintenance, changing from steam to diesel-electric locomotive
power, and replacing rolling stock with modern equipment. Bridge girders
resting in the gaps suffered little damage and could be raised and reused. Since
the bridges were of a standard design, damaged sections were easy to replace.
By 1959 through service between Saigon and Dong Ha resumed, but the expedient nature of the repairs required much slower train speeds. Through service
only lasted for a year and the government postponed further improvements.
In September 1961, service between Saigon and Loc Ninh was suspended for
security reasons. In the first six months of 1962, the Viet Cong blew up or
derailed sixteen trains, machine-gunned thirteen others, and sabotaged the
line at some thirty places. By April, all-night trains ceased operations. Viet
Cong sabotage and the destructive typhoons of 1964 reduced the rail system
to a little more than 180 miles. That year alone, the insurgents initiated 650
incidents of sabotage. In total, 620 bridges and over 236,000 feet of track were
destroyed. Locomotives were damaged nearly 400 times. Only short sections,

[4] Vietnam Transportation Study, pp. 83, 129; Lt. Col. Nelson P. Conover, "The Lines of
Communication Program in Vietnam, A Case Study," Research Paper, U.S. Army War College,
1973, p. 7 (hereafter cited as "LOC Program in Vietnam"), copy in Historians files, CMH.

*President Diem inaugurates railroad service from Saigon to Dong Ha on
7 August 1959.*

such as Saigon to Bien Hoa and Da Nang to Hue, remained operational;
the rest of the system remained too damaged to operate or unused. The rail
authority also dropped attempts to restore the Saigon to My Tho line in the
delta as impractical.[5]

Nevertheless, the South Vietnamese government considered the restoration of
the battered railway system paramount for the nation's development and stead-
fastly believed the railway a strategically important link. Government officials
believed that a strong Republic of Vietnam had to have an operating railroad to
serve as the backbone of the nation's mass transportation system. Besides plans
for restoring regular freight and passenger service, Vietnamese railroad officials
planned for new lines to haul coal, fertilizer, and cement from existing and newly
developed facilities. The railway service estimated that it would take six to eight
months of minimal repairs to restore all of the disrupted traffic.[6] (*Map 26*)

---

[5] Dunn, *Base Development*, pp. 108–09; "The Trans-Viet-Nam Railroad: Its History and
Significance," *News from Viet-Nam*, Press and Information Office, Republic of Vietnam
Embassy, 5, no. 14 (1 September 1959): 9–13; USOM, *Annual Report for the 1958 FY (July 1,
1957 to June 30, 1958)* (Saigon, n.d.), pp. 9–10, USOM, *Annual Report for FY 1961* (Saigon, 20
November 1961), p. 31, both in Historians files, CMH; Pinkepank, "Rails Through Vietnam—2,"
Trains 29 (April 1969): 38; *The Postwar Development of the Republic of Vietnam: Policies and
Programs*, 2 vols. (Saigon and New York: Joint Development Group, Postwar Planning Group,
Development and Resources Corporation, March 1969), 2:363, copy in CMH (hereafter cited
as *Postwar Development of Vietnam*).
[6] "The Trans-Viet-Nam Railroad," pp. 13–15; Vietnam Transportation Study, pp. 3, 132–
33; Mary E. Anderson, et al., *Support Capabilities for Limited War Forces in Laos and South
Vietnam* (Santa Monica, Calif.: Rand Corporation, 1962), p. 75.

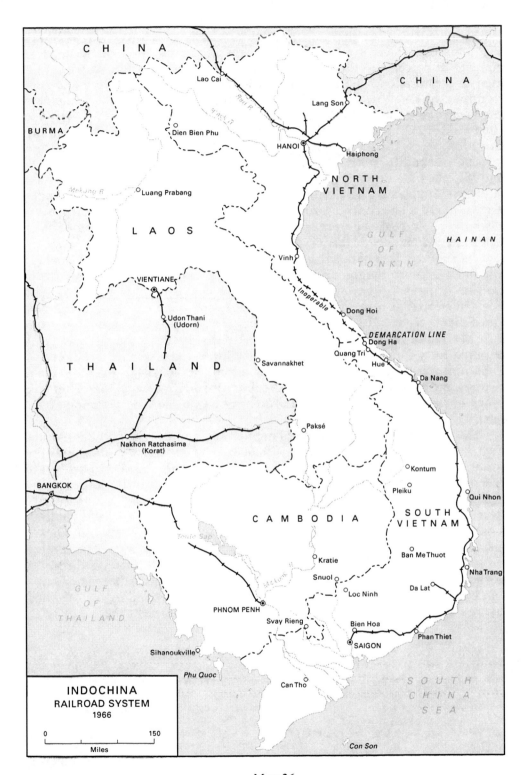

CHINA

Lao Cai

CHINA

BURMA

Dien Bien Phu

Lang Son

HANOI
Haiphong

NORTH
VIETNAM

HAINAN

Luang Prabang

GULF
OF
TONKIN

LAOS

VIENTIANE

Vinh

Udon Thani
(Udorn)

Inoperable

Dong Hoi

THAILAND

Savannakhet

DEMARCATION LINE
Dong Ha
Quang Tri
Hue

Da Nang

Paksé

Nakhon Ratchasima
(Korat)

Kontum

Pleiku

Qui Nhon

BANGKOK

CAMBODIA

SOUTH
VIETNAM

Tonle Sap

Kratie

Ban Me Thuot

GULF
OF
THAILAND

Snuol
Loc Ninh

Nha Trang

Da Lat

PHNOM PENH

Svay Rieng

Bien Hoa
SAIGON

Phan Thiet

Sihanoukville

Phu Quoc

Can Tho

SOUTH
CHINA
SEA

INDOCHINA
RAILROAD SYSTEM
1966

0   150

Miles

Con Son

MAP 26

Meanwhile, of all the U.S. nonmilitary assistance provided to South Vietnam under the U.S. Operations Mission's direction, most of the approximately 40 percent of the funds set aside for transportation was used on highway restoration. The U.S. Military Assistance Advisory Group considered roadwork important and closely coordinated projects with Vietnamese engineers and the operations mission and its American contractors. Between 1955 and 1962, the operations mission employed Capitol Engineering Corporation of Dillsburg, Pennsylvania, to do the engineering design and inspections, and Johnson, Drake, and Piper to do the construction. Starting in 1957, the construction firm built and improved over 180 miles of main highways. Highways 1A, 19, and 21 were designed and built to modern highway standards. The contractor also repaired other national and interprovincial routes and replaced damaged and demolished bridges with modern or temporary structures or ferries. A completely rebuilt four-lane Highway 1A and its eight bridges handled high-speed and heavy traffic over the twenty-one miles between Saigon and Bien Hoa and provided an impressive showcase of American highway engineering. The new Bien Hoa Highway also boasted two large concrete four-lane bridges spanning the wide Dong Nai and Saigon Rivers (1,489 and 3,234 feet, respectively). Descending from Ban Me Thuot to Ninh Hoa over its ninety-three-mile course, Highway 21 required forty-four bridges. The rebuilt Highway 19 between Qui Nhon and Pleiku also included forty new bridges within its 106-mile distance. Field surveys and construction plans proceeded for other major roads. After five years, however, the two firms left the country because of the worsening road security.[7]

The Vietnamese director general of highways, Ministry of Public Works, with U.S. financial and technical help, strived to maintain and rebuild the young nation's roads. The director's responsibility included managing the design, construction, and maintenance of the national and interprovincial highways, while local governments administered the provincial routes and city streets. To do this, the director relied on government workers, local contractors, and South Vietnamese Army engineer units. Because of President Diem's interest in developing the Central Highlands, the highway agency concentrated efforts on improving Highway 14, considered by the government as the transportation backbone of the region. The 114-mile road project, the agency's first major venture using mechanized equipment, included rebuilding the road and laying a double bituminous surface treatment, all to be done by the end of 1962.

---

[7] Vietnam Transportation Study, pp. 90–91; Conover, "LOC Program in Vietnam," pp. 8–9; Buttinger, *Vietnam: A Dragon Embattled*, 2:970–71; USOM, *Activity Report, 30 June 1954 to 30 June 1956*, pp. 38–40, USOM, *Annual Report (July 1, 1957 to June 30, 1958)*, pp. 5–8, USOM, *Annual Report for FY 1960* (Saigon, 1 October 1960), pp. 9–12, USOM, *Annual Report for FY 1961* (Saigon, 20 November 1961), USOM, *Annual Report for FY 1962* (Saigon, n.d.), pp. 21–23, all in Historians files, CMH; Memo, Chief, Temporary Equipment Recovery Mission (TERM) for Ch MAAG, 19 Jan 57, sub: Conference with USOM on Roads; Ltr, Dep Ch MAAG (Log) to Director, USOM, 10 Jun 57; DF, TERM Engineer to Ch MAAG, 11 Jun 57, sub: Road Priorities. Memo, Ltr, and DF in Samuel T. Williams Papers, MHI.

Another program included major bridge reconstruction. Designed jointly by Capitol Engineering and the director general of highways and assigned to local contractors, work began in 1961 with intent to replace bridges on Highways 1, 13, and 14. For replacement and repair of bridges destroyed or damaged by insurgent sabotage, the U.S. Operations Mission ordered large shipments of structural steel for fabrication into movable, military-type bridge sections.[8]

Meanwhile, South Vietnamese Army engineers focused their efforts on the "strategic roads" program in the highlands. Diem firmly believed that the new rural roads would open the remote area to settlement and support military operations. Lt. Gen. Samuel T. Williams, the chief of the military advisory group, fully supported the rural project because it provided good training and meshed nicely with U.S. contingency plans. One project involved Route 5 from coastal Mo Duc inland to Kontum. Completion of this project would have resulted in another strategic link between the highlands and the coast and possibly led to future commercial and strategic links to Laos and Thailand. The routing of a section through Pakse in Laos became the topic of many high-level discussions between U.S. and Vietnamese officials. While the highway department worked along the eastern one-third of the road, troops of the 4th Engineer Combat Group advanced from Kontum over the remaining and more difficult stretch. Then in late 1959, the Viet Cong launched several hit-and-run attacks along the road. The insurgents succeeded in setting fire to the wooden bridge decks (a frequent practice throughout the war), damaged some construction equipment, and captured an engineering work party. As the insurgents increased the number of attacks, American advisers were evacuated.[9]

Despite these setbacks, Saigon in early 1962 renewed the road-building plan for the entire country. Plans included more than 180 miles of military roads of minimum standards reaching to outposts in previously inaccessible mountainous areas near the Cambodian and Laotian borders. South Vietnamese Army engineers gave this effort high priority, and U.S. Seabee teams added some military roads at Special Forces camps. Several U.S. advisers considered the plan too ambitious, however, and in the end Saigon did not accomplish its goals. In May 1963, MACV headquarters reported that out of

---

[8] USOM, *Annual Report for FY 1961*, p. 26, and ibid., *FY 1962*, pp. 22–23; Dunn, *Base Development*, p. 99; Conover, "LOC Program in Vietnam," p. 8.

[9] Spector, *Early Years*, p. 307; "Communists Attack South Vietnamese Roadbuilders," Daily Intelligence Bulletin for 19 Nov 59, Asst CofS for Intel, DA, p. 4, Historians files, CMH; Memo of Conversation Between Diem and Williams, 20 Dec 57, in *Foreign Relations of the United States, 1955–1957*, vol. 1, *Vietnam* (Washington, D.C.: Government Printing Office, 1985), pp. 729, 889–93; Memo, Lt Gen Samuel T. Williams, Ch MAAG, for Amb Elbridge Durbrow, 14 Jan 59, in *Foreign Relations of the United States, 1958–1960*, vol. 1, *Vietnam* (Washington, D.C.: Government Printing Office, 1986), pp. 131–33. Both volumes contain numerous references to President Diem's interest in this program and highway restoration. On 9 May 1957, during his visit to the United States, the Vietnamese leader outlined his strategic highway program in the highlands to President Dwight D. Eisenhower. The American president, half jokingly, said there was an old adage that roads sometimes were a "golden bridge for your enemies." For more on the Eisenhower comment, see *Foreign Relations of the United States, 1955–1957*, vol. 1, *Vietnam*, pp. 797–98.

nine hundred miles of roads programmed for improvement and construction, the government's military and civilian engineers succeeded in accomplishing only 20 percent of the roadwork.[10]

By then the director general of highways lacked the capability to improve much less restore the nation's road system. Much of the equipment left by the American contractors stood idle for lack of repair parts and adequate maintenance support, and the central repair facility filled rapidly with unusable and deteriorating machinery. Although the director issued a "Ten-Year Plan of Highway Improvement, 1965–1975" in 1964, in reality no capability existed to carry it out. The Vietnamese also persisted in using outdated and improper engineering practices learned under the colonial administration. Many officials recognized these defects, but supervisors and laborers fearing the possible loss of jobs strongly resisted modernization. To make matters worse, the U.S. Operations Mission's advisory staff had dropped to just four people in the agency's Roads Branch of its Public Works Division.[11]

Deteriorating road conditions persisted. Frequently, the Viet Cong sabotaged roads using the "piano keyboard" method. This old Viet Minh practice cut trenches several hundred yards along alternate sides of the roadway, thereby stopping vehicle traffic but leaving a small walkway for pedestrians and insurgents. Cutting roadways and destroying culverts and bridges reversed most of the earlier rebuilding efforts. Because the work crews lacked sufficient security on the job, nearly all work on the highway system stopped. The storms that swept over the country in late 1964 exacerbated the already deteriorating road system. During 1965, entire sections of key Highway 1 were closed to traffic in several provinces, and the Viet Cong and North Vietnamese regular units threatened to sever Highway 19, effectively dividing the country.[12]

The U.S. Agency for International Development, the operations mission's successor organization, urgently revived its transportation program for the hard-pressed Vietnamese nation. In April 1965, the Agency for International Development negotiated a contract study with a Washington firm to analyze South Vietnam's entire transportation system, particularly roads and the means to their improvement. By then the United States had made a large investment in the road system, over $57 million between 1955 and 1964. Some of this work greatly helped the U.S. military buildup. The Saigon to Bien Hoa Highway carried an increased volume of military traffic and served as the link to large bases built along its route. Despite its poor condition and security, Highway 19 still served as a vital link in the movement and resupply of U.S. forces in the Central Highlands. Still, the study, which was completed in June

---

[10] MFR, MAAG J–3, 1 Mar 62, sub: Briefing for William P. Bundy, Dep Asst Sec Def for International Security Affairs; Memo, Robert H. Johnson, National Security Council Staff, for Walt W. Rostow, Dep Spec Asst to the President, 14 Oct 61, sub: Subjects for Exploration in Vietnam; MAAG-Vietnam Military Assistance Plan/Program, FY 1964–1969, 1 May 1963. All in Historians files, CMH. For more on the Seabee teams' roadwork, see Marolda, *From Military Assistance to Combat*, pp. 192–200, 346–55; and *Helping Others Help Themselves*, pp. 29–43.

[11] Conover, "LOC Program in Vietnam," p. 9; Vietnam Transportation Study, pp. 110, 113–14.

[12] Conover, "LOC Program in Vietnam," p. 8; Vietnam Transportation Study, p. 91; MACV History, 1965, p. 122.

1966, showed that damage had reduced the capacity of the country's transportation network by at least 50 percent. The study declared the railway system almost entirely inoperative, the highways cut in many places with more than one thousand bridges and culverts damaged or destroyed, and the capacity of inland waterways blocked in many places. Even if the director general's crews could repair some of the road damage, they could not have sustained the level of maintenance needed to overcome the deterioration caused by the increased weight and volume of heavy military vehicles. Between 1957 and 1966, railway freight dropped nearly 50 percent. Only coastal shipping and civil air transport managed to expand service. As for solutions, the Agency for International Development study outlined logical tasks for postwar nation building, but these were insupportable because no one could foresee the extent or duration of the war.[13]

Initial Restoration Arrangements

Early in the U.S. buildup, land-line improvements had lagged behind logistical facilities and air and water lines of communication, but this changed as road usage increased. Initially, engineer troops used expedient methods such as filling craters, tactical bridging, and some road and bridge improvements to support tactical and logistic traffic. During 1966, road traffic increased at an accelerated rate. It was helped by General Westmoreland who urged commanders to use the roads instead of relying on helicopters. By recognizing that major improvements would have to be made to the ground transportation system, engineering staffs developed plans to restore the railway system and major portions of the road network leading to ports and major bases. A bypass to move convoys around congested Saigon was also considered essential. In April 1966, General Ploger proposed building a new direct road connecting Long Binh with the 25th Division's base camp at Cu Chi. The proposed two-lane paved road included fabricating a large bridge near Phu Cuong. Though the II Field Force commander was reluctant to provide road construction security, General Engler, the USARV commander, approved the highway and detailed planning began. Ploger also directed his group commanders to help South Vietnamese roadwork and to issue materials if bridges and roads were built to U.S. specifications. As far as committing U.S. engineers to the highways, which were deteriorating because of heavy traffic, he had to balance roadwork with all the priority tasks assigned to Engineer Command. In early December, Ploger set maximum and minimum guidelines for his commanders. A minimum of 10 percent of unit effort would be devoted to roadwork. In January 1967, Westmoreland reiterated to his field commanders that because of limited tactical airlift capability and the need to strengthen the nation's

---

[13] Conover, "LOC Program in Vietnam," pp. 9–10; *Vietnam Transportation Study*, p. 91; *MACV History*, 1965, p. 3; Dunn, *Base Development*, p. 99; *Postwar Development of Vietnam*, 2:364.

economic posture, the "overall strategy will include aggressive action to open secure, and use land and water lines of communication."[14]

Military planners understood that programs funded by the Agency for International Development had strategic military value, and they began to take steps to merge the efforts between U.S. and Vietnamese military and civilian agencies. MACV J–4 and the Directorate of Construction coordinated roadwork with Saigon's Joint General Staff through a Joint Roads and Bridges Committee. During a meeting held in April 1966, the committee discussed plans to reinforce bridges along Highway 15 to the port at Vung Tau. At the same meeting, the committee agreed to dispatch a joint team of military and civilian engineers to reconnoiter Highway 1 from Ninh Hoa to Phan Rang and from Qui Nhon to Bong Son. In May, the committee set priorities and classification standards for the selected routes. Class 50 roads with two-way Class 35/one-way Class 50 bridging became the norm for divisional loads. Following the Highway 1 reconnaissance, the South Vietnamese Army chief of engineers used these standards to decide which bridges should be replaced. The MACV director of construction then arranged the transfer of steel beams from U.S. Agency for International Development stocks to U.S. and South Vietnamese engineers to get the bridge work under way.[15]

Similar arrangements took place to restore the railroad. In March 1966, a Joint MACV/Agency for International Development Study Group reviewed the reconstruction and security efforts and considered the South Vietnamese government's proposal that the U.S. command rent the railroad system. MACV headquarters, which did not plan to operate a U.S. military rail service, declined the offer, preferring to have U.S. forces continue the customer relationship with the Vietnam National Railway Service. In April, Washington pledged additional commodity support provided the Vietnamese government carried out railway operations and reconstructed those sections damaged by floods and the war. Reconstruction efforts were coordinated through three standing committees composed of members of the Military Assistance Command, Vietnam, the U.S. Agency for International Development, and the South Vietnamese government. Primary responsibility rested with the South Vietnamese/U.S. Railway Reconstruction Committee. A joint U.S. Railroad Coordinating Committee consisting of MACV staff and civilian aid officials met for the first time in August and held regular monthly meetings. The Directorate of Military Construction and the Agency for International Development prepared to reconstruct some 375 miles of a 713-mile system by the end of 1967. The plan encompassed four phases, and some slippage was anticipated. This objective was in line with one of the goals set forth during a recent U.S.-South Vietnamese conference in Honolulu: to increase the open

---

[14] Quoted in MACV History, 1967, vol. 1, p. 327; Dunn, *Base Development*, p. 99; Ploger, *Army Engineers*, p. 115–16; Ltr, Ploger to COs 18th Engr Bde, 79th Engr Gp, 159th Engr Gp, 4 Dec 66, sub: Allocation of Engineer Effort, box 31, 77/051, RG 319, NARA; Msg, COMUSMACV MACCOC11 08709 to CINCPAC, 19 Mar 66, sub: Rail and Road Communications—I and II Corps, box 5, 69A/702, RG 334, NARA.

[15] Quarterly Hist Rpts, 1 Apr–30 Jun 66, MACDC, pp. 6–7, and 1 Jul–30 Sep 66, MACDC, p. 8.

roads and railroads from about 30 to 50 percent. Progress depended on the phased securing of areas through which the railroad passed. Both governments agreed that the Vietnam National Railway Service would rebuild the main lines and U.S. forces would fund and build spurs, sidings, and marshaling yards. The Agency for International Development furnished construction materials such as rails, ties, structured steel, bridge trusses, and equipment. In addition, the U.S. military ordered two hundred rail cars and spare parts to supplement the fleet of Vietnamese rolling stock for the handling of military cargo. The 1st Logistical Command would control the cars and make arrangements with the Vietnam National Railway Service for maintenance on a reimbursable basis. Of the $25 million programmed for restoration from 1966 through 1969, about two-thirds ($16.8 million) of the funding came from the U.S. Agency for International Development.[16]

The railway contributed to the allied war effort in several ways. For instance, the first shipment along the Qui Nhon to Phu Cat section, which reopened in July 1966, delivered nine hundred tons of construction materials to build Phu Cat Air Base. During the construction, the Vietnam National Railway Service also hauled rock from the RMK-BRJ quarry outside Qui Nhon. This undertaking alone kept many dump trucks from clogging Highway 1, by then the main supply route to Bong Son. When Phu Cat Air Base became operational, the railroad hauled supplies, ammunition, and fuel on a daily forty-three-mile round trip along a secure route patrolled by allied troops. Similar arrangements were made in October to haul rock for air base construction at Tuy Hoa. Since the new airfield would cross the rail line, the Air Force had to relocate the line. Even closed sections were put to use as temporary bypasses for allied vehicular traffic. During Operation THAYER in late 1966, the redecking of the Bong Son railroad bridge provided a critical link over the Lai Giang River since the highway bridge could not support required loads. The conversion of the railroad bridge for truck traffic turned out to be extremely timely after the Viet Cong destroyed two sections of the highway bridge. In I Corps, the marines removed sections of track between Da Nang and Chu Lai for vehicular use. After getting Saigon's approval, the marines were supposed to stockpile the tracks, but South Vietnamese troops appropriated some rail left next to the right of way. The marines did transport the remainder to the Da Nang rail yard.[17]

Highway and railway security and restoration goals incorporated in the 1967 Combined Campaign Plan met with varying degrees of success. Allied military

---

[16] MACV History, 1966, pp. 294–96; Heiser, *Logistic Support*, pp. 165–66; Dunn, *Base Development*, p. 109; U.S. Agency for International Development/Vietnam, Project Status and Accomplishment Report, July–December 1970, Project: Railway Rehabilitation, p. 209, Historians files, CMH; Quarterly Hist Rpts, 1 Apr–30 Jun 66, MACDC, pp. 6–7, and 1 Jul–30 Sep 66, MACDC, pp. 8–10; Msg, Lt Gen Jean E. Engler, DCG USARV, MAC 5568 to Besson, 4 Jul 66, Dunn Papers, CMH; MFR, MACJ03, sub: Railroad Briefing, 10 March 1966, 12 Mar 66, box 5, 69A/702, RG 334, NARA; Msg, COMUSMACV MACDC-PO 40427 to CG, USARV, COMNAFORV, and Cmdr, Seventh Air Force, 7 Sep 66, sub: Vietnam Line of Communications Requirements, box 5, 69A/702, RG 334, NARA.

[17] Dunn, *Base Development*, pp. 109–10; MACV History, 1966, p. 296; Quarterly Hist Rpts, 1 Apr–30 Jun 66, MACDC, p. 7, 1 Jul–30 Sep 66, MACDC, pp. 9–11, and 1 Oct–31 Dec 66, MACDC, pp. 1–4; Pinkepank, "Rails Through Vietnam—2," p. 42.

operations included eliminating enemy ambushes from vital roads and rail lines. Land clearing of a 100- to 300-yard-wide strip of vegetation along key roads reduced the enemy's ability to mount ambushes and reduced the effectiveness of those that did occur. The cleared strips also provided cleared right-of-ways for later route restoration and upgrading. This surge in operations and restoration work in 1967 increased the percentage of military essential roads classified as "secure" (Green). During 1966, the roads defined as secure—those that could be traveled during daylight hours without an armored escort—varied from a low of 698 miles (23 percent) to 742 miles (43 percent). The 1967 Combined Campaign Plan identified 1,296 miles of roads essential for military operations. During the year, roads classified Green rose from 609 miles in January (35 percent) to 1,063 miles (61 percent) in December. The goal of 50 percent secure was exceeded by almost 11 percent. Military engineers and contractors in 1967 upgraded 170 miles of highway and city streets to MACV standards. They also maintained 4,432 miles of roads and streets, constructed 18,218 feet of new bridging, and rebuilt or replaced 39,845 feet of destroyed or damaged bridges with tactical bridging. Highway tonnages increased approximately 100 percent over that moved in 1966. Specific achievements included the reopening of Highway 1 in southern II Corps and northeastern III Corps and the opening or securing of remaining stretches in I Corps. In early 1968, engineers temporarily reopened the road from Saigon all the way to the Demilitarized Zone, but Highway 1 still required major upgrading and many sections remained vulnerable to sabotage and ambushes.[18] (*Map 27*)

On the other hand, railroad restoration plans continued to lag in 1967. The number of sabotage incidents clearly showed the Viet Cong's determination to disrupt restoration of the system. Attacks even took place on the outskirts of Saigon. In April, satchel charges caused light to heavy damage to ten locomotives and a crane parked in a rail yard. A month later, a mine caused a train's derailment. Because of these attacks, the Vietnam National Railway Service restored less than sixty-two miles in the period December 1966 to June 1967. To repair the recently damaged lines, the railway service diverted equipment and work crews from reconstruction projects. Some U.S. officials now seriously doubted the value of railroad restoration. In July, Robert W. Komer, who had recently assumed the new position as Westmoreland's deputy for civil operations and revolutionary development support, or pacification, proposed eliminating or deferring railroad restoration. Komer asserted that available roads and coastal shipping voided the railroad's economic significance. He believed railroad security troops could be put to better use in pacification. MACV J–4 reevaluated the plans and

---

[18] MACV History, 1967, vol. 2, pp. 762, 764–66, 844–45; Msg, COMUSMACV MACDC-PO 32400 to CINCPAC, 2 Oct 67, sub: Highlights of Land Clearing RVN—Opn PAUL BUNYON, Westmoreland Message files, CMH; Sharp and Westmoreland, *Report*, p. 144. For more on the reopening of Highway 1, see Msgs, COMUSMACV MACJ322 19486 to DCG USARV, CG III MAF, CG I FFV, CG II FFV, 15 Jun 67, sub: Opening of Highway 1; CG II FFV AVFBC-P 70009 to COMUSMACV, 1 Jul 67, sub: Opening of Highway 1; CG III MAF to COMUSMACV, 4 Jul 67, sub: Opening of Highway 1; CG I FFV AVFA-GC-PL A1974 to COMUSMACV, 5 Jul 67, sub: Opening of Highway 1. All in box 3, 70A/782, RG 334, NARA.

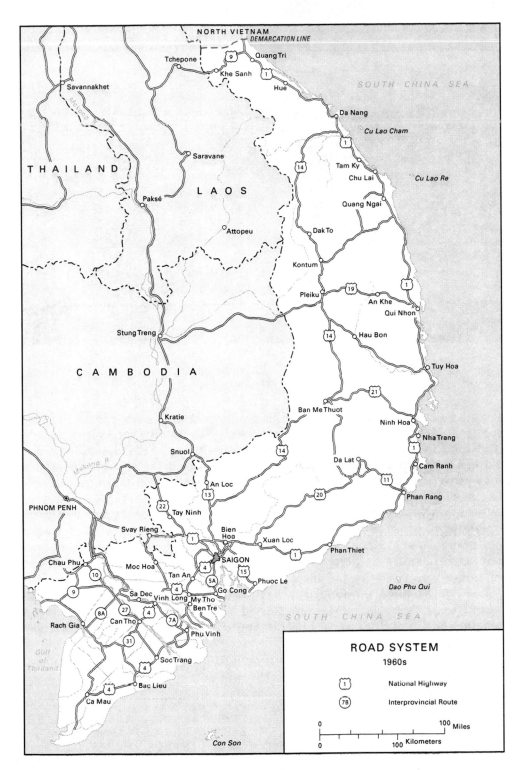

ROAD SYSTEM
1960s

 National Highway

 Interprovincial Route

0   100 Miles

0   100 Kilometers

MAP 27

*South Vietnamese railway workers repair tracks between Qui Nhon
and Phu Cat.*

facilities needed for U.S. logistical bases and recommended continuing restoration. General Abrams, acting for Westmoreland, advised Komer that "a decision to write off the rail system prematurely would also be undesirable. From a long-range standpoint, a balanced transportation system in Vietnam may prove valuable." By October, however, persistent sabotage and difficulties moving materials to work sites set back the railway service's scheduling by six months. At the end of 1967, 296 miles (39 percent) of the rail lines were operational, 61 miles (8 percent) were undergoing repair, and 413 miles were not operational (53 percent). Although the security goal of 55 percent was not reached, statistics alone did not present the complete picture. Movement of military tonnage by rail increased nearly threefold over that moved in 1966. The Vietnam National Railway Service also transported vegetables from Da Lat to coastal towns and started a daily passenger service between Saigon and Long Binh to ease the commute for the base's Vietnamese civilian workforce.[19]

[19] MACV History, 1967, vol. 2, pp. 766–69; Dunn, *Base Development*, p. 109; Memo, Robert W. Komer, Dep Civil Operations and Revolutionary Development Support, for Westmoreland, 9 Jul 67, sub: Vietnamese National Railroad System (VNRS) Restoration; Memo, Abrams for Komer, 22 Jul 67, sub: VNRS Restoration. Both in box 3, 70A/782, RG 334, NARA. For more on the restoration schedule, see Staff Study, MACJ45, 29 Jun 67, sub: Vietnamese National Railroad, box 3, 70A/782, RG 334, NARA.

Running the Programs

The mature Lines of Communication Program, robustly organized and decently funded, dates from the end of 1967. Progress had not been lacking the previous year as planners had begun turning their attention from ports and bases to land lines restoration and improvement. For one thing, the perennial tug-of-war over land transport financing had been resolved in the Agency for International Development/Department of Defense Realignment Program of November 1966, which recognized the Agency for International Development's fiscal limitations. Under this program, the Defense Department would fund certain civilian aid projects, including railway sabotage replacement and highway maintenance and upgrading, provided they were clearly war related or incurred more than 50 percent of their cost because of enemy attacks. Retroactive to 1 July 1966, the program tasked Department of the Army with programming, budgeting, and funding the highway maintenance program. The Army, in turn, passed the requirement to MACV headquarters and, in January 1967, the Construction Directorate assumed responsibility for developing the highway maintenance and upgrading portion of the realignment program.[20]

Even so, highway funding remained under pressure through 1967. Major shortfalls, including $35 million in Agency of International Development/Department of Defense funds and $50 million in military construction contingency money, affected the upgrading of Highway 1 north of Cam Ranh Bay and Highway 4 south of Vinh Long. The Construction Directorate feared that the holdup of contingency funds could cause work on several major bridges and bypasses to slip one year. As it is, Fiscal Year 1968 requirements had grown to $109.5 million, and the directorate estimated that about $130 million (including $50 million in military construction funds) would be needed each year through the close of Fiscal Year 1970.[21]

Into the breach stepped Robert Komer who, in November 1967, traveled to Washington to solicit support for the pacification and highway programs. There he presented a memorandum to Secretary McNamara, pointing out crucial funding needs. He contended that roads were not only militarily essential and saved airlift, but they were also vital for pacification. Only military engineers or U.S. contractors, he said, could do the job. And as the Agency for International Development did not have the money, the Defense Department

---

[20] MACV History, 1967, vol. 2, p. 763; Conover, "LOC Program in Vietnam," pp. 10–11; Quarterly Hist Rpts, 1 Oct–31 Dec 66, MACDC, pp. 42–43, and 1 Jan–31 Mar 67, MACDC, pp. 1, 53; MACDC-EBD Fact Sheet, [5 Dec 67], sub: Highway Restoration Program in Vietnam, box 3, 70A/782, RG 334, NARA. See also Msg, COMUSMACV MACDC-PO 08067 to Sec State and Sec Def, 9 Mar 67, sub: Upgrading of Highway System RVN, box 3, 70A/782, RG 334, NARA; and Memorandum of Understanding Between the Department of the Army and the Agency for International Development, 25 May 67, box 4, 74/167, RG 334, NARA.

[21] Conover, "LOC Program in Vietnam," p. 11; MACDC-EBD Fact Sheet, [5 Dec 67], sub: Highway Restoration Program in Vietnam. For more on funding problems during the period 17 April–26 June 1967, see backchannel messages between MACV's directors of construction and the Defense Department's chiefs of Southeast Asia Construction, Dunn Papers, CMH.

was his only recourse. McNamara scribbled a note in the margin of the memorandum directing his comptroller and installations and logistics chief to find
the money required. Komer later noted that the money was requested and
approved on military grounds for pacification purposes. "Here," he declared,
"was a case where the two went quite well together."[22]

From this point on, Washington recognized highway improvement work
as a distinct entity and always referred to it as the Lines of Communication
Program, or, in military terms the LOC Program, under the administration
of the Directorate of Construction. In December, the U.S. Mission Council
in Saigon agreed, and the formal transfer of military and civilian highway
programs to MACV headquarters took place on 1 February 1968. Early in
1968, the directorate's consolidated highway funding request for Fiscal Year
1969 again exceeded $100 million. Approved funding levels varied from year
to year, but by 1972, the funding level in all categories of restoration reached a
cumulative total of nearly $500 million.[23]

With the accent on organization and concentrated effort, the South
Vietnamese military also came on board. On 1 November 1967, recognizing the
inadequate coordination, program overlap, and duplication of work among the
two allies, the South Vietnamese Joint General Staff and MACV headquarters
agreed to consolidate all responsibilities for land lines of communication. A
Combined Central Highway and Waterways Committee consisting of representatives from Military Assistance Command, Vietnam, the U.S. Agency for
International Development, and South Vietnamese military and civilian agencies, and chaired by the chief of staff of the Joint General Staff, first met on
22 November to discuss and approve the committee's charter. Planning then
advanced to restore and upgrade approximately 2,532 miles of national and
interprovincial highways to support military operations and pacification and to
stimulate economic development. Highway construction standards applied to
all construction agencies: U.S., South Vietnamese, and allied engineers.[24]

---

[22] Conover, "LOC Program in Vietnam," pp. 20–21; Msg, Sec Def DEF 4127 to
COMUSMACV, 1 Dec 67, box 3, 70A/782, RG 334, NARA; Interv, Rand Corp. with Robert
W. Komer, Organization and Management of the New Model Pacification Program, 1966–1969,
7 May 1970, p. 122, Robert W. Komer Papers, CMH.

[23] Conover, "LOC Program in Vietnam," pp. 18–19, 21–23, 40; Memo, Lt Col James S.
Sibley, Chief, LOC Div, Engr Cmd, for CG, Engr Cmd, VN, 5 Jun 70, sub: Authority for LOC
Program, incl. 3, and Memo, Brig Gen William E. Bryan Jr., Acting CofS, MACV, for Rollins,
27 Dec 67, sub: Mission Council Action, incl. 4, 26th Mil Hist Det, LOC Study, 10 Feb 71,
VNIT 813, CMH. See also MACV Directive 415–6, 11 Mar 70, *Construction: Surface Lines of
Communication, Restoration and Maintenance of Highways and Bridges*, an. A, p. 2, Historians
files, CMH.

[24] Conover, "LOC Program in Vietnam," pp. 19–20; Quarterly Hist Rpt, 1 Oct–31 Dec 67,
MACDC, pp. V-13 to V-15; Dunn, *Base Development*, p. 100; Ploger, *Army Engineers*, p. 161;
MACV Directive 415–6, 11 Mar 70, an. A, p. 1. See also Charter for the Combined Central
Highway and Waterway Committee signed by Gen Westmoreland (COMUSMACV), Gen
Cao Van Vien (Chief, Joint General Staff, Republic of Vietnam Armed Forces), Mr Donald
G. MacDonald (Dir, USAID), and Mr Buu Don (Ministry of Public Works), incl. 6, 26th Mil
Hist Det, LOC Study, 10 Feb 71, VNIT 813, CMH; and Lt. Gen. Dong Van Khuyen, *RVNAF
Logistics*, Indochina Monographs (Washington, D.C.: U.S. Army Center of Military History,
1980), p. 380.

Organizational changes took place at the Directorate of Construction on 1 January 1968 with the establishment of a separate Lines of Communication Division. The Construction Management Division, formerly responsible for the highway program, was deactivated with its remaining functions transferred to other divisions. The new division, headed by Col. Charles R. Clark, directed the planning and execution of all national and interprovincial highway programs, plus waterways, railways, and dredging done by military engineers, contractors (mainly RMK-BRJ), and American and Vietnamese civilian agencies. The Agency for International Development retained responsibility for secondary road projects in the provinces. Finally, on 1 February, the Lines of Communication Division took over the advisory mission, which included advising the director general of highways and his five highway district engineers and their staffs and the provincial public works chiefs on all aspects of the country's roads and ferries. A third branch, the Advisory Branch, which included several civilian aid technicians, was added.[25]

Sustaining Highway Improvement

Before long, highway improvement evolved into the single biggest construction program of the Vietnam War. The Combined Central Highway and Waterway Committee and the Directorate of Construction's Lines of Communication Division envisioned a highway program lasting through 1974, but General Westmoreland, recognizing its tactical and economic importance, directed an accelerated plan to complete most restoration work by the end of 1971. As a result, priorities were adjusted, with 2,199 miles earmarked for earlier completion, and the remaining 333 miles placed in deferred status. These figures, however, did not include upgrading and new road construction supporting tactical operations in areas not served by the major road networks or other local road programs sponsored by the Agency for International Development. In addition, the Philippine Civic Action Group worked on local roads and bridges in Tay Ninh Province. Railroad restoration along the main branch lines also got increased attention. The Lines of Communication Division mainly concerned itself with U.S. military rail spurs, and the division regularly participated in the meetings of the joint railroad coordinating committee.[26]

[25] Conover, "LOC Program in Vietnam," pp. 22–23; Dunn, *Base Development*, pp. 22–23, 100; Msg, COMUSMACV 10964 to CG, USARV, Cmdr Seventh Air Force, COMNAVFORV, 18 Apr 68, sub: Transfer of Highway Advisory Functions to MACV, box 6, 71A/354, RG 334, NARA; Memo of Understanding, MACDC-LOC, 1 Feb 68, sub: HQ MACV Assumption of the Highway Civil Advisory Responsibility, in Booklet, MACDC-PO, Development of the Construction Directorate, 15 Jun 70, box 11, 72A/870, RG 334, NARA. See also 26th Mil Hist Det, LOC Study, 10 Feb 71, 10 Apr 68, sub: Transfer of Highway Advisory Functions to MACV, incl. 5, and above cited 1 Feb 68 Memo of Understanding, incl. 9, VNIT 813, CMH.

[26] Dunn, *Base Development*, p. 103; Conover, "LOC Program in Vietnam," p. 23; Tregaskis, *Building the Bases*, p. 391; Larsen and Collins, *Allied Participation in Vietnam*, p. 64; MACV History, 1968, vol. 2, p. 660; MACV History, 1969, vol. 2, pp. IX-90, IX-92; Memo, HQ MACV, MACJ45, sub: Minutes of Meeting, Joint MACV/USAID Railroad Coordinating Committee, 9 Feb 68, box 1, 71A/354, RG 334, NARA.

The centralized management and increased planning capability showed in other areas as well. Improved cost estimates and cost accounting made funding, if not easier, at least more orderly. More explicit construction standards set forth as the result of a clearly defined highway program enabled the Navy's Officer in Charge of Construction to hire specialized architect-engineer firms to solve design problems. Nearly all of the Navy's design firms were now involved in highway projects. Their diverse and sophisticated skills ranged from developing computerized bridge designs using prestressed concrete beams to photogrammetric designs of highway layouts. By 1969, the architect-engineer effort expanded to $15 million in design contracts alone. The road builders also anticipated the large bridge-building requirement, and the design firms developed a series of standard designs for bridge decks, beams, prestressed concrete piles, pier protection, security lighting, and abutments. As a result, design costs averaged a low 1.5 percent of bridge construction, reducing design costs, simplifying construction, and saving time.[27]

Army engineers relied on civilian consulting firms to provide engineering expertise and methods to improve quality control. For instance, the U.S. Army Engineer Construction Agency, Vietnam, hired Quinton-Budlong Company to design bridges, to review road and bridge designs prepared by the units, and to give advice on equipment operation and maintenance, safety, and other engineering services. Included among the more than one hundred employees in Vietnam were experts in quarry and asphalt plant operations, well drilling, welding, mechanics, electrical designs, and licensing equipment operators. Formally referred to as "technical representatives," several employees functioned at work sites and unit headquarters—including sixteen at 18th Engineer Brigade. Leo A. Daly Company concentrated on architectural engineering tasks, and Technical Services Corporation provided four asphalt equipment experts. The U.S. Army Construction Agency, Vietnam, trained quality assurance personnel who checked the quality of crushed rock, earth compaction, and asphalt production and gathered other quality control data. Since quarries and asphalt plants were the key to carrying out the highway program, commanders and staffs looked for ways to improve operations. In 1970, Brig. Gen. Henry C. Schrader, the 18th Brigade's commander, started organizing engineer industrial control centers at group headquarters. The centers, consisting of at least one officer and several noncommissioned officers and enlisted men from each battalion running a quarry and asphalt plant, served as a centralized agency in the planning, installation and construction, and operation for all industrial complexes in each group. Modern soil-testing devices such as nuclear dosimeters were also used, and a central materials testing laboratory was established at Long Binh. General Schrader personally played a major role in obtaining and evaluating the nuclear dosimeters.[28]

---

[27] Conover, "LOC Program in Vietnam," pp. 24, 48; Tregaskis, *Building the Bases*, p. 420; Interv, Richard Tregaskis with Frank S. McGarvey, Morrison-Knudsen International Company, Inc., Boise, ID, 19 Jun 70, p. 11, Richard Tregaskis Papers, Naval Facilities Engineering Command Hist Ofc, Port Hueneme, Calif.

[28] Msg, CG, USAECAV to CINCUSARPAC, 11 Jul 68, sub: Engineering Contract Support, box 2, 74/167, RG 334, NARA; Maj. Henry A. Stearns, Maj. Rudolph E. Abbott, and Capt.

Highway construction standards followed U.S. practices as much as possible. These standards, based on criteria established by the American Association of State Highways Officials, ranged from a Class A highway (23 feet wide and 8-foot shoulders) to a Class F road (23 feet wide and 2-foot shoulders). The Directorate of Construction incorporated the Class F road to make the most use of the French-constructed highway system's alignments on existing embankments. This class called for laying a rock or asphalt stabilized base over the existing road and widening the traveled way, or actual road less shoulders, to 23 feet and paving it with an asphaltic concrete surface. Design life of the rehabilitated highway system ranged from twenty years for a Class A road to ten years for Classes C or D. Design life did not apply to the Class F roads.[29]

Bridges were essential links along the stretches of restored highways. One inventory showed approximately 750 bridges in the road net's restoration and upgrading program. Over one-third of these bridges met satisfactory conditions or were under construction when the program began. Culverts would replace many bridges, and others would not have to be replaced for several years. Plans called for approximately 250 new bridges totaling 37,000 feet. Standard roadway widths for two-lane roads were set at 25 feet. Designs also provided for curbs and one-meter (3.3-foot) width sidewalks with guardrails. Short span bridges of 165 feet or less called for 40-foot roadway widths without sidewalks. Since bridges were the most vulnerable links of the highway system, engineers devised various methods of protection. Pier protection and security lighting systems were included in designs for major bridges. The chain-link fence stand-off pier protection system consisted of steel beams driven into the riverbed with fencing and concertina wire hung from the beams. In regions with extreme tidal variations, engineers built a flotation system known as a floating catwalk. This system could be fabricated easily, and the Styrofoam or steel balls were floated to the bridge for installation. Once emplaced and covered with wooden platforms, the catwalk was connected with chain-link fencing and the interior was filled with concertina. The floating catwalk also served as an observation post for guards, who could walk completely around the bridge pilings at water level. Antiswimmer devices and mine booms were developed to frustrate enemy demolition teams. Concertina wire was suspended from floating buoys and fastened on the river bottom. Changes in the tides, typical in most of Vietnam's waterways, caused the concertina to shift

James L. Campbell, "LOC Highway Restoration, Vietnam," *Military Engineer* 65 (March–April 1973): 85; 1st Lt. Doug Noble, "Quinton-Budlong, The Consulting Engineer," *Kysu'* 2 (Spring 1970): 9–10; Memo, Brig Gen Henry C. Schrader, CG, 18th Engr Bde, sub: Industrial Complex Control Center, 23 Jun 70, Henry C. Schrader Papers, MHI. See also Engr Cmd, Reg no. 415–30, Construction: Quinton-Budlong Consultants, 8 Apr 71, box 2, 74/571, RG 334, NARA; Interv, Capt David D. Christensen, 26th Mil Hist Det, with 1st Lt Robert McKay, OIC [Officer in Charge], Central Materials Testing Lab, USAECAV, 2 Jan 71, VNIT 794, CMH; 1st Lt. Robert Hoyler, 18th Engr Bde, Report on the Evaluation of Nuclear Moisture-Density Gauges, 18 Apr 71, Vietnam Collection, MHI.

[29] Dunn, *Base Development*, pp. 102–03; History, LOC Program, incl. 3, tab X, OCE Liaison Officer Trip Rpt no. 22, 8 May 72, p. 183, OCE Hist Ofc. For more on highway standards, see MACV Directive 415–6, 11 Mar 70, an. A.

Floating balls being readied for installation at the Phu Cuong Bridge pier protection system

and agitate unpredictably, hindering swimmers from trying to pass through it. Mine booms constructed of heavy timber attached to steel cables and across the river upstream of the bridge were employed to stop or detonate floating mines. Usually several feet of chain-link fence were hung from the timbers to catch submerged mines.[30]

The highway program required the procurement and transportation of vast quantities of construction materials and equipment to the work sites. As roadwork moved farther away from the logistics bases, a greater burden was placed on the entire supply system, and to reduce the number of times material was handled planners sought ways to ship directly from supply points to the users. Stocks of cement, asphalt, and culvert material dwindled rapidly, and engineers and logisticians improvised or borrowed materials from other services. For instance, the 45th Group borrowed construction materials from the Seabees when it moved to I Corps in early 1968 and began to reopen Highway 1 through the Hai Van Pass. It also became clear that the standard equipment assigned to engineer units to do roadwork would not suffice.[31]

The short timetable for completing the roadwork prompted ideas for additional road-building equipment. Even if commanders shifted more engineer troops to the roads, the construction rate would not increase since road construction was limited by rock production and rock-hauling capability. In 1968 at the suggestion of the Lines of Communication Division, the USARV Engineer, the Army Engineer Construction Agency, Vietnam, and the 18th and 20th Brigades analyzed rock requirements, production capabilities, and the advantages of furnishing Army engineer units with high-capacity civilian construction equipment. The engineers concluded that additional rock-production equipment (rock crushers, rock drills, and large-capacity rock-hauling trucks) and road-working equipment (heavy-duty and hand compactors, asphalt pavers and distributors, dump trucks, and cement mixers) would be required to complete the road program by the specified time. Evidence showed that "straight commercial" equipment used in the United States and not designed to meet military requirements (such as the twenty-ton rough terrain crane) or modified for military use (like the Caterpillar 830MB and Clark 290M wheeled tractors and the Euclid and LeTourneau 18-cubic-yard earthmoving scrapers) would increase the troops' road-building ability to the rigorous standards required for a permanent highway system.[32]

[30] Ploger, *Army Engineers*, pp. 120–21; MACV Directive 415–6, 11 Mar 70, an. B; Capt. Edward D. Florreich, "Lines of Communication Restoration Program," *Kysu'* 2 (Spring 1970): 13. See also Bridge Protection in the Republic of Vietnam, Engr Adv Div, MACV, 1 Jan 68, MHI; and Capt. Robert J. Oldmixon, "Pier Protection," *Engineer* 2 (Spring 1972): 26–29. The 41st Port Construction Company worked on many pier protection systems in and around Saigon. For more on techniques and equipment, see Interv, Capt David D. Christensen, 26th Mil Hist Det, with Capt Harold W. Wagner Jr., CO, 41st Engr Co (Port Const), 4 Dec 70, VNIT 774, CMH.

[31] Dunn, *Base Development*, p. 102. For more on engineering materials, see Interv, Christensen with Lt Col John L. Moffat, Dir of Materiel, USAECV, 7 Apr 71, VNIT 879, CMH.

[32] Dunn, *Base Development*, pp. 103–06; Conover, "LOC Program in Vietnam," pp. 24–25; Purchase of Construction Equipment with MCA [Military Construction, Army] Funds, p. 1, incl. 3, tab K, OCE Liaison Officer Trip Rpt no. 13.

Purchase of commercial construction equipment totaling some $23 million from Military Construction, Army, funds began in December 1968. By July 1969, most of the more than seven hundred items of machinery in original factory colors arrived at the ports of Cam Ranh Bay, Qui Nhon, and Saigon. For easier identification, a purple roadrunner cartoon character running on a green circle with the letters *MCA-LOC* appeared as an emblem on all the new equipment. Among the most important items reaching the 18th and 20th Engineer Brigades were eight 250-ton-per-hour rock crushers and 226 twelve-cubic-yard dump trucks. The high-volume all-electric crushers, considered the key to the success of the highway program, replaced the lower-capacity 75-ton-per-hour plants. Just as portable as the smaller plants but easier to operate and maintain, the new crushers required fewer operators (one compared to three for the 75-ton-per-hour crushers), and produced at least three times as much rock. In addition, six 225-ton-per-hour crushers from depot stocks in the United States arrived before the 250-ton-per-hour models, thus making possible the rapid rate of road improvement in 1969. When the new crushers were put into operation, many of the 75-ton-per-hour crushers judged uneconomically repairable were turned in. Feeding the fourteen large crushers required the addition of thirty-six track drills, ground-level rock feeders, and 600-cubic-foot-per-minute air compressors to the equipment purchase list. Quarry operations also improved with the addition of 29 six-cubic-yard scoop loaders to load the 100 fifteen-cubic-yard Euclid dump trucks purchased in 1967 for quarry work. Far simpler to operate, faster, and easier to maintain, each of the new scoop loaders replaced two 40-ton crane shovels. The new twelve-cubic-yard dump trucks joined the fleet of standard five-ton Army dump trucks, by then hard-pressed to meet hauling needs. New earth compaction equipment included sixty heavy-duty compactors plus hand compactors to speed up culvert backfilling. Also twelve new backhoes for culvert placement and excavation in confined areas saved thousands of man-hours of digging and many hours of equipment time on crane-mounted shovels and clamshells. To speed up the paving effort, six asphalt pavers and fourteen distributors augmented the equipment on hand. Seven asphalt curb extruders were ordered. Their use in the Central Highlands helped redirect and channel monsoonal runoff and the manufacture of curb and gutter systems in villages and towns. Since bridges required about an average of 165 cubic yards of concrete for abutments, deck slabs, and approach slabs, the engineers requested additional cement mixers. Central batch plants and transit mixers replaced the small 16S cement mixers and allowed work on more than one bridge at a time, thus significantly reducing production and maintenance man-hours. Altogether, the new equipment increased the construction capability of the two brigades by as much as 50 percent.[33]

---

[33] Ploger, *Army Engineers*, pp. 120, 124–25; Dunn, *Base Development*, pp. 103–07, 122; Army Activities Rpt, 9 Jul 69, p. 41, CMH; Engr Cmd History, 8 Jul 72, p. 23, Historians files, CMH; Florreich, "Lines of Communication Restoration Program," p. 12. See also Interv, 1st Lt Raymond F. Bullock, 26th Mil Hist Det, with Lt Col Paul J. Kline, Chief of Supply Br, Supply and Maint Div, Engr Sec, USARV, 9 Mar 69, incl. 4, VNIT 445, CMH. For the Army's justification to purchase commercial construction equipment, see Memo, Ploger, OCE, for Asst Sec Army (Installations and Logistics), sub: Requirement for Heavy Construction Equipment—USARV, 19 Jul 68, Historians files, CMH. The merits of military and commercial construction

Manufacturers quickly provided the off-the-shelf construction equipment. In fact, this transaction was much faster than the normal procurement of standard military equipment. Since most of the commercial buy was new to the Army, the contract inserted provisions for factory representatives to prepare the equipment for use and to train the operators. Management of the equipment operators school came under the 169th Engineer Construction Battalion, which also picked up and processed all the equipment destined for 20th Brigade units. By early 1970, the Long Binh–based battalion had distributed 103 pieces of equipment to the brigade's units in III and IV Corps. The U.S. Army Mobility Equipment Command also awarded a $5 million contract to Dynalectron Corporation of Fort Worth, Texas, to provide maintenance and repair parts support. In May 1969, the firm began setting up shop at twenty-six locations in Vietnam, and its employees, numbering between 125 to 200 technicians, accomplished all unit-level maintenance and ran small repair parts distribution points at job sites for the 18th and 20th Brigade units. Vinnell Corporation provided backup engineer equipment and automotive maintenance support from its field maintenance shop at Cam Ranh Bay.[34]

In carrying out the highway restoration program, engineer units built base camps and industrial sites along the routes to be upgraded. This included building and maintaining quarries, asphalt plants, and access roads. During 1969, the 18th and 20th Engineer Brigades used more than 70,000 tons of rock every week, a rate that more than doubled the following year. The following year, the plants operated by engineer troops produced 340,000 cubic yards of rock and 60,000 tons of asphalt monthly to support the highway program. In addition, Army engineers relied on RMK-BRJ's crushers for 38 percent of the rock required monthly to sustain an annual construction rate of 177 miles as set forth in 1968. By mid-1970, the two brigades ran eighteen rock-crusher sites. Most rock ended up as landfill, base courses, and asphaltic concrete on the roads. Of the eleven asphalt plants set up by 1970, all but two operated in conjunction with quarries. Both plants were located near paving sites, and trucks hauled rock to the plants from quarries and stockpiled.[35]

The near nonexistence of rock deposits in the Mekong Delta, where finding dry ground could sometimes be a problem, meant transporting large quantities of rock into the delta. At that time there were only two large sources for

equipment are discussed in Maj. Gerald M. Tippins, "CCE—An Impossible Dream?" *Engineer* 2 (Spring 1972): 12–15.

[34] Dunn, *Base Development*, p. 122; ORLL, 1 Nov–31 Jan 70, 169th Engr Bn, 14 Feb 70, p. 11, Historians files, CMH; 1st Lt. Eugene Roberts, "The Great Race, New Equipment," *Kysu'* 1 (Summer 1969): 11; Stearns, Abbott, and Campbell, "LOC Highway Restoration, Vietnam," p. 85; Visit to Various Headquarters in the Republic of Vietnam, 1–22 May 1969, incl. 5, p. 7, OCE Liaison Officer Trip Rpt no. 14, 7 Jul 69, OCE Hist Ofc; Engr Cmd History, 8 Jul 72, p. 16; "Other Logistical Support Services: Field Maintenance," *Vinnell* 11 (May 1969): 12–14.

[35] Dunn, *Base Development*, pp. 102, 111; Ploger, *Army Engineers*, pp. 124, 162; Briefing for the Undersecretary of the Army, incl. 14o, p. 7, OCE Liaison Officer Trip Rpt no. 16, 30 Apr 70, OCE Hist Ofc. For more on quarrying operations, see Spec. Mike Barry, "The Rhythm of Tam Quan Quarry," *Kysu'* 1 (Summer 1969): 3–5; Memo, Capt Kurt E. Schlotterbeck, 26th Mil Hist Det, for Mr Warner Stark, CMH, sub: 94th Quarry Detachment, 30 Nov 69, VNIT 547, CMH. VNIT 547 includes interviews, diagrams, photos, and a large-scale map of the Vung Tau area.

RMK-BRJ industrial sites such as the Phu Cat quarry and asphalt plant provided rock and asphalt for the highway program.

crushed rock in the area—RMK-BRJ's University Quarry outside Saigon and the 159th Engineer Group's quarry at Vung Tau run by the 103d Construction Support Company's quarry platoon. In November 1966, the consortium added a 400-ton-per-hour crusher to the 250-ton-per-hour plant and aggregate production jumped to an average of 6,500 tons per day. In late January 1967, it reached a record production of more than 12,000 tons per day. Also that month, RMK-BRJ set up a hot-mix asphalt and concrete batch plant at its University Quarry site. Similarly, the Vung Tau quarry now operated by the 94th Engineer Quarry Detachment evolved into the Army's largest producer of crushed rock. This quarry produced most of the rock hauled to the 34th Engineer Group's projects in the delta. By late 1969, Vung Tau's production averaged 70,000 tons per month. Among the few outcroppings found in the northwest corner of IV Corps, two South Vietnamese quarries—an army quarry at Nui Sam and a Ministry of Public Works quarry at Nui Sap—produced rock for local highway and construction projects. By 1968, barges and sampans were transporting more than 150,000 tons of rock per month from these four quarries to off-loading sites in the delta. In January 1970, more than 200,000 tons of rock were shipped, and in August 1971 production exceeded 272,000 tons.[36]

[36] Dunn, *Base Development*, pp. 102–03; Ploger, *Army Engineers*, p. 124; Tregaskis, *Building the Bases*, pp. 249–50, 261; *Diary of a Contract*, pp. 255–56; Florreich, "Lines of Communication Restoration Program," p. 12; ORLL, 1 Feb–30 Apr 66, 159th Engr Gp, p. 2, 13

Transporting rock to the delta greatly increased following the establishment of the MACV Delta Transportation Plan. The plan called for the construction of twelve barge off-loading points—seven by the U.S. Army (including the RMK-BRJ site at Can Tho) and five by the South Vietnamese. One key problem facing the planners was the gathering, repairing, and movement of off-loading equipment—clamshell cranes, bulldozers, and conveyor belts—and getting them to the sites. The U.S. Agency for International Development and U.S. Army, Vietnam, helped by obtaining some equipment for the South Vietnamese. Army Engineer Hydrographic Survey Team Number 1, a ten-man team based at Long Binh, the only one of its kind in the Army, charted water routes along the delta's many canals.[37]

Expediting the distribution of crushed rock to where it was needed the most was charged to a small Delta Rock Agency formed within the Lines of Communication Division. This meant the three major customers in the delta—the U.S. Army, the South Vietnamese Army, and Vietnam's Ministry of Public Works—would integrate construction operations and distribute rock on a quid pro quo basis. Their rock needs were forecasted for certain periods and forwarded to the Delta Rock coordinator, a Directorate of Construction staff officer, who decided what portion of each forecast would be filled at the discharge points. Crushed rock was loaded aboard barges linked together and ferried by Military Sea Transportation Service tugs to the unloading points. During 1971 over 7.1 million tons of crushed rock had been distributed, and almost 1.2 million tons were shipped in 1972. That year, two of the quarries ended operations: Vung Tau in February and University Quarry in June. To make up for the loss of the two quarries, South Vietnamese engineers increased production at Nui Sam, and the U.S. Agency for International Development extended its contract with Vinnell Corporation to run Nui Sap through 1973.[38]

May 66, Historians files, CMH; Memo, Schlotterbeck for Stark, 30 Nov 69, sub: 94th Quarry Detachment, p. 1; MACV History, 1971, vol. 1, p. IX-9; Ltr, CO, 34th Engr Gp to CG, 20th Engr Bde, 19 Oct 67, sub: Rock Requirements in the Delta, Historians files, CMH.

[37] Quarterly Hist Rpts, 1 Jan–31 Mar 68, MACDC, p. IV-19, 1 Jul–30 Sep 68, MACDC, p. IV-7, 1 Oct–31 Dec 68, MACDC, p. IV-9, and 1 Jan–31 Mar 69, MACDC, 19 Apr 69, pp. IV-9 to IV-10, Historians files, CMH; Background Statement Leading to Formation of the Delta Rock Agency, incl. 3, tab E–1, OCE Liaison Officer Trip Rpt no. 13; Spec. Mike Barry, "Down a Not So Lazy River," *Kysu'* 1 (Winter 1969): 25–26; MACV History, January 1972–March 1973, vol. 1, p. E-32. For more on the U.S. Army Hydrographic Survey Team, see HQ, Engr Cmd, AVCC-MO-I, U.S. Army Topographic Support of Military Operations in the Republic of Vietnam, 25 Feb 71, pp. 58–82, Historians files, CMH; and Capt. Joseph Cascio, "The Water Mappers," *Kysu'* 1 (Spring 1969): 7–8.

[38] Quarterly Hist Rpts, 1 Apr–30 Jun 68, MACDC, pp. IV-6 to IV-7, 1 Jul–30 Sep 68, MACDC, p. IV-8, 1 Oct–31 Dec 68, MACDC, p. IV-10, and 1 Jan–31 Mar 69, MACDC, p. IV-10; Concept of Organization, incl. 3, tab E–2, OCE Liaison Officer Trip Rpt no. 13; Barry, "Down a Not So Lazy River," pp. 25–26; MACV History, January 1972–March 1973, vol. 1, p. E-32. For more on the Delta Rock Agency, see AAR, Development of the Delta, Phase I, 26th Mil Hist Det, 18 Jul 69, which includes Interv, Capt Raymond F. Bullock, 26th Mil Hist Det, with Capt John H. Morgan, Coordinator, Delta Rock Agency, 10 Apr 69, incl. 7, and MACDC-LOC, sub: Standard Operating Procedure Delta Rock Agency, n.d., incl. 9, VNIT 445, CMH; and, HQ, MACV Directive 55–15, 8 May 69, *Transportation and Travel, Rock Movements, and Operations in the Delta*, Historians files, CMH.

Barges haul rock to the delta.

At times mud and the scarcity of rock, particularly in the delta, compelled the road builders to develop base course stabilization expedients. Engineers found that lime caused the clay to coagulate into sand-sized particles, which adhered together, forming a much more stable base than the clay alone. Henceforth, they used this stabilization method for all delta road construction. The clay-lime stabilization process was not new. It had been used by road builders in many parts of the world and tested on selected airstrips in Vietnam. Paddy clay could be used to build a strong and durable subgrade, making possible faster construction and allowing the rock barged in at great expense to be used only for the final base course and paving. As true of almost all other horizontal construction in Vietnam, this road-building technique could not be done in the monsoon season. When the rains stopped, roadside paddies were diked and pumped out. Bulldozers and earthmovers scraped out clay to form or expand a roadbed several feet above the surrounding paddies. As the clay dried out, controlled amounts of lime and moisture were added and mixed into the clay with a rotary tiller or disk harrow and compacted with a sheepsfoot roller or a segmented compactor and allowed to cure. Soil stabilization machines later were used to mix the lime with the clay fill. The engineers repeated this process with several (usually four) eight-inch layers of subbase. An eight-inch layer of crushed rock or cement was added to the last course, and a double layer of asphalt topped off the road. The 34th Engineer Group accomplished most of the thirty-seven miles along Highway 4 in the IV Corps area by using this method.[39]

[39] Dunn, *Base Development*, p. 107; Ploger, *Army Engineers*, pp. 118–19; Conover, "LOC Program in Vietnam," pp. 42–43; Florreich, "Lines of Communication Restoration Program,"

Elsewhere in Vietnam engineers used similar stabilization methods. In June 1969, the 554th Engineer Construction Battalion, 79th Group, started upgrading thirty-one miles of Highway 13 from Lai Khe to An Loc using Portland cement instead of lime. In this case, no suitable laterite for the subbase could be found in sufficient quantity along the route. The only source of rock for the base course available in the area was located near An Loc. Furthermore, the battalion would not have a rock crusher on site until late fall. Tests of the local silty sand, however, revealed an unusually low level of plasticity and twice the strength of lateritic material previously used for subbase construction. The battalion commander, Lt. Col. Elbert D. H. Berry, ordered further tests to see if cement would serve as a bonding agent between the silty sand particles, thus increasing the bearing capacity. After many tests, the battalion found that six inches of the soil-cement would provide Class A strength for the MACV Class F highway. This method eliminated the need for a rock quarry at An Loc, but it meant building a completely new parallel road since traffic could not travel on the base course during the seven-day curing process. Work started in late December when Vietnamese civilian workers laid out the bags of cement and dumped the contents on the new eight-inch subbase lift of silty sand. A grader in conjunction with a rotary tiller and a water distributor then spread and scarified the cement evenly over the surface. A segmented embankment compactor followed. The soil-cement, like concrete, required a moist cure. Since this method was almost impossible during the dry season, the 554th accomplished this with less effort by thoroughly dampening the completed base course and coating it with an asphalt cutback to seal in the moisture. After the curing period, the battalion paved the surface with two and one-half inches of asphaltic concrete.[40]

The Builders

To achieve the escalated rate of roadwork by 1971, the Lines of Communication Division and the Combined Central Highway and Waterway

---

pp. 12–13; Interv, 1st Lt Raymond F. Bullock, 26th Mil Hist Det, with Lt Col Richard E. Leonard, CO, 36th Engr Bn, 20 Mar 69, VNIT 445, pp. 5–8, CMH. VNIT 445 also includes other interviews, documents, and a large map covering highway and other construction in the delta.

[40] Ploger, *Army Engineers*, pp. 119–20; Spec. Blanchard de Merchant, "Cement Stabilization of QL 13," *Kysu'* 2 (Spring 1970): 23–25. The 554th also faced a strange soil stabilization problem at a site of a just-completed dry span bridge on Route 8A near Phu Cuong, part of the main supply route from Cu Chi to Long Binh. Seven days after the bridge opened to traffic, the north lane of the east approach mysteriously dropped seven feet. More laterite and rock fill were added, but the approach continued to sink. A team of geologists and soils experts from the 20th Brigade discovered that this phenomenon was caused by the newly added fill that was slipping along a huge subsurface plane. The laterite and rock had displaced the silty, fluid paddy clay to the area of least resistance away from the "dam" created by the old roadbed along the south side of the new roadbed and the perpendicular bridge abutments. Having identified the problem, the engineers added blast rock and fill to the north side of each approach, thus creating sufficient force perpendicular to the subsurface plane and containing the clay. Capt. Gary D. McDonald, "The Terrible Temper of Rach Ba Bep," *Engineer* 2 (Spring 1972): 10–11.

Committee reordered priorities and switched responsibilities among the builders. Carrying out the highway program were the 20th Engineer Brigade in III and IV Corps, the 18th Engineer Brigade in II Corps and southern I Corps along Highway 1 to the Binh Son Bridge south of Quang Ngai, and the 3d Naval Construction Brigade in the rest of I Corps. RMK-BRJ supported the troops by providing rock and asphalt and upgrading assigned sections of road and building major bridges. South Vietnamese and allied military engineers participated to a smaller degree. In several areas, including Highways 19 and 1 west and north of Qui Nhon, the consortium and troops worked jointly. The troops of the 45th Group prepared the roadbed and RMK-BRJ did most of the paving. The centralized management system introduced in late 1967 eased shifting of resources under changing conditions. For instance, engineer troops took over insecure road sections from RMK-BRJ. In addition, this planning took into consideration added repair and maintenance work caused by enemy action and monsoon weather. Often distances were too great to upgrade to MACV standards during one construction season, and road sections were simply patched up to survive the monsoon.[41]

Nevertheless, a considerable amount of work had already been done by 1968. The 35th Engineer Group began paving operations at Cam Ranh and advanced both north and south along Highway 1. By early summer of 1967, the group also started fabricating twenty-foot reinforced-concrete spans. The 159th Group paved Highway 15 between Long Binh and Vung Tau and built a causeway near the port city. Paving along Highway 19 between Qui Nhon and An Khe and Highway 1 north to Bong Son by the 45th Group and RMK-BRJ was well under way. During 1968, the Combined Central Highway and Waterway Committee formally assigned 1,182 miles of the highway program to U.S. Army engineers, 218 miles to the Seabees, and 729 miles to RMK-BRJ. Early in the buildup, South Korean Army engineers of the Dove Unit, a task force devoted to local improvement programs based in Bien Hoa Province, began road and bridge work in the Saigon area. Later, South Korean engineers helped RMK-BRJ build a new North Capital Bypass around Saigon.[42]

As time passed, these figures changed because of funding, program reviews, and U.S. troop withdrawals. In 1969, South Vietnamese Army engineers formally entered the Lines of Communication Program and assumed responsibility for 103 miles of highways. By the end of 1969, the troops and contractor completed 997 miles. Work progressed slowly in 1970 adding another 464 miles, for a total of 1,585 miles or 65 percent of the highway program. A cutback by the House Armed Services Committee in construction funds that

[41] Conover, "LOC Program in Vietnam," pp. 23–25, 27; History, LOC Program, OCE Liaison Officer Trip Rpt no. 22, p. 184. For an example of road priorities shown on sketch maps corps by corps, see MACDC-LOC, Construction Bulletin No. 415–3–2, 15 Oct 69, sub: Lines of Communication Restoration Priorities, incl. 8, 26th Mil Hist Det, LOC Study, 10 Feb 71, VNIT 813, CMH.

[42] Ploger, *Army Engineers*, pp. 117, 156; Larsen and Collins, *Allied Participation in Vietnam*, p. 123; Stearns, Abbott, and Campbell, "LOC Highway Restoration, Vietnam," p. 86. For more on Highway 1, see Spec. Randy Hunter, "The Long Road: Asphalt Lifeline from the Delta to the DMZ," *Kysu'* 1 (Fall 1969): 17–22.

year primarily involving highway restoration caused the Secretary of Defense to direct a thorough review of the highway program. In September, he reduced the fund ceiling by $49 million from $496.7 million to an interim $447 million. This resulted in the deferral of about 300 miles to be constructed by the South Vietnamese government later. U.S. Army troop responsibility dropped from 1,566 to 1,151 miles, the Navy from 267 to 248 miles, and the contractor from 614 to 560 miles. South Vietnamese Army engineer responsibility, however, increased from 103 to 322 miles. The priority assigned to the program remained high with 42 percent of the U.S. engineer troop effort, 25 percent of contractor effort, and 25 percent of South Vietnamese engineer troop effort devoted to the road program.[43]

The highway program had become the U.S. Army Engineer Command's largest single project. As of early 1969, twenty-one of the 18th and 20th Engineer Brigades' twenty-eight battalions and more than half of the 33,000 troops had been directly involved in the road program at one time or another. That May sixteen battalions, nine in the 18th and seven in the 20th Brigade, were committed to highway upgrading, which in March absorbed 32.1 percent and 31 percent, respectively, of their total effort. By the spring of 1970, more than 11,000 men of the 26,000 in the two brigades were occupied in some aspect of the highway construction program. As U.S. combat forces withdrew from Vietnam, a larger proportionate share of engineer battalions stayed behind, with fourteen carrying out roadwork to the end of 1970. By then the 19th Engineer Combat Battalion completed its portion of Highway 20 near the II/III Corps border and departed Vietnam. In 1971, the remaining battalions finished their road projects, with the last three departing near the end of the year. The Army road builders consisted largely of construction battalions and their attached units, including port construction companies and detachments. Nonetheless, an average of three combat battalions augmented with light equipment, construction support, and dump truck companies dedicated their efforts to the highway program. Actually, most of the combat battalions helped the effort by maintaining assigned sections of road and doing specific tasks such as base course work and building bridges.[44]

[43] Conover, "LOC Program in Vietnam," pp. 34–36; History, LOC Program, OCE Liaison Officer Trip Rpt no. 22, pp. 184–85; Stearns, Abbott, and Campbell, "LOC Highway Restoration, Vietnam," p. 86; Dunn, *Base Development*, pp. 111–12; Ploger, *Army Engineers*, p. 117; MACV History, 1970, vol. 2, pp. IX-63 to IX-67; Army Activities Rpts, 5 Aug 70, p. 21; 26 Aug 70, p. 23; 16 Sep 70, p. 23, CMH. For more on roadwork between 1967–1970, see LOC sections: Quarterly Hist Rpts, 31 Mar 67 through 31 Dec 70, MACDC; OCE Liaison Officer Trip Rpts, nos. 6–19; MACV History, 1968, vol. 2, pp. 651–53; MACV History, 1969, vol. 2, pp. 91–97.

[44] Dunn, *Base Development*, p. 102; Engr Cmd, AVHEN, Fact Sheet, sub: Vietnamization of the LOC Effort, 18 Jan 72, box 2, 74/571, RG 338, NARA; Conover, "LOC Program in Vietnam," pp. 33–34; Army Activities Rpt, 9 Jul 69, incl. 5, p. 41, CMH; Visit to Various Headquarters in the Republic of Vietnam, 1–22 May 1969, p. 3, OCE Liaison Officer Trip Rpt no. 14; History, LOC Program, OCE Liaison Officer Trip Rpt no. 22, pp. 186, 189, 194. For more on road-building problems facing Army engineer troops, see Interv, Capt Wilbur T. Gregory, 26th Mil Hist Det, with Lt Col James S. Sibley, Chief, LOC Div, Const Dir, USAECV, 5 Oct 70, VNIT 740, CMH. See also History of the 18th Engr Bde LOC Program, 7 Apr 70, Schrader Papers, MHI; and S. Sgt. Matt Glasgow, "Roads to Peace: 20th Brigade Road Construction" *Kysu'* 2 (Fall 1970): 2–5.

It became apparent that the existing multipurpose nine-hundred-man U.S. Army construction battalion was not the best organization to do extensive roadwork. Normally construction battalions assigned projects to the three identical construction companies, which in turn assigned tasks to the single earthmoving and two vertical construction platoons. Additional specialized personnel and equipment were attached as necessary. Since roadwork is considered horizontal construction (earthmoving, paving, and drainage work involving much heavy equipment), battalions experimented by pooling personnel and equipment into functional-type companies, typically an earthmoving company with most of the heavy equipment, a dump truck company, and a vertical construction company consisting of tradesmen (carpenters, electricians, masons, plumbers, and steelworkers) to do structural or finish work above ground. In the delta, where there were a large number of bridges to be built, the 159th Engineer Group organized a bridge-building company and gave it equipment suited to that task. By centralizing personnel and equipment, including Vietnamese civilian workers and commercial construction equipment and attached troops and equipment from other units, local commanders believed these organizations allowed better use of equipment and resulted in improved road production.[45]

In I Corps, the 3d Naval Construction Brigade and Marine Corps engineers finished their share of the highway program in 1971. Earlier, in mid-1968, when the number of Seabee battalions increased to twelve, about 30 percent of the Navy's engineer troop effort already had shifted to the highway program. One important project was the completion by Naval Mobile Construction Battalion 8 in September of the seven-and-a-half-mile road between Hue and the new permanent port facility at Tan My. In the spring of 1969, all five Marine Corps engineer battalions (including the two divisional battalions) and all ten Seabee battalions concentrated on Highways 1 and 9 in northern I Corps and Route 4 south of Da Nang. Starting at Lang Co, just north of the Hai Van Pass, the 3d Brigade rebuilt approximately eighty-seven miles of Highway 1 north to Dong Ha, then some twenty miles of Highway 9 west to the Vandegrift Combat Base at Ca Lu. Large supplies of crushed rock, estimated at 1.8 million tons, and asphalt were required. The 30th Naval Construction Regiment ran the quarries, crushers, and asphalt plants, and the five battalions of the 32d Naval Construction Regiment provided the road work crews. Road improvements undertaken during 1969 by the road builders in I Corps (including Army engineers of the 45th Engineer Group, South Vietnamese Army engineers, and RMK-BRJ) cut travel time between Da Nang and Hue from six hours to two, and between Da Nang and Dong Ha to four and one-half hours. Major bridge construction by the Seabees included the opening on 30 March of the new 825-foot, timber-piled-supported, concrete-decked Liberty Bridge at An Hoa, about nineteen miles southwest of Da Nang, which replaced the

---

[45] Stearns, Abbott, and Campbell, "LOC Highway Restoration, Vietnam," p. 85; Interv, Capt William A. Kunzman, 16th Mil Hist Det, with Col John W. Brennan, CO, 159th Engr Gp, 24 Feb 72, VNIT 1038, pp. 4, 8–9, CMH.

original bridge washed away by monsoon floods in late 1967. The new bridge not only cut travel time between Da Nang and An Hoa by half, but it also increased the allied capability to support tactical and pacification operations and economic development into the An Hoa basin. During the same month, the 1st Marine Engineer Battalion completed upgrading Route 4 to western Quang Nam Province. Before departing Vietnam, Seabee units completed road and shoulder work along with permanent and timber bridges on Highway 1 north and south of Da Nang. Naval Mobile Construction Battalion 5 resurfaced the steep, winding, and heavily traveled section through the Hai Van Pass. South of Da Nang, persistent enemy harassment failed to halt Naval Mobile Construction Battalion 62's improvement of Route 4. Naval Mobile Construction Battalions 62, 10, and 74 also completed three major concrete and steel highway bridges at Dong Ha, Quang Tri, and Cau Do (just south of Da Nang), respectively.[46]

In mid-1969, RMK-BRJ was completing a work-in-place rate of more than $6 million a month. By January 1970, the consortium had completed 375 miles with about 322 remaining. Overall, its share of the nearly $500 million highway program came to about $300 million. Normally, the assignment to construct the most difficult permanent bridges, including the manufacture of prestressed concrete bridge beams and pilings used in the highway program went to RMK-BRJ. For instance, from May through July 1971, the firm started building five major bridges near Saigon costing nearly $20 million. These included the 1,343-foot Tan An Bridge and the 1,738-foot Ben Luc Bridge along Highway 4 in the delta; the 2,654-foot Bien Hoa Bridge across the Dong Nai River leading to the Phu Cuong and North Capital Bypasses; the 1,819-foot Binh Loi Bridge over the Saigon River on Highway 13; and the 1,571-foot Saigon River Bridge on the North Capital Bypass. The consortium operated quarries and asphalt plants throughout the country, including the Saigon University industrial site, a major supplier to the Delta Rock Agency, and the Phu Cat site, which supplied rock and asphalt in the Qui Nhon region. Although normally assigned the less dangerous routes, the firm's crews still endured enemy harassment. Several incidents took place in early 1970 on Highway 1 between Qui Nhon and Nha Trang. In March, a project manager was killed by a mine north of Chi Thanh between Song Cau and Tuy Hoa. In that incident, a Vietnamese worker was killed and another badly injured. The same month, three South Korean soldiers were killed guarding the firm's Chi Thanh installation. In the next three months, three more supervisors became

---

[46] Charles R. Smith, *U.S. Marines in Vietnam: High Mobility and Standdown, 1969* (Washington, D.C.: History and Museums Division, Headquarters, U.S. Marine Corps, 1988), pp. 267–68; Tregaskis, *Building the Bases*, pp. 298–99, 412–14, 417–18; Lt. (jg) J. A. Schroeder, "The Col Co Road to Hue," *Military Engineer* 61 (May-June, 1969): 165–68. For more on Seabee road and bridge work, see Tregaskis, *Building the Bases*, pp. 409–22; I Corps highway construction summaries in Quarterly Hist Rpts, MACDC, for years 1969–1971; MACV History, 1968, vol. 1, pp. 475–80; MACV History, 1969, vol. 2, pp. IX-98 to IX-100; Lt. James M. Ramsey, USN, and Lt. (jg) John O'Blackwell, "The Cam Lo River Bridge," *Military Engineer* 62 (January-February 1970): 29–31; Lt. Cmdr. Gordon W. Callendar Jr., "Seabee Bridging at Hue," *Military Engineer* 62 (September-October 1970): 316–19.

casualties—one by another mine, the second in an ambush south of Tuy Hoa, and the third during a mortar attack at Chi Thanh. As for the thousands of Vietnamese workers, RMK-BRJ's prolonged stay in Vietnam afforded them the opportunity to learn new skills, which hopefully could be a vital source for future economic development after the war.[47]

U.S. military engineers also undertook several bridge projects of the scope usually assigned to the consortium. The 1,002-foot Phu Cuong Bridge spanning the Saigon River north of Saigon developed into one of the largest and most complex bridge projects undertaken by engineer troops in Vietnam. Begun in October 1967, and opened to traffic in late June 1968, the bridge completed a key link along the Phu Cuong Bypass. This main supply route bypassed Saigon and connected the depot complex at Long Binh with U.S. Army bases at Di An, Phu Loi, Cu Chi, and Lai Khe. Units of the 20th Brigade's 159th Engineer Group tasked to do the bridge included Company B, 92d Engineer Construction Battalion, and the 41st Engineer Port Construction Company. Construction consisted of fourteen 60-foot spans and two 81-foot navigational spans with approximately 20 feet of clearance above mean high water. The two abutments and five land-based piers were constructed of built-up steel members. Depths of penetration for the piles through the riverbed's organic silt averaged over 200 feet with a maximum depth of approximately 270 feet for several piles. Tidal fluctuation of eight feet in the river and the current caused tricky problems during the alignment of the steel jacket legs into which the piles were placed. Over six miles of steel piping alone were used in the piers. All circular piles and their prefabricated jackets were filled with concrete for additional rigidity. The two navigational spans consisted of eighteen 81-foot prestressed concrete beams, and the remaining fourteen spans held five steel stringers and precast concrete deck panels. Getting the 22-ton, 81-foot beams to the job site and in place turned into a major engineering operation. Trucks had to haul the beams some fifteen miles over the partially completed road directly onto the bridge. The 41st Engineer Company used Navy cube work barges mounting large cranes to lift the beams and move them from the point of pickup to the span being placed. This movement took place only during the slack tide because of the difficulty of controlling the barges maneuvered into place by a land craft, mechanized. Curbing was cast on the deck, which then was topped off with a four-inch asphaltic concrete wearing surface. An extensive lighting and pier protection system and other fixtures were completed in August.[48]

[47] Conover, "LOC Program in Vietnam," p. 31; Tregaskis, *Building the Bases*, pp. 418, 420–21, 429–30; Quarterly Hist Rpt, 1 Oct–31 Dec 71, MACDC, 8 Mar 72, pp. III-8 to III-9, Historians files, CMH.

[48] Capt. Thomas C. Weaver, "The Phu Cuong Bridge, Construction and Features," *Military Engineer* 61 (March-April 1969): 122–24; Quarterly Hist Rpt, 1 Apr–30 Jun 69, MACDC, 19 Jul 69, p. IV-5, Historians files, CMH; Unit History, 41st Engr Co (Port Const), n.d., p. 1, Historians files, CMH. See also Capt. John E. Schaufelberger, "Precast Concrete Deck Panels," *Military Engineer* 61 (March-April 1969): 124; 20th Engr Bde Press Release, Phu Cuong Bridge and Bypass, 11 Jun 68, Historians Files, CMH; and Interv, 1st Lt. Raymond F. Bullock, 26th Mil Hist Det, with Lt Col Robert L. Crosby, CO, 92d Engineer Battalion, 29 Nov 68, VNIT 336, CMH.

The Phu Cuong Bridge under construction

The construction of the Phu Cuong Bridge did not go unnoticed by the Viet Cong, who saw both military and propaganda value in its destruction. Mortar rounds frequently struck the bridge site during construction. During the Tet offensive, enemy troops launched a major attack on the South Vietnamese Army Engineer School a few hundred yards from the engineers' work camp. American engineers helped their South Vietnamese colleagues in a counterattack that succeeded in keeping the enemy from overrunning the school and bridge. On 6 November 1968, Viet Cong sappers swam to the bridge and expertly fastened a demolition charge that destroyed two spans and caused other major damage. The 65th Engineer Battalion, 25th Division, quickly installed an M4T6 floating bridge and reopened the critical bypass road to traffic within twenty hours. Soon afterward the 41st Port Construction Company returned to make permanent repairs. Capt. Thomas C. Weaver, the company commander, concluded that as little as fifty pounds of explosives had sufficed to do the damage because the original bridge design did not call for cross-bracing of the steel I-beams. An M48 tank sitting on the bridge at the time of explosion added its weight to the heavy weight of the bridge and multiplied the damage. The new design made it more difficult to unhinge the stringers if the concrete deck panels started breaking away. A major salvage operation followed. Divers, often working by feel alone and in changing tides, worked underwater clearing debris, checking piers, and locating damaged bridge components. Driving new piles became especially difficult because of the rubble, and the divers faced the extra hazard of placing charges on the bottom to break up the large pieces of concrete and steel blocking the way. Two piers were shifted several feet, which meant that ten 60-foot-length steel beams had to be extended to 65 feet. This splicing of five-foot extensions took two welders all day and about one hundred pounds of rod to complete the job. Time allowed for repair extended to four months. With additional support from the 159th Group, especially welders from other units and the 92d Engineer Battalion, the 41st beat its adjusted deadlines, completing the repairs on 14 February 1969.[49]

The Bong Son bridges crossing the Lai Giang River on Highway 1 some fifty miles north of Qui Nhon epitomized spans that underwent several changes during the war. During the Bong Son Campaign in 1966, the existing single-lane bridges, a multispan Eiffel truss highway bridge and a railway bridge covered with planking to carry heavier loads, were used. A third crossing during the dry season consisted of a combination causeway and M4T6 floating bridge. When the Viet Cong damaged the Eiffel bridge in October 1966, the 35th Engineer Combat Battalion raised the Eiffel spans and completed the reconstruction in late July several months before the next monsoon season. By late 1969, traffic became so congested that the 84th Engineer Construction Battalion prepared plans to build a new permanent bridge. The 1,634-foot-long span, claimed as one of the longest bridges to be built by American troops in Vietnam, with a dual-lane roadway twenty-four feet wide, evolved into one of the largest and most demanding projects undertaken by the battalion. An engineer task force consisting of Company B and the 536th Port Construction Detachment departed Qui Nhon and set up camp in a grove of coconut trees on the river bank. Although plans called for the removal of several trees, which represented a major portion of the local economy and sentimental value, the engineers relented and carefully fitted

---

[49] AAR, Repair of Phu Cuong Bridge, 26th Mil Hist Det, 4 Apr 69, VNIT 376, CMH; ORLL, 1 Feb–3 Mar 69, 92d Engr Bn, n.d., p. 8, Historians files, CMH; Unit History, 41st Engr Co (Port Const), pp. 1–2. VNIT 376 includes interviews, photographs, and design sketches.

the twenty-four tropical wooden buildings between the trees. This resulted in some of the trees jutting through the overhangs of the tin roofs, and several young trees were transplanted. (During work breaks, villagers harvesting the crops showed their appreciation by treating the Americans to fresh coconuts.) Site preparations also included setting up an industrial plant consisting of a precast deck slab yard, a rebar processing yard, and a stringer yard. Five bulldozers fashioned an earthen dike diverting water to the far shore so that a construction causeway could be built. Concurrent training took place on the operation of forty-ton cranes, welding machines, transit mixers, and special items of equipment. Bridge construction began in early 1970 and lasted until September. The bridge consisted of twenty-six 60-foot and one 70-foot steel spans. Each of the twenty-six piers included a reinforced-concrete cap resting atop ten H-piles. Because the I-beams came in random 30- to 40-foot lengths, all 162 stringers had to be spliced. Welding operations continued day and night. Once the stringers were in place and welded together with steel braces, a crane lifted eight-inch reinforced-concrete deck slabs into place. Four hundred and eighty-nine slabs went into the completed structure and three inches of asphaltic concrete surfaced the roadway. Handrails, curbs, lighting, and a pier protection system followed.[50]

During the American troop withdrawals, South Vietnamese engineers gradually took over more of the highway program. Before they assumed their part in the highway program, South Vietnamese Army engineers doing road and bridge work received support from U.S. engineer units in the form of materials as long as they followed U.S. specifications. While the Americans concentrated on roads used for tactical or logistical operations, the South Vietnamese, through their advisers and on-the-job training programs sponsored by neighboring U.S. Army engineer and Seabee units, gained experience in the more pacified areas. Over time, South Vietnam's engineering capability developed and expanded in size. Although South Vietnamese Army engineers were faced with many of their own increased combat support and construction missions, the South Vietnamese Army Chief of Engineers, Col. Nguyen Chan, agreed in late 1969 to participate fully in the highway restoration program. Initially, three construction battalions assumed responsibility for three sections totaling 103 miles along Highway 1 in II and III Corps and Highway 4 in IV Corps. This work included the building of fifty bridges totaling 9,445 feet long. In July 1970, U.S. troops transferred another 178 miles, including 65 miles of Highway 14 in II Corps; and sections of Route 7B (Cheo Reo to Highway 14 in central II Corps), Route 2 (Nui Dat to

---

[50] ORLLs, 1 Aug–31 Oct 69, 84th Engr Bn, 31 Oct 69, p. 3, 1 Feb–30 Apr 70, 84th Engr Bn, 30 Apr 70, p. 2, 1 May–31 Jul 70, 84th Engr Bn, 31 Jul 70, p. 2, and 1 Aug–31 Oct 70, 84th Engr Bn, 31 Oct 70, p. 2, all in Historians files, CMH; Capt. Roger L. Baldwin, "Long Bridge at Bong Son," *Military Engineer* 63 (September-October 1971): 334–35; 1st Lt. John Gamble and Spec. Newell Griffith, "Bridge Over Troubled Waters," *Kysu'* 2 (Fall 1970): 18–21. See also articles on the 577th Engineer Construction Battalion's bridge project on Highway 1 south of Tuy Hoa by 1st Lts. John P. Guthrie and Hubert C. Roche, "Cau Ban Thach Bridge," *Military Engineer* 61 (July-August 1969): 270–72; and the 46th Engineer Battalion's Bailey Bridge modification to a key bridge in Saigon by Maj. Maxim I. Kovel and Capt. Richard M. Goldfarb, "New Life for the Fish Market Bridge, Saigon," *Military Engineer* 60 (March-April 1968): 102–04.

Troops of the 84th Engineer Battalion build the new Bong Son Bridge.

Highway 1 in southern III Corps), Route 8A (Rach Gia to Route 27 in western IV Corps), and Route 2B (Ham Tan to Highway 1 in southern III Corps). At the end of 1970, the South Vietnamese commitment stood at 300 miles, of which 17.5 miles were completed.[51]

As the U.S. engineer presence decreased, South Vietnamese Army engineers took over quarries and asphalt plants and received more road-building equipment. In 1970, Vietnamese engineers took over operations at the Freedom Hill Quarry outside Da Nang. Later that year, they began on-the-job training at U.S. Army industrial sites, thus easing transfers at Phu Loc and Da Nang in I Corps, Weigt-Davis and Ban Me Thuot in II Corps, and

[51] Ploger, *Army Engineers*, pp. 116, 174–75; Tregaskis, *Building the Bases*, pp. 406–07; Conover, "LOC Program in Vietnam," p. 35; History, LOC Program, OCE Liaison Officer Trip Rpt no. 22, pp. 185, 187, 189. See also Spec. Peter Elliot, "Paving the Way to Vietnamization," *Kysu'* 2 (Fall 1970): 25–28.

South Vietnamese Army engineers became proficient in repairing and building bridges such as this panel bridge.

Nui Le in III Corps the following year. Also in 1971, construction units were authorized and equipped with 35 and later 75-ton-per-hour rock crushers, 5-ton dump trucks, and 150-ton-per-hour asphalt plants. In March 1972, the 6th Construction Group assumed responsibility for operations at the Dillard industrial work site south of Da Lat. By early 1973, the South Vietnamese had receipted for approximately 550 items of special heavy construction equipment. These transfers included much of the U.S. Army's commercial equipment, thus passing on the enhanced capability for high-quality road building. Dynalectron continued to supply the parts and repaired the equipment at the work sites.[52]

Bridge building became one task in which the South Vietnamese Army engineers excelled. Before entering the highway program, they completed numerous projects, including the rock-unloading piers for the Delta Rock Agency. In 1967 as the 45th Engineer Group upgraded Highway 1 to Bong Son, the South Vietnamese Army 20th Engineer Combat Group, using U.S. materials, worked on several bridges outside Qui Nhon. In late 1970 when Lt. Col. Francis R. Geisel's 554th Engineer Construction Battalion began upgrading thirty-two miles along Highway 20 near Bao Loc, the South Vietnamese 203d Engineer Combat Battalion already had completed three bridges. Geisel noted that the 203d did a good job despite the poor decking, which was not

---

[52] Conover, "LOC Program in Vietnam," pp. 35–36, 41, 44; MACV History, January 1972–March 1973, vol. 2, p. E-30; Khuyen, *RVNAF Logistics*, pp. 381, 384.

properly creosoted and placed and began rotting, and the shortage of some equipment. The 203d was also building all the bridges using U.S. materials along the 815th Construction Battalion's thirty-one miles of Highway 20 while the U.S. battalion concentrated on roadwork. Lt. Col. George K. Withers, the 815th's commanding officer, recalled in early 1971 that the 203d could get along just fine without his advice and was certain "that they can build bridges much better than we can." In late 1970, the 20th Group's 201st Engineer Combat Battalion finished one of the longest bridges in South Vietnam, the 3,281-foot Tuy Hoa Bridge on Highway 1.[53]

In late 1971, the U.S. troops' roadwork came to a close, RMK-BRJ's share neared completion, and the South Vietnamese took over the balance of the program. Earlier in the year, highway planners, facing the accelerating troop withdrawals, tried to resolve discrepancies and develop more realistic scheduling. By then the Seabees had completed nearly all of their assigned work in I Corps—188 miles, mostly sections of Highways 1 and 9, before departing Vietnam. U.S. Army Engineer Command's share was set at 1,094 miles, of which 751 miles had been completed. The planners deferred 279 miles of national and provincial highways, which would be completed by the Vietnamese. More adjustments followed as Washington hastened the departure of engineer troops and sought ways to close out the costly RMK-BRJ contract. Nevertheless, Maj. Gen. Robert P. Young, who replaced Brig. Gen. Robert M. Tarbox as the director of construction in December 1970, had to contend with transferring more highway and bridge construction to contractors. One approach included hiring Vietnamese contractors under so-called lump-sum contracts, which would have the added benefit of strengthening South Vietnam's construction industry, one of the nation-building goals for the country. This concept became known as the Clarke Plan, named for Lt. Gen. Frederick J. Clarke, who became Chief of Engineers in August 1969, and under whose direction the plan was devised. The Clarke Plan, put in final form in June 1971, called for the transfer of fifty-seven miles of highway in the Mekong Delta. A pilot contract for just over $1.6 million was awarded by the Officer in Charge of Construction to a local Vietnamese firm, and in November the contractor began to upgrade six miles of a badly deteriorated Route 27 north of Binh Thuy. In July, RMK-BRJ took over an additional 17.5 miles as the 20th Engineer Battalion prepared to stand down. The consortium's assigned work increased to 615 miles, with 595 miles completed at the end of the year. In addition, the contractor undertook major repairs, including repaving one hundred miles of Highways 13, 19, and 22 originally completed by the 18th and 20th Brigades. (Final reports still counted the one hundred miles as completed by the troops.) Base course and pavement failures along Highway 19 required the rebuilding of seventy-three miles of road east and west of An Khe. Work started in January with the construction of a base camp and installation of rock crushers and an asphalt plant. Roadwork got under way in March.

---

[53] Intervs, Capt David D. Christensen, 26th Mil Hist Det, with Lt Col Francis R. Geisel, CO, 554th Engr Bn, 17 Jan 71, pp. 2–3, 8–10, and Lt Col George K. Withers Jr., CO, 815th Engr Bn, 15 Jan 71, pp. 2–3, 9, both in VNIT 813, CMH; Khuyen, *RVNAF Logistics*, p. 381; Quarterly Hist Rpt, 1 Oct–31 Dec 70, MACDC, 17 Jan 71, LOC Sec, para 2d(1)(b), Historians files, CMH.

April saw a sharp increase in enemy action against the contractor. These attacks included direct rocket hits on the quarry, causing seven lost days of operation and damage to a generator shed, and an ambush in June that killed two Navy inspectors and wounded three employees. In December, U.S. Army engineers completed their commitment as the 35th Engineer Group's Task Force WHISKEY finished the final section of Highway 1 near Phan Thiet, bringing the total completed by Army troops to 905 miles. An Australian engineering squadron did subgrade repair and emplacement of base rock along Routes 2 and 23 in Phuoc Tuy Province, with paving carried out by RMK-BRJ and South Vietnamese Army engineers. The Directorate of Construction credited the Australians with completing 7.5 miles along Route 23. South Vietnamese Army engineers completed 127 of the 417 miles assigned to them.[54]

During 1972, RMK-BRJ completed its highway assignment, leaving the unfinished portion of the highway program to the South Vietnamese. In the Saigon region, the firm finished its assigned bridge and road projects. These included several bypass roads, most notably the North Capital Bypass in May, as well the realignment and rehabilitation of the eight miles of Highway 1 between Bien Hoa and the bypass. Traffic now could bypass Saigon's fast increasing congestion and converge with key roads outside the city. Also completed by midyear and transferred to the director general of highways were the bridges at Tan An and Ben Luc, the Binh Loi Bridge, the Saigon River Bridge on the North Capital Bypass, and the impressive concrete and steel Bien Hoa Bridge over the Dong Nai River. These immaculate, gleaming, off-white spans included precast concrete bridge beams and pilings manufactured by the firm. In the highlands, daylight attacks along Highway 19 forced RMK-BRJ to move its crews and equipment to Cam Ranh Bay, but they returned and finished the roadwork before the North Vietnamese launched their massive Nguyen Hue (Easter) offensive in April. The firm ended its Vietnam operations in May. Its construction accomplishments included 615 miles of road, many large bridges, and hundreds of small bridges.[55]

In 1972, South Vietnamese Army engineers' growing participation in the highway program resulted in the activation of two new construction groups. Elements from five construction battalions and one combat battalion were now committed to roadwork, and elements of seven additional battalions were tasked with bridge construction. Altogether, the equivalent of nine battalions was committed to the highway program. During the year, Vietnamese Army

---

[54] History, LOC Program, OCE Liaison Officer Trip Rpt no. 22, pp. 189–95; Stearns, Abbott, and Campbell, "LOC Highway Restoration, Vietnam," p. 86; Conover, "LOC Program in Vietnam," pp. 31–32, 34, 35–38; Tregaskis, *Building the Bases*, p. 421; Clarke Plan, incl. 2, tab X, LOC Program Summary, OCE Liaison Officer Trip Rpt no. 22; Army Activities Rpts, 17 Feb 71, p. 25; 19 May 71, p. 61; 9 Jun 71, p. 61; 20 Dec 72, p. 22, CMH; Msg, Maj Gen Robert P. Young, MACV Dir of Const, MAC 01763 to Brig Gen Richard McConnell, Asst Dir of Const Opns, Office of Asst Sec Def (Installations and Logistics), 8 Feb 71, sub: Switch of LOC Construction from Troop to Contractor; Msg, Young MAC 08609 to McConnell, 7 Sep 71, sub: VN Construction Industry Development. All messages in Robert P. Young Papers, MHI.

[55] Tregaskis, *Building the Bases*, pp. 420, 430–31; Stearns, Abbott, and Campbell, "LOC Highway Restoration, Vietnam," p. 86; Army Activities Rpt, 20 Dec 72, p. 22, CMH; Quarterly Hist Rpt, 1 Oct–Dec 71, MACDC, pp. III-8 to III-9.

engineers completed 58 miles of road and 3,400 feet of new bridging, bringing their totals to 181 of 417 miles of assigned roadwork and 24,842 of 37,650 feet of assigned bridge construction. Hanoi's offensive interrupted progress, but South Vietnamese Army engineers rallied to repair 150 bridges totaling 12,264 feet. Of that number, they permanently or temporarily repaired 122 with tactical bridges, bypasses, or rafts. The first Vietnamese lump-sum contractor completed three of the assigned six miles, and the Officer in Charge of Construction awarded two more contracts to Korean firms using Vietnamese subcontractors, one in March for thirty-five miles and another in June for twenty-seven miles. These contractors completed fifteen miles of road in 1972.[56]

Although the extensive development of roads, airfields, and water transportation overshadowed the restoration of the railroad and reduced the urgency of an operational railway system, the slight increases in rail service sustained basic civilian and military operations, particularly on short hauls. Railroad restoration made little progress in early 1968 due to the Tet offensive. In March, General Westmoreland became concerned that the closure of the Hue/Tan My or the Dong Ha/Cua Viet port complexes could seriously hamper operations in northern I Corps. With the influx of reinforcements into the area, he urged greater emphasis on reopening secure land lines of communication, both highway and rail. Planners estimated that opening the Da Nang to Hue rail line, which had been closed since early 1965, would increase the daily military resupply capability by 800 to 1,000 tons. Westmoreland also felt that reopening the railroad would be a tangible symbol to the Vietnamese people of the Saigon government's control of the area. As a demonstration of allied support, MACV assigned the 3d Naval Construction Brigade to do major bridge repairs along the line. In the spring of 1968, a Seabee detachment successfully raised and reconstructed a railway bridge just south of Phu Bai. In mid-July, Naval Mobile Construction Battalion 1 began rebuilding the three remaining demolished bridges on the southwest shores of Lap An Bay. Since the rail line and Highway 1 diverged southeast of the bay and no road ran parallel along this stretch, material and equipment were off-loaded from trucks and placed on rail cars for transport to the work camp and bridge sites. A Vietnam National Railway Service work train locomotive, however, would not be available for several months, and other attempts failed to push the loaded rail cars. The innovative Seabees then borrowed a small Marine Corps four-wheeled vehicle, called the Mule, which fitted exactly on the one-meter tracks and successfully moved the cars without difficulty. When possible, bridge sections were lifted back on to new piers. In November, enemy sappers dropped one side of the railroad bridge near Phu Loc, but the battalion raised and repaired the bridge

[56] MACV Construction Directorate, Lines of Communication Division Briefing, U.S. LOC Construction Progress, LOC Lump Sum Program, FY 1972 AID/DoD Program, Presented to the Combined Central Highway and Waterway Committee Plenary Session, 28 Jun 72, pp. 2, 5–7, Historians files, CMH; Stearns, Abbott, and Campbell, "LOC Highway Restoration, Vietnam," p. 86; Engr Cmd, AVHEN, Fact Sheet, sub: Vietnamization of the LOC Program, 18 Jan 72; MACV History, January 1972–March 1973, vol. 2, pp. E-30 to E-31; Khuyen, *RVNAF Logistics*, pp. 384, 386; Conover, "LOC Program in Vietnam," pp. 36–37.

*Bien Hoa Bridge, completed in 1972 by RMK-BRJ, was part of a new bypass
road north of Saigon.*

before the end of the year. Rail service between Da Nang and Hue resumed on
16 January 1969, and planning proceeded to expand service north from Hue
to Dong Ha.[57]

As the intensity of the Tet offensive lessened, repair trains began reopening other lines. The rail line between Saigon and Xuan Loc was restored, and
on 15 December 1968 passenger and cargo service started on the 200-mile
line between Qui Nhon and Phan Rang. Security continued to plague railway
progress. Yet, despite an average of fifteen interdictions per month, as compared to eight per month in 1967, operational trackage increased to 336 miles,
or about 45 percent of the railway system.[58]

Railway restoration made slow but steady progress. Enemy sabotage persisted, and the Vietnam National Railway Service had to contend with manpower and equipment shortages and the inevitable heavy rains. During 1969,

[57] MACV History, 1968, vol. 1, pp. 658–59; Lt. Theodore I. Harada, "Seabees on the
Railroad," *Military Engineer* 61 (July-August 1969): 262–66.
[58] MACV History, 1968, vol. 1, pp. 660–62.

the Seabees continued to help the beleaguered rail service restore its bridges, which included raising one span and replacing another at the Song Bo Bridge nine miles north of Hue. At U.S. logistical facilities, the construction of rail spurs and sidings by RMK-BRJ and the Vietnam National Railway Service neared completion. These included spurs from the military docks at Newport, Qui Nhon, and Da Nang. By early 1971, the railroad operated in three separate areas with approximately 60 percent, or 441 miles, of the 770 miles of the programmed main line and branch line system. The longest run, some 249 miles, ran from the Long Song River (south of Phan Rang) to Phu Cat. During 1970, cargo transported by rail climbed 15 percent over 1969 (from 595,190 to 685,030 short tons), and the number of passengers transported by rail increased a dramatic 40 percent over the same period (from 1.75 million to 2.4 million). In April 1970, President Thieu temporarily canceled restoration work, but he rescinded this order the following February. The railway service immediately resumed planning to close the remaining gap between Saigon and Phu Cat.[59]

A Legacy

With the return of all roadwork to the South Vietnamese government and the departure of the remaining U.S. troops, the Agency for International Development assumed the Directorate of Construction's responsibility for managing the highway program. Planning began in the fall of 1970 when MACV and the Agency for International Development agreed to set up a joint committee to study the return of the highway advisory function to the civilian agency. A contract consulting team, Booz Allen, was hired to recommend institutional changes that the director general of highways could use to increase its capability. In May 1971, the joint ad hoc committee submitted its detailed report, which evaluated the condition of the highway system, estimated maintenance requirements, appraised the director general of highways' capability, and proposed a timetable for the transfer of functions. One of the intermediate steps took place in July 1971 when the Directorate of Construction reestablished the Highways Engineering and Advisory Branch, as it existed before 1968. The advisory function was transferred on 1 July 1972. The Agency for International Development took over all MACV responsibility for managing the highway program on 1 January 1973, and the Delta Rock Agency in March. That month, the last American troops departed Vietnam in compliance with the Paris Peace Accords that were signed in late January. The only U.S. military connection during the ensuing armistice period continued through the small Engineer Branch, Defense Attaché Office, which provided

---

[59] MACV History, 1969, vol. 1, pp. 97–98; ibid., 1970, vol. 2, pp. IX-71 to IX-77; MACV History, January 1972–March 1973, vol. 2, p. E-22; Railroad Spur Construction, sec. III, Lines of Communication Division (MACDC-LOC), in Quarterly Hist Rpt, 1 Jan–30 Apr 71, MACDC, p. III-15, Historians files, CMH; Tregaskis, *Building the Bases*, p. 417; Dunn, *Base Development*, pp. 109–11.

Flatcars on front of a train south of Phu Cat were intended to detonate mines placed on the tracks by the Viet Cong.

some technical assistance and maintenance support through the Dynalectron contract for the commercial construction equipment.[60]

The director general of highways faced many uncertainties as a highway organization. Although U.S. forces and RMK-BRJ did transfer many major items of equipment, which included five former RMK-BRJ industrial sites and two prestressed concrete plants, it was recognized that the equipment would require a larger and technically skilled workforce. The Directorate of Construction identified the Vietnamese agency's problems earlier in a September 1968 study that emphasized the need for massive reforms in the Saigon government's procedures and policies. The director general of highways did increase its workforce to 6,000 employees in 1972, but the Construction Directorate considered a further expansion to 10,000 by 1975 necessary to maintain and use the equipment and facilities in its charge. Considering Saigon's requirements for military personnel and funding in other areas at the time, the Directorate of Construction doubted that the highway agency would get the needed expansion. Inevitably, the road network would require more major repairs. Problems facing the director general in 1972 included authority to hire an additional 4,000 workers, procuring over 1,500 pieces of equipment, rising costs, inadequate funding, the loss of over 300 engineers and trained

[60] MACV History, 1971, vol. 3, p. IX-18; ibid., January 1972–March 1973, vol. 2, p. E-30; USAID/Vietnam, Project Status and Accomplishment Report, July–December 1970, Project: Highways Improvements, p. 205; Khuyen, *RVNAF Logistics*, p. 381; Fact Sheet, Engineer Branch, Defense Attaché Office, 1 Apr 74, sub: The ARVN Lines of Communication Program, U.S. Army Engineer Programs, copy in OCE Hist Ofc. For more on the advisory transfer, see USAID/MACV Ad Hoc Committee Final Report, n.d., box 3, 72A/7061, RG 334, NARA.

technicians to the military draft since 1968, and the unfavorable competition with contractors for trained technicians.[61]

At best, the director general attempted to preserve the upgraded highway system, using its own personnel to do maintenance and minor repairs and relying on South Vietnamese Army engineers and contractors to do most major repairs and new construction. Some progress was made in carrying out its maintenance mission. In 1971, the highway department had 105 teams backed up by 16 support teams that prepared cold-mix patches, seal-coated road surfaces, and repaired wooden bridge decks. It also began preparations to train some 600 Regional and Popular Forces troops to do work on the rural roads. Between 1968 and 1971, highway crews repaired 491 miles of road and 28,215 feet of bridging. In 1971, the highway department seal-coated 149 miles of road and made plans to increase this effort to 435 miles. Also between 1968 and 1971, the director general of highways built 8,202 feet of new bridging and could have done more but lacked funds to utilize fully local contractors. It faced many repairs in I Corps because U.S. engineers in the rush to finish work did not upgrade all the bridges and culverts along Highway 1. In some sections, the enemy destroyed newly emplaced culverts, and the troops filled the gaps to keep the road open. As a result, severe flooding followed heavy rains. In 1971, the highway agency replaced twelve bridges and nine culverts and planned to continue this work through 1975.[62]

Following the departure of RMK-BRJ in mid-1972, South Vietnamese Army engineers carried out most of the remaining bridge and highway modernization tasks until the country's collapse in April 1975. Work progressed satisfactorily despite difficulties supplying spare parts and fuel and transporting long and heavy prestressed concrete beams to construction sites. There was also the continuing sabotage and fighting between 1973 and 1975. In 1974, the engineers began producing rock for all military and civilian highway projects in the delta. With help from Defense Attaché Office technicians, the quarry at Nui Sam was modernized with the addition of two 250-ton-per-hour rock crushers. By 31 July 1974, Vietnamese engineers reduced the remaining highway program to 146 miles of highway and 9,298 feet of bridging. In April 1975, only 70 miles of new roadwork and 6,600 feet of bridging remained to be completed. During 1974 and 1975, however, Communist forces intensified their sabotage efforts along the highways. Several sections of Highway 4 were blown up at one time, increasing the demand for more rock and tactical bridging. South Vietnamese engineers reacted by stockpiling rock and dirt along the road west of My Tho. The troops and the highway department increased

[61] MACV History, 1971, vol. 1, pp. IX-19 to IX-20; Conover, "LOC Program in Vietnam," pp. 45–46; For more on the director general of highways' projects, personnel, funding, equipment, and maintenance facilities, see USAID/MACV Ad Hoc Committee Final Report.

[62] MACV History, 1971, vol. 1, pp. IX-19 to IX-20; Briefing by director general of highways at June 28th, 1972 Combined Central Highway and Waterway Committee Meeting, box 1, 74/0089, RG 334, NARA. For more on the director general of highways' organization, its work along Highway 1 in I Corps, and status in late November 1971, see Briefing by MACV Construction Directorate Lines of Communication Division for General Frederick C. Weyand, Dep COMUSMACV, 29 Nov 71, box 4, 72A/7061, RG 334, NARA.

the tactical bridge reserves by recovering panel bridges used by U.S. engineers along secondary roads in the pacification program and substituted Eiffel bridges and steel-timber permanent and semipermanent structures in their place. Nearly 2,296 feet of panel bridging were recovered and placed in reserve stocks. Efforts were also under way to round out incomplete panel bridge and M4T6 floating bridge sets when the collapse occurred.[63]

The nearly completed highway system encompassed modern, paved highways stretching almost the entire length of the country. Of the program's 2,532 miles, the road builders had by late 1972 completed 1,911 miles, with 288 miles in progress or programmed, and 333 miles deferred. These figures did not include the many more local roads and bridges improved or built to support tactical operations and pacification or city streets funded under other programs. All the builders participated in one or more of these projects. In the case of local roads and bridges, the Combined Central Highway and Waterway Committee's standards did not apply, but the unpaved and lower weight-bearing bridges connected hamlets and villages to secondary roads, which in turn joined major roads. Funding allocations, including civilian and military assistance, for the Lines of Communication Program alone totaled $460.3 million, or more than $236,000 per mile. Altogether, thirty industrial sites, including quarries, rock crushers, and asphalt plants, were set up to supply rock and asphaltic concrete. Between 1968 and 1972, these sites provided over 27 million tons of rock and 2.5 million tons of asphalt. The improved roads included Highway 1 connecting Saigon with the coastal cities and towns north to Dong Ha and west to Go Dau Ha, where it connected with Highway 22 to Tay Ninh; Highway 19 from outside Qui Nhon to Pleiku; Highway 21 from Nha Trang to Ban Me Thuot; Highway 13 from Saigon to An Loc; Highway 20 from Da Lat to Highway 1 west of Xuan Loc; Highway 4 from Saigon to the delta, and connecting to other improved roads in this important rice-growing region; Highway 15 connecting Saigon to Vung Tau; and parts of Highway 14 in the highlands between Dak To, Pleiku, and Buon Blech. The bypasses around Saigon diverted traffic and helped improve the area's defenses. Besides the national highways, the highway program included lesser roads such as Routes 9, 10, and 27 in the delta. The improved roads were opened to two-way traffic with bridges of thirty-five-ton capacity and above. Save for some unfinished sections in the delta, the road system provided reliable, rapid surface routes to all the major population centers. The loss of life and vehicles to enemy mines dropped dramatically, and military convoys covered distances in one day what previously required two to three days. The heavy fighting starting with the Easter offensive damaged some roads and bridges, mainly the northern sections of Highway 1 nearest the Demilitarized Zone and Highway 13 near An Loc, but most of the road net survived intact. Since South Vietnam's armed forces depended much more on road transportation than U.S. units, the improved highway system became an important factor as Saigon moved forces to oppose the North Vietnamese. The nation-building value was very

---

[63] Khuyen, *RVNAF Logistics*, pp. 384, 386, 388–89.

evident as the volume of civilian truck traffic increased between major cities, and people settled along the new highways.[64]

Compared to the roads, the railroad system did not live up to its potential for moving military freight because it was so easily interdicted. Only five short portions operated with relative security: Saigon to Xuan Loc, Bien Hoa, and Long Binh; Nha Trang north to Tuy Hoa and south beyond Phan Rang; Qui Nhon to Phu Cat; and Da Nang to Hue. Nevertheless, the railroad was a valuable transportation asset for carrying newly landed cargo on short runs to the depots. One daily freight train from the Saigon port to Long Binh carried the equivalent of 6,800 truckloads a week, thus eliminating a large number of truck runs from Saigon's congested streets and the Bien Hoa Highway. Military cargo could also move by rail from Qui Nhon and Cam Ranh Bay to Phu Cat, Tuy Hoa, Nha Trang, and Phan Rang. The railroad also hauled large quantities of crushed rock from several quarries at low rates. Even though any value gained by rail transportation for U.S. forces diminished as the withdrawal gained momentum, the potential for commercial and military use by the South Vietnamese remained an important nation-building goal.[65]

[64] Conover, "LOC Program in Vietnam," pp. 40–41; Army Activities Rpts, 19 Jul 72, p. 24; 20 Dec 72, p. 22, CMH; MACV History, January 1972–March 1973, vol. 2, p. E-30; Ploger, *Army Engineers*, p. 118; Dunn, *Base Development*, pp. 107–08, 112; Khuyen, *RVNAF Logistics*, p. 379. See also "LOC," *Engineer* 1 (Spring 1971): 18–21; Florreich, "Lines of Communication Restoration Program," pp. 11–16; and HQ, Engr Cmd, AVCC-CC, Fact Sheet, sub: Tactical and Rural Roads Programs as of 29 Feb 72, incl. 6, OCE Liaison Officer Trip Rpt no. 22.

[65] Dunn, *Base Development*, p. 111; Khuyen, *RVNAF Logistics*, p. 76; Peter Braestrup, "Saigon is Restoring Railroad: Controversial Project Proceeds with U.S. Aid," *Washington Post*, 8 December 1968.

Year of Transition, 1969

The new year, 1969, brought a change in the direction of American policy in Vietnam as the United States committed itself to gradual withdrawal of its forces and to turning over greater responsibility to the South Vietnamese. After Secretary of Defense Melvin R. Laird toured the theater in March, planning began in earnest in Saigon and Washington for the execution of the opening phase of the withdrawals commencing in the summer and for accelerated training of the Vietnamese under the "Vietnamization" concept. The year thus became a time of transition for U.S. forces, including the engineers, as several units departed and others moved to new areas and missions.[1]

The State of the Construction Program

At the start of 1969, the $1.7 billion military construction program passed the 75 percent completion mark. Military engineers and contractors had built six deep-water ports at Da Nang, Qui Nhon, Cam Ranh Bay, Vung Ro Bay, Vung Tau, and Saigon with twenty-seven deep-draft berths providing a capacity of 600,000 short tons per month. In addition, shallow-draft facilities at nine other ports handled 800,000 short tons per month. Mooring buoys and unloading facilities capable of discharging 1.25 million barrels of fuel daily had also been constructed. Over 270 miles of pipeline moved fuel to major bases. Some pipelines were laid in secure areas such as the six-mile line from the Saigon River to Tan Son Nhut Air Base. Longer sections included the 109-mile pipeline running through contested territory between Qui Nhon and Pleiku. Eight jet-capable air bases with fifteen 10,000-foot concrete runway/taxiway systems and parking aprons were in service, supplemented by some eighty-three auxiliary airfields capable of handling C–123 or C–130 cargo planes. Together, these airfields supported over 5,750 aircraft of all types. The engineers had also built four major depot complexes at Long Binh, Da Nang, Qui Nhon, and Cam Ranh Bay. Most of the so-called hard-core logistical requirements such as ports, airfields, and depots were completed.[2]

The construction program remained a shared undertaking between contractors and troops. Contracts absorbed most funds for ongoing and approved projects (64 percent or $260 million against $149 million allocated for troop

---

[1] MACV History, 1969, vol. 1, p. II-3. For more on Vietnamization, see Clarke, *Final Years*, pp. 341–60.

[2] MACV History, 1969, vol. 2, pp. IX-87 to IX-88; Dunn, *Base Development*, pp. 129–30; Heiser, *Logistic Support*, p. 78; Quarterly Hist Rpt, 1 Oct–31 Dec 68, MACDC, pp. VI-4 to VI-5. See also HQ MACV, MACDC-BD, Construction Program South Vietnam (Complex Review), 1 Mar 69, Historians files, CMH.

The busy port at Cam Ranh Bay showing a tanker at a petroleum pipeline pier and ships along side of the four cargo piers.
(below) *Ship-to-shore pipeline at Cam Ranh Bay transferred large quantities of fuel daily.*

construction). However, the troops carried out more projects because their labor was not charged to the program. Although the RMK-BRJ contract was originally programmed to close out in late 1967, the highway program and the impending cutback in troop strength convinced planners to retain a good share of the consortium's construction capability. At the start of 1969, the firm's workforce consisted of 25,000 personnel, staffed and equipped to carry out about $22 million of work per month. (Planners used a rough figure of $1 million per month per one thousand employees.) To operate efficiently, RMK-BRJ required about $100 million of backlog work. It appeared certain at the time that some contract capability would be required as long as U.S. forces were stationed in Vietnam. The contract effort would continue at approximately a 40-30-30 distribution to the Army, Navy, and Air Force, respectively, until the latter two programs were completed. Then the consortium's effort would entirely support the Army. By calculating the firm's work-in-place rate and the backlogs, planners estimated that RMK-BRJ could finish its share of the Air Force and Navy projects by the beginning of August 1969 and January 1970, respectively. Ideally, the Army's backlog would be done by April 1970. Other firms continued to augment the construction effort. The facilities engineering contractors, Pacific Architects and Engineers (the largest next to RMK-BRJ with more than 20,000 employees) and Philco-Ford, also worked on minor new construction projects at bases. Vinnell Corporation, which built and operated several Army central power plants, received word to build three more power plants at Can Tho, Vinh Long, and Cha Rang outside Qui Nhon, raising the total to fifteen. There were smaller firms such as dredging companies and local contractors, the latter with a combined $1.5 monthly construction capability.[3] (*Table 1*)

Much work still remained to be completed by engineer troops. Army engineers, who had the largest share of troop projects, had finished only about 53 percent of their construction jobs. The Seabees were further along, finishing more than 61 percent of their projects. Together, Army, Navy, and Air Force engineers completed almost 56 percent of assigned work. From a fiscal standpoint, Army engineers had more than $104 million of work remaining, the Seabees $36.5 million, and Air Force engineers $8 million. The approximate backlog was thirty-one months.[4]

The Army's construction program came close to the total completion figure only because of the great amount of work done by contractors. By May 1969, the Army's $887.4 million program, managed in Vietnam by the U.S. Army Engineer Construction Agency, reached 72 percent completion. Army engineers finished $127.6 million or 57 percent of $224.6 million set aside for troop construction compared to the contractors $513.6 million or 77 percent of an assigned $662.8 million. More variation took place in highway construction. RMK-BRJ finished

[3] MACV History, 1969, vol. 2, p. IX-88; Army Troop Construction in the Republic of Vietnam, 1 Feb–1 Apr 69, incl. 6, pp. 1–3, OCE Liaison Officer Trip Rpt no. 14; PA&E History, FY 1969, pp. 1, 3; "12 Additional Power Plants," *Vinnell* 11 (May 1969): 10, Historians files, CMH.

[4] MACV History, 1969, vol. 2, pp. IX-88 to IX-89.

TABLE 1—PERCENT OF WORK IN PLACE AT SELECTED LOCATIONS, JANUARY 1969

| | Percent by | |
Location	Troops	Contractor
Saigon	2	98
Tan Son Nhut	6	94
Bien Hoa	21	79
Tuy Hoa	22	78
Can Tho	22	78
Cam Ranh Bay	29	71
Phan Rang	33	67
Nha Trang	37	63
Phu Cat	40	60
Vung Tau	46	54
Da Nang	48	52
Chu Lai	57	43
Long Binh	59	41
Qui Nhon	67	33
Pleiku	71	29
Phu Bai	73	27
An Khe	75	25
Cu Chi	92	8
Total	40	60

Sources: Military Assistance Command, Vietnam, *Construction Status Report, South Vietnam*, 28 Feb 69; JLRB, Monograph 6, *Construction*, p. 125.

42 percent of its share of roadwork compared to only 8.1 percent by the troops. This latter percentage, however, did not reflect the nearly $24 million in Military Construction, Army, funds used to buy commercial construction equipment. The Army received nearly $35 million in construction funds for the highways, but the troops expended only $2.8 million as the approximate value of work in place. Also, the MACV Construction Directorate planned to transfer as much work as possible to RMK-BRJ to enable the troops to do more operational support missions.[5] (*Chart 5*)

In 1969, Army engineers numbered approximately 36,000 soldiers and comprised 10 percent of the total Army force in Vietnam. Of this number, over

[5] Army Troop Construction in the Republic of Vietnam, 1 Feb–1 Apr 69, incl. 6, pp. 1–3, OCE Liaison Officer Trip Rpt no. 14.

CHART 5—ARMY CONSTRUCTION IN PLACE BY CATEGORY, TROOPS, AND CONTRACTOR, JANUARY 1969

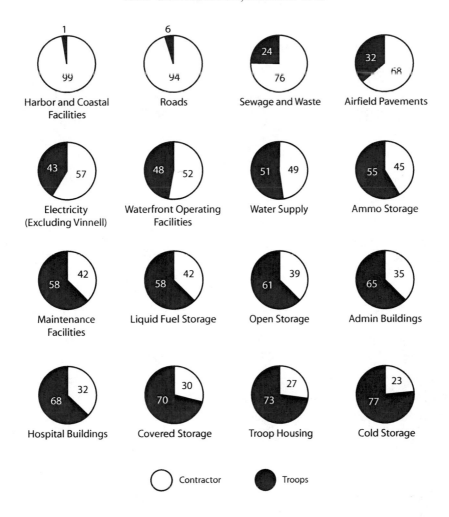

Sources: Military Assistance Command, Vietnam, *Construction Status Report, South Vietnam*, 28 Feb 69; JLRB, Monograph 6, *Construction*, p. 124.

28,000 served under General Parker. Most of the remaining engineers were assigned to the seven Army divisions, the four infantry and airborne brigades, and the armored cavalry regiment. Parker relied on the 18th and 20th Engineer Brigades, the U.S. Army Engineer Construction Agency, and a small planning staff at Headquarters, U.S. Army, Vietnam, to carry out his command's large construction and operational support responsibilities. Subordinate units comprised six groups, twenty-seven battalions, and many separate companies, teams, and detachments. The 18th Brigade under Brig. Gen. John H. Elder Jr., and after May 1969 Col. John W. Morris, reported 13,310 personnel on its rolls (95.5 percent of authorized strength). Col. Harold R. Parfitt's larger 20th Brigade stood near full

447

Aerial view of Cam Ranh Bay from the south and the port to the left and the peninsula filling in with depots and troop billets

strength (99.1 percent) with 13,798 troops. Brig. Gen. Elmer P. Yates' Engineer Construction Agency, which included utilities detachments, reported 1,106 troops on hand, or 87.2 percent of 1,268 authorized personnel. Critically short military occupational specialties ranged from plumbers (231 of 591, or 39 percent of authorized personnel on hand) to auto repair parts specialists (86 of 127, or 67.7 percent). On the other hand, there were almost double the number of carpenters authorized, and almost all the authorized number of quarrymen were present.[6]

A review of the 18th Brigade's projects in the spring of 1969 illustrates the variety and scope of construction carried out by its three groups, thirteen construction and combat battalions, and smaller specialized units. In I Corps, the 45th Engineer Group's four combat battalions took on several projects. The 14th Engineer Battalion expanded a surgical hospital at Camp Evans, and the 27th Engineer Battalion built an 11,520-square-foot UH–1 maintenance hangar, hardstand, and sixty-foot observation tower at Phu Bai. Major road-work carried out by the 19th and 39th Engineer Battalions included upgrad-

---

[6] Visit to Various Headquarters in the Republic of Vietnam, 1–22 May 69, incl. 5, p. 3, OCE Liaison Officer Trip Rpt no. 14. See also tabs H through J for station lists of all units down to company level, tab G listing critical personnel shortages, and tabs K through L for Letters of Instruction issued in January to commanding generals, Engineer Troops [Engineer Command], Engineer Construction Agency, and 18th and 20th Engineer Brigades.

ing Highway 1 between Quang Ngai and Bong Son. In northern II Corps, the 937th Engineer Group's 815th Engineer Construction Battalion at Pleiku completed air-conditioned communications buildings, a 33,450-square-foot aircraft maintenance hangar, maintenance shops, and troop housing. The 20th Engineer Combat Battalion worked on an 8,800-square-foot wood frame post exchange and a 12,800-square-foot wood frame theater for the 4th Division at nearby Dragon Mountain. At Qui Nhon, the 84th Engineer Construction Battalion finished two similar communications facilities, a prisoner of war hospital, two cold-storage warehouses, a prefabricated metal 120-by-200-foot warehouse, and seven 120-by-200-foot two-story wood frame buildings. In the port area, the 536th Port Construction Detachment built an ammunition off-loading facility, and the 84th Engineer Battalion and the 643d Engineer Company (Pipeline) extended two eight-inch petroleum pipelines eight miles to a tank farm. Extensive roadwork under way by the 84th and 815th Engineer Battalions, and the 299th Engineer Combat Battalion progressed along Highways 1, 14, 19, and Route 7B near Cheo Reo. In southern II Corps, the 35th Engineer Group's 864th Engineer Construction Battalion completed another communications facility and a seawall and hardstand at Nha Trang. The 577th Engineer Construction Battalion finished a 10,500-barrel tank farm and cantonment facilities at Tuy Hoa and worked on a 4,500-barrel tank farm at Da Lat. At Phan Thiet and Bao Loc, the 116th Engineer Combat Battalion built base camp facilities. Ongoing projects included the 589th Engineer Construction Battalion's 24,000 square feet of warehouses at Dong Ba Thin and the 87th Engineer Construction Battalion's 200,000-barrel tank farm for the Air Force at Cam Ranh Bay (with help from the 497th Port Construction Company and 643d Engineer Pipeline Company). All five battalions of the 35th Group were improving twenty-five MACV advisory camps, which stood at 48 percent complete and included new billets, mess halls, wells, water storage and sewage systems, and latrines and shower facilities. Also, all of the group's battalions were doing major roadwork along Highways 1, 11, 20, and 21.[7]

Likewise the 20th Brigade accomplished a variety of construction in III and IV Corps. Newly completed base construction at Long Binh by the 159th Engineer Group's 46th and 169th Engineer Construction Battalions included a stockade complete with prefabricated steel administrative buildings and wood frame barracks, mess hall, latrines, administrative buildings, and guard towers. Cantonment and water supply facilities were added for the 1st Division at Lai Khe by the 34th Engineer Construction and 168th Engineer Combat Battalions, and a new wood frame operations building for the Royal Thai Army Aviation Company was finished at Long Thanh North by the 92d Engineer Construction Battalion. Major work under way by the 46th and 92d Engineer Battalions consisted of eight prefabricated steel, open-sided Pascoe warehouse buildings for the rear echelons of the 1st Cavalry and 101st Airborne Divisions at Bien Hoa. The 92d and 168th Engineer Battalions

---

[7] Principal Construction Projects, 35th, 45th, and 937th Engr Gps, 1 Feb–15 Apr 69, incl. 6, tabs A through C, OCE Liaison Officer Trip Rpt no. 14; Ploger, *Army Engineers*, p. 163. See also PA&E History, FY 1969 and FY 1970.

were adding cantonment facilities at Long Thanh and Lai Khe. Also at Lai Khe, the 714th Power Distribution Detachment was building an electrical power distribution system, and at Cogido near Long Binh the 497th Port Construction Company was working on an ammunition off-loading wharf. Roadwork extended along Highway 13 (34th Engineer Battalion), the Phu Cuong Bypass (46th Engineer Battalion), and Highway 20 (169th Engineer Battalion). The 159th Group's 41st Port Construction Company also began the Phu Cuong Bridge pier protection system. In western III Corps, the 79th Engineer Group's 554th Engineer Construction Battalion completed a wood frame communications center, a maintenance hangar, and an aircraft fire station for the 25th Division at Cu Chi. The 588th Engineer Combat Battalion built dog kennels for the division at Tay Ninh, and the 554th Engineer Battalion finished MACV adviser facilities at Bao Trai. Ongoing projects included asphalt paving and an 11,520-square-foot hangar at Cu Chi airfield by the 554th Engineer Battalion and upgrading MACV adviser facilities by the 31st and 588th Engineer Combat Battalions. Highway improvements by the 554th and 588th Engineer Battalions progressed along Highways 1 and 22. In the delta, the 34th Engineer Group finished wood frame cantonments and water and power distribution systems for the 9th Division at Dong Tam (69th and 93d Engineer Construction Battalions), a 1,000-barrel petroleum tank farm at Vi Thanh (69th Engineer Battalion), and airfield paving at Vinh Long (36th Engineer Battalion). Roadwork by the group's 34th, 36th, and 69th Engineer Battalions and the 35th Engineer Combat Battalion covered approximately seventy-three miles.[8]

The Seabees continued supporting U.S. forces in I Corps. At the start of 1969, the 3d Naval Construction Brigade had ten naval mobile construction battalions, which included two battalions of reservists, divided among two construction regiments in I Corps. In addition, two construction battalion maintenance units, one in I Corps and the second at Cam Ranh Bay, detachments from an amphibious construction battalion, and fifteen Seabee teams carried out construction tasks. Battalions continued to rotate during the year, peaking at twelve at midyear and five at the end of the year when the 3d Marine Division departed. The drop in the number of Seabee battalions prompted the departure of the 30th Naval Construction Regiment in December. Highway upgrading made progress despite the added effort required to repair monsoon-damaged roads. Improved stretches extended along Highways 1 and 9. The Liberty Bridge over the Ky Lam River connecting An Hoa and Go Noi Island to Da Nang was completed in March, and the Hue highway bridge opened to two-way traffic in September. Projects for the Navy and Marine Corps in the Da Nang area included 176,000 cubic feet of refrigerated storage, over 300,000 square feet of covered storage, a data processing center, and concrete aircraft shelters at Da Nang Air Base and Marble Mountain Air Facility. Also completed or under way were five hangars and airfield facilities at Quang Tri, sixteen wood frame hangars for the U.S. Army at Camp Evans and Camp Eagle,

[8] Principal Construction Projects, 34th, 79th, and 159th Engr Gps, 1 Feb–15 Apr 69, incl. 6, tabs D through F, OCE Liaison Officer Trip Rpt no. 14; Ploger, *Army Engineers*, p. 163.

a 200-bed hospital for the 1st Marine Medical Battalion in Da Nang, concrete aircraft shelters at Chu Lai, and improvements at Marine Corps combat bases. At the start of 1969, the naval construction forces faced backlogs of twenty-four months of vertical construction and thirty months of horizontal construction. When the troop withdrawals gained momentum in 1969, the backlogs dropped dramatically when the naval command canceled or deferred several projects. By the end of the year, the programmed backlog fell to under four weeks of horizontal construction and under five weeks of vertical work, sufficient for the planned employment of five battalions.[9]

Marine Corps engineers comprised the third cohort of U.S. military engineers in I Corps. The two divisional battalions, two heavy engineer battalions, two bridge companies, and one line company from another battalion swept roads for mines, built firebases, maintained and improved roads, and ran water supply points. They also did some construction. The fifth battalion, the 9th Engineers at Chu Lai (headquarters and three companies) and Tam Ky (part of a fourth line company) focused on maintaining and improving Highway 1 north of Chu Lai, including some operational support for the Army's American Division. With the departure of the 3d Marine Division, the 3d and 11th Engineer Battalions halted all construction projects. Before leaving Vietnam in late November, the 11th Engineer Battalion dismantled two combat bases and helped the 3d Battalion prepare the Cua Viet and Dong Ha Combat Bases for transfer to the South Vietnamese. The 1st and 7th Engineer Battalions stayed with the 1st Marine Division in the Da Nang area, and the 9th Battalion and one bridge company remained at Chu Lai.[10]

In 1969, the five Air Force Red Horse squadrons of the 1st Civil Engineering Group concentrated on the aircraft shelter program, air base improvements, and emergency repairs. Between 10 August 1968 and 31 May 1969, the 554th Civil Engineering Squadron based at Phan Rang Air Base completed forty-one construction projects costing $1.2 million, and included an armament and electronics shop, a flight line fire station, a concrete access taxiway, six troop barracks, two 20-man officers quarters, a 49,000-square-yard asphalt hardstand, and a base theater. The unit also built more than the equivalent of one mile of reinforced-concrete revetment wall to protect Air Force barracks. During the same period, the 819th Civil Engineering Squadron at Phu Cat Air Base finished thirty-two projects totaling $1.8 million. Some of these projects included a 250,000-gallon water storage tank, a 5,000-foot chain-link security fence, new officers quarters, replacement aircraft ramps, and a communications facility. A sewer line and an officers dining hall were among the projects completed at Pleiku. The three other Red Horse squadrons—the 555th at Cam Ranh Bay, the 820th at Da Nang, and the 823d at Bien Hoa—carried out similar work. Red Horse units also built, improved, and manned perimeter defenses. At Phan Rang, teams deployed fourteen times in response

---

[9] MACV History, 1969, vol. 1, p. IV-13, vol. 2, pp. IX-98 to IX-100; Smith, *U.S. Marines in Vietnam, 1969*, pp. 267–69; Quarterly Hist Rpt, 1 Oct–31 Dec 69, MACDC, 19 Jan 70, p. II-3, Historians files, CMH.

[10] Smith, *U.S. Marines in Vietnam, 1969*, pp. 267–70, 377, 382.

to mortar, rocket, and sapper attacks. (The 589th Construction Battalion and other Army tenant units also helped guard their part of the base's perimeter.) At Da Nang, emergency repairs were severely tested on 27 April 1969 when a grass fire destroyed ammunition and bomb storage and fuel storage areas two miles southwest of the air base. Continuous concussions from the blasts also damaged many buildings. Red Horse work crews began repairing operations buildings and the passenger terminal, and within a month most of the damage was repaired. The Air Force also reduced some of its engineering force in 1969. In December, the 555th Civil Engineering Squadron at Cam Ranh Bay was inactivated, and the personnel and equipment were distributed among the four remaining squadrons.[11]

In 1969, aircraft shelters became a high-priority item, especially for the Air Force and Navy. The 1968 Tet attacks revealed the weakness of uncovered revetments when twenty-five Air Force aircraft alone valued at $94 million were destroyed and another 251 were damaged by high-trajectory rockets and mortars. Now faced with an urgent requirement to protect its planes, the Air Force adopted an existing double-corrugated steel arch building design, the "Wonder Arch" named after the building's manufacturer, the Wonder Trussless Building Company. Sections were bolted together and topped off with unreinforced concrete (fifteen inches thick on ridges and twenty-nine inches thick in valleys) which provided good protection. An added freestanding backwall gave equal protection and included an opening to let out jet exhaust. A few shelters could also be fitted with a front closure device. The Air Force program, called Project CONCRETE SKY, originally figured on 408 shelters costing $8.775 million at six air bases. However, sixteen shelters for Phu Cat Air Base were canceled, reducing the number of shelters to 392 and the price to $7.992 million. Production of materials began in mid-1968, and by October the 820th Civil Engineering Squadron began building the first Wonder Arches at Da Nang Air Base. Work got off to a slow start. Instructions were missing, many metal parts had been damaged during shipment, and the arches were larger than the spaces for them in the existing revetments. Despite these and other problems, Red Horse engineers cut the erection time for one shelter from one week to one day. All ninety-eight shelters at Da Nang were completed fifty-two days ahead of schedule. Elsewhere, Air Force engineers built 75 shelters at Bien Hoa, 62 at Tan Son Nhut, 40 at Phu Cat, 61 at Phan Rang, and 56 at Tuy Hoa. Since military engineers lacked enough concrete batch plants for the concrete covers, the MACV Construction Directorate committed RMK-BRJ's large-capacity plants and its fleet of truck-mounted transit mixers. The firm completed 53 concrete covers at Da Nang, 68 at Bien Hoa, and all 62 at Tan Son Nhut. Air Force engineers did the rest. All the shelters were completed by 13 January 1970 at a final cost of $15.7 million, well worth the investment in deterring

[11] Project RED HORSE, pp. 1–2, 7–9, 11, 13, 15–16, 19–25, 29–30; Roger P. Fox, *Air Base Defense in the Republic of Vietnam, 1961–1973* (Washington, D.C.: Office of Air Force History, United States Air Force, 1979), p. 115; Smith, *U.S. Marines in Vietnam, 1969*, pp. 263–64; Quarterly Hist Rpt, 1 Oct–31 Dec 69, MACDC, p. II-4.

Seabees pouring concrete for Wonder Arches at Da Nang Air Base

future damage to expensive aircraft. A comparable Navy program planned 299 shelters at five bases, but this number was eventually reduced to 122 at Da Nang, Marble Mountain, and Chu Lai. Elements of the 3d Naval Construction Brigade began building shelters in July 1969. By mid-1970, the brigade would complete 114 shelters and place concrete covers on all the shelters except 45 at Da Nang, which were poured by RMK-BRJ.[12]

Though the Wonder Arches proved cost-effective for expensive fighter aircraft, ensuring the same degree of protection for all aircraft did not appear warranted. The cost of a completed shelter, including the concrete, came to about $30,000, remarkably cheap when compared to saving a $2 million to $4 million aircraft. When the last shelter at Tuy Hoa Air Base was capped in early 1970, the Air Force had a combined total of nearly 1,400 protective structures, including approximately 1,000 revetments. Such cocoon-type shelters, however, seemed impractical for the large number of Army helicopters in Vietnam. Even flying skid-mounted

[12] Dunn, *Base Development*, p. 65; Tregaskis, *Building the Bases*, pp. 388–90; Fox, *Air Base Defense in the Republic of Vietnam*, pp. 70–71; MACV History, 1969, vol. 2, pp. IX-100, IX-102; ibid., 1970, vol. 2, p. IX-53; Project RED HORSE, pp. 6–7, 14, 18, 94–96; Nalty, *Air War Over South Vietnam*, pp. 35–36; Capt. Paul Y. Thompson, "Aircraft Shelters in Vietnam," *Military Engineer* 61 (July-August 1969): 273–74.

Wonder Arches like these at Bien Hoa Air Base could not be used for all types of aircraft, and the Army used less expensive revetments.

helicopters in and out of uncovered revetments was considered hazardous. A September 1969 study by the Engineer Strategic Studies Group, Office of the Chief of Engineers, reported that to shelter all 4,210 Army aircraft in Vietnam would cost a prohibitive $300 million. Just covering some of the Army's high-cost aircraft—the fixed-wing OV–1 Mohawk surveillance airplane, and CH–47 and CH–54 helicopters—would cost $54 million. Considering losses of Army aircraft in Vietnam, the study concluded it was not cost-effective to undertake a covered shelter program.[13]

Since some protection could be achieved with revetments alone, they were used and given high priority in 1969. The Army typically used the parallel and L-shaped revetments for helicopters and the U-shaped revetment for fixed-wing aircraft. Since aviation units moved frequently, each move required new revetment construction. Many helicopters had revetments in two or more locations. Designs ranged from revetment bins five feet high and four feet thick to portable revetments that could be easily assembled, disassembled, and moved. Precast concrete yards were already manufacturing bridge decking for the highway program, and conversion to movable concrete revetments was easily accomplished. The first concrete revetments were installed at Phu Loi in mid-1970. From late 1970 on, the precast concrete revetment was adopted for wide use and offered the advantage of long life and portability. Also, in early 1970, Army engineers

---

[13] Tregaskis, *Building the Bases*, p. 389; Fox, *Air Base Defense in the Republic of Vietnam*, pp. 70–71; Dunn, *Base Development*, p. 67; Engineer Strategic Studies Group, OCE, Sep 69, An Analysis of the Requirement for Army Aircraft Shelters, pp. ix, H-5.

L-shaped helicopter revetment

developed a practical revetment consisting of two walls of M8A1 runway matting with one foot of compacted earth between them. In March, four-sided revetments were specified for aircraft costing $1 million or more. A movable M8A1 revetment placed on a warehouse cart was developed as the fourth side to be connected and towed away from U-shaped revetments. Still, protection for most Army aircraft consisted of practical expedients such as earth- or rock-filled corrugated steel bins and fifty-five-gallon drums.[14]

Troop Withdrawals and Realignments

On 10 June 1969, President Richard M. Nixon announced that 25,000 U.S. troops would be withdrawn from South Vietnam starting within thirty days. Included in this number were 15,400 Army troops, 8,400 Marines, and 1,200 Navy personnel. Phase I redeployment involved most of the 9th Infantry Division from northern IV Corps and a regimental landing team in northern I Corps from the 3d Marine Division. MACV led off the redeployment with first-rate combat units to make the reduction credible to Hanoi, the American public, and the South Vietnamese government. MACV's other criteria were operational: it would

[14] Dunn, *Base Development*, pp. 67–68; Ploger, *Army Engineers*, pp. 162–63; *Department of the Army Historical Summary, Fiscal Year 1969* (Washington, D.C.: U.S. Army Center of Military History, 1973), p. 112; Capt. Glen E. Brisbine, "Combat Aircraft Revetments—A Review," Military Engineer Field Notes, *Military Engineer* 64 (March–April 1972): 92–94.

reduce the U.S. presence in areas where strong South Vietnamese units, backed by U.S. tactical air, artillery, and helicopter support, could take up all or part of the ground fighting; where progress in pacification and military operations was being made; and where nearby U.S. forces could readily reinforce in emergencies. Since Saigon had to be protected at all costs, III Corps would be exempt from the early reductions. On 16 October, the president announced Phase II reductions of 40,500 men, and a Phase III redeployment of 50,000 men on 15 December, to be completed by 15 April 1970. By then, U.S. military strength in the theater would drop to 434,000.[15]

The first withdrawals involved several engineer units. Most of the 9th Division, including its 15th Engineer Battalion left in September. Company C, however, remained with the 3d Brigade at Tan An in southern III Corps and became the 571st Engineer Combat Company. On 19 June, the 86th Combat Battalion at Camp Viking near My Tho received orders to depart in August. In the past, the 86th had supported the 9th Division and carried out road and bridge construction in III and IV Corps. Attached units (595th Light Equipment Company, 523d Port Construction Company, and 67th Dump Truck Company) were reassigned. Personnel with less than ten months' service in Vietnam were transferred to other engineer units. Line companies in the 86th were reduced to zero strength, and remaining troops were transferred to headquarters company, which departed with the battalion headquarters on 15 August. Most of the battalion's equipment was transferred to other engineer units, and the 93d Construction Battalion took over responsibility for the construction materials in the supply yard.[16]

By the end of the year, two more units had also departed Vietnam. In November, the 70th Combat Battalion, which had served in the highlands for more than four years, left. The 70th had moved from Pleiku to the Ban Me Thuot area in September 1968, where it built several firebases and helicopter revetments for an air cavalry unit moving into the area. About one-half of the engineers were reassigned to other units, equipment was either turned in or transferred to other units, and the main base camp (Camp Jerome) was torn down. On 19 November, the 70th transferred its remaining projects to the 864th Construction Battalion. Company B at the Hot Rocks Quarry east of Ban Me Thuot became Company C, 19th Engineer Combat Battalion. In December, Company C, 307th Engineer Battalion, returned to Fort Bragg, North Carolina, with the 3d Brigade, 82d Airborne Division.[17]

---

[15] MACV History, 1969, vol. 1, pp. IV-10 to IV-13, IV-16, IV-18. See also Clarke, *Final Years*, pp. 346–59.

[16] MACV History, 1969, vol. 1, pp. IV-13, IV-15 to IV-19; Quarterly Hist Rpt, 1 Oct–31 Dec 69, MACDC, p. II-8; Ploger, *Army Engineers*, p. 158; ORLLs, 1 Nov 68–31 Jan 69, 18th Engr Bde, 31 Jan 69, pp. 2–3, and 1 May–31 Jul 69, 86th Engr Bn, 9 Aug 69, pp. 1–2, 11, both in Historians files, CMH. For more on engineer units selected for redeployment, see Msg, Abrams MAC 7181 to Adm John S. McCain, CINCPAC, 6 Jun 69, sub: Force Planning, Abrams Papers, CMH.

[17] ORLLs, 1 Aug–31 Oct 68, 70th Engr Bn, 31 Oct 68, pp. 3, 5, and 1–29 Nov 69, 70th Engr Bn, 29 Nov 69, pp. 2, 5. Both in Historians files, CMH.

The year 1969 also saw the scheduled departures of the two National Guard engineer units sent to Vietnam in 1968. In August, nearly one thousand men of the 116th Engineer Combat Battalion from Idaho and the 131st Light Equipment Company from Vermont reached the end of their tours in southern II Corps. Both moved to nearby transfer stations and to their home states for demobilization ceremonies. The 116th spent most of its tour in the Bao Loc area upgrading or maintaining Highways 1 and 20 and Route 8B. The battalion also provided operational support for Task Force SOUTH, an American and South Vietnamese Army task force operating in southeast II Corps. Other duties involved maintaining Phan Thiet and Bao Loc airfields and improving MACV adviser camps. Civic action projects included providing water to the outlying Montagnard villages and improving drainage and village roads. To the north, the 131st Light Equipment Company, under the 70th Engineer Battalion's control, ran Hot Rocks Quarry, helped rebuild fifty-seven miles of Highway 21 to the juncture with Highway 1 north of Nha Trang, and built several airfields. The 131st also did extensive civic action work. One major project involved relocating three thousand Montagnards from seventeen villages to a more secure central village. Work included clearing and leveling a one-square-mile area for the consolidated village, digging wells, improving roads, transporting people and animals, and even moving houses on low-bed trailers and reassembling them at the new site.[18]

The 19th Engineer Combat Battalion filled the void left by the departing 116th Battalion. This was a logical move since the 19th was nearly done upgrading eighteen and a half miles of Highway 1 from the I/II Corps border to Duc Pho. Control of the battalion passed from the 45th Group to the 35th Group for the second time in sixteen months. On 5 August, an advanced party moved to Bao Loc and began the process of taking over the 116th's projects and equipment. Some of the 19th's equipment and all tactical bridging and construction supplies were either transferred to nearby engineer units or turned in to the Qui Nhon depot. Base camps were transferred or dismantled, including the main camp at Landing Zone LOWBOY. Several units, however, remained behind for a short time to complete the roadwork.[19]

After the 19th Engineer Battalion moved to southern II Corps, it inherited two units formerly attached to the 116th, the 572d Light Equipment Company and the 547th Asphalt Platoon at Camp Smith. After settling in, the 19th proceeded to upgrade fifty miles of Highway 20. Other units also moved.

[18] MACV History, 1969, vol. 1, pp. IV-16 to IV-17; ORLL, 1 Aug–31 Oct 69, 35th Engr Gp, 31 Oct 69, p. 7, Historians files, CMH; Lt. Col. Harvey L. Latham, "The Longest Weekend," *Kysu'* 1 (Summer 1969): 7–9; Ltr, CG, 18th Engr Bde, to CG, Engr Trps, VN, 18 Sep 69, sub: Evaluation of National Guard Units, pp. 6–7, 37–38, Historians files, CMH; AAR, Engineer National Guard Units, 26th Mil Hist Det, 17 Jul 69, pp. 2–3. The AAR also contains interviews from Capt. Raymond F. Bullock, CO, 26th Mil His Det, with Engineer National Guard Units to LOC Program, Vietnam, 35th Engr Gp, 70th Engr Bn, 116th Engr Bn, and 131st Engr Co, Jun–Jul 69, VNIT 446. The evaluation and report contain much information concerning the readiness of the two units, statistics comparing the units with Regular Army engineer units in Vietnam, commanders' evaluations, and interviews.

[19] ORLLs, 1 Aug–31 Oct 69, 35th Engr Gp, p. 7, and 1 Aug–31 Oct 69, 19th Engr Bn, 13 Nov 69, pp. 1, 5–6. Both in Historians files, CMH.

The 497th Port Construction Company, which had been attached to the 20th Brigade, returned to the 18th Brigade and resumed work at Cam Ranh Bay and Qui Nhon. The Earthmoving Platoon, Company C, 93d Construction Battalion, left the 35th Group and rejoined the battalion in the delta.[20]

Vietnamization

If Vietnamization was not a new idea, what was new were the tempo, timing, intensity, and relationship with pacification and the U.S. troop withdrawals. By 1967, the number of Army engineer advisers had increased several times over from around 50 in the early 1960s to around 150. The previous year, the MACV Construction Directorate had taken on advisory responsibilities for South Vietnamese engineer construction units, engineer equipment and maintenance, and the Engineer School at Phu Cuong. The Directorate's Engineer Advisory Division headed by a colonel was authorized nearly sixty officers and enlisted men. The rest of the engineer advisers were assigned to MACV corps and division advisory detachments with the advisers working with corps engineer staffs, combat engineer groups, and divisional and nondivisional engineer battalions.[21]

Training programs also expanded. The Engineer School tripled its student capacity to three thousand to handle the growing population of student officers, officer candidates, noncommissioned officers, and enlisted soldiers. In addition, a selected number of officers and enlisted men were sent to the U.S. Army Engineer School at Fort Belvoir to attend engineering courses. The number steadily increased from 34 in 1965 to 105 in 1968 and 218 in 1969.[22]

When the Americans started to get pacification on its feet in mid-1967, General Abrams, as General Westmoreland's deputy, was charged with upgrading South Vietnam's armed forces. By this time, the South Vietnamese Army engineers were organized to support a national army of ten infantry divisions. Organized along American lines, each division had one engineer battalion of over 400 men. Backing the divisional engineers, each corps now had a combat engineer group consisting of several combat engineer battalions (540 men each) and separate dump truck, float bridge, panel bridge, and light equipment companies. From north to south, the 10th, 20th, 30th, and 40th

[20] Report of Visit to Various Headquarters in the Republic of Vietnam, 1–21 Nov 68, incl. 3, p. 7, OCE Liaison Officer Trip Rpt no. 13; ORLLs, 1 Aug–31 Oct 69, 35th Engr Gp, p. 7, 1 Nov 69–31 Jan 70, 35th Engr Gp, 31 Jan 70, pp. 5–6, 1 Nov 69–31 Jan 70, 19th Engr Bn, 31 Jan 70, p. 1, and 1 Nov 69–31 Jan 70, 864th Engr Bn, 31 Jan 70, pp. 2, 7, 9. ORLLs in Historians files, CMH.

[21] Joint Table of Distribution, Construction Directorate, MACV, 1 May 1966; Talking Paper, sub: MACV Advisory Structure, 23 Jul 67, tab C, pp. 6, 10, and tab E, p. 5, Historians files, CMH. For more on the advisory effort, see Clarke, *Final Years*, pp. 49–78, and Cosmas, *Years of Escalation, 1962–1967*, pp. 288–90.

[22] In late 1964, six South Vietnamese captains and the author attended the same officer career course at Fort Belvoir. Unpublished Monograph, "The Development and Training of the South Vietnamese Army," CMH, 1973, pp. 16–17; Ploger, *Army Engineers*, pp. 167–71; Lt. Gen. Dong Van Khuyen, *The RVNAF*, Indochina Monographs (Washington, D.C.: U.S. Army Center of Military History, 1980), pp. 171, 206.

TABLE 2—ENGINEERS IN SOUTH VIETNAMESE ARMED FORCES
IMPROVEMENT AND EXPANSION PROGRAM, 1968–1973

	End of June					
Major Engineer Elements	*1968*	*1969*	*1970*	*1971*	*1972*	*1973*
Combat Engineer Battalion (Division)	10	10	10	10	11	12
Combat Engineer Battalion (Separate)	13	13	12	12	12	12
Combat Engineer Group Headquarters	4	4	4	4	4	4
Engineer Battalion Construction	8	10	13	17	17	17
Engineer Construction Group Headquarters	2	2	3	4	4	4

Source: Khuyen, *RVNAF*, p. 19

Groups supported I Corps, II Corps, III Corps, and IV Corps, respectively, with twelve combat battalions and several separate companies and platoons. There were two construction groups, the 5th supporting the two southern corps and the 6th in the two northern corps. Combat service support engineers were found in depot, maintenance, post engineer, topographical, utilities, and sawmill units. Altogether, there were over 14,000 soldiers in engineer combat units and nearly 3,000 troops in the engineer construction units.[23] (*Table 2*)

By 1969, the enlargement of the South Vietnamese Army engineer force had gained momentum. In MACV's Phase I expansion plan, the 71st Construction Battalion, the first of the four new construction battalions envisioned in the plan, was organized at Can Tho in June 1968. By the end of September, U.S. advisers reported the new battalion had 85 percent of its authorized personnel and 10 percent of its authorized equipment. Pending the formation of the 7th Construction Group, the 71st trained under the tutelage of the South Vietnamese 40th Engineer Combat Group and operated independently in the Bac Lieu area of southern IV Corps.[24]

For Phase II, MACV planned to add one nondivisional combat battalion, five construction battalions, two construction group headquarters, and three land-clearing companies. By 1971, the Phase II program would result in an engineer force of ten divisional battalions, twelve combat battalions in four

[23] South Vietnamese Army divisional engineer battalions were numbered the same as the parent division, and nondivisional engineer combat units were numbered in a pattern matching respective corps and groups such as the 30th Group in III Corps with the 301st and 302d Engineer Battalions; separate engineer companies also had a similar three-digit unit designation; construction groups were numbered 5th through 8th and their battalions began with the same numbers such as 5th Group and its 51st and 52d Battalions. Force Structure Plan, Republic of Vietnam Armed Forces, FY 1967–1969, Military Assistance Program Directorate, 10 Jan 67, pp. 9, 12; Fact Sheet, MACDC-PP, 19 May 71, sub: ARVN Engineer Troop Strength, both in Historians files, CMH; Quarterly Hist Rpt, 1 Apr–30 Jun 67, MACDC, p. II-11; Quarterly Hist Rpt, 1 Oct–31 Dec 68, MACDC, p. II-8.

[24] MACV History, 1968, vol. 1, pp. 250–51; Quarterly Hist Rpt, 1 Jul–30 Sep 68, MACDC, p. II-4.

groups, and seventeen construction battalions in four groups. This organization, MACV believed, would be able to provide more than enough combat and general support and allow the South Vietnamese Army engineers to take on a meaningful share of the highway program. American planners also developed a smaller force structure, called Phase III, in case of a lesser North Vietnamese and Viet Cong threat. A force of six or eight divisions instead of ten would mean a matching reduction of divisional battalions, and the number of nondivisional battalions would be reduced to twelve. The number of construction battalions would remain the same because of the highway improvement program.[25]

Initially MACV planners envisioned unit-to-unit equipment transfers from departing U.S. units, but in 1969 the transfer process included newer equipment from depot stocks. This decision was prompted by General Abrams who, on 31 May, established criteria designed to ensure that all used U.S. Army equipment was safe to operate, serviceable, and operable to the extent required for its intended purpose. Replacements for any missing or defective parts had to be on order for delivery to the Vietnamese armed forces before a piece of equipment was considered transferable. Equipment not available from the depots came from U.S. units. An engineer unit slated for redeployment or inactivation continued its mission as long as possible, then turned over responsibilities, bases, and equipment to Vietnamese units. In June 1969, the 18th and 20th Brigades received instructions to begin transferring equipment to Vietnamese engineer units on a massive scale. These demands were not excessive since Vietnamese engineer units were authorized less equipment than their American counterparts. During August and September 1969, the 18th Brigade transferred approximately 193 separate major items of equipment to the 40th Base Depot, the only engineer base depot in the Vietnamese Army. In late October, the brigade was tasked to transfer more equipment, this time to the new 805th Heavy Equipment Company at Da Nang and to the 63d Construction Battalion at Nha Trang. The 20th Brigade accomplished similar transfers, including equipment destined for the 63d Construction Battalion. A single U.S. battalion, the 92d Construction at Long Binh, received, stored, staged, and transferred some two hundred pieces of equipment for the brigade.[26]

Meanwhile, South Vietnamese Army engineer units gained valuable experience through training programs sponsored under Project BUDDY by U.S.

---

[25] MACV History, 1969, vol. 2, pp. VI-16 to VI-17, VI-19 to VI-20; Quarterly Hist Rpts, 1 Oct–31 Dec 68, MACDC, p. II-8, and 1 Jan–31 Mar 69, MACDC, p. II-11; Fact Sheet, U.S. Army Engr Cmd, AVHEN-MO, 29 Nov 69, sub: ARVN Engineer Improvement and Modernization, Historians files, CMH; Msg, Abrams MAC 7005 to Stanley R. Resor, Sec Army, 2 Jun 69, sub: Release of Information Pertaining to ARVN Force Levels, Abrams Papers, CMH.

[26] MACV History, 1969, vol. 2, pp. VI-20 to VI-21; Ploger, *Army Engineers*, pp. 172–74; Ltr, Col Harry A. Griffith, CO, 35th Engr Gp, EGA-CO to Brig Gen John W. Morris, CG, 18th Engr Bde, 22 Nov 69; DF, MACJ462, sub: Condition Criteria for Transfer of US Equipment to RVNAF, 7 Nov 69, incl. 12b, and Chart: Construction Equipment Comparison for U.S. and RVNAF Units, incl. 12d, OCE Liaison Officer Trip Rpt no. 16; ORLLs, 1 Aug–31 Oct 69, 18th Engr Bde, 31 Oct 69, pp. 7–8, and 1 Aug–31 Oct 69, 159th Engr Gp, 14 Nov 69, p. 26. Ltr, DF, and ORLLs in Historians files, CMH.

A South Vietnamese soldier operates a grader under the watchful eye of a U.S. soldier.

Army engineer units. Between August and December 1969, the 84th and 815th Construction Battalions trained 167 South Vietnamese soldiers from the 6th Construction and 20th Combat Groups to operate D7E bulldozers, 290M scrapers, twenty- and forty-ton cranes, scoop loaders, and asphalt plants. The 35th Group arranged the training and equipment transfers for the newly activated 63d Construction Battalion. In December, the 20th Brigade reported its units had trained 426 South Vietnamese soldiers, with another 176 undergoing training. On-the-job training sessions included courses on soils analysis by the 36th Construction Battalion, asphalt plant assembly by the 36th Battalion and the 544th Construction Support Company, and quarry operations by the 92d and 169th Construction Battalions and the 94th Quarry Detachment. Through such methods, the Vietnamese quickly learned to operate newer and complex machines. Besides these programs, the 100th Float Bridge Company between May and October provided on-the-job training to the 301st Combat Battalion, 30th Combat Group, on assembling the M4T6 floating bridge.[27]

---

[27] Ploger, *Army Engineers*, pp. 170–71; Dunn, *Base Development*, p. 139; Fact Sheets, 937th Engr Gp, sub: Engineer Training of ARVN Forces, 9 Feb 70, and 35th Engr Gp, sub: Current ARVN Affiliation, 12 Feb 70, both in incl. 15c, and Fact Sheet, 20th Engr Bde, sub: RVNAF OJT Program, 16 Feb 70, incl. 16c, OCE Liaison Officer Trip Rpt no. 16. For more on the

Joint projects, especially highway upgrading, increased, encouraging South Vietnamese engineers to assume more responsibility. In I Corps, the South Vietnamese expanded minesweeping, bridge repair, and land-clearing operations, and began to take on more road maintenance responsibilities. In II Corps, the 18th Engineer Brigade's efforts to help South Vietnamese engineer units develop a greater horizontal construction capability (quarry, road and airfield, and paving operations) began to bear fruit. After South Vietnamese Army engineers joined the highway improvement program in 1969, they began work on the long concrete and steel Highway 1 bridge at Tuy Hoa. The Vietnamese 20th Engineer Group built the bridge, and the U.S. 937th Group provided the materials and technical advisers. In the highlands, 20th Group units provided twenty dump trucks to support U.S. efforts along Highway 14. The 61st Construction Battalion, 6th Construction Group, assumed responsibility for upgrading thirty-four miles of Highway 1 between Phan Rang and Phan Thiet. In early 1970, the 20th Brigade began making plans to transfer the Gia Ray industrial site located midway between Saigon and the II/III Corps border. Sometime after 1 May, South Vietnamese engineers were expected to take over the production of all crushed rock and asphalt concrete for Highway 1 roadwork between Gia Ray and the II/III Corps border. In IV Corps, South Vietnamese engineers assumed a larger share of airfield maintenance and began building major highway bridges. The Vietnamese 40th Engineer Combat Group also took over quarry operations at Nui Sam.[28]

Operation SWITCHBLADE teamed U.S. and South Vietnamese land-clearing units under Project BUDDY. Originally MACV called for the activation of three land-clearing companies beginning in July 1970 and six-month training cycles for each under the tutelage of the 62d Engineer Battalion (Land Clearing). Lt. Gen. Julian J. Ewell, the II Field Force commander, considered land clearing one of his most valuable assets, and he proposed accelerating the activations and training by six months. MACV concurred and the 318th Land Clearing Company, 30th Engineer Combat Group, was activated on 1 December 1969. Lt. Col. Paul C. Driscoll, the 62d's commanding officer, named the training program Operation PLOWSHARE. However, he quickly changed the name to SWITCHBLADE after learning the same code word was used in the program for the peaceful use of atomic energy to make excavations. On 15 December, the 62d welcomed the first group of seventy-five trainees to Long Binh and assigned them to the 501st Land Clearing Company and Company A, the battalion's direct-support maintenance unit. The program consisted of classes followed by on-the-job training with the 62d's officers, tractor-trailer drivers, Rome plow operators, and mechanics acting as instructors on a one-to-one basis. Special seats were installed in the

administrative arrangements of Project BUDDY, see Memo, USARV Engr AVHEN-MO for Dep Engr, 22 Apr 69, sub: Utilization of US Facilities for OJT of ARVN Maintenance Personnel—"Project BUDDY," Historians files, CMH.

[28] MACV History, 1969, vol. 2, pp. VI-147 to VI-148; Fact Sheet, EGC-CS, sub: Engineer Training of ARVN Forces, 9 Feb 70, and Fact Sheet, 35th Engr Gp, sub: Current ARVN Affiliation, 12 Feb 70, both in incl. 15c; Ltr, 20th Engr Bde to CG, USA Engr Cmd, VN, sub: Review of the LOC Restoration Program, 19 Feb 70, incl. 16d, OCE Liaison Officer Trip Rpt no. 16.

cabs for the Vietnamese engineers to sit on and observe the American operators. Later in the training program, the Americans occupied the observers'
seats. From this vantage point, the trainees began to understand the intrica-
cies of land clearing and the importance of daily operational maintenance.
On 15 January 1970, thirty trainees accompanied the 501st to War Zone
C for forty-five days to support a 1st Cavalry Division operation near the
Cambodian border. Several large ten-ton tractors hauling trailers carrying
the Rome plows and their American and Vietnamese operators in the cabs
were driven by Vietnamese soldiers. In late January, a second cycle of thirty
Vietnamese soldiers began training with the 984th Land Clearing Company.
This group spent its forty-five days in the field supporting the 11th Armored
Cavalry Regiment in War Zone C. They were followed by thirty more trainees
who began training with the 60th Land Clearing Company. By the middle of
March, the 318th had almost sixty qualified plow operators, and the training
of support personnel was in high gear.[29]

The next stage of Operation SWITCHBLADE saw the emergence of the
318th as an operational land-clearing company. An earlier than expected inactivation of the 501st Land Clearing Company in April 1970 resulted in the
transfer of ten Rome plows and the formation of the 318th's first operational
land-clearing platoon. A joint provisional land-clearing company consisting
of American and Vietnamese personnel was formed and employed in a cutting
operation in the jungle east of Tay Ninh. The former trainees enthusiastically
manned the lead plows with the Americans following behind. Capt. Nguyen
Van Tich, the 318th's company commander, guided the cut from a low-flying
OH–58 observation helicopter. Tich, who attended the U.S. Army Engineer
School at Fort Belvoir, exuded confidence. When asked if he envisioned problems for the first operation, he replied, "My men are ready. I think we should
be able to match the American engineer's performance, at least on this first
operation." By early May, the operation was proclaimed a success, and the
convoy of trucks and lowboys, which returned the provisional company to
Long Binh, brought an end to direct U.S. supervision. Most of the remaining equipment, including twenty-five additional Rome plows, was transferred
from the 62d Engineer Battalion and other U.S. units during the fifteen-day
stand-down period. MACV's Engineer Advisory Division also assigned one
officer and two maintenance noncommissioned officers as advisers. On 25
May, the 318th began its first forty-five-day cycle. Captain Tich controlled

---

[29] Lt. Gen. Julian J. Ewell and Maj. Gen. Ira A. Hunt Jr., *Sharpening the Combat Edge: The
Use of Analysis to Reinforce Military Judgment,* Vietnam Studies (Washington, D.C.: Department
of the Army, 1974), p. 218; Ltr, HQ, 20th Engr Bde, AVRI-OS, to CG, USARV, 21 Jan 70,
sub: ARVN Land Clearing; Memo, Dep Senior Adviser, III Corps, MACCZ-IIIEN, for Senior
Adviser, III Corps, 7 Feb 70, sub: Training of ARVN Land Clearing Companies; Ltr, HQ, 62d
Engr Bn, Lt Col Driscoll to CO, 79th Engr Gp, 6 Nov 69, sub: Proposed Plan for Activation
of ARVN Land Clearing Companies, all in Historians files, CMH; ARVN Land Clearing
Training (Operation SWITCHBLADE) Feb 70, incl. 16g, OCE Liaison Officer Trip Rpt no. 16;
Interv, Capt Kurt E. Schlotterbeck, CO, 26th Mil Hist Det, with Lt Col Paul C. Driscoll, CO,
62d Engr Bn (Land Clearing), 22 Jun 70, VNIT 707, pp. 50–52, CMH. See also Spec. Bob Dart,
"Vietnamization of Land Clearing—The Beginning," *Kysu'* 2 (Spring 1970): 5–8; and Spec. Neil
Gaston, "Vietnamization of Land Clearing—The Final Phase," *Kysu'* 2 (Summer 1970): 12–16.

the company's operations, and only a few advisers accompanied the unit. The 318th operated like a U.S. land-clearing company, with the 30th Group's S–3 and S–4 sections working closely with their U.S. counterparts. Requisitions for repair parts were submitted to the Area Logistical Command to establish demands on the South Vietnamese supply system. The 62d continued to provide some parts and maintenance support, with the expectation that the South Vietnamese logistics system would take over this responsibility by 1 January 1971. After completing the cutting cycle on 9 July, the 318th reverted to the control of the 30th Group, but the company remained at Long Binh until the group could find a permanent home. This proved beneficial. Neither the 62d nor the 318th by themselves had enough ten-ton tractors and twenty-five-ton trailers on hand to move a company's thirty Rome plows to the field at once. As a simple solution, the two units pooled their tractors and trailers.[30]

Meanwhile, the 62d continued its training program. In late February 1970, soldiers from the 218th Land Clearing Company, 20th Engineer Combat Group in II Corps, started training. The 118th Land Clearing Company, 10th Engineer Combat Group in I Corps, began its training cycle the following month. The methods were the same. Almost every skill required of the U.S. unit was included during the on-the-job training program. Rome plow operators, heavy truck drivers, mechanics, and medical personnel trained and worked together. Vietnamese cooks, however, were not included. Colonel Driscoll recalled, "It wouldn't be too helpful to train them to cook our stuff when they're not going to be able to get it." By midyear, he expressed satisfaction on the progress of the training program. Maintenance skills, however, required more effort than operating the bulldozers because some trainees expected to come in from a cut, park their equipment, and go to bed. They quickly learned that the American operators took pride in making sure the equipment was ready to go the following morning. Driscoll did not see language as a major barrier. "Just put a GI and an ARVN in a cab and in a couple of days," he noted, "the ARVN's got it. It's rather amazing, really. But that was the key to the whole concept of the program, was the ability of the GI to be motivated to get the lesson across." Driscoll noted interesting differences between the South Vietnamese companies. Most of the 318th's soldiers had just completed basic training and were "cooperative and pliable," but the 118th consisted of older, experienced soldiers who required more motivation. This attitude may also have stemmed from the fact that the 118th would not be working with the 62d very long, and the Vietnamese may have considered the program as just another training exercise. However, in the field their attitude changed. The 118th developed a competitive spirit, and its soldiers did not hesitate to help man the berm line defenses when the Americans came under attack. Driscoll

---

[30] Capt. Nguyen Van Tich quoted in Gaston, "Vietnamization of Land Clearing—The Final Phase," pp. 12–16. Also, MFR, HQ II FFV, AVFB-EN, Col John Perkins III, II Field Force Engr, sub: Operation of 318th ARVN Engineer Land Clearing Company, 22 Apr 70, Historians files, CMH; Interv, Schlotterbeck with Driscoll, 22 Jun 70, pp. 52–54.

did not have any qualms about the 118th's future as a successful land-clearing company.[31]

Certain organizational differences allowed the South Vietnamese land-clearing companies to operate more independently than their U.S. counterparts. The Vietnamese companies included a direct-support maintenance section with a third-echelon maintenance capability, unlike the U.S. companies that depended on battalion maintenance. Also, the South Vietnamese planned to assign more tractors and trailers to their land-clearing companies. Colonel Driscoll recalled just before his departure in July 1970 that the Vietnamese land-clearing companies would have twenty 10-ton tractors with low-bed trailers compared to only ten in the U.S. companies. Nevertheless, the Vietnamese companies were not 100 percent mobile. Both the U.S. and Vietnamese companies depended on other engineer units to move thirty Rome plows at one time. Still, Driscoll believed that the South Vietnamese had "a real good TO&E [table of organization and equipment]."[32]

A memorandum of understanding between MACV and the U.S. Army Engineer Command, Vietnam, formalized the training and affiliation programs. Signed on 26 March 1970 by Brig. Gen. John A. B. Dillard, who replaced General Parker in December 1969 as the commander of Engineer Troops, and Maj. Gen. Raymond C. Conroy, MACV's J–4, the memorandum defined the procedures governing training assistance. The document also confirmed that the affiliation program supplemented the Vietnamese training system and MACV's advisory role. General Dillard, who was promoted to major general on 1 April 1970, aggressively pushed his command to accelerate assistance. Most of the command's groups and battalions worked closely with nearby Vietnamese Army engineer units. For instance, the 159th Engineer Group at Long Binh sponsored the 5th Construction Group at Hoc Mon. The 31st Combat Battalion supported the 301st Combat Battalion. Several U.S. units were associated with more than one Vietnamese unit. In I Corps, the 45th Group worked with the 8th Construction and 10th Combat Groups. In the delta, the 35th Combat Battalion sponsored four Vietnamese engineer battalions, the 402d and 403d Combat, the 73d Construction, and the 21st Combat of the 21st Infantry Division. Dillard felt that the South Vietnamese chief of engineers' decision to take on part of the highway restoration program and the Vietnamese engineers' willingness to seek training were indicative of real progress.[33]

[31] Interv, Schlotterbeck with Driscoll, 22 Jun 70, pp. 55–61; Gaston, "Vietnamization of Land Clearing—The Final Phase," p. 16.

[32] Fact Sheet, HQ, USA Advisory Group, III CTZ, MACV, MACCZ-IIIEN, 5 Feb 70, Historians files, CMH; Interv, Schlotterbeck with Driscoll, 22 Jun 70, p. 61.

[33] Ltr, Maj Gen John A. B. Dillard, CG Engr Cmd and USARV Engr, AVCC-CG to Ploger, OCE, sub: RVNAF Engineer Training, 14 Apr 70; Memorandum of Understanding Between MACV J–4 and CG, U.S. Army Engr Cmd, VN, sub: RVNAF Training, 26 Mar 70; 18th and 20th Engr Bdes ARVN Affiliation Program, Associated Organizations, incl. 2 to Ltr, Dillard AVCC-CG to Ploger, 14 Apr 70. All in Historians files, CMH. See also Spec. Curt Nelson, "U.S.-ARVN Training; Benefits on Both Sides," *Kysu'* 1 (Winter 1969): 21–24; and Elliot "Paving the Way to Vietnamization," pp. 25–28.

Turning Over Bases

As the U.S. withdrew more troops from Vietnam, many bases, ports, and other facilities became excess to American needs. Some departing U.S. units either tore down their camps or transferred them to other units. Prefabricated buildings were often disassembled and moved to other bases or returned to the United States. Sometimes new construction requirements arose when U.S. units not leaving Vietnam moved from a transferred base to another base. Although all remaining bases and facilities would eventually be given to the Vietnamese, MACV developed a priority system to allow other U.S. forces and agencies to have a last chance to use them.[34]

These transfers took place under broad guidelines established by MACV and the Joint General Staff. MACV wanted to turn over surplus bases as soon as possible, with priority given to the South Vietnamese armed forces. The command's directive outlined assurances that the U.S. occupants would provide for a continuity of operations and maintenance responsibility. Also, South Vietnamese personnel were supposed to get adequate training in the operation and maintenance of the facility's equipment and systems. Similar directives issued by the Joint General Staff and Central Logistics Command complemented the MACV directive. In November 1969, MACV and the Joint General Staff established a combined U.S.-Vietnamese committee for coordinating base and facility transfers along with similar committees in the four corps.[35]

When it came time to transfer a base to the Vietnamese, it was expected to be usable with a functional utilities system. If sophisticated equipment was essential to American operations at a base, but not for the incoming South Vietnamese, it would be removed. Usually this meant replacing high-voltage systems with low-voltage systems and removing air conditioners except in critical facilities such as hospital operating rooms. Generally, air conditioners were removed with the stipulation that windows would be provided. Still, Lt. Gen. Dong Van Khuyen, the commanding general of the Central Logistics Command, asserted that some U.S.-built bases were "too spacious and consisted of too many buildings and types of structures, all equipped with a complex array of utilities." Frequently, the bases were too big for the Vietnamese units to fill all the barracks, to man all the guard and defense positions, and to maintain all the facilities. Too, departing U.S. units usually left the installations in increments, failing to guard all the unoccupied areas, which gave rise to theft and destruction. "Sometimes," General Khuyen recalled, "entire mobs broke into a few bases and openly dismantled and took away everything in sight in complete defiance of the few PF [Popular Forces] guards." This situation would worsen at several bases as the withdrawal of U.S. forces accelerated.[36]

[34] Ploger, *Army Engineers*, pp. 171–72; Dunn, *Base Development*, p. 142.

[35] MACV History, 1969, vol. 2, pp. VI-123 to VI-124; ibid., 1970, vol. 2, p. IX-36; Khuyen, *RVNAF Logistics*, pp. 179–80. For more on base transfer policies, see MACV History, 1969, vol. 2, pp. IX-102 to IX-105; MACV Directive 735–3, 3 Nov 69, *Disposal of Excess US Armed Forces Real Property and Related Property in the Republic of Vietnam*, Historians files, CMH.

[36] Ploger, *Army Engineers*, p. 172; Khuyen, *RVNAF Logistics*, pp. 181, 184–85 (first quote, p.184, second, p.185); MACV History, 1970, vol. 3, pp. IX-36, IX-40.

The departure of the 9th Division in mid-1969 set the pattern for the departure of tactical units. Redeploying from a combat zone required the division's maneuver units to phase their stand-down and departures while keeping pressure on the enemy. For the 9th Division, this process started in July with the infantry battalions standing-down within their firebases while still conducting security operations before turning over the bases to the South Vietnamese. The units then moved to Dong Tam, where they were billeted aboard barracks ships. Tactical coordination with the South Vietnamese 7th Infantry Division, which operated in the same tactical area, took place without difficulty. As the 9th Division's operations drew to a close, engineering support also phased out. Projects in progress relating to pacification or civic action continued to completion. Not all of the 9th Division, however, departed. The 3d Brigade was reorganized as a separate brigade consisting of four infantry battalions and other elements totaling over 5,800 soldiers.[37]

Dong Tam marked the first major disposal of real property in Vietnam and foreshadowed problems with future base transfers. A joint Army–Pacific Architects and Engineers inventory of the base reported 1,001 buildings and facilities valued at $9.2 million. Some items—such as air conditioners, lighting fixtures, and ceiling fans—were removed and shipped out. Several buildings were also looted and damaged, probably by both departing Americans and arriving Vietnamese, despite the posted Off Limits signs. A Pacific Architects and Engineers real estate specialist complained about South Vietnamese Army transfer officials arriving late and delaying the signing of transfer forms that the facilities engineering contractor's clerks had worked around the clock to prepare. Despite these difficulties, Dong Tam was transferred for South Vietnamese occupancy on 1 September, but the unfamiliar procedures and delays held up the complete transfer until late October. It became obvious that training programs for the South Vietnamese covering certain skills such as operating and maintaining large generators had to be set up. One solution included temporarily assigning personnel from the incoming unit for on-the-job training before the transfer. Another and better approach called for Pacific Architects and Engineers to teach these skills. The firm already had in place an excellent training program for its Vietnamese employees.[38]

In mid-September, Pacific Architects and Engineers began training South Vietnamese soldiers. The first course trained engineers to operate and maintain Dong Tam's central power plant's four 500-kilowatt generators and thirty-nine miles of high-voltage distribution lines. Three American and one Vietnamese instructor assisted by three interpreters taught mechanical and electrical facets of high-voltage generation and distribution. Each course lasted seventeen weeks and a second cycle followed. The firm's electrical supervisor later commented that the Vietnamese soldiers were good students, learned fast, and were

---

[37] MACV History, 1969, vol. 3, pp. D-1 to D-2, D-5, D-13. For more on the redeployment of the 9th Division, see MACV History, 1969, vol. 3, an. D.

[38] PA&E History, FY 1970, pp. 12–13, and app., PA&E Memo, Froilan V. Udtujan, Real Estate Mgmt Spec, for Acting Mgr, Engr Dept, sub: Trip Report to Dong Tam Installation, 9 Sep 69; Quarterly Hist Rpt, 1 Oct–31 Dec 69, MACDC, p. V-6; MACV History, 1970, vol. 2, pp. IX-35 to IX-36, IX-40.

interested in the work. The engineers completed the second cycle at the end of May 1970, fully qualified as high-voltage technicians. Later, the contractor's program moved to Nha Trang Air Base where the firm taught 150 Vietnamese airmen comprehensive facilities engineering courses that ranged from eight to twelve months.[39]

The departure of the 9th Division and the transfer of Dong Tam generated new construction requirements for the 34th Group at a time when there were fewer units to do the work. The 3d Brigade, 9th Division, headquarters complex at Tan An required expansion; a new battalion-size base camp at Can Giuoc had to be built; and aviation and support units had to be moved elsewhere, which required new facilities. Little time was allowed despite the imminent departure of the 9th Division's 15th Engineer Battalion and the group's 86th Engineer Combat Battalion. The 34th Group also faced a critical shortage of construction materials. This problem was reduced when engineers removed bunkers from closed firebases and collected all unused lumber at Dong Tam. Materials held as a contingency reserve to repair damaged bridges were also used. The group diverted units from other projects, and work began at the 3d Brigade's headquarters and the base camp at Can Giuoc. Expanding the Tan An airfield and adding helicopter and fixed-wing parking aprons and revetments at Vinh Long and Can Tho also got under way. While the 93d Construction Battalion concentrated its efforts at Tan An, the 36th and 69th Construction Battalions worked on the other aviation projects.[40]

More base transfers followed. In October 1969, Blackhorse, forty miles east of Saigon and former base camp for the 11th Armored Cavalry Regiment, was transferred to the South Vietnamese 18th Division. This time no significant damage took place at the 700-acre camp valued at $1.9 million. Camps in III Corps occupied by the 3d Brigade, 82d Airborne Division, which departed Vietnam in December, were either dismantled or turned over to the Vietnamese. At Vung Tau, several U.S. Army buildings, mostly barracks, worth $3 million were turned over to the Vietnamese national police in March 1970. In February 1970, a portion of the 1st Division's camp at Dau Tieng was transferred, but that part of the camp included Pacific Architects and Engineers facilities. The firm had to move operations to the other side of the camp. Dau Tieng's new status also meant reduced manning. Consequently, the contractor moved the equipment and maintenance shop to a new parent installation at Cu Chi. In addition, 160 tons of supplies and 498 pieces of equipment no longer needed at Dau Tieng were redistributed to other installations in Vietnam. In I Corps, the South Vietnamese 1st Division accepted portions of the 3d Marine Division's base at Dong Ha, and in III Corps the Vietnamese Air Force received former RMK-BRJ facilities at Bien Hoa Air Base. Also, U.S. Army units inherited several vacated camps by the end of the year. The 3d Marine Division's base at Quang Tri became the home base for the 1st Brigade, 5th Infantry Division

---

[39] PA&E History, FY 1970, pp. 76–78. For more on the contractor's training program, civilian and military, see PA&E History's Appendix A, "Training Accomplishments."

[40] ORLL, 1 Aug–31 Oct 69, 34th Engr Gp, 1 Nov 69, pp. 5–8, Historians files, CMH.

(Mechanized). A portion of the departing Philippine Civic Action Group's camp at Tay Ninh was transferred to elements of the 1st Cavalry Division.[41]

Supporting Combat

As in 1968, a major share of the allied effort in 1969 remained committed to finding, fixing, and destroying North Vietnamese and Viet Cong formations and their base areas. In the northernmost provinces of I Corps, the 3d Marine Division, the 1st Brigade, 5th Division, and the 101st Airborne Division guarded the Demilitarized Zone. In southern I Corps, the 1st Marine and Americal Divisions carried out small-unit patrolling and security operations. In II Corps, the 4th Division screened the Central Highlands along the Cambodian border and the 173d Airborne Brigade conducted most of its security operations in Binh Dinh Province. In III Corps, the 1st Cavalry and 25th Divisions defended the area along the Cambodian border. The 1st Division, the 11th Armored Cavalry Regiment, the 199th Infantry Brigade, the 3d Brigade, 82d Airborne Division, and the 3d Brigade of the 9th Division shielded the Saigon approaches. The South Vietnamese armed forces, nearly one million strong, continued their security mission and participated in joint operations with U.S. forces. Allied forces also carried out security missions in their areas of responsibility: two South Korean army divisions protected the coastal lowlands in II Corps; the Korean marine brigade operated south of Da Nang; and the Australian and Thai forces secured the region southeast of Saigon and guarded the approaches from that direction.[42]

During the year, ground combat in South Vietnam lessened significantly, with occasional violent contacts between American and North Vietnamese units. In January, XXIV Corps began a series of offensive operations aimed at North Vietnamese base areas near the Laotian border. A reinforced U.S. Marine regiment struck an enemy stronghold north of the A Shau Valley in Operation DEWEY CANYON and uncovered large caches. In March, the 101st Airborne Division launched a series of operations in the A Shau Valley itself in an effort to destroy the redoubt and to interdict supply routes built by North Vietnamese engineers. One heavy and violent contact between American and North Vietnamese units occurred when the division's 3d Brigade and South Vietnamese Army elements launched a heliborne operation, APACHE SNOW, into the valley. On 11 May, one of the brigade's battalions collided with a sizable enemy force entrenched on Dong Ap Bia (Hill 937), resulting in a bloody battle better known as Hamburger Hill. Three more battalions were thrown into the fray before the crest finally fell. In the United States, the apparent waste of American lives in taking another enemy position and then

---

[41] MACV History, 1969, vol. 2, p. 105; ibid., 1970, vol. 2, pp. IX-35, IX-37; Quarterly Hist Rpt, 1 Oct–31 Dec 69, MACDC, pp. V-4 to V-5; PA&E History, FY 1970, pp. 13–14.

[42] MACV History, 1969, vol. 1, pp. II-13, V-3, vol. 2, p. VI-10; Stanton, *Rise and Fall of an American Army*, pp. 295, 303, 308, 319.

leaving it sparked more criticism on the conduct of the war. Washington sent out word to hold casualties down.[43]

Organic engineer units carried out minesweeping and demolition support and general support in their parent organizations' tactical areas. Divisional elements such as the 8th Engineer Battalion supported the 1st Cavalry Division's border operations by building and improving tactical roads, firebases, landing zones, and forward airfields. In September, the 8th began resurfacing and expanding the vital Bu Dop airstrip, but monsoonal rains and mud hampered building the additional four hundred feet of runway. Frequent enemy mortaring left craters pockmarking the strip, adding more repair work to the project. Tragedy struck the airmobile engineer battalion when a helicopter carrying the battalion commander, Lt. Col. Andre G. Broumas, and members of his staff during a flight from Quan Loi and Bu Dop was shot down, killing all aboard.[44]

A major accomplishment of the 1st Division's 1st Engineer Battalion in 1969 was the opening of nearly fifty-six miles of a direct all-weather supply route between Phuoc Vinh and Song Be. Before 1969, this stretch of Route 1A, actually a crude path, had been closed for five years and supplies had to be airlifted to bases at Dong Xoai, Bunard, and Song Be. The joint U.S.-South Vietnamese Army task force under the coordinating control of Lt. Col. Robert Segal's 1st Engineers got under way in January. As the battalion committed equipment, bridging, minesweeps, and its land-clearing section to the operation, the South Vietnamese 301st Engineer Combat Battalion concentrated on the roadwork and bridge construction. Land-clearing companies of the 62d Engineer Battalion cut 200-yard-plus-wide swaths along both sides of the road. The road was opened and secured in June, and civilian commercial traffic soon took advantage of the improved route.[45]

The 101st Airborne Division's Operation KENTUCKY JUMPER, a division-wide operation that included Operation APACHE SNOW, illustrated the operational support role of the new airmobile engineer battalion. Between 1 March and 31 July, the 326th Engineer Combat Battalion built nine new firebases. The 326th also opened over ninety landing zones, and on 13 June reopened the abandoned 1,500-foot C–7A Ta Bat airstrip on the floor of the A Shau Valley. At Camp Eagle, the division's base camp, and brigade base camps improvements were made to tactical operation centers, fire direction centers, helipads, bunkers, billets, and interior roads. Helicopter airlift of the 326's equipment

---

[43] Davidson, *Vietnam at War*, pp. 612–15. For more on Operation APACHE SNOW, see MACV History, 1969, vol. 2, pp. V-56 to V-57; ORLL, 1 May–31 July 69, 101st Abn Div, 20 Aug 69, pp. 4–6, Historians files, CMH; HQ, 22d Mil Hist Det, Narrative Operation APACHE SNOW, Historians files, CMH; Stanton, *Rise and Fall of an American Army*, pp. 299–302; and Samuel Zaffiri, *Hamburger Hill, May 11–20, 1969* (Novato, Calif.: Presidio Press, 1988).

[44] *Memoirs of the 1st Air Cavalry Division, August 1965–December 1969*, p. 185. See also Spec. Ken Hammond, "The Divisional Engineer," *Kysu'* 1 (Winter 1969): 6–8; and 14th Mil Hist Det, 1st Cav Div, Construction of a Fire Base, 10 Oct 69, Historians files, CMH.

[45] *The First Infantry Division in Vietnam, 1969* (n.p.: 1st Infantry Division, n.d.), "June 1969" and "1st Engineer Battalion" sections; AAR, Opn TOAN THANG pp. 2–64, Phase I, 1st Engr Bn, 20 Feb 69, 1st Engineer Battalion in Vietnam, vol. 2, pp. VIII-19 to VIII-32, CMH; ORLL, 1 May–31 Jul 69, 1st Inf Div, 29 Aug 69, pp. 5–6, Historians files, CMH. See also 1st Engr Bn, Fundamentals of Engineer Support, [1969], Historians files, CMH.

jumped 30 percent over the previous quarter. Together, the airlifts included 144 CH–47 Chinook and 192 CH–54 Flying Crane sorties. Techniques developed to disassemble equipment into transportable sections paid off in early June when the battalion rebuilt the Ta Bat airstrip. Four D5A bulldozers, two M450 bulldozers, two TD6 bulldozers borrowed from the Seabees, four graders, two backhoes, five 3/4-ton dump trucks, three sheepsfoot rollers, a thirteen-wheel roller, and other assorted equipment moved by air from Camp Eagle in several stages. Upon completion of the clay-surfaced airstrip and the first landing and takeoff by a Caribou, the equipment was dismantled for airlift back to Camp Eagle the next morning. Three miles south of Camp Eagle, the battalion reopened and improved an abandoned firebase (later renamed ARSENAL) to counter North Vietnamese rocket attacks. Roadwork to the firebase, which involved making side hill cuts and placing several culverts and the reconstruction of the firebase served as excellent training models for the battalion.[46]

During APACHE SNOW and the bitter fighting for Dong Ap Bia, the 326th airlifted teams often under heavy enemy fire to clear landing zones. The teams, encumbered with chain saws and rucksacks loaded with plastic explosives, rappelled through holes in the dense forest canopy and cut down trees to make landing zones for helicopters to bring in more troops and supplies and evacuate casualties. Engineer teams also opened holes in the canopy so that the infantry's 81-mm. mortars could fire.[47]

In terrain suitable for armored and mechanized infantry operations, attached engineers provided equally important support. In mid-March 1969, the 1st Brigade, 5th Division, formed Task Force REMAGEN around elements of an armored battalion and a mechanized infantry battalion and other supporting arms, including a reinforced engineer platoon. The task force advanced along Highway 9 from Ca Lu to the Khe Sanh plateau. It then turned south, screening the northern flank of the 3d Marine and 101st Airborne Divisions operating in the valleys to the southeast. Company A, 7th Engineer Battalion, led the task force down the dirt road, clearing mines, building bypasses, and spanning streams with armored vehicle launched bridges, which were immediately lifted back onto their carriers after the column crossed. Task Force REMAGEN roamed the area until the end of April, encountering only light resistance and some mortar fire directed from Laos.[48]

In III Corps, the 919th Engineer Company sent its platoons with each of the 11th Armored Cavalry Regiment's three squadrons on their wide-ranging sweeps. Additional support was also available with the activation of a fourth platoon in early 1969. In June, the 4th Platoon moved with the regimental headquarters to Quan Loi. There it made repairs to the headquarters area, improved base defenses, served as a ready reaction for perimeter defense, and escorted convoys along Highway 13. During one ambush along the road, the

---

[46] ORLL, 1 May–31 July 69, 101st Abn Div, pp. 1, 4, 39–42; Bowers, *Tactical Airlift*, p. 493.

[47] AAR, Opn APACHE SNOW, 3d Bn, 187th Inf, 20 Jun 69, p. 1, Historians files, CMH; Zaffiri, *Hamburger Hill*, pp. 66–67, 104–06.

[48] Starry, *Mounted Combat*, pp. 153–54; Smith, *U.S. Marines in Vietnam, 1969*, pp. 64, 66; Stanton, *Rise and Fall of an American Army*, pp. 297–98.

engineers responded by firing their combat engineer vehicle's demolition gun against enemy troops for the first time.[49]

In 1969, the 18th and 20th Engineer Brigades committed nearly one-half of their total effort to operational support missions. Combat and construction engineers worked on airfield and bridge repairs, minesweeps, base camp defenses, aircraft revetments, tactical roads, and land clearing. In I Corps, the 45th Group assigned the 14th Combat Battalion to upgrade five miles of Route 8B to an all-weather two-lane compacted dirt road for the 3d Marine Division, the 39th Combat Battalion to restore an ammunition supply point for the American Division, and the 27th Combat Battalion to build a new helicopter hardstand for the 101st Airborne Division at Camp Eagle.

In II Corps, the 937th Group's 20th Combat Battalion stabilized the subbase and base course and added four inches of new asphalt at the Cheo Reo Airfield. Also in the highlands, the 299th Combat Battalion upgraded Route 512 between Dak To and Ben Het to one-lane all-weather capability. The 20th Combat and 84th and 815th Construction Battalions erected high chain-link stand-off fences around tank farms at An Khe, Qui Nhon, and Pleiku. In the Bong Son area, the 19th Combat Engineers repaired artillery gun pads and helipads for the 173d Airborne Brigade. In southern II Corps, the 35th Group directed the 70th Combat Battalion's building of an underground medical bunker at Ban Me Thuot. The 116th Combat Battalion built nine L-shaped and twenty-one parallel aircraft revetments at Phan Thiet, and the 577th and 589th Construction Battalions constructed gun pads, guard towers, and bunkers at firebases. At Nha Trang, the 864th Construction Battalion resurfaced an aircraft ramp.

In III Corps, the 159th Engineer Group (46th, 92d, and 169th Construction and 168th Combat Battalions) improved defenses at base camps, built bridge pier protection systems, and constructed aircraft parking aprons and revetments. A major undertaking by the 169th was the construction of a 3,800-square-foot reinforced-concrete underground tactical operations center for MACV and III Corps headquarters next to the Bien Hoa Air Base. Also in III Corps, the 79th Group's 31st Combat Battalion completed bunkers and defenses for an advisory team at Song Be, and the 554th and 588th Construction Battalions made airfield improvements for the 1st Cavalry and 25th Divisions.

In IV Corps, the 34th Group's 86th Combat Battalion built gunpads for the 9th Division's firebases, and the 36th Construction Battalion improved perimeter defenses at Vinh Long. At Vi Thanh Airfield, the 69th Construction Battalion upgraded the facility to C–130 status, and the 93d Construction Battalion added aircraft revetments at Dong Tam. Also, the 35th Combat Battalion built 96 bunkers, 9 guard towers, and 7 tactical operations centers

---

[49] Starry, *Mounted Combat*, pp. 154–55; ORLLs, 1 Nov 68–31 Jan 69, 11th Armd Cav Regt, 10 Feb 69, p. 21, and 1 May–31 Jul 1969, 11th Armd Cav Regt, 18 Aug 69, p. 18. Both in Historians files, CMH.

and laid over 15,750 feet of concertina barbed wire at Binh Thuy and Soc Trang base camps.[50]

Combat engineers often formed their own infantry units and ran reconnaissance patrols and small search and clear operations. Such organizations and operations became more prevalent as infantry units withdrew or were committed elsewhere. For instance, the 19th Engineer Battalion endured incessant harassment while working on Highway 1 between Bong Son and Duc Pho. The most serious incident occurred on 22 July 1968 when a convoy of the attached 137th Light Equipment Company was ambushed near Tam Quan, resulting in the deaths of 12 engineers and major damage to vehicles and equipment. During the quarter ending 31 January 1969, the 19th Battalion recorded 202 incidents, including small arms and grenade attacks and mine and booby trap explosions that caused 5 deaths and 31 wounded and damage to vehicles and equipment. Explosives also destroyed twelve culverts, and sappers frequently cut trenches and built barricades of bamboo, wire, and metal along the road. In reaction, the 19th began platoon-size reconnaissance and combat operations, often at night, near base camps. In one instance, the engineer-infantrymen, using night-vision devices, detected an enemy company-size force moving near the perimeter. A call for supporting artillery and fire from a dual 40-mm. "Duster" air defense cannon destroyed the column. In May, during one period of heightened enemy activity, Companies A and D reorganized as infantry to protect work parties and to carry out sweeps, night ambushes, and patrols. One platoon from each company returned to their engineering duties, but the bulk of the two companies continued their infantry role for several months. The battalion continued to carry out infantry missions after moving to the Bao Loc area, where there were no U.S. combat units nearby. Its previous experiences with the enemy on Highway 1 proved valuable for the men of the 19th because they still had to deal with roadblocks, booby traps, ambushes, and sniper fire.[51]

Another notable combat engineer feat in 1969 was the 27th Engineer Battalion's extension of a main supply route deep into the A Shau Valley. Previous expeditions into the valley had found both the enemy and adverse weather almost insurmountable barriers. To retain control permanently would

---

[50] Ploger, *Army Engineers*, p. 163; Principal Construction Projects, 159th Engr Gp, 1 Feb–15 Apr 69, incl. 6, tab F, Engineer Operational Support in the Republic of Vietnam, 1 Feb–15 Apr 69, incl. 7, OCE Liaison Officer Trip Rpt no. 14. For more on the nondivisional combat engineers during this period from the perspective of their commanders, see Interv, Schlotterbeck with Lt Col James E. Hays, CO, 70th Engr Bn, 26 Oct 69, VNIT 492, CMH; Interv, Schlotterbeck with Lt Col George N. Andrews, CO, 31st Engr Bn, 11 Jan 70, VNIT 573, CMH; and Interv, Col Robert D. Arrington, with Lt Col James E. Caldwell, CO, Company B, 31st Engr Bn, 1985, MHI.

[51] ORLLs, 1 Nov 68–31 Jan 69, 19th Engr Bn, 31 Jan 69, pp. 3, 6, 1 May–31 Jul 69, 19th Engr Bn, 11 Aug 69, pp. 2, 14, and 1 Nov 69–31 Jan 70, 19th Engr Bn, 31 Jan 69, p. 3. All in Historians files, CMH. For more on the 22 July 1968 ambush, see AAR, 13th Mil Hist Det, Ambush of the 137th Engineer Company's Convoy in Binh Dinh Province, 12 Aug 68, VNI 187, CMH. See also Interv, Schlotterbeck with S Sgt Phillip V. Hawkins, S–2 NCO, 19th Engr Bn, 30 Mar 70, VNIT 670, CMH; and Spec. Paul Grieco, "Engineer Groundpounders," *Kysu'* 2 (Summer 1970): 26–28.

require the movement of armor and heavy artillery and a large steady flow of supplies from the 101st Airborne Division's base at Camp Eagle across the coastal plain, through the mountains, and into the valley. Beyond Firebase BASTOGNE, Route 547, which was nothing more than a six- to eight-foot cart trail, proceeded westward thirty miles to the junction with Route 548 near the Ta Bat airstrip. When given the go-ahead, the 45th Engineer Group assigned the road project to the 27th Combat Battalion commanded by Lt. Col. Malcolm D. Johnson and after 26 June by Lt. Col. Stuart Wood. Additional units were attached to help. Besides his four line companies and headquarters company, Johnson controlled all or parts of eight other engineer companies. These included two light equipment companies, a panel bridge company, a land-clearing company, an earthmoving platoon from a construction battalion, and a company and platoon from another combat engineer battalion. A combat engineer company from the South Vietnamese 1st Division also joined the engineer task force. This conglomeration of engineer units, totaling some 1,500 men, was designated Task Force TIGER. Work on the all-weather twelve-foot-wide supply link began in March.[52]

The project, code-named Operation HORACE GREELY, featured nearly every facet of road construction. Daily minesweep teams headed out from the firebases and night defensive positions preceding the roadwork. While the teams were checking for mines, the battalion commander and his operations officer reconnoitered the road by air. Blasting and removal of hundreds of tons of rock were required to widen the roadway in areas that had steep drops of 200 to 300 feet with sheer bluffs rising directly above. Explosives were used to such an extent that Lt. Gen. Richard G. Stilwell, XXIV Corps commander, personally requested more when the engineers exceeded their allocation. The 59th Land Clearing Company did much of the cutting and clearing of vegetation and earthwork. Since the vegetation had been defoliated by chemicals for some time, visibility and safety for the advancing plows was eased. The bulldozers' protective cab covers also reduced the danger of large falling trees. Clearing areas bordering the roadway sometimes dropped off sharply to 60 percent grades. The engineers solved this problem by developing an unusual method of tandem dozing called yo-yo dozing. This method used two bulldozers, one positioned on the roadway as an anchor and the other hooked to the first, rear end to rear end, then lowered over the edge of the slope with its plow pointed straight down to clear a swath through the trees and brush. When the dozer with the plow reached the bottom of the grade, the two dozers reversed their winches and the anchor dozer drew the second dozer back up the grade to begin a new cut. This technique was repeated until the 250-yard strip was cleared. Culverts were assembled in the base camps and transported, dangling by hook and line from CH–47s, to the construction sites. Farther down the

[52] Ploger, *Army Engineers*, p. 152; Interv, Schlotterbeck with Lt Col Stuart Wood Jr., CO, 27th Engr Bn, 15 Jun 70, VNIT 680, pp. 6–7, 17, CMH; ORLL, 1 May–31 July 69, 27th Engr Bn, 7 Aug, 69, pp. 3–4, Historians files, CMH; Zaffiri, *Hamburger Hill*, pp. 47–48.

A CH–47 Chinook helicopter drops off preassembled culvert on Route 547.

road, other elements of Task Force TIGER built a 160-foot panel bridge with concrete abutments and piers over the Song Bo River.[53]

By mid-June, Task Force TIGER reached the juncture with Route 548. By working out of the firebases and night defensive positions and the 101st Airborne Division providing security, the task force steadily pushed west. By 18 June, a small trace had been cut opening the route to small tracked and wheeled vehicles. A month later, the engineers had widened the road to pass tanks. By the end of July, 12.5 of the 18.5 miles between Firebase BASTOGNE and Firebase CANNON had been upgraded to a Class 50 two-lane all-weather road. Three new bridges (a 160-foot Class 60 panel bridge, an 80-foot Class 50 one-lane timber trestle bridge, and a comparable 63-foot timber trestle bridge), ninety-five culverts, and four fording sites complemented the roadwork. Work on the route's remaining section did not get beyond pioneer road status, sufficient for the passage of five-ton trucks. In mid-September, the 101st Airborne Division completed another operation (MONTGOMERY RENDEZVOUS) to deny the enemy access into the valley. Before long, however, the withdrawal of American forces precluded future efforts to keep the North Vietnamese Army out of the A Shau. The 101st Airborne Division pulled back toward the coast, and the engineers ceased further roadwork. Again the allies had to concede control of the valley. When Colonel Wood later checked the road, he found it cut by eighty-six rock

---

[53] Ploger, *Army Engineers*, pp. 152–53; Interv, Schlotterbeck with Wood, 15 Jun 70, pp. 7–12; ORLL, 1 May–31 Jul 69, 14th Engr Bn, pp. 5–6. See also 1st Lt. J. P. Donahue, "547, The A Shau Expressway," *Kysu'* 1 (Fall 1969): 6–9.

Bridge under construction along Route 547

slides, three serious enough that it would have taken a bulldozer two days to clear.[54]

In 1969, land-clearing operations continued to prove their effectiveness in supporting tactical operations, and land-clearing units increased in size and number. Before December 1968, three land-clearing platoons (or teams at the time) operated in Vietnam, each equipped with thirty Rome plow blades and protective cabs mounted on D7E bulldozers. Two of the platoons were assigned to the 20th Brigade to carry out land-clearing missions for II Field Force, and the third was assigned to the 18th Brigade to support I Field Force. Other engineer units, particularly the divisional battalions, carried out land clearing on a smaller scale. The successful use of modified bulldozers as a tactical weapon spurred tactical commanders to request more units. Engineer leaders agreed that expansion was needed and began programming organizational changes. With only sixty-four men, existing land-clearing teams could do little more than oper-

---

[54] Interv, Schlotterbeck with Wood, 15 Jun 70, pp. 26–28; ORLL, 1 May–31 Jul 69, 27th Engr Bn, pp. 3–4; ORLLs, 1 May–31 Jul 69, 45th Engr Gp, 31 Jul 69, pp. 2–3, 1 May–31 Jul 69, XXIV Corps, 20 Aug 69, p. 19, and 1 Aug–31 Oct 69, XXIV Corps, 14 Nov 69, pp. 10, 24. ORLLs in Historians files, CMH.

ate the equipment, and their parent units experienced a severe drain on their maintenance capabilities.[55]

Expansion began in December 1968 when the three teams expanded to three companies with the same number of Rome plows and more administrative and maintenance capabilities. One month later, in January 1969, three additional land-clearing companies were activated, bringing the total to six. These were evenly divided between the two brigades, which took different approaches to their command arrangements and locations. The 18th Brigade's area of responsibility encompassed the widespread and restrictive geography of I and II Corps and their road nets. Consequently, it assigned each of its three companies, the 59th, 538th, and 687th, to a group. On the other hand, the 20th Brigade's area of operations included more level expanses in III Corps and road nets adequate for the large convoys of heavy tractor-trailers from a central base to the cutting areas. The assignment of its three companies to an existing battalion seemed the most practical choice. The 62d Construction Battalion based at Long Binh, one of the units programmed for inactivation or personnel cutbacks in 1968, was selected for this purpose. In January 1969, its three lettered construction companies (B, C, and D) were inactivated and replaced by the 60th, 501st, and 984th Land Clearing Companies. Company A, the equipment and maintenance company, was modified for the battalion's new mission. The 62d's authorized personnel strength dropped from 905 to 647, but the actual number hovered around 700, which included many volunteers from the lettered construction companies. Also that month, the brigade reassigned the 62d from the 159th Group to the 79th Group, which directly supported II Field Force operations.[56]

The 62d became the first land-clearing battalion ever to be assembled. Its immediate concerns focused on assigning troops to the right jobs and ensuring an adequate stock of repair parts. Soldiers from the construction battalion were reassigned to suitable specialties or retrained to fill vacancies. Arrangements were made with the 1st Logistical Command to order repair parts to build up stockages. A new commander also arrived. On 13 January, Lt. Col. Valentine E. Carrasco, who guided the battalion's transition from a construction to a land-clearing organization, transferred the colors to Lt. Col. Maximiano R. Janairo. In turn, Colonel Janairo led the new battalion for six months before transferring command to Colonel Driscoll. Driscoll commanded the battalion for a full year instead of the normal six months. When asked why during his end of tour interview, he replied: "I don't know. Nobody told me why it was. When my six months was drawing to a close, I just kept my mouth shut. It could have been that my date was overlooked. It could have been that maybe I had so many skeletons buried in the closet that they didn't feel like foisting them off on someone else." He surmised that the real reason to leave him in

---

[55] Ploger, *Army Engineers*, p. 99.

[56] Ibid., pp. 99–100; ORLLs, 1 May–31 Jul 68, 62d Engr Bn, 13 Aug 68, p. 2, 1 Feb–30 Apr 69, 62d Engr Bn, 10 May 69, p. 2, and 1 Nov 68–31 Jan 69, 159th Engr Gp, 14 Feb 69, p. 2, all in Historians files, CMH; Land Clearing Battalion Chronology, p. 1, incl. 10 to Rpt, 26th Mil Hist Det, 16 Jul 69, sub: Land Clearing, VNIT 522, CMH.

command was to carry out the training at a critical time for the new South Vietnamese land-clearing companies.[57]

During 1969, the 62d incorporated new procedures and refined land-clearing doctrine. At company level, the key to successful operations was close coordination with its security element. Platoon leaders worked closely with their infantry or armored cavalry escorts. Maintenance, a constant problem because of the wear and tear caused by extended stays in the field and dangerous operating conditions, was improved thanks to a more responsive repair parts system. An authorized float of twenty-seven D7E tractors also compensated for any shortage of repair parts. Heavily cratered areas caused by B–52 strikes added to the many hazards involved in clearing operations, especially after vegetation around the crater grew back. It typically took several hours to retrieve a D7E that had crashed through the vegetation and fallen in. Pulling out the larger D9 bulldozer became a major effort. As for the danger from enemy fire, Colonel Driscoll recalled that a soldier assigned to his battalion had a one in three chance of being a casualty and a two in three chance if he was assigned to a land-clearing company.[58]

The transformation to a land-clearing battalion proceeded quickly and was soon followed by clearing operations. Reorganizing one of the existing teams into the 60th Land Clearing Company was not feasible until the team returned from the field. On 14 January, the 60th deployed for the first time. It cleared two hundred yards on each side of Route 1A between Phuoc Vinh and Dong Xoai as part of the 1st Engineer Battalion's road project. Similarly, the 501st Land Clearing Company deferred its reorganization from a team until it returned from the Trang Bang area where it cut hedgerows and destroyed bunker complexes in support of the 25th Division. In January the new 984th Land Clearing Company began organizing. The number of D7E Rome plows increased from nine to fifteen by the end of the month. By then one platoon was ready and was sent to help the 60th along Route 1A. Later operations bore out the effectiveness of company-size land clearing. The 60th cut nearly 5,000 acres and destroyed forty-five bunkers and fighting positions for the 199th Infantry Brigade in the Gang Toi and Hat Dich areas east of Saigon. During this six-week operation, the 60th lost three D7Es and two men were wounded. In mid-January, the 501st moved to War Zone C in support of the 1st Cavalry Division and cleared over 3,700 acres in two weeks. The 984th cut over 3,000 acres in the Trapezoid area for the 1st Division, but the company lost three D7Es, and thirty-five engineers were wounded.[59]

As 1969 progressed, II Field Force wore down the enemy and pushed him back toward his Cambodian base areas. General Ewell exploited these successes by pushing teams of armored cavalry and Rome plows to open roads

[57] Land Clearing Battalion Chronology, VNIT 522, p. 1; Interv, Schlotterbeck with Driscoll, 22 Jun 70, pp. 4–5 (quoted words).

[58] Land Clearing Battalion Chronology, VNIT 522, p. 8; Interv, Schlotterbeck with Driscoll, 22 Jun 70, pp. 15, 66.

[59] Land Clearing Battalion Chronology, VNIT 522, p. 1; Land Clearing Operations (1 Nov 69–31 Jan 70), pp. 1–2, incl. 16f, OCE Liaison Officer Trip Rpt no. 16. Inclosure 16f also has seven overlays depicting the operations during this period.

in War Zone C. Cuts on each side of the roads were reduced to about fifty yards, which automatically sped up operations fourfold. The 100-yard-plus swath allowed adequate aerial observation and did not unduly expose vehicle to ambushes. Also, instead of completely clearing large areas, the 62d adopted a lane-clearing technique, initially a 1,000-yard-plus lane, which would expose an enemy base area. Again much time was saved, and the lanes were further reduced to a little more than 300 to 600 yards, again increasing the savings. This technique was successfully used south of the Michelin Rubber Plantation and east of the Saigon River. By the end of 1969 and in early 1970, the 62d's three companies had spread out in III Corps taking advantage of the first half of the dry season. They cut large areas for the 1st Infantry and 1st Cavalry Divisions, the 199th Infantry Brigade, and South Vietnamese and Thai forces.[60]

The 18th Engineer Brigade's land-clearing operations met with equal success. Soon after its activation on 1 January 1969, the 59th Land Clearing Company moved north to the 45th Group. It began clearing operations for the 3d Marine Division in "Leatherneck Square" near Dong Ha, where it cleared nearly seven thousand acres in the first full-scale land-clearing operation in I Corps. After a one-week maintenance break, the company cleared another one thousand acres for the marines before joining the 27th Engineer Battalion in Operation HORACE GREELY. Besides the work along Route 547, the 59th cleared sections along Route 548 and other areas near firebases and the A Shau airstrip. Later platoons joined the 14th, 27th, and 39th Engineer Battalions in clearing operations. In northern II Corps, the 937th Group's 35th Land Clearing Team expanded into the 538th Land Clearing Company. It cleared sections along Highways 1 and 14 and Route 6B near Qui Nhon. The 18th Brigade's third land-clearing company, the 687th, was formed from the inactivated 87th Engineer Construction Battalion and attached to the 35th Group. Upon completion of its training, the new company moved to southern II Corps and began clearing operations along Highways 11, 20, and 21A. In April, the 2d Platoon was sent to Qui Nhon to clear the area around the petroleum tank farm. In June, the platoon deployed to I Corps to take part in a Marine Corps operation. Other elements of the 687th completed the clearing of almost seven thousand acres. After a maintenance break, the 687th on 30 October resumed clearing operations along Highway 1 and in the Cam Ranh Bay area. One month later, the company moved to Ban Me Thuot to clear both sides of local Route 1.[61]

Elsewhere, engineers supported clearing operations with ordinary bulldozers, some of which were equipped with Rome plow kits. Using Rome plows to destroy bunker complexes proved much more thorough when compared to the usual method of employing demolitions. Often demolitions only destroyed the beams. The holes remained, and the enemy could easily rebuild the bunker. If the engineers blasted the roof in, the enemy could burrow even deeper

[60] Ewell and Hunt, *Sharpening the Combat Edge*, p. 220; Land Clearing Operations (1 Nov 69–31 Jan 70), pp. 1–2, incl. 16f, OCE Liaison Officer Trip Rpt no. 16.

[61] Smith, *U.S. Marines in Vietnam, 1969*, p. 24; ORLLs, 1 Nov 68–31 Jan 69, 18th Engr Bde, 31 Jan 69, p. 2, 1 Feb–30 Apr 69, 18th Engr Bde, 30 Apr 69, p. 3, 1 May–31 Jul 69, 18th Engr Bde, 31 Jul 69, pp. 3–4, 1 Aug–31 Oct 69, 18th Engr Bde, 31 Oct 69, pp. 3–4, and 1 Nov 69–31 Jan 70, 18th Engr Bde, 31 Jan 70, pp. 2–3. All in Historians files, CMH.

and continue to use the area. On 9 June 1969, the 4th Division dispatched an infantry battalion and two 26-man engineer teams from the division's 4th Engineer Battalion to destroy a large bunker complex in the western highlands. Equipment consisted of two D7E bulldozers mounting Rome plow kits, two D7Es with standard bulldozer blades, one combat engineer vehicle with a bulldozer blade, one M48A3 tank, and a trailer-mounted air compressor. The D7Es, especially those with the Rome plow blades, could rip off the tops of bunkers, fill in the holes, and clear the area of vegetation. The combat engineer vehicle and tank provided additional security. During the clearing operation, the compressor blew out the bulldozers' radiators three or four times a day to remove dirt and debris churned up while stripping the jungle. Working closely with the infantry companies, the teams destroyed 1,350 bunkers and cleared potential landing zones. Another method of land clearing was the continued use of anchor chains dragged by a team of two bulldozers. A third D7E followed the chain to help clear excessive debris collected in front of the chain. The 65th Engineer Battalion, 25th Division, complemented the 62d Land Clearing Battalion's work by using this method. Using six D7Es, the 65th daily knocked down an average of 125 acres of trees from four to eight or twelve inches in diameter.[62]

In the delta, with its inundated rice paddies and poor drainage, Rome plows had been considered impractical until, early in 1969, a joint U.S. and Vietnamese Army engineer operation changed that view. One Viet Cong-dominated area between Sa Dec, Vinh Long, and Can Tho, excluding segments next to Highway 4, became known as the Y base area because of its canal network. Rice paddies covered 80 percent of the Y, the rest being cultivated wood lines. Four main canals formed the Y with many others interspersed throughout the area. Thick vegetation around the canals provided the enemy cover and concealment. Removing this vegetation became the major objective in destroying this sanctuary. The area was considered too large for hand clearing, and it was decided to try the Rome plows. After the South Vietnamese infantry cleared out enemy main force units, an infantry security force and the South Vietnamese 9th Engineer Battalion, 9th Infantry Division, moved in by land. Simultaneously, the U.S. 595th Light Equipment Company arrived by water in landing craft. Conventional blade D7E bulldozers built a dirt access road parallel to the main canal far enough to cover a day's land clearing. Underbrush was also cleared some thirty to fifty yards from the canal. From this road, perpendicular branches were extended out to the paddy land to provide access for the Rome plows. This pattern was repeated, the bulldozers building the roads and the Rome plows cutting the trees. Not all the wood line could be cut, and chain and hand saws were used to cut the remaining trees. During the initial thirty-seven-day period, three Rome plows from the company's land-clearing detachment cut five miles along both sides of one canal. Over the

---

[62] MACV Combat Experiences 4–69, MACJ3-053, 3 Nov 69, sec. 1, pp. 1–11, Historians files, CMH; Capt. Robert A. Adams, "Anchor Chain Land Clearing," *Military Engineer* 63 (July-August 1971): 259. See also Capt. David R. Fabian, "Bunker-busting Operation," Military Engineer Field Notes, *Military Engineer* 62 (March-April 1970): 102–03.

next seventy-five days, another 1,833 acres were cleared, linking the northern secure areas with those in the south. The detachment also destroyed over two thousand bunkers along with living areas and a small hospital and captured large quantities of ammunition, equipment, and documents. It succeeded in removing all traces of enemy activity in a fifty-square-mile area. During the operation, the small U.S. contingent suffered nearly 50 percent casualties, and mines disabled one-half of the bulldozers.[63]

At times Army and Marine land clearers pooled their Rome plows into provisional organizations in support of operations. In June, the 2d Platoon of the 687th Land Clearing Company moved to I Corps to take part in the 1st Marine Division's Operation PIPESTONE CANYON, joining personnel and equipment from the 7th and 9th Marine Engineer Battalions to form a provisional land-clearing company operating under the control of the III Marine Amphibious Force. PIPESTONE CANYON took place in two notorious areas, Dodge City and Go Noi Island, approximately six to twelve miles south of Da Nang and four to twelve miles west of Hoi An. For several years, the Viet Cong and North Vietnamese had used both areas as havens and staging areas for attacks into the coastal lowlands and on Da Nang. During the 164-day operation, the combined force transformed Go Noi Island from a densely vegetated tract to a cleared area free of tree lines and other cover long used by the enemy to conceal his movement across the island. Nearly twelve thousand acres were cleared before the platoon rejoined the 687th at Phan Thiet on 1 September. In August, a provisional land-clearing platoon from the 39th Engineer Combat Battalion joined the 9th Marine Engineer Battalion to clear areas for the 1st Marine and American Divisions. One operation swept across an enemy staging area known as Barrier Island about twelve miles south of Da Nang. A joint operation, this effort included an amphibious landing on the seaward side of the island with security provided by South Korean Marines. Another operation cleared a coastal area just north of Duc Pho. In late November, the platoon returned to the battalion's base at Chu Lai for maintenance, and fifteen days later made another amphibious landing along the coast just south of Da Nang. After clearing more than two thousand acres, the platoon returned to Chu Lai on 31 December.[64]

The cumulative effects of land-clearing operations in Vietnam had a decided impact on the enemy. He increasingly had to adjust to the disappearance of his base areas and the unveiling of his trails. Land clearing enabled friendly forces to observe, shoot, and move through hundreds of thousands of acres of territory he considered his domain. Land clearing also went hand in hand with pacification. Traffic flowed on hundreds of miles of roads made safer by pushing back the jungle growth that had provided excellent concealment. Newly cleared land

[63] Capt. Stephen E. Draper, "Land Clearing in the Delta," *Military Engineer* 63 (July–August 1971): 257–59.

[64] MACV History, 1969, vol. 1, p. V-58; ORLLs, 1 May–31 Jul 69, 18th Engr Bde, p. 3, 1 Aug–31 Oct 69, 18th Engr Bde, p. 3, 1 May–31 Jul 69, 45th Engr Gp, p. 3, and 1 Nov 69–31 Jan 70, 39th Engr Bn (Cbt), 31 Jan 70, pp. 1–3, 13, all in Historians files, CMH; Smith, *U.S. Marines in Vietnam, 1969,* pp. 174–87, 270; 1st Lt. J. P. Donahue, "The Clearing of Barrier Island," *Kysu'* 2 (Spring 1970): 26–28.

also had the potential for agriculture and settlement. The tactical effectiveness of the Rome plows continued in 1970 as preparations were being made to deprive the North Vietnamese Army from the immunity of retreat into sanctuaries long extant in Cambodia.[65]

[65] Ploger, *Army Engineers*, p. 103.

15

Year of Cambodia, 1970

During 1970, American policy stayed on course as U.S. forces hastened their withdrawal from Vietnam. For the most part, military operations were characterized by dispersed and highly mobile spoiling attacks, geared to exploiting progress in Vietnamization and pacification. But though the withdrawal proceeded rapidly, the war also took a dramatic upswing with a multidivision operation into Cambodia during the spring.

The redeployments in 1970 took place in three increments. At the beginning of the year, U.S. ground elements included some 330,000 soldiers and 55,000 marines out of a total U.S. force of 475,000. By 15 April, however, the 1st Division, the 3d Brigade of the 4th Division, a Marine Corps regimental landing team, and an Air Force fighter group had departed. As of 15 October, the 3d Brigade of the 9th Division, the 199th Light Infantry Brigade, another Marine Corps regimental landing team, and three Seabee battalions had gone. By the end of December, the rest of the 4th Division and two of the three brigades of the 25th Division returned to the United States. When 1970 ended, U.S. troop strength had dropped to 335,000, including 250,000 soldiers and 25,000 marines.[1]

More U.S. Army engineers also left, but this hardly lessened their workload, and the engineers organized accordingly. The engineer command structure under U.S. Army, Vietnam, which had reached a peak strength of 33,000 in 1969, dropped to about 22,000 by the end of 1970. By then the six engineer groups had dropped to five and the twenty-four battalions to twenty. Inevitably, the command reorganized, as the USARV Engineer and the U.S. Army Engineer Construction Agency consolidated.[2]

Engineer Consolidation

The policy of entrusting to one senior Army engineer the dual responsibilities of USARV staff engineer and command of all Army engineer units that were not organic to other commands had continued since the early days of the buildup. After April 1968, the engineer's dual role titles were Commanding General, Engineer Troops, Vietnam (Provisional), and USARV Engineer. The Engineer Troops, Vietnam, consisted of three subordinate commands—the 18th and 20th Engineer Brigades and the construction agency. Under this

---

[1] MACV History, 1970, vol. 1, pp. I-5, IV-7.
[2] Military Personnel Strength Summary, incl. 6A, OCE Liaison Officer Trip Rpt no. 18, 2 Nov 70, OCE Hist Ofc; ORLLs, 1 Nov 70–29 Apr 71, 18th Engr Bde, 30 Apr 71, p. 2, and 1 Nov 70–15 Apr 71, 20th Engr Bde, 25 Apr 71, p. 3, both ORLLs in Historians files, CMH.

organization, the two brigades commanded all nondivisional engineer combat and construction units while the construction agency centralized control over the Army's construction program and facilities engineering.[3]

In October 1969, USARV headquarters directed its major commands, including the engineers, to start consolidating or eliminating headquarters. The engineers considered several options before deciding to combine the headquarters of the Engineer Troops/USARV Engineer with the Engineer Construction Agency into a new consolidated engineer command and staff headquarters. The consolidation made sense. The Engineer Troops/USARV Engineer and the construction agency were located at Long Binh. There was already overlap and duplication of functions between the USARV Engineer's Facilities Engineering Division and the construction agency's Real Property Management and Real Estate Divisions. There was also uncertainty over responsibility for staff supervision and control and management of troop construction projects between the USARV Engineer's Construction Division and the construction agency's Engineering Division. The USARV Engineer did not have the technical engineering capability to review construction requests, contractor designs, and to do its own designs. This capability existed only in the construction agency's Engineering Division. Finally, the loss of twenty-three spaces in the USARV Engineer required some internal reorganization, and the planners anticipated that more personnel savings would be possible if the headquarters of the Engineer Troops/USARV Engineer and the Engineer Construction Agency were consolidated.[4]

The organization of the new U.S. Army Engineer Command, Vietnam (Provisional), received USARV approval on 7 January 1970 and was formalized as a command on 1 February. General Dillard, the commander, continued to serve concurrently as the USARV Engineer. The Engineer Command staff functioned as the USARV Engineer staff, and consisted of the Directorate of Construction, Directorate of Facilities Engineering, Military Operations Division, Materiel Division, and administrative and special staff offices. With the consolidation of the construction agency as part of the Engineer Command's staff, two major subordinate commands remained, the 18th and 20th Brigades. The authorized strength of the Engineer Command stood at over 26,000 men, making it the second largest subordinate element of USARV after the 1st Logistical Command. General Tarbox, formerly commanding general of the construction agency, became the deputy commander and the deputy USARV engineer.[5]

---

[3] Ploger, *Army Engineers*, p. 179.

[4] Debriefing, Parker, 14 Oct 69, p. 7, Senior Officer Debriefing Program, DA, Historians files, CMH; Interv, OCE Hist Ofc with Maj Gen David S. Parker, 7 Jan 85, pp. 271–72, OCE Hist Ofc; Memo, Maj Gen David S. Parker for OCE Hist Ofc, Dec 1977, pp. 2–3, Historians files, CMH; Briefing, USARV Engineer Reorganization, n.d., incl. 14P, pp. 1, 4–8, 10–11, OCE Liaison Officer Trip Rpt no. 16. See also Briefing, USARV Engineer, 7 Jan 70, Historians files, CMH.

[5] Briefing, USARV Engineer Reorganization, incl. 14P, pp. 11–14, OCE Liaison Officer Trip Rpt no. 16; "Engineer Command Reorganized," *Castle Courier*, 9 February 1970, p. 3. See also Ltr, Lt Gen Frank T. Mildren, AVHEN, to CG, USAECV(P), sub: Letter of Instruction (LOI), 17 Feb 70, incl. 14J, OCE Liaison Officer Trip Rpt no. 16.

Two new positions, the Director of Construction and the Director of Facilities Engineering, were established. Colonels occupied both, and they directed two of the command's most pressing missions, the highway program and facilities engineering services involving base transfers and closures. The Director of Construction oversaw several components carrying out the Army's construction program. The Engineering Division planned, programmed, and designed the Army's construction and highway improvement projects in Vietnam, and the Construction Division supervised and coordinated the accomplishment of these projects. Also, the Director of Construction's Electrical Systems Office managed the Army's high-voltage program in Vietnam. Both divisions maintained close ties and coordinated RMK-BRJ's

General Dillard

army construction projects with the Navy's Officer in Charge of Construction. The Director of Facilities Engineering supervised the Real Estate Division, the Installations Division, and three District Engineer offices. Through its local area offices, the Real Estate Division managed over 1,100 land-use concurrences with the Vietnamese government and over 400 leases totaling about $19 million a year. In 1970, the only Women's Army Corps officer authorized to wear the Corps of Engineers insignia, Lt. Col. Margaret M. Jebb, served as deputy chief of the Real Estate Division. Coordination and supervision of the Pacific Architects and Engineers' facilities engineering contract, now amounting to about $100 million per year, came under the Installations Division. Similarly, the division supervised the high-voltage power operation contract with Vinnell Corporation. In addition, the Installations Division inherited the Base Development Branch, formerly in the construction agency's Engineering Division. Though emphasis on base development had declined, the branch played a key role in base transfers and new requirements when U.S. forces moved from one base to another. The engineer districts remained much as they were under the construction agency except the Greater Saigon Engineer District, now incorporated into the Southern Engineer District. This district had its headquarters in Saigon and oversaw all U.S. Army installations in III and IV Corps. The Central Engineer District at Cam Ranh Bay carried out the same role in II Corps, as did the Northern District at Da Nang for I Corps.[6]

---

[6] Briefing, USARV Engineer Reorganization, incl. 14P, pp. 11–14, OCE Liaison Officer Trip Report no. 16; "Engineer Command Reorganized," p. 3; Spec. Larry Crabtree, "Only WAC Wearing Engineer Brass Receives Ph.D. of Philosophy," *Castle Courier*, 13 July 1970.

The Military Operations Division and Materiel Division, also headed by colonels, made up the third and fourth major staff elements. The Military Operations Division handled a wide range of operational functions, including coordinating engineer support for tactical operations and preparing plans for the disposition of engineer forces and contingency plans. It served as the central agency for all USARV mine and countermine activities and carried out mapping and intelligence missions. This division also had the responsibility for all USARV engineer activities related to Vietnamization and the training of South Vietnamese Army engineer units. Finally, the Materiel Division coordinated the movement and distribution of all engineer equipment and construction materials with other USARV staff agencies and engineer units.[7]

General Dillard believed in the reorganization. On 23 February, he wrote to General Clarke, Chief of Engineers, noting improved responsiveness and control over the command's programs and improvement in morale. The reorganized engineer command and staff arrangement was even serving as a model for the consolidation of the 1st Logistical Command and the USARV logistical staff. His main concern about the proposed logistical reorganization was the possibility that his command would be placed under a deputy commanding general for logistics. This was an old worry of engineers because Army doctrine had in the past called for engineer construction commands to operate in rear areas under the theater army logistical commands. Although he and Clarke preferred to be entirely separate or included in the operational chain, both senior engineers did not foresee any problems or lack of responsiveness in operating this way. Anyhow, the engineer command and staff structure continued to operate as a major staff agency and major command within U.S. Army, Vietnam. Within six months, the provisional command received Department of the Army approval as a table of distribution organization.[8]

Toward the Sanctuaries

For years the North Vietnamese and Viet Cong had taken advantage of sanctuary in Laos and Cambodia, where refitting, resupplying, and training could be carried out without interference. These staging areas and supply points continued to pose a danger to South Vietnam, especially the large base areas and support depots within short striking distance of Saigon. As the North Vietnamese and Viet Cong increased their hold over the Cambodian border region, it became increasingly clear to American officials that, despite his currently reduced strategy of small-scale attacks, the enemy could stage from these bases to launch an invasion after the Americans withdrew.[9]

---

[7] Briefing, USARV Engineer Reorganization, incl. 14P, p. 13, OCE Liaison Officer Trip Report no. 16; "Engineer Command Reorganized," p. 3.

[8] Ltr, Maj Gen John A. B. Dillard, CG, Engr Cmd and USARV Engr, to Maj Gen Frederick Clarke, Ch Engr, 23 Feb 70; Ltr, Clarke to Dillard, ENGME-P, 20 Mar 70, both in Historians files, CMH; FM 5–1, *Engineer Troop Organizations and Operations*, September 1965, p. 1–2; Interv, Capt David D. Christensen, 26th Mil Hist Det, with Maj Gen Charles C. Noble, CG, USAECV, 25 Jul 71, VNIT 919, p. 7, CMH.

[9] MACV History, 1970, vol. 1, p. III-77; Davidson, *Vietnam at War*, p. 623; Stanton, *Rise and Fall of an American Army*, p. 336.

Allied operations after the Tet offensive of 1968 had destroyed many of the enemy bases in South Vietnam, making the sanctuaries even more critical to Hanoi. By early 1969, the enemy was sortieing regularly from the cross-border redoubts. These forays were typically preceded by logistical buildups inside Vietnam, which General Abrams called a logistical nose. The allied countertactic was to cut off this logistical nose and thus frustrate the attack that was intended to follow. A variety of forces were deployed to prevent these buildups and to seal the borders of South Vietnam. The 1st Cavalry Division and its engineers were at the forefront of this campaign.[10]

Beginning in the fall of 1968, the 1st Cavalry Division had positioned most of its combat units along the Cambodian border. The division assumed responsibility along an arc some ninety miles wide and thirty miles deep. The division's area of responsibility comprised the provinces of Phuoc Long, Binh Long, Tay Ninh, and Binh Duong and straddled the enemy's lines of communication leading southward from the border to Saigon. As the 1st Division redeployed from Vietnam, the 1st Cavalry Division moved in, setting up operations, working with local Vietnamese forces, and learning about the enemy. The North Vietnamese and Viet Cong attempted to reestablish their logistical system, with the aim of repeating the attacks of Tet 1968. Throughout 1969, the 1st Cavalry Division fought a series of skirmishes along these supply lines, thwarting strong North Vietnamese forces attempting to position themselves closer to the capital. Although enemy attempts in the early months of 1970 were weaker, he still tried to operate in force in these critical areas. In several instances, the cavalrymen uncovered base camps and fair-sized caches. Apparently, cutting the trails was causing a backlog of supplies in Cambodia. This was an ideal time to strike at the Cambodian supply base.[11]

The 1st Cavalry Division's 8th Engineer Battalion continued to improve base camps and support tactical operations. Battalion headquarters and headquarters company joined the division headquarters at Phuoc Vinh; Company A at Tay Ninh supported the 1st Brigade; Company B helped the division settle in at Phuoc Vinh then moved to Lai Khe with the 2d Brigade; and Company C supported the 3d Brigade at Quan Loi. In the field, the battalion concentrated on building and upgrading landing zones and firebases, improving base defenses, and giving minesweep and demolition support to the infantry battalions. As the division uncovered enemy supply caches, the engineers stored and distributed captured rice to nearby villages.[12]

As in the past, the 20th Engineer Brigade's 79th Engineer Group reinforced the divisional engineers in III Corps north and west of Saigon. Much of the priority work went to improving the forward airfields, main supply

---

[10] Starry, *Mounted Combat*, pp. 138–39.

[11] Tolson, *Airmobility*, pp. 218–20; *Memoirs of the First Team*, p. 42. For more on the 1st Cavalry Division's operations along the Cambodian border, see J. D. Coleman, *Incursion* (New York: St. Martin's Press, 1991) pp. 34–211; Keith William Nolan, *Into Cambodia: Spring Campaign, Summer Offensive, 1970* (Novato, Calif.: Presidio Press, 1990), pp. 3–65; Shelby L. Stanton, *Anatomy of a Division: The 1st Cav in Vietnam* (Novato, Calif.: Presidio Press, 1987), pp. 158–77.

[12] *Memoirs of the First Team*, pp. 184–85.

routes, and base camps. The 31st Engineer Combat Battalion with the help of the 557th Light Equipment Company completed the resurfacing of the Phuoc Vinh airfield and built helicopter and fixed-wing revetments at Phuoc Vinh and Song Be. Work at the Phuoc Vinh airfield involved removing the M8A1 matting, preparing the base course, putting down a twelve-inch layer of soil-cement, and topping that off with a three-inch layer of asphaltic concrete. With the additional 400 feet of runway, the field was upgraded to a 4,100-foot Type II C–130 airfield. Upon completion of this work in February 1970, the battalion shifted to upgrading taxiways and parking ramps. Vertical construction involved building a new sixty-foot control tower, a helicopter refueling point, and other facilities for the 1st Cavalry Division. The 31st also upgraded several forward airstrips scattered across northern III Corps, including the strips at Tong Le Chon, Katum, Bu Dop, Dong Xoai, and Loc Ninh Special Forces camps. The 588th Engineer Battalion, which included the 362d Light Equipment and the 544th Construction Support Companies, covered some 6,400 square miles and maintained 110 miles of roads. Other support for the 1st Cavalry and 25th Divisions included building new or improving artillery gun pads, aircraft revetments, bunkers, air control and counter-mortar fire towers, and clearing fields of fire. In the Lai Khe area, the 168th Engineer Combat Battalion concentrated on operational support and base camp construction for the 1st Division. Elements of the battalion supported the 1st Cavalry Division at Quan Loi and Phuoc Vinh, where one platoon began work on the foundation and floor for a hangar being moved from Blackhorse. In September 1969, when enemy forces were detected approaching Lai Khe, the battalion served as a base defense force under the 1st Division. The 168th departed Vietnam about the same time as the 1st Division in April 1970.[13]

As the 79th Group's only construction battalion, the 554th Engineer Battalion was committed to base construction and highway improvement work. Most of this effort took place at the 25th Division's Cu Chi camp and along Highways 1 and 22. At Cu Chi, improvements included building bunkers and revetments for the 12th Evacuation Hospital. Working with Pacific Architects and Engineers, the 554th also reconstructed a concrete and asphalt helipad for the hospital's medical evacuation helicopters. Another job involved paving interior roads at the base. In late 1969, the 554th began shifting operations to Lai Khe and Highway 13 upgrading.[14]

The overthrow of the Cambodian chief of state, Prince Norodom Sihanouk, in March 1970 hastened the events leading to the U.S. attack into Cambodia. Sihanouk's successor, Prime Minister Lon Nol immediately demanded that the North Vietnamese and Viet Cong leave Cambodia,

[13] Annual Hist Sum, Aug 69–Sep 70, 31st Engr Bn, pp. 6–7; ORLL, 1 Feb–30 Apr 70, 31st Engr Bn, 9 May 70, pp. 5–9; ORLLs, 1 Feb–30 Apr 69, 588th Engr Bn, 14 May 69, pp. 1–2, 1 May–31 Jul 69, 14 Aug 69, 588th Engr Bn, pp. 1–2, 1 Nov 69–31 Jan 70, 588th Engr Bn, 15 Feb 70, pp. 1–2, 1 May–31 Jul 69, 168th Engr Bn, 11 Aug 69, pp. 2, 4–8, and 1 Aug–31 Oct 69, 7 Nov 69, 168th Engr Bn, pp. 2, 4–11, all in Historians files, CMH.

[14] ORLLs, 1 May–31 Jul 69, 554th Engr Bn, n.d., pp. 2, 4–9, 1 Aug–31 Oct 69, 554th Engr Bn, 14 Nov 69, pp. 2, 4–7, 1 Nov 69–31 Jan 70, 554th Engr Bn, 14 Feb 70, pp. 2, 3–5, and 1 Feb–30 Apr 70, 554th Engr Bn, 14 May 70, pp. 2, 4–7, all in Historians files, CMH.

and threatened to block the flow of supplies coming by sea through the port of Sihanoukville. Almost at once, the North Vietnamese Army and *Khmer Rouge* (the Communist insurgents in Cambodia) launched attacks in the Cambodian border areas to retain control over all routes leading to their bases. The small, inexperienced Cambodian Army was no match for the North Vietnamese, who drove west and threatened the capital, Phnom Penh. It became evident that without outside help, Lon Nol and his pro-Western government would be overthrown, the port of Sihanoukville would be reopened to the Communists, and all of Cambodia could become a major base area outflanking allied forces in South Vietnam. The collapse of Cambodia would be a disastrous blow to South Vietnamese hopes for survival, imperil the time needed to complete Vietnamization, and jeopardize the safe withdrawal of U.S. forces.[15]

Lon Nol's appeal for help prompted the South Vietnamese to mount large-scale operations across the border. In mid-April, III Corps units launched Operation TOAN THANG (TOTAL VICTORY) 41, a three-day thrust into the Angel's Wing area of Cambodia, a key stronghold west of Saigon. Because U.S. forces were not allowed in Cambodia, American advisers stayed behind, although a brigade of the U.S. 25th Division supported the operation by screening the border. Before the end of the month, the South Vietnamese launched two short thrusts into Cambodia from IV Corps, basically confidence builders for the South Vietnamese Army. Although the allies declared the attacks a success, the North Vietnamese avoided contact and knew about the operations in advance. The short durations of the operations prevented a careful search of the base areas.[16]

Although U.S. engineers did not accompany the South Vietnamese across the border, they lent considerable help. In preparation for TOAN THANG 41, III Corps engineer advisers arranged with the 20th Brigade to provide forty complete M4T6 floats, cranes, and bridge boats on 30 March to an assembly area near Go Dau Ha. Elements of the South Vietnamese 30th Group assembled the raft sections, which were then transported by CH–54s to the Ben Soi Bridge site west of Tay Ninh. Other 30th Group engineers launched the sections into the Vam Co Dong River and connected them into a floating bridge, which was completed the following day. A 600-foot access road and a 90-foot panel bridge leading to the floating bridge were also built by 30th Group engineers. In early April, the group's 303d Combat Battalion built a M4T6 dry span and

[15] Davidson, *Vietnam at War*, pp. 624–25; Palmer, *25-Year War*, pp. 98–99. See also, MACV History, 1970, vol. 3, pp. C-1 to C-13; Brig. Gen. Tran Dinh Tho, *The Cambodian Incursion*, Indochina Monographs (Washington, D.C.: U.S. Army Center of Military History, 1979), pp. 1–17, 29–32; Starry, *Mounted Combat*, pp. 166–67; Lt. Gen. Sak Sutsakhan, *The Khmer Republic at War and the Final Collapse*, Indochina Monographs (Washington, D.C.: U.S. Army Center of Military History, 1980), pp. 32–83.

[16] Davidson, *Vietnam at War*, p. 625; Palmer, *25-Year War*, p. 99; Starry, *Mounted Combat*, pp. 167–68. For more on the South Vietnamese reaction and April cross-border operations, see Tho, *Cambodian Incursion*, pp. 32–50. See also MACV History, 1970, vol. 3, pp. C-14 to C-35; Rpt, Opn TOAN THANG 41, 14–17 Apr 70, U.S. Army Advisory Gp, III Corps, 7 Jun 70, Historians files, CMH.

a panel bridge along the approaches to the Angel's Wing. As the operation neared, the III Corps Engineer on 10 April asked the 20th Brigade to provide three 45-foot M4T6 dry spans. Four days later, another request asked for 200 feet of 24-inch culvert. The 20th Brigade did not have this size culvert in stock, but supply personnel found the required size and amount in the 25th Division's 65th Engineer Battalion and delivered it to Go Dau Ha that night. Early planning called for the dry spans to be assembled at Go Dau Ha and lifted by CH–54s, but weather and other problems prevented using the helicopters. Two of the bridges were loaded on trucks and the third was partially assembled so that it could be lifted by a Chinook on the morning of 14 April if required. During TOAN THANG 41, the 30th Group attached a platoon from the 301st and 303d Combat Battalions to each of the three attacking South Vietnamese task forces and assigned one company from the 301st in general support. One task force also had an attached platoon of engineers from the South Vietnamese 25th Division's 25th Engineer Battalion. Other work done by the 30th Group included opening and keeping open the main supply route between Bien Hoa and the Cambodian border and maintaining roads on the Vietnamese side of the border. On the Cambodian side, the South Vietnamese installed dry spans and one culvert and found the roads to be in good condition, although overgrown with grass in places. As for the Vietnamese engineer effort, the III Corps advisory group's report of the operation succinctly noted: "Planning was timely and complete. Advance preparation prevented delays in execution. Provision was made to cope with any contingencies. Execution was aggressive and effective."[17]

Main Attacks

Meanwhile, President Nixon and his advisers concluded that U.S. forces would have to intervene in Cambodia to ensure the success of the South Vietnamese raids. On 28 April, Nixon made the decision to send U.S. forces across the border in combined operations with the South Vietnamese. The objectives were to relieve pressure on the ragtag Cambodian Army, to destroy supplies and enemy troops in the base areas, and to seize the elusive enemy headquarters for southern Vietnam, *COSVN*. The main attack consisted of operations by South Vietnamese forces from III and IV Corps and by elements of II Field Force. Weather was a major concern, for the monsoon season would begin in late May. Good weather conditions were expected in early May, and the area chosen for the first attack was flat, with few natural obstacles to cross-country movement. For the Americans, timing and distance were critical factors. To allay the expected outcry in the United States, the Americans would be in and out of Cambodia by the end of June. Nixon also limited the U.S. invasion to no more than thirty kilometers (eighteen and a half miles) into Cambodia.[18]

---

[17] Tho, *Cambodian Incursion*, p. 48; Rpt, Opn TOAN THANG 41, pp. 7-1 to 7-2.

[18] Davidson, *Vietnam at War*, p. 625–26; Palmer, *25-Year War*, p. 99–100; Starry, *Mounted Combat*, pp. 168, 178; MACV History, 1970, vol. 3, p. C-32. For an overview and operational

The major South Vietnamese part of the offensive, Operation TOAN THANG 42 (also called ROCKCRUSHER), began on the morning of 29 April when III Corps forces entered Cambodia's Svay Rieng Province. The mission included opening and securing Highway 1 to allow the evacuation of Vietnamese refugees and to help the Cambodian Army regain control of its territory. Three South Vietnamese infantry-armor task forces moved south and west in the Angel's Wing, meeting stubborn resistance as the Viet Cong and North Vietnamese attempted to remove supplies and equipment. Nonetheless, the South Vietnamese quickly seized the first objectives and, on 1 May, swept past the provincial capital of Svay Rieng, opening Highway 1 east to Vietnam. South Vietnamese helicopters quickly moved captured troops and materiel back across the border, while larger caches were guarded, pending removal, or destroyed. TOAN THANG 42 evolved into several phases as forces attacked south into the Parrot's Beak, and the advance continued westward along Highway 1 to clear the southern part of Svay Rieng. On 3 May, III Corps forces linked up with IV Corps units advancing from the south. After seizing more weapon and supply caches, the South Vietnamese advanced west toward Phnom Penh, opening the road for Vietnamese residents who wanted to flee Cambodia. Other forces attacked north to the Chup Rubber Plantation near Kompong Cham, a provincial capital located on the Mekong River twenty-one miles northeast of Phnon Penh, and west from Vietnam along Highway 7. U.S. ground participation involved sending elements of the 3d Brigade, 9th Division, into the Parrot's Beak not far from its area of operations in Long An Province.[19]

While the South Vietnamese were busy in the Angel's Wing, the TOAN THANG 43 attack started at dawn on 1 May into the Fishhook area of Cambodia. To spearhead the attack, Lt. Gen. Michael S. Davison, the II Field Force commander, put together a highly mobile force around the 1st Cavalry Division. All together, about 12,000 Americans and nearly 5,000 South Vietnamese formed the assault echelons, making the Cambodian incursion the biggest operation since JUNCTION CITY in 1967. This force, commanded by Brig. Gen. Robert L. Shoemaker of the 1st Cavalry Division, included for its airmobile elements the 3d Brigade, 1st Cavalry Division, and the South Vietnamese 3d Airborne Brigade. The 11th Armored Cavalry Regiment; the 2d Battalion, 34th Armor, of the 25th Division; the 2d Battalion (Mechanized), 47th Infantry of the 9th Division; and the South Vietnamese 1st Cavalry Regiment provided ground reconnaissance and assault capabilities. Following intensive B–52 and tactical air and artillery strikes, the airborne brigade air-assaulted into an area north of the Fishhook to seal off escape routes. As air cavalry screened to the north, American armor and mechanized infantry, meeting some resistance, advanced across the border from

---

summaries of the Cambodian Incursion in III and IV Corps, see MACV History, 1970, vol. 3, pp. C-35 to C-86, C-93 to C-97; II FFV Commander's Evaluation Report, Cambodian Operation, 31 Jul 70, Incl to ORLL, 1 May–31 Jul 70, II FFV, 14 Aug 70, Historians files, CMH.

[19] Starry, *Mounted Combat*, pp. 167–68, 176–77; Tho, *Cambodian Incursion*, pp. 51–70, 83–90, 127. For more on the 3d Brigade, 9th Division's participation, see AAR, Opn TOAN THANG 500, 3d Bde, 9th Inf Div, 3 Jun 70, Historians files, CMH.

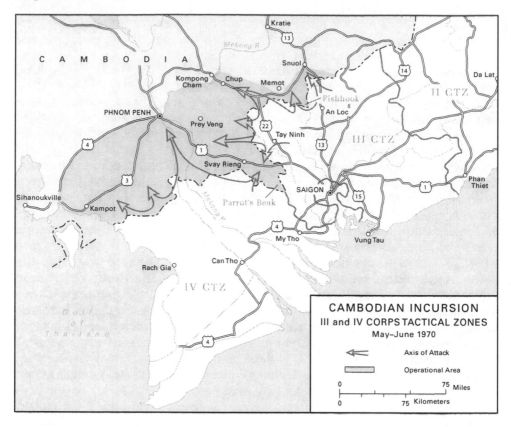

CAMBODIAN INCURSION
III and IV CORPS TACTICAL ZONES
May–June 1970

←⎯ Axis of Attack

▭ Operational Area

0 ⎯⎯⎯⎯⎯ 75 Miles

0 ⎯⎯⎯⎯⎯ 75 Kilometers

MAP 28

the south toward the suspected location of *COSVN*, hoping to envelope the enemy flank to the west. All first-day objectives were taken by dark.[20] (*Map 28*)

Task Force SHOEMAKER advanced steadily over the next few days. On 2 May, the armored column cut the road east of Memot and linked up with the South Vietnamese airborne forces. The following day, elements of the 1st Cavalry Division's 2d Brigade reinforced the task force. North Vietnamese forces avoided contact, sprinkling the roads with antitank mines. They left behind fully stocked depots as the allies probed their positions and uncovered caches. By early afternoon on the fourth, the 11th Armored Cavalry broke out of the jungle. The armored column proceeded north twenty-five miles through large rubber plantations using Highway 7 as the axis of advance toward Snuol and its important road junction. After taking Snuol on 5 May, the cavalrymen discovered an improved road, large enough for trucks and

[20] Starry, *Mounted Combat*, pp. 170–71; Tho, *Cambodian Incursion*, pp. 70–75; Palmer, *Summons of the Trumpet*, pp. 233–34; Tolson, *Airmobility*, pp. 222–24. For more on the operation, including day-to-day events, see AAR, Cambodian Campaign, 1st Cav Div, 18 Jul 70; AAR, Opn DONG TIEN II and TOAN THANG 43, 11th Armd Cav Regt, 9 Dec 70, both in Historians files, CMH. See also Coleman, *Incursion*, pp. 234–48; Nolan, *Into Cambodia*, pp. 85–183.

*An armored vehicle launched bridge being emplaced in the Fishhook
area of Cambodia*

carefully hidden under the jungle canopy. Advancing along this road they
found and destroyed an abandoned truck convoy laden with supplies. Snuol
turned out to be the hub of an extensive logistical operation. It included
a fully equipped motor park, complete with grease racks and spare parts,
and a large storage area containing tank gun ammunition. When President
Nixon on 7 May proclaimed the cross-border operation a success and said
that U.S. troops would be withdrawn from Cambodia by 30 June, the search
efforts intensified.[21]

On 5 May, the 1st Cavalry Division dissolved Task Force SHOEMAKER and
took direct control of TOAN THANG 43. During the period 6 to 13 May, 1st
Cavalry Division troopers found a huge complex dubbed the City. The City
covered about one square mile consisting of a well-organized storage depot
capable of receiving and issuing large quantities of supplies. Most of the storage bunkers contained large quantities of weapons and munitions. Storage
bunkers held clothing, food stocks, and medical supplies. The large number
of mess halls and training facilities showed that the complex was also used to

---

[21] Starry, *Mounted Combat*, pp. 172–74; Tho, *Cambodian Incursion*, pp. 74–76; Tolson,
Airmobility, pp. 224–25.

Soldiers of the 1st Cavalry Division uncovered the huge complex known as the City.

provide refresher military and political training to recent replacements from North Vietnam.[22]

Meanwhile, on 6 May other allied forces launched three simultaneous attacks against other bases straddling the border. The 1st Brigade, 25th Division, deployed two mechanized and two straight infantry battalions in the Dog's Head area northwest of Tay Ninh City and attacked toward Krek in Operation TOAN THANG 44 (also called BOLD LANCER). On 11 May, the brigade found a large cache fourteen miles west of the border, and the operation ended three days later. The next day, the 1st Brigade deployed to the Fishhook and relieved the 1st Cavalry Division. Elements of the 1st Cavalry Division's 2d Brigade and the South Vietnamese 5th Division launched the second and third cross-border operations, TOAN THANG 45 and TOAN THANG 46, from Binh Long and Phuoc Long Provinces. On the second day, the air cavalry troopers found an important weapons cache, nicknamed Rock Island East, which eventually yielded 329 tons of munitions. This cache was so large that it was decided to build a pioneer road linking the base area with Highway 14 across the border. Meanwhile, the division moved the 1st Brigade from War Zone C to the area around the Cambodian town of O'Rang, east of the 2d Brigade not far from the II/III Corps boundary in South Vietnam. A bat-

---

[22] Tho, *Cambodian Incursion*, pp. 76–78; Tolson, *Airmobility*, pp. 226–29.

talion of the 199th Infantry Brigade joined the foray, and this brought U.S. combat forces in the area to eight battalions. By the end of May, contacts with enemy forces increased along with the discovery of more weapons and supply caches. By early June, the entire division moved into the area. On 5 June, a hospital complex with surgical facilities was discovered northwest of Bu Dop. During the second week of June, units of the 3d Brigade unearthed an underground shelter containing more communications equipment and a shop with parts and accessories. Sites located by the South Vietnamese in Base Area 350 included a hospital complex capable of providing medical treatment for five hundred men. The 1st Cavalry Division began withdrawing its forces from Cambodia on 20 June. It closed the last firebase on 27 June, and by 29 June all U.S. units had returned to South Vietnam.[23]

During the eight-week incursion, around-the-clock support, much of it from Brig. Gen. Edwin T. O'Donnell's 20th Brigade, came from the engineers. Besides the attacking units' organic engineers, five nondivisional battalions, and several separate companies and detachments, provided direct and general support. The bulk of the units came from Col. Ernest J. Denz's 79th Group. Task Force EAST, composed of the 31st Combat and 554th Construction Battalions and the 362d Light Equipment Company, supported the 1st Cavalry Division. Task Force WEST, composed of the 588th Combat Battalion, attached elements of the 92d Construction Battalion, and the 557th and 595th Light Equipment Companies supported the 25th Division. Units in general support included the 62d Land Clearing Battalion; the 79th Provisional Bridge Company; a mixture of separate companies, platoons, and detachments; and twenty-one tractors and trailers from the 34th Group and the 18th Engineer Brigade. Priority went to opening main supply routes over long-unused roads from Tay Ninh and An Loc. Deteriorating forward tactical airfields were also repaired and upgraded. Priority also went to building logistic facilities at the airfields and upgrading the surrounding roads to all-weather. Organic engineers, the 1st Cavalry Division's 8th Engineer Battalion, the 25th Division's 65th Engineer Battalion, and the 11th Armored Cavalry Regiment's 919th Engineer Company, concentrated on opening, maintaining, and closing firebases and attack routes and helped with the removal and destruction of caches. Together, the equivalent of three nondivisional battalions supported the 1st Cavalry Division. This was a threefold increase compared to engineer effort for the division during normal operations. With only two days' notice, supporting engineers during the three days from 29 April through 1 May opened nine firebases for Task Force SHOEMAKER. Over forty-five miles of laterite roads and two dirt airstrips had to be upgraded

---

[23] Starry, *Mounted Combat*, pp. 174–75; Tolson, *Airmobility*, p. 229; Tho, *Cambodian Incursion*, pp. 78–83. See also AAR, Cambodian Campaign, 1st Cav Div; AAR, Cambodian Operations, 25th Inf Div, 19 Jul 70; AAR, Opn BOLD LANCER I, 1st Bde, 25th Inf Div, 18 Jul 70, all in Historians files, CMH; *The U.S. 25th Infantry Division, October 1969–October 1970*, Yearbook (n.p.: Tropic Lightning Association, n.d.), pp. 80–81; Nolan, *Into Cambodia*, pp. 205–402.

on the same two days' notice to handle the division's large resupply effort channeled through Tong Le Chon and Katum.[24]

The 11th Armored Cavalry Regiment's rapid advance to Snuol along Highway 7 relied on hasty bridging. After the lead tanks broke out of the jungle on the early afternoon of 4 May, they raced to the first of three bridges destroyed by the retreating North Vietnamese. The cavalry secured the site, placed an armored vehicle launched bridge, and went on. Resuming the advance the next day, the cavalry laid another armored vehicle launched bridge at a second crossing site. However, plans for the third crossing site called for a M4T6 dry span. Lt. Col. Scott B. Smith, the 1st Cavalry Division engineer and commander of the 8th Engineer Battalion, had reconnoitered the bridges with the cavalry regiment's intelligence officer and found the third site too wide for the armored vehicle launched bridge. There were no other nearby launching sites. Smith requested the 79th Bridge Company at Quan Loi to preassemble the dry span for helicopter lift when the 11th Armored Cavalry secured the area. On the morning of 5 May, Smith airlifted a team of engineers—a platoon from Company C, 31st Engineer Battalion, the bridge sections, and a bulldozer dismantled in two sections—to the original site. Once assembled, the bulldozer began leveling the embankments. The engineers assembled the two 30-foot ramp sections and a 15-foot center span delivered later in the morning by a CH–54 in three loads. Although the engineers worked hard, it became obvious that the bridge would not be ready until late afternoon. Anxious not to lose the momentum of the attack, Col. Donn A. Starry, the 11th's commanding officer, set out on foot with an armored vehicle launched bridge. He found a place where the span could be used. After carefully testing several places, the vehicle's operator let down the bridge with a few inches to spare across the gap and a cavalry troop tried it out. By 1300, the regiment's column of M48 Patton tanks and M551 Sheridan armored cavalry assault vehicles resumed the advance. Working in waist-deep mud and fighting the Flying Crane's hurricane-like winds most of the day, the engineers wrestled the last of the dry span sections into place and completed the 75-foot bridge later that afternoon.[25]

More engineers soon crossed the border. On 14 May, the 31st Engineer Battalion, which had already assigned Company C to maintain Highway 13 from An Loc to Snuol, ordered the company to Snuol to build bypasses around the heavily damaged town. This work consisted primarily of placing a laterite cap on existing light roads through the rubber plantation. Culvert, preassembled by Company D in Song Be, was airlifted to crossing sites. On 5 June, Company C, 588th Engineer Battalion, joined 11th Armored Cavalry elements at the western edge of the Memot Plantation. The company began

[24] Ploger, *Army Engineers*, p. 176; AAR, Cambodian Campaign, 1st Cav Div, p. G-1; AAR, Opn Toan Thang 43, 20th Engr Bde, n.d., pp. 1–3, 9; AAR, Opn Toan Thang 43, 79th Engr Gp, 15 Aug 70, pp. 3–4, 7, all in Historians files, CMH. For more concerning 79th Group operations, see Intervs, Capt David D. Christensen, 26th Mil Hist Det, with Col Ernest J. Denz, CO, 79th Engr Gp, and staff officers, 26 Nov 70, VNIT 773, CMH.

[25] Starry, *Mounted Combat*, pp. 172–73; AAR, Opn Dong Tien II and Toan Thang 43, 11th Armd Cav Regt, pp. 10–11; AAR, Opn Toan Thang 43, 20th Engr Bde, p. 10; AAR, Opn Toan Thang 43, 79th Engr Gp, p. 23; 31st Engr Bn, Annual Hist Sum, Aug 69–Sep 70, 31st Engr Bn, p. 7; Coleman, *Incursion*, p. 243; Nolan, *Into Cambodia*, pp. 135–36.

A Sheridan of the 11th Armored Cavalry crosses an armored vehicle launched bridge.

to upgrade existing trails to help the regiment's search for the enemy and his caches. The onset of the rains, which fortunately held off for three weeks in May, slowed progress, but this did not hamper the momentum of the armored cavalry. However, by mid-June, heavy rains and traffic caused steady deterioration of the two launched bridge crossings. Company C, 31st Engineer Battalion, moved to nearby Firebase COLORADO to make repairs and complete a 110-foot triple-single panel bridge. Although the rains caused problems for the engineers, a larger concern was a sudden increase in North Vietnamese activity in the area. Early one morning during heavy rains, the enemy struck the crowded base with rocket-propelled grenades and small-arms fire, wounding thirteen engineers and damaging several vehicles. Thirty dump trucks from the 104th Dump Truck and 79th Bridge Companies that reached the base with their loads of rock the previous day had to park in partially exposed positions. On the morning of 17 June, work parties en route to the bridge sites were taken under fire and were forced to remain within the base for the remainder of the day. The next day, eight engineers were wounded on their way to a bridge site.[26]

At this time, the incursion reversed, and the U.S. troops began to withdraw from Cambodia. On the nineteenth, disassembly of the panel bridge and

[26] AAR, Opn TOAN THANG 43, 20th Engr Bde, pp. 11, 18, 21, 26; AAR, Opn TOAN THANG 43, 79th Engr Gp, pp. 9–10, 12; AAR, Cambodian Campaign, 1st Cav Div, pp. G-6 to G-7. For more about the 31st and 588th Engineer Battalions during the Cambodian operation, see Interv, Capt Kurt E. Schlotterbeck, 26th Mil Hist Det, with Lt Col Gwynn A. Teague, CO, 31st Engr Bn, 6 Jul 70, VNIT 728; Interv, Schlotterbeck with Lt Col Thomas A. Stumm, CO, 588th Engr Bn, 24 May 70, VNIT 668, both in CMH.

497

*A CH–54 Flying Crane transports fixed-span bridging during the
Cambodian incursion.*

75-foot dry span began. Two separate sections of the dry span were easily
hauled away by helicopter. The third section, however, was so deeply mired
in the mud that it could not be removed. It had to be destroyed in place. The
110-foot panel bridge was dismantled and carried away in trucks. On 26 June,
all U.S. units in the area returned to Vietnam through Katum.[27]

Throughout the operation, the engineers worked hard on the main supply
routes. In the western sector, the 588th Engineer Battalion upgraded and maintained forty miles of Route 4 from Tay Ninh to Memot. The main effort involved
building a new road from Katum. Work on the extension just south of the border
began on 5 May. It took the 588th, with the help of the 595th Light Equipment
Company's earthmoving equipment and the 984th Land Clearing Company's
Rome plows in the lead, three days to gouge out a tactical road through the
jungle. This allowed tracked vehicles to move to Firebase X-RAY just across the
border. Augmented by five 290M scrapers and a roller from the 595th Light
Equipment Company, the 588th Battalion's Company A pushed on, completing the eleven-mile road on 16 May. Meanwhile, Company D, 92d Engineer
Battalion, joined Company D, 588th Engineer Battalion, in a joint effort to
maintain Highway 22 from Tay Ninh to Cambodian Route 78, and on to its

---

[27] AAR, Opn TOAN THANG 43, 20th Engr Bde, p. 11; AAR, Opn TOAN THANG 43, 79th Engr
Gp, pp. 12–13.

juncture with Highway 7 at Krek. One of two major spans built along the way involved erecting a 180-foot panel bridge consisting of a 140-foot triple-double span and a 40-foot double-single span supported by an intermediate pier.[28]

Also busy in the western sector were the 25th Division's engineers. One of the first tasks facing the division's 65th Engineer Battalion was to blaze a road to the Rach Cai Bac, a stream along the border some five miles west of Thien Nhon. While an infantry battalion made a helicopter assault to secure a bridgehead, a few miles on the Cambodian side, Companies B and E began installing a M4T6 floating bridge. Bulldozers cleared away brush on the river banks, and the bridge trucks loaded with bridge decking, each with an inflatable float set on top, pulled up at intervals. The men of the combat line and bridge companies worked past dark, completing the job forty-five minutes before midnight. At dawn on 7 May, armored vehicles began crossing the 225-foot river into Cambodia. Meanwhile, Company A, which routinely supported the 1st Brigade, helped prepare the support base at Thien Nhon. Five Rome plows from the 984th Land Clearing Company began to widen and lengthen the airstrip and clear fields of fire around the base.[29]

In the eastern sector, the 31st Engineer Battalion opened routes to handle the heavy loads needed to supply the 1st Cavalry Division and return captured supplies. Besides its work along Highway 13, the 31st on 26 May began upgrading Highway 14A, which was little more than a trail, from Loc Ninh to Bu Dop. By 1 June, Companies A and D reinforced with equipment from the 557th Light Equipment Company and 554th Construction Battalion, with protection provided by two cavalry troops, completed the upgrading. On 11 May, a platoon from Company B, a team of Rome plows from the 62d Land Clearing Battalion, and a panel bridge platoon from the 79th Bridge Company started working north along the heavily mined stretch from Bu Dop to Firebase BROWN inside Cambodia. Encountering an 80-foot gap along the way, the engineers used bulldozers to push out causeways from each bank. This narrowed the gap and allowed fitting the 38-foot M4T6 dry span airlifted from Quan Loi. Once across the gap, Rome plows opened the road to Firebase BROWN. Without the plows, the road would have taken several additional days to open. Operating from Firebase BROWN, and then Firebase MYRON, the engineer task force added three panel bridges. These included an 80-foot span to replace the dry span, another standard dry span, and several culverts.[30]

In addition to the various supply routes, the engineers built access roads to the cache sites. Initially the 1st Cavalry Division planned to remove the captured stores at the City by helicopter. However, after finding more weapons and munitions, the division ordered the engineers to build an access road to evacuate the supplies by vehicle. Aerial reconnaissance revealed an unused trail leading east across the border to Company C, 31st Engineer Battalion's, camp

---

[28] AAR, Opn TOAN THANG 43, 20th Engr Bde, pp. 12, 17–18; AAR, Opn TOAN THANG 43, 79th Engr Gp, pp. 9–11; AAR, Cambodian Campaign, 1st Cav Div, pp. G-6 to G-7; S. Sgt. Matt Glasgow, "Thrust into Cambodia," *Kysu'* 2 (Fall 1970): 13.

[29] The U.S. 25th Infantry Division, October 1969–October 1970, Yearbook, pp. 80, 82; Tho, *Cambodian Incursion*, pp. 78–79; Nolan, *Into Cambodia*, pp. 220–21.

[30] AAR, Opn TOAN THANG 43, 20th Engr Bde, pp. 21, 26; AAR, Opn TOAN THANG 43, 79th Engr Gp, pp. 13–14; AAR, Cambodian Campaign, 1st Cav Div, p. G-6.

near Highway 13. On 9 May, Company C and the 557th Light Equipment Company moved to the foot of the trail and began ripping a narrow channel through the thick jungle. With a D7E bulldozer in the lead, and a hovering helicopter keeping the work party on course, Rome plows widened the swath while road construction equipment shaped a new road. Late that day, the bulldozers broke through the trees at one side of the cache. Engineer and transportation vehicles quickly began hauling several thousand tons of supplies back to Vietnam over the 13.5-mile "Cache City Road." Fortunately, rainfall had not yet become a factor in the operation, and the dry weather enabled the engineers to use hasty fords, thus eliminating the need for culverts and large amounts of fill. In late May, the 31st's Company B helped the 1st Cavalry Division remove several hundred tons of supplies from Rock Island East. Rome plows widened a three-mile trail, formerly a North Vietnamese supply route, through the jungle. The improved route allowed engineer and transportation units to haul the captured material to Highway 14 and back to Vietnam.[31]

Engineers also made headway at the forward airfields, adding more offloading and logistical facilities. At the heavily used Bu Gia Map airstrip, the 8th Engineer Battalion added a new 250-foot square soil-cement parking ramp, a helicopter refueling facility, two gunship rearm points, and a 15,000-gallon-per-day water supply point. Daily maintenance of the laterite runway surface and parking ramp kept the fair-weather airfield open despite heavy daily rains. The 31st Engineer Battalion did similar work at Bu Dop and Loc Ninh. Planners looked to Loc Ninh as a dependable backup to Bu Dop. Similar improvements were made at Thien Nhon, Katum, and Tong Le Chon. Company D, 588th Engineer Battalion, and the 362d Light Equipment Company were already committed to upgrading the Thien Nhon airfield when the Cambodian operation began. The large number of aircraft sorties and the resultant failures in the M8A1 matting kept the 588th busy patching the runway and welding the matting's joints. Reinforcing elements of the 92d Engineer Battalion from Long Binh arrived on the eleventh to begin work on a brigade-sized logistics base, which was completed within six days. Thien Nhon was unusual in that a section of Highway 22 served as the runway. At Katum, Company C, 588th Engineer Battalion, and elements of the 362d Light Equipment Company started extensive maintenance work on 2 May. The airfield could not be closed during daylight hours, so the engineers worked at night, using soil-cement stabilization covered with a MC70 asphalt coating. On 11 May, Company D moved to Katum from Cu Chi to place some 8,000 square yards of M8A1 matting on the 1,200-foot runway. Company A joined the effort and built a logistics complex, an airfield bypass road, a staging area, a road in the ammunition storage area, a berm to protect parked tractor-trailers, and an aerial resupply hook-out pad to sling load cargo. At Tong Le Chon, rain and heavy air traffic caused several soft spots on

---

[31] AAR, Opn TOAN THANG 43, 20th Engr Bde, pp. 21, 26, 32; AAR, Opn TOAN THANG 43, 79th Engr Gp, pp. 13–15; Glasgow, "Thrust into Cambodia," pp. 13, 16. See also Spec. Harry Huntington, "The Cache Road," *79th Pioneer* 2 (30 May 1970): 3.

The Thien Nhon airstrip used a section of Highway 22 as a runway.

the airfield. On 19 June, a platoon from the 557th Light Equipment Company moved to the base and completed repairs in ten days.[32]

Land-clearing units again displayed their ability to support combat operations. Teams probed suspected bunker and cache areas and cut pioneer roads for tracked and wheeled vehicles to haul away captured supplies. For the first time since the activation of the 62d as a land-clearing battalion, smaller task forces or platoons instead of complete companies moved to the cutting sites. For instance, between 10 May and 5 June, the 2d Platoon, 60th Land Clearing Company, supported the 31st Engineer Battalion by opening roads to cache sites. The platoon also helped the armored cavalry in reconnaissance-clearing missions. During the cuts, plows frequently unearthed more caches. Both sides of Highway 14A were also cleared back to about twenty-five yards for a distance of about six miles. On 18 May, another task force of Rome plows from the 984th Land Clearing Company joined the 588th Engineer Battalion at Thien Nhon to clear suspected ambush sites along Highway 22 and Cambodian Route 78. The thirteen plows also cleared fields of fire around the base. The first plows entered Cambodia on 26 May, and soon one plow detonated a mine. No one was hurt and the plow withstood moderate damage. On 4 June, the task force moved west of the base along Route 20 and cleared both sides of the road to the border. It finished this job on 6 June. In mid-May,

---

[32] AAR, Cambodian Campaign, 1st Cav Div, p. G-5; AAR, Opn TOAN THANG 43, 20th Engr Bde, pp. 40, 43, 46; AAR, Opn TOAN THANG 43, 79th Engr Gp, pp. 10–11. For more on earlier work at Thien Nhon by the 588th Engineer Battalion, see Capt. Francis L. Smith Jr., "Combat Engineers at Thien Nhon," *Military Engineer* 62 (November-December 1970): 392–93.

501

it became evident that if the armored cavalry was to maintain momentum around Snuol, routes of travel would have to be cleared through the jungle. A task force from the 60th Land Clearing Company, consisting of twenty-eight men and nine Rome plows, joined the cavalrymen. Although ambushed several times, in seven days the land clearers cut some 155 acres.[33]

Due to the small number of plows available and the extremely dense bamboo growth in the Snuol area, the land clearers developed several cutting techniques. The standard echelon formation was not workable. Instead, a lead plow pushed ahead some fifty to one hundred yards into the jungle with the remaining plows dividing and simultaneously pushing to the left and right. Once the direction of the cut was decided, the lead plow rejoined the others. When the plows neared the point of the deepest penetration, the process was repeated. This method allowed the plows to double and triple up in removing the bamboo and left a thirty-to-forty-yard-plus corridor free of downed trees. Although daily production dropped, this method had the advantage of giving the accompanying armored cavalry a much wider and faster reaction time to enemy attacks. The cavalry dropped the normal procedure of keeping the armored vehicles in cleared areas, and instead moved some of their forces ahead and on both flanks of the plows. This was a slow process. The armored vehicles had trouble forcing their way through the jungle, but this method did break up potential ambushes and gave the plows better protection. Firing at random intervals, "recon by fire," apparently proved effective, for the enemy did not test the advance. Colonel Driscoll, the 62d's commanding officer, however, was convinced that the recon by fire method was not really all that good. He recalled: "We've been hit within two or three seconds after gunships have worked the precise area in which we get hit." In addition, Driscoll did not see much of a deterrence in using "shotguns" on the plows. "There is a drawback to putting a second man in the cab of the plow," he noted, "because there's no buddy seat in there . . . and he sits up high and he's exposed an awful lot more than the operator. If they hit a mine, he's almost a cinch to get some fragmentation from it." Finally, the cavalry-engineer teams used "automatic ambushes" to protect the land clearers in the cutting areas and in their night defensive positions. At the end of the cutting day, security forces placed trip wires across trails connected to a series of five or six claymore mines. This helped reduce enemy activity, but it also required the delicate and time-consuming task of deactivating the weapons the next day.[34]

In early June, both land-clearing companies entered the Fishhook. There they worked with the 11th Armored Cavalry to find and destroy enemy fortifications, hiding places, and supplies and equipment. Until 1 June, the bulk of the 60th Land Clearing Company had been cutting near Minh Thanh northwest of Lai Khe. The 984th had been working in the Bearcat area south-

[33] AAR, Opn Toan Thang 43, 20th Engr Bde, pp. 46–52; AAR, Opn Toan Thang 43, 79th Engr Gp, pp. 15–16.

[34] AAR, Opn Toan Thang 43, 20th Engr Bde, pp. 59–60; Interv, Schlotterbeck with Driscoll, 22 Jun 70, pp. 25–26.

east of Long Binh. North Vietnamese resistance had been increasing in the Fishhook, and the land clearers encountered the heaviest enemy attacks of the Cambodian campaign. In its first day of cutting, the 60th's security force discovered fifty bunkers, some weapons and munitions, and twenty tons of rice. Two days later, the cavalry-engineer force found another bunker complex about one square mile in area. On 7 June, they unearthed a large complex of seventy-five bunkers with two feet of overhead cover and fifty huts. During the search, more weapons and medical supplies were found. On the ninth, the land clearers and their cavalry escort came under fire several times, but they did find and destroy another 150 bunkers and huts. Such finds continued until the close of the operation on 24 June. Enemy attacks were frequent, from small arms, mortar rounds, and rocket-propelled grenades. On the sixteenth, the enemy made some seventy individual attacks against the cutting force. Three operators suffered wounds and two plows received light damage. One Rome plow struck a mine, wounding the operator and damaging the plow. The next day, the lead plow was struck by some one hundred rounds of automatic rifle fire and a rocket-propelled grenade. This attack killed one engineer and wounded another and left one plow moderately damaged. The 984th, which arrived in Cambodia in mid-June, also had its share of enemy contact. On the night of 19 June, rocket and mortar attacks against the company's night defensive position killed three and wounded eighteen engineers. Harassing attacks on the twenty-second and twenty-third wounded another eleven engineers. Still, between 4 and 24 June, the two companies cleared about 1,700 acres of jungle and destroyed more than 1,100 enemy structures. Supplies uncovered included X-ray equipment, medical supplies, communications equipment, various weapons, 10,000 rounds of small-arms ammunition, and 500 mortar rounds. Both units departed Cambodia via Katum and were among the last units to cross the border.[35]

The Cambodian operation illuminated the methods and hazards of land clearing. During the June cuttings, the mission of the company-size operations was to support the armored cavalry to locate enemy caches. The 60th and 984th made their cuts as rays from night defensive positions. In the Fishhook, II Field Force usually assigned one land-clearing company to support one armored cavalry troop. The land clearers, however, found one platoon of ten plows sufficient to support a troop and recommended this ratio for future operations. If, during their probing cuts, the land clearers found they could not finish the cut that day, they could expect enemy contact the following day. To allay this problem, they bypassed all large trees and continued the cut as fast as possible within a seven- to nine-hour period. After returning to the night defensive position, operators still had daily maintenance to do. This meant sharpening blades, changing oil, and cleaning their machines. In addition, each plow required sixty to seventy gallons of gas and seventy

---

[35] AAR, Opn Toan Thang 43, 20th Engr Bde, pp. 52–59; AAR, Opn Toan Thang 43, 79th Engr Gp, pp. 17–22; Starry, *Mounted Combat*, pp. 175–76; Glasgow, "Thrust into Cambodia," pp. 16–17; Spec. Dave Massey, "Land Clearing Team, Rome Plows on the Border," *Hurricane* (September 1970): 34–37.

gallons of water daily. Of the two land-clearing companies in Cambodia, the 984th suffered more casualties, mostly by indirect fire against night laagers. In the attack that caused three deaths and eighteen wounded, the engineers had inherited the position used the previous day by the armored cavalry. Typically, engineers tried to complete overhead cover before nightfall while bulldozers formed a berm around the position. If the land clearers could not finish the fortifications, they slept under their bulldozers or heavy equipment. The 60th made this a standard practice. Although not as comfortable or as protective as a bunker, the bulldozers gave ample protection.[36]

Maps and terrain intelligence played important roles in the Cambodian operation. Four years earlier, the Engineer Command had established two map depots in Vietnam. Due to the large troop withdrawals before Cambodia, only the 547th Map Depot Platoon at Long Binh remained in Vietnam. The Cambodian offensive severely tested the map supply system. The depot received no advance notice of the operation, and apparently II Field Force did not prepare map supply plans. At first, stocks of the border region seemed adequate for limited needs, but as the operation grew in scope restockage became necessary. On 1 May, II Field Force made the first emergency requisition for 58,000 Cambodian large-scale sheets. The 66th Engineer Company (Topographic) (Corps), a unit under the control of II Field Force, sent the request directly to the 29th Engineer Battalion (Base Topographic) in Hawaii. The maps reached the depot on the fourth. That day the 66th forwarded another request to the 29th, this time for 600,000 Cambodian and Laotian map sheets. The twenty-eight tons of maps reached Vietnam on 11 May. Between 1 and 10 May, the 66th's Reproduction Platoon printed over 45,000 map sheets of Cambodia to meet initial demands. From 1 May to 2 June, the depot issued over 685,000 maps, almost three times the average issued over the same period. Cartographers also produced two uncontrolled mosaics of Cambodian airfields. Altogether, the 66th Topographic Engineer Company issued almost one million maps to support the Cambodian operation. The 517th Engineer Terrain Detachment in the Engineer Section at II Field Force headquarters furnished additional topographic information. There, expert terrain analysts looked at contours, vegetation, soil-bearing capacities to support armored vehicle movement, and information on likely enemy storage sites and supply routes. Given the poor secondary roads and the reliance on motor vehicles, especially in areas like the Dog's Head and Fishhook, terrain and road knowledge before the attack became essential.[37]

South Vietnamese engineers continued to do a creditable job of supporting their combat forces. At the start of TOAN THANG 42, the 30th Group placed one platoon in direct support and one company in general support of each

---

[36] AAR, Opn TOAN THANG 43, 20th Engr Bde, p. 61; Massey, "Land Clearing Team, Rome Plows on the Border," p. 37. For more on a commander's views of the Cambodian operation, see Interv, Schlotterbeck with Driscoll, 22 Jun 70, pp. 23–39.

[37] AAR, TOAN THANG 43, 20th Engr Bde, p. 64; U.S. Army Topographic Support of Military Operations in the Republic of Vietnam, pp. 100–101, 117–18; ORLL, 1 Feb–30 Apr 70, II FFV, 14 May 70, p. 42, Historians files, CMH. See also Spec. Larry Mayo, "Battle Maps," *Kysu'* 1 (Summer 1969): 26–28.

Supplying enough maps for the Cambodian operation put a strain on the remaining map depot in Vietnam.

corps task force. Before the operation, the group again borrowed M4T6 floats and equipment from the 20th Brigade. These floats made possible the crossing by two task forces over the Vam Co Dong. Flying Crane helicopters picked up finished raft sections from the assembly site and dropped them off at the bridge site, where the South Vietnamese 301st and 303d Engineer Combat Battalions and the 317th Float Bridge Company tied the sections together into a bridge. By early evening of the twenty-fifth, the bridge was finished well before the armored task forces began their move into Cambodia. On 1 May, direct support grew to company-size operations. South Vietnamese combat engineers opened and repaired supply routes along the axis of advance, built passable roads out of trails, and erected M4T6 dry spans and panel bridges or bypasses for the armor and wheeled columns to follow. Other tasks included building supply points protected by solid earthen revetments and helipads to evacuate the wounded. Also, on 30 May the 303d Engineer Combat Battalion finished a forward airstrip with a laterite runway. To the south in IV Corps, the 40th Combat Group built a C–7 Caribou airstrip at Neak Luong. In mid-June, an enemy mortar attack destroyed six floats at the critical Ben Soi Bridge, but the 301st Engineer Combat Battalion repaired the bridge in four hours. At Svay Rieng, the 303d Engineer Battalion took four days to replace a destroyed bridge with a 160-foot double-double panel bridge. Later in the operation, the 301st Engineer Battalion used four D7E bulldozers to clear vegetation on both sides of Highway 7 near the Chup Rubber Plantation. In June and July, as some forces withdrew, unguarded bridges were disassembled and stocked. Not to be outdone by enemy sappers, 30th Group engineers used demolitions

505

to destroy a 272-foot Eiffel bridge to deny its use to the enemy. The North Vietnamese Army had used the bridge spanning the Kompong Spean River at Kompong Trach for several years. Work shifted to improving defensive positions at bridges and bypasses as enemy forces increased their attacks. Engineer camps were also attacked. On 4 July, 120 rounds of 82-mm. mortar fire struck in and around the perimeter of Company A, 301st Engineer Battalion, at Krek. The attack killed five engineers and damaged several pieces of equipment. American advisers reported that the Vietnamese engineers gained tremendous confidence during the operation. This was the first time that battalion-size engineer units had to live and work in a combat environment, making sure that bridges and roads were completed on time.[38]

The Cambodian operation confirmed basic engineer principles in supporting a corps in the attack. Detailed planning at the 20th Brigade began only two to three days before the campaign and then continued for the next three weeks. General O'Donnell held nightly command and staff meetings, including representatives from groups and battalions, at brigade headquarters just outside Bien Hoa Air Base. At this time, orders were issued for the following day. O'Donnell stopped this practice after the units completed their moves, adequate bridge stocks and building materials reached the forward areas, and the scope of the operation became apparent. Task force organizations worked without difficulty since the brigade and groups routinely allocated resources based on the extent of the assignment. As in the past, engineer units quickly moved out of their base camps and firebases. They easily switched from regularly assigned construction jobs to the hasty, constantly changing main supply routes and combat support missions. Working relationships did not pose any problems since engineer units worked with the same supported combat unit. The brigade and group continued the practice of allowing the two battalion task force commanders the flexibility to accept combat and operational support missions directly from the divisions. This helped speed up reaction time. Tactical commanders assigned priorities, and the 20th Brigade and 79th Group allocated the needed resources.[39]

Despite the short lead times and some setbacks during the incursion, U.S. engineers could look back on an impressive list of accomplishments. They built 35 miles of new roads and opened, repaired, or upgraded another 221 miles, and built or reinforced fifteen bridges. Roads were made passable by improving drainage, patching holes, and capping the surface with rock or laterite. Consequently, locating laterite borrow pits became critical. When placed on a dry base, laterite caps held up throughout the monsoon season. The engineers also built enough turnouts on the many single-lane roads to handle passing vehicles. Large amounts of tactical bridging were brought forward and used, over 345 feet of M4T6 dry span and 710 feet of panel bridging. Several delays in building the panel bridges occurred, however, because of missing parts and inexperience. Whenever possible, work crews used culverts or the large mul-

---

[38] AAR, Opn TOAN THANG 42, U.S. Army Advisory Gp, III Corps, pp. 4-3, 14-2 to 14-4; Khuyen, *RVNAF Logistics*, pp. 193–94.
[39] AAR, Opn TOAN THANG 43, 20th Engr Bde, pp. 69–70.

tiplate arch to conserve bridging and make the crossings less vulnerable to destruction.[40]

The key to opening and maintaining forward tactical airfields lay in the building of bypass roads where the main roads also served as runways. Early construction of the bypasses at Thien Nhon and Katum permitted their continued use. The engineers placed six hundred feet of M8A1 matting at each end of the two airfields in an attempt to ensure runway usage when the rains came. However, poor drainage under the matting made continuous maintenance necessary. When necessary, work took place at night to allow daylight landings. Patching potholes with soil-cement, laterite, and hot- and cold-mix asphalt continued throughout the operation. The cold mix proved to be the best material because it could be stockpiled at each end of the field. In case it became necessary, sufficient matting for one complete airfield in Cambodia was stockpiled at Quan Loi.[41]

As in the past, road mines posed major threats to armored columns and supply convoys. During the Cambodian operation, II Field Force used the only available vehicle-mounted mine detonator, the M48 Tank-Mounted Clearing Roller. This roller was obtained through the ENSURE (Expediting Non-Standard Urgent Requirements for Equipment) management system. Through this program, items were developed and procured either for evaluation purposes or for operational requirements. To speed up clearing operations for this operation, two tanks and one combat engineer vehicle mounted the rollers. The rollers cut clearing times by 50 percent when compared to using hand-held detectors. One precaution was necessary. If the roller only made one pass, all vehicles behind it had to follow in its exact path. Also, when the rollers set off mines, one or more of the replaceable wheels and sometimes the wheel arm assemblies had to be replaced. Mines over sixty pounds usually caused enough damage to require replacement of the entire arm assembly. Replacement typically took four men forty-five minutes. The 65th Engineer Battalion tried mounting the roller on one of its four combat engineer vehicles. Although designed for the M48A3 tank chassis it did, after removal of the bulldozer blade and some minor modifications, fit the combat engineer vehicle's chassis. However, to replace the rollers and assembly arms, a wrecker was required because the engineer vehicle's boom was too short for the job. After using the roller for several days and after setting off three mines, the engineers realized that enemy sappers were trying to figure the minimum size needed to destroy the mine roller. Combat engineers judged it best to use the roller infrequently along the same road, because daily usage encouraged the enemy to increase the number and size of mines. They argued it was better to use the roller with the tank since mounting the roller on the combat engineer vehicle took a specialized piece of equipment away from its primary mission. No matter how they were mounted, the rollers were considered better than the expedients of "running" or "busting" the road with tanks or dump trucks.[42]

---

[40] Ibid., pp. 70–71.

[41] Ibid.

[42] Ibid., p. 12; AAR, Opn TOAN THANG 43, 79th Engr Gp, p. 33; AAR, Cambodian Operation,

*Rollers placed in front of M48 tanks and this combat engineer vehicle
expedited clearing roads of mines.*

The cross-border operation took its toll on engineers. Casualties in 20th Brigade units totaled 7 killed and 132 wounded. Mortar and rocket attacks accounted for most casualties, and most of these losses took place between 16 and 19 June in the Fishhook. On 16 June, a work party en route from Firebase COLORADO to a bridge site was ambushed and suffered eight wounded. It was during this period that the 584th Land Clearing Company lost twenty-one soldiers. Other casualties were scattered in sporadic attacks or mines throughout the campaign. Tragedy also struck at the higher commands. On 12 May, General Dillard's helicopter was shot down while he was reconnoitering Route 509 about ten miles southwest of Pleiku. Others killed in the incident included Col. Carroll T. Adams, commanding officer, 937th Engineer Group; Lt. Col. Fred V. Cole, commanding officer, 20th Engineer Combat Battalion; Capt. William D. Booth, aide de camp; Cmd. Sgt. Maj. Griffith A. Jones of the Engineer Command; and five others. Cmd. Sgt. Maj. Robert W. Elkey of the 937th Group was the sole survivor.[43]

25th Inf Div, pp. 11–12; Heiser, *Logistic Support*, p. 53. For more on the tank roller evaluations in armored operations, see Final Report, Army Concept Team in Vietnam (ACTIV), sub: Optimum Mix of Armored Vehicles for Use in Stability Operations, ACTIV Proj. No. ACG 69F, 27 Jan 71, pp. III-8, G-33, Historians files, CMH.

[43] AAR, Opn TOAN THANG 43, 20th Engr Bde, p. 68; Ploger, *Army Engineers*, pp. 176–77; "MG Dillard, 9 Others Die in Crash," *Castle Courier*, 18 May 1970; Nolan, *Into Cambodia*, p. 201.

Secondary Assaults

The secondary attacks from II Corps were designed to draw North Vietnamese forces north and to cut their logistics lifeline north of the main battle. U.S. and South Vietnamese forces spearheaded the operations, known as BINH TAY I through IV, from Kontum in the north to Ban Me Thuot in the south. Headquarters I Field Force and II Corps received little warning of the attack. When orders were received on 2 May, units of the U.S. 4th Division and the South Vietnamese 22d Infantry Division were operating in enemy Base Area 226 in northwest Binh Dinh Province. The 173d Airborne Brigade in northern Binh Dinh Province was the only other American tactical unit of any size in II Corps. The 4th Division had already lost a brigade to redeployment.[44]

On 6 May, elements of the 4th Division reinforced by the South Vietnamese 40th Regiment, 22d Division, commenced Operation BINH TAY I. The task force made air assaults into Base Area 702 west of Kontum and just south of the tri-border area of Laos, Cambodia, and South Vietnam. South Vietnamese forces began the BINH TAY II and III operations on 14 and 20 May into Base Areas 701 and 740 west of Pleiku and Ban Me Thuot, respectively. BINH TAY IV, carried out by the South Vietnamese in late June, was not directed toward destruction of enemy forces or bases but toward the evacuation of Cambodian and Vietnamese refugees.[45]

Although preceded by B–52 strikes, the 4th Division's attacks ran into trouble, and the division's foray across the border only lasted ten days. An infantry battalion on loan from the 101st Airborne Division failed to reach its intended landing zone because of heavy fire. Other units of the 4th Division also met resistance at their landing zones and were forced to withdraw. But after this initial opposition, the North Vietnamese avoided contact throughout the operation. By 7 May, the 1st Brigade's three battalions reached the objective area, followed by the 2d Brigade's three battalions. When the troops landed, helicopters dropped off 105-mm. howitzers to provide close fire support. Heavier artillery such as 8-inch howitzers and 175-mm. guns provided support from bases on the South Vietnamese side of the border. On 9 May, the airborne battalion uncovered a rice depot containing about 500 tons. Two days later, elements of other battalions discovered a North Vietnamese dispensary complete with surgical equipment and medicine and a twenty-bed hospital. On the thirteenth, most of the 2d Brigade returned by helicopter to Camp Radcliff. Once at their home base, they began standing down pending redeployment to the United States. The 1st Brigade began pulling out of Cambodia on 15 May, completing its withdrawal the following day. South Vietnamese units stayed behind, continuing the search in the base area. BINH TAY I ended on 25 May.[46]

The number of training areas, base camp complexes, and other facilities discovered during BINH TAY I revealed the scope and permanency of the North

---

[44] MACV History, 1970, vol. 3, p. C-88; Tho, *Cambodian Incursion*, p. 90.

[45] MACV History, 1970, vol. 3, pp. C-87 to C-91; Tho, *Cambodian Incursion*, pp. 90–100.

[46] MACV History, 1970, vol. 3, pp. C-87 to C-88; Tho, *Cambodian Incursion*, pp. 90–95. For more on BINH TAY I, see AAR, Opn BINH TAY I, 4th Inf Div, 21 Jul 70, Historians files, CMH.

Vietnamese forces in this stronghold. By 13 May, the 4th Division had uncovered over 15 tons of munitions, 500 tons of food, large amounts of medical supplies, over 800 crew-served weapons and small arms, and a variety of other military equipment. The South Vietnamese found still more. Given the effort it would take to reestablish Base Area 702 during the rainy season, the North Vietnamese had suffered a blow.[47]

During BINH TAY I, the 4th Division's 4th Engineer Battalion and elements of the 937th Engineer Group, 18th Engineer Brigade, rendered support. The I Field Force Engineer Section planned and coordinated the construction effort and operational support tasks. Divisional engineers carried out the usual missions: one line company in direct support of each brigade (Company B less one platoon to the 1st Brigade and Company C to the 2d Brigade) and one line company (Company D) in general support. (Company A had redeployed with the 3d Brigade.) Company A, 299th Engineer Combat Battalion, provided the third platoon to accompany the 1st Brigade's attached airborne battalion. In turn, each line company doled out one platoon to each infantry battalion while the company headquarters remained with the brigade headquarters. On the morning of 4 May, the 4th Engineer Battalion's forward headquarters moved from Camp Radcliff to Engineer Hill outside Pleiku. On 11 May, the battalion forward deployed with the division headquarters to New Plei Djereng. At the end of the operation on 16 May, the 4th Engineer Battalion returned to Camp Radcliff with the division headquarters. In the midst of BINH TAY I, the 4th Engineer Battalion and 937th Group changed commanders. On 13 May, Lt. Col. Richard L. Curl replaced Lt. Col. John R. Brinkerhoff, who had commanded the 4th Battalion for one year. Curl would depart with the division in December. At the 937th, Col. James C. Donovan replaced Colonel Adams, who died when General Dillard's helicopter was shot down. Adams had commanded the group for five months.[48] (*Map 29*)

The engineers provided the normal close support for the infantry battalions assaulting the base area. Primary tasks included opening landing zones, clearing mines and booby traps, and building fortifications. Typically, rappelling teams, with their chain saws and explosives, dropped from hovering helicopters and cleared the landing zones. They usually were followed by a Case 450 airmobile bulldozer to help build battalion-size firebases. On the Vietnamese side of the border, Company D, reinforced with D7 bulldozers and graders from Company B, repaired Route 509 from the juncture with Highway 19 to New Plei Djereng. At New Plei Djereng, one platoon along with elements of Company E, the bridge company, built field fortifications for the 4th Division's forward logistical base. Company E anticipated its bridging mission by moving two armored vehicle launchers with three bridges to New Plei Djereng. It also kept a bridge platoon on alert with two sets of M4T6 floating bridge at Engineer Hill. When the bridge mission was discontinued on 14 May, the platoon returned the bridging to Camp Radcliff; the trucks

[47] MACV History, 1970, vol. 3, pp. C-87 to C-88.
[48] AAR, Opn BINH TAY I, 4th Inf Div, pp. 17–18; ORLL, 1 May–31 Jul 70, I FFV, 15 Aug 70, p. 45, Historians files, CMH; Ploger, *Army Engineers*, pp. 221, 226.

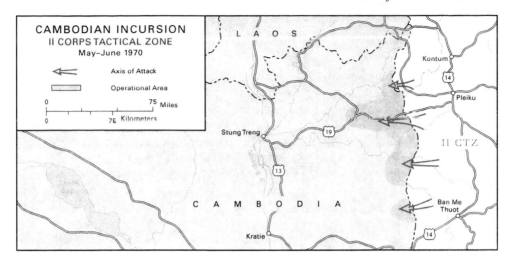

CAMBODIAN INCURSION
II CORPS TACTICAL ZONE
May–June 1970

⟵ Axis of Attack

▭ Operational Area

0   75 Miles
0   75 Kilometers

LAOS
Kontum
14
Pleiku
Stung Treng
19
II CTZ
13
Ban Me
Thuot
C A M B O D I A
Kratie
14

MAP 29

returned to New Plei Djereng and transported elements of an infantry battalion back to Camp Radcliff. At New Plei Djereng, the 4th Engineer Battalion built helicopter pads and refueling points. Two asphalt distributors also spread peneprime in an attempt to reduce the acute dust problem at the forward base. Reinforcing the 4th Engineer Battalion, Company B, 20th Engineer Combat Battalion, maintained Highway 19 from Highway 14 to Duc Co. At the Duc Co forward base, the company built berms around the fuel storage point, an ammunition supply point, and a rearm point. On 16 May, the company took over the Route 509 maintenance from Company D and continued this task until 29 May when it returned to Engineer Hill.[49]

Engineers supporting BINH TAY I experienced several annoying problems. Due to the shortage of ten-ton tractors and trailers, it took four days to shuttle all the bulldozers and graders to the Route 509 work sites. Heavy air and road traffic at New Plei Djereng forced the engineers to spread peneprime only at night. Also, the 4th Engineer Battalion's limited equipment and supply of peneprime did not completely overcome the dust problem. On the other hand, heavy rains during the latter part of the operation hampered overland supply. Without nearby sources of good rock, engineers resorted to using the local basaltic soil, which when used as a surface course turned into a soupy mud under heavy rains and traffic. Had the operation continued longer with no change in the weather, resupply by road would have been drastically reduced. The 4th Engineer Battalion also had mixed results using the combat engineer vehicle with expendable rollers. During dry weather, the engineers could clear the main supply route within an hour, but the onset of heavy rains reduced the rollers' effectiveness. Running at slower speed helped. Still, mud collected in the road wheels and caused excessive wear to the mine roller's bearings.[50]

[49] AAR, Opn BINH TAY I, 4th Inf Div, pp. 17–19.
[50] Ibid., pp. 19–20.

511

The cross-border incursions hurt the North Vietnamese Army. By the end of June, U.S. and South Vietnamese forces captured or destroyed almost 10,000 tons of materiel and food. This amounted to enough rice to feed more than 25,000 enemy troops a full ration for an entire year, individual weapons to equip fifty-five full-strength battalions, crew-served weapons to equip thirty-three full battalions, and recoilless rifle ammunition for more than 9,000 typical attacks against allied forces. In all, 11,362 enemy soldiers were killed and over 2,000 captured. For the 1st Cavalry Division, the operation in Cambodia exceeded all expectations. Maj. Gen. George W. Casey, who took command of the 1st Cavalry Division on 12 May from Maj. Gen. Elvy B. Roberts, summed up the results as "impressive." In a letter to his troops on 6 July, he wrote that the division "killed enough of the enemy to man three NVA [North Vietnamese] Regiments; captured or destroyed individual and crew-served weapons to equip two NVA Divisions; and denied the enemy an entire year's supply of rice for all his maneuver battalions in our AO [area of operations]." For the next fourteen months, there were almost no North Vietnamese and Viet Cong operations in South Vietnam other than sporadic ambushes and attacks by fire.[51]

Operations After Cambodia

Besides the Cambodian incursion, allied military operations in 1970 focused on enemy base areas inside South Vietnam, on protecting the population, and on exploiting progress in Vietnamization and pacification. In some areas, the Regional and Popular Forces assumed greater responsibilities for security. On 1 July, allied forces launched a countrywide Summer Campaign to intensify the pacification efforts, with the emphasis on eliminating the Viet Cong underground and guerrilla forces. The Fall Campaign began on 1 October, and it concentrated allied forces against contested border areas. With these operations and the hundreds of small-unit actions, Saigon's control over the countryside expanded.[52]

In I Corps, which faced the heaviest enemy threat, XXIV Corps took over command from the Marine Corps, and the South Vietnamese assumed greater responsibility. At the beginning of the year, U.S. ground forces consisted of some 50,000 Army troops and 55,000 marines, down from 79,000 marines a year earlier before the departure of the 3d Marine Division. In Quang Tri Province, the 1st Brigade, 5th Division, continued to guard the invasion routes across the Demilitarized Zone, and the 101st Airborne Division protected Hue. The III Marine Amphibious Force with its 1st Marine Division guarded Da Nang. During the summer, the marines made one last sweep through the Que Son Valley, keeping the area quiet until the division departed. In southern I Corps, the Americal Division continued security operations in Quang Tin

---

[51] Quoted words from Tolson, *Airmobility*, pp. 232–33; Starry, *Mounted Combat*, pp. 179–81; MACV History, 1970, vol. 1, p. I-3; Stanton, *Rise and Fall of an American Army*, p. 342. For more on the results of the Cambodian operation, see MACV History, 1970, vol. 3, pp. C-103 to C-108.

[52] MACV History, 1970, vol. 1, pp. I-1, I-4, V-6; Stanton, *Rise and Fall of an American Army*, p. 343.

and Quang Ngai Provinces. South Vietnamese forces assumed responsibility in western Quang Tri Province from the U.S. Marines and took control of several firebases south of the Demilitarized Zone. Meanwhile, U.S. and South Vietnamese forces carried out combined and separate operations to keep pressure on the enemy and deny him access to the population.[53]

In II Corps, the allies continued to pursue the main forces, attack the Viet Cong underground, and protect the population. Enemy activity was generally light. After the 4th Division redeployed in October and November, the South Vietnamese 22d Division assumed responsibility in the highlands. By then, the South Vietnamese controlled most of the corps area except parts of the coast. The 173d Airborne Brigade in northern Binh Dinh Province and the two Korean divisions to the south carried out pacification operations.[54]

An example of engineers supporting pacification was the completion of a bridge and causeway in the village of Tam Quan in northern Binh Dinh Province. The bridge and causeway stretched nearly 800 feet and provided fishermen from an off-shore island with a more rapid link to mainland markets, allowing them to make a quick sale of their daily catch. The project fell to the men of the 2d Platoon, Company B, 39th Engineer Battalion. Only two of the squad leaders and platoon sergeant were experienced in building timber bridges, and this one would be entirely over water. According to Sp4c. Richard Del Gaudio, "Before we moved down there, they told us we'd be working on an island paradise—ideal conditions, beautiful setting and plenty of beach." They did find a quiet village on the mainland separated by a tidal river from the off-shore island. The men wanted to live there, but they had to settle for nearby Landing Zone ENGLISH NORTH and commute to the bridge site.[55]

Each day's early morning hours were spent sweeping for mines in the quarry area north of Tam Quan and the one-and-a-half-mile road from Highway 1 through the village to the job site. "We found eleven mines altogether," 1st Lt. John W. Erdman noted, "Nine using the detectors and two the hard way." One mine was pointed out by a child as the engineers were sweeping through the village. Dump trucks from Companies A and D at Chu Lai were sent south on loan to haul the large quantity of rock from the Tam Quan quarry to complete the first section of the causeway. A crane from the 137th Light Equipment Company was then dispatched to drive bridge piling. While Sp4c. James Bradford operated the crane, the 39th Battalion's surveyor, Sp4c. Ronald Rauch ensured that the bridge remained in line. "This proved to be quite a hassle, considering the rocky bottom conditions the piling had to be driven through," recalled Rauch. Work progressed on the timber bridge, and the engineers figured they would be finished within a week. One morning during the minesweep, however, the men sensed something was wrong. No

[53] MACV History, 1970, vol. 1, pp. I-5, V-7; Graham A. Cosmas and Lt. Col. Terrence P. Murray, USMC, *U.S. Marines in Vietnam: Vietnamization and Redeployment, 1970–1971* (Washington, D.C.: History and Museums Division, Headquarters, U.S. Marine Corps, 1986), pp. 2, 17, 84.

[54] MACV History, 1970, vol. 1, pp. I-5, V-8. For more on some of the allied operations in II Corps, see ORLL, 1 May–31 Jul 70, I FFV, 15 Aug 70, pp. 10–15.

[55] Spec. Willis Meeuwsen, "Bridge for the People," *Kysu'* 2 (Summer 1970): 6–7.

mines were found, and as the mine team rounded the last curve they found the bridge was in flames. The bridge was too far gone to be saved. Lieutenant Erdman questioned the villagers through an interpreter and learned that after a brief firefight with the Popular Forces night security, the Viet Cong had set fire to the abutment and substructure of several pile bents. The fire spread quickly upward along the freshly creosoted piles to the superstructure. By noon all the spans had dropped into the water. Six weeks of work seemed lost.[56]

Col. William R. Wray, commanding officer of the 45th Engineer Group, and Lt. Col. Hugh G. Robinson, the commander of the 39th Battalion, flew to the job site. After viewing the damage, Colonel Wray suggested the bridge be rebuilt using the unburnt portions of the piles for support. That evening pencil sketches were made and discussed, and by morning the idea had taken shape. Brig. Gen. John W. Morris at 18th Engineer Brigade sent a message emphasizing the importance of the project. "Everyone was down in the dumps," Erdman observed, "but it didn't take much push after it was decided what had to be done." With infantry from the 173d Airborne Brigade setting up a guard position and additional Popular Forces security now on the scene, the engineers responded by working late hours and sometimes through the night. "We had to work at night to cut off the burnt piles, because only then was the tide low enough," said Specialist Del Gaudio. After cutting off the piles, a 10-by-12-inch cap was bolted and scabbed onto them. This provided a substructure for the modified timber trestle bridge. The troops also developed a way to prefabricate the twenty-foot bents at Landing Zone ENGLISH NORTH and haul them to the bridge site. By using the sawed-off piles and prefabricated bents, the platoon rebuilt the bridge in only twenty-two days. To complete the second section of causeway, the 511th Engineer Company (Panel Bridge) contributed thirteen dump trucks and a platoon of men to help the 39th's engineers. Four days later, the entire project was completed without further incident.[57]

In III and IV Corps, the allies consolidated their gains while enemy activity remained light to moderate. Later in the year, following the departure of 25th Division headquarters and two of its brigades, the 9th Division's 3d Brigade, and the 199th Infantry Brigade, the South Vietnamese Army assumed more responsibility for the border and over 50 percent of the territory of III Corps. Allied plans included South Vietnamese forays into Cambodia and battalion and smaller size operations to strengthen government control over the countryside. The 1st Cavalry Division, the 25th Division's 2d Brigade, and the 11th Armored Cavalry Regiment, the only U.S. ground combat units in III Corps, interdicted enemy supply routes, located and destroyed caches, and supported pacification. Australian and Thai forces continued to support pacification east and southeast of Saigon. The South Vietnamese had already assumed responsibility for ground operations in IV Corps when the 9th Division departed. A strong U.S. advisory effort, naval forces, helicopter units, and some combat support elements, including the 34th Engineer Group, remained in the delta.[58]

---

[56] Ibid., pp. 7–8
[57] Ibid., p. 8.
[58] MACV History, 1970, vol. 1, pp. I-5, V-8 to V-10; ORLL, 1 Aug–31 Oct 70, II FFV, 14 Nov

Although Cambodia represented the year's operational highlight, 1970 still saw the engineers active in other areas. They began or completed construction projects, continued operational and combat support missions that included land clearing, improved MACV advisory facilities, and made progress in the highway program. Some units prepared to stand down and others moved to new locations to replace departing units. In I and II Corps, the 18th Engineer Brigade commanded by General Morris and then by his successor, General Schrader, who assumed command in May, still controlled three groups consisting of eleven battalions, altogether almost 13,000 men. In I Corps, the three battalions of the 45th Group (14th, 27th, and 39th Combat) carried out operational support tasks for XXIV Corps and highway projects. Work included building bunkers, aircraft revetments, security towers, access roads to bases, and repairing the Mai Loc Special Forces airstrip in Quang Tri Province. Seabee units, which went from five battalions early in the year to two battalions by year's end, worked on construction projects, including aircraft shelters. Three Marine Corps engineer battalions carried out similar tasks for the 1st Marine and American Divisions. Another Marine engineer task included dismantling prefabricated buildings in the Force Logistics Command's depot outside Da Nang for shipment to Okinawa. By the end of the year, only the 1st Marine Engineer Battalion and one reinforced company of the 7th Marine Engineer Battalion remained.[59]

In northern II Corps, the 937th Group's four battalions (20th and 299th Combat and 84th and 815th Construction) and in southern II Corps the 35th Group's four battalions (19th Combat and 577th, 589th, and 864th Construction) completed bunkers, aircraft revetments, and repairs to several airfields. Much effort went to the highway program along with some construction. The 864th Battalion was so spread out that it formed two large task forces, Task Force 21 outside Ban Me Thuot and, about 125 miles to the south, Task Force WHISKEY not far from Phan Thiet. The battalion headquarters remained at Nha Trang, over 60 miles by air from Ban Me Thuot and almost 125 miles from Phan Thiet. To prepare for the turnover of Camp Enari near Pleiku, the 937th disassembled two aircraft hangars and ten Pascoe buildings and reassembled them at Camp Radcliff.[60]

70, pp. 23–29, Historians files, CMH.

[59] ORLLs, 1 Feb–30 Apr 70, 18th Engr Bde, 30 Apr 70, pp. 2–4, 1 May–31 Jul 70, 18th Engr Bde, 31 Jul 70, pp. 3–4, and 1 Aug–31 Oct 70, 18th Engr Bde, 31 Oct 70, pp. 3–4, all in Historians files, CMH; Cosmas and Murray, *U.S. Marines in Vietnam, 1970–1971*, pp. 324–27; Combat-Operational Support, I/II CTZ (Corps Tactical Zone) (18th Engr Bde), incl. 15b, OCE Liaison Officer Trip Rpt no. 16. See also Significant Construction Projects and Operational Support Activities of 45th Engr Gp, incl. 10H, OCE Liaison Officer Trip Rpt no. 17, 30 Jul 70, OCE Hist Ofc, and incl. 7G, OCE Liaison Officer Trip Rpt no. 18.

[60] ORLLs, 1 Feb–30 Apr 70, 18th Engr Bde, pp. 2–4, 1 May–31 Jul 70, 18th Engr Bde, pp. 3–4, 1 Nov 69–31 Jan 70, 35th Engr Gp, p. 5, and 1 Nov 69–31 Jan 70, 864th Engr Bn, pp. 2–13; Combat-Operational Support, I/II CTZ (18th Engr Bde), incl. 15b, OCE Liaison Officer Trip Rpt no. 16. See also Significant Construction Projects and Operational Support Activities of 35th and 937th Engr Gps, incls. 10G and 10I, OCE Liaison Officer Trip Rpt no. 17, and incls. 7F and 7H, OCE Liaison Officer Trip Rpt no. 18.

In III and IV Corps, the 20th Brigade headed by Brig. Gens. O'Donnell and Kenneth B. Cooper, who took command in November, also controlled three groups made up of twelve battalions, some 12,000 men. In northern and western III Corps, the 79th Group's four battalions (31st and 588th Combat, 554th Construction, and 62d Land Clearing), besides supporting the Cambodian operation, worked on highways, some construction, airfield improvements, and a variety of operational support missions. The 159th Group and its four construction battalions (34th, 46th, 92d, and 169th) focused efforts in the Saigon–Long Binh area and the highway program. In IV Corps, the 34th Group's four battalions (35th Combat and the 36th, 69th, and 93d Construction) supported the remaining U.S. units in the delta and sought to complete its share of the highway program.[61]

The 326th Engineer Battalion's work at Firebase RIPCORD typified the experience of engineers working where the only access was by helicopter. In April, the 101st Airborne Division established Firebase RIPCORD atop a hill twenty-four miles west of Hue to help prevent the North Vietnamese Army from moving into populated coastal regions of Quang Tri and Thua Thien Provinces. On 11 April, a small engineer party equipped with hand tools, power saws, and demolitions landed and fanned out and cut fields of fire and established fighting positions while the infantry searched the surrounding area. The engineers next cleared a rough landing zone. With enough space opened, a CH–54 lifted in a Case 450 tractor followed by a CH–47 carrying the blade and a drum of diesel fuel. Within minutes the engineers manhandled the blade to the tractor, which then cleared the hilltop of debris and excavated the first gun pit for the artillery. Another helicopter dropped off a Case 580 combined scoop loader and backhoe. Ammunition berms were pushed up and more gun pits were carved into the hilltop. Although poor weather held off the arrival of the artillery pieces for several days, construction continued into July. By then, the North Vietnamese began to subject the base to daily attacks by mortars, recoilless rifles, and rocket-propelled grenades. Artillery fire, air strikes, and ground sweeps failed to drive off the determined North Vietnamese, who appeared to be preparing for a full-scale attack. Rather than face another siege like Khe Sanh, the division decided to close the firebase. The evacuation, which included the engineer equipment, took place on 23 July under heavy enemy fire, and several Chinook helicopters were heavily damaged. In August and September, this story was repeated at Firebase O'REILLY some five miles north of RIPCORD. O'REILLY was abandoned in early October. Clearly the North Vietnamese were determined to protect their most important base areas and supply routes, especially in northern I Corps, and their

---

[61] ORLLs, 1 Feb–31 Apr 70, 20th Engr Bde, 26 May 70, pp. 3–4, 7–8, 1 May–31 Jul 70, 20th Engr Bde, 13 Aug 70, pp. 2–3, 6, and 1 Aug–31 Oct 70, 20th Engr Bde, 15 Nov 70, pp. 3–4, 6–8, all in Historians files, CMH; Fact Sheet, 20th Engr Bde, Operational Support and Combat Support Missions, 1970, incl. 16b, OCE Liaison Officer Trip Rpt no. 16. See also Significant Construction Projects and Operational Support Activities of 34th Engr Gp, 79th Engr Gp, and 159th Engr Gp, incls. 11E to 11G, OCE Liaison Officer Trip Rpt no. 17, and incls. 8C to 8E, OCE Liaison Officer Trip Rpt no. 18.

pressure on Firebases RIPCORD and O'REILLY indicated they still had enough strength to exploit allied weakness.[62]

The steady departure of engineer units continued in 1970. Two engineer combat battalions, the 1st and 168th, left in April when the 1st Division returned to Fort Riley, Kansas. The 168th returned to Fort Lewis, Washington, where it was inactivated. On 15 August, a color guard detachment representing the 35th Engineer Combat Battalion, which had served in I, II, and IV Corps, headed to Fort Lewis for inactivation. A color guard detachment representing the 4th Engineer Battalion left on 6 December for the 4th Division's new home base at Fort Carson, Colorado. Two days later, the 65th Engineer Battalion less one company returned with the 25th Division headquarters and two brigades to their home base in Hawaii. Company B remained behind in III Corps with the 2d Brigade. On 15 December, color guards from the 19th and 588th Engineer Battalions returned to Fort Lewis for inactivation ceremonies. Also by the end of the year, the first of the six engineer group headquarters, the 79th at Long Binh, had left the country. Engineer units of the other services also returned to home bases in the Pacific or United States. In February, the Air Force's 819th and 820th Red Horse Squadrons departed for the United States. The U.S. Navy's Mobile Construction Battalions 1, 5, 7, 10, 62, 74, and 121 completed their tours and returned to their home bases. Most were not replaced. At the end of the year, only Naval Mobile Construction Battalions 3 and 74, the latter returning to Vietnam for its fourth and final tour, and Construction Battalion Maintenance Unit 302 remained in Vietnam.[63]

Facilities Upgrades and Transfers

Though base construction had declined, it remained a major factor in engineering planning and execution. Strict controls still applied to new construction, and only emergency requirements such as the aircraft shelter program were considered. This was emphasized by General Abrams, who in early 1970 advised his commanders to use existing facilities, and for more temporary needs "make do or do without." In April and October 1970, the Army's funding allocation for Vietnam projects increased by only $20.2 and $15 million, respectively, for urgently required work. This brought the Army's total construction program to $992 million. During the year, the Navy and Air Force programs reached $446.6 million and $405.3 million, respectively. In dollar figures, the three services' military construction programs at the end of 1970 totaled $1.774 billion. Work in place proceeded at a steady rate, from $15 million per month in the second half of 1969 to $13 million per month through April 1970. The MACV Construction Directorate reviewed the remaining projects and assigned work to be carried out by troops and

[62] MACV History, 1970, vol. 3, pp. G-1 to G-6; Cosmas and Murray, *U.S. Marines in Vietnam, 1970–1971*, pp. 84–85; Stanton, *Rise and Fall of an American Army*, pp. 344–46.

[63] Quarterly Hist Rpts, 1 Jan–31 Mar 70, MACDC, 18 Apr 70, p. II-2, 1 Apr–30 Jun 70, MACDC, 24 Jul 70, p. II-3, 1 Jul–30 Sep 70, MACDC, 23 Oct 70, p. II-7, and 1 Oct–31 Dec 70, MACDC, 17 Jan 71, p. II-9 to II-10. All in Historians files, CMH.

RMK-BRJ. Some savings were made when RMK-BRJ freed some $30 million after reducing overhead, amortization, and other cost efficiencies, releasing the funds to fulfill unfinanced requirements. Facilities engineers also began to face sizable cuts in operations and maintenance funds after the transfer of bases. Washington only provided $108 million for Fiscal Year 1971 instead of the projected $139.2 million. As a result, Pacific Architects and Engineers and Philco-Ford faced a reduction in services and personnel.[64]

During 1970, the emphasis on Vietnamization continued, with most funds released by the Department of Defense slated for improving and modernizing Vietnamese facilities and for the highway program. The total funded and unfunded dollar of work remaining came to $379 million for the Improvement and Modernization Program, U.S. forces, and the highways and railways program. Also, during the first half of the year, MACV and the service components carefully reviewed unfunded requirements for possible deletions. As a result, $214.5 million in unfunded requirements dropped to $81.8 million by the end of the year. Maximum use was to be made of vacated facilities to fill South Vietnamese requirements. South Vietnamese Army project groupings included building and upgrading hospitals, training centers and schools, communications facilities, depots, and company-size camps for the regular forces and regional forces. By the end of 1970, many of these facilities were completed or under way.[65]

Notwithstanding the reductions in funding, engineer troops and contractors made progress in upgrading bases, roads, and rail lines. The emphasis on roadwork shifted to quality control instead of miles paved, with more than half of the program completed. Security concerns temporarily halted main rail line restoration, but work proceeded on the spurs. The Air Force completed its aircraft shelter program, and the Navy's program neared completion. Other new work consisted primarily in improving forward airfields, upgrading staging areas at ports and marshaling yards, and improving MACV adviser facilities.[66]

Rigid health standards imposed by the Department of Agriculture and the U.S. Public Health Service caused an upgrading of the staging areas for returning equipment and material. The facilities included wash racks, parking areas, grease racks, maintenance shops, and lighting systems for twenty-four-hour operation to clean, prepare, and stage vehicles and equipment for shipment to the United States. By the end of the year, all the required facilities at Newport,

---

[64] MACV History, 1970, vol. 2, pp. IX-45, IX-49; Ltr, MACDC-PP to CINCPAC, sub: CPAF [Cost Plus Award Fee] Contractor, RVN, 22 Oct 70, Historians files, CMH; Ltr, Abrams, MACDC-PO to CG, USARV, 5 Feb 70, sub: Utilization of Facilities and Construction Resources, Historians files, CMH; Report of Liaison Visit to the Office of the USARV Engineer, incl. 6, p. 1, OCE Liaison Officer Trip Rpt no. 18.

[65] MACV History, 1970, vol. 3, pp. IX-49, IX-51.

[66] Ibid., pp. IX-52 to IX-53, IX-66, IX-73; Quarterly Hist Rpts, 1 Jan–31 Mar 70, MACDC, pp. VI-3 to VI-4, and 1 Jan–31 Mar 71, MACDC, p. II-1; Report of Liaison Visit to Office of the USARV Engineer, incl. 6, p. 1, OCE Liaison Officer Trip Rpt no. 18.

Long Binh, and Da Nang were completed. The last retrograde facility at Qui Nhon was completed on 1 February 1971.[67]

The upgrading of MACV advisory facilities, which began in 1968, neared completion. Of 244 sites selected for vertical construction and water supply projects, troops and contractors completed work at 191. Ongoing work continued at 27 sites, while 26 awaited the start of construction. During the first quarter of 1971, 9 more sites would be added, with 228 completed by 31 March, 12 in progress, and only 13 programmed to start construction.[68]

For the South Vietnamese, the construction of dependent shelters remained an important program that increasingly drew American backing. This program dated from 1961 when 64,000 family units were planned for servicemen below the grade of sergeant. An eight-year long-range program developed jointly by MACV and the Joint General Staff estimated a total requirement of 200,000 shelters. Over the years, several problems hampered construction. These included a lack of contractor interest, difficulty in obtaining land, shortages of transportation, and in 1964 the diversion of material caused by floods. Between 1961 and 1963 the government completed only 18,700 units; about 6,500 in 1965; under 10,000 for active military and a separate Regional Forces program in 1966; 2,408 units in 1967; and 4,951 units in 1969. In 1966, the Defense Department tried a pilot program for self-help dependent housing in III Corps by providing supplies and materials to build over 9,000 units. However, due to tactical operations, combat units could not find the time to do the work. In late 1968, Saigon decided to set aside some money for self-help, but only about 2,400 shelters were finished. Starting in 1969, U.S. and South Vietnamese funds were also allocated for the Popular Forces. In November 1969, MACV and the Joint General Staff established the Dependent Shelter Program Group to manage the jointly funded program. By April 1970, the joint group determined the need for 240,000 shelters, taking into account that 40,000 families could be sheltered in vacated U.S. bases. After almost ten years, the requirement for new housing still stood at 200,000 units to be built in eight years.[69]

The plans for 1970 identified about 16,000 shelters for construction. U.S. funds, provided from the services' accounts, totaled $4.3 million, with the Vietnamese government spending the equivalent of $5 million. Several designs were examined, and the approved standards were functional and austere, in deference to the living standards of the Vietnamese. Under these standards, the cost for each unit would be held to $600. The United States contributed cement, lumber, and corrugated metal, and the Vietnamese government provided the rest of the materials and military engineers, contractors, and troop units to do the work. American funding was expected to grow to $6 million

[67] MACV History, 1970, vol. 3, p. IX-54, Quarterly Hist Rpt, 1 Jan–31 Mar 71, MACDC, p. II-1.

[68] Ibid. See also 1st Lt. Daniel L. Campbell, "Builders for the Advisors," *Kysu'* 2 (Summer 1970): 20–22.

[69] MACV History, 1970, vol. 3, pp. IX-54 to IX-57; Quarterly Hist Rpt, 1 Jan–31 Mar 70, MACDC, p. II-3. For more on the dependent shelter program, see Khuyen, *RVNAF Logistics*, pp. 389–95.

for each of the following seven years. Costs were expected to increase, with additional funds required for community facilities such as schools and dispensaries. However, deliveries of U.S. materials for 1969 and 1970 fell behind schedule, and it took until mid-1970 before materials began to arrive in large amounts. In May, President Thieu wrote President Nixon asking for more help to complete the program in four years. Nixon responded by promising financial support to build 20,000 shelters per year for five years. By the end of the year, workers completed 7,909 shelters with 5,739 under construction. In addition, 1,121 military families lived in excess U.S. military facilities. Because of Thieu's interest in getting more dependent housing as soon as possible, his military planners considered committing the equivalent of four engineer construction groups in 1971. Problems continued to trouble the housing program. One difficulty included Saigon's timely release of funds. Another involved theft by corrupt officials and officers of the U.S.-furnished materials.[70]

Base transfers gained momentum in 1970. By the end of the year, seventy-seven transfers ranging from firebases to large base camps were completed with thirty-eight still in process. In I Corps, several Marine Corps facilities, including the Da Nang Force Logistics Command compound and 7th and 9th Engineer Battalion camps, were declared excess. They were transferred in January and May. In II Corps, the 4th Division transferred Camp Enari/Dragon Mountain (over 1,400 buildings totaling over 2 million square feet of floor space) to the South Vietnamese 22d Division in April. Camp Radcliff would be transferred after the departure of the 4th Division. Major transfers in III Corps included the 1st Division's Lai Khe base camp (over 500 buildings totaling 445,000 square feet of floor space) to the South Vietnamese 5th Division. The United States also turned over the 25th Division's base camp at Cu Chi (over 3,200 buildings amounting to over 2.3 million square feet of floor space) to the South Vietnamese 25th Division in December.[71]

Transfers also extended to ports. In November 1969, a joint U.S.-Vietnamese committee announced schedules for complete or partial turnovers at Can Tho, Saigon, Vung Tau, Nha Trang, Qui Nhon, and Da Nang. In January 1970, the South Vietnamese began to move cargo through the shallow-draft ports of Hue, Dong Ha, and Chu Lai. The ports of Can Tho and Saigon (excluding Newport) were transferred in March and June. Not everything would be left behind. Planning was under way to disassemble some prefabricated buildings, airfield matting, and DeLong piers. Transfer problems persisted, with the lack of trained Vietnamese facilities engineers among the major obstacles. To make up for these deficiencies, U.S. forces continued on-the-job training programs

[70] MACV History, 1970, vol. 2, pp. IX-59 to IX-63; Quarterly Hist Rpts, 1 Apr–30 Jun 70, MACDC, pp. II-3 to II-IV, 1 Jul–30 Sep 70, MACDC, pp. II-5 to II-6, and 1 Oct–31 Dec 70, MACDC, pp. II-2.

[71] MACV History, 1970, vol. 1, p. I-5, vol. 2, pp. IX-36 to IX-39; Quarterly Hist Rpt, 1 Jan–31 Mar 70, MACDC, pp. V-4 to V-5; Khuyen, *RVNAF Logistics*, p. 182; For more transfer lists, see Quarterly Hist Rpts, 1 Apr–30 Jun 70, MACDC, p. V-3, 1 Jul–30 Sep 70, MACDC, p. III-3, and 1 Oct–31 Dec 70, MACDC, pp. V-3 to V-5; PA&E History, FY 1970, app. Summary of Transfer and Abandonments of Real Property; PA&E History, FY 1971, pp. 82, 84, Historians files, CMH.

and using contract instructors. The Army's facilities engineering contractor, Pacific Architects and Engineers, continued to play a major role in training Vietnamese personnel as well as administering base transfers.[72]

By the end of 1970, the future of South Vietnam looked hopeful thanks to pacification and Vietnamization. While U.S. forces had cut their manpower by almost one quarter, the South Vietnamese military, including the Regional and Popular Forces, hovered around one million. Improved security made some of the fruits of the construction effort more meaningful to the people of Vietnam. The highway program, for example, had completed 1,461 miles or 65 percent of the 2,226 miles of main roads slated for improvement. At the start of 1970, 1,613 miles of road had reached the Green security status. By the end of the year, the allies claimed 71.4 percent Green and 97 percent in the less secure Amber status along 2,745 miles of road, which included the Green miles. The security status for the railway system at the end of the year only reached 44 percent Green, but the lower-priority rail system managed to support essential and military operations. Security goals along the vital waterways especially in the Mekong Delta reached 80 percent Green. MACV considered the Cambodian operation an important military success. Perhaps the clearest indication of success was the lowering of the U.S. profile as South Vietnamese units assumed more of the burden of fighting and supporting its forces. As the 1970 MACV History declared: "If progress in Vietnamization and pacification was the keynote in 1970, the theme of 1971 would be testing the viability of our progress." This would be seen with the forthcoming South Vietnamese foray in Laos in February.[73]

---

[72] MACV History, 1970, vol. 2, pp. IX-36 to IX-43; Quarterly Hist Rpt, 1 Jan–31 Mar 70, MACDC, p. V-4; DF, USARPAC, GPCO-ME, to CofS Army, sub: Management Study of DeLong Barges Used as Piers, 29 Jun 70, Historians files, CMH. For more on PA&E's facilities engineering and training efforts, see PA&E History, FY 1970, pp. 35–54, 69–80; PA&E History, FY 1971, pp. 37–86. Both in Historians files, CMH.

[73] MACV History, 1970, vol. 1, p. I-6, vol. 2, pp. VII-1 to VII-3, IX-65 to IX-66, IX-69 to IX-72, IX-78 to IX-79. Green status roads were physically open and traffic could move during daylight hours with relative freedom. Amber status roads were physically open but security measures, such as armed escorts, were required. MACV noted that for measurement purposes, total roads in Amber status included miles in Green status. Red was the unsafe status, and the roads were closed by enemy control or by explosive damage. The same security codes applied to railroads.

16

Last Battles and Departure

The years 1971 to 1973 saw American and other allied forces complete their withdrawal from Vietnam and the signing of the Paris Agreement, which was supposed to bring the war to a close. These years were highlighted by a South Vietnamese cross-border thrust into Laos, an operation which showed that Vietnamization still needed work, and the large-scale North Vietnamese Easter offensive one year later, which showed that Vietnamization might be succeeding after all. For the military engineers, it was a time of wrapping up the construction program, making headway on highway improvements, transferring facilities and responsibilities to the South Vietnamese, and providing combat operational support to the dwindling allied combat force.

Laos: The Plan

Although an attack plan had long been on the books, the trigger for the allied raid into Laos was evidence of a North Vietnamese buildup across the border for what was shaping up to be a major dry-season offensive in northern I Corps in early 1971. On 23 December 1970, President Nixon approved the campaign in principle, subject to final review. After further coordination between Washington and Saigon, the president approved the operation in detail on 18 January 1971. President Thieu concurred and named the operation LAM SON 719 after the birthplace of Le Loi, a national hero of antiquity. The 719 denoted the year (1971) and the objective (Highway 9). The attacking ground forces would be solely South Vietnamese because the U.S. Congress after Cambodia had prohibited the expenditure of funds for any American ground troops campaigning outside South Vietnam.[1]

LAM SON 719 was to be a spoiling attack, not to seize terrain but to upset North Vietnamese plans and deter future offensives. American and South Vietnamese commanders planned to cut supply and infiltration routes, destroy logistical complexes, and inflict losses on enemy forces. The incursion would take place within a narrow corridor some fifteen miles on either side of Highway 9 and go no farther in Laos than Tchepone, a logistical center on the Ho Chi Minh Trail about twenty-five miles west of the border. Allied commanders envisioned a four-phase operation that would last at least ninety days, or until the onset of the Laotian rainy season in early May.

---

[1] MACV History, 1971, vol. 2, pp. E-15, E-17; Davidson, *Vietnam at War*, pp. 637, 639, 641; Palmer, *25-Year War*, p. 109. For more background on LAM SON 719, see Maj. Gen. Nguyen Duy Hinh, *Lam Son 719*, Indochina Monographs (Washington, D.C.: U.S. Army Center of Military History, 1979), pp. 32–35.

In the first phase, Operation DEWEY CANYON II, starting on 30 January, U.S. XXIV Corps would seize the approaches inside South Vietnam up to the border. Advancing on D-Day, a task force under the control of the 1st Brigade, 5th Infantry Division (Mechanized), would occupy the Khe Sanh area and clear Highway 9 for the advancing South Vietnamese. Farther to the south, the 101st Airborne Division would conduct diversionary operations in the A Shau Valley. In Phase II, South Vietnamese I Corps forces, mainly the Airborne Division reinforced by the 1st Armored Brigade, artillery, and engineers, would attack along Highway 9 in a series of leapfrogging air assaults and armored advances. In the third phase, the Airborne and 1st Infantry Divisions, after occupying Tchepone, would expand search operations to destroy enemy bases and stockpiles while rangers manned blocking positions to the north. Phase IV would involve withdrawing the South Vietnamese from Laos, ending the operation on or about 6 April. During all phases of LAM SON 719, U.S. forces would provide artillery fire, helicopter airlift, and tactical and strategic air support.[2]

For the engineers, the operation began on 12 January, eighteen days before D-Day, when Col. Kenneth E. McIntyre, commanding officer of the 45th Engineer Group, was summoned to meet with the XXIV Corps G–2 in Da Nang. Formerly at Phu Bai, the XXIV Corps in March 1970 had moved to the former III Marine Amphibious Force's headquarters at Camp Horn, Da Nang; the 45th Group headquarters had also moved from Phu Bai on 1 December 1970 and was comfortably quartered at the former Seabee Camp Haskins at Red Beach just west of Da Nang. The XXIV Corps engineer, Col. John H. Mason, was out-of-country, and Lt. Gen. James W. Sutherland, the XXIV Corps commander, wanted to have a senior engineer involved in the planning. McIntyre was asked to take charge of engineer planning until Mason's return two days later. Critical information was needed as soon as possible. Could Highway 9 be opened for wheeled vehicles to Khe Sanh by D plus 1 and to the Laotian border by D plus 7? Could the Khe Sanh airstrip be opened to handle C–130 aircraft? McIntyre responded that he would immediately get his staff on the job and reconnoiter the road and airfield. The G–2 told him, however, that planning was restricted to a small circle of men, and there could be no ground reconnaissance or unusual activity in the area. Fortunately, Brig. Gen. John G. Hill, the commander of the 1st Brigade, 5th Division, had a ready-made solution since the brigade constantly operated west along Highway 9. When the G–2 agreed that some reconnaissance was necessary, Hill arranged for McIntyre to accompany an aerial reconnaissance carried out under the guise of a helicopter raid.[3]

---

[2] MACV History, 1971, vol. 2, p. E-25; Davidson, *Vietnam at War*, pp. 642–43; Hinh, *Lam Son 719*, pp. 35–40; Palmer, *Summons of the Trumpet*, pp. 239–40; Palmer, *25-Year War*, p. 111. See also Cosmas and Murray, *U.S. Marines in Vietnam, 1970–1971*, pp. 195–96.

[3] Brig. Gen. Kenneth E. McIntyre, "The Magnificent Sight," *Army Engineer* 2 (November-December 1994): 30; Col. Kenneth E. McIntyre, "Secret Planning," *Kysu'* 3 (Fall 1971): 15; Interv, Capt David D. Christensen, 26th Mil Hist Det, with Col Kenneth E. McIntyre, CO, 45th Engr Gp, 8 Mar 71, VNIT 891, pp. 3–4, CMH; ORLL, 1 Feb–30 Apr 71, 45th Engr Gp, 30 Apr 71, p. 1, Historians files, CMH.

A few hours later, the 45th Group commander, seated aboard a 101st Airborne Division AH–1 Cobra gunship and accompanied by another Cobra and two UH–1 Hueys, took off from Camp Eagle and headed west for Khe Sanh and the Laotian border. In between rocket firings, the pilot and Colonel McIntyre searched for a trace of Highway 9, which for long stretches was difficult to find, especially west of Khe Sanh, where the road was almost completely overgrown along the remaining fourteen miles to the border. In this stretch alone, all ten bridges had been destroyed, and the road had been cut in several places by bomb and shell craters. All but three of the thirty-three bridges along Highway 9 between the abandoned Firebase VANDEGRIFT, twelve miles east of Khe Sanh, and

Colonel McIntyre

a staging area for the operation, and the Laotian border were down. Flying over the former 3,700-foot Khe Sanh airstrip, McIntyre saw that the AM2 aluminum matting was riddled with hundreds of rocket and shell holes, and a huge crater overlapped the strip near its eastern end.[4]

Hoping he sounded more confident than he felt, Colonel McIntyre reported to the G–2 that the job could be done. To do this, the engineers would have to repair, replace, or bypass thirty bridges, clear about twenty-two miles of road, and virtually rebuild several sections of roadway. The only way to meet the deadlines would be to transport tactical bridging or culverts by helicopter to each stream crossing to join up with engineer teams and equipment that had moved overland to their work sites. As far as Khe Sanh airfield was concerned, McIntyre asked that the 101st Airborne Division's 326th Engineer Battalion land there on D-Day to begin repairing the runway and apron and to prepare two nearby sites for tactical bridging. The G–2 approved this request, and McIntyre continued his solitary planning. Although the only engineer involved thus far in DEWEY CANYON II, McIntyre later stated, "I'm sure what made it a success from a engineering viewpoint is that we were brought in on it early enough so that engineer aspects could influence the total plan." This allowed him to mesh tactical planning with engineer planning and gave him enough

---

[4] McIntyre, "The Magnificent Sight," pp. 29–31; Interv, Christensen with McIntyre, 8 Mar 71, pp. 4–6; Interv, Christensen with Col Kenneth E. McIntyre, CofS, Engr Cmd, 20 Jul 71, VNIT 923, pp. 19–20, CMH.

lead time to identify long-range equipment requirements. What he considered most important was getting heavy-lift helicopter support that might not have been available if he had joined the planning team at a later date.[5] (*Map 30*)

McIntyre still needed more data on Khe Sanh airfield and the bridge sites. Some limited aerial photo coverage, including a low-level movie film of Highway 9, proved helpful. He also arranged with General Hill for additional reconnaissance. On 13 January, ranger and engineer reconnaissance teams from the 1st Brigade, 5th Division, were inserted at Khe Sanh and the Bridge 36 site a few miles east of the airfield. After spending about ninety minutes on the ground, the team at the airstrip verified what McIntyre had seen from the air. At the bridge site, the team found that both embankments needed work before the gap could be bridged. Although the teams gathered the required information, they did not know of the impending operation.[6]

When Colonel Mason returned, he and McIntyre began preparing the engineer plan. Normally the 45th Group would receive missions from the 18th Engineer Brigade and Engineer Command, but General Sutherland thought there would be undue strain on security if all levels of the engineer command were made aware of the operation. As a result, the 45th Group was placed under the operational control of XXIV Corps effective 18 January. Then, about ten days before the operation, McIntyre received permission to bring the commanders of the 14th, 27th, and 326th Engineer Battalions into the planning group. Eventually the engineer planning team grew to eight officers from the 45th Group and two from the XXIV Corps Engineer Section. Still, the security restrictions on DEWEY CANYON II and LAM SON 719 presented problems. Colonel McIntyre recalled that even the XXIV Corps G–4 was not involved in the planning until a few days before D-Day. As for the engineers' major supplier, the U.S. Army Support Command at Da Nang, only the commanding general, his deputy, and another officer participated in the planning. The few planners did not have the experience and information they needed to get all the details right. Although group and battalion operations officers were included in the planning, company commanders and platoon leaders were not briefed until D minus 2. The troops were informed on D minus 1, 29 January, and then confined to their bases.[7]

During the preparations, cover stories and deceptions were devised. When, a few days before the operation, bridging was moved to Quang Tri Combat Base, the troops were told that Highway 1 was in danger of being cut north of the Hai Van Pass. When the 27th Engineer Battalion at Camp Eagle began

---

[5] Interv, Christensen with McIntyre, 20 Jul 71, pp. 22–23 (quotation, p. 22); McIntyre, "The Magnificent Sight," p. 31; Interv, Christensen with McIntyre, 8 Mar 71, pp. 7–8, 20–21.

[6] McIntyre, "Secret Planning," p. 15; Interv, Christensen with McIntyre, 8 Mar 71, p. 5; Intervs, Christensen with Maj Richard A. Miles, Bde Engr, 1st Bde, 5th Inf Div, Capt George B. Schoener, CO, Co A, 7th Engr Bn, and 1st Lt Allen D. Ackerman, Asst Bde Engr, 1st Bde, 5th Inf Div, 10 May 71, VNIT 887, pp. 4–7, CMH.

[7] McIntyre, "The Magnificent Sight," p. 32; Interv, Christensen with McIntyre, 8 Mar 71, pp. 7, 15, 20; Interv, Christensen with Col John H. Mason, XXIV Corps Engr, 8 May 71, VNIT 892, pp. 3, 9–11, 17–18; Interv, Christensen with Lt Col Russell L. Jornes, CO, 27th Engr Bn, 21 Mar 71, VNIT 897, pp. 3–4. All VNITs in CMH.

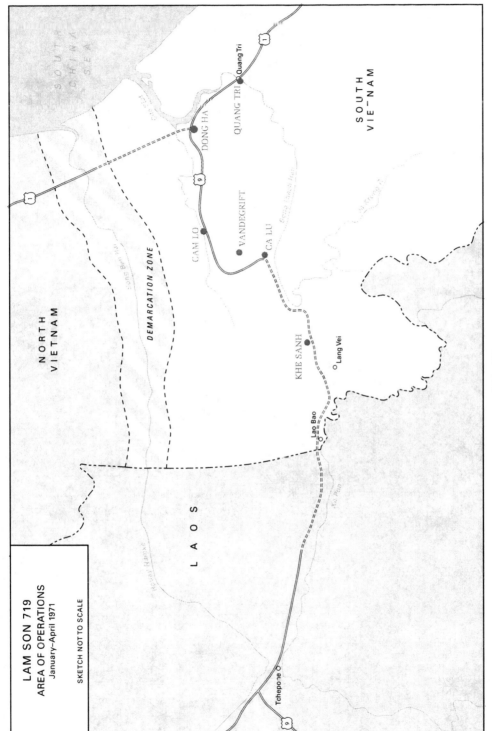

LAM SON 719
AREA OF OPERATIONS
January–April 1971

SKETCH NOT TO SCALE

NORTH
VIETNAM

SOUTH
CHINA
SEA

DEMARCATION ZONE

Song Ben Hai

1

Cam Lo

CAM LO

Dong Ha

DONG HA

QUANG TRI

Quang Tri

9

VANDEGRIFT

CA LU

Song Thach Han

Song Quang Tri

SOUTH
VIET-NAM

KHE SANH

°Lang Vei

Lao Bao

LAOS

Xepon River

Ka Dong

Tchepone

9

MAP 30

stockpiling culvert and practicing putting down and taking up of aluminum matting, the story was that the unit was about to enter the A Shau Valley.[8]

As for the Khe Sanh airstrip, several surprises arose during the planning. Since AM2 repair kits were unavailable, the replacement of damaged panels would have to be improvised, and restoring the runway would take longer than originally predicted. Fearing a serious delay, McIntyre recommended a parallel assault strip, 3,200 by 60 feet, be constructed and used until the old runway could be repaired. He considered this one of his better decisions, for the repair of the existing runway would take more than four weeks. Even this approach gave some concern, for the parallel dirt airstrip was sited based on a topographic map of the area: it could have been right on top of a minefield or old underground bunkers. In addition, McIntyre learned that the soil was almost entirely clay, which mandated topping the assault strip with matting. At first he considered requisitioning the light-duty M8A1 matting because it was readily available, but he realized it might not hold up on the clay. Since the AM2 matting was not available in the quantities required, he used MX19 aluminum matting.[9]

When the 45th Group received its mission statement, it divided the engineering tasks into six parts. In the first, Lt. Col. Robert H. Carpenter's 14th Engineer Combat Battalion based at Quang Tri would open Highway 9 to Khe Sanh. In the second, the battalion would work with Company A, 7th Engineer Battalion, in building an alternate parallel road north of Highway 9. The third task entailed building logistical support facilities at the reopened VANDEGRIFT base. Lt. Col. Russell L. Jornes' 27th Engineer Battalion, which would move overland on D plus 1 to Khe Sanh, was given the fourth and fifth tasks: open Highway 9 west of the base and repair the old Marine airstrip. For the sixth task, Task Force 326, comprised of six platoons totaling about 230 men from the 326th Engineer Battalion under the command of its executive officer, Maj. Gene A. Schneepeck, and under the operational control of the 45th Group, would be airlifted to Khe Sanh and begin constructing the assault airstrip. The task force would also prepare nearby bridge sites for tactical bridging. When the 27th Engineer Battalion reached Khe Sanh, it would assume operational control of the airborne engineers and provide heavier construction equipment as they continued working on the assault airstrip.[10]

One unknown was the weather. Ground transportation for the corps-size operation of about 35,000 troops depended on a single long, narrow, winding

---

[8] McIntyre, "The Magnificent Sight," p. 32; Interv, Christensen with Jornes, 21 Mar 71, pp. 9–12, 22–23; Interv, Christensen with Mason, 8 May 71, p. 11; Interv, Christensen with Maj Gene A. Schneepeck, XO, 326th Engr Bn, 18 Mar 71, VNIT 893, pp. 15–19, CMH.

[9] McIntyre, "The Magnificent Sight," pp. 31–32, 35; Interv, Christensen with McIntyre, 8 Mar 71, pp. 5–6, 10–12; Interv, Christensen with Mason, 8 May 71, p. 4–6; ORLL, 1 Feb–30 Apr 71, 45th Engr Gp, p. 18.

[10] McIntyre, "The Magnificent Sight," p. 32; Interv, Christensen with McIntyre, 8 Mar 71, pp. 8–9; Interv, Christensen with Mason, 8 May 71, pp. 9–10; ORLL, 1 Feb–30 Apr 71, 45th Engr Gp, p. 22; Interv, Christensen with Schneepeck, 18 Mar 71, p. 2; Interv, Christensen with Jornes, 21 Mar 71, pp. 28–29; Interv, Christensen with Lt Col Robert H. Carpenter, CO, 14th Engr Bn, 7 Mar 71, VNIT 886, pp. 9–10, CMH.

road that hardly existed. Air transportation would have to rely on uncertain airstrips at Khe Sanh. Heavy rains could wash out most of the culvert bypasses and effectively close Highway 9 and possibly the airstrips. These concerns were expressed when Colonels McIntyre and Mason met with General Sutherland, seven days before the operation. Acknowledging these concerns, Sutherland replied, "I am confident that you engineers will do whatever is necessary to make this operation a success."[11]

By D minus 1 the engineers were ready. Two panel bridge sets had been moved from Chu Lai to the Quang Tri Combat Base. All M4T6 bridging in depot contingency stocks at Da Nang to be used as dry spans also arrived at Quang Tri. Four armored vehicle launched bridges and three launchers from the 1st Brigade, 5th Division's Company A, 7th Engineer Battalion, and the 1st Battalion, 77th Armor, were made available to cross gaps along Highway 9. In the last few days before the operation began, the 14th and 27th Battalions assembled 4,700 feet of 36-inch corrugated metal culvert and rigged them in bundles of four 20-foot sections for helicopter lift. The engineers also assembled ten 38-foot M4T6 dry gap spans for CH–54 Flying Cranes. One day before the start of DEWEY CANYON II, the 45th Group Headquarters (Forward), some thirty men, moved to the Quang Tri Combat Base. Over half of the forward consisted of aviators, crew chiefs, and maintenance troops. Altogether, the group would assemble nearly 1,100 troops for the operation, including the men from Task Force 326.[12]

The Operation

DEWEY CANYON II began under the guise of an artillery raid. At 0400 on 29 January, D minus 1, an armored task force accompanied by Company D, 14th Engineer Battalion, left Quang Tri City northward along Highway 1 and turned west on Highway 9, reaching Firebase VANDEGRIFT in midmorning. Other elements of the 14th Battalion moved to Ca Lu. After arriving at VANDEGRIFT, Company D began upgrading the approaches to the firebase. At day's end, the company went through the motions of bedding down for the night. Shortly after midnight, two columns advanced west toward Khe Sanh, setting in motion DEWEY CANYON II. The main effort, Troop A, 3d Squadron, 5th Cavalry, with elements of the 14th Engineer Battalion and the 59th Land Clearing Company working together as an armored cavalry-engineer team, moved from VANDEGRIFT along Highway 9. The secondary effort, the remainder of the 3d Squadron, 5th Cavalry, and elements of Company A, 7th Engineer Battalion, moved from Firebase ELLIOTT north of VANDEGRIFT to build the alternate road to Khe Sanh.[13]

---

[11] McIntyre, "The Magnificent Sight," p. 33 (quotation).

[12] Ibid., p. 32; Interv, Christensen with McIntyre, 8 Mar 71, p. 31; Interv, Christensen with Carpenter, 7 Mar 71, pp. 10–11.

[13] McIntyre, "The Magnificent Sight," p. 33; Robert P. Miller, "The Road to Laos," *Army* 40 (January 1990): 27–29; AAR, Opn LAM SON 719, XXIV Corps, 30 Jan–6 Apr 71, 14 May 71, p. 4, Historians files, CMH; Interv, Christensen with Carpenter, 7 Mar 71, pp. 13–16.

The main column on Highway 9 set out at one minute after midnight, 30 January. At the lead were a Rome plow, dismounted cavalry troops and engineer minesweeping teams, a Sheridan tank equipped with an infrared light, and an armored cavalry assault vehicle. Engineer bridge and culvert work crews, Rome plows, bulldozers, tanks, and armored cavalry assault vehicles down the line followed. The pace during the first hour in the darkness and no moonlight was much too slow to suit 1st Lt. Robert P. Miller, one of the platoon leaders from Company D, 14th Engineer Battalion, near the front of the column. Miller later recalled it was like driving down a black tunnel, and he asked and received quick approval from the troop commander to turn on the lead Rome plow's headlights. The minesweep also slowed the advance, and it was discontinued because the operators could not get even a partial swing of their mine detectors in the high elephant grass in the road. After advancing about four hundred yards, the Sheridan became mired at the first stream crossing. The Rome plows passed the tank and pushed on, scraping the road clear of vegetation and widening it to some degree. On this stretch the road lay in the narrow Song Thach Han River valley, with the river on the left and hills on the right. Not able to stray far from the original roadbed, the column forded streams as it advanced. The Rome plows with their angled blades did a good job clearing vegetation, but a straight line D7 bulldozer had to be brought up to cut the steep banks to bypass the downed bridges. Although none of the first nine intermittent streams were flowing, a bulldozer did bog down in a wide muddy stream bed, and winching it out cost the column more precious minutes. In the darkness, the column had advanced only two and a half miles, but the advance increased speed as daylight spread over the valley.[14]

As the day dawned, the sky began to fill with helicopters ferrying troops, howitzers, small bulldozers, culvert sections, and supplies of all kinds to Khe Sanh. Joining Major Schneepeck's Task Force 326 were three infantry battalions that would secure the base and begin operations in the area. Sling-loaded CH–47 and CH–54 helicopters carrying M4T6 dry span bridges and culvert met teams of engineers at bypassed bridge sites. The preassembled frame of each bridge was carried by a Flying Crane followed by a Chinook with a load of deck balk slung underneath. Chinook after Chinook brought four twenty-foot sections of preassembled culverts dangling at the end of a fifty-foot sling. An impressive 116 helicopter sorties flew in support of 45th Group missions that first day, 84 by Chinook and Flying Cranes. During the first two days, Army and Marine Corps helicopters also moved 425 tons of engineer supplies and equipment to Khe Sanh in one of the most ambitious airlifts in support of engineer operations in the war.[15]

---

[14] Starry, *Mounted Combat*, p. 187; McIntyre, "The Magnificent Sight," p. 33; Miller, "The Road to Laos," pp. 27–29; ORLL, 1 Feb–30 Apr 71, 45th Engr Gp, p. 19; AAR, Opn LAM SON 719, XXIV Corps, p. 4; Interv, Christensen with Carpenter, 7 Mar 71, pp. 13–16. See also Steven J. Rogge, "A Combat Engineer Squad Leader in Vietnam," *Army Engineer* 2 (August-September 1994): 32; Keith W. Nolan, *Into Laos: The Story of DEWEY CANYON III/LAM SON 719, Vietnam 1971* (Novato, Calif.: Presidio Press, 1986), pp. 64–99.

[15] McIntyre, "The Magnificent Sight," p. 33; AAR, Opn LAM SON 719, XXIV Corps, p. 4; AAR, Opn LAM SON 719, 101st Abn Div, 30 Jan–9 Apr 71, n.d., p. 12–2, Historians files, CMH.

A dump truck and Rome plow begin clearing and widening Highway 9 during the opening phase of LAM SON *719.*

As the cavalry-engineer column pushed westward along Highway 9, Task Force 326 engineers began preparing bridge sites near Khe Sanh for tactical bridging. At 0830, shortly after infantry units had landed on nearby hills, a platoon of Task Force 326 landed between Bridges 33 and 34 and teams moved the short distance to work on the abutments at both sites. At Bridge 33, an abutment was prepared for an armored vehicle launched bridge when the cavalry-engineer column reached the bridge site later that day. The airmobile engineers also placed a helicopter-delivered M4T6 dry span at Bridge 34 and built a culvert bypass at Bridge 35, clearing mines and booby traps along the way. Frequently their mine detectors could not be used because of the heavy vegetation, especially the very sharp-edged ten- to twelve-foot elephant grass, and they had to probe for mines with bayonets. From Bridge 34 to 35, the teams visually swept the road followed by two small M450 bulldozers that sheared off the top two to three inches of soil and rolled off antitank mines to the side. Near the end of the day, the bulldozers reached Bridge 36 and work crews began working on the bridge abutments. The cavalry-engineer column reached Bridge 36 by nightfall, only two bridges short of Khe Sanh. Before D-Day was over, the 14th Engineer Battalion had placed three armored vehicle launched bridges and seven M4T6 dry spans and constructed thirteen culvert bypasses, thereby meeting the D-Day objectives. Colonel McIntyre attributed this remarkable progress to the massive helicopter support and "the stellar performance of platoon leaders and squad leaders whose aggressive leadership, flexibility and good engineering judgement carried the day. The baton had been successfully passed from the planners to the doers."[16]

Some mix-ups did occur. While watching the progress of the Chinooks and Flying Cranes from his own helicopter, Colonel McIntyre noticed some returning with their loads of bridges and culvert. He followed them back to their pickup points and discovered some of the helicopter crews had not been able to communicate with the engineers on the ground. All radio frequencies had been changed on D minus 1, but some of the crews had not been given the new ones. McIntyre spent the next several hours passing out slips of paper with the correct frequencies. He also developed a simple arrangement by having large numbers painted on bed sheets and informing the helicopter pilots to deliver materials to the matching bridge site on the ground.[17]

The next day, 31 January, the road was open to Khe Sanh. That morning a M4T6 dry span was in place at Bridge 36, and the first armored units reached Khe Sanh at 1230. By the end of the day, tracked and wheeled vehicles were flowing into the combat base. Roadwork west of the base to the Laotian border was continued by Colonel Jornes' 27th Engineer Battalion. Two Rome plows from the 1st Platoon, 59th Land Clearing Company, now attached to the 27th Battalion, blazed the way for the 1st Squadron, 1st Cavalry, all the way to the border by the end of D plus 2. About two hundred yards short of the border, the engineers put up a road sign warning American troops not to go any farther. Highway 9 was opened to traffic on D plus 5, two days ahead of schedule. Together, the 14th and 27th Battalions and Task Force 326 crossed a total of 39 streams with 12 M4T6 dry spans, 8 armored vehicle launched bridges, 2 eighty-foot panel bridges, and 84 forty-foot sections of culverts. All the bridges would be removed at the end of the operation.[18]

Work continued along Highway 9 throughout the operation. Except for brief maintenance periods and when wet weather caused very slick conditions, the road remained open twenty-four hours a day. On 2 February, a tank retriever towing an M48A3 Patton tank damaged an eighty-foot double-single panel bridge at Bridge 10 several miles north of VANDEGRIFT. Traffic continued to move over the bypass during the construction of a new bridge, which opened four days later. With one exception two-way wheeled traffic could traverse all the way from Dong Ha to the Laotian border. The switchback crossing at Bridge 36 was a major bottleneck early in the operation. Heavy traffic had caused the M4T6 dry span bridge to settle fast, and it was replaced by an armored vehicle launched bridge. A sharp turning radius and steep slopes

[16] McIntyre, "The Magnificent Sight," pp. 33–34 (quotation); Miller, "The Road to Laos," p. 29; ORLL, 1 Feb–30 Apr 71, 45th Engr Gp, p. 19; AAR, Opn LAM SON 719, 101st Abn Div, p. 12–1; Interv, Christensen with Schneepeck, 18 Mar 71, pp. 9, 31–32, 38–49.

[17] McIntyre, "The Magnificent Sight," p. 34; Interv, Christensen with Carpenter, 7 Mar 71, pp. 18–19.

[18] McIntyre, "The Magnificent Sight," pp. 34–35; ORLL, 1 Feb–30 Apr 71, 45th Engr Gp, p. 19 and incl. 5; Interv, Christensen with McIntyre, 8 Mar 71, pp. 22–23; Interv, Christensen with Jornes, 21 Mar 71, pp. 21–22. See also AAR, Opn LAM SON 719, 101st Abn Div, p. 12–1.

An armored vehicle launched bridge at Bridge 33

on both sides of the bridge caused some trucks and trailers and low-beds to miss the turn. A bypass could not be built, and the engineers resorted to blasting some thirty-five feet of rock from the northern approach, making entry easier. On 8 March, 45th Group engineers installed a second armored vehicle launched bridge with higher abutments and a better approach. The following day, rains in the mountains to the north caused a sudden rise in the water. Rushing water caused some damage to the abutments of the original bridge. This approach was further improved, and Bridge 36 suffered no other problems. On the peak day some 2,100 vehicles used the road. Security west of Khe Sanh improved with fifty-yard-wide strips on both sides of the road cleared by Rome plows.[19]

Along the alternate route to Khe Sanh, Company A, 7th Engineer Battalion, began clearing a tracked vehicle road, called Red Devil Road, in honor of the unofficial nickname for the 5th Division. Building this second route had been considered for several years, and past reconnaissances had shown this was possible. Moving in a southwesterly direction, bulldozers began cutting a tank trail along a rising ridge line, which ran about four and one-half miles and required fewer stream crossings and side hill cuts as well as better drainage and a minimum of work. This took two and one-half days. Then instead of following the ridge to the top of the mountain, the bulldozers proceeded down the hill with a steep side hill cut, making several switchbacks,

[19] McIntyre, "The Magnificent Sight," pp. 34–35; Miller, "The Road to Laos," pp. 30, 33; ORLL, 1 Feb–30 Apr 71, 45th Engr Gp, pp. 19–20; Interv, Christensen with Carpenter, 7 Mar 71, pp. 28–29.

and pushed across a valley about 300 to 400 yards wide. Here the trail had to be gouged out of dense vegetation and one stream crossed on the fourth day. The steep banks of the stream made fording impossible, and a thirty-foot section of forty-eight-inch culvert was brought up, rolled into the stream, and topped off with dirt. On the other side of the valley, the bulldozers moved up a steep ridge with grades 30 to 40 percent in a number of spots. The next stretch extended across a flat region known as the Punchbowl, crisscrossed by several small streams and more than one mile of dense jungle on the eastern end. Bangalore torpedoes were used to help cut through the heavy vegetation, and the streams were either forded or temporary box culverts made out of logs were built. On 8 February, the column met a 14th Engineer Battalion column pushing out from Khe Sanh. Little contact was made with the enemy, and opening Red Devil Road was accomplished in much less time than anticipated. Improvements continued throughout the operation. These involved installing more culverts and upgrading the pioneer road to a fair-weather wheeled vehicle road.[20]

During LAM SON 719, the 1st Brigade, 5th Division, exploited the use of armored cavalry-engineer teams to build an additional fifty miles of pioneer trails and fair-weather roads in the Khe Sanh area. Such team efforts usually consisted of a mechanized engineer platoon with three armored personnel carriers and one or two bulldozers from Company A, 7th Engineer Battalion, and an armored cavalry platoon. This type of organization afforded a highly armed road-building team that quickly bulldozed pioneer roads through rugged terrain and steep grades. Later the grades were reduced and the road surfaces improved for wheeled vehicles. Steep hills surrounded the Khe Sanh Plain, and the new roads provided added mobility for the armored and mechanized units and ground access to firebases. These measures kept most of the enemy's mortar fire out of the Khe Sanh area and rocket fire to a minimum. After some ten days into the operation, a cavalry-engineer team built another route, called Red Devil Drive, some nine miles along a ridgeline from Khe Sanh to a ranger group headquarters near the border. Built in four days as a tank trail, this road was soon improved to handle truck traffic and allowed the movement of 175-mm. guns and ammunition to a firebase supporting South Vietnamese forces inside Laos. A tank trail was also blazed north of Red Devil Road to block possible North Vietnamese threats from that direction. Other trails followed, linking areas with Highway 9 and Red Devil Road.[21]

The most difficult stretch of road was built south of Red Devil Road during the latter stages of LAM SON 719. About three miles long, this road served two purposes. It was to link up with Firebase CATES overlooking Highway 9 and also to confuse North Vietnamese forces in that area as to what the next allied course of action would be. Rugged triple-canopy jungle barred the way, and

---

[20] Interv, Christensen with Miles, et al., 10 May 71, pp. 10–11, 22–25, 32, 45–49; ORLL, 1 Feb–30 Apr 71, 45th Engr Gp, p. 22.

[21] AAR, Opn LAM SON 719, XXIV Corps, pp. 92–93; ORLL, 1 Feb–30 Apr 71, 1st Bde, 5th Inf Div (Mech), 11 Jun 71, pp. 12, 21–22, Historians files, CMH; Interv, Christensen with Miles, et al., 10 May 71, pp. 29–31, 34–36.

A bucket loader and bulldozer at work on Red Devil Road

the steep grades were just about the maximum for bulldozers to traverse. This was the only place that the bulldozers of Company A, 7th Engineer Battalion, received rocket-propelled grenade fire. One bulldozer was hit twice, badly damaging the engine, and it had to be dragged out by another bulldozer.[22]

Meanwhile, at Khe Sanh work proceeded on the two airstrips. On D-Day, Major Schneepeck's Task Force 326 started around-the-clock work on the assault airstrip, and the 1st Platoon, Company A, 27th Engineer Battalion, began to remove damaged matting from the existing airstrip. Once the engineers were on the ground, it became obvious that they could not complete the assault airstrip by the end of D plus 3. Mines laid by the marines and the lack of minefield records restricted movements in and around the base. The clayey soil coupled with a steady drizzle and heavy ground fog made it difficult for the soil to dry enough to get proper compaction. Even with the 27th Engineer Battalion's D7E bulldozers and scrapers, the task force did not complete the dirt strip until D plus 5. The first C–130 landed at 1600 the next day, 4 February, opening Khe Sanh to cargo aircraft for the first time in three years. However, the plane got mired in seven-inch ruts. Although a safe takeoff was narrowly managed, the Air Force determined that the packed earth airstrip was unsatisfactory for prolonged

---

[22] Interv, Christensen with Miles, et al., 10 May 71, pp. 35–36.

At Khe Sanh Airfield, a CH–54 Flying Crane drops off sectionalized engineer equipment.

use during poor weather. Meanwhile, another C–130's wheels bumped into a concrete slab about an inch under the runway's surface, and the aircraft had difficulty maneuvering around this obstacle. The engineers quickly excavated the entire slab. On 7 February, a twenty-sortie C–130 effort hauled 250 tons of MX19 matting from Bien Hoa to Quang Tri. In turn, heavy-lift helicopters began delivering the matting to Khe Sanh. Despite intermittent rain, the engineers again compacted the assault strip using heavy rollers, and on 12 February they began laying the first of nearly 17,000 panels of the aluminum matting. The matted assault strip opened on the fifteenth, receiving its first C–130 that afternoon. On that day transports began regular flights into Khe Sanh. In addition, the battalion cleared, graded, compacted, and laid M8A1 matting for a new 350-by-850-foot parking apron near the west end of the assault strip. Other airfield work included repairs to an old parking apron between the two runways, turnarounds, and overruns. On 20 February, Major Schneepeck and his Task Force 326 were released from operational control of the 27th Engineer Battalion and returned to its home at Camp Eagle.[23]

[23] "Lam Son 719," *Engineer* 1 (Summer 1971): 27; AAR, Opn LAM SON 719, 101st Abn Div, pp. 12-2 to 12-4; ORLLs, 1 Feb–30 Apr 71, 45th Engr Gp, pp. 20, 22, and 1 Feb–30 Apr 71, 27th Engr Bn, 30 Apr 71, pp. 1, 4, 10, Historians files, CMH; Bowers, *Tactical Airlift*, pp. 513–14; Nalty, *Air War Over South Vietnam*, pp. 257–58; Interv, Christensen with Schneepeck, 18 Mar 71, pp. 54–78; Interv, Christensen with Jornes, 21 Mar 71, pp. 18–21, 23–24, 26–28. See also Nolan, *Into Laos*, pp. 100–15.

Repair of the old AM2 runway proved even more difficult. Removing and replacing the damaged panels was a very slow process, because the subgrade had failed in many places, and there were scores of dud rounds buried under the matting. The engineers used several thousand pounds of explosives to clear old minefields and the dud rounds. Furthermore, the expedient anchorage system of U-shaped pickets driven into the ground and bent over the matting was giving way, and the touchdown area at the eastern edge of the runway had failed. In one instance, the gust created by a helicopter's blades lifted up a section of matting and flipped it over. Since the standard anchorage system was not on hand, the expedient method was effectively modified. Two eight-foot U-shaped pickets were welded to a three-foot angle iron. The pickets were driven into the ground with the angle irons holding the edge of the mats in place. It took a month for the 27th Engineer Battalion to restore the runway for C–130 use. It opened on 1 March. Nearly two thousand AM2 panels were used in the repair work. If AM2 repair kits had been available, repairs would have proceeded more quickly. "In retrospect," Colonel McIntyre later wrote, "it would have been faster to have pulled the AM2 matting off to one side and to have started over from scratch."[24]

The Air Force's decision to enlarge the refuel and rearm area and add other improvements compounded the engineers' workload. Task Force 326 had to move its bivouac site to make room, and when the airborne engineers started clearing the new refuel and rearm area they uncovered huge amounts of buried debris, including fifty-five-gallon drums, timbers, old matting, and parts of bunkers. Bulldozer operators had to be especially careful in scraping away the dirt since the bulldozer blades could easily rip open barrels containing aviation fuel or diesel. More reinforced-concrete slabs were also found and had to be removed. Improvements made by the 27th Engineer Battalion's attached 591st Light Equipment Company included a 538,200-square-foot helicopter staging area, improved roads within the base, and berms for the helicopter rearm pad. The company used over 90,000 gallons of peneprime to stabilize the helicopter area and interior roads. Another task assigned to the light equipment company required digging up and removing the remains of over one hundred bunkers built by the Marine Corps.[25]

Despite the many problems at Khe Sanh, the airfield readily handled the forty C–130 flights planned for each day. This figure was reached on 19 February for the first time with the arrival of 350 tons of supplies, including 57,000 gallons of aviation fuel. Supply buildup was rapid, and a two-day stockage was attained in critical 105- and 155-mm. ammunition. On the peak day, 28 February, sixty-two C–130s flew into Khe Sanh. From 15 February until the end of the operation, one or both of the runways were available for service. On 9 March, however, a C–130 lost its right rear wheel while landing,

[24] Quote from McIntyre, "The Magnificent Sight," pp. 34–35; "Lam Son 719," p. 27; Nalty, *Air War Over South Vietnam,* p. 258; ORLLs, 1 Feb–30 Apr 71, 45th Engr Gp, pp. 20–21, 1 Feb–30 Apr 71, 27th Engr Bn, p. 9, and 1 Feb–30 Apr 71, 27th Engr Bn, 30 Apr 71, pp. 1, 4, 9, Historians files, CMH; Interv, Christensen with Schneepeck, 18 Mar 71, pp. 56–57.

[25] Interv, Christensen with Schneepeck, 18 Mar 71, pp. 78–79; Interv, Christensen with Jornes, 21 Mar 71, p. 17; ORLL, 1 Feb–30 Apr 71, 27th Engr Bn, p. 8.

Aerial view of reopened Khe Sanh Airfield

and the bare axle ripped the matting for some two thousand feet. Lacking enough replacement matting, repair crews quickly patched the damaged area with epoxy cement mixed with river run sand hauled in from the Ca Lu area. This method proved much better than soil-cement because of its quicker curing time. In fact, weather became more of a limiting factor than the condition of the runways.[26]

Besides the road and airfield work, Army engineers carried out routine operational support tasks during the cross-border operation. Company A, 7th Engineer Battalion, made minesweeps along Highway 9 between Dong Ha and Cam Lo and minesweeps and assault bridging for the mechanized brigade's artillery raids around VANDEGRIFT. The company also provided a 1,500-gallon-per-hour water point and built fortifications at Khe Sanh. The 326th Engineer Battalion built bunkers, fortifications, and tactical operations centers for the 101st Airborne Division and helped close down the Khe Sanh base. Units of the 45th Group built operational and logistical facilities at VANDEGRIFT, Khe Sanh, and along the juncture of Highway 9 and the Laotian border. During the operation, group units continued to provide engineering support to other U.S. units throughout I Corps. For example, the 27th Engineer Battalion's other three line companies and the attached 591st Light Equipment Company built dog kennels at Phu Bai for the battalion's

---

[26] McIntyre, "The Magnificent Sight," p. 35; Bowers, *Tactical Airlift*, p. 514; ORLLs, 1 Feb–30 Apr 71, 45th Engr Gp, p. 20, and 1 Feb–30 Apr 71, 27th Engr Bn, p. 10; Interv, Christensen with Jornes, 21 Mar 71, pp. 34–36.

mine detection dogs, constructed new fighting positions and bunkers at Camp Eagle, and reopened and maintained thirty miles of Route 547 into the A Shau Valley.[27]

Once the engineers reopened Highway 9 on the Vietnamese side of the border, South Vietnamese armored and airborne forces crossed into Laos at 1000 on 8 February and advanced west five and a half miles toward Tchepone. Although intelligence reports had indicated that the terrain along the Laotian part of Highway 9 was favorable for armored vehicles, a road net did not exist. Highway 9 was a neglected forty-year-old single-lane dirt track, with high shoulders on both sides and no maneuver room. Huge bomb craters, undetected earlier because of the dense grass and bamboo, restricted the armored vehicles to the road. Helping to lead the way, the South Vietnamese 101st Engineer Battalion used bulldozers to fill in craters and ditches. Some parts of the road had been destroyed beyond quick repair, and the engineers built bypasses. As the armored units moved along Highway 9, elements of the South Vietnamese Airborne Division, 1st Infantry Division, and rangers made assaults into landing zones north and south of the road. U.S. Marine Corps helicopters lifted small D4 bulldozers to open the firebases. Meanwhile, U.S. Army helicopters and artillery had moved to Khe Sanh to provide fire support, and air cavalry reconnaissance teams searched for North Vietnamese forces.[28]

Enemy reaction to the border crossing was swift and violent. Soon South Vietnamese units were bogged down by heavy North Vietnamese resistance and bad weather. Elements of some five North Vietnamese divisions plus tanks and large numbers of field and antiaircraft artillery faced the South Vietnamese, and the drive toward Tchepone stalled. After several days' delay, the South Vietnamese air-assaulted into heavily damaged Tchepone, but by then the North Vietnamese counterattacked with Soviet-built T54 and T55 tanks, heavy artillery, and infantry. They struck the rear of the South Vietnamese forces strung out on Highway 9, blocking their main avenue of withdrawal, and overwhelmed several South Vietnamese firebases, depriving the troops on the road of critically needed flank protection. The result was a near disaster. U.S. Army helicopters trying to rescue South Vietnamese soldiers from their besieged hilltop firebases encountered intense antiaircraft fire. Panic soon ensued, and desperate troops even clung to helicopter skids to reach safety. In one instance, U.S. CH–54 helicopters airlifted two small bulldozers to help set up a crossing point for armored vehicles on the steep-banked Xepon River in order to reach Highway 9. The last elements of the 1st Division left Laos on 21 March, and the remaining forces withdrew into South Vietnam over the next few days. Operation LAM SON 719 officially ended on 6 April.[29]

---

[27] ORLLs, 1 Feb–30 Apr 71, 1st Bde, 5th Inf Div, p. 12, 1 Feb–30 Apr 71, 45th Engr Gp, p. 18, and 1 Feb–30 Apr 71, 27th Engr Bn, pp. 4–8; AAR, Opn LAM SON 719, 101st Abn Div, pp. 12-4 to 12-5.

[28] Starry, *Mounted Combat*, pp. 190–92; Tolson, *Airmobility*, pp. 240–41; Hinh, *Lam Son 719*, p. 65–68; Davidson, *Vietnam at War*, p. 656; Interv, Christensen with Mason, 8 May 71, p. 27. See also Cosmas and Murray, *U.S. Marines in Vietnam, 1970–1971*, pp. 196–99.

[29] Starry, *Mounted Combat*, pp. 192–96; Tolson, *Airmobility*, pp. 241–44; Hinh, *Lam Son 719*, pp. 117–18.

Although the South Vietnamese reached Tchepone, it was of little consequence. Their stay there was brief, and the supply caches discovered were disappointingly small. Operations along the Ho Chi Minh Trail were hardly disrupted. Actually, the North's infiltration reportedly increased during the operation as Hanoi shifted traffic to roads and trails farther to the west in Laos. In addition to heavy personnel losses, the South Vietnamese lost large amounts of equipment, and all of the combat engineer equipment, including thirty-one bulldozers, in their disorderly withdrawal. Over 100 U.S. Army helicopters were also destroyed and more than 600 damaged, the highest number in any one operation of the war. LAM SON 719 vividly illustrated that Vietnamization still had a long way to go and that the South Vietnamese were still heavily dependent on U.S. advisers. There were some successes. During the incursion, South Vietnamese forces inflicted heavy casualties on the North Vietnamese, forestalling a spring offensive in the northern provinces. LAM SON 719 also helped delay major operations by Hanoi for the remainder of 1971 and early 1972. The operation, however, failed to sever the Ho Chi Minh Trail for any appreciable time. The trail was in full operation within a week.[30]

By 25 March, U.S. forces had begun making preparations to pull back from their forward bases. Concern now centered on whether North Vietnamese forces would push across the border toward Khe Sanh. Already, sappers had penetrated the VANDEGRIFT logistical base and succeeded in blowing up 10,000 gallons of aviation fuel. The enemy had also harassed Khe Sanh with mortar, artillery, and sapper attacks. As a result, General Sutherland ordered the removal of the airfield matting at Khe Sanh several days before the planned 1 April date. Company A, 326th Engineer Battalion, began to clean up and recover salvaged construction materials. Two additional engineer platoons were airlifted to the base and, by 1 April, the day the airfield officially closed, 524 bundles of MX19, 179 bundles of AM2, and 34 bundles of M8A1 matting were ready to be put aboard departing C–130s. All the MX19 and AM2 matting emplaced at the Khe Sanh airfield was removed with the exception of the panels used to repair the old AM2 runway. On 5 April, a platoon from Company C, 326th Engineer Battalion, augmented with four M450 bulldozers arrived by air and began tearing down the base. Equipment that was not salvageable was buried, and bunkers and munitions left behind were destroyed. The engineers departed the next day, the same day the last South Vietnamese and U.S. forces boarded helicopters at the base.[31]

During the campaign, the 45th Group suffered personnel and equipment losses. Six engineers were killed and eleven wounded in action, seven seriously enough to be medically evacuated. Five of the deaths and three seri-

[30] MACV History, 1971, vol. 2, pp. E-33 to E-34, E-45; Starry, *Mounted Combat*, p. 197; Tolson, *Airmobility*, p. 252; Hinh, *Lam Son 719*, pp. 127, 131, 139; Davidson, *Vietnam at War*, pp. 650–51.

[31] William M. Hammond, *Public Affairs: The Military and the Media, 1968–1973*, United States Army in Vietnam (Washington, D.C.: U.S. Army Center of Military History, 1996), p. 483; Hinh, *Lam Son 719*, pp. 109, 117, 119, 125; Bowers, *Tactical Airlift*, p. 518; ORLL, 1 Feb–30 Apr 71, 45th Engr Gp, pp. 21–22; AAR, Opn LAM SON 719, 101st Abn Div, 30 Jan–9 Apr 71, p. 12–5.

ously wounded were members of the 59th Land Clearing Company during an artillery attack late in the operation near the Laotian border. Several days later, a road grader operator from the 14th Engineer Battalion was killed in an attack along Highway 9. Two other engineers were wounded. The next day, a 14th Engineer Battalion convoy heading toward Khe Sanh was ambushed: one engineer was wounded and two 10-ton tractors were damaged. Major equipment losses during the operation included six bulldozers.[32]

Attacks along Highway 9 were mostly harassing small arms, mortar, and rocket-propelled grenade fire on work parties and convoys. Minesweep teams easily detected the poorly buried mines. While rocket, mortar, and artillery fire did some damage to the Khe Sanh airfield, repairs were usually made within a few hours. During the sapper attack, some 230 rounds were directed at the base, but the air facility suffered no damage. No injuries were inflicted upon 45th Group personnel and no equipment was damaged. Of the forty sappers who tried to enter the base, eighteen were killed or captured.[33]

Redeployments and Reorganizations

LAM SON was the last major engineer operation of the war, for the redeployments in 1971 and 1972 saw the departure of nearly all engineer troops and major organizational changes. In the engineer advisory effort, corps-level field force headquarters were consolidated with corps advisory elements. The forming of the Third Regional Assistance Command in III Corps and the Second Regional Assistance Command, later Second Regional Assistance Group, in II Corps joined together the former field force engineer sections and corps engineer advisory teams. In III Corps, for example, the II Field Force Engineer Section at Plantation became the Third Regional Assistance Command Engineer Section. Engineer advisers at III Corps headquarters at Bien Hoa and the 30th Engineer Combat Group at Hoc Mon now reported to the command engineer. Engineer advisers with South Vietnamese construction units, such as those with the 5th Engineer Construction Group at Hoc Mon, continued to report to the MACV Engineer Advisory Division in Saigon.[34]

After much study, Engineer Command, Vietnam, decided its share of the Increment VI drawdown in early 1971 would be the 18th and 20th Brigade headquarters, the 937th Group headquarters, and the 46th and 589th Construction Battalions. Maj. Gen. Charles C. Noble, the commanding general from June 1970 to August 1971, noted that the ability to get around by aircraft and modern communications allowed him to forego the intermediate brigade headquarters. "Intermediate headquarters," he wrote in his debriefing report, "if not essential for command or control, only serve to slow down the action, to dilute the force of policies and directives, and to fritter away

[32] ORLL, 1 Feb–30 Apr 71, 45th Engr Gp, p. 21; Hinh, *Lam Son 719*, p. 130.
[33] ORLL, 1 Feb–30 Apr 71, 45th Engr Gp, pp. 21–22.
[34] MACV History, 1971, vol. 1, pp. II-20 to II-21.

precious high quality manpower resources." As far as he was concerned, the restructuring of Engineer Command "has tightened up the 'outfit.' "[35]

Col. William J. Schuder's 937th Engineer Group at Phu Tai outside Qui Nhon underwent several unit changes before it departed in early April 1971. When the 19th Engineer Combat Battalion left the Ban Me Thuot area in December 1970, its area of responsibility was divided. The 20th Engineer Combat Battalion took over responsibility for the Ban Me Thuot area in Darlac Province. Task Force SIERRA (610th Construction Support Company; Company D, 84th Engineer Construction Battalion; and one platoon of the 299th Engineer Combat Battalion) was formed to continue the upgrading of Highway 21. Meanwhile, the group's 84th Battalion prepared to move to Da Nang and 45th Group control, but serious flooding in Binh Dinh Province delayed the move until January 1971 because of the need to reopen washed-out roads. In March, the 937th's headquarters reduced its strength to zero, and its remaining units and area of responsibility in northern II Corps were transferred to the 35th Engineer Group.[36]

The 18th and 20th Engineer Brigades left soon afterward. Among the major functions transferred from the brigades to Engineer Command were deputy engineer, aviation officer, chaplain's office, and communications section. Special advisers, some not previously assigned at Engineer Command level, included officers assigned to other organizations. The chief of the Plans Division, Military Operations Directorate, took on the additional duties as inspector general; the surgeon in the 92d Engineer Battalion became the medical adviser; and the headquarters commandant's duties were taken over by the executive officer. Although the size of the headquarters slightly increased, the departure of the two brigades resulted in almost three hundred spaces. On 18 April, General Schrader saw his 18th Brigade furl its flag at Dong Ba Thin. At the ceremony, the brigade was awarded the Meritorious Unit Commendation, Second Oak Leaf Cluster, a culmination of its service since arriving in Vietnam on 3 September 1965. Two days later, a similar ceremony was held at Bien Hoa where Lt. Gen. William B. McCaffrey, deputy commanding general of the U.S. Army, Vietnam, saluted the men of the 20th Brigade. The brigade's commanding officer, General Cooper, assumed new duties as the deputy commanding general of Engineer Command. Upon the departure of the two brigade headquarters, Engineer Command assumed direct control of the four remaining engineer groups.[37]

---

[35] Debriefing, Maj Gen Charles C. Noble, 6 Aug 71, p. 1 (quotation), p. 2 (quoted words), Senior Officer Debriefing Program, DA, Historians files, CMH; Reorganization of HQ, USAECV, Briefing, Commanding Officers Conference, Engr Cmd, 18–19 May 71, p. 1, Historians files, CMH.

[36] ORLLs, 1 Feb–24 Mar 71, 937th Engr Gp, 24 Mar 71, pp. 2–3, and 1 Feb–30 Apr 71, 35th Engr Gp, 30 Apr 71, pp. 3–4, both in Historians files, CMH. For more about the 937th Group commander's experiences during this period, see Intervs, Christensen with Col William J. Schuder, CO, 937th Engr Gp, 10 Dec 70, VNIT 784, and 5 Apr 71, VNIT 878, CMH.

[37] ORLLs, 1 Feb–30 Apr 71, Engr Cmd, n.d., p. 1, 1 Feb–29 Apr 71, 18th Engr Bde, 30 Apr 71, p. 7, and 1 Nov 7–15 Apr 71, 20th Engr Bde, 25 Apr 71, p. 3; AARs, Standdown After Action Rpt, 18th Engr Bde, 27 Apr 71, p. 1, and Standdown After Action Rpt, 20th Engr Bde, 1 May 71, p. 1. All in Historians files, CMH. See also Spec. James Lohre and Spec. Albert Gore, "18th & 20th Drawdown," *Kysu'* 3 (Fall 1971): 20–27.

Despite the large Increment VI withdrawals, Engineer Command was still a large organization. Assigned strength on 30 April 1971 stood at 17,500, some 2,500 below authorization strength and down from 23,000 the previous November. By 30 September, the command's military strength dropped to 14,400, but Vietnamese workers increased 15 percent between October 1970 and April 1971, from about 4,100 to over 4,700. In April, the command's four groups commanded eighteen battalions (six combat, eleven construction, and one land clearing) and assorted companies and detachments. Engineer Command assigned each group engineering responsibility for a corps area: 45th Group, I Corps; 35th Group, most of II Corps; 159th Group, III Corps and part of southern II Corps; and 34th Group, IV Corps. Training of South Vietnamese Army engineers gained momentum for programmed transfers of roadwork and industrial sites. At the time, the command's productive effort averaged 46 percent highway work, 45 percent combat and operational support, and 9 percent construction. The Northern, Central, and Southern Engineering Districts, although reduced by 26 percent because of base closings and decreased projects, continued to supervise contract construction and the facilities engineering contracts.[38] (*Chart 6*)

When Col. John S. Egbert's 35th Group headquarters at Cam Ranh Bay took charge of the 937th Group's missions and units, its area of responsibility expanded to all II Corps except Lam Dong Province bordering III Corps. Egbert now commanded six battalions and assorted smaller units. Highway restoration remained the group's major effort (63 percent), with combat and operational support and some construction consuming the remaining capability (28 and 9 percent, respectively). In late April, the 589th Battalion completed thirty-four miles of Highway 1 and then stood down. Task Force SIERRA continued roadwork along Highway 21 through June. In August, another task force, WHISKEY, was organized to complete Highway 1 between Phan Thiet and the II/III Corps border. Construction included refurbishing convalescent hospital buildings at Cam Ranh Bay for a religious retreat center, disassembling several prefabricated aircraft hangars at the An Khe and Tuy Hoa airfields and warehouses at the Long My depot outside Qui Nhon for return to U.S. stocks, and civic action projects. Enemy action against the engineers diminished. While an average of five incidents was reported between April and July, the number dropped to one over the next three months. Most attacks usually amounted to one or two mortar or rocket rounds causing little damage. Despite the heavy convoy traffic, road ambushes averaged just over one per month. Beginning in August, the group began cutting its strength, from a high of 5,200 down to just over 1,700 at the end of October. By then the 20th and 864th Engineer Battalions and Task Force SIERRA had been redeployed or inactivated. The 815th Battalion

---

[38] ORLL, 1 Feb–30 Apr 71, Engr Cmd, pp. 2–3, 12–13; U.S. Army Engineer Support Organizations, South Vietnam, 3–29 Apr 71, tab A, incl. 2, OCE Liaison Officer Trip Rpt no. 19, 28 Jun 71; HQ, Engr Cmd, Periodic Report of Engineer Command Activities, 20 Nov 71, p. 2; Org Chart, Engr Cmd, 1 Dec 71. Last two in Historians files, CMH. For a listing of engineer unit redeployments, see HQ, Engr Cmd, sub: Engineer Command History, 8 Jul 72, app. G-1, Historians files, CMH.

CHART 6—U.S. ARMY ENGINEER COMMAND ORGANIZATIONS,
OCTOBER 1971

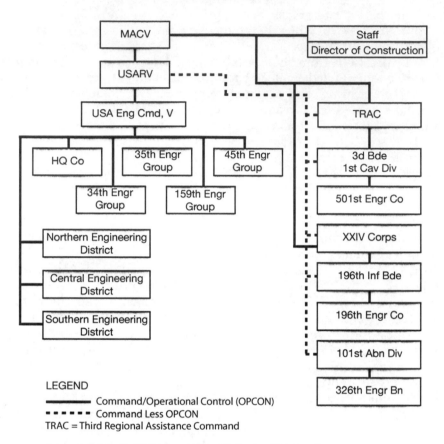

LEGEND

━━━━━━ Command/Operational Control (OPCON)
▪▪▪▪▪▪ Command Less OPCON
TRAC = Third Regional Assistance Command

Source: Tab A, Incl 1, OCE Liaison Officer Trip Rpt no. 22, 8 May 72

was transferred to the 159th Group. In early November, the 35th Group headquarters began closing down and departed on the twenty-first, almost six and one-half years after arriving at Cam Ranh Bay on 9 June 1965. The 299th Battalion also departed that month, ending more than six years of service in Vietnam. As 1971 drew to a close, only two troop units, the 577th Engineer Battalion with a reduced strength of 252 personnel and Task Force WHISKEY, remained in II Corps doing highway work under Engineer Command's direction.[39]

The 34th Group at Binh Thuy in the delta also departed in November. Col. James A. Johnson and his successor Col. John F. McElhenny, who assumed command in May, commanded four construction battalions, the 34th, 36th,

---

[39] ORLLs, 1 Feb–30 Apr 71, 35th Engr Gp, 30 Apr 71, pp. 1, 3–4, and 1 Aug–31 Oct 71, 35th Engr Gp, 21 Nov 71, pp. 1–2, both in Historians files, CMH. See also Interv, Christensen with Col John S. Egbert, CO, 35th Engr Gp, 6 Jan 71, VNIT 813, CMH.

69th, and 93d and their attached units. The group's main tasks included highway work and the training of South Vietnamese engineer units to take over the roadwork in IV Corps. Despite heavy road and bridge work, units began to depart that summer. Combat and operational support dropped to zero, and the engineers experienced only minor contact with the enemy. The most significant incident happened in July when an explosive charge was set off next to a work barge of the 523d Port Construction Company at a bridge site along Route 7A in Vinh Binh Province. Eight engineers were wounded, and the barge suffered minor damage. An ambush of a convoy along Highway 4 the following month took the life of a driver of a ten-ton tractor with low bed. Meanwhile, the 93d Battalion had dropped to zero strength on 31 July and was inactivated. By the end of September, only one full battalion, most of another, and part of a third remained, less than two thousand men. On 15 October, the 34th Battalion completed its drawdown to a color detachment, which flew to the United States a few days later. Also on 15 October, the 69th Battalion began its drawdown scheduled for completion by 15 November. Simultaneously, the group headquarters began to stand down, and the sole remaining battalion in the delta, the 36th Construction, was transferred to 159th Group control on 1 October. Upon the departure of the 34th Group, Colonel McElhenney moved to the deputy engineer position at Engineer Command.[40]

The 45th Group in I Corps continued operations until early 1972. In April 1971, Col. Walter O. Bachus assumed command, replacing Colonel McIntyre, who moved up to Engineer Command to assume duties as chief of staff. At this time, the group consisted of the 14th, 27th, and 39th Combat and the 84th Construction Battalions and assorted smaller units, some 3,500 troops. The 14th Battalion was the only engineer unit permanently located along the Demilitarized Zone, but the Quang Tri–based battalion departed in August. The other units were located in the Hue–Phu Bai, Da Nang, and Chu Lai areas, with the 39th Battalion supporting the Americal Division and the 27th Battalion supporting the 101st Airborne Division. Some of the work accomplished by the group included a 5,000-square-foot hardstand at Marble Mountain Airfield (84th Battalion); nine miles of fair-weather roads in the Quang Tri base area (14th Battalion); airfield improvements and repairs at Chu Lai (39th Battalion); 2,500 square feet of M8A1 matting for the Phu Bai airfield terminal (27th Battalion); and improvements at Firebases VEGHEL, BASTOGNE, RAKKASAN, and SALLY (27th Battalion). During the three-month period ending in November, the group reported fifty-six projects or operations requiring one platoon week of effort. Serious setbacks in work, however, occurred in late 1971 as a result of Typhoon Hester. The storm and its fourteen inches of rain in a twenty-four-hour period on 24 October damaged buildings, knocked out several bridges, and washed out parts of Highway 1.

[40] ORLL, 1 Aug–31 Oct 71, 34th Engr Gp, 13 Nov 71, pp. 1–4, Historians files, CMH; Engr Cmd Statistical Summary, 30 Sep 71, tab DD, incl. 2, OCE Liaison Officer Trip Rpt no. 21, 4 Jan 72. See also Intervs, Christensen with Col John F. McElhenny, CO, 34th Engr Gp, 14 Sep 71, VNIT 966; Lt Col Francis A. Sarnowski, CO, 34th Engr Bn, 13 Sep 71, VNIT 965. Both in CMH.

A section of the Pohl Bridge built by the Seabees in 1968 and a major link to the firebases along Route 547 was carried away as thousands of tons of logs and debris accumulated against the concrete pillars. Work crews from the 26th and 27th Engineer Battalions and the South Vietnamese 127th Float Bridge Company rushed to the scene to clear debris and begin rafting operations. At Chu Lai, widespread damage to buildings took place at the time the 39th Engineer Battalion's strength was about 50 percent due to redeployment. By 8 November, the 39th Battalion at Chu Lai had dropped to company strength with the mission to support the 196th Infantry Brigade. By then tactical operations had diminished and so did the group's casualties. From May through October, 9 engineers were killed in action, 8 died from nonbattle causes, and 34 were wounded, with all the battle deaths and most of the wounded taking place between May and July. No casualties were suffered between November 1971 and January 1972 when the combat battalions began to stand down. As the 101st Airborne Division began preparing for its departure, the 27th Battalion started redeploying to Fort Bragg, North Carolina, on 31 December. The group headquarters began to stand down on 15 January 1972. This left the 84th Construction Battalion at Da Nang with one company and a utilities detachment at Phu Bai and Company A, 39th Battalion, at Chu Lai as the only nondivisional engineer units in I Corps.[41]

The 159th Engineer Group at Long Binh, the last of the six group headquarters in Vietnam, stayed until April 1972. Col. John W. Brennan, who replaced Col. Levi A. Brown in late May 1971, assumed control of the 31st Combat; the 62d Land Clearing; the 92d, 169th, and 554th Construction Battalions and their attached units; and the 66th Topographic Company, altogether just under 3,900 troops. The stand-downs, including the 46th Construction Battalion in April and the 62d Land Clearing Battalion's headquarters in July, were offset in October when the 36th and 815th Construction Battalions, formerly 34th and 35th Group units, joined the 159th Group. The number of troops jumped to more than 6,700, but diminished with each unit stand-down. During 1971 and early 1972, the 36th, 169th, 554th, and 815th Battalions concentrated on the highway program, the 92d Battalion on base development work, and the 31st Battalion and the attached land-clearing companies on combat and operational support missions. The group's area of responsibility included all of III and IV Corps and Lam Dong Province in southern II Corps, where the 554th Battalion tried to complete thirty-one miles of Highway 20. Several unusual but priority vertical construction projects, all at Long Binh, involved constructing a confinement facility (92d Battalion), a post exchange warehouse complex (169th Battalion and later 92d Battalion), a medical retention facility for drug testing (92d Battalion and Pacific Architects and Engineers), and a structure containing three handball courts at nearby Plantation (46th Battalion and 92d Battalion). These and several other projects of a more permanent nature at

[41] ORLLs, 1 Aug–30 Nov 71, 45th Engr Gp, 30 Nov 71, pp. 1, 3, 7–13, and 1 Dec 71–28 Jan 72, 28 Jan 72, pp. 1–3, both in Historians files, CMH; "Hester Cuts Priorities" and "Debris Rams Bridge, Span Lost to Current," *Castle Courier*, 15 November 1971.

Long Binh came about because of the possibility that the base might house a residual force for an indefinite period.[42]

As a result of these stand-downs, Engineer Command reduced manning and streamlined its organization. Because of the reduced requirement for engineering services, the Engineer Directorate, responsible for designs, design reviews, and cost estimates, was reduced to a division and placed under the Construction Directorate on 1 August 1971. Programming and review functions for Military Construction, Army, projects were transferred to the Operations Division, Facilities Engineering Directorate, and responsibility for monitoring Military Construction, Army, funds moved to the Program and Budget Office. Meanwhile, senior engineers in Washington and Saigon pondered the practicality of one general heading both Engineer Command and the MACV Construction Directorate. General Clarke, the chief of engineers, proposed the idea, and General Noble and General Young, who had been the director of construction since December 1970, fully agreed. In Washington, General Dunn and General Raymond, the first two directors, saw no difficulty in one general filling both positions with a deputy general at Engineer Command. The dual-hatted responsibility passed to General Young on 5 August upon the departure of General Noble. During the Increment X drawdown in January 1972, the Engineer Command headquarters dropped from 334 to 213 personnel. In March, General Young departed Vietnam and General Johnson, the deputy commanding general, assumed the two positions.[43]

During the Increment XI drawdown that took effect in April 1972, the Engineer Command was replaced by a smaller U.S. Army Engineer Group, Vietnam. By the end of that month, U.S. troop strength dropped a further seventy thousand, and the engineer headquarters was reduced by an additional one hundred spaces. As a result, Engineer Command designed an engineer force consisting of 903 military, augmented with Vietnamese laborers and facilities engineering contractors, to provide limited operational support and minor construction and facilities engineering services. Earlier, on 1 January 1972, subordinate provisional "engineer regions," one in each corps area, or military region, were formed. The company-size engineer regions, commanded by lieutenant colonels, assumed all engineer support missions by 15 April. As the engineer regions became operational and assumed facilities engineering responsibilities, the district engineer offices were inactivated. In Military Region 1, personnel and equipment from the 84th Engineer Construction

---

[42] ORLLs, 1 Feb–30 Apr 71, 159th Engr Gp, 20 May 71, pp. 1–2, 5, 35–7, and 1 Aug–31 Oct 71, 159th Engr Gp, 20 Nov 71, pp. 1–3, 4–5, both in Historians files, CMH. See also Interv, Capt William Kunzman, 46th Mil Hist Det, with Col John W. Brennan, CO, 159th Engr Gp, 24 Feb 72, VNIT 1038; Interv, Christensen with Lt Col Charles E. Eastburn, CO, 92d Engr Bn, 29 Jun 71, VNIT 900. Both in Historians files, CMH.

[43] Engr Cmd History, 8 Jul 72, p. 1; Msg, Maj Gen Charles C. Noble, CG, Engr Cmd, ARV 1574 to Maj Gen Frederick J. Clarke, 1 May 71; Msg, Brig Gen Curtis W. Chapman, Dir Mil Engineering, OCE, 232 to Maj Gen Robert P. Young, 4 May 71, sub: Engineer Organization, both in Young Papers, MHI; Memo, Col N. C. Manitsas, CofS, Engr Cmd, AVCC-C/S, for Dirs of Const, Materiel, and Mil Ops, sub: Reorganization of Headquarters, Engineer Command, 26 Feb 71; Ltr, HQ, Engr Cmd, ACCC-E, sub: Limited Reorganization of Headquarters, Engineer Command, 28 Jul 71. Memo and letter in Historians files, CMH.

Battalion at Da Nang, the last engineer battalion to stand down in the region, were formed into Engineer Region, Military Region 1. The region engineers moved to the eastern side of the Da Nang River to Camp Horn, the headquarters for the First Regional Assistance Command. Also retained for a while were a port construction platoon and a well-drilling detachment. In II Corps, the 497th Port Construction Company, which completed its stand-down on 15 April, manned and outfitted Engineer Region, Military Region 2, at Pleiku. At Long Binh, the 92d Engineer Construction Battalion served as the nucleus for Engineer Region, Military Region 3, before drawing down in April. A well-drilling detachment also remained. Engineer Region, Military Region 4, at Binh Thuy received personnel and equipment from the 36th Construction Battalion when it stood down in January. On 30 April, Col. Alfred L. Griebling, former chief of staff of Engineer Command, assumed command of Engineer Group, Vietnam. On the same day, the new headquarters became operational and took control over the engineer regions and assumed the engineer staff role in U.S. Army, Vietnam, headquarters. By then all battalion-size engineer units had redeployed, inactivated, or stood down, including the 159th Group on 30 April. Key staff elements carried over from Engineer Command included Operations, Administrative, Facilities, and Materiels Directorates, which became divisions, and a Programs and Budget Office.[44] (*Chart 7*)

There was, however, some concern whether this smaller engineer force had the capability to support the 69,000 U.S. troops still in country. In his March 1972 debriefing, General Young noted that the engineer group would be mostly facilities engineering oriented. There would be platoon-size troop organizations that could undertake minor construction and some combat support. However, Young pointed out, "This austere approach to engineers in the total force structure cannot provide the level of engineer support, particularly combat engineer support, dictated by experience and doctrine." He also warned, "It is a reasonable, calculated risk which will succeed if no major combat action or natural disaster is experienced." The facilities engineering contractors could provide some additional construction capability, but South Vietnamese engineers would have to be called to do additional combat and operational support.[45]

Nearly all the other services' engineers had departed by this time. In I Corps, the marines transferred or demolished their bases. In September and October 1970, Marine Corps engineers leveled much of An Hoa Combat Base, dismantling or demolishing 340 buildings and flattening fortifications, leaving intact only the airfield, the industrial complex, and a small portion of the base for the South Vietnamese. Base demolition accelerated in early 1971. Ironically, Firebase RYDER in the Que Son Valley had been rehabilitated

---

[44] Engr Cmd History, 8 Jul 72, pp. 1–2; Debriefing, Young, 5 Aug 71–15 Mar 72, 6 Nov 72, p. 16, Senior Officer Debriefing Program, DA, Historians files, CMH. The term *military region* emerged in July 1970 when the South Vietnamese government, in an administrative move that incorporated the territorial forces into the regular army, redesignated the corps tactical zones as military regions. I Corps Tactical Zone, for example, became Military Region 1, or MR 1.

[45] Debriefing, Young, 6 Nov 72, p. 16, Senior Officer Debriefing Program, DA, Historians files, CMH.

CHART 7—U.S. ARMY ENGINEER SUPPORT ORGANIZATIONS, MAY 1972

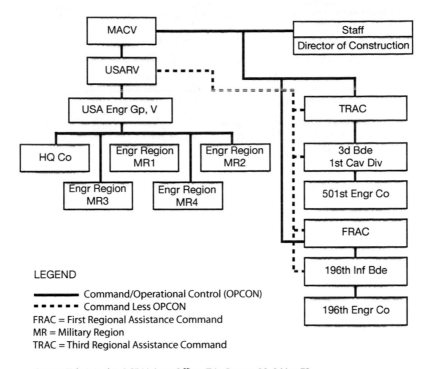

LEGEND

━━━━━━ Command/Operational Control (OPCON)
■ ■ ■ ■ ■ Command Less OPCON
FRAC = First Regional Assistance Command
MR = Military Region
TRAC = Third Regional Assistance Command

Source: Tab A, Incl 1, OCE Liaison Officer Trip Rpt no. 22, 8 May 72

only the previous September. During February and March, the 1st Engineer Battalion redeployed, leaving its Company A with Company A, 7th Engineer Battalion, as elements of the stay-behind 3d Marine Amphibious Brigade to guard the Da Nang area. During the final months of redeployment, Marine engineers provided combat support, operated water points, and leveled numerous camps and firebases near Da Nang. In May, the 3d Marine Amphibious Brigade ceased combat operations and turned over operations to the 196th Infantry Brigade. The last elements of the 3d Marine Amphibious Brigade and its engineers departed in late June.[46]

Meanwhile, the Seabees had consolidated headquarters and had withdrawn almost all units. In March 1970, the 32d Naval Construction Regiment assumed the functions of the 3d Naval Construction Brigade, and the Officer in Charge of Construction, Vietnam, took over the duties of brigade commander. The latter part of 1970 saw the 32d Regiment headquarters moving to Saigon and three construction battalions redeploying to their home ports and the arrival in Da Nang and Bien Hoa of two replacement battalions. Both battalions returned to their home bases in April 1971, and Naval Mobile Construction Battalion 5 arrived at Bien Hoa for its sixth and final tour in

---

[46] Cosmas and Murray, *U.S. Marines in Vietnam, 1970–1971*, pp. 243–44, 246, 325, 327.

549

Vietnam. In November, Naval Mobile Construction Battalion 5 completed its assigned coastal radar sites and departed for Guam. The brigade headquarters stood down at the same time. Construction Battalion Maintenance Unit 302, which helped build dependent shelters in I Corps, left its Cam Ranh Bay base the following February and moved to Subic Bay, Philippines. By March 1972, only four Seabee teams remained, three in the delta and one in III Corps, mostly doing civic action projects. This was quite a drop from eight teams in 1967 and fifteen teams in 1968. The team at Go Cong on the coast southeast of Saigon finished its work in April and transferred its camp to Regional Forces engineers. By the end of the month, the teams and controlling detachment headquarters left Vietnam, thus ending over nine years of service in South Vietnam.[47]

The few remaining Air Force civil engineers continued to concentrate on improvements and repairs at the major air bases. Of note was support of Projects ENHANCE and ENHANCE PLUS when, beginning in May 1972, Washington supplemented combat losses during the Easter offensive with additional equipment for new South Vietnamese armor, artillery, and air force units. The increased aircraft inventory of the rapidly expanding South Vietnamese Air Force resulted in new projects totaling over $3 million in Military Assistance funds for facilities improvements at eight air bases. Work included a depot overhaul at Bien Hoa Air Base, installation of additional petroleum storage tanks at several air bases, and reorganization of the civil engineering school. Construction of the petroleum storage tanks at five of the bases was to be awarded to contractors in April 1973 and completed in August. The Bien Hoa construction and the remainder of the base projects were scheduled to start in March 1973 and be completed in July. Several of the prefabricated airstrips and buildings, however, would be retrograded and returned to stock at overseas bases and stateside depots. In late 1972, civil engineers dismantled several prefabricated barracks buildings at Da Nang Air Base. The Seventh Air Force also recovered almost 8 million square feet of AM2 aluminum matting from Phan Rang, Cam Ranh Bay, and Tuy Hoa.[48]

The few U.S. Army engineer units in South Vietnam departed before the end of 1972. The 501st Engineer Company, formed from one of the line companies of the 1st Cavalry Division's 8th Engineer Battalion, supported the stay-behind 3d Brigade, 1st Cavalry Division, at Bien Hoa. At Da Nang, one of the line companies of the 26th Engineer Battalion, Americal Division, became the 196th Engineer Company, 196th Infantry Brigade. Increment XII withdrawals, which saw the departure of 20,000 more troops by the end of June, left only 49,000 U.S. military personnel in Vietnam. The two brigades were reduced to battalion-size task forces, and the engineer companies were reduced to platoons. In August, the two task forces departed, marking the end

---

[47] Tregaskis, *Building the Bases*, pp. 300–301, 417–18, 436; MACV History, 1971, vol. 1, pp. V-39 to V-41.

[48] Clarke, *Final Years*, pp. 452–53; MACV History, 1972–1973, vol. 2, p. E-60; Fact Sheet, 8 May 72, sub: Status of Retrograde of Aluminum Airfield Matting, incl. 1, p. 321, OCE Liaison Officer Trip Rpt no. 22.

of the Army's ground combat role in the war. The next two increments ending in November further reduced U.S. forces to 27,000. By then Engineer Group, Vietnam, now down to some 221 soldiers, no longer had a troop capability, and only supervisory military personnel remained in the engineer regions.[49]

Operational Support After Laos

Following LAM SON 719, U.S. ground forces concentrated on defending critical bases while the South Vietnamese assumed primary responsibility for the ground war. The days of massive U.S. offensive operations were over. In III Corps, South Vietnamese Army units began replacing U.S. units along the border in early 1971 and carried out operations trying to block the return of enemy main force units and launched counterinfiltration operations into Cambodia. Although U.S. ground forces could not enter Cambodia, many continued to support South Vietnamese cross-border operations until their redeployment. Ground activity then centered on patrolling and sweeping the rocket belts, strips of land from which the enemy could fire barrages at bases and cities. This security role was dubbed dynamic defense, and by the end of the year all U.S. ground units had shifted into it. After LAM SON 719, the 101st Airborne Division gradually disengaged from direct contact with North Vietnamese Army units in the jungled western regions in I Corps and concentrated on dynamic defense in its assigned coastal region. The Americal Division fulfilled a similar role in southern I Corps. Occasionally the enemy lashed out at the defensive positions. The most severe attack occurred in March 1971 when sappers overran Firebase MARY ANN southwest of Tam Ky killing and wounding over one hundred American soldiers. In November, the newly organized South Vietnamese 3d Infantry Division took up positions along the Demilitarized Zone formerly manned by the 1st Brigade, 5th Division.[50]

As 1972 opened, only sixteen U.S. maneuver battalions remained in country, and most soon departed. In I Corps, only the 196th Infantry Brigade was left guarding Da Nang. Allied ground operations in II Corps revolved around the two South Korean divisions carrying out security missions and combat operations in their assigned areas. In III Corps, only the 3d Brigade, 1st Cavalry Division, and the 2d Squadron, 11th Armored Cavalry, remained in positions along the rocket belt surrounding the Saigon–Bien Hoa–Long Binh complex. No U.S. ground combat force had operated in IV Corps since August 1969.[51]

Seven of the thirteen nondivisional combat engineer battalions had departed by early 1971. Only the 31st Engineer Battalion in III Corps, the 20th and 299th Battalions in II Corps, and the 14th, 27th, and 39th Battalions in I Corps remained in South Vietnam. Over the next year, these units carried

---

[49] Chart, U.S. Army Engineer Support Organizations, South Vietnam, 1 May 72, tab A, incl. 1, OCE Liaison Officer Trip Rpt no. 22, 8 May 72; Redeployments of U.S. Forces from the RVN, Army Activities Rpt, 8 Nov 72, p. 3, CMH; Clarke, *Final Years*, app. C, U.S. Troop Redeployments, p. 524; ORLL, 1 Aug–31 Oct 72, Engr Gp, Vietnam, 31 Oct 72, p. 3, OCE Hist Ofc.

[50] MACV History, 1971, vol. 1, pp. I-5, IV-3, IV-21, IV-31, IV-35.

[51] Ibid., 1972–1973, vol. 1, p. 9.

out combat and operational support missions primarily geared to support the remaining combat forces as they shifted into their dynamic defense roles. All but the Long Binh–based 31st Battalion would be gone by early 1972, and it would depart in mid-March.[52]

The 31st Engineer Battalion and its attached light equipment company typified these operations. During the first half of 1971, the battalion centered its efforts on building firebases and on local road upgrades in III Corps. In February, a platoon-size force, while at work improving and maintaining the local road net and building gun pads for eight-inch howitzers at Firebase BLUE near the Cambodian border, helped defend the base when it was attacked in the early morning hours of the twenty-third. The engineers took up arms, put out fires, helped root out several sappers holed up in bunkers, and after the attack disposed of unexploded munitions. Miraculously, the platoon suffered only minor injuries and no fatalities. Over the next several months, the battalion worked on other firebases near the border. In June, Company B and the 1st Platoon, 557th Light Equipment Company, completed Firebase PACE on Highway 22. Work included building four heavy gun pads, an interior road net, hardstand and storage areas, a berm with four 40-mm. Duster positions, and an access road to the base. The following month, the company and light equipment company elements completed the rehabilitation of Firebase ELSENBERG just off Route 13 southwest of Tay Ninh. At the same time, the 557th Light Equipment Company upgraded Highway 22 between Firebase PACE and Tay Ninh to an all-weather road, with a twelve-foot roadway and four-foot shoulders. To the east at Song Be, Company A improved Firebase BUTTONS, several local routes in the area, and airlifted elements to Bu Gia Map and Katum to carry out assorted tasks. Other Company A elements removed a M4T6 floating bridge and returned it to stock at Long Binh, repaired the runway at Loc Ninh, and improved the adjacent firebase. Meanwhile, Company C labored to upgrade two legs of the so-called Phu Hoa Triangle Road in Binh Duong Province southwest of Phu Cuong, giving access to rice paddies unused since the 1968 Tet offensive. However, the onset of the rainy season and other priorities halted this work for several months.[53]

At midyear, the 31st Battalion's emphasis shifted to improving base defenses, installing aircraft revetments, participating in the highway improvement program, and other operational support tasks. In July, Company C began installing prefabricated bunkers, erecting towers, clearing fields of fire, and building 875 yards of berm and a berm road at Phu Loi. Late that month, the company started building a hardstand and erecting aircraft revetments. Heavy rains caused problems in compacting the hardstand over a saturated subgrade, but this work was completed in September and the perimeter work by the end of the year. Later in the year, Company B completed Bridge 4 on Highway 20. This project involved two 60-foot spans, a center pier with

---

[52] ORLL, 1 Feb–30 Apr 71, Engr Cmd, incl. 1.
[53] ORLLs, 1 May–31 Oct 71, 31st Engr Bn, 4 Nov 71, pp. 1–4, and 1 May–31 Oct 71, 159th Engr Gp, 20 Nov 71, pp. 5–6, both in Historians files, CMH; Spec. Albert Gore, "Threatened Overrun of FSB BLUE Thwarted by Quick Reactions," *Castle Courier*, 5 April 1971, pp. 4–5.

a steel pile concrete cap, steel pile and concrete cap abutments, and concrete slab decks. The battalion employed its M48 mine-rolling tank on uncharted or suspected minefields. Between May and July, nearly five thousand antipersonnel mines were detonated on the perimeter at Di An. This was followed by the rolling of a suspected minefield on the southwest corner at Phu Loi Post to allow the clearing of fields of fire. Another task involved the construction of a large bunker sheltering four Signal Corps teletype vans.[54]

When the 62d Land Clearing Battalion headquarters stood down, its two land-clearing companies, maintenance company, and attached bridge company remained intact and were transferred to the 31st Combat Battalion. In more than four years, the Rome plows had cleared more than 215,000 acres in III Corps, and the 62d Battalion cleared some 60,000 acres since the Cambodian incursion. The affiliated South Vietnamese 318th Land Clearing Company also cleared 22,000 acres in III Corps as well as considerable land in Cambodia. Other cuts included 3,300 acres cleared in January and February 1971 by the 984th Land Clearing Company for the Royal Thai Army forces east of Bearcat base and 5,500 acres cut from mid-January to early March by the 60th Land Clearing Company around the village of Vo Dat in Binh Tuy Province. The 60th Land Clearing Company then moved to Hau Nghia Province to clear 18,000 acres. Several of the larger D9 bulldozers being evaluated for land-clearing operations were formed into a team for smaller land-clearing projects. Although the D9 was found excellent for cutting first traces, it quickly outpaced the security force's armored cavalry assault vehicles and D7 bulldozers. Land clearing continued to be dangerous work. During the period 1 November 1970 to 30 April 1971, there were 76 casualties, including 6 killed in action, a rate of about 15 percent in a unit averaging 500 men. Security forces, especially infantry on foot, were also more prone to suffer casualties as the Rome plows crashed through the jungle. In June, Lt. Col. Robert P. Monfore, who had commanded the 62d Battalion for nearly one year, turned over command to Lt. Col. Walter P. Hayes. Following the closeout of land-clearing battalion headquarters, Colonel Hayes became the Engineer, Third Regional Assistance Command.[55]

In anticipation of the quickening redeployments, commanders kept up the high pitch of land-clearing operations during 1971. In July and August, the 984th Land Clearing Company and the 318th Land Clearing Company cut over 9,000 acres in the Boi Loi Woods resulting in the virtual destruction of an enemy sanctuary in lower Tay Ninh Province. In August, the 60th Land Clearing Company began a cut in the Tan Uyen rocket belt north of Bien Hoa in an attempt to keep the enemy from launching rockets at the bases in the area. After clearing more than 1,400 acres, the company was pulled from this mission and moved by sea to I Corps in a high-priority mission with the

---

[54] ORLL, 1 May–31 Oct 71, 31st Engr Bn, 4 Nov 71, pp. 4–5.

[55] ORLLs, 1 Nov 70–30 Apr 71, 62d Engr Bn, 14 May 71, pp. 1–8, 17, Historians files, CMH, and 31st Engr Bn, 1 May–31 Oct 71, p. 1; Spec. Ray Smietanka, "Rome Plows," *Hurricane* (April–June 1971): 42–44; Interv, Maj John L. Hitti, 17th Mil Hist Det, with Lt Col Walter P. Hayes, CO, 62d Engr Bn, and Third Regional Assistance Command (TRAC) Engr, 17 Oct 71, VNIT 983, pp. 2, 10–11, CMH.

45th Group's 59th Land Clearing Company to clear the Da Nang rocket belt. Meanwhile, the 984th and 318th Land Clearing Companies continued the Tan Uyen cut and cleared another 3,500 acres. In October, the 984th Company deployed north to clear strips along Highway 13 between An Loc and Loc Ninh.[56]

In II Corps, the 35th Group's 538th Land Clearing Company completed clearing jungle around isolated villages in Tam Quan District near Bong Son. While clearing one area for a night defensive position, the land clearers discovered they had chosen a North Vietnamese bunker complex. The enemy soldiers apparently made a hasty retreat from the area, since fires were still burning, ice was found, and laundry was left to dry. During the next forty days, the land clearers found and destroyed hundreds of tunnels and bunkers. Returning to its base camp at Cha Rang Valley near Highways 1 and 19, the company relaxed, recuperated, and repaired bulldozers before moving out for another forty-five to sixty days in the field, this time clearing 200-yard strips of thick vegetation that had grown on both sides of Highway 19 in the An Khe Pass. December, however, saw the departure of the 59th, 538th, and 984th Land Clearing Companies. U.S. land-clearing operations came to an end on 14 December when the 60th Land Clearing Company finished its cut in the Ham Tan region in the southeastern corner of III Corps and departed the following month. All land clearing now passed to the South Vietnamese.[57]

The building of firebases, which guarded the approaches to the U.S. bases and protected the Saigon and Da Nang military complexes, was among the last major operational support missions for U.S. Army engineers. In III Corps, this work involved building firebases for the 3d Brigade, 1st Cavalry Division; the 2d Squadron, 11th Armored Cavalry Regiment; and the 23d Artillery Group. Construction of these bases was the culmination of the experiences gained over the years. Extensive land clearing, earthwork, construction of berms, and excavation for fire direction centers, personnel shelters, and materials were required. In late February 1972, the 3d Brigade, recently reduced from four to three infantry battalions, began building six firebases northeast of Saigon to provide an artillery fan over the rocket belt and its approaches. Among the firebases were Bunker Hill, Grunt II, Ennis, and Crossed Sabers. Earlier, the 2d Squadron moved to its new firebase, Fiddler's Green, north of Saigon. From these more elaborate and permanent firebases, the cavalrymen and airmobile infantry launched operations to keep enemy rocketeers off balance. The firebases also served as rest and recuperation centers for the company-size units rotating in and out of the field and boasted sound defenses, better living conditions, and even wells at three of the bases. Intended to serve as models for the South Vietnamese,

---

[56] ORLLs, 1 May–31 Oct 71, 159th Engr Gp, pp. 7–8, and 31st Engr Bn, 1 May–31 Oct 71, pp. 5–6; Spec. David L. Myers, " 'Rocket Pocket' Gets Plowed," *Castle Courier*, 4 October 1971.

[57] Spec. David L. Myers, "538th LC Co. Strips Treacherous An Khe Pass," *Castle Courier*, 15 November 1971; Engr Cmd History, 8 Jul 72, app. G-1; HQ, TRAC, Semiannual Written Review, 23 Jan 72, p. 64, Historians files, CMH.

*Excavating a tactical operations center at one of the new firebases north of
Bien Hoa Air Base*

the firebases would be transferred to the South Vietnamese when the U.S.
units departed.[58]

Illustrative of the firebases was BUNKER HILL located on a squat laterite
hill seventeen miles northeast of Saigon and nine miles north of Bien Hoa.
Construction plans were developed by Maj. Peter J. Offringa, the air cavalry brigade's engineer, who directed the 3d Brigade's 501st Engineer Company operations and coordinated additional engineer support with the Third Regional
Assistance Command Engineer and the 159th Group. Soon a task force under
the direction of the 3d Platoon, 501st Engineer Company, and earthmoving
equipment from the 557th Light Equipment Company began cutting the initial
swath through the scrubby undergrowth covering the hill. Timing was critical,
for the light equipment company only had fifteen days before standing down.
A vertical construction platoon from the 92d Construction Battalion rounded
out the task force. For the next two weeks, an average of four Caterpillar 830s
hauled their 20-cubic-yard loads from 0730 to 1730, leveling the area and forming a perimeter berm. Once the scrub was removed, the familiar triangular
shape of another 1st Cavalry Division firebase became evident. The scrapers
also dug slots for fighting and living bunkers under the berm while bulldozers
dug slots for the tactical operations center, mess hall, ammunition and supply

[58] HQ, TRAC, Engr Adviser, Fact Sheet for G–3, n.d., sub: Engineer After Action Rpt, 15
Dec 1971 Thru 6 Aug 1972, p. 1, Historians files, CMH; Maj. Peter J. Offringa, "Bunker Hill:
Anatomy of a Firebase," *Engineer* 3 (Summer 1973): 19–20.

storage areas, medical bunker, base defense and fire direction center, and a post exchange bunker. Helicopter pads were positioned close to the entrance of the base and accessible to the aid station.[59]

As the earthwork progressed, other engineers began working on the structures. The vertical construction platoon built the framework for the tactical operations center and the mess hall and supervised Vietnamese laborers pouring concrete floors for both structures. Modular design was incorporated whenever possible. Conex containers were widely used. Dug eight feet into the ground, the sturdy tactical operations center was covered by M8A1 matting and the equivalent of seven layers of sandbags. An underground tunnel connected the operations center to the base defense/fire direction center bunker. The 105-mm. howitzers were placed on a 200-foot equilateral triangular pad raised six feet to provide direct fire over the berm. Individual mortar pits were sandbagged to a height sufficient to protect the crews. The mess hall consisted of a 20-by-40-foot storage area and two 10-by-20-foot serving areas. Even though it was eight feet underground and covered with a roof, the mess hall was well ventilated through a series of vents consisting of M8A1 matting and 72-inch culverts. Troops were not exposed to enemy fire while being served, and the eating areas were dispersed. Wide flange beams were used to support the ammunition storage point's roof, which was topped off with M8A1 matting, plastic sheeting for waterproofing, three feet of soil, and coated with peneprime. The basic design incorporated the majority of the living and fighting bunkers as part of the eight-foot-high triangular earth berm in order to take full advantage of its protection. Conex containers, with the doors removed, were used as two-man living bunkers and positioned into trenches before the berms were finished. Twelve-foot-wide rolls of plastic were used for waterproofing, and five feet of earth protected the top, front, and rear. The design of the fighting bunkers emphasized primary and supplemental bunkers. Corner bunkers were constructed as a three-unit module with the center Conex oriented with the apex, and the units to the left and right were positioned to provide coverage to the front and the flanks. Above each corner arose fifteen-foot observation towers with radar towers on the roofs. Extensive rows of barbed-wire obstacles reinforced by claymore mines, trip flares, and fougasse (a combination of napalm and C4 demolitions) surrounded the base.[60]

The thirty-five-day construction effort resulted in a comfortable and efficient battalion-size operating base. Water was pumped from a nearby river, treated, and pumped to a sixteen-foot-high tower. Gravity-feed distribution led to showers, drinking water tanks, mess hall, and mess hall wash rack area. Hauling water by truck was only needed to fill fifty-five-gallon washing barrels in the vicinity of the living and latrine areas. This home away from home also boasted three temporary volleyball courts, horseshoe pits, and an outdoor stage used for movies, USO shows, and a briefing area. BUNKER HILL was the

---

[59] Offringa, "Bunker Hill: Anatomy of a Firebase," pp. 20, 25.
[60] Ibid., pp. 21, 24–25.

Aerial view of Firebase BUNKER HILL

last major firebase built by Army engineers, and it epitomized the evolution of firebase construction during the Vietnam War.[61]

Allied commanders anticipated a significant enemy offensive in 1972, most likely again during Tet, but no one expected the blitzkrieg-like invasion of the South that began on 30 March. This 1972 assault, known as the Easter offensive or Nguyen Hue offensive, broadly resembled the Tet offensive of 1968, except the North Vietnamese Army, not the Viet Cong, bore the brunt of combat. With nearly all U.S. combat troops gone and South Vietnamese military capacity still lacking, North Vietnam sensed an opportunity to demonstrate the failure of Vietnamization and hasten eventual victory. Total U.S. military strength in country was about 95,000, of which only 6,000 were combat troops. Countering the offensive fell almost exclusively to the South Vietnamese and their U.S. advisers, and allied air power. Attacking on three fronts North Vietnamese Army regulars poured out of the Demilitarized Zone and Laos and captured Quang Tri. The provincial capital was retaken in September after a bloody battle. In the Central Highlands, North Vietnamese units moved into Kontum Province, forcing the South Vietnamese to give up several border posts before being halted. In III Corps, the enemy took Loc Ninh on 2 April and surrounded An Loc, where a three-month-long battle ensued until the siege was broken. Often massive firepower provided by U.S. air and naval forces helped decide victory or defeat for the South Vietnamese. At An Loc, resupply airdrops for the besieged city by U.S. Air Force C–130

[61] Ibid., pp. 24–25.

cargo planes became regular and efficient, and the defenders were kept well supplied.[62] (*Map 31*)

During the Easter offensive, U.S. ground forces, including supporting engineers, stayed in their defensive roles near major installations. The largest incremental drawdown was entering its final month when the enemy offensive began. It continued unabated. Redeployments remained on schedule with eight U.S. maneuver battalions standing down by 30 June, leaving only the two battalion task forces at Bien Hoa and Da Nang. April and May saw over 30,000 U.S. troops leave Vietnam. Another 20,000 would depart by 1 July, reducing the number in Vietnam to 49,000. Allied forces—Australians, Thais, and South Koreans—had departed or were in the process of leaving. While maintaining redeployment and installation transfer schedules, the United States reacted to the enemy attack by bringing in additional fighter-bomber, airlift, and gunship aircraft to help the beleaguered South Vietnamese. The enemy offensive also brought about an acceleration of the Improvement and Modernization Program, and more armor, aircraft, and ships were delivered. Occasionally, American installations at Da Nang and Bien Hoa, where U.S. Marine Corps fighter-bombers deployed to beef up tactical air support, received light rocket attacks but no ground assaults. As for the fighting, aviators bore the brunt of combat, flying bombing, aerial resupply, and gunship missions. Advisers with South Vietnamese combat units directed the air strikes, which helped decimate Hanoi's forces.[63]

Although constantly dwindling in numbers during this period, the regional engineers carried out operational support missions, minor construction projects, and other tasks. In I Corps, a detachment of Military Region 1 engineers working on bunkers at Quang Tri was trapped for two days until rescued. The last rescue helicopter also carried out a Pacific Architects and Engineers generator operator. Later, the same soldiers built a bunkered forward tactical operations center (TOC) (dubbed Super TOC) for the First Regional Assistance Command at Hue and placed 4,500 mines around the communications facilities at Phu Bai. During the height of the enemy attacks near Pleiku, Military Region 2 engineers replaced missing matting at the Camp Holloway airfield to allow C–130s to land and deliver supplies and evacuate personnel. Military Region 3 engineers built revetments for medical evacuation helicopters and ready rooms for the crews at Tan Son Nhut and improved defenses at Long Binh and Third Regional Assistance Command at nearby Plantation. In the delta, engineers built live-in bunkers and replaced revetments and worked

[62] MACV History, 1972–1973, vol. 2, p. L-9; Lt. Gen. Ngo Quang Truong, *The Easter Offensive of 1972*, Indochina Monographs (Washington, D.C.: U.S. Army Center of Military History, 1980), pp. 127, 134, 171–72. For more on the Easter offensive and South Vietnamese counterattacks, see MACV History, 1972–73, vol. 1, pp. 33–101, vol. 2, annexes J to L; Maj. Charles D. Melson and Lt. Col. Curtis G. Arnold, *The U.S. Marines in Vietnam: The War That Would Not End, 1971–1973* (Washington, D.C.: History and Museums Division, Headquarters, U.S. Marine Corps, 1991), pp. 2–184; Dale Andrade, *Trial by Fire: The 1972 Easter Offensive, America's Last Vietnam Battle* (New York: Hippocrene Books, 1995).

[63] MACV History, 1972–1973, vol. 1, pp. 33, 35, 37, 46, 49; Clarke, *Final Years*, p. 482; Melson and Arnold, *U.S. Marines in Vietnam, 1971–1973*, pp. 157–58, 162, 164.

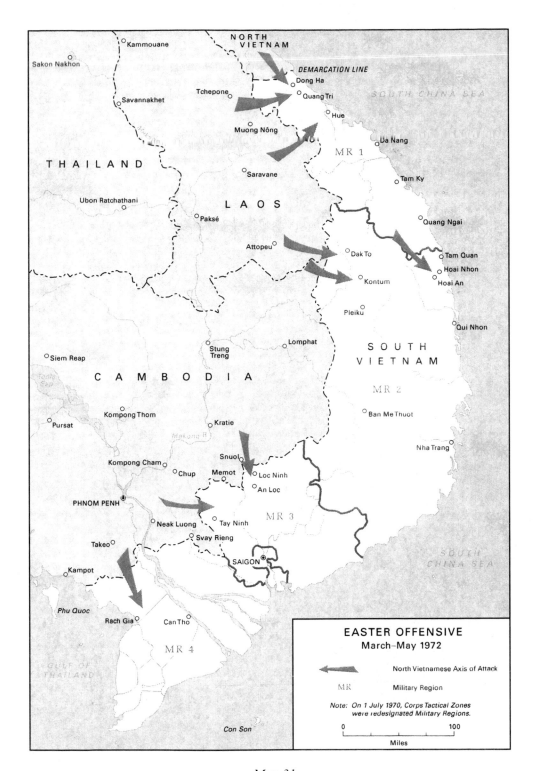

Kammouane

Sakon Nakhon

NORTH VIETNAM

DEMARCATION LINE

Dong Ha

Tchepone

Quang Tri

SOUTH CHINA SEA

Savannakhet

Hue

Muong Nông

Da Nang

MR 1

THAILAND

Saravane

Tam Ky

Ubon Ratchathani

LAOS

Paksé

Quang Ngai

Attopeu

Dak To

Tam Quan
Hoai Nhon
Hoai An

Kontum

Pleiku

Qui Nhon

Lomphat

SOUTH
VIETNAM

Stung
Treng

MR 2

Siem Reap

CAMBODIA

Kompong Thom

Ban Me Thuot

Pursat

Kratie

Snuol

Kompong Cham

Memot

Loc Ninh

Chup

An Loc

Nha Trang

MR 3

PHNOM PENH

Neak Luong

Tay Ninh

Takeo

Svay Rieng

Kampot

SAIGON

SOUTH
CHINA SEA

Phu Quoc

Rach Gia

Can Tho

MR 4

GULF
OF
THAILAND

Con Son

EASTER OFFENSIVE
March–May 1972

North Vietnamese Axis of Attack

MR   Military Region

*Note: On 1 July 1970, Corps Tactical Zones
were redesignated Military Regions.*

0   100

Miles

MAP 31

on perimeter defenses at Can To Airfield. Regional engineer troops also substituted for Vietnamese workers who stayed home during the attacks.[64]

A dramatic engineer-related mission during the Easter offensive involved the destruction of bridges spanning the 500-foot-wide Mieu Giang River at Dong Ha near the Demilitarized Zone. On 2 April, Easter Sunday, a North Vietnamese tank column approached from the north. The Seabee-built steel girder wooden decked highway bridge, capable of bearing sixty-ton loads, seemed to be the logical crossing for the drive down Highway 1 to Quang Tri and Hue. A South Vietnamese marine battalion and army tanks took up defensive positions on the southern bank, while gunfire from U.S. warships helped slow the enemy approaching the highway bridge and an old railroad bridge upstream. Although one of the railroad bridge's spans had been destroyed, foot soldiers could still get across. South Vietnamese aircraft joined in and destroyed several tanks nearing the highway bridge. Meanwhile, confusion reigned among South Vietnamese commanders and U.S. senior advisers whether or not to blow the highway bridge. At the scene were two American advisers, Marine Corps Capt. John W. Ripley with the marines and Army Maj. James E. Smock with the armored unit. When it appeared that the North Vietnamese would make a crossing in overwhelming force, the advisers requested and received permission to destroy the bridge. Engineers from the 3d Division had hauled five hundred pounds of TNT blocks and C4 plastic explosives to the base of the highway bridge. Ripley, who had received extensive demolitions training, realized that the explosives still had to be properly placed in the supports. A high chain-link fence topped by sharp steel concertina wire prevented easy access to the girders. Once Ripley cleared the fence, he had Smock push the boxes of explosives over the fence to him. He then crawled hand over hand dragging the boxes one by one along the underside of the southern span and placed them in the channels formed by the six supporting I-beams. All this took some two and one-half hours. At first the North Vietnamese watched but then started shooting at him. The tanks could roll across the bridge at any moment, but for some reason they did not. While Ripley completed preparations at the main highway bridge, Smock hastily placed demolitions on the railroad bridge. Meanwhile, Ripley had prepared both a time fuse and an electrical detonation. The electrical charge failed to set off the explosives, and the fuse seemed to take forever. At this time, a South Vietnamese colonel and his American adviser emerged, and they felt the bridge should not be destroyed. Suddenly there was an explosion. The span was neatly severed from the abutment, and it plunged into the river. It was not certain whether the fuse finally set off the charges. A previously requested air strike in the area may have set off a sympathetic detonation of the demolitions. In any case, the enemy armor advance toward Quang Tri was halted at Dong Ha, at least for the time being. The determined North Vietnamese tried get-

---

[64] General comments, Gerald E. Galloway, Historians files, CMH; ORLL, 1 Aug–31 Oct 72, Engr Gp, Vietnam, pp. 5, 7.

Aerial view of the Third Regional Assistance Command headquarters at Plantation with the engineer buildings to the right of the entrance road

ting tanks across at a bridge farther west at Cam Lo, but naval gunfire and air strikes halted this advance. Air strikes later destroyed this bridge.[65]

Fewer U.S. engineer advisers and smaller engineer staffs remained at the end of the April drawdown. In the Third Regional Assistance Command Engineer Section, all military personnel but the plans and operations officer had departed. Colonel Hayes, a captain, and two noncommissioned officers returned to the United States. All the advisers with the 30th Engineer Combat Group also departed. The remaining member of the section, Maj. Adrian G. Traas, was transferred to the G–3 to carry out engineering and other duties in the section. He continued as the sole link to the III Corps Engineer and 30th Group commander, Col. Le Van Nghia, often arranging helicopter transport and materials and accompanying Colonel Nghia on his visits to the field. The three division advisory teams in III Corps retained one engineer officer as engineer battalion advisers, several advisers remained with the 5th Engineer Construction Group, and a lieutenant colonel served as the III Corps Civil Operations and Revolutionary Development Support engineer adviser. The

---

[65] MACV History, 1972–1973, vol. 2, pp. L-8 to L-9; Starry, *Mounted Combat*, p. 208; Melson and Arnold, *U.S. Marines in Vietnam, 1971–1973*, pp. 50–58, 60–61, 64; Andrade, *Trial by Fire*, pp. 90–94. There are several versions on the destruction of the bridges at Dong Ha. See Col. Gerald H. Turley, *The Easter Offensive* (Novato, Calif.: Presideo Press, 1985), pp. 132–58, 177–91; John G. Miller, *The Bridge at Dong Ha* (Annapolis: Naval Institute Press, 1989); Vicki Vanden Bout, "Ripley at the Bridge," *Leatherneck* 69 (February 1986): 16–19.

Colonel Nghia inspects repaired bridge at Go Dau Ha, which was completed by the 302d Engineer Battalion on 18 July after being destroyed during the Easter offensive.

Military Region 3 Engineer, Lt. Col. Ralph A. Luther, former commander of the 92d Construction Battalion, reported to Engineer Group, Vietnam. In the other corps, engineer positions were consolidated. In I Corps, the Military Region 1 Engineer, Lt. Col. Gerald E. Galloway, assumed additional duties as the First Regional Assistance Command staff engineer. In II Corps, the adviser to the 20th Engineer Combat Group assumed the duties as Second Regional Assistance Command Engineer, and a similar assignment was given to the 40th Combat Group adviser in IV Corps.[66]

By this time, South Vietnamese Army engineers had taken over almost all operational support missions. In III Corps, months before the Easter offensive, Colonel Nghia's 30th Group concentrated on supporting cross-border operations in Cambodia and road and bridge work. The group reopened and upgraded Highway 22 from Tay Ninh to the border, placed a laterite cap on Route 13 from Chon Thanh to Minh Thanh, opened a rural road in the Angel's Wing between Highway 1 west of Go Dau Ha and Ben Cau, upgraded Highway 1 from Xuan Loc to the II/III Corps border, and improved local roads in Tay Ninh and Binh Duong Provinces near the Cambodian border. In early 1972, the group assumed contingency responsibility from U.S. engi-

[66] HQ, TRAC, Semiannual Written Review, 23 Jan 72, pp. 36–37; OCE Liaison Officer Trip Rpt no. 22, 8 May 72, pp. 4–8.

neers for thirty-two critical bridges in the corps area. Arrangements were made by the engineer adviser, Major Traas, to transfer contingency stocks of panel and M4T6 bridges along with excess bridge parts stored at Long Binh. Responsibility for all forward airfields within the corps was also assumed by the 30th Group. During and after the Easter offensive, the group reopened sections of main supply routes cut by the enemy. Examples of this work included replacing a blown culvert on Highway 1 near Trang Bang and rebuilding a blown bridge across the Vam Co Dong River at Go Dau Ha. Attempts to relieve surrounded An Loc involved moving the group's 318th Land Clearing Company to Highway 13 where it cleared growth and enemy bunkers on each side of the road north of Lai Khe.[67]

The Easter offensive was a grave challenge for South Vietnam, and for the most part South Vietnamese troops resisted with determination. Well-planned U.S. air strikes, particularly the massive B–52 raids, caused many of the 100,000 North Vietnamese and Viet Cong casualties and destroyed at least one-half of their tanks and large-caliber artillery. To some degree, it appeared that Vietnamization was working, provided that massive U.S. airpower, logistical support, and advisers were available. The halting of the North Vietnamese offensive allowed the United States the appropriate amount of time, a "decent interval," to bow out of Vietnam. Nevertheless, North Vietnam still had gained considerable territory along the Laotian and Cambodian borders as well as the area just south of the Demilitarized Zone. These occupied areas would serve as the launch points for the final Communist assault in 1975.[68]

Winding Down the Construction Program

During 1971 and 1972, the massive military construction program neared the end. In 1972, for the first time since the 1965 buildup, a fiscal year Defense Department budget contained no request for authorizations or appropriations for military construction to support operations in Southeast Asia. Thus far, more than $1.8 billion for military construction had been appropriated. Still, much work needed to be done, particularly in the highway and modernization and improvement programs. Although the Department of the Army did approve $36 million for Fiscal Year 1970, this was disapproved by the Defense Department, and further funding by Congress became difficult. Command guidance stressed making do and doing without. Even more rigid construction limitations were established in September 1971. New project starts costing over $50,000 had to get Defense Department approval prior to the obligation of funds, and project overruns would have to be financed from funds already assigned. In terms of dollars spent and man-hours, 1971 was another busy year, but the downward trend was unmistakable and inevitable. As the year

[67] HQ, TRAC, Semiannual Written Review, 23 Jan 72, pp. 36–37.
[68] Truong, *Easter Offensive*, pp. 158, 172, 176, 179–80.

drew to a close, about $97 million in construction funding previously appropriated still remained to be completed.[69]

Rather than saying construction was coming to an end in Vietnam, the MACV Construction Directorate declared that the program was entering a new phase, and emphasized a new and higher standard of construction to complete outstanding projects. Buildings still needed to be erected or improved as units moved to other locations. A great majority of the camps were designed for a five-year life. The temporary standard two-story wooden buildings, some going back five or more years, showed signs of the ravages caused by monsoons and termites. Despite the drastic restrictions, new construction carried out by the U.S. Army Engineer Command, Vietnam, continued at a surprisingly high rate. General Noble reported that 10 to 15 percent of the Engineer Command's effort was still devoted to construction. In May 1971, the MACV Construction Directorate received $25 million in Fiscal Year 1971 Military Construction, Army, funds, boosting the Army's share of military construction in South Vietnam to $968.4 million. This figure changed little the following year, rising to $970.2 million and dropping to $954.1 million later in the year. For Fiscal Year 1973, the directorate forecasted the need for $58 million in Military Construction, Army, funds to support the facilities needed by residual forces and the modernization and improvement program for the South Vietnamese. Almost $41 million of this amount would be used to upgrade the facilities at Tan Son Nhut and Long Binh. More work was also programmed at Can Tho and Da Nang.[70]

Attention increasingly turned to improving living standards. A concerted effort had been under way for some time to improve the advisory camps. General Abrams deemed better housing and recreation facilities at base camps necessary to keep troops away from the less wholesome attractions of Vietnamese society, while at the same time lowering the American profile in Vietnam. As the war wound down, disciplinary problems, especially drug use, had become major command concerns. In May 1971, Abrams reiterated "the necessity for creative and imaginative programs to meet the diversified needs and interests of our personnel during the period of redeployment." Emphasis was placed on intramural sports programs and building sports facilities on a self-help basis using operations and maintenance or nonappropriated funds. Troops also spent spare time improving barracks and game rooms. General Noble told his commanders that their troops "must do something about these needs, be it the repair of a movie projector or the construction of a basketball court."[71]

---

[69] MACV History, 1971, vol. 1, pp. IX-1, IX-13 to IX-14; Construction Directorate Briefing for Mr. Barry J. Shillito, Asst Sec Def (Installations and Logistics), 13 Dec 71, p. 2, Historians files, CMH.

[70] MACV History, 1971, vol. 1, pp. IX-13 to IX-14; Construction Directorate Briefing for Mr. Shillito, 13 Dec 71, pp. 3–4; Quarterly Hist Rpt, 1 Oct–31 Dec 71, MACDC, 8 Mar 72, p. II-8, Historians files, CMH; Army Activities Rpts, 21 Jan 72, p. 2, 16 Aug 72, p. 19, and 20 Dec 72, p. 21; Debriefing, Noble, 6 Aug 71, p. 6, Senior Officer Debriefing Program, DA, Historians files, CMH; Engr Cmd History, 8 Jul 72, p. 29.

[71] Quotes from Ltr, MACJ12, Abrams to DCG, USARV, sub: Morale and Welfare Activities, 7 May 71, with first Indorsement, Lt Gen William B. McCaffrey, DCG, USARV, 25 May 71,

Indicative of these requirements, Engineer Command reported the completion of 103 major base construction and repair projects between February 1970 and April 1972. This work, valued at $4 million in materials, included new recreation facilities at Long Binh, a confinement facility at Long Binh (better known as LBJ or Long Binh Jail), a religious retreat center at Cam Ranh Bay, an equipment retrograde facility at Da Nang, and a post exchange warehouse complex at Long Binh. Among the high-priority projects with short suspense dates were the retrograde facilities at ports such as the one at Da Nang. All were completed on time. Physical security projects—the construction of perimeter lighting, chain-link fencing, and concrete revetments—became a continuous operation. During the same period, contract work valued at $21 million resulted in the completion of forty-seven major projects, including the Long Binh and Newport rail spurs, a petroleum pipeline at Qui Nhon, a cinder block cantonment for the 8th Radio Research Unit at Phu Bai, and an indoor theater and outdoor pool at Long Binh completed by the summer of 1972. Another $1.4 million was used by troops and contractors to improve MACV adviser sites. Engineer troops worked on ten port construction and repair projects, expending some $600,000 in materials, at Da Nang, Qui Nhon, Cam Ranh Bay, and Newport. Also, in late 1971 damage caused by Typhoon Hester diverted much engineer activity in I Corps to repair work. The port of Da Nang required extensive repairs caused not only by the typhoon but also heavy usage by ships involved in retrograde operations. At Qui Nhon, port construction troops were called on to repair a failing wharf. On the operational side, the command directed over two thousand taskings, ranging from airfield repairs to firebase construction.[72]

During 1971, Seabee units with help from RMK-BRJ kept up a high pace of work. Major efforts centered on completing assigned sections of the highway program in I Corps, building logistical facilities at new and existing naval bases throughout the country, completing a network of sixteen coastal radar surveillance sites covering the entire South Vietnamese coast, and helping build dependent housing for South Vietnamese naval personnel. Work at naval bases and radar sites was part of the Navy's Accelerated Turnover to the Vietnamese—a time-phased transfer of equipment, operational and logistical bases, and responsibilities to the South Vietnamese Navy. The first of the radar sites, which were designed to complement MARKET TIME operations by detecting Hanoi's seaborne infiltration, commenced operations on 1 July at Vung Tau. One month later, a second site atop Monkey Mountain overlooking Da Nang harbor was turned over to the South Vietnamese. By December, the seventh site had been turned over and the remaining nine stations would follow by mid-1972. Seabee teams also drilled wells, repaired bridges, and helped the U.S. Agency for International Development on various civic action projects. From Soc Trang in the delta to Xuan Loc in III Corps, detachments worked on roads, schools, a maternity clinic, and other projects to raise living standards

and second Indorsement, Noble, 2 Jun 71, to subordinate commanders and staffs, Historians files, CMH; Construction Directorate Briefing for Mr. Shillito, 13 Dec 71, p. 4.

[72] Engr Cmd History, 8 Jul 72, pp. 7–8.

in outlying areas. In the delta, teams found time to install playground equipment at local hospitals and orphanages, grade local roads and install culverts, and repair water systems.[73]

Construction for South Vietnamese modernization and improvement began to show fruition. This work accomplished under the Military Assistance Service Funded (MASF) category of military construction included depot upgrades. The depot program was designed to provide the South Vietnamese armed forces with a rebuild capability for most engineer, signal, and ordnance equipment. The 40th Engineer Base Depot in Saigon was one of the beneficiaries. Initiated in 1969, the depot upgrade program was almost complete in 1972. Large equipment deliveries brought about by Projects ENHANCE and ENHANCE PLUS, however, prompted an urgent need for a greater combat vehicle rebuild capability. In December, almost $2 million was added by the Defense Department for new rebuild facilities. A part of this money was slated for more upgrades for the 40th Engineer Base Depot. During 1972, much work also went into building new wards for provincial hospitals. Work ranged from $231,000 for rehabilitation work in Qui Nhon to over $2.3 million for a 450-bed hospital in Saigon.[74]

The South Vietnamese dependent housing effort still had its ups and downs. On the upward side, the U.S. made a commitment in 1970 to furnish $6 million to purchase building materials over a five-year period from 1971 to 1975 to build an additional 100,000 housing units. Also most South Vietnamese divisions were taking over the large bases left behind by U.S. forces and making sufficient land available for housing. Construction was enhanced when Pacific Architects and Engineers transferred a cement block–making machine to the 5th Engineer Construction Group, which from its rear base at Hoc Mon began making cement blocks for the program in III Corps. In addition, the South Vietnamese chief of engineers designed and built thirty-three manual and electric operated cement block–making machines for issue to engineer units in the other corps areas. Shortages in other building materials were made up by diverting materials from other projects. Construction carried out by Vietnamese troops and small local contractors progressed satisfactorily until 1972 when the Easter offensive damaged a number of houses already built or under construction. The program was nearly stopped as contractors abandoned work sites, the supply of materials slowed down, and engineer units devoted more effort to support military operations. Work resumed in late 1972 but was interrupted again when the U.S. Congress cut assistance funds. With only a limited amount of assistance funds forthcoming, Saigon had to suspend the program after 75 percent of the housing had been completed.[75]

During 1971, the time finally arrived to phase out the construction contract with the joint venture contractor, Raymond, Morrison-Knudsen, Brown and

---

[73] Marolda, *By Sea, Air, and Land,* pp. 318–19; Tregaskis, *Building the Bases,* pp. 417–18; MACV History, 1971, vol. 1, pp. V-6 to V-7, V-37 to V-41.

[74] MACV History, 1971, vol. 1, p. VIII-57; ibid., 1972–1973, vol. 2, pp. E-7 to E-8, E-23; Clarke, *Final Years,* pp. 452–53. See also Khuyen, *RVNAF Logistics,* pp. 307–26.

[75] MACV History, 1972–1973, vol. 1, pp. E-32, E-34; Khuyen, *RVNAF Logistics,* pp. 394–95.

Root, and J. A. Jones. In early 1970, the Naval Facilities Engineer Command in Washington decided the best way to end the cost-reimbursable contract was to do it in stages. During Phase I, RMK-BRJ and the Officer in Charge of Construction in Saigon reviewed all completed work totaling some $1.3 billion. This phase, which was completed in May 1971, proved so successful that a Phase II analysis covering another $227 million was completed in November. Completion of about $200 million still remained, and a final goal of nearly $1.9 billion was set for 1 July 1972. In order to meet this date, all projects had to reach the Officer in Charge of Construction not later than 1 May 1971 to allow time to forward notices to proceed to the contractor by 1 June. Planning also started for the orderly disposal of construction equipment, depots, and industrial sites. There were over 7,500 major items of equipment alone. With only eleven months to go, $38 million worth of materials, supplies, and spare parts had to be used and any residual inventory passed on to another agency.[76]

Despite the impact of the North Vietnamese 1972 Easter offensive, RMK-BRJ completed most of its tasks ahead of schedule in mid-May. As soon as the jobs were finished, the contractor reduced manning and closed out sites. The Officer in Charge of Construction worked closely with MACV for the transfer of the contractor's camps, shops, and other facilities to U.S. and Vietnamese government agencies, with the U.S. Agency for International Development accepting many for final transfer to the Vietnamese. A system was developed to distribute a complete listing of usable equipment and materials worldwide. U.S. military organizations could request any item, paying only the cost of shipping. As part of the Vietnamization effort, about three thousand items of equipment were transferred to the U.S. Agency for International Development for use in a program to foster the growth of a construction industry in Vietnam. Equipment not claimed, about one-half, was transferred to the U.S. Army Property Disposal Agency, Vietnam, for sale. Nearly all the surplus construction materials were purchased from the contractor or transferred to other agencies in Vietnam. On 3 July 1972, the Officer in Charge of Construction held a ceremony in Saigon to note the end of the RMK-BRJ contract, which had run nearly eleven years. Accounting and administrative closeouts followed, and the Officer in Charge of Construction, Saigon, office closed its doors in early October. At that time, no disputes remained between the Navy and RMK-BRJ, a remarkable achievement for a contract relationship of this scope.[77]

RMK-BRJ's part in the wartime building effort was striking. The firm accounted for more than 60 percent of the construction in Vietnam, with the rest by military engineers and other contractors. RMK-BRJ employed over 200,000 people from 1962 to 1972, with Vietnamese comprising at least 80 percent of the workforce. To sustain the large construction load, equipment valued at $208 million was shipped across the Pacific and distributed to seventy-five construction camps and hundreds of smaller work sites. Along with the massive amount of airfield, port, depot, cantonment, and roadwork,

---

[76] Tregaskis, *Building the Bases*, pp. 423–29.
[77] Ibid., pp. 423–29, 431–33, 436.

some of RMK-BRJ's significant building feats included MACV headquarters, much of Long Binh Post, the U.S. Embassy, the deep-draft Newport and its railroad spur connecting line to Long Binh, and five major bridges and a new Saigon bypass. Left behind to be operated by the Vietnamese government were several industrial facilities such as quarries, concrete prestress plants, dredging, and heavy construction equipment and maintenance repair shops. It was anticipated that South Vietnam's future economic growth would greatly benefit through the marketable skills learned by more than 150,000 Vietnamese men and women. Women's roles in construction, thanks to RMK-BRJ and other U.S. firms, had expanded to heavy equipment operators, welders, and electricians.[78]

The closeout of the RMK-BRJ contract did not end the Navy's involvement in administering construction contracts. During the late stages of the RMK-BRJ program, the Officer in Charge of Construction awarded an increased number of contracts to Vietnamese and non-Vietnamese firms. In early 1972, the construction agency awarded a major road-building project in the delta to a Vietnamese firm and a $3.5 million contract to a Vietnamese joint venture for a combined arms school near Long Binh. As the RMK-BRJ contract was nearing an end, the Naval Facilities Engineering Command in Washington made plans for a new organization to replace the Officer in Charge of Construction. On 1 September, that part of the agency administering the remaining contracts in Vietnam was reconfigured as the Director of Construction, Republic of Vietnam. The new office reported to the Officer in Charge of Construction, Thailand, thus returning to the organizational structure that existed before 1 July 1965.[79]

Modifications were also made for facilities engineering services. At the beginning of Fiscal Year 1971 (1 July 1970), Pacific Architects and Engineers, Philco-Ford Corporation, and Vinnell Corporation employed over 25,100 workers carrying out facilities engineering and high-voltage power production costing $108.6 million. Some 1,100 military personnel assigned to Engineer Command's Facilities Engineering Directorate, the three district engineer offices, and engineer detachments supervised the contracts. In addition, a small organization, Facilities Engineer, Saigon, composed of direct-hire U.S. and Vietnamese civilians, provided facilities engineering services to leased facilities in Saigon. The Army determined it was in the best interest of the government to negotiate sole-source contracts for facilities engineering requirements. Cost-plus-award-fee contracts were again used and cost-incentive features were added as part of the award fees with the expectation that the two facilities engineering contractors would perform at a high level and at the same time reduce costs to increase earnings. In July 1970, Pacific Architects and Engineers began its eighth year in South Vietnam with an $82 million contract (down from $100 million in Fiscal Year 1970) and 21,000 employees. The contract renewal charged the firm to continue buildings and grounds

[78] "Summary of Vietnam Construction," *Military Engineer* 64 (September-October 1972): 56–58; Merdinger, "Civil Engineers, Seabees, and Bases in Vietnam," p. 235.

[79] Tregaskis, *Building the Bases*, pp. 436–37.

maintenance, power production, potable water and ice supply, sanitation, air conditioning and refrigeration, and fire protection services at U.S. and allied installations in II, III, and IV Corps, and in I Corps north of the Hai Van Pass. During the year, however, the number of major bases dropped from twenty-nine to sixteen, the contract was reduced another $16 million, and the workforce dropped to 14,000. Following the turnover of logistic support in I Corps to the Army in the summer of 1970, Philco Ford continued to carry out facilities engineering support at Da Nang and Chu Lai. By 30 June 1971, base transfers and budgetary constraints reduced the three contractors' workforce by 7,200 and the contracts to $90.6 million.[80]

In July 1970, land-based high-voltage power requirements were separated from facilities engineering requirements and opened to competitive bidding. A contract was awarded to Pacific Architects and Engineers to operate and maintain generators and distribution systems at thirty-two sites. Vinnell Corporation still operated and maintained eleven T–2 tankers as power barges (four at Cam Ranh Bay, two at Nha Trang, three at Qui Nhon, and two at Vung Tau), an interrelated land site recently completed at Cam Ranh Bay, and land-based power plants at Da Nang, Phu Bai, and Chu Lai under cost-plus-incentive terms. Plans were under way since 1969 to begin removing the tankers when the land-based plants came online. The estimated cost of returning the ships to the United States for disposal, including towing, came to $140,000 per ship. Since the ships were old and obsolete and the U.S. Navy no longer had requirements for their future use, the Maritime Commission agreed to sell the ships to competitive bidders as scrap at their present locations. On 20 August 1970, the first two barges were removed from the system at Cam Ranh Bay, and in March and April 1971 all three barges at Qui Nhon were removed from power generation. Vinnell then took over operation of the newly completed land-based power system for the remainder of the fiscal year.[81]

During Fiscal Year 1972, the Army renewed the sole-source facilities engineering and power barge contracts and again put the land-based high-voltage contracts out to competitive bidding. While the facilities engineering contracts with Pacific Architects and Engineers and Philco-Ford remained a cost-plus-award-fee contract with an incentive feature, the power barge contract with Vinnell was changed to a fixed price. As for the land-based high-voltage requirement, Pacific Architects and Engineers competed with six other firms and again was awarded a contract to operate most of the plants, including Qui

[80] PA&E History, FY 1970, p. 2; ibid., FY 1971, pp. 1–3; Summary of Facilities Engineering Support for FY 71, tab T, pp. 1–3, OCE Liaison Officer Trip Rpt no. 20, 2 Sep 71; Engr Cmd History, 8 Jul 72, p. 10; AAR, Opn COUNTDOWN, HQ, USARV, 4 Jun 73, vol. 2, p. D-2-1, Historians files, CMH; Col. Warren S. Everett, "Contractors in the Combat Zone," *Military Engineer* 64 (January-February 1972): 37–38. For more on facilities engineering, see Intervs, Christensen with Col Marion M. Wood, Northern Dist Engr and later Dep Engr, Engr Cmd, 11 Aug 71, VNIT 963; Lt Col Pierpont F. Bartow, Northern Dist Engr, 28 Jul 71, VNIT 962; and Lt Col Paul D. Matthews, Central Dist Engr, 19 Aug 71, VNIT 964. All in CMH.

[81] Summary of Facilities Engineering Support for FY 71, pp. 3–4, OCE Liaison Officer Trip Rpt no. 20, 2 Sep 71; Engr Cmd History, 8 Jul 72, pp. 11–12; MFR, USARPAC G–4, 1 Mar 69, sub: Replacement of T–2 Ships RVN; Ltr, Dir Const, USARV Engr, to USARPAC, sub: Disposition of Power Barges in RVN, 12 Jun 70, both in Historians files, CMH.

Nhon. Vinnell continued to operate high-voltage plants at Da Nang, Phu Bai, Chu Lai, and Cam Ranh Bay. By July 1971, Pacific Architects and Engineers' facilities engineering contract dropped to $51.68 million and its workforce to 12,444 (788 U.S., 860 third-country national, and 10,796 Vietnamese). With the departure of U.S. Marines and the consolidation of remaining forces in I Corps, Philco-Ford's contract expanded slightly to $14.46 million and its workforce to 3,206 (134 U.S., 447 third-country national, and 2,625 Vietnamese).[82]

By late 1972, nearly all facilities engineering and power generation were consolidated under one contractor, Pacific Architects and Engineers. Philco-Ford decided to close out its Vietnam operation at the end of September; Facilities Engineer, Saigon, phased out in February 1973 with most employees transferring to Pacific Architects and Engineers; and the firm was selected for a new consolidated contract awarded on 1 October for the remainder of Fiscal Year 1973. Pacific Architects and Engineers also became responsible for facilities engineering at Da Nang and Phu Bai, the remaining high-voltage barges at Nha Trang and Tuy Hoa, and increased management responsibilities in the Military Assistance Service Funded program. Major installations offices remained at Nha Trang, Camp Holloway outside Pleiku, ROK Valley outside Qui Nhon, Tan Son Nhut, Long Binh, and Can Tho. Clustered under these installations were other bases and MACV advisory team sites. For the Nha Trang and ROK Valley installations, South Korean forces became the facilities engineering firm's major customer. In order to comply with the new contract terms, Pacific Architects and Engineers made organizational changes, including the creation of a Da Nang office and a Military Assistance Service Funded Program Office, and incorporated the power barges and land-based power plants under the installation offices. By helping the South Vietnamese to assume ever-increasing responsibilities for former U.S. installations, the firm's Military Assistance Service Funded Program Office expanded on-the-job training and supplies for the many low-voltage generators at transferred installations and signal sites. To this end, the office dispatched six contact teams to carry out training and to help maintain and overhaul the power plants and distribution systems.[83]

---

[82] Summary of Facilities Engineering Support for FY 72, pp. 1–2, and Facilities Engineer FY 72 Proposals, pp. 1–2, both in tab T, OCE Liaison Officer Trip Rpt no. 20, 2 Sep 71; Status of Land Based High Voltage Power Plants, tab GG, incl. 1, p. 255, OCE Liaison Officer Trip Rpt no. 22, 8 May 72. For more on facilities engineering from the contractor's perspective, see Intervs, Christensen with Mr Charles N. Leightner, General Manager's Representative, Military Region 1 (MR 1), PA&E, 29 Jul 71, VNIT 959; Mr William F. Lane, General Manager's Representative, MR 2, PA&E, 18 Aug 71, VNIT 960; and Mr Lawrence Farnum, Installation Engr, Philco-Ford, 31 Jul 71, VNIT 961. All in CMH.

[83] PA&E, General Manager's Special Instructions, Contract DAJB11–73–0026, Revised Organizational Structure, September 1972, Historians files, CMH; AAR, Opn COUNTDOWN, HQ, USARV, vol. 2, p. D-2-1.

Last Transfers

The accelerated redeployments in 1971 and 1972 dramatically increased base transfers and reductions in leases. Compared to the transfer of 75 bases in 1970, the number in 1971 more than doubled to 173, including 105 former Army facilities. By the end of the year, a total of 278 bases, adviser sites, and industrial complexes, or 33 percent, had been turned over to the South Vietnamese since the transfers began in 1969. Military leases were reviewed to see if leaseholders could be moved to rent-free facilities as units and activities redeployed. As a result, the number of leases decreased steadily from 420 in July 1970 to 316 by 1 February 1971, reducing leasing costs from $11.8 million to $10.4 million. This savings, however, did not correspond to the drop in the number of leases since the costs of renewed leases increased. Savings in Military Assistance Program funds were achieved by substituting U.S. facilities slated to be closed instead of proceeding with planned improvement and modernization projects. The Construction Directorate, working closely with the MACV J–4, identified eighty-four projects valued at $11.4 million for cancellation. The directorate also closely worked with the U.S. Agency for International Development to determine what facilities not needed by the South Vietnamese armed forces could be used by civilian agencies such as the Ministry of Education, which expressed interest in using excess facilities for schools.[84]

By then several South Vietnamese military units and civilian agencies had moved into the former U.S. bases. Larger operational bases, composed chiefly of tin-roofed wooden barracks and prefabricated metal buildings used for offices and warehouses, along with surfaced roads, electrical-generating systems, and deep-well water supply systems, were usually turned over to South Vietnamese divisions. Despite their advantages, these bases caused certain problems. Some were too spacious even for an entire South Vietnamese division. Often units taking over the bases found themselves short of personnel to fill all the barracks and to man all the guard posts and defense positions and incapable of properly maintaining all the facilities. As a result, South Vietnamese units gradually shrank their living perimeter and cannibalized buildings no longer in use. New major South Vietnamese bases included Dong Tam (occupied since 1969 by the 7th Division); Vinh Long Airfield (9th Division); Blackhorse eight miles south of Xuan Loc (18th Division training center); Bearcat (Armor School and the site for a newly constructed Infantry School); Cu Chi (25th Division and several logistical units); Lai Khe (5th Division); Di An (Marine Division and the Railway Service); Camp Enari outside Pleiku (47th Infantry Regiment); Freedom Hill outside Da Nang (3d Division); and Camp Eagle (1st Division). The more permanent logistical bases usually accommodated logistical and field support units. Many expedient battalion and brigade bases, however, were usually destroyed or abandoned. Large or small, abandoned bases, including Camp Radcliff at An Khe, were often looted. Engineer units

---

[84] MACV History, 1971, vol. 1, p. VIII-61; Ltr, HQ, MACV, MACDC-RPM, to CINCPAC, sub: Planning for Disposal of Facilities, 13 Mar 71, Historians files, CMH.

were too preoccupied to participate in salvage operations and were usually used to recover prefabricated metal buildings only.[85]

The redeployment of two U.S. brigades characterized the stand-downs of separate infantry brigades and their camps. Before ending operations in late April 1971, the 2d Brigade, 25th Division, closed six firebases and transferred three others to the South Vietnamese and Thai forces in the area, transferred its main base camp (Camp Frenzell-Jones) near Long Binh to other U.S. units, and consigned two base camps at Xuan Loc to the South Vietnamese 18th Division. At the closed firebases, the brigade's 54th Engineer Company leveled berms, swept the areas for mines, and hauled away usable materials. Along the Demilitarized Zone, the 1st Brigade, 5th Division, made arrangements with Pacific Architects and Engineers and the 101st Airborne Division to transfer the bases at Quang Tri and Dong Ha to the South Vietnamese 1st Division prior to its departure in August. The South Vietnamese were slow to move in, however, and the lack of security led to looting.[86]

The American Division's stand-down was centered at the large 18,000-man Chu Lai base camp. The base was served by a water system consisting of sixteen wells and five lakes capable of producing 1.3 million gallons of potable water per day. A 7,000-kilowatt high-voltage electrical system with one main plant and three substations served about 85 percent of the base, the remainder being served by a low-voltage system. The $50 million installation contained some four thousand buildings of temporary and preengineered construction, an airfield with three runways, a dock area with five ramps along the beach, and a thirty-three-mile perimeter lighting system. Planning began in early August 1971 to transfer the entire base to the South Vietnamese 2d Division. In this case, detailed instructions and close coordination with the installation engineer and well-controlled South Vietnamese quartering parties prevented excessive looting. Units were not cleared to depart until all trash had been moved to a sanitary fill. On 2 October, Typhoon Hester caused extensive damage and a virtual stop to the policing campaign. About one thousand buildings were repaired mostly through unit self-help, and South Vietnamese officials declared their requirements would be satisfied without further repairs. The 2d Division requested that storm debris be left in place and no attempt be made to clear collapsed buildings. Additionally, the quartering parties requested a halt to the trash removal because trash was very often of value to the South Vietnamese soldier. In the final days, the police emphasis was placed on recovering abandoned equipment and materials. Although the base was transferred on 27 November, some U.S. contractor personnel remained for

[85] Khuyen, *RVNAF Logistics*, pp. 180–85.

[86] AAR, Opn Keystone Robin (Charlie), 2d Bde, 25th Inf Div, 28 Apr 71, pp. 109–10, 117; AAR, Opn Keystone Oriole (Bravo), 1st Brigade, 5th Inf Div, 19 Aug 71, pp. 90–92. Both in Combined Arms Research Library (CARL), Fort Leavenworth, Kans. For more on other stand-downs in late 1970, see AAR, Opn Keystone Robin, 3d Bde, 9th Inf Div, 8 Oct 70; AAR, Opn Keystone Robin, 199th Inf Bde, 12 Oct 70; and AAR, Opn Keystone Robin, 25th Inf Div, 15 Dec 70, all in CARL.

awhile to reconfigure the electrical system to South Vietnamese needs and to assist and train the new tenants to operate the system.[87]

During 1972, the remaining major bases were transferred or dismantled. Most of the real estate and facilities at Cam Ranh Bay were turned over by 1 May. Only the Navy's MARKET TIME installation at the southern end of the peninsula, the Sea-Land transport facility, and the property disposal yard remained under U.S. control. At Qui Nhon, the port facilities and the Phu Tai complex had been turned over in November 1971, followed by the incremental transfer of the camps in and around the city. At Da Nang, the bulk of facilities were transferred in May and June, with Camp Horn, the 95th Evacuation Hospital, a covered storage area near Camp Horn, the air base, and China Beach still in U.S. hands. Other major transfers included Phu Cat Air Base in December 1971, Phan Rang Army and Air Base during the period March to May 1972, Tuy Hoa Air Base on 22 February 1972, and Bien Hoa Air Base in June 1972. In late October, as the peace negotiations appeared to be nearing fruition, the Joint Chiefs of Staff directed MACV to title transfer all remaining bases and adviser sites. Procedures were streamlined, and by 10 November the title transfer was completed with the understanding that U.S. forces would continue to occupy facilities in use until no longer needed. This transfer also conveniently preempted the provisions in the cease-fire agreement concerning the mandatory dismantling of U.S. bases in South Vietnam. Joint physical inventories followed over the next sixty days. On 12 November, the South Vietnamese took formal possession of the Long Binh complex, estimated at $107 to $120 million in value. Extensive prior planning had taken place, and as the South Vietnamese units (the 3d Area Logistics Command and other logistical units, a 175-mm. artillery battalion, an armor squadron, an airborne battalion, and three ranger groups) moved in, U.S. units were consolidated and moved to another part of the post. By November, the recently reorganized U.S. Army, Vietnam/MACV Support Command headquarters, including Engineer Group, Vietnam, completed the move of its 2,900 member staff to Tan Son Nhut.[88]

Several bases or portions of bases were taken over by civilian agencies. At Long Binh, the Ministry of Economic Affairs took over a one-mile stretch bordering the Bien Hoa Highway to expand the Bien Hoa Industrial Zone. The ministry also took over several other camps, including RMK-BRJ's facilities at Cam Ranh Bay for industrial development. At Phu Loi, a former 1st Cavalry Division camp, the Ministry of War Veterans used the base as a vocational and handicraft training for veterans and disabled service men. The Ministry of National Education converted a former South Korean logistics base at Nha Trang into a community college.[89]

With the drawdowns and base transfers, large quantities of construction equipment and materials were retrograded to depots in the United States and

---

[87] AAR, Opn KEYSTONE ORIOLE (CHARLIE), Americal Div, 15 Dec 71, pp. F-1 to F-3, CARL.

[88] Turn Over Plans, p. 235, OCE Liaison Officer Trip Rpt no. 22, 8 May 72; MACV History, 1972–1973, vol. 2, pp. E-28 to E-29, F-31; AAR, Opn COUNTDOWN, HQ USARV, vol. 1, pp. A-3-I-3-1 to A-3-I-3-2, vol. 2, pp. D-1, D-10, D-3-1; Khuyen, *RVNAF Logistics*, pp. 183–86.

[89] Khuyen, *RVNAF Logistics*, pp. 183–84.

overseas or transferred to the South Vietnamese. Many of the Army engineer units were inactivated in country and their equipment was turned in. The Engineer Command's Materiel Directorate monitored the turn-ins and reviewed requests by MACV and USARV logisticians for lateral transfers to U.S. and South Vietnamese engineer units. As a result, about 90 percent of the equipment issued to South Vietnamese engineer units under the Improvement and Modernization Program came from Engineer Command's assets rather than depot stock. Most of this equipment was used in the highway restoration program and included the commercial equipment and industrial plants. In February 1972, the MACV Engineer Advisory Division assumed responsibility from Engineer Command for administering the Dynalectron maintenance contract for the commercial equipment. Similarly, much of RMK-BRJ's equipment and industrial complexes were transferred to the South Vietnamese. What could not be used or cared for by the South Vietnamese, such as prefabricated buildings and AM2 matting, was disassembled and shipped out. Equipment and materials not transferred to the South Vietnamese or considered too expensive or too much trouble to retrograde were turned in to property disposal channels to be sold to the highest bidders.[90]

Some of the high-dollar equipment returned to U.S. stocks included high-voltage generators and DeLong piers. These were major items in the retrograde program involving equipment valued at over $75 million. Shipment of the 5,000-kilowatt generators offered no real problems, but the large 15,000-kilowatt generators, each almost the size of a locomotive, required heavy lift cranes and heavy transport vehicles to be moved along carefully reconnoitered roads and bridges capable of supporting the heavy loads to the ports. Returning these generators also depended on the hookup of replacement low-power generators or connection of existing loads to commercial power. By 1 November 1972, only 38 of the 129 generators had been returned to the United States, but the potentially successful cease-fire talks spurred the shipment of 57 more valued at over $9.6 million by 28 March 1973. The rest were shipped out by MACV's successor Defense Attaché Office.[91]

DeLong piers at five ports were steadily reduced. The two A-type barges comprising the DeLong pier at Da Nang were removed by RMK-BRJ in June and towed to Subic Bay. Between July and November, DeLong Corporation prepared the two A-type barges of the pier at Vung Ro Bay for shipment to the United States. In both instances, U.S. Navy tugboats towed the sections. In July, DeLong Corporation began preparing the two A-type barges of Pier Number 1 at Cam Ranh Bay for shipment the following June. By November 1972, five DeLong piers, valued at some $60 million, were still in country: one at Qui Nhon consisting of four A-type barges, one at Vung Tau consisting of seven B-type barges, and three at Cam Ranh Bay consisting of four A-type and

[90] Engr Cmd History, 8 Jul 72, p. 14–15; Fact Sheets, MACJ46-AE, sub: Dynalectron Maintenance Contract, n.d., tab U, and RVNAF Engineer Participation in the Lines of Communications Program, 2 Mar 72, OCE Liaison Officer Trip Rpt no. 22, 8 May 72.

[91] AAR, Opn COUNTDOWN, HQ USARV, vol. 1, p. 77, vol. 2, pp. D-3 to D-4, D-4-1 to D-4-20.

*One of the DeLong piers being towed to Cam Ranh Bay for return to the
United States*

six B-type barges. In December, the ammunition pier (six B-type barges) at Cam Ranh Bay was released for retrograde and by February 1973 was disassembled and en route to the United States. In June 1973, dismantling of another pier at Cam Ranh Bay and the remaining pier at Qui Nhon was nearing completion. Dismantling of the last DeLong pier at Cam Ranh Bay was programmed to begin in June 1974. All that remained of the five piers at Cam Ranh Bay was the permanent pier built by RMK in 1964.[92]

Bowing Out

In late October 1972, the prospect of the cease-fire negotiations appeared to be nearing a successful conclusion. General Weyand, who replaced General Abrams in June, began planning for successor agencies once a treaty was signed. Three new organizations would assume MACV's responsibilities: a successor headquarters in Thailand, a U.S. element of a four-party joint military commission putting the cease-fire machinery in place once the cease-fire went into effect, and the Defense Attaché Office to monitor military activities and provide technical assistance to the South Vietnamese. The cease-fire began on 28 January

[92] Fact Sheet, AVCC-CC, sub: DeLong Piers, RVN, 17 Mar 72, incl 10, OCE Liaison Officer Trip Rpt no. 22, 8 May 72; Periodic Rpt of Engr Cmd Activities, 17 May 72, p. 7, OCE Hist Ofc.

1973, and the war ground to a halt, at least temporarily. In the sixty days that followed, over 58,000 foreign troops departed South Vietnam, including about 23,000 Americans and 35,000 South Koreans. MACV headquarters dissolved on 29 March, and the three new agencies took over its remaining functions.[93]

The continuing drawdown of U.S. forces resulted in more changes for the few remaining U.S. military engineers. Major reorganizations of the U.S. Army Engineer Group, Vietnam, occurred in September and December 1972. The final organization resulted in a staffing of 100 military, 40 Department of Army civilians, and 674 Vietnamese employees. By this time, all the region engineers had ceased functioning as separate elements, and their personnel were assigned to the group's Headquarters and Headquarters Company. On 28 December the group, the USARV Engineer Section, the MACV Directorate of Construction, and the Engineer Advisory Division were consolidated into one organization under Col. Russell J. Lamp, who became the MACV Command Engineer. The USARV Engineer Section and U.S. Army Engineer Group merged to direct projects and plans and administration. A Construction Division took care of construction and engineering and real estate. The combining of the MACV, USARV, and Engineer Group real estate offices significantly reduced processing time for leasing and base transfers. A newly established Engineer Branch, Defense Attaché Office, took over functions that would provide some support to the South Vietnamese after the last U.S. troops departed. These duties consisted of facilities maintenance, the highway improvement program to include technical assistance and maintenance support for the commercial equipment, the retrograde of equipment, support for large generators and air conditioners at certain communications sites, dependent shelters, the Military Assistance Service Funded program, and construction materials. Region engineers, who had stood their troop elements down, also reported to the Engineer Branch. By then the region engineers had been reduced to a few personnel. As the region engineers stood down, the Engineer Group formed a mobile engineer platoon at Tan Son Nhut and platoon-size forces of local nationals in each region. Pacific Architects and Engineers and Facilities Engineer, Saigon, provided the bulk of facilities engineering support.[94]

In late January 1973, following the signing of the armistice agreement, the MACV Command Engineer accomplished one final project. The engineers were tasked to help provide facilities for the Four-Party Joint Military Commission and the International Commission of Control and Supervision overseeing the cease-fire terms. Lacking any planning guidance, this final construction requirement came as a surprise. Within twenty-four hours, local national employees and Pacific Architects and Engineers began building or renovating offices, living quarters, and conference facilities. About two weeks later, all but two of the fourteen regional sites had been completed. By the time all U.S. and allied forces departed on 29 March, most of the fifty-two team sites and twelve control points were done.

---

[93] MACV History, 1972–1973, vol. 2, pp. G-1, H-2; Clarke, *Final Years*, pp. 495–96.

[94] MACV History, 1972–1973, vol. 2, pp. E-30, G-12; AAR, Opn COUNTDOWN, HQ USARV, vol. 1, pp. 30–31, 75–76, 78–79, vol. 2, pp. D-1 to D-2, D-8, D-1-1, D-7-1.

Pacific Architects and Engineers continued unfinished work under the direction of the Engineer Branch, Defense Attaché Office.[95]

By then all that remained of the engineer infrastructure was the small Engineer Branch, in the Army Division of the Defense Attaché Office. Altogether, this division consisted of three military and 277 U.S. civilian personnel. The Defense Attaché Office comprised five attachés representing the three services. Pacific Architects and Engineers continued to provide facilities engineering and limited construction services to the Defense Attaché Office until the final collapse of South Vietnam in 1975.[96]

---

[95] AAR, Opn COUNTDOWN, HQ USARV, vol. 1, pp. 33–34, vol. 2, pp. D-4 to D-5, D-11 to D-12.
[96] MACV History, 1972–1973, vol. 2, p. G-12.

17

Conclusion

There is no doubt that the accomplishments of the military and civilian engineers in Vietnam were numerous, varied, and significant. Army engineers and their military and civilian partners opened new ports along the coast, constructed and improved airfields in the dense jungle, built and paved roads, erected logistical facilities and housing, and provided support for tactical operations to counter an attempted Communist takeover of South Vietnam. The number of Army engineer units grew from the 35th Group's initial contingent of two construction battalions to an engineer command consisting of two brigades, six groups, twenty-eight battalions, forty-two separate companies, and various teams and detachments. Another seven engineer battalions were assigned to the divisions and eight companies to the separate brigades and regiment. During the war, over 200,000 soldiers served in engineer units, on engineer staffs assigned to major commands, in other staff and command positions, and as engineer advisers. In early 1969, when the Army reached its peak strength in Vietnam, there were over 40,000 troops serving with Army engineer units. During the course of the war, over 1,500 soldiers (engineers and non-engineers), including 143 officers, were killed or died of injuries while serving in engineer units and in other engineer assignments.[1]

The construction effort in South Vietnam allowed the United States to deploy and operate a modern 500,000-man force in a far-off underdeveloped land. Ground combat troops were able to fight the enemy from well-established bases, which gave U.S. and allied forces the opportunity to concentrate and operate when and where they wished. Although most of the construction was temporary, more durable facilities, including airfields, port and depot complexes, headquarters buildings, communications facilities, and an improved highway system, were intended to boost South Vietnam's defensive capabilities and developing economy.

Control and Management

In Vietnam, the employment of the engineers required adjustments to Army doctrine to cope with the nature of the war. There were no front lines; the enemy was anywhere and everywhere, often intermingled with the population.

---

[1] Galloway, "Essayons," pp. 252, 298; Dunn, *Base Development*, p. 26; Computer Reports by Military Occupational Specialty, 26 Jan and 20 Apr 2005, Information Technical Management Division, Washington Headquarters Services, Department of Defense; Casualty Information Sheet, n.d., Historian, U.S. Army Engineer School, Fort Leonard Wood, Mo. See also "U.S. Army Hostile Deaths by Combat Arms Branch," *VFW Magazine* 90 (June-July 2003): 22.

Consequently, construction units were not assigned to a communications zone to the rear, nor were combat engineer units to a combat zone to the front. Rather, engineers not assigned to divisions and separate brigades were placed under a single central command starting with the 18th Engineer Brigade and eventually U.S. Army Engineer Command, Vietnam.[2]

Centralization violated the long-held doctrine of giving the corps commander control of engineer resources, but this was always more theoretical than real in Vietnam, since the commander had merely to pick up the phone, and any nearby units would be his as long as he needed them. In a letter to General Duke, General Larsen of I Field Force had nothing but praise for the cooperation his headquarters received from the 18th Engineer Brigade. As far as he was concerned, the response of the brigade's units surpassed all expectations to the point that operational control was never an issue. General DePuy, a former commander of the 1st Infantry Division, later remarked that Engineer Command forces supported him so effectively that he was unaware for some time that they were not under the control of II Field Force. When asked of the desirability of assigning construction units to his 1st Logistical Command, General Eifler replied that the separate engineer disposition served many requirements and he was satisfied whenever he received his fair share of construction support.[3]

What did engineers themselves think of the centralized system? Engineer general officers supported it. Duke, Parker, Noble, and Young emphasized the high priority given to operational support and the flexibility in the use of engineer units. In late 1967, a survey conducted by Maj. Gerald A. Galloway, a student at the U.S. Army Command and General Staff College at Fort Leavenworth, Kansas, of past and present engineer brigade, group, battalion, and separate company commanders, found a two-to-one margin of approval. The minority position came from seven of nine division engineers, who were concerned about being saddled with base development projects, and one of three field force engineers.[4]

Like the nondivisional engineers, construction resources were also centrally managed, in this case by the MACV director of construction. Until the advent of the directorate, control at the service level had led to competition among the services' construction agencies for limited resources and a variety of construction programs responsible to different heads in Saigon and Washington. The first year of the buildup, when construction projects mush-

---

[2] Galloway, "Essayons," pp. 254–63. See also FM 100–10, *Field Service Regulations, Administration*, July 1963, para. 2-6 to 2-7; FM 100–15, *Field Service Regulations, Larger Units*, March 1966, para. 2-6 to 2-12; and FM 5–1, *Engineer Troop Organizations and Operations*, September 1965, para. 1-1 to 1-10.

[3] Ploger, *Army Engineers*, pp. 140–41; Galloway, "Essayons," pp. 276–78 (interviews with Generals DePuy and Eifler); Debriefing, Young, 15 Mar 72, p. 1, Senior Officer Debriefing Program, DA, Historians files, CMH; Ltr, Lt Gen Stanley R. Larsen, CG, I FFV, to Brig Gen Charles M. Duke, CG, 18th Engr Bde, n.d., copy in Rpt of Visit to Various Headquarters in Vietnam, 15–30 Sep 67, incl. 7, tab A, OCE Liaison Officer Trip Rpt no. 9.

[4] Galloway, "Essayons," p. 278; Debriefings, Duke, 14 May 68, p. 1, Parker, 5 Dec 69, p. 7, Noble, 2 Jul 71, p. 2, and Young, 15 Mar 72, pp. 2–3, Senior Officer Debriefing Program, DA. All debriefings in Historians files, CMH.

roomed with no central planning or direction, proved especially difficult in this regard. Unification under MACV brought clarity to project assignments, priority of effort, and construction standards. Reporting directly to the MACV commander, the office worked well enough that a later Defense Department logistical review recommended that future contingency plans establish the composition and role of a construction directorate on the staff of every joint forces commander.[5]

Materiel and Maintenance

To ensure that projects were carried out efficiently, the engineers depended on having the right equipment and receiving their supplies on time. One of the innovations of the Vietnam War was the purchase of commercial equipment to speed up the highway program. Senior engineers were pleased with the results. General Roper, a former 18th Engineer Brigade commander, believed military construction equipment manufactured to Army specifications was many years behind that used in the construction industry. He became a strong advocate of procuring commercial equipment without the restriction of military specifications, noting that several ten-ton dump trucks could be bought for the price of one standard military five-ton dump truck. Commercial vehicles operated in almost every environment that the military vehicle faced and outproduced the military dump truck by a factor of two to one. In justifying the equipment, Roper had his staff review the highway program to see how long it would take to finish it with available equipment. The answer was eight years, but incorporating additional, mostly commercial construction equipment, cut the time to four years or less without an increase in manpower. This result led to the request in 1968 for commercial equipment using Military Construction, Army, funds. General Parker, who commanded all nondivisional engineer troops in 1968 and 1969, conceded that the commercial equipment was not as rugged as the military, but it had performed well in the year since it was procured.[6]

Although engineers usually made do with the equipment on hand, modifications often resulted in better performance. A good example was the D7E bulldozer, equipped with a larger-than-normal Rome plow blade, sharpened each day, and its stinger projection used to split large trees before the blade sliced them off. Adding this powerful blade along with a protective cab, thereby creating the famous Rome plow, made the standard bulldozer an effective tactical weapon in clearing jungle growth.[7]

Lighter engineer equipment also arrived on the scene, initially with the 1st Cavalry Division. Before the division deployed, the air assault tests at Fort Benning had been promising with respect to engineer organization, training, operational concepts, and light equipment. But the proof was in combat, and

---

[5] Dunn, *Base Development*, pp. 18–19,137; JLRB, Monograph 6, *Construction,* pp. 87–89, 91–92.

[6] Debriefings, Parker, 14 Oct 69, p. 11, and Roper, 7 Oct 68, p. 4, Senior Officer Debriefing Program, DA. Both in Historians files, CMH.

[7] Roberts, "Trends in Engineer Support," pp. 34–35; Ploger, *Army Engineers*, pp. 96, 103.

the division's battle experience in the Central Highlands and along the coast exceeded the most optimistic forecasts. Again and again, light equipment of the 8th Engineer Battalion—small dump trucks, bulldozers, and scrapers— moving with the heliborne assault elements, cleared landing zones and forward airstrips and helped to elevate the tempo of operations. Later in the war, Colonel Malley, the engineer who took the 8th Battalion to Vietnam, thought that all combat engineers, including the nondivisional, should be issued airmobile equipment. A start was made in 1968 when ten airmobile sets arrived in Vietnam, seven going to units, among them the 101st Airborne Division, which was being reorganized as an airmobile division, and three to equipment pools under the control of Field Force (Corps). When additional sets arrived a year later, along with an infusion of repair parts, there was finally enough light equipment on hand to meet the airmobile requirements of operations.[8]

Airfield matting, on the other hand, caused headaches throughout the war. T17 membrane airfields required constant monitoring and repairs before ruptures caused by landings and takeoffs allowed water to get through to the subgrade. Experience showed that if T17 matting was used, it always required repairs. M8A1 steel matting was not much better. Crews struggled to connect panels made by different manufacturers and keep the panels aligned. Manufacturers apparently did not follow the same specifications. Once this problem was reported, Army Materiel Command insisted that the manufacturers meet its criteria. Commanders were more favorably inclined to AM2 and MX19 aluminum matting. The problem with the AM2 matting was the need for an expedient anchoring system. Later, accessory kits contained systems that correctly secured the matting. Better yet, the MX19 matting was initially provided with anchoring kits and was reported to be easy to place and remove. The trend was definitely toward using the MX19 as the standard airfield matting.[9]

The single greatest deficiency in engineer operations during the war, however, was poor equipment maintenance. General Parker believed that this defect needed to be corrected in the military school system. Added to the maintenance issue was the variety of makes and models of engineering equipment. To correct these problems, Parker pushed the idea of employing command maintenance inspection teams and established a school at Long Binh for repair parts clerks. These steps helped lower deadline rates, but equipment still on deadline was often the most needed on construction projects. Assigning more light equipment companies to every combat battalion helped, but the additional units outstripped the ability of the logistical system to provide backup maintenance. Parker suggested giving a third-echelon maintenance capability to light equipment companies similar to that of the construction battalions. Generals Morris and Chapman, in reviewing their experiences in

[8] Malley, "Engineer Support of Airmobile Operations," pp. 15–17; Logistics Review, USARV, vol. 7, Engineering Services System, pp. T-58 to T-60, U-34 to U-36, V-8, V-12 to V-14.
[9] Roberts, "Trends in Engineer Support," pp. 40–43.

commanding the 18th and 20th Engineer Brigades respectively, proposed the same capability for the combat battalions.[10]

Engineer Technique

In Vietnam, engineer reconnaissance teams played key roles in examining future work sites and determining if construction material was available nearby. Often, however, the teams had to travel over insecure roads. Security arrangements for the teams varied. When an engineer battalion carried out a reconnaissance operation in a division's area of operations, the division provided security. For example, when the 19th Engineer Battalion began road work between Bong Son and Duc Pho in 1967, south of the I/II Corps boundary, the 1st Cavalry Division protected the work parties. North of the boundary, Task Force OREGON provided protection. But in many cases, the engineers were on their own. Helicopters helped speed up reconnaissance, but there was always a question of availability and the danger of being shot down by enemy fire. When doing ground reconnaissance, many engineers preferred the armored personnel carrier (M113) to the 3/4-ton truck. With their mobility, protection, and firepower, M113s gave engineers entry into some of the most difficult and dangerous terrain in the country.[11]

To deny cover and concealment to the enemy, the engineers made important breakthroughs in counter-tunneling and land clearing. Through trial and error, engineers found that the acetylene method supplemented with satchel charges did a better job at collapsing tunnels than standard shaped charges. Rome plows came to be regarded as first-class tactical weapons and a trend of the future. Throughout the war, the Engineer School gave land-clearing classes to the officers in the advanced course under the reasonable assumption that future counterinsurgencies fought in jungles would require land-clearing units to spearhead advances and deny terrain to the enemy. But when U.S. forces withdrew from Vietnam and the Army turned its attention to conventional scenarios in Europe and elsewhere, the land-clearing classes vanished from the school curriculum. The land-clearing experience survived, if tenuously, in the few lessons-learned tracts that were published during the war.[12]

The enemy's use of mines and booby traps was a significant problem in Vietnam and resulted in a large proportion of the battle casualties suffered by U.S. soldiers. Mines alone caused two-thirds of all combat losses of armored personnel carriers and tanks. All sorts of munitions, many of U.S. origin, were turned into lethal mines. In 1969, the Army established the Mine Warfare Center at U.S. Army, Vietnam, headquarters, and began to disseminate reports on enemy techniques and countermeasures. While the losses to mines

[10] Debriefings, Brig Gen John W. Morris, CG, 18th Engr Bde, 13 Jul 70, p. 13, Parker, 14 Oct 69, pp. 10–11, Chapman, 30 Oct 69, pp. 8–9, and Brig Gen Harold R. Parfitt, CG, 20th Engr Bde, 1 Nov 69, pp. 6–7, Senior Officer Debriefing Program, DA. All in Historians files, CMH.

[11] Roberts, "Trends in Engineer Support," pp. 50–52; ORLLs, 1 Feb–31 Mar 67, 14th Engr Bn, p. 12, 1 Aug–31 Oct 67, 20th Engr Bn, p. 22, 1 Nov 66–31 Jan 67, 27th Engr Bn, p. 15.

[12] Roberts, "Trends in Engineer Support," pp. 35–36, 54–55; Debriefing, Parker, 14 Oct 69, p. 2, Senior Officer Debriefing Program, DA, Historians files, CMH.

remained serious, General Parker believed that the center's efforts helped to improve the ratio of mines detected to undetected. Using mine-detecting dog teams also showed promise, and improved models of metallic and nonmetallic mine detectors began to arrive in Vietnam. Despite this progress, mine warfare still favored the forces that placed the mines and put those forces that had to detect them or neutralize their effects at a continuing disadvantage.[13]

Another major innovation during the Vietnam War was the firebase in whose design and construction engineers played key roles. By late 1966, U.S. tactical units made it a common practice to build firebases within range of impending operations with artillery and infantry, headquarters elements, medical facilities, and other support. Typically, engineer bulldozers cleared fields of fire on slightly elevated land. Taking drainage into consideration, protective berms were pushed up from the outside, which would allow water to be drained and create a natural obstacle such as a moat. Drained artillery pads followed. Sixty-inch culvert or Conex containers topped off with pierced-steel planking and roofing paper and about two to three feet of earth or sandbags could be fitted in the berms as troop living and fighting positions. By piecing together the materials, tactical operations and fire-direction centers could be built. Fuel storage bladders could also be positioned in berms pushed up by the bulldozers. Firebases were constantly improved while the infantry maneuvered and the artillery provided fire support.[14]

Mapmaking provided another arena for engineer innovation in Vietnam. Up-to-date maps and topographic information were key ingredients to military operations in Vietnam, especially the placement of artillery fire. During the early stages of the war, artillery units normally supported ground units from fixed positions into which ground control had been extended. Surveys enabled the artillery to ensure the accuracy of fire, but as artillery units moved to more remote areas it became more difficult to support friendly units because surveys were lacking. In early 1967, Lt. Col. Arthur L. Benton, the former chief of the Mapping and Intelligence Division of the Engineer Section, U.S. Army, Vietnam, who had returned to Vietnam on temporary duty from the Army Map Service in Washington, D.C., developed a system known as photogrammetric positioning. By tying aerial photographs to base maps, artillery surveyors could readily obtain azimuth and location of firing positions. Working with the photograph and overprint of a map, aerial observers could give accurate references to targets. Tests proved favorable, and a system was in place after Operation CEDAR FALLS.[15]

[13] Hay, *Tactical and Materiel Innovations*, pp. 130–36; Debriefing, Parker, 14 Oct 69, pp. 3–4, Senior Officer Debriefing Program, DA, Historians files, CMH. For more on mine countermeasures, see Ewell and Hunt, *Sharpening the Combat Edge*, pp. 136–47.

[14] Hay, *Tactical and Materiel Innovations*, pp. 97–106; Ltr, 20th Engr Bde to CG, USARV, 9 Aug 69, sub: Fire Support Base Construction Experience, Historians files, CMH. See also Ltr, 14th Mil Hist Det to Office Ch of Mil Hist, sub: The Construction of a Fire Base in the 1st Cavalry Division (Airmobile), 10 Oct 69, Historians files, CMH; Maj. Geoffrey A. Fosbook Jr., "Clearing Artillery Fire Support Bases," *Military Engineer* 61 (March-April 1969): 87–89.

[15] ORLLs, 1 Aug–31 Oct 66, II FFV Artillery, 14 Nov 66, pp. 12–13, 1 Feb–30 Apr 67, II FFV Artillery, n.d., pp. 8–9; Ltr, Col Arthur Benton to author, 25 Aug 98; Citation, Incl in

Finally, the engineers fought as infantry when needed. All engineer soldiers were trained to fight as infantry, and engineer units were organized and equipped to defend themselves. From the earliest days of the buildup, engineers in units like the 173d Engineer Company, 173d Airborne Brigade, were assigned infantry duties to defend firebases and run patrols. The 1st Engineer Battalion had its M48 tankdozers acting as armor to protect infantrymen and engineers. It also organized a platoon made up of armored personnel carriers mounted with flamethrowers, tankdozers, scoop loaders, and air compressors to destroy tunnel complexes. In the highlands, the 4th Engineer Battalion combined its flame platoon's four flamethrower tracks with the four tankdozers from the four line companies and a tank retriever to form an engineer armored task force to protect convoys hauling sand between Dragon Mountain and Kontum City, reducing the dependence on infantry and armor for security. In the delta, combat engineers attached to the riverine force frequently filled out infantry units that were short of riflemen. During the Tet offensive, the retaking of Pleiku City was helped by a task force from the 4th Engineer Battalion. Clerk typists, carpenters, draftsmen, surveyors, and mapmakers of the 79th Engineer Group headquarters joined the troops on the perimeter to help defend II Field Force headquarters at Plantation when it came under attack. Outside Qui Nhon, the 84th Engineer Construction Battalion defended the approaches to the city.[16]

The Soldiers

Despite the improvements in organization, equipment, and technique, it was still the engineer soldier who had to carry out the mission under trying conditions. Almost all senior officers looking back as the war wound down commented on the dedication and courage of their engineers. These same officers, however, were struck by the youth and inexperience of the men in the ranks. Because the bulk of the reserves were not called up, the Army had to create an entire new corps of junior officers and noncommissioned officers on a crash basis through candidate schools, faster promotions, and direct appointments. New, inexperienced junior noncommissioned officers were dubbed "shake and bake" to denote how quickly they got their new ranks. General Noble in particular believed that something was missing from their training. They seemed to lack initiative and confidence and, more pointedly, leadership training. He wrote that modern-day training seemed to be focused on the classroom, not on the practical art of handling men. He considered it regrettable that the Army had cut back on close-order drill and marching in formation. "In close order drill," he wrote, "the chain of command, from the squad leaders on up, exercised positive, man to man and voice leadership."

Recommendation for Award, Lt. Col. Arthur L. Benton, 26 Jul 67, Engr Cmd. All in Historians files, CMH.

[16] FM 5–135, *Engineer Battalion: Armored, Infantry, and Infantry (Mechanized) Divisions*, November 1965, p. 2–1; FM 5–1, *Engineer Troop Organizations and Operations*, September 1965, pp. A-B-11, A-B-15, A-B-48, A-B-52, A-B-71; Debriefing, Duke, 14 May 68, p. A-12, Senior Officer Debriefing Program, DA, Historians files, CMH.

The Vietnam-era Army lost that time-honored means of nurturing effective leaders, and there were consequences for performance in the units.[17]

The vagaries of the one-year tour in Vietnam also seriously affected unit performance. The threat was particularly insidious on the one-year anniversary of a unit's arrival in the theater, for on that date the unit, in theory, would undergo a 100 percent turnover in officers and men. To reduce this "rotation hump," the Army tried various expedients, one of which, in the 45th Engineer Group, required that there be no manpower loss larger than 25 percent in any thirty days during a unit's second year in Vietnam. Although the one-year tour was considered a morale booster, the constant rotation of troops caused turbulence as the replacements took time to become proficient. The long duration of the war also meant second and sometimes third tours for career soldiers. By 1970, frequent family separations caused some captains and majors to leave the service.[18]

The engineer school system and on-the-job training were supposed to provide all the skills that the soldier needed, but both sometimes fell short. General Morris believed that mechanics and specialized equipment operators running asphalt pavers, rock crushers, and well drillers should have been trained in the United States and sent to Vietnam as teams. Noble remarked that commanders never seemed satisfied that incoming troops were adequately prepared. Even a few more weeks of practicing on equipment at the Fort Leonard Wood training center did not seem to produce the proficiency to operate graders, cranes, paving machines, and scrapers. Proficiency only came after months on the job, but most soldiers either left the service after completing their tours or were promoted. In either event, the operator was lost and new ones had to be trained. The highest rank for equipment operators then authorized was specialist fifth class. A possible solution, according to Noble, was to authorize the next higher rank (E–6) to operate the more complex construction equipment. This solution could have resulted in significant savings to the Army in the long run.[19]

Regardless of the concerns expressed by the senior leadership, the engineer soldier demonstrated the ability to adapt to the many challenges thrown in his path. Though plagued by acute deficiencies in engineering skills, the first engineers in Vietnam quickly responded to the missions placed upon them. In some cases, only one noncommissioned officer may have had the experience needed to get the job done, but within a short time the engineer soldiers learned from their sergeant to carry out the new tasks with confidence. Their efforts were evident in the struggle against a relentless enemy and adverse physical conditions. Accompanying the infantry, engineer soldiers ferreted out and destroyed enemy mines and tunnel complexes. Engineers manning Rome plows cleared paths through the jungle. Engineers cleared defensive perimeters

[17] Debriefings, Noble, 6 Aug 71, p. 10 (quoted words), and Chapman, 30 Oct 69, p. 9, Senior Officer Debriefing Program, DA. Both in Historians files, CMH.

[18] Debriefings, Chapman, 30 Oct 69, p. 10, and Morris, 13 Jul 70, p. 9, Senior Officer Debriefing Program, DA. Both in Historians files, CMH.

[19] Debriefings, Morris, 13 Jul 70, p. 9, and Noble, 6 Aug 71, p. 39, Senior Officer Debriefing Program, DA. Both in Historians files, CMH.

at base camps and firebases and built and often manned the bunkers and fighting positions. In their builder role, they completed or restored a multitude of airfields, roads, and bridges. The combat troops had base camps they could return to, not as comfortable as they may have liked, but constantly being improved. All these achievements were made possible through the endeavors of the engineer soldier and his officers.[20]

---

[20] Ploger, *Army Engineers*, pp. vii, 183.

Bibliographical Note

This account of engineer operations in South Vietnam draws on a wide range of sources. These include published and unpublished accounts, command histories, unit reports and histories, messages, personal papers, interviews, memoirs, studies, and research papers. Of special value in preparing this volume were the monographs *U.S. Army Engineers* by Maj. Gen. Robert R. Ploger and *Base Development* by Lt. Gen. Carroll H. Dunn, both in the U.S. Army Vietnam Studies series; the Naval Facilities Engineer Command's *Southeast Asia: Building the Bases* by Richard Tregaskis; and, "Essayons: The Corps of Engineers in Vietnam," a thesis prepared by Brig. Gen. Gerald E. Galloway while a student at the U.S. Army Command and General Staff College. These publications and thesis include many useful maps, charts, tables, glossaries, and appendixes. The companion combat volumes in the United States Army in Vietnam series fully depict the combat operations and provide more detailed maps of the operations. Altogether, these works provided the framework in this study's account of construction and engineer combat support.

Other sources were valuable in fleshing out the engineer story. Most of the engineer-related interviews were gleaned from the Center of Military History Vietnam Interview Collection. Personal accounts, memoirs, and articles helped fill in gaps, and like the interviews added a human element missing in the official reports. Interviews of senior officers conducted later were also consulted. At higher levels, the Military Assistance Command, Vietnam (MACV) commanders' and senior engineers' collections provided backchannel messages, memorandums, and journals. Likewise, official debriefing reports, some with enclosures, gave frank assessments of the war and engineer efforts. Various reviews, trip reports, and studies on the construction program, engineer combat support, and engineer organizations were of great value in preparing this study. The trip reports prepared by the Office of the Chief of Engineers liaison officers provide not only narratives of engineer operations in South Vietnam but also informative annexes dealing with topics of interest and numerous color slides used by the liaison officer in his briefings.

Unpublished Sources

Among the unpublished sources were quarterly command reports and the operational reports – lessons learned (ORLLs) prepared by engineer units at the battalion level and higher. Periodic after-action reports (AARs) prepared by the units provided accounts of specific operations. These sources vary in quality from unit to unit and are heavy in production figures that show the effectiveness of the support carried out by the engineers. A few units produced histories that also highlighted unit accomplishments and added some human interest. Similarly, interviews conducted in the field added a personal touch to engineer operations. Staff papers and reports put together by higher headquarters such as the Construction Directorate and the Office of the Chief of Engineers provide information on decisions affecting engineer operations, organization, logistics, and personnel matters. Special reports, studies, and investigations carried out at higher levels by the Defense Department, various headquarters, special study groups, and agencies depicted pros and cons of the construction effort. In all, the blending of these sources helped to achieve a balanced appreciation of the engineer endeavor in South Vietnam.

The ORLLs that replaced the command reports in late 1965 recorded a unit's activity over a three-month span (November–January, February–April, May–July, and August–October), arranged into sections on intelligence, operations, training, personnel, and logistics. Usually enclosures included an after-action account of an operation, statistics, maps, and design sketches. Engineer activities were also included in ORLLs prepared at logistical and support command, corps, division, and separate brigade levels. The ORLL was most useful in describing what the unit did during the period, innovative techniques, and avoiding future mistakes. In general, ORLLs lacked sufficient detail of individual operations and specific accomplishments by engineer soldiers in successfully carrying out unit missions. For this, the researcher has to turn to the combat after-action report, or AAR, interviews, and public affairs articles.

Unfortunately, AARs were not prepared for the bulk of engineer operations. Usually only significant operations merited AARs. Like ORLLs, engineer contributions to operations were included in AARs prepared by division and separate brigades. The AARs prepared by engineer units described the support rendered to tactical operations and specific projects such as constructing forward airfields. The majority of engineer AARs were prepared at the battalion level, but a few covering major operations such as the Cambodian incursion were prepared by engineer group and brigade headquarters. In a division-size operation, the division engineer prepared accounts depicting operations by the organic engineer battalion and other supporting engineers. Typically, the same format was used with topical headings, including task organization, intelligence, mission, concept of operation, chronology, execution, logistics and administration, and conclusions and recommendations. Sometimes AARs included maps, photographs, and design sketches. In some cases, AARs included oral interviews conducted by Engineer Command's military historians. The value of the AARs, besides the usual record-keeping account, was encompassed in the lessons on which the Army could base future doctrine, techniques, and possible corrective action.

National Archives and Records Administration

Many of the documents cited in this volume are in the custody of the National Archives and Records Administration (NARA). Records dealing with operations are housed in its facility at College Park, Maryland. Four record groups—Record

Group (RG) 319, Records of the Army Staff; RG 330, Records of the Department of Defense; RG 334, Records of Interservice Agencies; and RG 338, Records of the Army Commands—contain most of the ORLLs, AARs, correspondence and messages, and annual histories. Recently, NARA transferred the Vietnam War material in RGs 334 and 338 to a new record group—RG 472, Records of the U.S. Forces in Southeast Asia, 1950–1975. The documents consulted in RG 330 are the backup papers used in the Defense Department's report by the Joint Logistical Review Board dealing with construction and advanced base facilities maintenance. Since RGs 334 and 338 citations are reflected in this volume, researchers need to rely on the archivists at NARA to locate the documents in RG 472.

U.S. Army Center of Military History

The U.S. Army Center of Military History (CMH) located at Fort Lesley J. McNair in Washington, D.C., houses records that were critical in researching the engineer story. A photocopied set of the papers of General William C. Westmoreland assembled during his tour of duty as MACV commander, contains his backchannel messages and journal, as well as additional material that deals with engineer matters such as the messages between Westmoreland and Washington and Honolulu dealing with the establishment of a construction czar. Some engineer-related material can also be found in the photocopied collection of General Creighten W. Abrams. Critical, however, are the collection of papers of Lt. Gen. Carroll H. Dunn, the first MACV director of construction; the backchannel messages of Maj. Gen. Charles M. Duke, the commander of the 18th Engineer Brigade and later as the Engineer, U.S. Army, Vietnam; and copies of correspondence by Maj. Gen. Robert R. Ploger, the first commander of 18th Engineer Brigade and Engineer Command.

The CMH Vietnam Interview Collection, altogether some two thousand interviews, included over two hundred interviews with engineer commanders, staff officers, and soldiers. Many of the engineer-related interviews have been transcribed. Summaries of the interviews and occasionally AARs are contained in the files. Most of the interviews were with departing battalion commanders covering the period of their command. There are also interviews with commanders and staff officers at command, brigade, and group levels. Only a few interviews were held with company commanders and enlisted soldiers and contractor representatives. The interviews with AARs that stand out are with the commander and staff of the 35th Engineer Battalion (Combat) after it reopened the Hai Van Pass; the commander and soldiers of Company A, 70th Engineer Battalion (Combat), following the evacuation of Kham Duc; the commanders and staff of the 34th and 35th Engineer Groups who discussed the highway improvement program in their areas of responsibility; and the commander of the 62d Engineer Battalion (Land Clearing) following the Cambodian incursion. Most valuable are the interviews that took place soon after an operation or event because details were still fresh in the minds of the participants. The end-of-tour interviews were of limited value in this volume because the questions and answers mostly dealt with trends such as training, discipline, and morale but little on specific operations during the six-month or one-year tour.

Other useful holdings are the copies of command histories prepared by U.S. Military Assistance Command, Vietnam (MACV); the Commander in Chief, Pacific (CINCPAC); U.S. Army, Vietnam (USARV); U.S. Army, Pacific (USARPAC); the histories of the Joint Chiefs of Staff; and the twelve-volume Defense Department Study on U.S.-Vietnam Relations commonly known as the Pentagon Papers. The Historians files collection, which are maintained in the working files of Histories Division or the research files of the Historical Resources Branch, include copies of AARs, ORLLs, interviews, letters, messages, memorandums, reports, briefings, studies, unit histories such as a two-volume unpublished history of the 1st Engineer Battalion in Vietnam, and other documents. Also available in the Historical Resources Branch are the published unit yearbooks of several divisions and separate brigades and the 1st Engineer Battalion in Vietnam; contemporary field manuals and tables of organization and equipment (TOEs); the papers of Thomas B. Thayer (with some references dealing with construction, base transfers, land mines, and booby traps) assembled by the former director of the Southeast Asia Office in the Office of the Assistant Secretary of Defense for Systems Analysis; the report by the Joint Logistics Review Board, which includes volumes on construction and facilities engineering; a volume dealing with engineering services in the multivolume *Logistics Review, U.S. Army, Vietnam*; and engineer-related items in the personal papers of Robert W. Komer, the MACV Deputy for Civil Operations and Revolutionary Development Support (DEPCORDS). Copies of the weekly Army Buildup Progress Report (11 August 1965 to 20 December 1972), prepared by the Office of the Chief of Staff of the Army (and which became the Army Activities Report: Southeast Asia on 26 March 1969), include charts, graphs, and maps depicting base development progress and engineer unit dispositions as well as disclosing engineer problems. A collection of senior officer debriefing reports portray the views of several engineer generals before their departure from Vietnam. Some of the reports contain engineer subject annexes. Also in the files are copies of engineer-related documents, MACV complex reviews, Brig. Gen. Daniel A. Raymond's Observations on the Construction Program, development of the MACV Construction Directorate, backup documents for the MACV command histories, and annual histories prepared by Pacific Architects and Engineers.

U.S. Army Military History Institute

The U.S. Army Military History Institute (MHI) at Carlisle Barracks, Pennsylvania, has large holdings of Vietnam War documents, varying from unit-level to high-level conduct of the war. In many instances, these documents duplicate holdings in the National Archives. MHI also holds the papers of many senior Army officers, including Corps of Engineers generals. These include the papers and messages of Generals Robert P. Young, Charles C. Noble, and Henry C. Schrader. The Senior Officers Oral History Program files at MHI include an extensive interview with Maj. Gen. Robert R. Ploger. (A copy of the Ploger interview is also available at CMH.) In addition, MHI's Senior Officer Oral History Program contains interviews with Army War College students who served as engineer company commanders during the Vietnam War. MHI also houses the

Senior Officer Debriefing Reports, which include debriefings with senior Corps of Engineers officers. MHI is the main repository of newspapers and magazines published by commands in Vietnam. Engineer Command's biweekly newspaper, *Castle Courier*, and a quarterly magazine, *Kysu'*, contain a variety of articles on engineer operations, construction, highway restoration, and many human interest stories.

Office of History, Headquarters, U.S. Army Corps of Engineers

The Office of History, U.S. Army Corps of Engineers, is located at the Humphreys Engineer Center in Alexandria, Virginia, near Fort Belvoir. The Office of History's extensive Research Collection includes a wide variety of engineer-related material. Holdings contain many ORLLs and AARs prepared by engineer units in Vietnam, General Raymond's Development of the Construction Directorate and observations on the construction program, documents, reports, personal papers, photographs, interviews, and field and technical manuals. The office also contains periodicals such as the Engineer Command's biweekly newspaper, *Castle Courier,* and quarterly magazine, *Kysu',* and bimonthly *Military Engineer* published by the Society of Military Engineers.

The Office of History received copies of all the Office, Chief of Engineers (OCE) Liaison Officer Trip Reports, which were prepared to brief the Chief of Engineers and Washington officials. Besides a narrative of his visit to Army headquarters in Honolulu and Saigon and engineer units in South Vietnam, the OCE Liaison Officer assembled a wealth of information on unit dispositions, operations, construction, and engineer logistics and personnel. The narratives typically included candid comments during his meetings with senior engineer officers. Altogether, twenty-two reports were heavily mined in the preparation of this volume.

The Office of History has conducted a series of oral history interviews with senior engineer officers that cover their careers and service in Vietnam. This series called Engineer Memoirs includes interviews with Lt. Gen. Frederick J. Clarke, Chief of Engineers during the Vietnam War; Lt. Gen. Carroll H. Dunn, MACV Director of Construction, MACV J–4, Director of Military Construction at the Office of the Chief of Engineers, and Deputy Chief of Engineers; and other senior officers. The full citation for the memoir used in this volume is as follows:

Engineer Memoirs, Lieutenant General Carroll H. Dunn. Alexandria, Va.: Office of History, U.S. Army Corps of Engineers, 1998.

The Office of History, U.S. Army Corps of Engineers, and engineer divisions have published histories of their organizations and activities. For this volume, the author consulted an overview of the Corps' history and a history prepared by the Pacific Ocean Division. Full citations follow:

Thompson, Erwin N. *Pacific Ocean Engineers: History of the U.S. Army Corps of Engineers in the Pacific, 1905–1980.* Honolulu: Pacific Ocean Division, n.d.

U.S. Army Corps of Engineers. *The U.S. Army Corps of Engineers: A History.* Alexandria, Va.: Office of History, U.S. Army Corps of Engineers, 2007.

The Engineer Studies Group, later the Engineer Studies Center, a former field operating activity of the Corps of Engineers, did a broad spectrum of studies to assist OCE, the Army Staff, and the Defense Department and the Joint Chiefs of Staff, in making decisions on base development and strategic forces planning. The Office of History prepared a history of the Engineer Studies Center. A bibliography of the Engineer Studies Center publications is in the historian's files. The Engineer Studies Group and Center files are now in the National Archives. Complete citations used in this volume are as follows:

An Analysis of the Requirement for Army Aircraft Shelters. Engineer Strategic Studies Group, Office, Chief of Engineers, Department of the Army, September 1969.

Baldwin, William C. *The Engineer Studies Center and Army Analysis: A History of the U.S. Army Engineer Studies Center, 1943–1982.* Fort Belvoir, Va.: U.S. Army Corps of Engineers, 1985.

Bibliography of Publications, U.S. Army Corps of Engineers, Engineer Studies Center. N.p., n.d.

U.S. Navy Seabee Museum

The U.S. Navy Seabee Museum at Port Hueneme, California, houses the reports prepared by Seabee units following their tours in Vietnam and the Richard Tregaskis Papers. Tregaskis, noted for his book *Guadalcanal Diary*, had written *Southeast Asia: Building the Bases* for the Naval Facilities Engineering Command. *Building the Bases* was a chief source for sketching the work done by Seabee units and the joint-venture contractor, Raymond, Morrison-Knudsen, Brown and Root, and J. A. Jones (RMK-BRJ). Backup materials, including interviews used in the preparation of *Building the Bases,* are held in the Tregaskis Papers. Although Tregaskis concentrated on the Seabees and RMK-BRJ, he did include some work done by Army engineers. Most important were his interviews with senior Army and Navy engineers plus an interview with General Westmoreland in the Pentagon after he became the Chief of Staff of the U.S. Army. Also available at the U.S. Navy Seabee Museum is Diary of a Contract, a history prepared by RMK-BRJ. The Seabee unit reports should be of interest to researchers who want to delve into the accomplishments of the naval construction battalions.

Other Holdings

Several other repositories were useful in the preparation of this volume. The Office of Air Force History at Bolling Air Force Base in Washington, D.C., maintained files used in the preparation of John Schlight's *The War in South Vietnam: The Years of the Offensive, 1965–1968*, which cites Air Force documents and histories on air base construction and Red Horse engineers. The holdings of the U.S. Naval History and

Heritage Command at the Navy Yard in Washington, D.C., and the Marine Corps History and Museums Division at Quantico, Virginia, were checked for materials on Navy and Marine Corps engineers. The library at the U.S. Army Command and General Staff College at Fort Leavenworth, Kansas, has large holdings of ORLLs and AARs on the Vietnam War and copies of the MACV Complex Reviews. Copies of the MACV Complex Reviews were also found at the Army War College's library at Carlisle Barracks, Pennsylvania. In addition, many of the ORLLs, AARs, and Senior Officer Debriefing Reports are in the holdings of the Defense Technical Information Center at Fort Belvoir. The command historian of the Engineer School at Fort Leonard Wood, Missouri, has some holdings similar to the Corps of Engineers Historical Office. These include ORLLs, AARs, and OCE Liaison Officer Trip Reports.

Student theses and essays done by engineer officers attending the Command and General Staff College and Army War College provided additional background on engineer construction and combat support. Those consulted are as follows:

Conover, Lt. Col. Nelson P. "The Lines of Communication Program in Vietnam, A Case Study." Research paper, U.S. Army War College, 1973.

Galloway, Maj. Gerald E. "Essayons: The Corps of Engineers in Vietnam." Master of Military Art and Science thesis, U.S. Army Command and General Staff College, 1968.

Malley, Lt. Col. Robert J. "Engineer Support of Airmobile Operations." Student essay, U.S. Army War College, 1967.

Roberts, Col. Charles R. "Trends in Engineer Support." Student thesis, U.S. Army War College, 1969.

Published Primary Sources

Among the published primary sources are the histories prepared by major commands and the Joint Chiefs of Staff. Sections of the histories cover engineering activities. Complete citations for the histories used in this volume are as follows:

Arsenal for the Brave: A History of the United States Army Materiel Command, 1962–1968. Washington, D.C.: Historical Office, U.S. Army Materiel Command, 1969.

Headquarters, Commander in Chief, Pacific. "CINCPAC Command History 1961." Honolulu: Deputy Chief of Staff for Military Assistance, Logistics and Administration, 1962.

_____. "CINCPAC Command History, 1962." Honolulu, 1963.

_____. "CINCPAC Command History, 1963." Honolulu, 1964.

_____. "CINCPAC Command History, 1964." Honolulu, 1965.

Headquarters, U.S. Army, Pacific (USARPAC). History of U.S. Army Operations in Southeast Asia, 1 January–31 December 1964. Honolulu: Military History Division, Office of the Assistant Chief of Staff, G–3, 1965.

Historical Division, Joint Secretariat, Joint Chiefs of Staff. "The Joint Chiefs of Staff and the War in Vietnam, 1960–1968," Part 2, 1965–1966. Washington, D.C., 1970.

Military History Branch, Headquarters, United States Military Assistance Command, Vietnam. "Command History, 1964." Saigon, 1965.

_____. "Command History, 1965." Saigon, 1966.

_____. "Command History, 1966." Saigon, 1967.

_____. "Command History, 1967." Volumes 1–3. Saigon, 1968.

_____. "Command History, 1968." Volumes 1–2. Saigon, 1969.

_____. "Command History, 1969." Volumes 1–3. Saigon, 1970.

_____. "Command History, 1970." Volumes 1–4. Saigon, 1971.

_____. "Command History, 1971." Volumes 1–2. Saigon, 1972.

_____. "Command History, 1972–1973." Volumes 1–2. Saigon, 1973.

Procurement Support in Vietnam, 1966–1968. U.S. Army Procurement Agency, Vietnam. Japan: Toshio Printing Co., n.d.

Weinert, Richard P. *The Role of USCONARC in the Army Buildup, FY 1966*. Fort Monroe, Va.: U.S. Continental Army Command, 1967.

Admiral U. S. G. Sharp and General William C. Westmoreland directed the preparation of an overview of their commands covering the period from 1964 to 1968. Sharp's part of the report includes a section covering construction in Vietnam and Thailand, and Westmoreland's part presents a command account of the war in South Vietnam, including several appendixes. One of the appendixes outlines base development and construction in South Vietnam. The complete citation is: Sharp, Admiral U. S. G., and General William C. Westmoreland. *Report on the War in Vietnam (As of 30 June 1968)*. Washington, D.C.: Government Printing Office, 1969.

Two sets of studies published under Department of the Army auspices were used in the preparation of this volume. The first series, the Vietnam Studies, consists of twenty-one monographs by senior officers who served in Vietnam. The monographs cited in this volume are:

Collins, Brig. Gen. James L. Jr. *The Development and Training of the South Vietnamese Army, 1950–1972*. Washington, D.C.: Department of the Army, 1973.

Dunn, Lt. Gen. Carroll H. *Base Development in South Vietnam, 1965–1970*. Washington, D.C.: Department of the Army, 1972.

Eckhardt, Maj. Gen. George S. *Command and Control, 1950–1969*. Washington, D.C.: Department of the Army, 1974.

Ewell, Lt. Gen. Julian J., and Maj. Gen. Ira A. Hunt Jr. *Sharpening the Combat Edge: The Use of Analysis to Reinforce Military Judgment*. Washington, D.C.: Department of the Army, 1974.

Fulton, Maj. Gen. William B. *Riverine Operations, 1966–1969*. Washington, D.C.: Department of the Army, 1973.

Hay, Lt. Gen. John H. Jr. *Tactical and Materiel Innovations*. Washington, D.C.: Department of the Army, 1974.

Heiser, Lt. Gen. Joseph M. Jr. *Logistic Support*. Washington, D.C.: Department of the Army, 1974.

Kelly, Col. Francis J. *The U.S. Army Special Forces, 1961–1971*. Washington, D.C.: Department of the Army, 1973.

Larsen, Lt. Gen. Stanley R., and Brig. Gen. James L. Collins Jr. *Allied Participation in Vietnam.* Washington, D.C.: Department of the Army, 1975.

Pearson, Lt. Gen. Willard. *The War in the Northern Provinces, 1966–1968.* Washington, D.C.: Department of the Army, 1975.

Ploger, Maj. Gen. Robert R. *U.S. Army Engineers, 1965–1970.* Washington, D.C.: Department of the Army, 1974.

Rogers, Lt. Gen. Bernard W. *Cedar Falls–Junction City: A Turning Point.* Washington, D.C.: Department of the Army, 1974.

Starry, General Donn A. *Mounted Combat in Vietnam.* Washington, D.C.: Department of the Army, 1978.

Tolson, Lt. Gen. John J. *Airmobility, 1961–1971.* Washington, D.C.: Department of the Army, 1973.

A second set of studies published by the Center of Military History in limited quantities, the Indochina Monographs, was done by contract and authored by senior South Vietnamese, Cambodian, and Laotian officers. The series consists of twenty monographs covering subjects of special interest to the authors who consulted available records. The series also provides personal commentary and experience, thereby giving them the character of primary sources. The following monographs were used:

Hinh, Maj. Gen. Nguyen Duy. *Lam Son 719*, Washington, D.C.: U.S. Army Center of Military History, 1979.

Khuyen, Lt. Gen. Dong Van. *The RVNAF.* Washington, D.C.: U.S. Army Center of Military History, 1980.

_____. *RVNAF Logistics.* Washington, D.C.: U.S. Army Center of Military History, 1980.

Lung, Col. Hoang Ngoc. *The General Offensives of 1968–69.* Washington, D.C.: U.S. Army Center of Military History, 1981.

Sutsakhan, Lt. Gen. Sak. *The Khmer Republic at War and the Final Collapse.* Washington, D.C.: U.S. Army Center of Military History, 1980.

Tho, Brig. Gen. Tran Dinh. *The Cambodian Incursion.* Washington, D.C.: U.S. Army Center of Military History, 1983.

Truong, Lt. Gen. Ngo Quang. *The Easter Offensive of 1972.* Washington, D.C.: U.S. Army Center of Military History, 1980.

Several published unit histories were consulted. Typically, divisions and brigades prepared histories in the form of yearbooks depicting operations. Unit histories usually have sections on the activities of organic engineer units. The 1st Engineer Battalion, 1st Infantry Division, also published several histories. These published histories typically include photographs of commanders and troops, photographs of operations, and maps. The histories used in this publication and available at the Center of Military History are as follows:

Always First: A Pictorial History of the 1st Engineer Battalion, 1st Infantry Division, October 1965–March 1967. 1st Engineer Battalion, n.d.

Always First: 1st Engineer Battalion, 1967–1968. 1st Engineer Battalion, n.d.

Coleman, Maj. J. D., ed. *1st Air Cavalry Division: Memoirs of the First Team, Vietnam, August 1965–December 1969*. Tokyo, Japan: Dai Nippon Printing Co., n.d.

The First Three Years: A Pictorial History of the 173d Airborne Brigade (Separate). Brigade Information Office, n.d.

The 1st Infantry Division in Vietnam, 1969. N.p.: 1st Infantry Division, n.d.

The 25th's 25th . . . in Combat, Tropic Lightning, 1 October 1941–1 October 1967. The 25th Infantry Division, 25th Division Public Affairs Office, n.d.

The U.S. 25th Infantry Division, October 1969–October 1970, Yearbook. N.p.: Tropic Lightning Association, n.d.

Vietnam, the First Year: Pictorial History of the 2d Brigade, 1st Infantry Division. Tokyo, Japan: Brigade Information Office, n.d.

The MACV Directorate of Construction prepared periodic reports of the construction program known as Complex Reviews, which show a broad picture of facilities requirements as determined by the force structure in South Vietnam. Sketch maps of major base complexes are included. Related is the 1964 base development plan for Southeast Asia published by U.S. Army, Pacific. Full citations are as follows:

Base Development Plan No. 1–64, Vol. VI, Southeast Asia. Headquarters, U.S. Army, Pacific. Fort Shafter, Hawaii: U.S. Army, Pacific, n.d.

Construction Program South Vietnam (Complex Review). Saigon: Directorate of Construction, MACV, 1 December 1966.

Construction Program South Vietnam (Complex Review). Saigon: Directorate of Construction, MACV, 1 April 1967.

Construction Program South Vietnam (Complex Review). Saigon: Directorate of Construction, MACV, 15 January 1968.

Construction Program South Vietnam (Complex Review). Saigon: Directorate of Construction, MACV, 1 March 1969.

In 1969, the Department of Defense established the Joint Logistics Review Board and asked it to address the entire construction process. Two monographs cover base development planning, funding, standards for facilities, construction resources, and facilities engineering. The studies include numerous charts and tables. A similar study was prepared by U.S. Army, Vietnam. Full citations for studies used in this publication are as follows:

Joint Logistics Review Board. *Logistics Support in the Vietnam Era*. Monograph 1, *Advanced Base Facilities Maintenance*. Washington, D.C.: Department of Defense, n.d.

_____. *Logistics Support in the Vietnam Era*. Monograph 6, *Construction*. Washington, D.C.: Department of Defense, n.d.

_____. *Logistics Support in the Vietnam Era*. Monograph 12, *Logistics Planning*. Washington, D.C.: Department of Defense, n.d.

The Logistics Review, U.S. Army, Vietnam, 1965–1969, vol. 7, *Engineering Services System*. Headquarters, U.S. Army, Vietnam, n.d.

What became known as the Pentagon Papers were classified histories of Defense Department policy-making on Vietnam from 1945 through early 1968, prepared at Defense Secretary Robert S. McNamara's direction. They were leaked to the press in 1971 and commercial editions were published. The narrative in these volumes is supplemented by extracts and reproductions of many high-level documents, including several dealing with a deployment of engineer troops. The two publications used in this volume are as follows:

United States-Vietnam Relations, 1945–1967: Study Prepared by the Department of Defense, 12 vols. Washington, D.C.: Government Printing Office, 1971.

The Pentagon Papers: The Defense Department History of United States Decisionmaking on Vietnam. Senator Gravel Edition, 4 vols. Boston: Beacon Press, 1971.

The Department of Defense and Department of the Army have published annual histories, which include discussions of military construction in Vietnam. Those consulted and cited in this study are as follows:

Annual Report of the Department of Defense for Fiscal Year 1965. Washington, D.C.: Government Printing Office, 1967.

Annual Report of the Department of Defense for Fiscal Year 1967. Washington, D.C.: Government Printing Office, 1969.

Annual Report of the Department of Defense for Fiscal Year 1968. Washington, D.C.: Government Printing Office, 1971.

Department of the Army Historical Summary, Fiscal Year 1969. Washington, D.C.: U.S. Army Center of Military History, 1973.

Reports published by the U.S. Operations Mission (USOM) in South Vietnam and studies sponsored by the U.S. Agency for International Development (USAID) were useful in providing background to economic assistance, such as improvements to airports, ports, waterways, railroads, and roads. Complete citations for USOM reports and the USAID study used in this volume are as follows:

The Postwar Development of the Republic of Vietnam: Policies and Programs. 2 vols. Saigon and New York: Joint Development Group, Postwar Planning Group, Development and Resources Corporation, March 1969.

United States Operations Mission (USOM). *Activity Report of the Operations Mission to Vietnam, 30 June 1954 to 30 June 1956.* Saigon, n.d.

USOM. *Annual Report for the 1958 Fiscal Year (July 1, 1957 to June 30, 1958).* Saigon, n.d.

USOM. *Annual Report for Fiscal Year 1960.* Saigon, 1 October 1960.

USOM. *Annual Report for Fiscal Year 1961.* Saigon, 20 November 1961.

USOM. *Annual Report for Fiscal Year 1962.* Saigon, n.d.

Vietnam Transportation Study, Based on Field Studies Conducted April 1965 through May 1966 for the Government of Vietnam and the U.S. Agency for International Development. Washington, D.C.: Transportation Consultants, Inc., June 1966.

Also used were contractor-prepared histories by RMK-BRJ and Pacific Architects and Engineers. These were published in limited quantities and include photographs. Copies are in Historians files at the Center of Military History. These are:

Diary of a Contract, NBy (Navy Bureau of Yards and Docks) 44105, January 1962–June 1967. Raymond, Morrison-Knudsen, Brown and Root, and J. A. Jones (RMK-BRJ). Port Hueneme, Calif.: U.S. Navy Seabee Museum, July 1967.

Johns, Eric D. *History PA&E, Pacific Architects and Engineers Incorporated: Repairs and Utilities Operations for U.S. and Free World Military Forces in the Republic of Vietnam, 1963 to 1966.* [Saigon], n.d.

Johns, Eric D. and Frank A. Lerney. *History PA&E,* Calendar Year 1967. [Saigon], n.d.

_____. *History PA&E*, January–June 1968. [Saigon], n.d.

_____. *History PA&E*, Fiscal Year 1969. [Saigon], n.d.

_____. *History PA&E,* Fiscal Year 1970. [Saigon], n.d.

History PA&E, Fiscal Year 1971. [Saigon], n.d.

Frequently consulted were U.S. Army Field Manuals (FMs) for this period covering engineer missions, organizations, field data, and a reference for staff officers. For example, FM 5–34 is a pocket-size manual containing field data on explosives and demolitions, bridging, field fortifications, land mines, airfields and helipads, conversion tables, and other topics for noncommissioned officers and officers at the platoon level. Full citations are:

FM 5–1, *Engineer Troop Organizations and Operations.* Washington, D.C.: Government Printing Office, May 1961.

FM 5–1, *Engineer Troop Organizations and Operations.* Washington, D.C.: Government Printing Office, September 1965.

FM 5–34, *Engineer Field Data.* Washington, D.C.: Government Printing Office, 1 December 1969.

FM 5–135, *Engineer Battalion: Armored, Infantry, and Infantry (Mechanized) Divisions.* Washington, D.C.: Government Printing Office, November 1965.

FM 100–10, *Field Service Regulations, Administration.* Washington, D.C.: Government Printing Office, July 1963.

FM 100–15, *Field Service Regulations, Larger Units.* Washington, D.C.: Government Printing Office, March 1966.

FM 101–10–1, *Staff Officers' Field Manual.* Washington, D.C.: Government Printing Office, July 1972.

Several official histories from the perspective of the enemy have been published in Hanoi by the Socialist Republic of Vietnam. Used primarily as references in the combat operations volumes in this series, one has been used as a source for this work. The volume has been translated under the auspices of the Center of Military History. The histories and translations may be used by researchers who visit the Center. The citation for the publication used in this book is:

Luc Luong Vu Trang Nhan Dan Tay Nguyen Trong Khang Chien Chong My Cuu Nuoc [*The People's Armed Forces of the Western Highlands During the War of National Salvation Against the Americans*]. Hanoi: Nha Xuat Ban Quan Doi Nhan Dan [People's Army Publishing House], 1980.

Other Primary Publications

Indo-China, Geographical Handbook Series, Great Britain. Cambridge, England: Naval Intelligence Division, December 1943.

Military Operations, Lessons Learned: Land Clearing. Department of the Army Pamphlet 525–6. Washington, D.C.: Government Printing Office, 16 June 1970.

Special Operational Report–Lessons Learned (ORLL), 1 January–31 October 1965. Department of the Army, 23 August 1966.

Published Official Histories

U.S. Army

The Center of Military History's U.S. Army in Vietnam series currently consists of eleven published works, including a pictorial volume, two volumes on MACV, two advice and support volumes, two public affairs volumes, two combat volumes, one volume on communications, and *Engineers at War*. Additional volumes in progress cover combat operations, logistics, and advice and support. Those used in this volume are:

Bergen, John D. *Military Communications: A Test for Technology.* Washington, D.C.: U.S. Army Center of Military History, 1986.

Carland, John M. *Combat Operations: Stemming the Tide, May 1965 to October 1966.* Washington, D.C.: U.S. Army Center of Military History, 2000.

Clarke, Jeffery J. *Advice and Support: The Final Years, 1965–1973.* Washington, D.C.: U.S. Army Center of Military History, 1988.

Cosmas, Graham A. *MACV: The Joint Command in the Years of Escalation, 1962–1967.* Washington, D.C.: U.S. Army Center of Military History, 2006.

Hammond, William M. *Public Affairs: The Military and the Media, 1962–1968.* Washington, D.C.: U.S. Army Center of Military History, 1988.

_____. *Public Affairs: The Military and the Media, 1968–1973.* Washington, D.C.: U.S. Army Center of Military History, 1996.

MacGarrigle George L. *Combat Operations: Taking the Offensive, October 1966 to October 1967*. Washington, D.C.: U.S. Army Center of Military History, 1998.

Spector, Ronald H. *Advice and Support: The Early Years, 1941–1960*. Washington, D.C.: U.S. Army Center of Military History, 1983.

In addition, the author consulted other published works of the Center of Military History. These are:

Cash, John A. "Fight at Ia Drang, 14–16 November 1965." In *Seven Firefights in Vietnam* by John A. Cash, John Albright, and Allan W. Sandstrum, pp. 3–40. Washington, D.C.: Office of the Chief of Military History, United States Army, 1970.

Dod, Karl C. *The Corps of Engineers: The War Against Japan*, United States Army in World War II. Washington, D.C.: Office of the Chief of Military History, United States Army, 1966.

U.S. Air Force

Several Air Force histories have been used in this study. The Office of Air Force History has published several volumes on air operations in Southeast Asia, and their full citations are:

Bowers, Ray L. *Tactical Airlift*, Washington, D.C.: Office of Air Force History, United States Air Force, 1983.

Buckingham, William A. Jr. *Operation RANCH HAND: The Air Force and Herbicides in Southeast Asia, 1961–1971*. Washington, D.C.: Office of Air Force History, United States Air Force, 1982.

Fox, Roger P. *Air Base Defense in the Republic of Vietnam, 1961–1973*. Washington, D.C.: Office of Air Force History, United States Air Force, 1979.

Futrell, Robert F. *The Advisory Years to 1965*. Washington, D.C.: Office of Air Force History, United States Air Force, 1981.

Nalty, Bernard C. *Air War Over South Vietnam, 1968–1975*. Washington, D.C.: Air Force History and Museums Program, United States Air Force, 2000.

Schlight, John. *The War in South Vietnam: The Years of the Offensive, 1965–1968*. Washington, D.C.: Office of Air Force History, United States Air Force, 1988.

In addition, the Seventh Air Force prepared a history on airfield construction in South Vietnam. A copy is in the historian's files. The full citation is as follows: Martin, Jean, S. Sgt. Douglas W. Stephens, and S. Sgt. Robert F. Jakob. *USAF Airfield Construction in South Vietnam, July 1965–March 1967*. Historical Division, Directorate of Information, Headquarters, Seventh Air Force, n.d.

U.S. Navy

Naval histories covering the accomplishments of the Seabees and RMK-BRJ were extensively used. Those published by the Naval Historical Center are:

Hooper, Edward B. *Mobility, Support, Endurance: A Story of Naval Operational Logistics in the Vietnam War, 1965–1968.* Washington, D.C.: Naval Historical Division, Department of the Navy, 1972.

Marolda, Edward J. *By Sea, Air, and Land: An Illustrated History of the U.S. Navy and the War in Southeast Asia.* Washington, D.C.: Naval Historical Center, Department of the Navy, 1994.

Marolda, Edward J., and Oscar P. Fitzgerald. *From Military Assistance to Combat, 1959–1965.* United States Navy and the Vietnam Conflict. Washington, D.C.: Naval Historical Center, Department of the Navy, 1986.

The Naval Facilities Engineering Command contracted with Richard Tregaskis (*Guadalcanal Diary*) to write a history of the Seabees and RMK-BRJ on the construction effort in Southeast Asia, which was frequently cited in this volume: Tregaskis, Richard. *Southeast Asia: Building the Bases: The History of Construction in Southeast Asia.* Washington, D.C.: Government Printing Office, 1975.

In addition, the Seabee headquarters in Hawaii prepared a publication on the accomplishments of the Seabee Technical Assistance Teams sent to South Vietnam and Thailand. This publication covers the accomplishments of each team that supported the Army Special Forces Detachments early in the war and civic action projects for the U.S. Operations Mission: COMCBPAC Reports, Seabee Teams, October 1959–July 1968. Commander Naval Construction Battalions, U.S. Pacific Fleet, 1969.

U.S. Marine Corps

The Marine Corps and Museum Division has completed publication of its chronological series on U.S. Marines in South Vietnam. The following publications were used in the preparation of this study:

Cosmas, Graham A., and Lt. Col. Terrence P. Murray. *U.S. Marines in Vietnam: Vietnamization and Redeployment, 1970–1971.* Washington, D.C.: History and Museums Division, Headquarters, U.S. Marine Corps, 1986.

Melson, Maj. Charles D., and Lt. Col. Curtis G. Arnold. *The U.S. Marines in Vietnam: The War That Would Not End, 1971–1973.* Washington, D.C.: History and Museums Division, Headquarters, U.S. Marine Corps, 1991.

Shore, Capt. Moyers S. II. *The Battle for Khe Sanh.* Washington, D.C.: History and Museums Division, Headquarters, U.S. Marine Corps, 1969, reprinted 1977.

Shulimson, Jack. *U.S. Marines in Vietnam: An Expanding War, 1966.* Washington, D.C.: History and Museums Division, Headquarters, U.S. Marine Corps, 1982.

Shulimson, Jack, and Maj. Charles M. Johnson. *U.S. Marines in Vietnam: The Landing and the Buildup, 1965.* Washington, D.C.: History and Museums Division, Headquarters, U.S. Marine Corps, 1978.

Shulimson, Jack, and Lt. Col. Leonard A. Blasiol, Charles R. Smith, and Capt. David A. Dawson. *U.S. Marines in Vietnam: The Defining Year, 1968.* Washington, D.C.: History and Museums Division, Headquarters, U.S. Marine Corps, 1997.

Smith, Charles R. *U.S. Marines in Vietnam: High Mobility and Standdown, 1969*. Washington, D.C.: History and Museums Division, Headquarters, U.S. Marine Corps, 1988.

Telfer, Maj. Gary L., Lt. Col. Lane Rogers, and V. Keith Fleming Jr. *U.S. Marines in Vietnam: Fighting the North Vietnamese, 1967*. Washington, D.C.: History and Museums Division, Headquarters, U.S. Marine Corps, 1984.

Secondary Works

Anderson, Mary E., et al. *Support Capabilities for Limited War Forces in Laos and South Vietnam*. Santa Monica, Calif.: Rand Corporation, 1962.

Andrade, Dale. *Trial by Fire: The 1972 Easter Offensive, America's Last Vietnam Battle*. New York: Hippocrene Books, 1995.

Baldwin, Capt. Roger L. "Long Bridge to Bong Son." *Military Engineer* 63 (September-October 1971): 334–35.

Berry, F. Clifton Jr. *Sky Soldiers*. The Illustrated History of the Vietnam War. New York: Bantam Books, 1987.

———. *Gadget Warfare*. The Illustrated History of the Vietnam War. New York: Bantam Books, 1988.

Buttinger, Joseph. *Vietnam: A Dragon Embattled*, 2 vols. New York: Praeger, 1967.

Coleman, J. D. *Incursion*. New York: St. Martin's Press, 1991.

Davidson, Lt. Gen. Phillip B. *Vietnam at War: The History, 1946–1975*. Novato, Calif.: Presidio Press, 1988.

Fink, Col. George B. "Engineers Move to I Corps." *Military Engineer* 60 (September-October 1968): 358.

"Floating and Land Based Power Plants," *Vinnell* 11 (Spring 1969): 6–11.

Fulton Lt. Col. Taylor R. "Conglomerate Tactical Bridging." *Military Engineer* 59 (September-October 1967): 323.

Garland, Lt. Col. Albert N., ed. *Infantry in Vietnam*. New York: Jove Books, reprint of 1967 Infantry edition, 1985.

Gelb, Leslie H., and Richard K. Betts. *The Irony of Vietnam: The System Worked*. Washington, D.C.: Brookings Institution, 1979.

Karnow, Stanley. *Vietnam: A History*. New York: Viking Press, 1983.

Kiernan, Lt. Col. Joseph M. "Combat Engineers in the Iron Triangle." *Army* 17 (June 1967): 42–45.

Kouns, Maj. Darryle L. "Combat Engineers in Operation DUCHESS." *Military Engineer* 62 (May-June 1967): 173–76.

Malley, Lt. Col. Robert J. "Forward Airfield Construction in Vietnam." *Military Engineer* 59 (September-October 1967): 318–22.

Mangold, Tom, and John Penycate. *The Tunnels of Cu Chi*. New York: Random House, 1985.

Mayo, Sp4c. Larry. "Battle Maps." *Kysu'* 1 (Summer 1969): 26–28.

McIntyre, Brig. Gen. Kenneth E. "The Magnificent Sight." *Army Engineer* 2 (November-December 1994): 29–35.

———, Col. "Secret Planning." *Kysu'* 3 (Fall 1971): 15.

Merdinger, Capt. Charles J. "Civil Engineers, Seabees, and Bases in Vietnam." In *Vietnam: The Naval Story*, edited by Frank Uhlig Jr., pp. 228–253. Annapolis: Naval Institute Press, 1986.

Meyerson, Joel D. "War Plans and Politics: Origins of the American Base of Supply in Vietnam." In *Feeding Mars: Logistics in Western Warfare From the Middle Ages to the Present,* edited by John A. Lynn, pp. 281–87. Boulder, Colo.: Westview Press, 1993.

"MG Dillard, 9 Others Die in Crash," *Castle Courier*, 18 May 70.

Miller, John G. *The Bridge at Dong Ha*. Annapolis: Naval Institute Press, 1989.

Moore, Lt. Gen. Harold G., and Joseph L. Galloway. *We Were Soldiers Once . . . and Young: Ia Drang—The Battle That Changed the War in Vietnam*. New York: Random House, 1992.

Murphy, Edward F. *Dak To*. Novato, Calif: Presidio Press, 1993.

Nolan, Keith William. *Battle for Hue: Tet, 1968*. Novato, Calif.: Presidio Press, 1983.

_____. *Into Laos: The Story of DEWEY CANYON III/LAM SON 719, Vietnam 1971*. Novato, Calif: Presidio Press, 1986.

_____. *Into Cambodia: Spring Campaign, Summer Offensive, 1970*. Novato, Calif.: Presidio Press, 1990.

_____. *The Magnificent Bastards: The Joint Army-Marine Defense of Dong Ha, 1968*. Novato, Calif: Presidio Press, 1994.

Oberdorfer, Don. *Tet!* Garden City, N.Y.: Doubleday, 1971.

Offringa, Maj. Peter J. "Bunker Hill: Anatomy of a Firebase." *Engineer* 3 (Summer 1973): 19–20.

Palmer, General Bruce Jr. *The 25-Year War: America's Military Role in Vietnam*. Lexington: University Press of Kentucky, 1984.

Palmer, Col. Dave R. *Summons of the Trumpet: U.S.-Vietnam in Perspective*. San Rafael, Calif: Presidio Press, 1978.

Rhodes, Lt. Col. Nolan C. "Operation DUKE." *Military Engineer* 61 (September-October 1969): 332–33.

Schneebeck, Maj. Gene A., and Capt. Richard E. Wolfgram. "Airmobile Engineer Support for Combat," *Military Engineer* 59 (November-December 1967): 397–98.

Scigliano, Robert G. *South Vietnam: Nation Under Stress*. Boston: Houghton Mifflin Co., 1963.

Shandler, Herbert Y. *The Unmaking of a President: Lyndon Johnson and Vietnam*. Princeton: Princeton University Press, 1977.

Sheehan, Neil. *A Bright Shining Lie: John Paul Vann and America in Vietnam*. New York: Random House, 1988.

Son, Lt. Col. Phan Van, and Maj. Le Van Duong, eds., *The Viet Cong Tet Offensive, 1968*, trans. J5/Joint General Staff Translation Board. Saigon: Printing and Publications Center, Republic of Vietnam Armed Forces, 1969.

Spector, Ronald H. *After Tet: The Bloodiest Year in Vietnam*. New York: Free Press, 1993.

Stanton, Shelby L. *Anatomy of a Division: The 1st Cav in Vietnam*. Novato, Calif.: Presidio Press, 1987.

_____. *The Green Berets at War: U.S. Army Special Forces in Southeast Asia, 1956–1975*. Novato, Calif.: Presidio Press, 1985.

_____. *The Rise and Fall of an American Army: U.S. Ground Forces in Vietnam, 1965–1973*. Novato, Calif.: Presidio Press, 1985.

"Summary of Vietnam Construction." *Military Engineer* 64 (September-October 1972): 56–58.

Taylor, Maxwell D. *Swords and Plowshares*. New York: W. W. Norton and Co., 1972.

"Tet Truce Offensive, 1968," and "In Defense of Plantation, 1968." *Hurricane*, II Field Force, Vietnam, Magazine (April–June 1971): 2–4.

Thompson, Capt. Paul Y. "Aircraft Shelters in Vietnam." *Military Engineer* 61 (July-August 1969): 273–74.

Trainor, Capt. Francis E. "Tunnel Destruction in Vietnam." *Military Engineer* 60 (September-October 1968): 341.

Turley, Col. Gerald H. *The Easter Offensive: Vietnam, 1972*. Novato, Calif.: Presidio Press, 1985.

"Vinnell Makes History at Long Binh." *Vinnell* 11 (May 1969): 8–10.

Walker, Maj. Bryon G. "Construction of a Delta Base." *Military Engineer* 60 (September-October 1968): 333–35.

Weaver, Capt. Thomas C. "The Phu Cuong Bridge, Construction and Features." *Military Engineer* 61 (March-April 1969): 122–24.

Westmoreland, William C. *A Soldier Reports*. Garden City, N.Y.: Doubleday and Co., 1976.

Yens, 1st Lt. David P., and Capt. John P. Clement III. "Port Construction in Vietnam." *Military Engineer* 59 (January-February 1966): 20.

Zaffiri, Samuel. *Hamburger Hill, May 11–20, 1969*. Novato, Calif.: Presidio Press, 1988.

Glossary

Aggregate

Crushed rock used with cement to form concrete.

AM2 Matting

Rectangular aluminum matting used on jet fighter bases and selected Army airfields.

Armored Vehicle Launched Bridge

A sixty-three-foot scissors-type bridge launched from an armored carrier for spans up to fifty-seven feet and capable of supporting Class 60 loads, that is, the approximate weight in tons of vehicles, equipment, and armor.

Bailey/Panel Bridge

The Bailey bridge, also called a panel bridge, is a portable preengineered tactical truss bridge from World War II vintage named after its British designer. The bridge, which can be easily and quickly constructed, is supported by two main trusses formed from ten-foot steel panels that are coupled with steel pins.

Balk Decking

Hollow aluminum decking used on M4T6 floating bridges and fixed spans. The deck balk pattern distributes the load over more than one pontoon or float. (*See also* M4T6 Bridge.)

Bent

A structural frame similar to a tower used to support a bridge or railroad trestle.

Butler Warehouse

A large prefabricated metal building.

CINCPAC	Commander in Chief, Pacific. Traditionally a naval commander heading the U.S. Pacific Command (PACOM); controls all service forces in the Pacific area, including Military Assistance Command, Vietnam (MACV), and reports directly to the Joint Chiefs of Staff.
Class (30, 50, 60)	Classification of bridges that indicate the safe weight-bearing capacity; small yellow classification signs are affixed to the front and sides of vehicles showing their safe loads.
Class 60 Floating Bridge	Pneumatic floats covered with two deck-tread panels, curbs, and filler panels that can support division loads as a float bridge or raft; assembling the bridge requires cranes and air compressors.
Compacted	Soil compressed by rolling to reduce surface volume in preparation for construction.
Conex	A reusable metal container used for shipping military equipment.
COSVN	*Central Office for South Vietnam*
Creosoted Timber	Piles and timbers treated with wood-tar distillate as a preservative.
D7E Bulldozer	New standard military bulldozer introduced in 1966 in the D7 series of bulldozers dating back to World War II.
Decomposed Granite	Granite deposits often marked by outcroppings of granite boulders.
DeLong Pier	A prefabricated pier developed by the DeLong Corporation used for rapid building of port quays and ship berths.

Dry Span

A portable bridge for crossing short gaps under forty-five feet or longer spans using trestle bents; made up of M4T6 floating bridge superstructure components such as trestles and balk decking. (*See also* Balk Decking.)

Eiffel Bridge

A French designed and manufactured bridge for carrying light traffic.

Fair-Weather Road

Unpaved roads normally suitable for traffic during dry season but susceptible to damage during rainy season.

Float Bridge

Tactical bridge designed to float on water on inflated rubber floats or metal pontoons. Types of float bridges included a foot bridge, light tactical raft, M4T6 bridge and raft, and Class 60 bridge and raft.

Hardstand

Hard-surfaced area for parking aircraft, equipment and vehicles, and supplies.

Hopper Dredge

A self-propelled and seaworthy dredge using hydraulic suction pipes to excavate bottom material into internal hoppers as opposed to separate barges. The dredge can carry its waste to a dumping site.

Horizontal Construction

Engineering term used to differentiate construction at or near ground level as compared to vertical construction, which is associated with structural work above ground; included is use of equipment for earthmoving, paving, and drainage work to build roads, canals, airfields, and open-storage areas. (*See also* Vertical Construction.)

Laterite	A red to brown, porous soil rich in secondary oxides of iron and aluminum. Laterite is unique in tropical regions such as Vietnam and is used widely as subgrade material for roads, airfields, and open-storage areas.
LCM	Landing craft, mechanized
LCU	Landing craft, utility
LOC	Lines of communication
LST	Landing ship, tank
M4T6 Bridge	Portable bridge used to build rafts, float bridges, and short fixed spans with hollow-tubed decking (also known as balk decking).
M8 Matting	Pierced steel landing matting distinguished by round holes in the matting developed in World War II.
M8A1 Matting	Solid steel airfield matting widely used on forward airfields and parking areas.
MAAG	Military Assistance Advisory Group
MACV	Military Assistance Command, Vietnam
MAP	Military Assistance Program
MCA	Military Construction, Army. Funds appropriated for Army construction.
MILCON	Military Construction. Funds appropriated for major construction.
MUST	Medical unit, self-contained, transportable
MR	Military Region
MX19 Matting	Square aluminum matting with honeycombed interior introduced in 1966 and proved preferable for heavily used forward airfields.

OICC	Officer in Charge of Construction, Republic of Vietnam. Naval Facilities Engineer Command's officer and office charged with supervising RMK-BRJ (Raymond, Morrison-Knudsen, Brown and Root, and J. A. Jones) in South Vietnam.
OMA	Operations and Maintenance, Army. Funds appropriated for repair and maintenance of facilities.
Operation MOOSE	Move Out Of Saigon Expeditiously
OPLAN	Operation Plan
PA&E	Pacific Architects and Engineers. California engineering firm under contract by the Army to provide facilities engineering services and minor construction.
PACOM	Pacific Command. Joint command under the CINCPAC who commands all service forces in the Pacific, including MACV.
Peneprime	A rapidly placed, temporary dust-control material with an asphalt base that was sprayed on a leveled, graded, and compacted area, such as helipads, to bond sand particles.
Philco-Ford Corporation	Facilities engineering contractor working for the Navy and later the Army in I Corps.
Pioneer Road	Hastily built roads supporting tactical operations; the roads may be subsequently improved or abandoned upon completion of an operation or the closing of a firebase.
Prime Beef	Air Force engineering teams derived from nickname (Prime) and acronym (Base Engineering Emergency Force).

Red Horse	Air Force engineering squadrons derived from acronym Red (Rapid Engineering Deployable) and Horse (Heavy Operational Repair Squadron, Engineering).
RMK-BRJ	Raymond, Morrison-Knudsen, Brown and Root, and J. A. Jones. A joint venture of four American contractors under the direction of the OICC, Republic of Vietnam, hired to do large construction projects for the armed forces and U.S. agencies in South Vietnam.
ROK	Republic of Korea
Rome Plow	A D7E bulldozer equipped with a heavy-duty protective cab and a special tree-cutting blade manufactured by the Rome Plow Company of Cedartown, Georgia. Used to clear vegetation and deny cover and concealment to enemy forces.
RVN	Republic of Vietnam, also known as South Vietnam.
Sheepsfoot Roller	A towed roller with attached feet used to compact loose soil as contrasted to a motorized pneumatic-tired and steel-wheeled roller.
T–2 Tanker	A tanker used as an electrical power ship.
T17 Membrane	Rubberized fabric used as an expedient airfield surfacing material on runways, taxiways, and parking areas.
TOE	Table of organization and equipment
Transphibian Tactical Tree Crusher	A land-clearing device that was only marginally effective in felling trees and vegetation.
Turnkey	Air Force construction contract package used at Tuy Hoa without Army or Navy engineering resources.
USARPAC	U.S. Army, Pacific

USARV	U.S. Army, Vietnam
USAID	U.S. Agency for International Development. State Department agency reorganized and renamed in 1961 to carryout long-range economic assistance programs. (*See also* USOM.)
USOM	U.S. Operations Mission. A field agency under USAID and its predecessor organization charged with carrying out economic and technical assistance in underdeveloped nations.
USNS	U.S. Navy Ship
Vertical Construction	In contrast to horizontal construction, vertical construction is more labor intensive and requires the use of hand tools; also includes efforts by carpenters, masons, steelworkers, plumbers, electricians, and other building tradesmen to build aboveground structures. (*See also* Horizontal Construction.)
Vinnell Corporation	California firm awarded a contract with the Army to convert T–2 tankers to generate electricity and to build, operate, and maintain land-based power distribution systems.
Wonder Arches	Steel aircraft shelters, reinforced with concrete that were designed to protect Air Force and Marine Corps fighter aircraft from bomb and rocket damage.

Map Symbols and Terms

Military Units

Function

Airborne Infantry . . . .

Airmobile Infantry . . . .

Armored Cavalry . . . .

Engineer . . . .

Infantry . . . .

Marine Corps . . . .

Size Symbols

Battalion or Armored Cavalry Squadron   II

Regiment or Group   III

Brigade   X

Division . . . .   X X

Corps . . . .   X X X

Examples

70th Engineer Battalion (Combat) . . . .

35th Engineer Group (Construction) . . . .

3d Brigade, 25th Infantry Division . . . .

1st Cavalry Division (Airmobile) . . . .

Boundary between the 1st and 25th Infantry Divisions . . . .

Geographic Terms

Nui	Mountain
Song	River

Acronyms and Abbreviations

Arty	Artillery
ARVN	Army of the Republic of Vietnam
Aust	Australian
BEQ	Bachelor Enlisted Quarters
BOQ	Bachelor Officers' Quarters
Bn	Battalion
CAV	Cavalry
CTZ	Corps Tactical Zone
Evac	Evacuation
FFV	Field Force, Vietnam
HQ	Headquarters
Inf Div	Infantry Division
LCM	Landing Craft, Mechanized
LCU	Landing Craft, Utility
Log	Logistics
LST	Landing Ship, Tank
LZ	Landing Zone
MACV	Military Assistance Command, Vietnam
MAF	Marine Amphibious Force
MR	Military Region
Ops	Operation
Ord	Ordnance
PA&E	Pacific Architects and Engineers
POL	Petroleum, Oils, and Lubricants
Repl	Replacement
Rgt	Regiment
RMK-BRJ	Raymond, Morrison-Knudsen, Brown and Root, and J. A. Jones
ROK	Republic of Korea
TF	Task Force
USAF	U.S. Air Force
USARV	U.S. Army, Vietnam

Index

629